Carpenter's Neurophysiology

Sixth Edition

Carpenter's Neurophysiology

A Conceptual Approach

Dunecan Massey

Nick Cunniffe

Imran Noorani

CRC Press
Taylor & Francis Group
Boca Raton London New York

CRC Press is an imprint of the
Taylor & Francis Group, an **informa** business

Sixth edition published 2022
by CRC Press
6000 Broken Sound Parkway NW, Suite 300, Boca Raton, FL 33487-2742

and by CRC Press
2 Park Square, Milton Park, Abingdon, Oxon, OX14 4RN

CRC Press is an imprint of Taylor & Francis Group, LLC

Library of Congress Cataloging-in-Publication Data
Names: Massey, Dunecan, author. | Cunniffe, Nick, author. | Noorani, Imran, author. | Carpenter, R. H. S. (Roger H. S.), 1945–2017. Neurophysiology.
Title: Carpenter's neurophysiology : a conceptual approach / Dunecan Massey, Nick Cunniffe, Imran Noorani.
Other titles: Neurophysiology
Description: Sixth edition. | Boca Raton : CRC Press, 2021. | Preceded by Neurophysiology : a conceptual approach / Roger Carpenter, Benjamin Reddi. 5th ed. 2012. | Includes bibliographical references and index.
Identifiers: LCCN 2020057776 | ISBN 9780367340674 (hardback) | ISBN 9780367340605 (paperback) | ISBN 9780429323720 (ebook)
Subjects: MESH: Neurophysiology
Classification: LCC QP360 | NLM WL 102 | DDC 612.8--dc23
LC record available at https://lccn.loc.gov/2020057776

ISBN: 978-0-367-34067-4 (hbk)
ISBN: 978-0-367-34060-5 (pbk)
ISBN: 978-0-429-32372-0 (ebk)

Typeset in Palatino and Futura
by Chernow Editorial Services, Inc.

Printed in the UK by Severn, Gloucester on responsibly sourced paper

Roger Carpenter
2 September 1945–27 October 2017
(Photograph courtesy of Professor Yao Liang, reproduced with permission)

CONTENTS

PREFACE TO THE SIXTH EDITION

Roger's death in 2017 marked the loss of a world-leading neuroscientist, extraordinary teacher and dear friend. Our aim in updating this book was to keep some sense of Roger's personality alive in his uniquely conversational textbook of neuroscience, which we have now termed *Carpenter's Neurophysiology*; may it join the ranks of *Gray's Anatomy* and Guyton's *Physiology* for perpetuity. We are all former students of Roger and are now supervisors (the Cambridge term for academic Tutor; Tutor at Cambridge means something else entirely) in our own right – like Roger, we have learned to write upside down when explaining challenging concepts over the supervision table with pen and paper to our students. For it is the reduction of increasingly complex reference material into core concepts that has always been the strength of the Cambridge supervision system, and which is the key aim of this book.

The fifth edition had undergone such a significant update that relatively minor editing has been necessary to bring the book up to date. For the most part, we have corrected minor factual errors and updated terms that had become anachronistic, while maintaining the essence of Roger's writing voice. Clinical boxes, which had been deployed in the fifth edition to bridge the gap between preclinical science and clinical practice, have undergone significant revision. In places, we have re-combined figures to make them clearer and easier to understand. We have also replaced and updated the figure legends which Roger chose to omit from the fifth edition.

We have finally dispensed with the increasingly clunky *Neurolab* (additional electronic material) as it is unplayable on contemporary operating systems and we are not convinced it was ever a key element of the book, the value of which was always in the clarity of writing, such that we saw no purpose in having the software re-written. We wonder how many former purchasers of *Neurophysiology* ever even opened it!

As ever, we remain grateful to our own students, past, present and future, for whom this book is written and who provide a consistent inspiration.

Dunecan Massey, Nick Cunniffe and Imran Noorani
Gonville and Caius College, Cambridge, 2021

PREFACE TO THE FIFTH EDITION

Since the last edition, the internet has arrived in earnest. Does this mean the death of the textbook? Are students no longer to read? No; but a very positive consequence of all this is that reading becomes less of a reflex activity. No longer will heads be bowed over turgid texts as line after line is dutifully enhanced with fluorescent highlighter; it is now permissible to *think* a little before deciding whether, and what, to read. Time is precious: one should never pick up a book without asking a preliminary, and rather fundamental, question: Why am I doing this? Is it to look up a specific fact? Is it to understand an elusive concept? Or is it to gain an overall perspective of a topic on a more expansive canvas? So much of the mind-numbing cramming that medical students in particular have to endure is, quite literally, meaningless. But it need not be like that.

To be sure, from the point of view of not killing people, nothing could be more important than memorising the proprietary names of pharmaceutical substances. But from the point of view of developing one's intelligence and nurturing one's imaginative powers – of learning how to be a *producer* rather than a consumer of knowledge – nothing could matter less. The internet is only a threat to the kinds of books that merely list facts. *Neurophysiology* has never been a conventional book; its special qualities mean that it is needed more than ever as an adjunct to what is so easily available online. But the facts are still of course there, complemented now by new sections that explicitly relate science and clinical practice; these are supplemented by a comprehensive set of practice MCQs.

We believe passionately that – like scientists – some doctors at least should be permitted to *think*, and it is this prejudice that permeates the pages of this book. Since thinking is also quite fun, once you get used to it, we hope that you will enjoy the experience of reading it.

It is a pleasure to acknowledge our indebtedness to the students who have provided valuable feedback about previous editions of the book as well as identifying those particularly sticky areas that they needed us to address; to Dr Deborah Field and Dr Danielle Reddi for clinical comment; to Dr Joanna Koster, Stephen Clausard and Amina Dudhia at Hodder; to Dr Sanjay Manohar; to Lotika Singha; and to the various authors and publishers where mentioned for permission to reproduce material.

Roger Carpenter and Benjamin Reddi
Cambridge and Adelaide, 2012

PREFACE TO THE FOURTH EDITION

The rapid march of technology means a CD-ROM rather than diskette, and therefore very much more in the way of goodies. NeuroLab is still there, in an improved form that – thanks to it having been translated into Java – no longer insists on being run on a PC: Macs welcome! But in addition we have full-colour illustrations, short video clips of neurological interest, a complete brain atlas of histological sections and scans, with interactive labels and the ability to do self-tests, a selection of audio material relating to sound perception and disorders of speech, a somewhat experimental three-dimensional interactive brain, and of course a complete version of the text, with the advantage that you can search it for individual words. Meanwhile the book itself has had a spring-clean, with much rewriting and additional material and redrawing of the illustrations, a new Appendix on methods of studying the brain, and a large number of completely new figures.

Much of this has been due to the hard work of a number of present and former students: I would like to thank Robin Marlow, Oliver Sanders and especially Sanjay Manohar for their brilliant translation of NeuroLab into Java, Dr Dunecan Massey and Dr Chris Allen for the video material, Alice Miller for her work on the brain atlas, and Ruaraidh Martin, Atman Desai and Atanu Pal for their contributions to the rotatable brain. Lastly, I must acknowledge a huge debt to Ben Reddi, whose trenchant criticisms and suggestions have immeasurably strengthened both text and figures.

R. H. S. Carpenter
Montalcino, 2001

PREFACE TO THE THIRD EDITION

Most of neurophysiology is concerned with dynamics, with sensory coding, with feedback, with plasticity and stability. These are easy concepts to teach to a few students round a table, with a plentiful supply of paper to scribble on, less easy to convey in a book. But the coming of age of the personal computer has changed all that, and with NeuroLab you have the opportunity – the first of its kind – to experiment with model systems and see for yourself how they respond, as well as having the chance to try out experiments and demonstrations on yourself. It has been fun to develop, and I am certain that you will find it fun to use. Meanwhile, the text has been thoroughly revised and updated, with some changes of emphasis, and a more extensive use of supplementary notes, that are indicated with the following symbol in the text: . The increasingly molecular approach to much of neurophysiology is an unwelcome trend in many ways, not least for the medical student who faces the 'anatomisation' of yet another area of study – how soon will it be before they are required to memorise stretches of DNA? But it has also brought with it some unifying simplifications which have helped to bring a little more tidiness to certain areas.

Once again, to my students huge thanks for their stimulation and support, particularly in fine-tuning NeuroLab to meet their needs.

R. H. S. Carpenter
Studland, 1995

PREFACE TO THE SECOND EDITION

A second edition has provided an opportunity to remedy some defects in the first. One of which I was particularly conscious was that in trying to achieve a connected and coherent narrative, I had sometimes neglected to be sufficiently exhaustive in enumerating the factual detail that is so greatly enjoyed by both students and examiners. This has now been rectified by the use of boxes containing tables and other systematic information outside the text itself. In addition, a number of topics receive a wider coverage; these include: pain, subcortical visual mechanisms, eye movements, central auditory mechanisms and the hypothalamus. Finally, an attempt has been made in the last chapters to distil from the preceding ones some kind of answer to that troublesome question that students so frequently ask: what principles govern the processes that convert patterns of sensory information into patterns of behaviour? What, in short, does the brain do?

As always, I owe a special debt to my long-suffering students for acting as guinea pigs for certain lines of approach, and for their encouragement and criticism.

R. H. S. Carpenter
Talloires, August 1989

PREFACE TO THE FIRST EDITION

The supervision system practised at Cambridge and elsewhere brings many benefits both to teacher and taught: not least, that lecturers are brought face to face with the results of deficiencies in their own teaching in a peculiarly immediate and painful way. What has seemed to many supervisors a most worrying trend over the last 10 years or so is the extent to which a student may come away from a series of lectures on (let us say) the circulation, with an impressive amount of detailed information, including perhaps the minutiae of experiments published only a month or two previously, yet with little sense of what might be called function: of what the circulation really does, of how it responds to actual examples of changed external conditions, and how it relates to other major systems. And in the case of the central nervous system things seem even worse: a student may acquire an immensely detailed knowledge of the anatomical intricacies of the motor system, yet not be able to tell you even in the broadest terms what the cerebellum actually does, or have the slightest feel for what kinds of processes must be involved in such an act as throwing a cricket ball. The result is much knowledge, but little understanding, and very little sense of ignorance.

I believe this to be the result of two factors. The first is, paradoxically, that over the last decade or so, Universities and Teaching Hospitals have quite rightly begun to take teaching much more seriously than once was the case, and consequently a perfectly laudable sense of competition has developed amongst lecturers to gain the approval of their audiences. But students – at least in the short term – tend to form judgements rather on the basis of the number of 'facts' that they have succeeded in copying down in the course of a lecture: the more recent these facts are, the better they are pleased. Lecturers naturally respond to this by filling their lectures with increasing amounts of detail, at the expense of fundamental principles. The students' notebooks swell with quantities of undigested information, but they are bewildered – even resentful – when asked simple but basic questions like 'how does a man stand upright?'. This change in emphasis has made physiology less enjoyable either to study or teach than it used to be, as well as less educational in the broadest sense: there is no time and little motivation to ask questions of oneself, and all is reduced, in the end, to rote-learning.

The second factor that has debased the intellectual quality of much of our teaching is the increasing emphasis that is put on mechanism instead of function. More time is often spent in talking about the detailed physics of nerve conduction than in discussing exactly what information is being carried by nerves, how it is coded, and how the nervous system is actually used. Again, lecturers' fear of instant student opinion is perhaps partly the cause: most students get immediate and easy satisfaction (of a limited kind) by seeing the detailed steps that cause a particular phenomenon; and if all can be reduced to a series of biochemical reactions, then so much the better. To understand whole systems and their interactions requires rather more effort of thought, and one can never be sure one is right. But in the long run, and most particularly for medical students, it is precisely the large-scale functioning of physiological systems that is important. A doctor needs to have a feel for what is likely to be the consequence of chronic heart failure in terms of problems of fluid balance, or for what may happen if his asthmatic patient decides on a holiday in the Andes. Whether the cardiac action potential is due mainly to calcium or to sodium, and whether or not the substantia nigra projects to the red nucleus, are for him matters of singularly little interest or significance.

This book is an attempt to go counter to this trend by starting from the premise that a more satisfactory way to teach physiology is to build a scaffold of general principles on which factual details may later be hung as the need arises, and to prefer to consider what systems do rather than how they do it. However, this is largely a matter of emphasis and organisation rather than of content, and the reader will find details of mechanism if they are required. Above all, the aim has been to recreate something of the intellectual excitement to the study of physiology that has been lost sight of in recent years, and to encourage the student to think and to question. If it is at all successful in this, the thanks should go not to me but rather to those past and present students of mine for whose intellectual stimulation I am – as all teachers must surely be – deeply indebted.

R.H.S. Carpenter
Cambridge, 1984

It is a pleasure to acknowledge my indebtedness to Dr Susan Aufgaerdem, Professor George L. Engel, Mr Austin Hockaday, Dr J. Keast-Butler, Dr Richard Kessel, Dr Peter Lewis, Dr J. Purdon Martin, Dr N. R. C. Roberton, Dr T. D. M. Roberts and Mr Peter Starling for their help in providing illustrations; to the editors and to the staff of Hodder Arnold for their criticisms and support; to various authors and publishers where mentioned for permission to reproduce material; and above all to the late Dr R. N. Hardy, who died so tragically while the book was in its final stages: his friendly encouragement, and his qualities of wisdom and humanity are sadly missed by all who knew him.

PART 1

NEURAL MECHANISMS

CHAPTER 1

THE STUDY OF THE BRAIN

This book is about trying to understand the brain; we might begin by asking whether such an aim is not, in fact, hopelessly ambitious. Your brain is a machine whose complexity far exceeds anything constructed by human hands. It is made up of units – cells called *neurons* – that provide both the pathways by which information is transmitted within the brain and also the computing machinery that makes it work. There are quite a lot of these neurons: about thirty times as many as the entire population of this planet. Another not very useful fact to impress your friends is that every cubic inch of your cerebral cortex has 10 000 miles of nerve fibre in it. A typical neuron is wired up to a thousand or two of its neighbours, and it is the *pattern* of these connections that determines what the brain does. All this makes the brain a fantastically complex structure, a thicket of twisting, interweaving fibres.

Our brains *need* to be complex: part of what they do is to embody a kind of working model of the outside world, that enables us to imagine in advance what will be the result of different courses of action. It follows that the brain must be at least as complicated as the world we experience. So the study of the brain is like the study of human society: for a society can only be fully understood if we comprehend not just the behaviour of isolated individuals, but also the interactions that each of them makes with others. Understanding the brain is a task as daunting as trying to comprehend the behaviour of the entire human race, its politics, its economics, and all other aspects of what it does; in fact, about thirty times more difficult. As a result, the study of the brain has in some respects a closer affinity with 'arts' subjects like history than it does with much conventional science. For many people that is its attraction.[1]

The need for neurons

How has such complexity come about? The evolutionary history of the brain is not well understood but in broad outline was certainly something like the account that follows.

For a single-celled organism such as an amoeba, coordination is essentially *chemical*: its brain is its nucleus, acting in conjunction with its other organelles.[2]

But a multicellular organism clearly needs some system of communication between its cells, particularly when, as in the primitive invertebrate *Hydra*, they are specialised into different functions: secretion, movement, nutrition, defence and so on. Communication between cells practically always means chemical communication. One cell releases a chemical substance; somewhere else another cell is waiting for this chemical, which tells it to do something. Nearly always, the target cell has receptor proteins in its cell membrane that recognise the particular chemical. Often these form part of what are technically called *ligand-gated channels* in the membrane, opening in response to the chemical and often allowing ions such as calcium, which act as messengers, to enter the cell. This kind of chemical communication is most familiar as *hormones*, operating within the body, perhaps to cause widespread changes: a good example is adrenaline, released in dramatic circumstances and alerting more or less the whole body. The chemical is not directed at any particular cell: it is *broadcast* throughout the whole body.

Diffusion time

In small and slow creatures, this works fine. The one shown in **Figure 1.1** has sensory cells (S) that control motor cells (M) by releasing a chemical transmitter or hormone into the common fluid space. But as an organism gets bigger, things get difficult, in two ways. First of all *time:* how long it takes for the message to get from the sender cell to its recipient depends on diffusion. It depends rather dramatically, because diffusion time is proportional to the *square* of the distance travelled – Einstein's Law (**Fig. 1.2**) – so it takes a disproportionately longer time for a chemical signal to travel over increasing distances: 1 μm takes 0.5 ms, but 10 mm takes 15 hours. If speed of response is not particularly important, this kind of communication may still be satisfactory even in very large organisms: our own hormonal control systems are of course precisely of this kind. Circulation helps, but even so your blood takes about a minute to go once round the body. If a tiger suddenly burst into your room, although

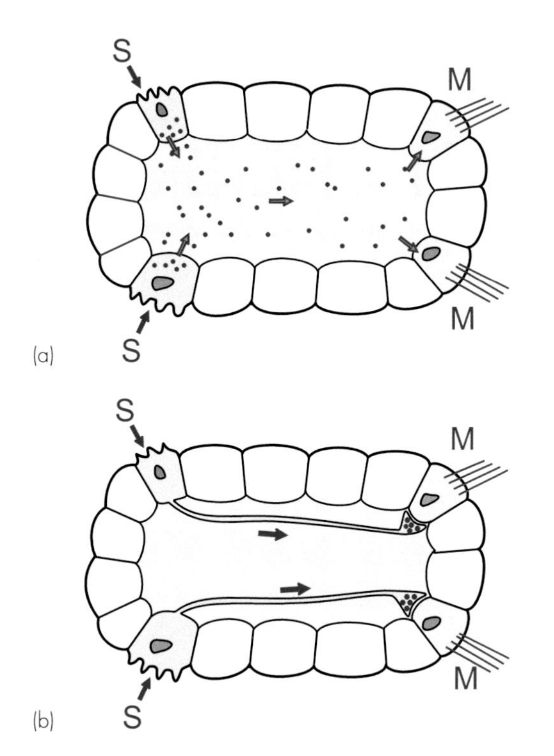

Figure 1.1 (a) A hypothetical multicellular organism with sensory cells (S) that control motor cells (M) by releasing a chemical transmitter or hormone into the common fluid space. (b) Direct connections between sensory and motor cells by means of nerve axons, providing communication that is both far quicker and more specific. Their ultimate action on the motor cells is still chemical.

you would instantly release a burst of adrenaline into your bloodstream, it would take about a minute for it to reach its target tissues. By that time you might well have been devoured.

Specificity

The other problem is *specificity*: the little creature in **Figure 1.1** can do just two things, but we can do much more. Each of us has well over 1200 separate muscles, for a start. And as we get more complicated with lots of different receptor cells and lots of different muscle cells, we want to do more specific things in response to specific circumstances. In a hormonal system, specificity can only come about by having a range of different chemicals, with target cells responsive to some but not others. In fact there is a very long list of hormones in the body, running into hundreds – but still not enough for the countless actions we might want to undertake. What is the solution?

In hormonal systems, a message is, in effect, shouted throughout the whole body. One way of being more specific is, as it were, to whisper it confidentially into the ear of the target cell. And this precision is where neurons come in. Neurons are simply cells with special elongated output processes called *axons*, which can often be very long indeed – the longest neurons measure about a metre long. The axons grope out towards their targets and make physical contact with them at regions called *synapses* (**Fig. 1.3**). They still release their chemical at the synapse, but

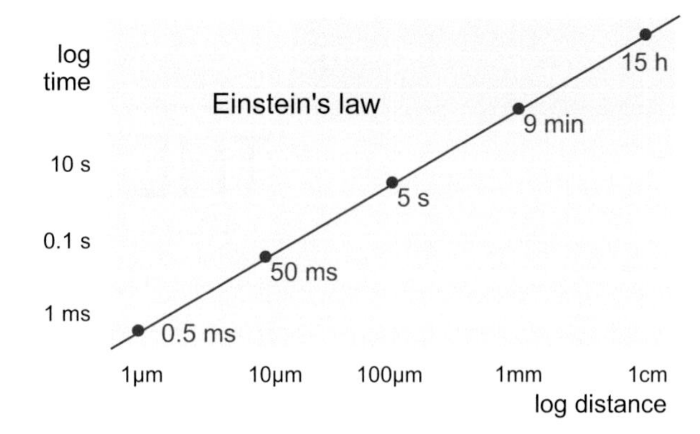

Figure 1.2 Diffusion time is a function of distance: note the logarithmic scales.

this time it only affects the cells it makes contact with and not the others, so we no longer need lots and lots of different transmitters. For instance, every muscle in your body is controlled by the *same* chemical substance, acetylcholine. Though specific, neurons can still be widespread in their actions: because axons can branch, they are capable of influencing thousands of target cells that can be scattered over a very wide area. These axons end either on muscles (where the synapse is called a *neuromuscular junction*), on the cell body (*soma*) of a target neuron, or often on the tree-like *dendrites* that enormously increase the surface area of many neurons.

It is still *chemical* communication, but we call the chemicals *transmitters* rather than hormones. There are several dozen known transmitter substances, falling into three main groups: amino acids, amines and peptides. In fact, knowing the name of a transmitter doesn't tell you much, because transmitters don't have to do the same thing at every site. The receptor molecules on the target cell translate arrival of the transmitter into some specific action, but different cells may use different receptor proteins and therefore translate the event differently. Thus there is no necessary logical connection between what the transmitter *is* and what it *does*. For example, acetylcholine excites skeletal muscles but inhibits heart muscle: the receptor proteins in the heart muscle cell membrane are different

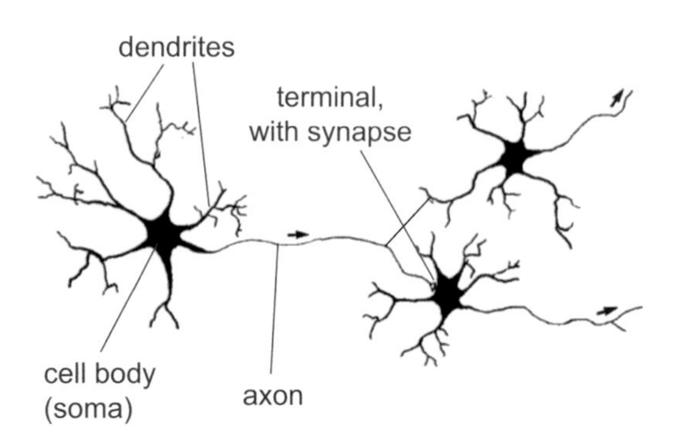

Figure 1.3 Schematic representation of a 'classical' central neuron synapsing with two others, in one case on a dendrite and in the other on the cell body or soma.

Clinical box 1.1 Neurotransmitters as drug targets

The issue of specificity is one of the first hurdles facing the clinician. As noted above, there are only a limited number of transmitter chemicals to bear a near infinite number of bits of information. The evolutionary solution has been the neuron, giving precision to the signal using spatial coding by delivering the transmitter to a particular recipient. Rather as the order 'fire' is one of a limited number of commands a sniper might expect to hear in the field, the general accomplishes specificity by ensuring the command is issued to the correct person; confusing his or her cables could clearly deliver a suboptimal result. The problem the clinician faces when trying to restore brain function by using administered analogues of neurotransmitters is that of targeting the signal to the correct recipient.

For example, morphine acts on a family of receptors and mimics endogenously generated signals called *opioids*. We know morphine as a rather powerful painkiller, and, not surprisingly, it acts on neurons that block pain signalling. Unfortunately, there are other morphine receptors located elsewhere in the body, the activation of which has rather less desirable characteristics; for example, neurons exist in ganglia near the bowel, which, when activated by morphine, reduce gut motility. Indeed, one of the most distressing side effects of morphine is constipation. The body gets around this problem by targeting its opioid release to those sites where it is needed – pain pathways when pain is disabling, bowel ganglia when gut activity is excessive. The clinician does not have the option of targeting individual neurons in this way, and most side effects of neurological drugs stem from this lack of specificity (see treatment of Parkinson's disease and the use of benzodiazepines discussed in Chapter 3, p. 51).

from those in skeletal muscle. The message is the same, but the meaning is different.

So nerves provide a system that is very specific both in what it does and where it does it. But what about speed? As we have seen, diffusion down long neurons would be hopelessly slow for distances of more than a few microns. Yet the fastest neurons can actually convey information at over 100 m/s. They are able to do this because they use a much quicker physical process than diffusion, namely electricity and the flow of current: these processes are discussed in the next chapter.

The brain

The basic ground-plan of the brain

In vertebrates, neurons develop from the *ectoderm*, the outer layer of embryonic germ cells that also forms skin. Some remain in epithelia as *sensory receptors* that are sensitive to mechanical or chemical stimuli, to temperature or to electromagnetic radiation. Others migrate inward and become specialised as *interneurons*, which respond only to the chemicals released locally by sensory neurons or other interneurons. They in their turn release transmitter at their terminals, which form synapses either with interneurons or with effectors such as muscles or secretory cells. So interneurons provide the communication channels by which information is passed rapidly from one part of the central nervous system to another, through mechanisms that will be discussed in Chapters 2 and 3. In *Hydra*, for example, we find a network of such intercommunicating neurons, making contact on the one hand with sensory cells on its surface that respond to touch and chemical stimuli, and on the other with muscle cells and secretory glands. *Hydra*'s brain is thus spread more or less uniformly throughout its body (**Fig. 1.4**), with only a slight increase in density in the region of its mouth: yet even such a relatively undifferentiated structure can generate well co-ordinated, even 'purposeful' behaviour.

The next step in the evolution of the nervous system came with the increasing specialisation of sensory organs, particularly of *teloreceptors* such as eyes and olfactory receptors. For an animal that normally moves in one particular direction, as in the earthworm in **Figure 1.4b**, such organs tend to develop at the front end, and the result of the consequent extra influx of sensory information to a localised region is an increased proliferation of interneurons in the head. In *Planaria* we see the first true brain of this kind, a dense concentration of neurons close to the eyes and sensory lobes of the head, giving rise to a pair of nerve cords that run down the body and send off side branches connecting with other neurons and effector cells. In segmented animals like the earthworm, the nerve cords

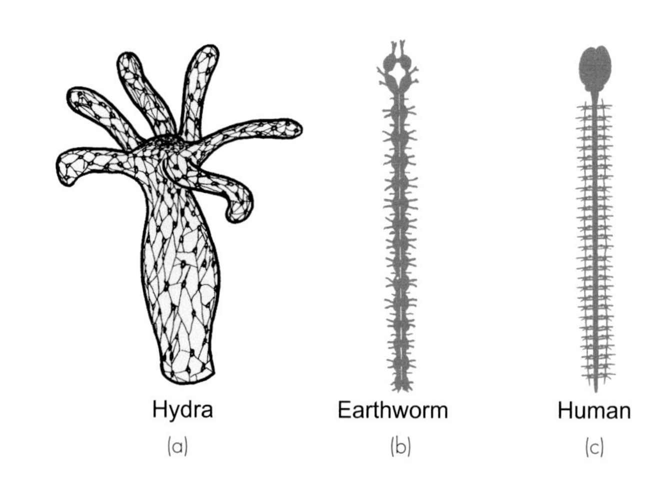

Figure 1.4 Schematic representation of Hydra nerve-net (a) and central nervous systems of earthworm (b) and human (c). (Partly after Buchsbaum, 1951; permission pending.)

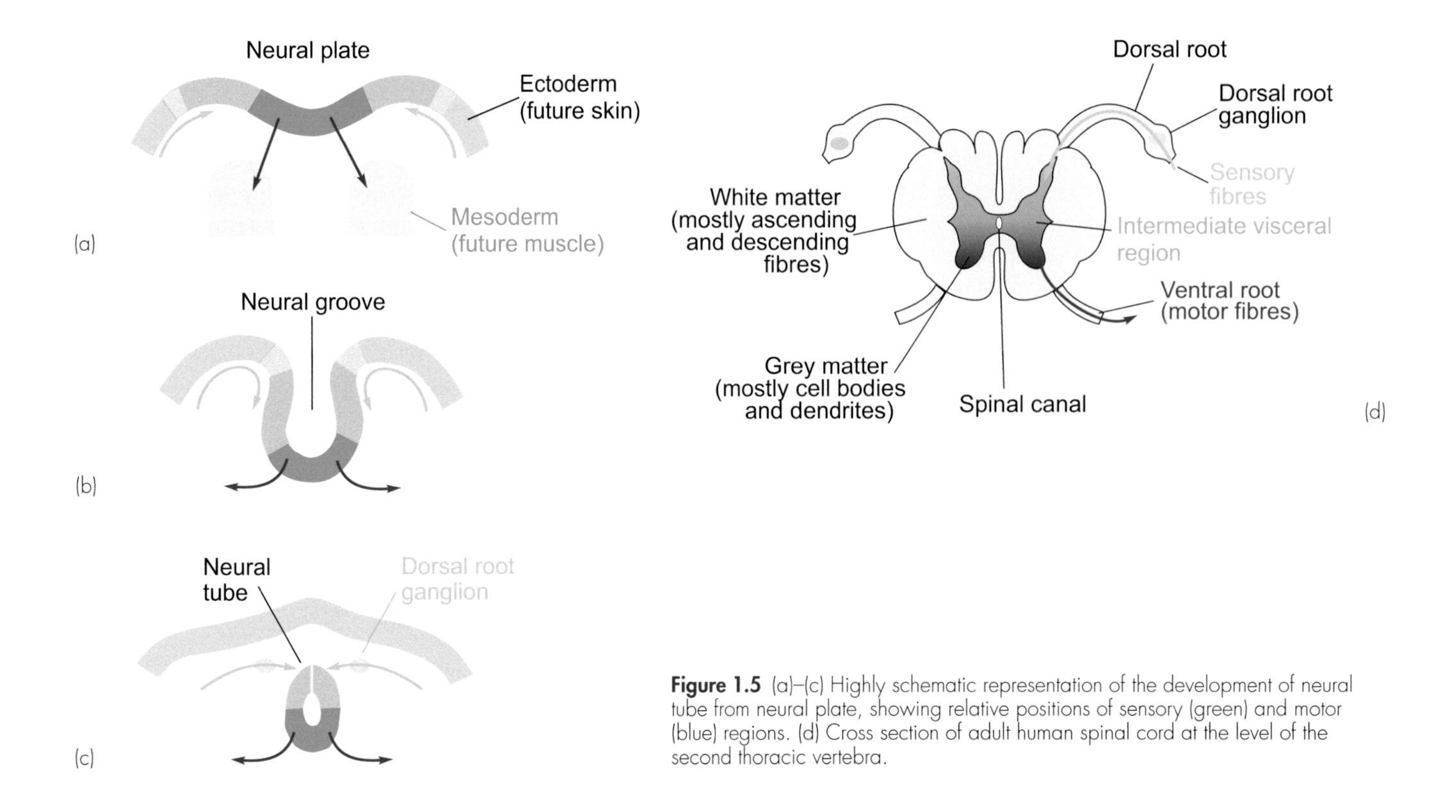

Figure 1.5 (a)–(c) Highly schematic representation of the development of neural tube from neural plate, showing relative positions of sensory (green) and motor (blue) regions. (d) Cross section of adult human spinal cord at the level of the second thoracic vertebra.

show a series of swellings or ganglia, one to each segment. Each is a kind of brain in its own right, and a decapitated earthworm is still capable of many kinds of segmental and intersegmental co-ordination. Though our bodies are not of course segmented, our nerve cord, the *spinal cord*, still shows some segmental properties, particularly in the organisation of the incoming and outgoing fibres, and in the existence of corresponding chains of ganglia along each side (**Fig. 1.4c**). We shall see in Chapter 10 that our spinal cord is also capable of a limited degree of brain-like activity. The primitive nerve net has not been altogether superseded, but survives as an adjunct to the central nervous system in the diffuse networks near the viscera that control movements of the gut and some other visceral functions (for example, stretching the gut causing contractions) and in the ancient diffuse core of the brain, the *reticular formation*.

The subsequent development of the brain is rather more complex, and not well understood. By looking at its evolutionary history in conjunction with the sequence of its growth in foetal development, one can postulate a framework that may help to relate the primitive nervous system to the more intricate structure of the adult human brain.[3]

The central nervous system is derived from a narrow strip of ectoderm, the *neural plate*, which runs down the middle of the vertebrate embryo's back (**Fig. 1.5a**). The centre of this strip becomes depressed into a trough or groove, and eventually its edges come to meet in the middle to form a closed structure, the *neural tube* (**Fig. 1.5b**; sensory regions shown green, motor regions, blue). Quite a common defect, *spina bifida*, occurs as a result of failure of this process of closure in the spinal cord; higher up, a similar defect produces, fatally, *anencephaly*. It is natural for sensory fibres from the skin to enter at the margins of the neural plate, and for motor fibres to the more medial musculature to leave the plate nearer the midline, and as a consequence one finds that it is in the dorsal half of the neural tube that the *sensory* fibres terminate (their cell bodies lying in the *dorsal root ganglia* on each side of the tube, **Fig. 1.5c**), while the cell bodies of the efferent *motor* fibres lie in the ventrolateral part of the neural tube: in between are afferents and efferents to the viscera. This dorso-ventral arrangement can be traced right up to the highest levels of the brain, but is particularly evident in the adult human spinal cord. In **Figure 1.5d**, one can see in cross-section (at the level of the second thoracic vertebra) the *ventral and dorsal horns,* consisting of masses of *grey matter* (mostly cell bodies), a less prominent central region concerned with the neural control of visceral function, and a surrounding sheath consisting of *white matter,* mainly bundles of nerve fibres running longitudinally up and down the cord. If one visualises a highly schematised hypothetical creature like the one in **Figure 1.6**, the topological reasons for these structural relationships become obvious.

At the head end, the process of cephalisation mentioned earlier causes the basic plan of the neural tube to be modified. The central fluid-filled canal, which is very small in the spinal cord, widens out at two separate points to form hollow chambers or *ventricles*; at the same time it migrates back to the dorsal surface of the neural tube, so that the ventricles are open on the dorsal side: as a result, the motor areas are most lateral, the sensory areas most medial, and visceral areas in-between. The surface of the ventricle is covered by the *choroid* membrane, the site of production of the cerebrospinal fluid that fills the canals

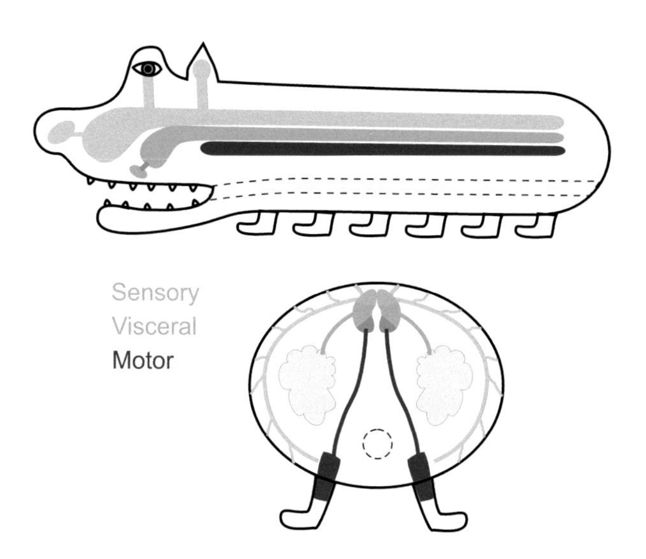

Figure 1.6 A highly schematised, hypothetical creature showing the relationships between sensory and motor regions of the spinal cord.

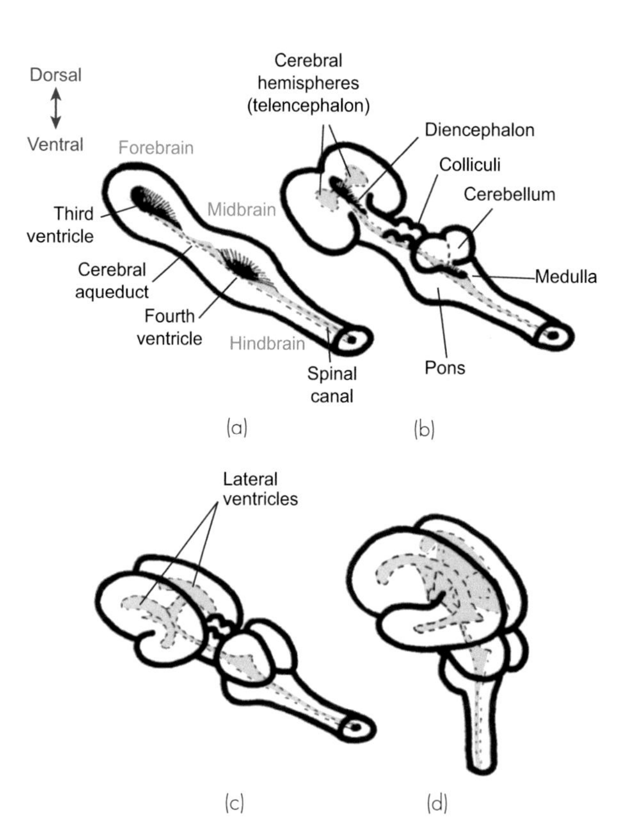

Figure 1.7 The notional steps leading from neural tube to human brain. (a) The opening of the canal to and from the third and fourth ventricles. (b) and (c) The growth of the cerebellum and of the cerebral hemispheres with their associated lateral ventricles. (d) Human brain, showing greatly enlarged cerebral hemispheres and flexion of the neural tube.

and ventricles of the brain. The more caudal of the ventricles is called the fourth ventricle, and the region around it is the *hindbrain* or *rhombencephalon*; it is connected to the more rostral third ventricle by the *cerebral aqueduct*. The region round the aqueduct is called the *midbrain* or mesencephalon, and that round the third ventricle is the *forebrain or prosencephalon* (**Fig. 1.7a**). The hindbrain is likewise divided into two regions: the caudal part is called the *medulla*, and the rostral part of it (*metencephalon*) is marked by the outgrowth of the *cerebellum* over the dorsal surface, and a massive bundle of fibres associated with it, the *pons*, on the ventral surface. Subsequently, the third ventricle produces a pair of swellings at the front end, which become inflated into the *lateral ventricles:* the neural tissue surrounding them forms the *cerebral hemispheres* (*telencephalon*) while the rest of the forebrain is called the *diencephalon* (**Fig. 1.7b**).

Table 1.1 Some important structures within the three divisions of the primitive brain

Hindbrain	Myelencephalon	Medulla
		Vestibular nuclei
		Medullary reticular formation
	Metencephalon	Pontine nuclei and reticular formation
		Cerebellum
Midbrain	Mesencephalon	Tectum (colliculi)
		Red nucleus
		Substantia nigra
		Mesencephalic reticular formation
Forebrain	Diencephalon	Thalamus
		Hypothalamus
		Septum
	Telencephalon	Corpus striatum
		Hippocampus
		Amygdala
		Cerebral cortex

In less developed species such as fish all these structures are easily recognisable without dissection, but in humans not only has the extraordinary ballooning growth of the cerebral hemispheres engulfed the other surface features (leaving only the cerebellum and medulla peeping out at the back), but the massive fibre tracts needed to connect the cerebral hemispheres to each other and to the rest of the brain have tended to elbow older structures out of the way and have often considerably distorted their shape. Another factor that makes the anatomy of the human brain somewhat confusing is that the neural tube has become bent forward: while the axis of the hindbrain is near vertical, that of the forebrain is horizontal (**Fig. 1.7d**).

An overview of the brain's structure

The most important areas of the human brain are shown in the sagittal and coronal sections of **Figure 1.8**.[4]

The *cerebellum* is an important co-ordinating system for posture and for motor movements in general (see Chapter 12), that arose originally as an adjunct to the vestibular apparatus, a sensory organ concerned with balance and the detection of movement (see Chapter 5). Important landmarks on the dorsal surface of the midbrain are the four bumps (corpora quadrigemina) formed by the *superior* and *inferior colliculi* ('colliculus' = 'little hill'), primitive sensory integrating areas for vision and hearing respectively; in higher vertebrates their function is mainly that of organising orienting reactions and other semi-reflex responses to visual and auditory stimuli. In the diencephalon, on each side of the third ventricle, lie the two halves of the *thalamus*, a dense group of nuclei whose neurons partly act as relays for fibres that project upward to the cerebral hemispheres. Close to them but more lateral is the *corpus striatum*, an old area that is concerned with the control of movements (see Chapter 12). Also in the diencephalon, but lying on the floor of the third ventricle, is the *hypothalamus*, the brain's interface with the hormonal and autonomic systems that control the body's internal homeostasis (see Chapter 14). More laterally, various nuclei, fibre tracts and other areas form a loosely defined system called the *limbic system*, which connects with the hypothalamus, with the olfactory areas, and with many other regions of the brain; they are concerned with such functions as emotion, motivation and certain kinds of memory (Chapter 14). Finally, there is the *cerebral cortex* ('cortex' = 'rind'), which covers the lateral ventricles and is deeply convoluted and furrowed in higher vertebrates, enabling a large superficial area of tissue to be crammed into a relatively small volume. Its role in carrying out some of the most complex things we do is discussed in Chapter 13.

The divisions of the brain described so far are very gross ones, and for the most part quite obvious to the naked eye. Finer anatomical distinctions can only be made by looking at the neurons themselves, and the way their populations vary from one region to another.

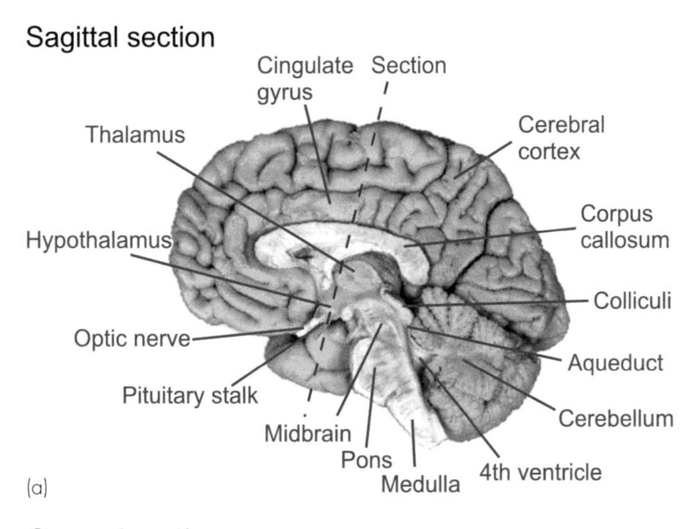

Figure 1.8 Sections of the human brain: (a) sagittal, showing medial aspect; (b) coronal, In the plane of the dashed line in (a).

Central neurons

Neuronal morphology

Neurons from different parts of the nervous system show a wide range of shapes and sizes (**Fig. 1.9**). What they have in common is a compact cell body containing the nucleus, and a number of projecting filaments that generally show extensive branching: these *dendrites* form the input pathways by which information from different sources is

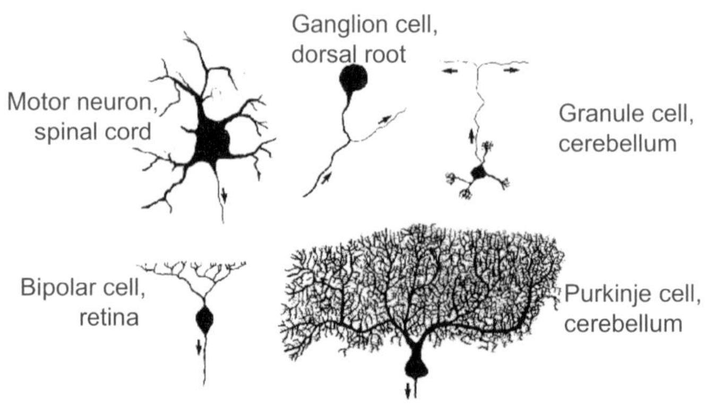

Figure 1.9 Some typical and somewhat idealised central neurones: motor neuron, spinal cord; ganglion cell, dorsal root; granule cell, cerebellum; bipolar cell, retina; Purkinje cell, cerebellum. The arrows indicate the axon.

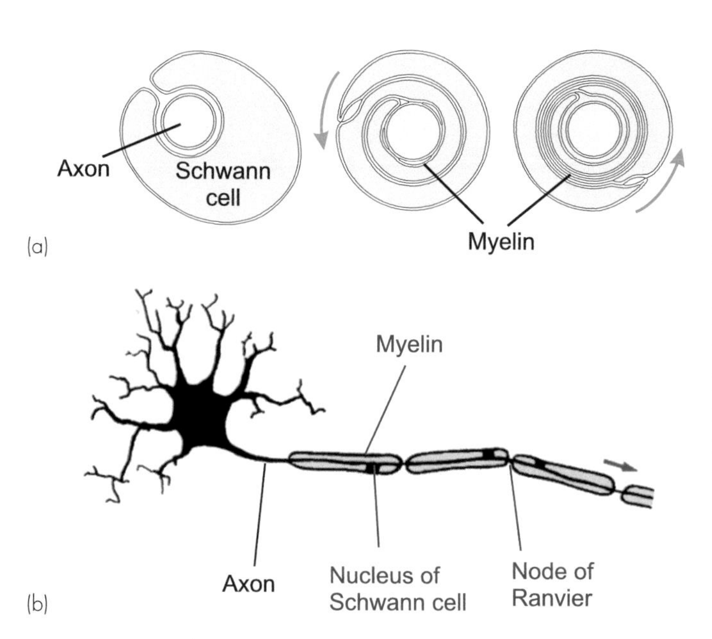

Figure 1.10 (a) Schematic cross section of myelinated axon, showing the layers of myelin formed by the Schwann cell wrapping itself round and round. This is performed by oligodendrocytes in the central nervous system. (b) A neuron with myelinated axon (not to scale: the distance between the nodes of Ranvier is typically of the order of 1 mm, and the cell body might be 20–80 µm in diameter).

gathered together by the neuron, and then transmitted in turn to some other region by the axon. However, there are many exceptions to this arrangement: cells in the dorsal root ganglion, for instance, have no dendrites at all and the cell body simply lies to one side of a single continuous axon, which travels from sensory receptors in the periphery to destinations in the spinal cord; and in many sites within the brain the dendrites are known to act as outputs as well as inputs.[5]

The axon is specialised for carrying information rapidly over long distances, and may often be very long indeed – as, for example, those from motor neurons, that carry muscle commands all the way from spinal cord to the end of a limb. Apart from very short ones, their information is carried in the form of propagated *action potentials*, discussed in Chapter 2. The larger axons are frequently swathed in layers of *myelin*, a lipid substance that improves the conduction of action potentials by acting as an electrical insulator. These layers are the result of accessory cells – *oligodendrocytes* in the brain and spinal cord and *Schwann cells* in the peripheral nervous system – sending out myelin-rich processes that wrap themselves round and round the axons to form a kind of Swiss roll (**Fig. 1.10a**). Oligodendrocytes are a specialised form of the *glial cells* that make up the bulk of the remaining nervous tissue, whose functions include regulation of the brain's ionic environment and possibly more complex functions as well: they are discussed further in Chapter 3. The myelin is interrupted at regular intervals, at the *nodes of Ranvier*. The picture in **Figure 1.10b** is not to scale – the distance between the nodes of Ranvier is typically of the order of 1 mm, while the cell body might be 20–80 µm in diameter.

Within neurons are found the usual intracellular organelles and other components, including 10 nm *neurofilaments* extending linearly along axons and dendrites as well as within the cell body, and the larger microtubules (*neurotubules*) that seem to be associated with the transport of organelles such as vesicles and mitochondria to and from nerve terminals, at over 10 mm per hour, and also proteins, more slowly. The far end of the axon is usually branched, and its terminals make synaptic contact either with the dendrites or the cell bodies of other neurons, with secretory cells, or (in the case of *motor neurons*) with muscle cells. At this point the signal – previously electrical – makes the terminal release a tiny quantity of chemical transmitter, which then acts on the postsynaptic cell.

Neural networks and complexity

Conceptually, therefore, a neuron is quite simple: responding to particular patterns of afferent input by communicating information – in the form of changes in potential – along its length to alter the release of neurotransmitters. Brains, however, are not. On the one hand, we have all the unspeakable wonders of our minds, of which we are so inordinately proud; on the other hand, when we open up the skull and peep inside all we see is a porridgy lump containing millions and millions of seemingly untidy neurons. The fundamental problem of neuroscience is that of linking these two scales together: can we trace the relationship between molecular and cellular mechanisms all the way to what was going on in Michelangelo's head as he painted the Sistine Chapel?

Very nearly, and the trick is to force yourself to think of the brain as a *machine* that carries out a well-defined job. That job is to turn patterns of stimulation, *S*, into responses, *R*: the sight of dinner into attack and jaw-opening; a page of music into finger movements. How it does this is clear, in principle at least. The brain is a sequence of neuronal levels, successive layers of nerve cells that project on to one another (**Fig. 1.11**). At each level, a pattern of activity in one layer gets transformed into a different pattern in the next. Thus an incoming sensory pattern S is transmitted from level to level, modified at each stage until it becomes an entirely different pattern of response R at the output. The network is able to generate specific patterns of response to particular patterns of input because, on a smaller scale, each of the neurons that makes up a particular level is *itself* a sort of miniature computer and responds only to a particular pattern of activity amongst the neurons of the level immediately in front of it (**Fig. 1.11b**). By joining together billions of units that are each *quite* intelligent, we end up with something that is *astonishingly* intelligent. Similar 'neural' networks embodied in computers can carry out such apparently high-level tasks as responding to individual voices or faces, or suggesting the names of books you might want to read.[6]

Each neuron is influenced by an extensive set of neurons in the preceding layer, yet this set forms only a fraction of that layer. But because this convergence is repeated

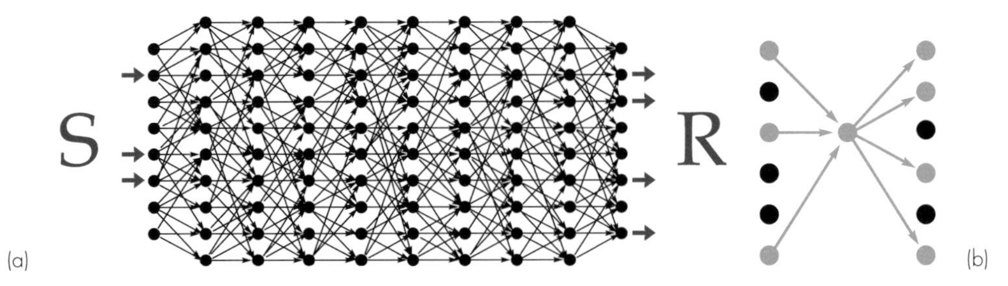

Figure 1.11 (a) Representation of the nervous system as a series of neuronal levels, through which sensory patterns (S) are transformed into patterns of response (R). It works because each neuron itself responds to particular patterns amongst the neurons of the preceding layer (b).

at every layer, over and over again, neurons in deeper layers can respond to more and more complex and wide-ranging aspects of the sensory world. This is particularly obvious in the case of the visual system. Receptors in the eye convey information about only a miniscule part of the retinal image, in effect a single pixel; but after a few levels have been passed, in the visual cortex, we find units that are able to respond to a specific type of stimulus, such as a moving edge, over wide areas of the visual field. We shall see later that as one gets to the deepest levels of the brain, neurons increasingly tend to be multimodal, responding not just to one kind of sense such as vision, but to others, such as sounds and to stimulation of the skin, as well. And of course if we consider the final level of all, the muscles that move our bodies, it is clear that they are *utterly* multimodal, since we can obviously learn to activate any particular muscle in response to any pattern of sensation, involving any or all of our sensory modalities.

Thanks to the work of biophysicists who have studied the way in which different synaptic inputs to a neuron interact with one another, we have a fair idea of the general rules that determine whether or not a cell will respond to a given pattern of excitation. In particular, we know that these rules depend very critically on the shape of the dendritic tree and the distribution of synapses upon it. For instance, two synapses interact in a very different way if they lie near one another on a single branch, rather than far apart on separate branches (discussed in Chapter 3). It follows from this that a full understanding of the function of a single neuron requires detailed knowledge of the size and shape of its dendrites, and of the origin of all the thousands of afferent fibres synapsing with it. So if we want to understand the behaviour of the whole thing in detail, we have a massive problem – the immense size and complexity of the whole thing. The network in the figure above looks pretty complex, but it only has 96 neurons; in the human brain there are some 10^{12} neurons, each smothered in synapses: so, more important, the *connections* that actually determine the brain's behaviour must be something on the order of 10^{15}.

It is extraordinarily difficult to grasp just how complex this structure is. Seventy years ago, when the first computers – 'electronic brains' – began to make their appearance, people were awed by their complexity; now of course they are orders of magnitude more complex, but still not nearly complicated enough to serve as an adequate metaphor for the brain (**Fig. 1.12**). A hundred years ago people were equally thrilled by what seemed to them the astonishing complexity of the telephone exchanges then being developed – the massive incoming cables like nerve trunks, the plugs and sockets used for switching like little synapses. What seemed then an exciting piece of technology was often used as a model for the brain. We'll see later that the idea of the plug-board, forming associations between elements, and programming the system in exactly the same way as the earliest computers were programmed, is indeed quite a powerful metaphor for what the cortex does. Today, the world's entire Internet is the most complex man-made device that exists, with over half a million miles of interconnected cabling, but it is only as complex as a *single* human brain – and nowhere near as intelligent.

So it really is a completely hopeless task to try and start with the biophysical properties of single neurons and hence somehow work out from that how 10^{12} such neurons connected together will behave: the problem turns out to be utterly uncomputable before we reach even a dozen of them. It is not so much that we don't know enough about the behaviour of each individual element, but that we don't understand how they act as a whole. The brain is composed of elements that are individually quite simple, but they are *joined together* in extremely complex ways. In that respect at least it does have certain

Figure 1.12 ENIAC (Electronic Numerical Integrator and Computer), the first electronic general-purpose computer. (Source: http://i.imgur.com/vccoLnX.jpg)

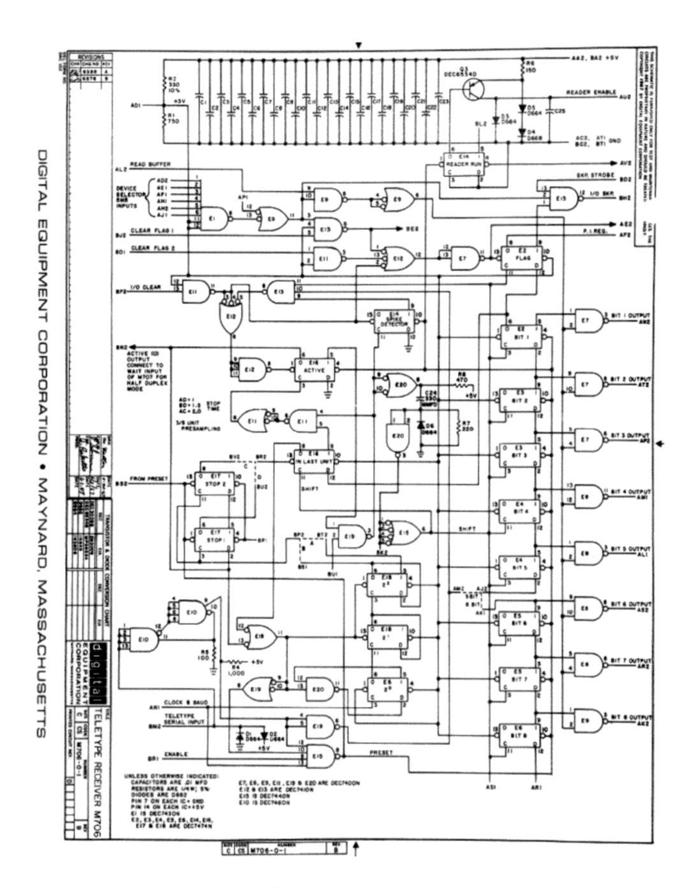

Figure 1.13 Logic diagram of a very minor part (the keyboard receiver) of a very small computer. (Source: Courtesy of Digital Equipment Corporation.)

similarities with a computer. The circuit diagram in **Figure 1.13** is just a tiny part of a very small and simple computer, of the circuits that perform the relatively trivial task of getting data from the keyboard. Without having to know very much about logic circuits, you can see that there are a lot of elementary units in the circuit (the logic gates), and that most of them are identical: it is clear that what the circuit does as a whole must be a matter of how these gates are connected to one another. A full understanding of the details of this circuit's operation requires a considerable intellectual effort; to master the computer of which it forms a part might take years of study; yet the brain is thousands of millions of times larger in scale.

In other words, even if our anatomical techniques were perfect, and we had the patience to identify each of the 10^{15} or so individual pathways in the central nervous system, ending up with something like a wiring diagram of the brain, would we really be much the wiser as a result? It is true that we could *in theory* then apply our knowledge of the biophysical properties of individual neurons to calculate how the whole thing would react to any given pattern of stimulation at the sensory receptors: the problem is of course the almost inconceivable difficulty of ever actually carrying out such a calculation. With modern computers, the accurate simulation of the behaviour of even a few dozen neurons connected together in a realistic network is about the limit of what can be achieved within a reasonable period of time. Thus it is not so much that we know too little of the behaviour of individual neurons, but rather that we lack the conceptual techniques for analysing their behaviour as a whole.

What, in fact, *would* it be like to know how the brain works? Would it, for instance, be like knowing how a computer works? Most of us are aware that a PC comprises peripherals (keyboard, screen, mouse etc.), short-term memory or RAM, a hard disk drive and some sort of central processor that reads instructions from the memory, shunts information around and does arithmetic and so on. Some of us may understand the language of these instructions and be able to predict how the machine will behave in response to a given program. Computer enthusiasts will know about how these programs relate to the actual hardware: the functions of the various microchips on the motherboard and the information highways that connect them to each other. This is what is meant by 'understanding the computer'. However, delving further – into the quantum physics that describes the action of the millions of gates that make up each microchip – is *not* something that helps to answer questions about how a computer works. In the same way, 'understanding the brain' can never be a solely molecular matter, a question of the properties of single channels, of configurational changes in single molecules, of biochemistry and biophysics; these are merely a diversion from the demanding task of trying to understand the brain at its own level.

Holism versus reductionism

Another example of a highly complex system made out of vast numbers of quite simple elements is provided by substances like foam rubber (**Fig. 1.14**). Here, too, the shape and behaviour of each individual element – each bubble – is unique, yet determined by relatively simple physical laws. But it would clearly be a hopeless task to attempt to work out the overall properties of a slab of such foam – its elasticity, for example – by painstakingly considering one bubble at a time and calculating its effects on each of its neighbours.[7] An alternative to a bottom-up or *reductionist* approach of this sort is to step back a little from the system, until the differences between each of its elements become blurred, and then try to derive some description of their average behaviour *en masse* without the necessity of considering each one in detail.

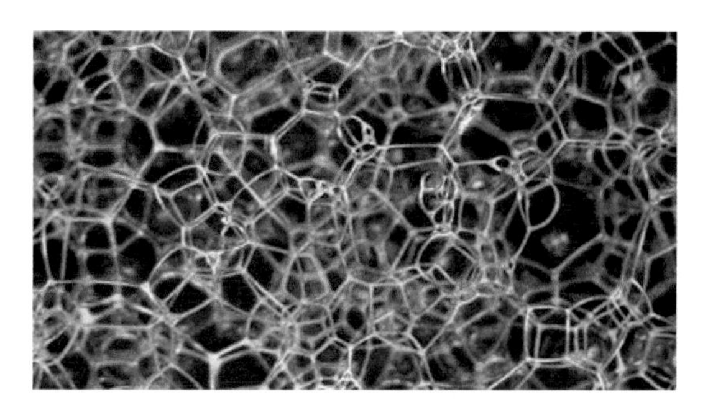

Figure 1.14 Microphotograph of foam rubber.

A well-known example of this kind of top-down or *holistic* approach is the study of the physical properties of gases. Gases exert a pressure because their molecules are constantly rushing about and bumping against the walls of their containers, generating an average force that depends on their velocity or temperature and on how many of them there are. If you knew the exact position and velocity of every single molecule of such a gas, then in principle you could compute each of their individual trajectories and hence predict their collisions with the walls and with each other: you would then have a perfect understanding of the gas. But this is clearly out of the question. A more tractable approach is to turn the large numbers involved into an actual advantage, by considering only the 'lumped' properties of the molecules, their average behaviour as a whole, and then derive statistical descriptions of the way in which pressure depends on volume and temperature, through the methods of *statistical mechanics* (**Fig. 1.15**).[8]

Might we hope that – by analogy – we might be able to develop a sort of statistical mechanics for neuronal populations? Unfortunately, the one thing we cannot do with neurons is to take averages: it is precisely the *differences* between them – the patterns of their activity – that are significant. These patterns arise because different cells have different sensitivities. Whereas uniform pressure will affect all the bubbles in the foam rubber more or less equally – when, for instance, we shine a uniform blue light on the retina – some of the cells will be more excited than others, quite apart from the spatial and temporal patterns being generated all the time by our interaction with the outside world. The last thing we want is to average all this out. It is as unhelpful to talk about the average behaviour of a sheet of cells in the retina or cerebral cortex as it would be to talk about an average page of this book, or an average Beethoven piano sonata. In fact we shall see again and again that at every level of the brain there are special mechanisms of *lateral inhibition* whose specific function is to exaggerate differences of activity between neighbouring neurons and thus to enhance or amplify any patterns that may exist, and such mechanisms make the whole system more uncomputable than ever.

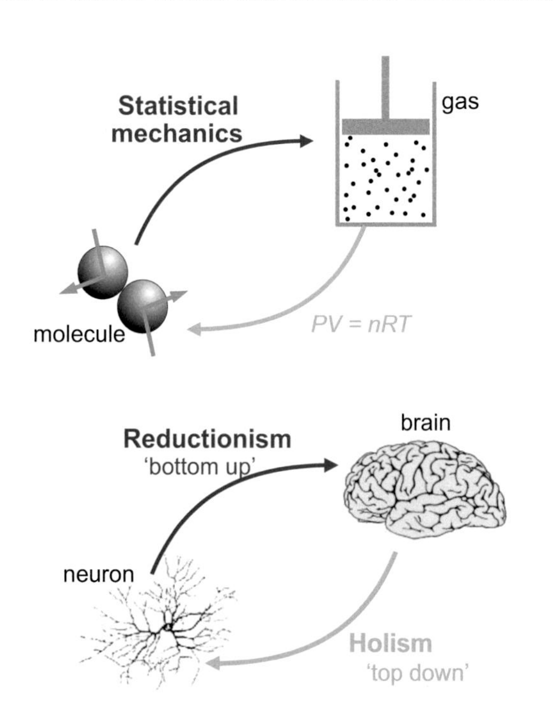

Figure 1.15 Some possible approaches to studying the brain: Statistical mechanics, holism and reductionism.

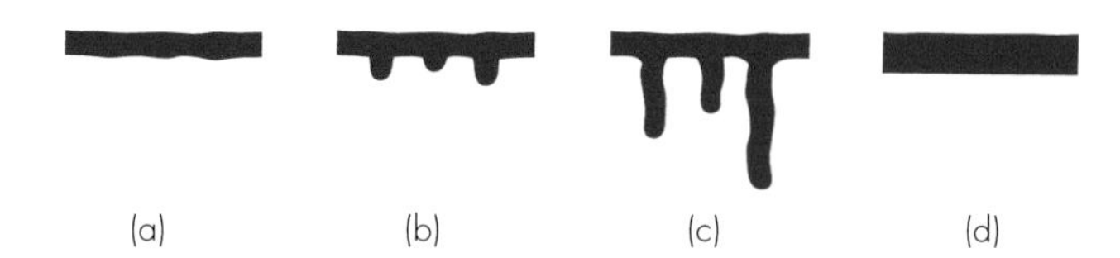

Figure 1.16 A system that amplifies patterns. We make a horizontal dribble of ink on a vertical card (a). Drops begin to form at points where there happens to be more ink (b), and in doing so they draw ink away from neighbouring regions, thus exaggerating the original non-uniformity (c). A prediction of the behaviour of the ink based only on averages (d) is the one result that is never observed!

A simple example may help to illustrate why this is. Suppose we take a piece of card, dribble a line of ink across it, and then upend it (**Fig. 1.16a**). The ink starts to run down, in an irregular way that is partly a function of the amount of ink at different points along the line (**Fig. 1.16b**). But as soon as a drop starts to form at the lower edge, it begins to draw ink off from neighbouring regions, thus reinforcing itself at their expense. We end up with a series of discrete trickles (**Fig. 1.16c**), in which the original pattern of 'activity' (the quantity of ink at different places) has been automatically amplified. A description in terms of average behaviour would be meaningless: if we repeated this a thousand times, the average descent of the ink would be perfectly uniform (**Fig. 1.16d**), the one thing that is *never* actually observed. Here, and in the brain, we are dealing with one of those fashionably chaotic systems in which the tiniest initial perturbations may generate incalculably huge effects.[9]

Another reason why we cannot take averages is that one person's neuronal connections are different from another's, both at the level of description of individual nerve cells, but also to a certain extent at grosser levels. One example concerns a part of the cerebral cortex called the motor area, where – as we shall see in Chapter 12 – the movements of various parts of the body are represented in a systematic pattern. These 'motor maps' differ quite markedly not only from one individual to another, but even in the same individual when measured on different occasions, partly as a function of use and disuse. This is very far from being an isolated example: we shall see later that connections in many parts of the brain are to a large extent determined by experience, and that it is precisely this *plasticity* of function that enables us to learn to adapt our behaviour to circumstances. So the functioning of the brain is probably as much a matter of how it has been programmed as of its hardware – of the general outline of its wiring specified by genetic instructions – and these programs are necessarily as varied as the experiences of the brain's owner. The same computer processor chip may be used for writing a book or for accessing Facebook: no amount of peering at it down a microscope will tell us *which,* for it is a matter of how it is programmed. So even if we did somehow manage to make some sense of the wiring diagram of A's brain, it is not at all obvious that it would throw much light on the functioning of B's.

Limitations of experimental approaches

So there are many levels at which one may try to investigate the brain, and corresponding to them a number of distinct branches of the neurosciences have emerged. To pursue the computer analogy a little further, biophysicists investigate the properties of the individual logic gates; neuroanatomists trace the connections that link one unit to another; psychologists describe the programs in the machine and how they got there; and pharmacologists study the colours of the wires. The task of *neurophysiologists* has generally been to try to bridge the levels at which the other disciplines perform their investigations,[10] by trying to correlate anatomical structure with patterns of neuronal activity and neural events with overt behaviour and sensation. In practice, one cannot hope to do much more than to try to identify neurons that are typical of a particular area, and form some idea of their general properties – for instance, that they are all involved in the control of eye movements. We are helped in this by the fact that neurons are to a large extent grouped in homogeneous communities called *nuclei*, groups of similar cells projecting to the same area of the nervous system whose afferent fibres likewise have common origins. The existence of nuclei implies that of the *tracts* that join them, and much of the work of neuroanatomists is to identify the inputs and outputs associated with particular nuclei. Some of the techniques they use are to trace pathways from one area of the brain to another and to attempt to correlate them with particular functions. These techniques fall into four basic classes: the purely anatomical tracing of pathways, and the three physiological techniques of recording, stimulating and making lesions. These techniques are discussed in more detail in the Appendix, but it is helpful at this stage to think a little about how their intrinsic limitations constrain our understanding. We have already seen that pure anatomy in itself can tell us little, but it turns out that the other approaches suffer from conceptual limitations of their own.

Recording

Neuronal recording, for instance, has to be related to something the animal experiences or does. Consequently it is of most use at the input, sensory end of the brain, where we have the closest control over the incoming stimuli (**Fig. 1.17**). All we need do is record from single neurons at the level we are interested in, perhaps the visual cortex, whilst applying a variety of visual patterns. We will then discover which kinds of patterns the neurons do or do not respond to, and can then hope to deduce the nature of their functional connections. This task is made easier for us because we have a pretty good idea in advance of what *sort* of patterns to try, namely those that actually occur commonly in real life. Thus in the visual system it is not unreasonable to start with things like straight lines and edges: a search for neurons in a cat's cortex that specifically responded to letters of the Greek alphabet might take quite a long time. But as we go further in to the brain, neurons get fussier and fussier about what they respond to; so in an experiment of finite length, we may never discover what a neuron is interested in. A consequence of this is that large areas of the cortex seemed to early investigators to be 'silent', apparently not responding to anything much at all. It is now clear that there are actually plenty of responses from these areas, but often to such specialised and peculiar stimuli or situations that one may not happen to hit on them. Because experimenters are guided by what the animal normally receives as stimuli, this in itself introduces a degree of bias: if you only test for what you *expect*, then that is what you *find*.

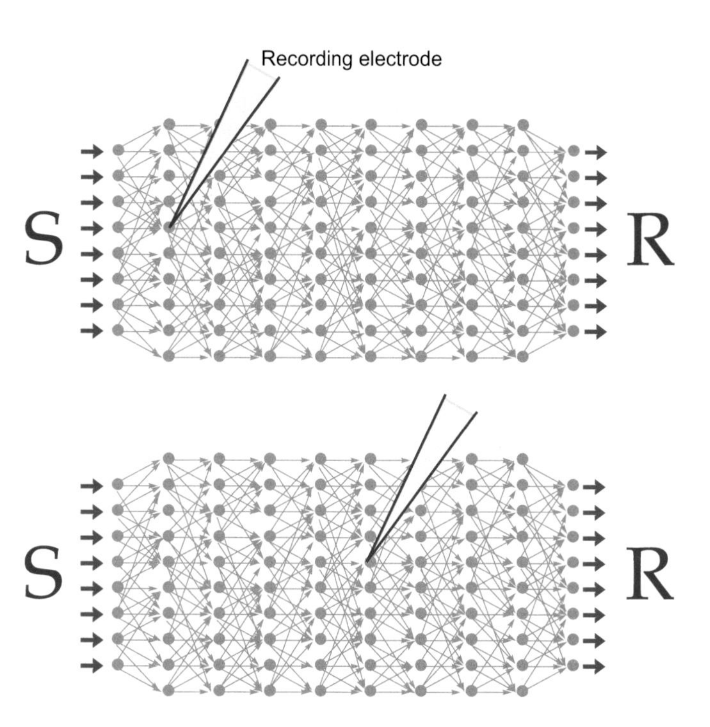

Figure 1.17 Recording from different parts of the brain: the further 'into' the layers of the brain recordings are taken (bottom, akin to recording from the higher centres), the more likely it is that nothing will be recorded at all.

Stimulation

As we move further towards *R*, in the motor system, the problem is turned inside out. We now have to *stimulate* the neurons individually and see what patterns of movement are generated as a result. But there is a snag. To get any response at all, we have to provide a pattern of stimulation that will be recognised at the next level down the chain and result in activation of the muscles. Quite apart from the extreme technical difficulty of actually generating any such pattern in the first place, there is a further difficulty. What patterns should we apply? Which of them are 'physiological' in the sense that they commonly occur in real life, and therefore likely to arouse a response in the next layer? Unlike the situation in sensory systems, we do not have the patterns that we see in the outside world as a guide. Consequently, when trying to stimulate the motor system, particularly in its more central regions, we find that stimulating single units seldom produces any effects at all, and that when movements are achieved, they are usually the result of using such large currents spread over such large areas that the specificity of the neurons is swamped, resulting in responses that are diffuse and unphysiological.

Lesions

Lesions pose the greatest problems of interpretation of all. Suppose we notice that destruction of a specific area *X* of the brain results in the loss of some function. One might think it reasonable to conclude that the function is localised in *X*. The trouble is, however, that there are many other ways in which such a result might be explained. The function might in fact be localised somewhere else altogether, but either with connecting fibres that merely happen to run through the region *X* (**Fig. 1.18a**), or perhaps requiring some kind of tonic permissive influence from *X* (**Fig. 1.18b**); or a lesion at *X* may simply interfere with the blood supply to the true area of localisation of the function

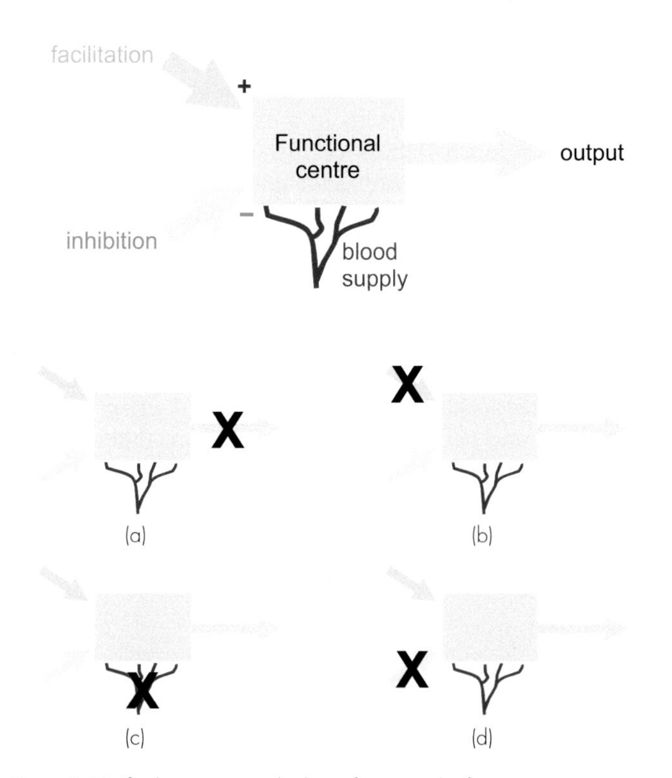

Figure 1.18 If a lesion causes the loss of a particular function, it may indeed be because it has directly interfered with the area responsible for that function; equally, it may be that it (a) has merely interrupted fibres of passage, (b) has abolished a tonic 'permissive' input or (c) has interfered with blood supply. A lesion that abolishes a source of tonic inhibition may give rise to (d) 'release', the appearance of new and abnormal reactions.

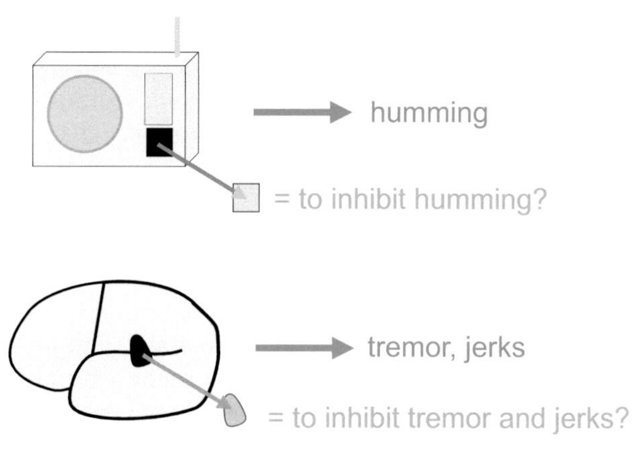

Figure 1.19 If removing a particular bit of a complex device creates complex behaviour, does that mean the function of that bit was to inhibit the behaviour?

(**Fig. 1.18c**). Paradoxically, one is on safer ground if one finds that a lesion in area *X* has absolutely *no* effect whatever on the function, for then one can be fairly sure (unless the function is localised at two independent sites that somehow work in parallel) that it is not localised in *X*.[11]

Sometimes the result of a lesion is not the abolition of a function, but the sudden appearance of new behaviour not previously seen, a phenomenon called *release* (**Fig. 1.18d**). Again, it is easy to jump to erroneous conclusions. If you remove one of the circuit boards from your radio and it suddenly starts making a humming noise, is it fair to conclude that the function of the board you removed was to inhibit humming (**Fig. 1.19**)? Yet it is still commonly said that because the effect of lesions in certain parts of the corpus striatum is a jerkiness in moving, and tremor at rest, the function of those areas is to smooth out movements and inhibit tremor!

The differences between peripheral and central parts of the brain, which we noted in the case of recording and stimulation, are even more problematic for lesions. Near the periphery there is usually a simple relationship between the lesion and what goes wrong – retinal damage causes local blind spots, and spinal injury causes local paralysis. But interpretation becomes much more difficult with lesions in higher areas of brain: the higher you go, nearer the centre of **Figure 1.20**, the *vaguer* the functional effects become, the bigger the occurrence of *release* and of long-term *recovery* as alternative routes through neighbouring regions start to be used instead. Part of the problem is undoubtedly that the whole notion of localisation in centres is too crude and simple-minded, a hangover from the days of phrenology, with its picture of the brain divided up into a number of discrete little compartments with highly specialised functions (**Fig. 1.21**). The relative ease with which beautifully coloured pictures of localised cortical activity can be obtained with modern brain scanning techniques has unfortunately led to a revival of this simple but deeply unsatisfying approach.[12] In the early days of surgical neurology, clinicians prided themselves on being

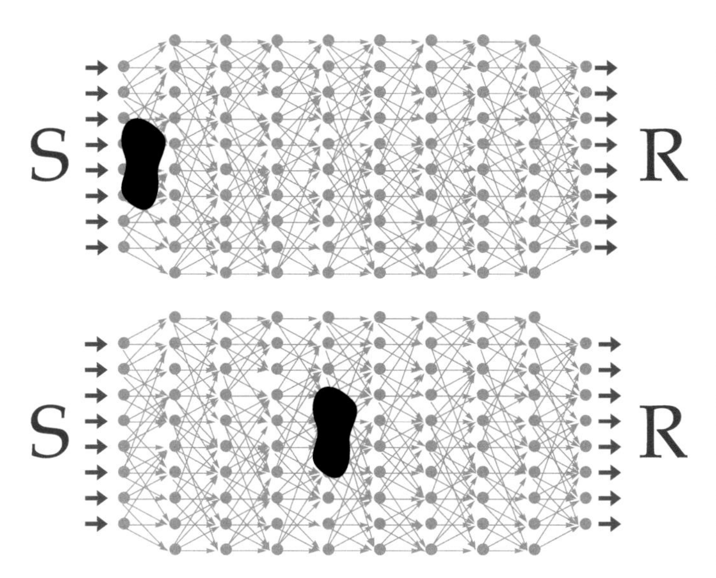

Figure 1.20 Lesion studies: akin to recording studies (**Fig. 1.17**), the further 'into' the layers of the brain lesions are created, the more unclear the effects are to the experimental observer.

able to predict which part of the brain had gone wrong by making intelligent observations of behaviour; nowadays it is much easier – though less interesting – to obtain the same information by means of a scan. The legacy of this approach has been an obsession with localisation of brain function. But *where* is not in the end a very interesting question, and a great deal less demanding than *how*. It also encourages a naïvely thoughtless approach to function that other branches of biology have long grown out of: to look for a centre for happiness in the brain is precisely as silly as to search for the happiness gene.

Some classical experiments once performed on rats by the psychologist Karl Lashley cast doubt on such a view of localisation. He made lesions of different sizes in the cerebral cortex, avoiding the regions specialised as primary areas of input and output, and found that the defect of performance in a task like learning to run a maze depended more or less on the quantity of cortex removed and surprisingly little on the region where it was taken from. Although when we record at one particular time from a neuron in a particular region of the cortex it may seem to be doing something very specific, it is likely that large areas of the cortex are not in fact rigidly committed to specialised tasks, but may change their function as the result of experience or of damage to neighbouring areas. For instance, the map of the human body found in the sensory cortex readjusts itself in a matter of days after amputation of a limb. The situation is reminiscent of that in the old electromechanical telephone exchanges, where automatic switches called selectors had the job of connecting one subscriber to another in response to signals from the telephone dial.[13] But there was not one set of selectors for each subscriber: if so, the total number of selectors required would have become astronomical. Instead, they were lumped together in a common pool: on lifting your receiver, a device called a line-finder hunted through the available bank of equipment until it found one free, and then gave you the dialling tone. Being shared, any one selector could thus be used by a large number of subscribers. As a result, if we were to drop a bomb into the exchange, blowing up a certain proportion of the selectors, then service would get worse for *all* subscribers, in that they would be more likely to find all the lines engaged. This kind of *plasticity* – the ability of one area to take over the function of another that is out of action – is often found at higher levels of the brain (as, for example, in recovery from stroke) and makes deductions from lesions more difficult still.

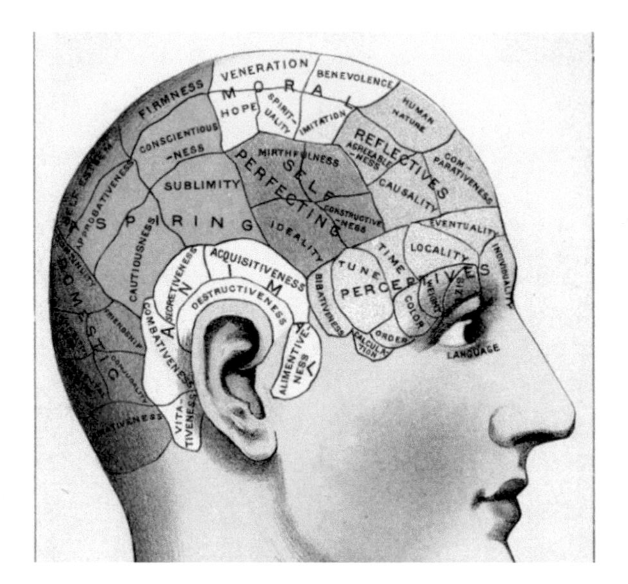

Figure 1.21 Phrenological head: an early attempt at cerebral localisation. (From The Household Physician, 1921; Source: https://www.flickr.com/photos/crackdog/14866775264)

Thinking about brain function

Thus there are profound intellectual and technical difficulties with all the methods of investigation currently in use, and the history of brain science has not been a simple linear progression. It is as if we tried to find out how our mobile phone works by sticking pins in it and measuring the voltages on them by passing large currents through it or by drilling holes into it at different points to see what happens. It is astonishing, in fact, that we know anything at all about the brain.[14]

Nevertheless, do not despair. Although in trying to understand the brain one may sometimes feel like a lone explorer hacking his way through impenetrable jungle, with no sense either of how far he has come or what – if any – is the destination, there are one or two tricks that can be immensely helpful both in making your machete strokes as effective as possible and in getting a sense of where you are and where you are heading.

Think big. Knowledge of most subjects is tree-shaped: a central trunk of fundamental concepts divides to form first large branches, then smaller ones and finally twigs and individual leaves. Students tend to approach too close to the tree, become obsessed by the leaves, and set about laboriously memorising them one by one. At best the result is patchy, and the work is made even more laborious by not realising how one factoid relates to another. Standing back to see the shape of the tree as a whole is not just intellectually more enlightening, it actually makes the memorisation easier and quicker.[15] If you really understand a system, you don't memorise, you calculate. Think of arithmetic: once you understand the rules of multiplication you are spared the necessity of *memorising* the fact that 315 × 457 = 143 955.

Think backwards. At first sight it may seem logical to start by considering the inputs to the system and what they connect to, and then what *they* connect to, and so on. The problem is that – because of divergence – very soon the number of things affected by the input gets out of control. If you start at the output, you can ask a simpler question: what caused this to happen? And what caused *that* to happen? Mazes are often easier to solve by starting in the middle and working your way out.

Think functionally, not structurally. Students often use structure as a crutch: it seems tangible and memorisable. But structure follows function and is meaningless without it.[16] As you approach a new system in the brain, ask yourself first what this system is *for*, in the simplest possible terms: establish the trunk of the tree of knowledge. Play God: ask yourself how *you* would design a system to carry out that function – what the difficulties are likely to be, what the physical constraints are. You can then turn to the real system with a much better idea of what it is all about.

Think quantitatively: Lord Kelvin, the great nineteenth-century physicist, once wrote: '*When you can measure what you are speaking about, and express it in numbers, you know something about it*'.[17] Whenever possible, think about sizes and numbers. How many fibres are there in my optic nerve? Could this limit what I see? How long are my reaction times? What could be filling up that time-gap between stimulus and response?

Think . . .

Notes

1. Brain mirroring the world Santiago Ramón y Cajal (1852–1934), whose contributions to neuroanatomy won him the Nobel Prize in 1906, wrote in *Charlas de Café: 'As long as our brain is a mystery, the universe, the reflection of the structure of the brain, will also be a mystery'.*

2. The brains of single-celled organisms A hugely stimulating recent discussion of this is in Bray, D. (2009). *Wetware: A Computer in Every Living Cell.* (New Haven, CT: Yale University Press).

3. Evolution of the brain Some aspects of this are discussed speculatively in Eccles, J. C. (1989). *Evolution of the Brain: Creation of the Self.* (London: Routledge).

4. Neuroanatomy There is an impressive range of excellent textbooks of neuroanatomy to choose from. A very full text with much reference to medical material is Brodal, A. (1981). *Neurological Anatomy in Relation to Clinical Medicine.* (Oxford: Oxford University Press), or his more recent (2004). *The Central Nervous System: Structure and Function.* (Oxford: Oxford University Press); Nauta, W. J. H. & Feirtag, M. (1986). *Fundamental Neuroanatomy.* (New York, NY: WH Freeman) is slimmer but has fine illustrations and a synoptic approach that helps conceptual understanding. Another thoughtful account is Jones, E. G. (1983). *The Structural Basis of Neurobiology.* (New York, NY: Elsevier). For absolutely stunning pictures in full colour, England, M. A. & Wakely, J. (1991). *A Colour Atlas of the Brain and Spinal Cord.* (London: Wolfe) is highly recommended. If you are lucky enough to find a copy of Chandler, E. H. (1963). *Textbook of Neuroanatomy.* (London: Pitman), turn to and admire the extraordinary clear diagrams of the development of the ground-plan of the brain, which have not been bettered in more recent books. Brown, M., Keynes, R. & Lumsden, A. (2001). *The Developing Brain.* (Oxford: Oxford University Press) is a more recent book that also illustrates development with particular clarity. A resolutely clinical approach is adopted in Haines, D. (2011). *Neuroanatomy.* (London: Lippincott William & Wilkins), with unusually large and clear diagrams and excellent illustrations of scans and histology; it is particularly good at presenting various ways of looking at brain structure side-by-side.

5. The neuron Two very clear accounts of the cell biology of the neuron, with good illustrations, are Levitan, I. B. & Kaczmarek, L. K. (2001). *The Neuron.* (Oxford: Oxford University Press), and Fain, G. (1999). *Molecular and Cellular Biology of the Neuron.* (Cambridge, MA: Harvard University Press).

6. Neural programming The field of neural networks is now a hugely popular one, with many books aimed at different levels of technical expertise. An approachable introduction to man-made neural networks is Gurney, K. (2009). *An Introduction to Neural Networks.* (London: University College). Something for the more mathematically inclined is Hassoun, M. H. (1993). *Associative Neural Memories: Theory and Implementation.* (Oxford: Oxford University Press). Another book designed specifically to link the computational approach to the way the brain actually works is Churchland, P. S. & Sejnowski, T. J. (1994). *The Computational Brain.* (Cambridge, MA: MIT Press). Edelman, G. M. (1987). *Neural Darwinism: The Theory of Neuronal Group Selection.* (Oxford: Oxford University Press) is an account of a rather specific proposal for how the general rules of neuronal programming might be implemented. Grillner S., Selverston, A. I., Stein, P. S. & Stuart, D. G., eds. (1997) *Neurons, Networks and Motor Behaviour.* (Cambridge, MA: MIT Press). is an attractive, biologically oriented account, and Cotterill, R. M. (1998). *Enchanted Looms: Conscious Networks in Brains and Computers.* (Cambridge: Cambridge University Press) has an outstandingly clear account of neural networks, among other things: a stimulating and wide-ranging book.

   Finally, an example plucked almost at random of how 'neural' networks in man-made computers may be set to work to carry out tasks of extreme complexity: Svärdström, A. (1993). Neural network feature vectors for sonar targets classification. *Journal of the Acoustic Society of America* **93** 2656–2665. Or a book that might make you a millionaire: Dhamija, A. K. (2009). *Forecasting Exchange Rate: Use of Neural Networks in Quantitative Finance.* (Saarbrücken: VDM Verlag).

7. Foam The photograph is courtesy of Simon Thomas, a sculptor whose work is informed by systems such as foams that are deeply complex yet the result of simple laws: see www.simonthomas-sculpture.com.

8. Reductionism and Holism A recurring theme in Hofstadter, D. R. (2000). *Gödel, Escher, Bach: An Eternal Golden Braid.* (New York, NY: Harvester), a brilliant tour de force and one of the great popular scientific works of the last few decades. (This is the twentieth anniversary edition.)

9. Pattern generation and amplification A readable, clear and beautiful account of the formation of natural patterns – the generation of complexity through simple interactions – is Stevens, P. S. (1976). *Patterns in Nature.* (Harmondsworth: Peregrine).

10. Neurophysiology No book better encapsulates the particularly physiological approach to brain function than Walsh, E. G. (1964). *Physiology of the Nervous System.* (London: Longmans) – sadly, long out of print. If you find a copy, snap it up. At one time, 'Neuroscience' was a trendy and exciting word; but as Frank Harary once remarked, subjects that have 'science' as part of their title are usually not very scientific (quoted in Weinberg, G. M. (2001). *An Introduction to General Systems Thinking.* (New York, NY: Dorset House Publishing)).

11. Interpretation of lesions There is an apocryphal story of the man who demonstrated his findings about the auditory system of the grasshopper at a meeting of the Royal Society. *'Gentlemen,'* he said, *'I have here a grasshopper that I have trained to jump whenever I clap my hands.'* He claps, and the grasshopper jumps. *'Now, gentlemen, I shall cut off its legs – so – and you will now observe that when I clap my hands it no longer jumps. It is therefore clear that grasshoppers hear with their legs.'* Perhaps the logical flaw is obvious here – the fact that grasshoppers *do* hear with their legs is beside the point! Yet hardly less glaringly flawed deductions have quite often found their way into the literature.

12. Brain scans – dynamic phrenology? See, for instance, Raichle, M. E. (1994). Visualizing the mind. *Scientific American* **April** 36–42. Also Legrenzi, P. & Umilta, C. (2011). *Neuromania: on the Limits of Brain Science.* (Oxford: Oxford University Press). In that thoughtful but challenging work, *Man on His Nature* (1940). (Cambridge: Cambridge University Press), the Nobel Laureate Sir Charles Sherrington wrote: *'Facts rebut the over-simplified conceptions such as to ascribe to separate small pieces of the roof-brain, wedged together like a jigsaw puzzle, separate items of highly integrated behaviour. A special place for comprehension of names, a special place for arithmetical calculation, a special place for musical appreciation, and so on. . . . Rather, we may think, the contributions which the roof-brain, in collaboration with the rest of the brain and spinal cord, make toward integrated behaviour will, when they are ultimately analysed, resolve into components for which at present we have no names. To state the organisation of the mind in terms of roof-brain activities is a desideratum not in sight.'*

13. The telephone exchange as a metaphor of the brain There has been a tendency throughout history for the brain to be likened to the latest piece of technology. We saw that telephone exchanges, with their thousands of incoming and outgoing wires and their awesomely intricate and ever-changing patterns of connections, were as popular in the early years of this century as images of 'how the brain works' and computers in the 1970s. Now the boot is on the other foot, for neural network computers were based on neurophysiology rather than the other way round.

14. Despair about the brain Has been a constant phenomenon through the ages. Some examples: Sir Thomas Browne, *Religio Medici* (1642). *'In our study of Anatomy there is a mass of mysterious philosophy . . . for in the brain, which we term the seat of reason, there is not anything of moment more than I can discover in the crany of a beast: and this is no inconsiderable argument of the inorganicity of the soul, at least in the sense we generally so receive it. Thus we are men, and we know not how.'* Albrecht von Haller (1765): *'Our present knowledge does not permit us to speak with any show of truth about the more complicated functions of the mind . . . hypotheses of this kind have in great numbers reigned in the writings of physiologists from all time. But all of them alike have been feeble, fleeting, and of short life.'* R. W. Gerard, 1949: *'It remains sadly true that most of our present understanding would remain as valid and useful if, for all we knew, the cranium were stuffed with cotton wool.'* An admirably complete, thoughtful and readable account of the development of ideas about the brain, with a wealth of illustrations, is Finger, S. (1994). *Origins of Neuroscience: A History of Explorations into Brain Function* (Oxford: Oxford University Press). A more general, and engaging, exploration of the underlying issues is Gregory, R. L. (1981). *Mind in Science.* (London: Weidenfeld & Nicolson).

15. Think big *'My advice to any young man at the beginning of his career is to try to look for the mere outlines of big things with his fresh, untrained and unprejudiced mind.'* Selye, H. (1956). *The Stress of Life.* (New York, NY: McGraw-Hill).

16. Structure follows function *'Form is a diagram of forces'*: D'Arcy Wentworth Thompson (1917). *On Growth and Form.* (Cambridge: Cambridge University Press).

17. Lord Kelvin From a lecture delivered in 1883; published in *Popular Lectures* **I** 73.

CHAPTER 2

COMMUNICATION WITHIN NEURONS

Although it had been known for nearly 2,000 years that nerves served to communicate between the body and the brain, the question of *how* they did it was settled only a matter of decades ago. Nervous conduction is now one of the best-understood processes in the whole field of physiology, and its elucidation has been a major triumph for that branch of the subject known as biophysics – the application of purely physical methods to biology. However, before we delve into how nerves are able to conduct signals so precisely over long distances, it is important to remember that their sole *function* is to release neurotransmitter; all that follows is simply a means to an end.

Neurons and electricity

That there was some link between *electricity* and nervous and muscular action had been sensed as far back as the end of the eighteenth century, with Galvani's celebrated observation of the twitching of frogs' legs when in contact with certain combinations of metals.[1] Further understanding had to wait for the development of galvanometers sensitive enough to register the passage of very small electrical currents. By the end of the nineteenth century it was clear not only that nerves and muscle could be activated by electrical stimulation, but conversely that their normal activity was always accompanied by changes in electrical potential. It did not follow, of course, that these electrical changes were the *cause* of neural communication, since activity of many other kinds of tissue also gave rise to electrical effects. It was not until 1939 that it was finally established that electrical currents were not only generated as a side effect of nervous conduction, but were also *necessary* for conduction to take place at all.

Figure 2.1 shows how to convince even the most hardened sceptic of this. Dissect out a short length of myelinated nerve fibre and lay it in a pool of saline solution on a glass slide. Stimulate one end while recording from the other: you pick up the action potentials that are transmitted down the axon. Now introduce a pool of (non-conductive) oil across the middle of the slide, dividing the saline into separate pools at each end. Stimulate again: there is no response. Yet we have done nothing to the *interior* of the fibre, so nerve conduction cannot be due to movements of fluids or of filaments within the axon. Next, join the two pools of saline together, with a copper wire. Suddenly the fibre starts conducting action potentials again. Something passing along the copper from one pool to the other must have enabled this to happen. Heat is too slow: the only possible candidate is electricity, completing the circuit, of which the other half lies within the axon. In other words, the current generated by an active nerve is not just an

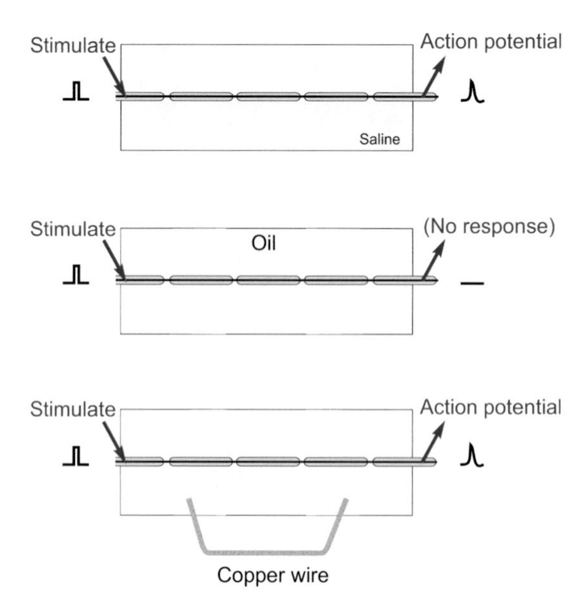

Figure 2.1 Knock-down proof that action potentials depend on electrical currents (see text).

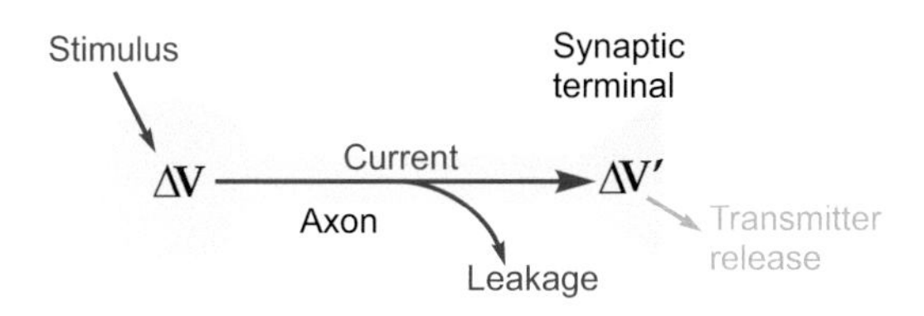

Figure 2.2 Idealised passive transmission of information by a neuron. A stimulus causes a change in potential; the resultant current flows down the axon but is subject to leakage, so that the voltage at the terminal is less than the original. Nevertheless, it causes the release of neurotransmitter from the ending.

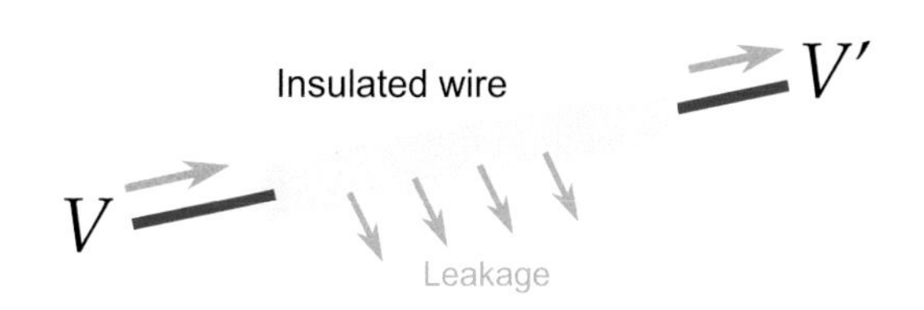

Figure 2.3 An axon is analogous to an insulated wire. Both consist of a central conductor and an insulated sheath, which nevertheless allows some current to leak out, so that the transmitted voltage is not the same as the original.

accidental by-product of transmission, like the noise from a car: it is an essential *determinant* of the entire process.

The flow of electrical current along nerves

Now nerve fibres, with their conductive central core of axoplasm surrounded by an insulated membrane often reinforced with extra non-conductive layers of myelin, are clearly very like ordinary insulated wires: if you apply a voltage at one end, current flows down the core and should make the other end change its voltage as well and trigger transmitter release (**Fig. 2.2**). Could action potentials simply be transmitted by this kind of passive conduction, in the same way that signals pass along a telephone wire? For most nerves, the answer is no. The snag is *leakage*. We're used to electrical wires and cables in which the core is a very good conductor – it has a low resistance – and the outside is a very good insulator – it has a very high resistance. So if you apply a given voltage ΔV at one end, you get essentially the same voltage $\Delta V'$ at the other end. But not exactly the same voltage, because a little of the current will have leaked out of the wire along the way, so the current leaving the far end will be slightly less than the current entering. With ordinary insulated wires like the ones used for domestic wiring very little current is lost, but nevertheless *some* is, and you notice it if you're trying to conduct over very long distances (**Fig. 2.3**). For the Victorian engineers who laid the first transatlantic telegraph cables, for example, this loss was a severe problem. You might apply quite large voltages to your cable in Porthcurno and find that by the time they reached Nova Scotia practically nothing was left – the electricity had just leaked away into the Atlantic Ocean (**Fig. 2.4**). You need to understand leakage a little more quantitatively, because it underlies how a nerve fibre actually works, and this involves a little physics: more precisely, something called its *cable properties*.

(a)

(b)

Figure 2.4 (a) A section of one of the first transatlantic telegraph cables, for which cable theory was originally developed. (b) A telegraph relay, analogous to the node of Ranvier in myelinated nerves (described below).

Nerve fibres as cables

Imagine the axon sliced up into a series of little imaginary compartments or units (**Fig. 2.5a**). For each unit, the current has a choice: it can either leak away across the membrane or carry on to the next unit. What it actually does depends both on how good its insulation is – the resistance of the axon membrane – and on how much resistance is offered to currents flowing longitudinally through the axoplasm. We can call the transverse, insulating, resistance of the membrane for one unit R_M, and the longitudinal resistance of the axoplasm per unit R_L. Joining lots of these units in series, we get a ladder-like network of resistors called the *equivalent circuit* of the nerve fibre. We can assume that the external medium offers no significant resistance to current flow (the justification for this assumption will become apparent later) and treat all the outer ends of the individual R_M's as if they were in electrical continuity.

Now imagine a potential V_0 applied to the end of this ladder. The current entering the first compartment has a choice: it can either flow out through R_M or continue through R_L to enter the next compartment. In fact, some fixed fraction of the total current takes the former route and the rest takes the latter: the bigger the resistance of the membrane, R_M, and the smaller that of the axoplasm, R_L, the greater will be the tendency of the current to carry straight on rather than leak away through the membrane. Unless the insulation is perfect, or the axoplasm infinitely conductive, the current entering the second cell will be smaller than that entering the first by some fixed ratio. In the same way, the current entering the third cell will be smaller than that entering the second, and so on all the way down the chain. The result is that the *potential* seen at each compartment will fall in a fixed ratio as one goes from compartment to compartment down the line, resulting in an exponential decline in voltage as a function of distance along the axon, at a rate that will depend on the ratio R_M/R_L (**Fig. 2.5b**). Thus V is given by $V_0e^{-x/\lambda}$, where e is the well-known constant whose value is about 2.718, and λ is a fundamental parameter called the *length constant* (sometimes referred to as the space constant) that describes how quickly the voltage declines as a function of the distance x. λ is in fact the distance you have to go before the voltage has dropped to $1/e$ (about 37 per cent) of its original value V_0. It turns out in fact that λ is actually equal to $\sqrt{R_M/R_L}$.

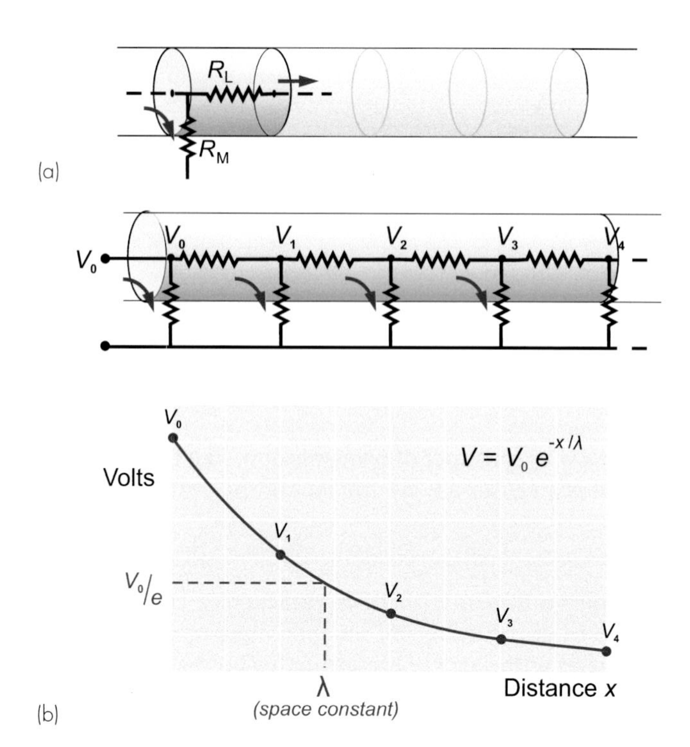

Figure 2.5 Passive spread of electrical current along an axon. (a) Each unit length of fibre can be thought of as having a longitudinal resistance, R_L, and a transverse or membrane resistance R_M. The whole axon may be represented by the ladder-like equivalent shown in (b). If a voltage V_0 is applied at one end, the voltage measured at different distances, x, along the axon will fall off exponentially because of current leakage through the membrane. The space constant, λ, is the distance for which the voltage falls by a factor e.

For telephone wires this length constant works out at some hundreds of miles and doesn't present much of a problem. But the performance of nerves, unfortunately, is a bit of a disaster. Consider, for instance, a large myelinated frog nerve fibre, say 14 µm in diameter, and assume for the moment that the myelin is uninterrupted along its length, without nodes of Ranvier. The axoplasm is intrinsically a vastly better conductor of electricity than myelin: their specific resistances are of the order of 100 ohm cm and 600 megohm cm respectively. But because the cross-sectional area of the axoplasm is so small, the ratio of R_M to R_L is not very great: for 1 mm, R_M comes out as about 250 megohm and R_L about 14 megohm, giving a space constant of some 4 mm or so. In other words, a potential generated at one end of the axon will have dropped to less than half at a distance of 4 mm, to about a tenth of its original value after a centimetre, and by two centimetres will only be some 1 per cent of the original stimulus and probably undetectable in the general background electrical noise. In other words, axons are quite incapable of acting as reliable passive conductors of electricity over distances of more than a centimetre or two at most. In many parts of the brain and in some sense organs, for example within the thin layer of the retina, a length constant of this size is fine. But if you're trying to get a message from your spinal cord to your feet it is obviously absolutely hopeless.

There are two reasons for this disappointing behaviour. The first is that the *materials* that nerve fibres have to be made of are not ideal. The core conducts about 50 times worse than it if it were made of, say, copper, while the insulating part, the cell membrane, is absolutely hopeless: compared with rubber its degree of insulation is about a *million* times worse. The other reason is that nerves are exceedingly *small* and the layer of insulation extremely thin. Size has a very important effect on the length constant: if you double the diameter of a cable, R_M drops by half (it depends on the surface area) but R_L drops by a quarter, because it depends on the cross-sectional area, which varies with the square of the diameter (**Fig. 2.6**). As a result, the ratio R_M/R_L is doubled, so the length constant

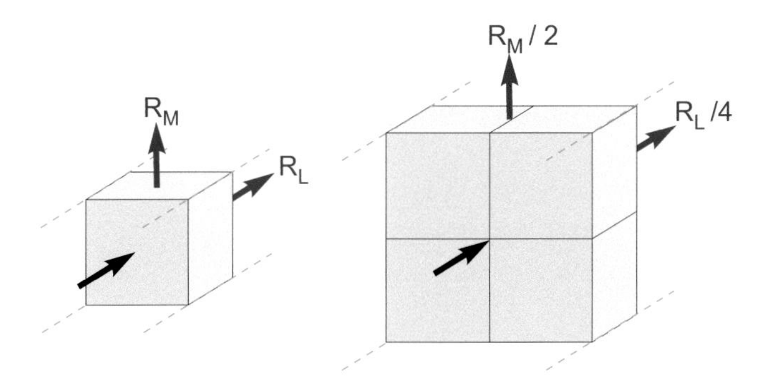

Figure 2.6 Doubling the diameter of an axon reduces R_M by a factor of 2, but R_L by a factor of 4. (The fact that axons are not usually square does not affect this result!)

increases by only the square root of two. In general the length constant is proportional to the square root of the diameter: the bigger the nerve, the further it can conduct. However, there are obviously constraints on how much you can improve nerve this way. Consider the sciatic nerve in your leg, whose large, 10 μm, fibres have length constants of a couple of mm or so. Even if we increased the diameter of each fibre to 1 cm, the length constant would still only be some 15 cm: messages would fade out before they got half-way down your thigh. You can calculate for yourself that such a fibre would need to be something like 9 cm across to be able to conduct reliably all the way down your leg, but the nerve would then be as big as the leg itself! (This analysis is not quite fair because we have assumed a constant thickness of myelin: but even if we allow this thickness to increase in proportion with the axon itself, the fibre will still need to be almost a hundred times bigger. Bearing in mind the fact that many important fibre tracts in the body contain millions of fibres, one must still conclude that passive conduction is not a practical possibility over long distances.)

For most nerves, then, passive conduction is not an option. How Victorian engineers solved the same problem in long telegraph cables was to have little amplifiers at intervals, called *repeaters,* which regenerated the signal, building the voltage up again each time to a full size. Since these repeaters used external energy to do their work, this was a process not of passive but of *active* conduction.

How does the *nerve* do it? It too has repeaters. In nerve fibres that have to conduct over long distances there are channels in the membrane that open in response to small changes in voltage across the membrane – called voltage-gated channels. When they open they trigger off a very large voltage burst or spike, with a fixed size of some 100 mV in amplitude but only 2–3 ms in duration: this dramatic and explosive event is called an *action potential.*

Action potentials

Regeneration

Many properties of action potentials demonstrate this process of regeneration. If we record the action potential from a single electrical stimulus at different points along a nerve fibre, we find that its amplitude does not in fact decrease at all as a function of distance, but stays at a constant value. Even more strikingly, the size of this action potential is not even a function of the size or nature of the stimulus that initiated it in the first place. So long as the strength of the stimulus is above a certain *threshold* value (below which no action potential is seen at all), neither the amplitude nor shape or speed of the action potential is in any way influenced by the nature of the original stimulus, a property known as the *all-or-nothing* law. These two features, the all-or-nothing law and the existence of a threshold, are never shown by passive conduction, where the energy for the transmitted signal must come entirely from the original stimulus.

A familiar example of this kind of behaviour is a burning sparkler. Here the combustion is continually regenerative: the heat of the burning region raises the temperature of the next region to the point where it too catches fire, and so on all along its length. In other words, there is a continuous cyclic process in which heat triggers combustion and combustion generates heat, this heat coming of course from the chemical energy stored in the sparkler's coating. The temperature and rate of advance of the burning region is not a function of the temperature of the flame which originally ignited it, so long as this was sufficient to light it at all (**Fig. 2.7a**).

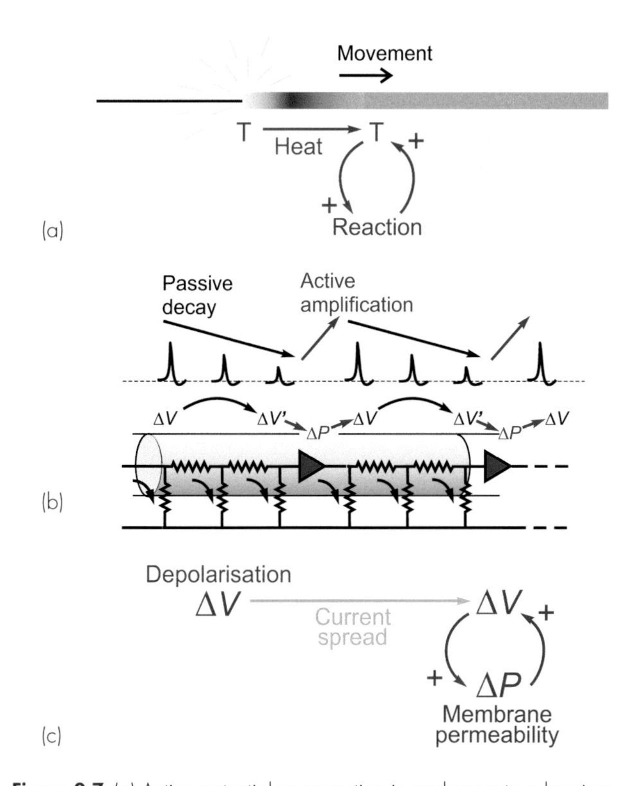

Figure 2.7 (a) Action potential regeneration is analogous to a burning sparkler. (b) The components of action potential propagation. A depolarisation, ΔV, at one point on the fibre results, through local current flow, in a smaller depolarisation, $\Delta V'$, some way down the axon. This triggers off a permeability change, ΔP, in the membrane, which in turn produces a voltage, ΔV, that is larger than $\Delta V'$. (c) The whole sequence – decay followed by regeneration – is repeated indefinitely.

Table 2.1 Ionic composition of three kinds of electrically active cell

	Na^+	K^+	Cl^-
Frog muscle (resting potential ca. –100 mV)			
Internal (mM)	10	124	1.5
External (mM)	109	2.3	78
Equilibrium potential (mV)	+65	–105	–100
Squid axon (resting potential ca. –60 mV)			
Internal (mM)	50	400	50
External (mM)	440	20	560
Equilibrium potential (mV)	+55	–75	–60
Mammalian Cell (resting potential ca. –90 mV)			
Internal (mM)	14	140	5
External (mM)	142	4	110
Equilibrium potential (mV)	+61	–94	–81

It turns out that this is a surprisingly close analogy to the mechanism of propagation of the action potential. What happens is that the original stimulus to the fibre, ΔV, makes local currents flow passively through the membrane, causing a spread of potential as in **Figure 2.7b**. It is sensed by neighbouring regions of the fibre and triggers a mechanism in the membrane, ΔP, that generates a voltage many times larger (thus introducing an amplification of the original signal), which in turn sets up local currents that cause a potential change still further down the axon . . . and so on, until the potential change has been transmitted from the point of stimulation to the end of the axon. This whole cyclical process is known as the *local circuit* mechanism of action potential propagation.

Each cycle consists of three distinct stages: first, there is the mechanism by which a potential at one point results in a passive flow of current and thus in depolarisation of regions further down the axon; second, the mechanism by which this depolarisation triggers off some change in the membrane; and third, the mechanism by which this change produces a new depolarisation that is much larger than what originally triggered it off (**Fig. 2.7c**).

Now, of these processes, the *passive* part of conduction – current flow – was understood very early on: it is simply a matter of physics, and in this respect nerve behaved in exactly the same way as the telephone and telegraph cables whose properties had been sorted out since the end of the nineteenth century, and can be described in terms of equivalent circuits. The other two processes obviously require identification of this mysterious change in the membrane that results in amplification of the voltage that triggers it. It turns out that this change is the product of a change in the *permeability*, ΔP, of the membrane to certain ions. If we measure the electrical resistance of the membrane, we find that during an action potential it drops enormously – with a time-course not very different from that of the action potential itself. To understand how a change in permeability can affect potential, we need first to understand the ionic composition of the interior of nerve fibres and the fluid by which they are surrounded (**Table 2.1**).

Ionic concentrations

From a physiological point of view, the body's function is ultimately to provide a stable environment for our genes. Genes require complex biochemical machinery to make them work, and this machinery in turn has quite stringent requirements in terms of the concentrations of ions and other substances that need to be around. And this of course is why we evolved *cells*; and in their membranes are all the mechanisms that carry out cellular homeostasis, particularly of their ion concentrations. But cells simply adrift in their environment are still at the mercy of environmental change. So the second major stage in evolution was the building of cells into tissues and bodies, which can then seal themselves off from the harsh world outside and create a sort of *inner* environment tailor-made to suit the cellular requirements. As a result, we have a nested series of compartments: *intracellular* containing intracellular fluid (ICF), *extracellular* with its extracellular fluid (ECF), and finally the real environment. Our genes are in a sense doubly buffered, like gold bullion locked in a safe inside a vault in Fort Knox. The brain is particularly well protected, through the blood–brain barrier (the tight junctions between astrocytes that separate cerebral capillaries from neurons) and buffering effect of the cerebrospinal fluid (CSF), which is in ionic equilibrium with brain ECF and whose composition is actively maintained by transport processes in the choroid plexus.

ICF differs from ECF in a number of ways, that vary slightly from one cell to another, and between species, but the underlying pattern is very similar. The single most important feature is that inside an axon we have a lot of potassium and not much sodium, whereas ECF is the opposite, with a lot of sodium and not much potassium. ECF or blood is in many ways remarkably like seawater – perhaps not so remarkable when you realise that it was in the ocean that cells first evolved into multicellular organisms: you are a walking bag of seawater, your cells still rocked in the cradle of the deep.[2]

These differences in concentration are maintained in two ways: by overall homeostatic mechanisms involving such things as the kidney and the regulation of intake, which determine the composition of the ECF, and by specific mechanisms within cell membranes that determine the ICF. What can be deduced from the striking differences between ICF and ECF in **Table 2.1**? For a start, the membrane of the cell must be essentially impermeable to all these ions: otherwise there would be nothing to stop Na and K from simply swapping with each other, so eventually there would be no difference in concentration at all. Second, the membrane must have set these *particular* concentrations up in the first place, presumably by manipulating the movement of ions across it.

The impermeability of the cellular membrane is a fundamental consequence of its construction as a bilayer of phospholipid molecules, with hydrophilic heads on the outside and hydrophobic tails pointing toward the centre. Such a structure ought not to let water or ions across. But floating in it are large specialised molecules – proteins or glycoproteins – that in effect provide passages or tunnels that let small molecules through, in a controlled and specific manner (**Fig. 2.8**). Now there are several different ways an ion can cross a cell membrane, but for the moment we shall just consider two major categories: *passive* and *active*. Given the chance, ions will always flow passively down a concentration gradient: in this passive diffusion, from high concentration to low, the ion necessarily loses energy. Conversely, if you see an ion moving up its energy gradient then it is gaining energy that must come from somewhere: this is called *active transport*, and uses ATP. From the nerve fibre's point of view, by far the most important of these membrane processes is what biochemically minded people think of as a sodium–potassium ATP-ase but which physiologists – loosely – refer to as the sodium pump (**Fig. 2.9**).

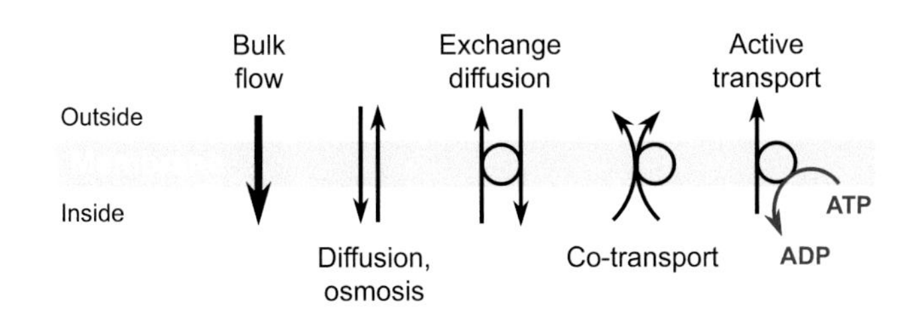

Figure 2.8 The major classes of mechanisms that transport ions across membranes

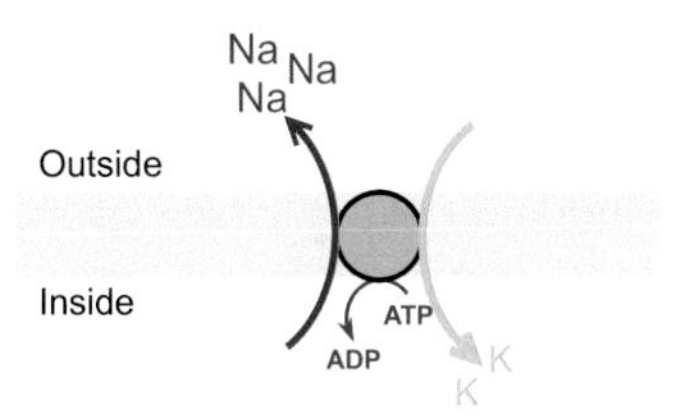

Figure 2.9 The 3:2 sodium:potassium pump, using adenosine triphosphate (ATP) to transport sodium out of the cell and potassium in, against their respective concentration gradients.

The sodium pump

The sodium pump swaps sodium ions on the inside with potassium ions on the outside (actually three sodiums for two potassiums). Because in each case this means moving ions against their concentration gradient, such pumps consume energy – hence the ATP-ase activity. (It is composed of a rather complicated protein straddling the membrane – it actually crosses it 10 times – and a protruding glycoprotein: the protein part has a number of functionally significant sites – a sodium-binding site, a potassium-binding site, and, inside, another region that acts as an ATP-ase.) The sodium pump is not just found in excitable cells like nerve and muscle, for its primary function is not really electrical at all. This can be seen by blocking the pump, for instance with a substance called *ouabain*, a cardiac glycoside found naturally in certain plants in Africa, where it has been used as an arrow poison. What then happens is that the normal background leakage of sodium into the cell is no longer baled out, and potassium is similarly gradually lost from the inside. Because the gain of sodium is faster than the loss of potassium, chloride moves with it to preserve neutrality, and water is drawn in osmotically. As a result, the cell begins to swell and eventually bursts. In other words, the pump is essential to maintain *osmotic equilibrium*.

What this and the many other varieties of pump *also* achieve is the creation of a *store of potential energy*. The cell membrane is like a car battery, charged by the dynamo provided by the sodium pump. This constant charging incurs a heavy metabolic cost – some 20 per cent of the resting energy consumption of the brain. It is rather a special energy store in two respects. First, unlike other stores in the body such as glycogen or fat or even ATP, this store is available for instant use – within a fraction of a millisecond. This makes it ideal for rapid nerve conduction. Second, it is specifically available at the *membrane* of the cell, where it can be used for tasks that are themselves localised in the membrane. Many cells use this energy for transporting other substances, through coupled transport: sodium's desire to enter the cell, down its concentration gradient, can for instance be coupled to the transport of glucose. Potassium is not used in this way, and in fact the axon membrane is characterised by being slightly permeable to potassium in its resting state, through one or more kinds of potassium leakage channels. We shall see later that these

Clinical box 2.1 The importance of the distribution of ions

The fluid surrounding cells is, as far as ions are concerned, in direct continuity with the blood. Hence, any change in the blood ionic composition leads to a change in the extraneuronal ionic concentration, and, correspondingly, the concentration gradient across the neuronal membrane. We know that the distribution of ions across the membrane determines the electrical properties of the neurons and it follows that disturbance in blood Na, K or Ca concentration will have profound neurological sequelae.

Hypokalaemia, by hyperpolarising the myocyte membrane (which has very similar properties to the neuronal membrane) causes muscle weakness which can be severe enough to cause paralysis and respiratory failure. Hyperkalaemia, in contrast, leads to depolarisation across cell membranes, yet this, as is described later in this chapter, promotes inactivation of voltage-gated sodium channels, which can block the propagation of action potentials by the process of accommodation. Thus, hyperkalaemia can also cause muscle weakness, though it is often a more pressing emergency as it typically impacts electrical conduction in the heart, causing dysrhythmia and death.

Severe hypernatraemia (perhaps from severe dehydration) can lead to cerebral depression and sedation, and further to coma and death. Conversely, patients with hyponatraemia (for example after success at water-drinking competitions) are prone to developing seizures. Unfortunately, restoration of extracellular sodium following prolonged hyponatraemia can be fatal as the rapid osmotic changes that ensue can lead to the irreversible osmotic demyelination syndrome. Doctors must therefore be particularly careful when prescribing intravenous fluids to those with hyponatraemia.

Hypercalcaemia is a curious case; calcium is essential for muscle contraction, neurotransmitter release and membrane channel function. One might therefore assume that hypercalcaemia is associated with seizures, tingling and neuronal hyperactivity. Instead the converse is true; hypercalcaemia is associated with weakness, drowsiness and reduced bowel motility. The reason for this is that calcium is predominantly an extracellular ion, and the membrane, having the properties of a capacitor, tends to accumulate these cations on the outer surface. This causes the local transmembrane electrical potential to be augmented, effectively hyperpolarising the cell membrane and reducing electrical transmission.

channels have an important role in determining the electrical properties of the axon at rest.

One might well wonder how it is possible to have a channel that is specifically permeable to potassium and not to sodium, an ion which is also singly positively charged and has a smaller hydrated size. Recent studies of the molecular structure of particular potassium channels have demonstrated a complex structure almost perfectly designed for this task (**Fig. 2.10**). Negative charges associated with the channel proteins attract positive ions, while carboxyl oxygens substitute for the water to which the ion is normally wedded: unhydrated sodium, being bigger than unhydrated potassium, is then excluded.

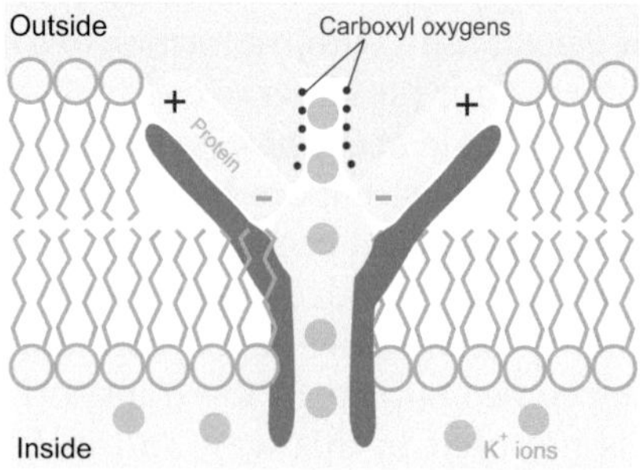

Figure 2.10 Postulated molecular structure of one class of potassium channel, showing how the presence of carboxyl groups and charged groups in proteins lining the channel enables it to distinguish between potassium and sodium ions. (After Doyle *et al.*, 1998; with permision from AAAS.)

This then is the ionic backdrop against which the axonal membrane operates. Much of its operation is common to all cells. As we shall see, they all share physical mechanisms that convert any changes in membrane permeability to changes in potential. What is unique about nerve and muscle cells is that they *also* possess special mechanisms by which changes in permeability are in turn triggered off by changes in potential, thus completing the cycle of three links by which the action potential is propagated over the membrane surface. Thus to understand how nerves work we need to be able to answer two questions:

How do ionic permeabilities affect membrane potential?
How do membrane potentials affect ionic permeabilities?

How membrane potential depends on ionic permeabilities

Imagine that we have a system of two compartments, A and B, and that initially A contains a strong solution of KCl, and B a weak one (**Fig. 2.11a**). Now suppose that the membrane separating them, initially impermeable, suddenly becomes permeable to potassium ions (**Fig. 2.11b**). Clearly there will now be a tendency for K^+ to diffuse through the membrane down the concentration gradient between A and B. Since the ions carry a positive charge, compartment B will become more and more positive with respect to A as they migrate in this way, setting up an electrical gradient that will tend to oppose the entry of further ions from A. Eventually there will come a point where the

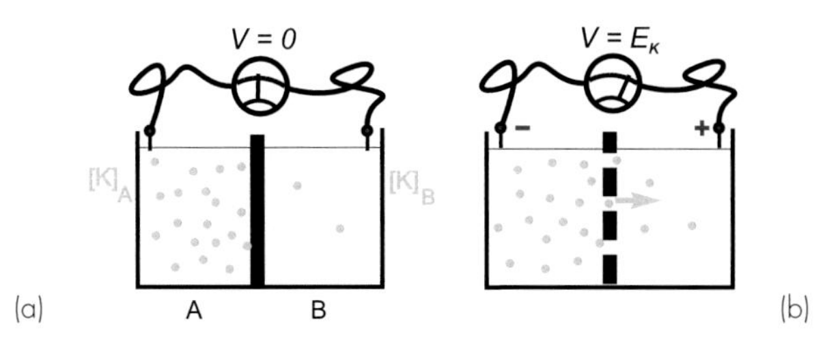

Figure 2.11 Two compartments, each containing potassium chloride at different concentrations. (a) The barrier separating them is impermeable to potassium, so there is no potential difference. (b) The barrier is made permeable to potassium ions (green dots), which therefore tend to diffuse down their concentration gradient. But in doing so they set up an electrical potential in the other direction which eventually grows to the point at which it prevents further ions moving across: this is the Nernst or equilibrium potential, E_K.

concentration gradient from A to B will be exactly equal and opposite to the *electrical gradient* from B to A, and the system will be in equilibrium, since there will be no net flow of ions across the membrane.

The resultant electrical potential between A and B is then called the *equilibrium potential* for potassium, E_K. To work out how big this potential will be, consider the energy involved in moving one potassium ion from A to B. The work done in moving it against the electrical gradient will be given by its charge e multiplied by the potential difference E_K; since the system is in equilibrium, this work must be exactly equal to the energy gained in moving down the concentration gradient, which can be shown to be

$$kT \ln \frac{[\mathrm{K}]_\mathrm{A}}{[\mathrm{K}]_\mathrm{B}}$$

where T is the absolute temperature, and k is Boltzmann's constant. So we can write:

$$eE_K = kT \ln \frac{[\mathrm{K}]_\mathrm{A}}{[\mathrm{K}]_\mathrm{B}}$$

or

$$\begin{aligned} E_K &= \frac{kT}{e} \ln \frac{[\mathrm{K}]_\mathrm{A}}{[\mathrm{K}]_\mathrm{B}} \\ &= \frac{RT}{F} \ln \frac{[\mathrm{K}]_\mathrm{A}}{[\mathrm{K}]_\mathrm{B}} \\ &\approx 58 \log_{10} \frac{[\mathrm{K}]_\mathrm{A}}{[\mathrm{K}]_\mathrm{B}} \text{ mv} \quad \text{(at 20°C)} \end{aligned}$$

(In the alternative form, derived from consideration of a mole rather than a single ion, R is the gas constant, and F is Faraday's constant, equal to $N.e$). This relationship (the *Nernst equation*) is true for any ion in equilibrium across a membrane to which it is freely (and solely) permeable, with the proviso that if the charge on the ion is not +1 (as for instance in the case of Cl^- or Ca^{2+}) we need to include this ionic charge z as well:

$$E_X = \frac{kT}{ze} \ln \frac{[X]_A}{[X]_B}$$

Only a tiny number of ions need to cross the membrane to set up such an equilibrium, so that the concentrations of the ions on each side remain effectively unchanged, and the equilibrium potential is set up virtually instantaneously after a rapid permeability change of this kind. Suddenly making the membrane permeable to potassium – perhaps by opening up little channels in it that allow K^+ through but nothing else – is rather like connecting a battery of voltage E_K across the equivalent circuit in **Figure 2.5**.

Now we noted earlier that the axonal membrane at rest is indeed slightly permeable to potassium. Since – taking the figures for the squid giant axon – the concentration of potassium inside is about twenty times greater than outside, we can calculate that there ought to be an equilibrium potential across the membrane of some 75mV, negative inside. If you actually put an electrode into a squid axon and measure the potential, you do indeed find a standing negative voltage, the resting potential, which is of the right order of magnitude but a little smaller (some –60mV). The discrepancy tells us that our equations need to be refined.

Membrane permeability to more than one ion

The limitation of the Nernst equation is that it represents a highly idealised – indeed hypothetical – state of affairs, because at any one moment, real membranes are in fact permeable to *more* than one ion. Let us go just one step further and imagine a situation of two compartments just as before, but this time instead of the membrane being permeable to just one of the ions, it is now permeable to both (**Fig. 2.12**). One thing we can deduce at once is that is that the overall potential, V, will not just be a function of the concentration differences of the ions. It must also depend on just *how permeable* the membrane is to each of them.

To see that this must be so, consider an extreme case – suppose that we make the permeability to sodium smaller

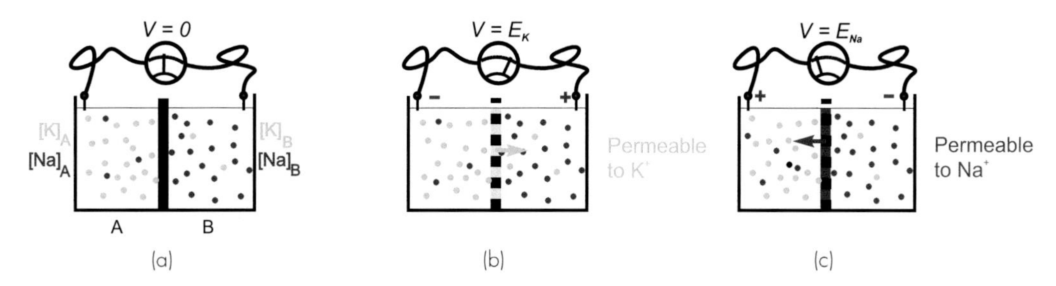

Figure 2.12 Two compartments (as in Figure 2.11) containing sodium and potassium chloride and separated by a barrier having a permeability P_K to potassium, and P_{Na} to sodium. (a) Both permeabilities are zero. (b) The barrier is permeable only to potassium (green dots), producing a potential of E_K. (c) The barrier is permeable to sodium (red dots), producing E_{Na}.

and smaller and smaller, so that eventually it reaches zero. Then V must be given by $E_K = 58 \log_{10} [K]_A/[K]_B$, or about –75mV. Equally, if we make the permeability to potassium smaller and smaller, to zero, then V will be given by $E_{Na} = 58 \log_{10} [Na]_A/[Na]_B$, or about +55mV. So the overall potential must also depend on the permeabilities of the two ions, which we can call P_K and P_{Na}. These are called *permeability coefficients*, and represent the ease with which the ion can pass through the barrier for a given concentration ratio.

Because of the P's, the resultant equation for the voltage V, the *Goldman constant field equation*, is a *little* more complicated than the simple Nernst equation, but not *very*.

$$E = \frac{kT}{e} \ln \frac{P_K[K]_A + P_{Na}[Na]_A}{P_K[K]_B + P_{Na}[Na]_B}$$

It has the same general form, but as you would expect we now have terms for both ions, and they are multiplied by their respective P's. So in a sense their weighting in the expression depends on how permeable they are. The more permeable the membrane is to potassium, the more the potassium concentration ratio will matter; in the limit, if we reduce one of the permeabilities to zero, something magical happens – it turns into the Nernst equation. So in a sense the Nernst equation is simply a special case of this one.

If you're not very happy even with this sort of elementary maths, there is a graphical representation of all this which most people find a helpful way of looking at it. In a sense, E_K and E_{Na} represent extreme values of the range that V can take up: it can't get more positive than E_{Na} or more negative than E_K. So we set up a vertical voltage axis and mark these two potentials on it. These voltages will be essentially fixed provided the cells are in good condition, because the concentrations are fixed, the temperature is fixed, and so are the other bits in the Nernst equation – k, e are universal physical constants. So these are like two rigid boundaries, and the actual voltage V at any moment must lie somewhere between them (**Fig. 2.13**). What the Goldman equation is saying is that V behaves as if it were under the influence of two forces: P_K pulls it towards E_K, and P_{Na} towards E_{Na}: where it ends up depends simply on the balance between the two. In other words, *changes in permeability cause changes in potential.*

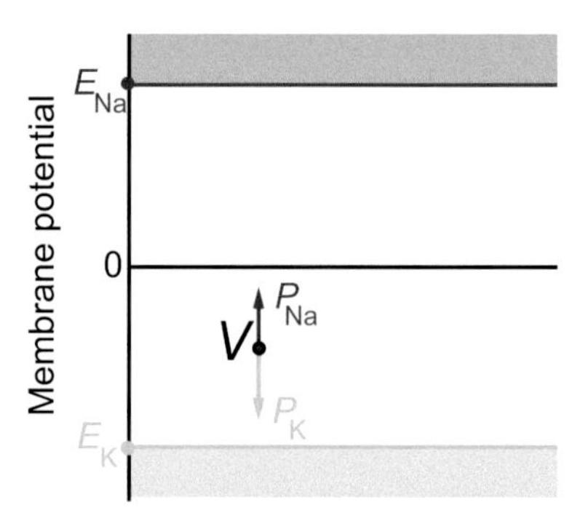

Figure 2.13 The membrane potential, E, can be thought of graphically as being an equilibrium between the pull of P_K towards E_K and P_{Na} towards E_{Na}.

Chloride

Haven't we forgotten *chloride*? In our simple model, we assumed that chloride ions were unable to diffuse across, so we were justified in omitting them from the constant field equation. But experiments show that real nerve and muscle membranes have significant chloride permeabilities, and it is obvious from the data that have been presented that there is a considerable imbalance in the concentrations of Cl^- on each side. Nevertheless, there are two reasons why this ion can, for the moment, be safely neglected. The first is that in practice the Nernst potential for chloride is usually very close to the equilibrium potential of the nerve membrane, so that changes in its permeability have negligible effects on the resting potential. What happens is that potassium and chloride are free to move together as KCl until the Nernst potentials for both chloride and potassium are equal: that is, until [K]out/[K]in = [Cl]in/[Cl]out. Because internal [Cl] is so very much smaller than [K], a shift of a given quantity of KCI has an enormously greater effect on the chloride ratio than on the potassium (since the external concentrations remain essentially unchanged). Thus chloride adjusts itself to a resting potential that is essentially determined by potassium. Second, it turns out that during the action potential no significant alterations in chloride permeability occur; as we shall see, this is in sharp contrast to what happens to sodium and potassium. However, when, in the next chapter, we look at synaptic mechanisms, we shall find that there are certain occasions when chloride cannot be neglected at all, and indeed most inhibitory synapses actually work through changes in chloride permeability.

The resting potential

With Goldman safely behind us we are now in a position to understand why the resting potential is *close* to E_K but not actually *at* it. The reason is that although at rest the permeability for potassium is much higher than for sodium, sodium permeability is *not* in fact zero: the ratio of the permeabilities is about 100:1, in frog muscle, at least. As a result, the resting potential is pulled a little more positive than would be expected for potassium alone, and the Goldman equation gives a pretty accurate prediction of the resting potential.

In fact, we can predict a lot more than that. If the equation is correct, then we should be able to predict V not just under resting conditions, but also when we deliberately mess about with the ionic concentrations. For example, we could alter the concentration of potassium in the external fluid: from the Nernst equation you can see that this will have the effect of altering the value of E_K.

In a classic experiment, Hodgkin and Horowicz did exactly this. They bathed a frog muscle cell in solutions with different concentrations of potassium and measured the resultant resting potentials. Now if the membrane had been *only* permeable to potassium, then it would have obeyed the Nernst equation, and the potential would be proportional to the log of the concentration outside. So

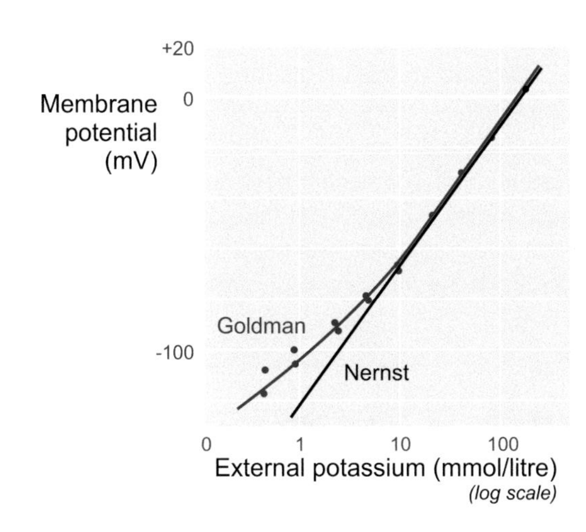

Figure 2.14 Measurement of membrane potential of frog muscle fibres (data points) in response to different external potassium concentrations. The black line shows what would be expected from the Nernst equation if the membrane were permeable only to potassium ions, whereas the blue line shows the expectation if it is about 1 per cent as permeable to sodium ions as it is to potassium. (After Hodgkin and Horowicz, 1959; with permission from John Wiley and Sons.)

(a)

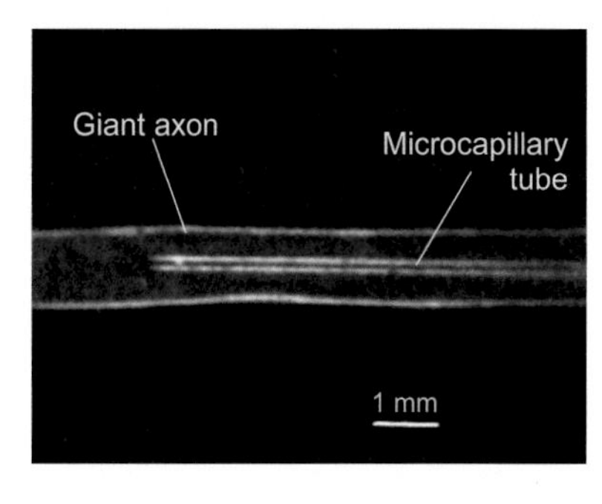

(b)

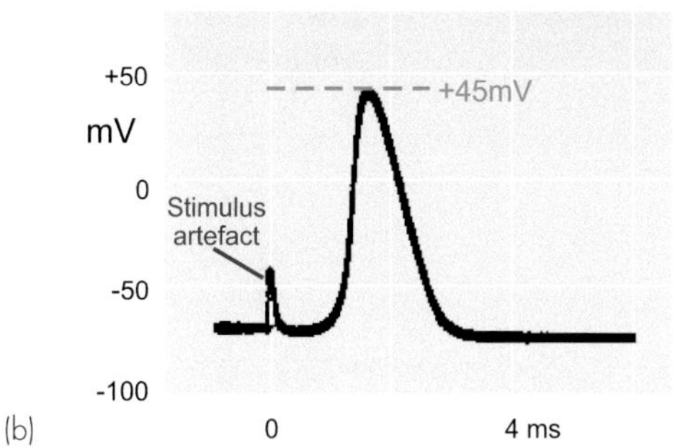

Figure 2.15 (a) Photograph of giant squid axon with a microcapillary tube introduced within its lumen. (b) Intracellular recording of action potential in squid giant axon (Hodgkin and Horowicz, 1956; with permission from John Wiley and Sons.)

plotting potential against the log of the concentration ratio would have given a straight line (**Fig. 2.14**). For large concentrations, this nearly worked, but as the concentration was lowered, the results increasingly deviated from the straight line. But by using the Goldman equation instead of the Nernst, and feeding in a ratio of about 100 for the permeabilities of potassium and sodium, the prediction was almost perfect.

Similar experiments have been done on the giant axons of squids, creatures that have in fact played a surprisingly large part in the discovery of how nerves function. The reason is that squids are shy, timid creatures: when alarmed, they contract their mantle to force water out of their siphon, propelling themselves backwards. These are very rapid responses, and need very fast conduction, using a special extra large fibre innervating the mantle, called a giant fibre. 'Giant' here is somewhat relative (somewhere between 0.1 and 1 mm across), but they are big enough that one can put glass tubes down them (**Fig. 2.15a**). This in turn means that one can do two things: one is to measure the electrical potential, for instance during an action potential; the other is to sample the cytoplasm inside (*axoplasm*) and see what it's made of. Indeed it is possible to squeeze their axoplasm out with a kind of miniature garden roller and replace it with fluids of different composition, thus altering the potassium concentration inside as well as outside.

The action potential

Now if we put a microelectrode inside a muscle fibre or squid axon and stimulate it, we find that the potential of the inside relative to the outside suddenly reverses from its resting –50mV (in the squid) to a peak of some +40mV, and then rapidly declines back to the resting potential again (**Fig. 2.15b**). This is the *action potential* or AP. (It is actually a *monophasic* action potential. You can also record action potentials with big electrodes outside the nerve fibre, and these are called *biphasic* action potentials. **Box 2.1** explains the difference.) What sort of permeability change could account for this? Not just a simple short-circuit, because then the potential would only tend towards zero, not reverse and become positive. In the squid the peak of the action potential is not far off at some +45mV, strongly suggesting that a sudden increase in Na permeability pulls the membrane potential temporarily towards E_{Na} (around +55mV). By experimenting with various concentrations of sodium ions inside and outside the axon it was possible to demonstrate directly that the peak of the action potential was indeed dependent on E_{Na}. If the concentration of external sodium was altered, by diluting the seawater with isotonic dextrose, although the resting potential didn't vary – or only very little – what *did* alter was the height of the action potential: the lower the external sodium concentration, the smaller the AP until eventually it is abolished altogether (**Fig. 2.16**). Another test was to change the sodium not outside but *inside*, an experiment that is only feasible with the squid axon because of its huge diameter. Once again, alteration of sodium altered the size of the AP, but the other way round: the *smaller* the internal Na, the *bigger* the AP. The peak of the action potential therefore depended critically on the ratio of $[Na]_{out}/[Na]_{in}$, in other words, on the Nernst potential for sodium. So the natural explanation was that the AP was caused by a brief increase in P_{Na}, driving the membrane potential towards E_{Na}.

So it looked as if there was an increase of P_{Na} at the start of the action potential, but what was setting it off? One attractive idea was that the depolarisation of the membrane by local currents from the previous bit of nerve

Box 2.1 Inside and outside nerve fibres: monophasic and biphasic action potentials

The potentials that nerve fibres use to convey information are potentials across the membrane. To measure them, you need one electrode on one side and one on the other – in other words, one (an *intracellular* electrode) penetrating the cell, and the other somewhere outside: if external resistance is low, it doesn't much matter where. With this arrangement, and the electrodes connected to a differential amplifier that amplifies the difference in voltage between them as an action potential (AP) passes down a fibre, you will record a trace that quite accurately reflects the true potential across the membrane at every moment.

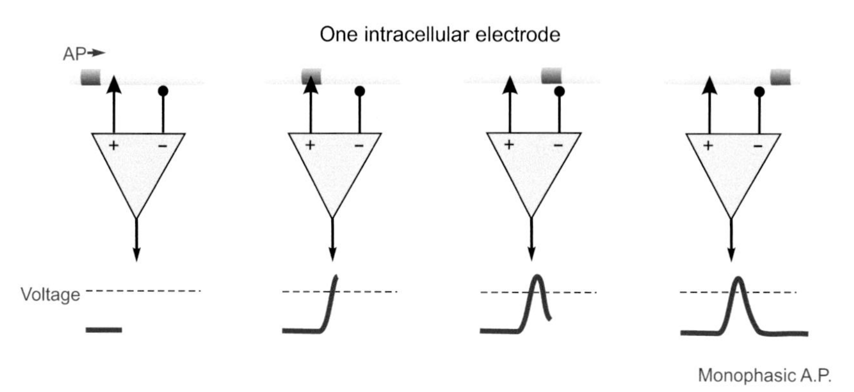

However, it is technically difficult to make electrodes small enough to penetrate axons, and pointless if all one wants to do is detect action potentials, rather than find out their exact shape. Instead, we can use a pair of extracellular electrodes, spaced a little apart along the nerve fibre, and connected as before to a differential amplifier. Now, as the action potential passes, each electrode in turn becomes more negative than the other while the membrane beneath it is depolarised. Because the amplifier is looking at the difference between the voltages, the recorded potential swings first one way, then the other, producing a *biphasic* action potential, as opposed to the *monophasic* one that you obtain with an intracellular electrode.

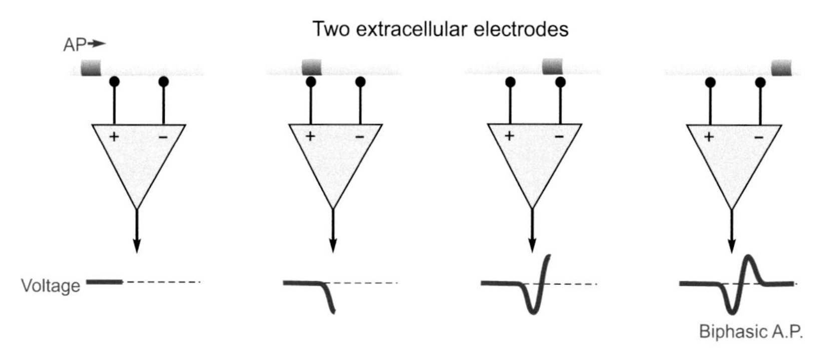

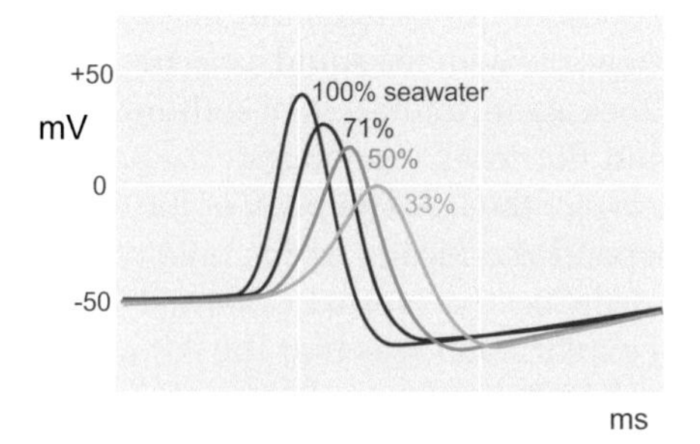

Figure 2.16 Action potentials in squid axon, showing the effect of different external sodium ion concentrations. The seawater was diluted with isotonic dextrose. (After Hodgkin and Katz, 1949; with permission from John Wiley and Sons.)

membrane might actually *cause* the increase in P_{Na}. This would work beautifully, because there would then be positive feedback – depolarisation gives increase in P_{Na}, which in turn causes more depolarisation – which would have exactly the kind of explosive regenerative effects that were needed (**Fig. 2.17**).

So at this stage, of the three pieces of the puzzle (local currents, the effect of permeability of voltage, the effect of voltage on permeability) the first two were well understood – simple physics and the Goldman equation – but the third was a complete mystery. It was Alan Hodgkin and Andrew Huxley in Cambridge who set about trying

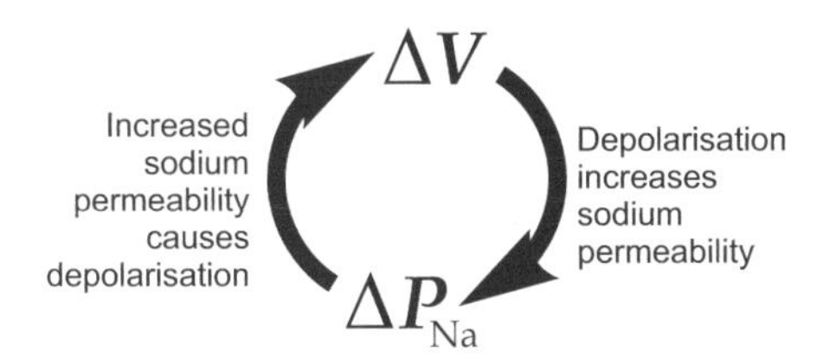

Figure 2.17 Positive feedback behaviour of voltage gated sodium channels, which initially open in response to membrane depolarisation.

to complete the puzzle by identifying the nature of this mysterious third process.[3]

How ionic permeabilities depend on potential

At first sight, the problem looks easy enough: what we want to know is how the permeability changes if the nerve fibre is depolarised. So why not simply use a microelectrode to pass different currents through the membrane, to set up different membrane potentials V, and see what happens to its conductance, g? We can work out g very easily: conductance is simply the inverse of resistance: resistance is V/I, so g is I/V – the current divided by the voltage. And we can assume that g is in turn simply proportional to the sum of all the permeabilities (**Fig. 2.18**).

That sounds simple, but unfortunately it cannot work. The reason is precisely *because* the system we are looking at has a feedback loop built into it. As soon as we alter the voltage, the conductance will change, and this will in turn mess up the voltage (the blue arrow in **Figure 2.18**), so it is no longer what we thought it was.

The voltage clamp

What we need to do is provide some way of setting the membrane potential at the level we want, and holding or *clamping* it there despite any changes in conductance that may be going on. The way that Hodgkin and Huxley did this was by using a negative feedback circuit (**Fig. 2.19**). We start, paradoxically, not by stimulating but by *measuring*. Using a long electrode inserted into the axon, we measure the actual voltage and compare this with the desired voltage V, the voltage we actually want the membrane to be at. The difference between the two represents the *error*, the amount by which the potential needs to be adjusted. One can then introduce a simple circuit – a current controller – that responds to this error signal by sending a current I through the membrane (using a second electrode inserted into the axon) in such a way as to move the actual voltage towards the value we actually intend. This amounts to clamping the membrane at the level we want, by hitting it very hard with a large injection of current as soon as it tries to wriggle free. All this can happen very fast, so we can put in different command voltages and see how the membrane responds. As before, if we measure the current we can calculate g by taking the ratio I/V. In principle one can use this technique with any sort of cell, but using the squid axon had particular advantages. Being so large, it was not *too* difficult to put long electrodes right down the middle, which increases the *total area* available and hence the size of the currents, which is technically an advantage.

So what did Hodgkin and Huxley find? The basic approach was to apply steps of depolarisation and measure the currents. If you suddenly reduce the potential from the resting potential of –65mV to say –50mV, the current changes in a characteristic and repeatable way that shows three main components (**Fig. 2.20a**).

- The first is a brief pulse or spike of current that is in a way a kind of artefact, due to having to charge the capacitance of the fibre up to the new potential. For the moment you can forget about it: we will be considering capacitance later (p. 41).
- Then there is a longer-lasting period during which current enters the axon, rising to a peak, then getting smaller again and in fact reversing to give:
- A current flowing out of the fibre, which lasts as long as the depolarisation is maintained. When the voltage is returned to the resting potential, this current drops relatively slowly back to zero.

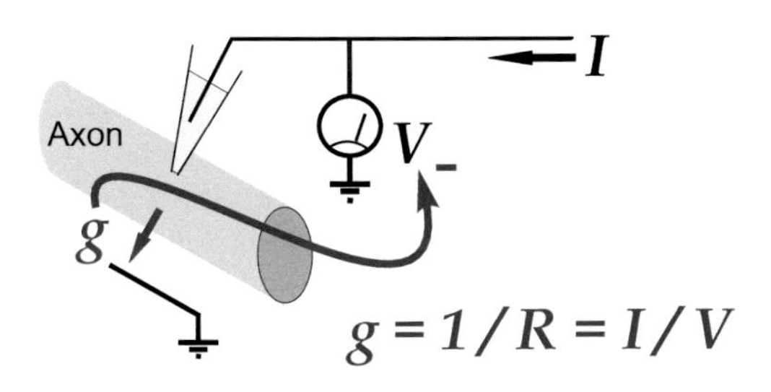

Figure 2.18 Trying to determine how the conductivity, g, depends on the potential, V. One might simply try injecting a current, I, to create a given potential, V, and then measure the ratio I:V to find the conductivity; but any change in g will *alter* the value of V (blue arrow), so the experiment cannot be done.

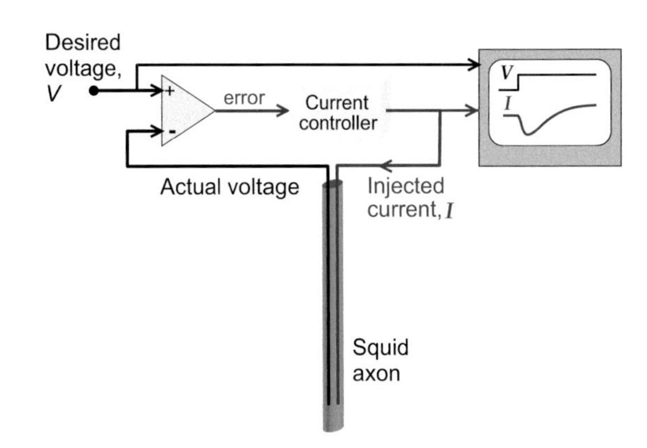

Figure 2.19 The principle of the voltage clamp. Two electrodes are inserted in the squid axon; the voltage measured by one of them (black) is compared with the 'desired voltage', V, and any difference between the two (error) automatically alters the current. I passes into the axon through the second electrode (purple). The time course of V and I are displayed on an oscilloscope; here, the current in response to a step change in V is shown (somewhat simplified and schematic).

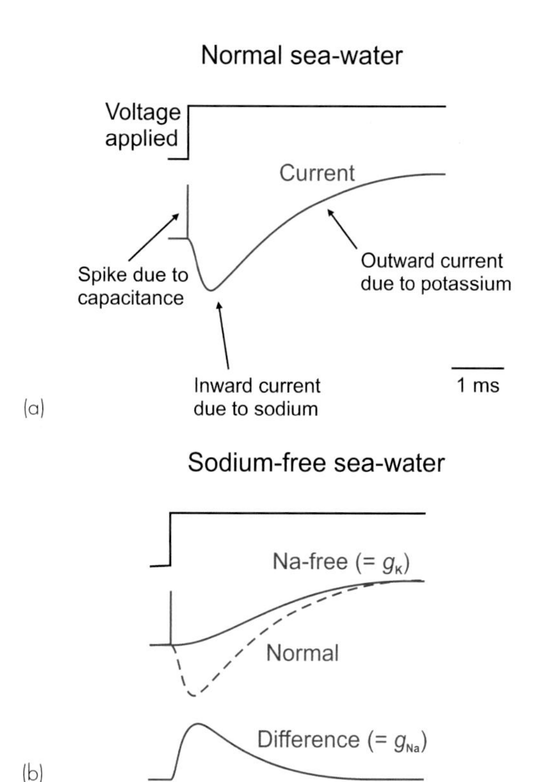

Figure 2.20 (a) Time course of current in response to a step depolarisation in a voltage clamp experiment on a squid axon, showing the three main components of the current. (b) By repeating the experiment in sodium-free seawater, the sodium component can be eliminated, leaving the potassium current behind; the difference can be presumed to be the time course of what the sodium current was originally. (Simplified, after Hodgkin & Huxley, 1952 (with permission from John Wiley and Sons); Hodgkin, 1958 (permission pending).)

What is going on? The fact that the second component is a current entering the fibre suggests that this could be the sodium current caused by the increase in P_{Na} that had been predicted. Similarly, the fact that in the third phase the current is outwards suggests that it is carried not by sodium but by potassium. One can test whether the second phase is indeed due to sodium by replacing all the external sodium with something else – Hodgkin and Huxley used choline, a positive ion much larger than sodium. They then found that the entry of current was completely abolished, leaving just the third component, which could be taken to represent the potassium current and thus the potassium permeability (**Fig. 2.20b**). Then, if one subtracts this potassium component from the total current curve, what is left must be sodium. This was confirmed in later experiments using two selective poisons that block the two channels. Tetrodotoxin (TTX) specifically blocks these sodium channels, and has the same effect as replacement of external sodium. Similarly, another substance called TEA (tetra-ethyl-ammonium) selectively blocks the potassium channels.[4]

So what we have now are curves showing how the P_{Na} and P_K vary after a depolarisation. Basically, P_{Na} starts to rise first, up to a peak, and then falls back. P_K also rises, but more slowly, and unlike P_{Na} it stays up for as long as the depolarisation is held, then decaying fairly leisurely to its base level. But something else happens to sodium that isn't immediately apparent in these records, which is that when the P_{Na} falls back to its resting level during the depolarisation, the channels have not in fact returned to their normal condition, for if one quickly returns the potential back to resting level and then depolarises again, there is virtually no response at all: the channels are in fact inactivated. They only recover when the membrane has been at the resting potential for a sufficient period of time. In other words, whereas the potassium channels exist in *two* possible states, open or closed, the sodium channels exist in *three* possible states: open, inactivated and closed.

Permeabilities during the action potential

By systematically measuring these permeability changes in response to steps of different size and starting from different initial voltages, Hodgkin and Huxley were able to derive equations expressing g_K and g_{Na} in terms of membrane potential, that summarised their data and made it possible to predict, in general, how the permeabilities would vary in response to *any* given pattern of depolarisation of the membrane. For once one knows how a system behaves in response to a small step input, then by breaking up any given voltage pattern into a series of small steps and adding the results together one can calculate the response to the whole thing. In particular, starting with the time-course of the intracellular action potential of the squid axon, one can work out in this way what permeability changes would result from it, ending up with the curves shown schematically in **Figure 2.21**. This shows that at the start of the action potential, P_{Na} rises abruptly, followed with a short delay by P_K; P_{Na} then starts to decline to its resting value while P_K is still quite high; the latter declines relatively leisurely to its resting state.

We have now come full circle, for if we know the time course of the permeabilities we should be able to calculate the shape of the action potential that should result from them. Even informally, it is obvious from **Figure 2.21** that at the start of the process, the sudden increase in P_{Na} will pull the action potential towards E_{Na}, but that it will fall again as P_K starts to overtake; and in the final phase, the prolonged undershoot, when the potential is hyperpolarised towards E_K, can be explained by the long time taken for P_K to return to normal.

But we can do this sort of thing much more quantitatively. If we feed the permeabilities into the Goldman constant field equation, and finish up with what we started with, the time-course of the action potential itself, then we know we have a complete description of how the action potential is regenerated. More precisely, the action potential represents the solution of the set of differential equations that embody the results of the voltage clamp experiment, the electrical properties of the membrane, and the constant field equation: the fact that this solution (shown in **Figure 2.22**, for a squid axon at 18.5°C) is so nearly identical to the shape of the actual action potential testifies to the completeness of Hodgkin and Huxley's

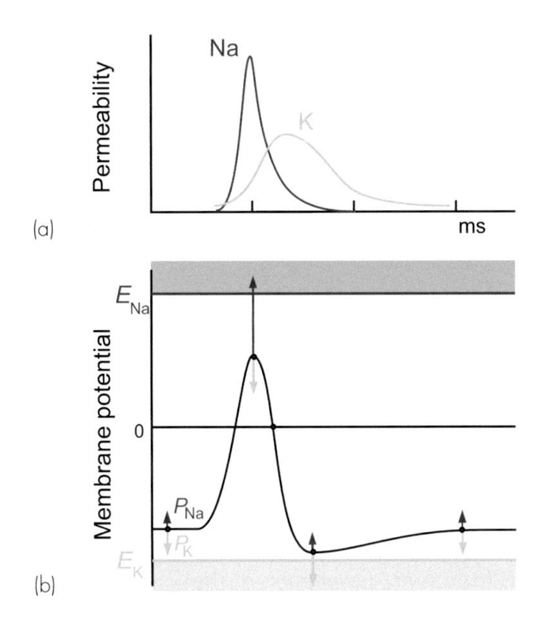

Figure 2.21 (a) Changes in potassium and sodium permeability associated with the action potential. (b) How these changes result in the form of the action potential itself. The arrows above and below the trace indicate roughly by their length the relative sizes of P_{Na} and P_K, pulling the potential respectively towards E_{Na} and E_K.

description of the way in which membrane permeability depends on voltage. This work was a landmark in biological science – the first time that a fundamental biological phenomenon had been entirely and completely described by a purely physical model.[5]

Patch-clamping

Curves like those in **Figure 2.22** look continuous: but it important to bear in mind that they are the result of the summation of thousands of single events (the opening and closing of channels) which are themselves *quantal*, or binary: a single channel is either open or shut, and the dynamics of overall permeability changes really reflect the way in which the *probability* of a channel being open varies with time and voltage. This can best be seen by using a refinement of the basic clamp technique called

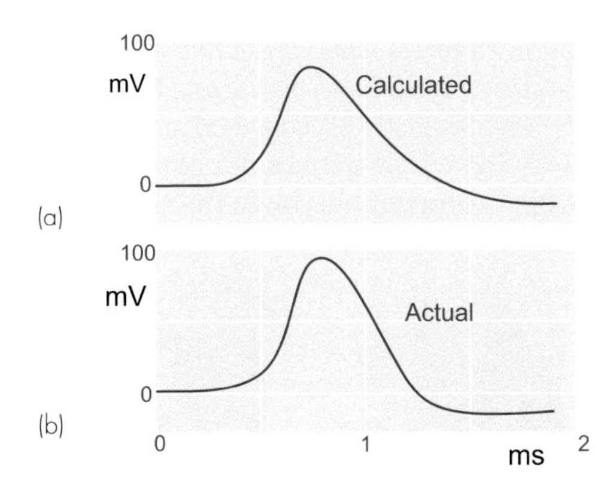

Figure 2.22 (a) Theoretical solution of the differential equations embodying the electrical properties of squid axon, the constant field equation, and the results of voltage-clamp experiments. (b) Actual action potential in squid axon at 18.5°C. (After Hodgkin and Huxley, 1952; with permission from John Wiley and Sons.)

patch-clamping (**Fig. 2.23**). The principle here is exactly the same as the ordinary voltage clamp, but *micro-miniaturised*. In voltage clamping, you measure the membrane potential, compare it with what you want it to be, and then if there's a difference or error, pass a current across the membrane to bring it back. In patch-clamping, as the name suggests, instead of clamping an entire cell, or most of it, what you do is operate on a *tiny part* of it. You take some of the membrane in question and simply suck it on to the end of a micropipette, then use the pipette as an electrode whose potential you measure and down which the clamping circuitry passes the currents necessary to do the clamping. Because the area of membrane is so small, there will typically be only a few channels in it – sometimes if you're lucky just one – so you can observe exactly what it does in response to depolarisation.

Under these conditions the isolated channels in fact behave very differently from what we saw in the whole squid axon, in two ways. First their opening and closing is stochastic: either they're open or they're closed, they cannot be in-between. Second, like vesicles at the neuromuscular junction they behave *probabilistically*: the effect of depolarisation is simply to alter the probability of their conducting rather than not conducting. These two things can be seen in the patch-clamping records of voltage-gated sodium channels in rat myotube membrane in **Figure 2.23**. The quantal nature of the responses is very obvious in the

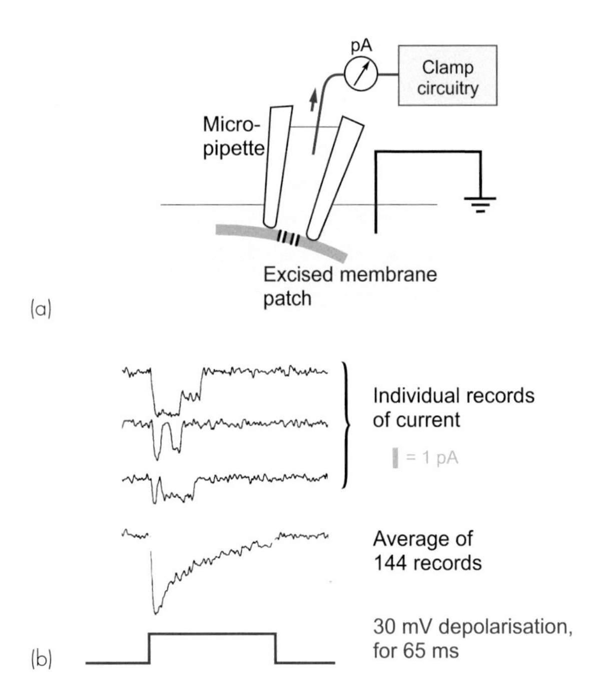

Figure 2.23 Patch-clamping. (a) Schematic view of the method: an excised patch of membrane is held tightly against a micropipette so that the potential across it can be clamped to various levels by passing current through the pipette. (b) Behaviour of individual voltage-gated sodium channels in rat myotubule membrane, as revealed by patch-clamping in single trials (three top traces), individual channels can be seen opening in response to depolarisation and shutting again spontaneously, their currents adding together when more than one is open. When many such records are averaged, the probabilistic summation leads to a curve similar to that seen for whole-fibre preparations (bottom trace). (Data from Pattak and Horn, 1982; with permission from Rockefeller University Press.)

three top records from single trials, and you can see that there must be at least two channels in this particular piece of membrane. You can also see how every time you apply the same potential it does something slightly different. But if you do lots of records and then average them (bottom trace), the average probability of their being open rather than closed then generates a continuous curve that is extremely similar to the kinds of curve that Hodgkin and Huxley measured originally in the whole squid axon.

So to summarise, what we learn from patch-clamping is that *individual channels* are stochastic, with a probability of opening that is a function either of voltage or of concentration of the transmitter. The smoothness of the overall response is simply because of the very large numbers of channels involved, that makes the random fluctuations almost invisible when measuring from whole cells.

Structure of voltage-gated channels

A quantitative description of a biological phenomenon can also often suggest the underlying mechanism. Hodgkin and Huxley showed that the changes in permeability in response to step depolarisations obey quite simple mathematical laws, with an equally simple mechanistic interpretation. The rise of potassium permeability, for instance, obeys the same kind of dynamics as a fourth-order reaction. In chemistry, an nth-order reaction is one where n molecules have to come together for the reaction to occur: if the probability of any *one* of them arriving is p, the probability of the whole reaction occurring is going to be p^n. High-order reactions have a number of characteristics: for instance, they tend to be more temperature-dependant than low-order ones, because p is often proportional to T, so any effect of temperature is amplified by the fourth power. They also tend to be slower than comparable reactions of lower order: p^4 is necessarily smaller than p^3 or p^2 or p. And in fact, by looking at the time-course of a reaction after suddenly doing something that increases p, you can tell from its shape what order the system is (**Fig. 2.24**).

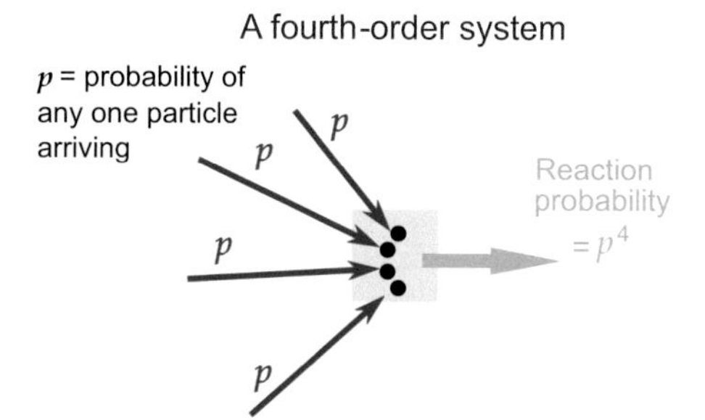

Figure 2.24 In a fourth-order system, four independent events must coincide for the outcome to occur; thus the probability of the outcome is the fourth power of the probability of each individual event.

This, in effect, is what Hodgkin and Huxley did, assuming that the probability p depended on the voltage at any moment. The particular interpretation they put on it was simply that the potassium channels were normally blocked by four independent particles. When the membrane is depolarised, there is simply an increased probability that any particular particle moves out of the way; but to unblock the channel, all four have to move out – hence the fourth order. The recovery is more rapid because it only takes one blocker to flip back for the channel to be blocked once again.

The sodium channel can be modelled in a similar way, but with two important differences: first, it obeys third-order – rather than fourth-order – dynamics (which is why sodium permeability rises more quickly), and second, once open it spontaneously closes again, entering the inactivated state. A plausible model is thus of three blocking particles that move aside when the membrane is depolarised, together with a fourth that does the opposite, moving in to inactivate the channel. What is absolutely extraordinary about all this, and shows the power of really exact quantitative analysis, is that several decades later when it became possible to look at the channels and sequence the proteins of which they were constructed, their hypothetical model turned out to be entirely correct. A sodium channel is a single protein that does indeed consist of exactly four domains, each very similar, composed of six alpha helices spanning the membrane. One of the alpha helices has a number of positively charged residues and seems to constitute the voltage-sensitive part of the complex; another part called the pore loop appears to make the channel selective for sodium rather than other ions. The four domains are believed to arrange themselves in the membrane as shown in **Figure 2.25** (the cylinders represent the alpha helices), and the idea is that when depolarised they tend to twist in such a way as to open the channel.[6]

Might it be helpful at this point to summarise what is known of the electrical propagation of the action potential? *A local depolarisation of a section of nerve gives rise, at first, to an increase in P_{Na} that causes the membrane to become still more depolarised as the potential moves towards E_{Na}. Meanwhile, however, P_K starts to rise, and the sodium permeability to fall, causing the potential to start to drop back towards the resting value. This in turn tends to shut off both the sodium and potassium channels; but because of the delayed response of potassium permeability, there is a period during which P_K is greater than in the resting state, and the membrane is hyperpolarised; eventually the resting potential is regained. Meanwhile, the currents generated by this process have spread to neighbouring regions of the fibre, causing them to depolarise and thus initiating, at a distance, the same sequence of changes all over again.*

In this way the whole pattern of potential and permeability changes is propagated down the fibre (**Fig. 2.26**).

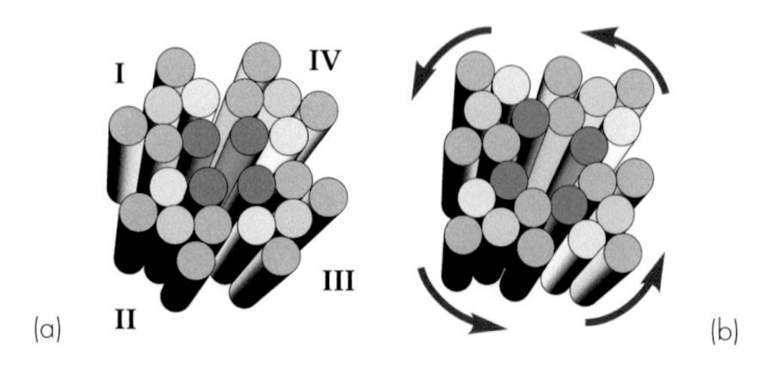

Figure 2.25 (a) Possible structure of a sodium channel, formed of four domains, each comprised of six cylindrical alpha helices. (b) Opening of the channel as a result of small rotations of each domain.

Clinical box 2.2 Broken channels

Knowing the molecular structure of these voltage-gated ion channels, we can now explain the actions of several neurotoxins and a number of genetic disorders affecting nerve conduction; of course, ion channels also come in a ligand-gated variety, which we consider in Chapter 3.

Neurotoxins acting at voltage-gated sodium or potassium channels typically either have pore-blocking properties or alter the gating kinetics and voltage-dependence of their conductance. The aforementioned puffer-fish-derived TTX (see the voltage clamp, p. 29), for example, physically occludes the pore to inhibit the passage of sodium. Meanwhile, scorpion venoms are a nasty cocktail of different components, one of which slows sodium channel inactivation, while another lowers the threshold for the sodium mechanism, so that the fibres fire paroxysmally and then block themselves. Some poisonous frogs secrete batrachotoxin – used in South America as an arrow poison – that lowers the sodium threshold and knocks out inactivation completely.

Inherited disorders of ion channel function – the 'genetic channelopathies' – are a rapidly expanding group of neurological disorders. For example, the most common mutation to cause Dravet syndrome (a severe childhood epilepsy associated with developmental and cognitive impairment), affects the SCNA1 gene to alter the pore-forming unit of the fast sodium channel $Na_V1.1$; this channel is particularly important in inhibitory (GABAergic) neurons, the loss of which results in hyperexcitability in the nervous system and, in turn, epilepsy. In addition to action potential propagation, voltage-gated ion channels are similarly essential for neurotransmitter release (Chapter 3): mutations of the CACNA1A gene, for example, impair the ability of the $Ca_V2.1$ channel to trigger neurotransmitter release, manifesting in paroxysms of unsteadiness, termed episodic ataxia.

Incidentally, a common misconception – actually taught in many schools – is that the action potential that follows the opening of voltage-gated sodium channels is due to a large increase in intracellular sodium concentration that is then corrected to restore resting membrane potential: 'Sodium ions rush in, and during recovery they are pumped out again by the sodium pump'. The clearest demonstration of the falsity of such a view is that after blocking the sodium pump with ouabain, a squid axon can carry several *thousand* action potentials before the internal sodium finally rises to the point where the axon can no longer conduct; this is because actual ion currents involved in an *individual* action potential are absolutely tiny. Furthermore the sodium pump and voltage-gated channels are operating on differing *timescales*; an action potential is over in a matter of milliseconds, whereas the pump continually chugs away in the background, maintaining sodium and potassium concentration gradients over the course of your entire lifetime.

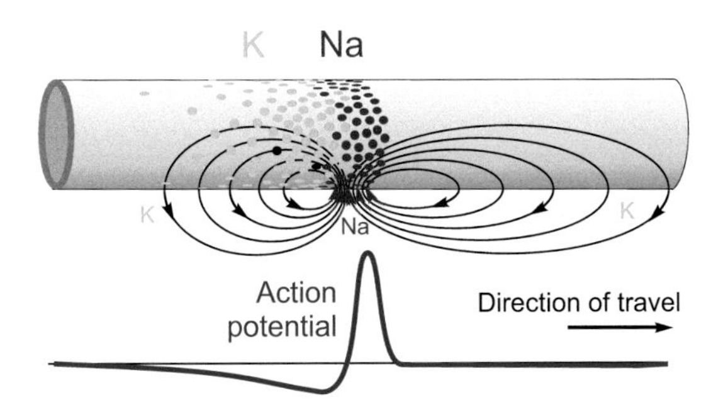

Figure 2.26 'Snapshot' of a nerve axon with an action potential travelling from left to right. The red holes represent the approximate relative density of open sodium channels, the green ones of potassium channels. Below the flow of current and distribution of potential along its length; the 'action potential' could be perceived from a different perspective as an 'action current'.

Table 2.2 Nerve conduction

Short distances	Long distances
Passive	Active
Unmyelinated	Myelinated or unmyelinated
Graded potentials	Action potentials
Cheap	Costly

The term 'action potential' is *itself* a reflection of the history of the study of nerve propagation, which is dominated by descriptions of how membrane voltage changes over time and distance. This is because it was simply easier to record voltage than current – voltmeters have very high electrical resistance and are the ideal 'impartial observer'; ammeters have very low resistance and effectively short out membranes if you try to directly record current flow across them. It took the later invention of the voltage clamp before currents flowing across membranes could be recorded indirectly. If we consider nerve propagation from the perspective of a continually regenerating, depolarising inward sodium current, we could just as easily call it an 'action current'.

Finally, splendid though the action potential is, it is not the *purpose* of nerves to carry action potentials, their purpose is to release neurotransmitters, and passive conduction, though only possible over short distances, is a very much more efficient way to transmit information.

Threshold properties

Once we understand the mutual relationship between membrane potential, on the one hand, and ionic permeabilities, on the other, we can easily explain many of the functional properties of action potentials that make them

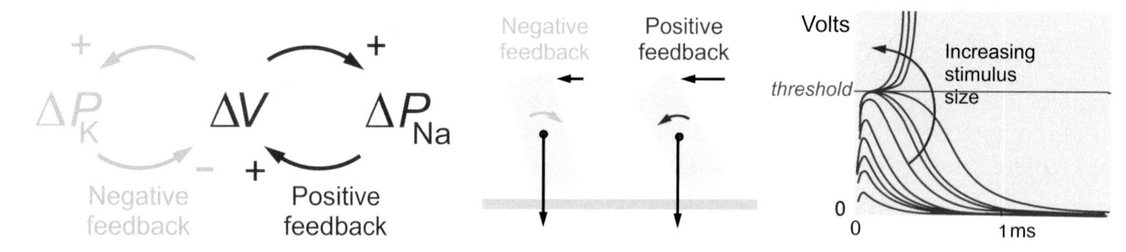

Figure 2.27 The relation between small depolarisations, ΔV, and changes in permeability to potassium ΔP_K and sodium ΔP_{Na} illustrating the existence of negative feedback in the former and positive in the latter. Whether the system as a whole shows negative or positive feedback depends on the relative size of the two components as in the brick shown in the middle. For a small push it shows negative feedback and is stable; for a larger push it shows positive feedback and topples over. The graph on the right shows response to stimulating currents of increasing size applied to crab nerve near the recording electrode, showing stability for small stimuli and instability (action potential generation) for larger ones: close to the threshold it teeters on the brink. (Partly after Hodgkin, 1938; with permission from The Royal Society (U.K.).)

behave so differently from simple passive conduction. In particular we can explain why there is a *threshold* and the existence of the *all-or-nothing law.*

The fundamental concept that underpins practically everything nerves do is the fact that there are feedback loops in the nerve membrane: more exactly, just two of them, one for sodium and one for potassium (**Fig. 2.27**). With potassium, a depolarisation causes an increase in P_K, which then tends to oppose the depolarisation by bringing the membrane potential nearer to E_K: a good example of a *negative feedback* system that tends to stabilise the membrane near its resting potential. Sodium is the exact opposite: if we depolarise the membrane, sodium permeability rises; and we know from the Goldman equation that if sodium permeability rises, this will depolarise the membrane even more. So what we have here is *positive* feedback – and a good thing too, since that is what underlies the membrane's regeneration of action potentials whose amplitudes have dropped because of the losses caused by passive conduction. But *uncontrolled* positive feedback is very bad news. In fireworks, heat stimulates reactions that generate more heat; in atom bombs a nuclear reaction occurs that generates neutrons that trigger more nuclear reactions. Indeed, if you look at any explosive process you will invariably find positive feedback going on.

Whether the system *as a whole* shows negative or positive feedback depends on the relative strengths of these two components. The reason that nerves are slightly less explosive than barrels of gunpowder is that the positive feedback of the sodium loop is tempered by the negative feedback of the potassium loop. Here a depolarisation – as with sodium – causes increased potassium permeability, but the big difference is that when P_K rises the nerve becomes less depolarised rather than more depolarised. So this is not explosive at all, but the reverse: potassium has a stabilising effect. What matters in nerve is the balance between the hysterical sodium response and the calming influence of potassium, in other words whether overall the feedback is positive or negative. Luckily, Nature has arranged things so that in the resting state, at the resting potential, the potassium effect is actually stronger than the sodium one: there is therefore net negative rather than positive feedback, at least for small displacements of potential. But if you push a little harder, with bigger and bigger depolarisations, there comes a point where the response to sodium overtakes potassium, so that there is net positive feedback, and this is what sets the fibre off and generates an action potential. So the threshold is in effect simply the point at which the two effects are just balanced.

A close analogy for all this is a brick or domino being pushed over. Again, there is a balance between positive and negative feedback: for a small push it shows negative feedback and is stable: for a larger push it shows positive feedback and topples over. In between, near the threshold, the brick may teeter on the brink before toppling one way or the other. As you can see in **Figure 2.27**, nerve fibres do exactly the same sort of thing: this is a crab nerve fibre receiving stimuli of increasing size, near the recording electrode, showing the stability for small currents and instability – action potentials – for large; close to threshold it teeters on the brink.

So any factor that favours the potassium mechanism rather than the sodium one will tend to raise the membrane threshold. Two important instances of this occur in the *refractory period* and in *accommodation*.

Refractory period

If we try to stimulate a nerve with a pair of shocks, gradually reducing the interval of time between them, we find that there comes a point when the threshold for the second shock begins to rise relative to that for the first. Eventually, as we go on decreasing the time between the stimuli, we find that we cannot activate the nerve a second time at all, no matter how large the current we use (**Fig. 2.28**). This period, during which it is impossible to stimulate the nerve for a second time, is known as the *absolute refractory period:* the period during which it can be stimulated, but only by using a larger current than usual, is called the *relative* refractory period. The latter corresponds quite well with the period just after the peak of the action potential during which P_K is still raised relative to its resting level, thus tending to stabilise the membrane potential.

The absolute refractory period seems to be due mostly to a property of the sodium channels. We saw earlier that in the voltage clamp experiments the sodium permeability rose quickly in response to a step of depolarisation, and

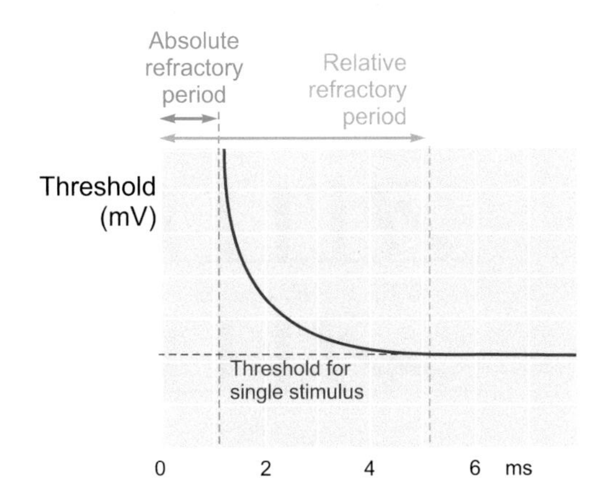

Figure 2.28 Refractoriness of nerve. The voltage, *V*, required to stimulate an axon at different times after a previous suprathreshold stimulus, showing the absolute refractory period and the relative refractory period.

then declined spontaneously, leaving the channels in an inactivated condition which lasts as long as the voltage is maintained. We noted that even when the voltage is returned to its original value, it takes a certain period of time for the sodium channels to revert from their inactivated state to one in which they can once again respond to changes of voltage. Thus after the peak of the action potential has passed, there is a period of recovery during which the sodium mechanism is unresponsive, making the membrane absolutely stable to stimuli of any size. The existence of the refractory period is of considerable functional importance, since this is what prevents the action potential from being conducted in both directions at once. Because the local currents flow almost equally both ahead of the action potential and behind it, it is essential that the region over which it has just passed should not be reactivated all over again; its refractoriness prevents this happening.

Accommodation

Finally, we saw earlier that sodium is quick off the mark but potassium responds more slowly. One consequence of this is that rapid depolarisations are more effective at stimulating the nerve fibre than slow ones, because they get at sodium as it were before potassium has time to rise. This phenomenon has a special name, *accommodation*. It is most obvious if you depolarise not with a step but with what is technically called a *ramp*, rising at different rates. The slower the rate of depolarisation, the higher the threshold; and if the ramp is slow enough it never fires at all, however much you depolarise it – it is said to have accommodated itself to the rising voltage (**Fig. 2.29**).

This peculiar phenomenon can easily be understood if we think in terms of the balance between the sodium and potassium mechanisms. In the voltage clamp experiments, we saw that a sustained step of depolarisation gave rise to an immediate but transient increase in sodium permeability and a delayed but sustained increase in that of potassium. Thus there is only a short period during which

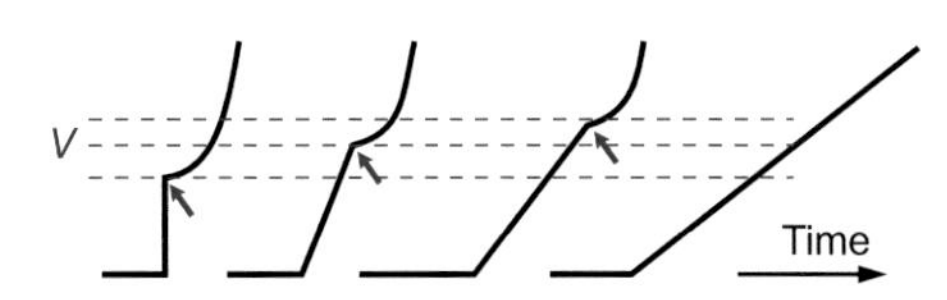

Figure 2.29 Accommodation. The threshold for generating an action potential (arrows) depends on the rate of depolarisation: if this is too slow, the fibre may never fire at all, however much it is depolarised.

the sodium mechanism dominates: time is on the side of stability. Suppose, for example, we were to stimulate a nerve not with one large step of depolarisation, but with a staircase-like sequence of little ones (**Fig. 2.30**). It is clear that whereas P_K increases cumulatively with each new step, P_{Na} does not, since it is only transient; furthermore, the transient increase in P_{Na} will steadily decline with increasing depolarisation, because of the steadily increasing degree of sodium channel inactivation, and also because the potential is closer to E_{Na}. Thus the more gradually we depolarise a nerve fibre, the more we push the sodium/potassium balance in favour of potassium, and the further we need to depolarise it in order to reach the threshold; and if we depolarise it slowly enough, there will come a point where P_{Na} is never great enough relative to P_K for the nerve to fire at all, and the membrane will therefore completely accommodate. A paradoxical consequence of this is that if you slowly depolarise a nerve fibre to a steady level just under its normal threshold, you might think it would therefore be easier to stimulate. In fact it's actually *more* difficult to stimulate than it was originally. In fact the whole concept of 'threshold', which is after all just an experimental observation, is somewhat arbitrary: threshold is as much to do with the *rate of change* of the depolarising stimulus as the stimulus size itself.

However, there is one big difference between a domino and a nerve: the feedback is *changing* in strength and direction all the time in nerve, and as a result it is finally able to recover and right itself. Sodium is quick on the draw, but quickly gives up and in fact keels over altogether because it gets inactivated. Potassium on the other rises slowly but *inexorably* to its final value. And this is exactly as it should be, because it means that in the long run potassium always wins and the membrane is absolutely bound to return to its original excitable state. We shall see later that the mechanism of accommodation can sometimes be an important determinant of the way in which sensory receptors respond to slowly changing stimuli.

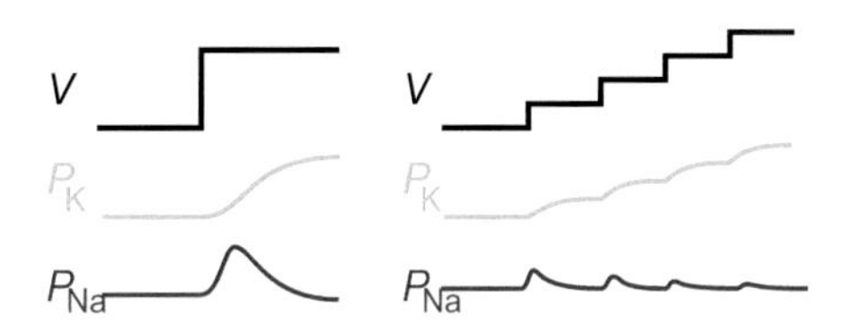

Figure 2.30 The changes in P_K and P_{Na} in response to a clamped step of voltage. In response to a series of small steps, approximating a slowly increasing depolarisation, potassium permeability increases steadily, while sodium permeability declines through inactivation.

The all-or-nothing law

We are now in a position to explain something of huge functional importance to nerve, the *all-or-nothing law*. When you record action potentials, one thing that becomes very obvious is that their size does not vary. In particular, it is not affected by the size of the stimulus that caused them. This something that's true of *any* system that relies on propagation through regeneration via positive feedback – think of our sparkler, or a barrel of gunpowder: it makes no difference whether you ignite the barrel with a match or a flame-thrower – the bang is just the same.

Any system with positive feedback will tend to behave in a manner approximating to all-or-nothing behaviour. If we nudge a domino over, ultimately it hits the ground with much the same force, whether the original nudge was large or small – much the same, but not *exactly* the same: clearly, if the energy of the push is appreciable in comparison with the domino's stored potential energy, the force with which it strikes the ground will be increased. More exactly, the energy released on falling over will be $P + E$, where E is the stored potential energy and P the energy imparted by the original push. In the case of nerves, the all-or-nothing law is not found to be strictly obeyed if one records close to the point of stimulation – within a length constant or two – since the stimulus energy then contributes in part to what is recorded. But as the action potential is propagated further and further away from its origin, this contribution becomes increasingly negligible, and it eventually settles down to its standard form. What we have, in effect, is not just one domino but a whole line of them (**Fig. 2.31**): when one falls, it imparts a fraction of its energy to the next, sufficient to knock it over, and so on in turn all the way down the line. Imagine for the sake of argument that one-tenth of a falling domino's energy is used in knocking over the next. Then the first domino imparts an energy $(P + E)/10$ to the second, which in turn imparts $((P + E)/10 + E)/10$ to the third, and so on: it is clear that the contribution of the original push, P, to the energy with which the nth domino hits the ground will get vanishingly small as n gets larger, and that this energy will in fact settle down at a constant level: the system as a whole will then obey the all-or-nothing law exactly. Thus the basic cause of all-or-nothing behaviour is the regenerative process that produces action potentials.

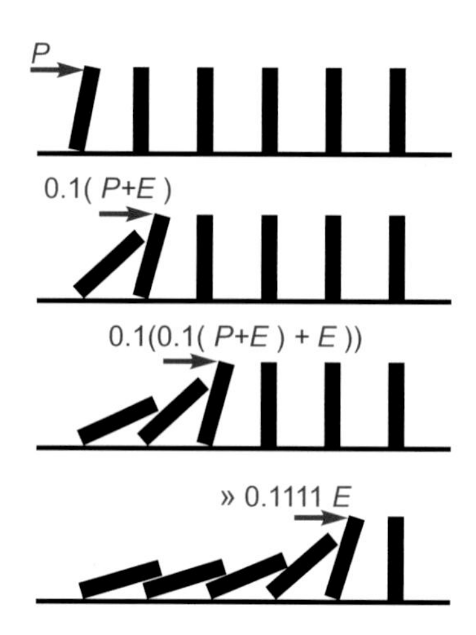

Figure 2.31 All-or-nothing behaviour of a row of falling dominoes.

Why all-or-nothing?

This law is of fundamental significance in the nervous system, and it is worth reflecting on its functional implications. Why should it have evolved? It clearly imposes very severe limitations on the kinds of messages that nerves can convey, prohibiting direct transmission of graded quantitative information (of the kind conveyed, for example, by the varying concentration of a hormone in the blood), the only messages permitted being of the binary 'yes/no' variety. The answer certainly lies in the problems of trying to send messages along cables that are so leaky that currents cannot be conveyed passively more than a matter of millimetres. An engineer faced with such a problem – as found on a somewhat larger scale in transatlantic submarine cables – would probably deal with it by introducing a series of booster amplifiers at intervals along the cable, to restore the losses caused by leakage. In the case of nerve axons, we have already seen that the length over which they are required to conduct is so vastly greater than the length constant that many thousands of such stages of amplification would be required; each node of Ranvier is in effect a booster of this kind.

What are the characteristics of a chain of amplifiers of this sort? All amplifiers, however good their quality, suffer from two defects: they introduce *noise*, and they create

Clinical box 2.3 Nerve conduction studies, a valuable clinical tool

Neurologists measure nerve conduction in the investigation of sensory and motor disorders that are thought to originate from pathology of the peripheral nerves. Typically, the clinician stimulates the nerve by placing a negative (depolarising) cathode and adjacent (hyperpolarising) anode over the nerve and then generating an electrical pulse between them. As the stimulus amplitude is increased fibres of greater threshold are stimulated; a supramaximal stimulus is used to ensure that all nerve fibres within the nerve are activated and that slight changes in stimulus intensity or relationship between nerve and surface electrode do not affect the recorded signals. Signals are then recorded from an electrode placed either over another, distant part of the nerve (for sensory nerves) or over the muscle innervated by a motor nerve. The recording electrode registers an initial upward (negative) deflection followed by a downward deflection. The recorded signals are the summed response from all the fibres in the nerve, called *compound motor action potentials* (CMAPs) or *sensory nerve action potentials* (SNAPs) depending on whether the nerve is predominantly motor or sensory.

The clinician then measures the signal amplitude (maximal height of the negative deflection), the area (area under the negative deflection) and the signal latency (time from stimulus to onset of negative deflection). The latency can then be used to calculate the conduction velocity. Values are then compared with those recorded from cohorts of normal individuals matched for age and sex.

Clearly, there is a problem calculating motor nerve conduction velocity in this way: the time for the stimulus to reach the recording electrode is affected not only by the nerve fibre, but also the relatively slow neuromuscular transmission (see Chapter 3) and the conduction in the muscle fibre. To tease out the pure motor nerve component the nerve is stimulated at two points along its course, and the latencies, or time from stimulus to recorded signal, are subtracted from each other; the difference in latency, therefore, reflects the time taken to conduct the distance between the two points of stimulation on the nerve. The conduction velocity is the spatial separation from the two points of stimulation divided by the difference in latency. In the case of SNAPs, conduction velocity is simply the spatial separation of the electrodes divided by the latency.

Interpreting the signal

There are three main pathologies affecting the peripheral nerve: axonal degeneration, disorders of myelination and conduction block. Nerve conduction studies (NCS) attempt to distinguish between these. While in reality this distinction is a little simple, and overlap exists, NCS still provide essential information, guiding diagnosis and treatment.

Axonal degeneration is characterised by reduced signal amplitude, as there are fewer functional fibres available to carry the action potentials that contribute to the summed response picked up by the recording electrode. Examples of axonal degeneration include forms of hereditary, chronically progressive sensorimotor neuropathies, or certain acquired forms of acute peripheral neuropathy such as axonal Guillain-Barré syndrome (GBS). The latter can progress from tingling in the toes to complete paralysis, often sparing only the eye movements, within 24–48 hours. This illness has autoimmune pathogenesis, and is usually provoked by an antecedent infection, such as with *Campylobacter jejuni;* it has also been described following Zika virus and SARS-CoV-2 infections. Chronic alcoholism can also lead to a much more insidious axonal neuropathy either directly or because of vitamin deficiencies.

Diseases caused predominantly by a loss of the myelin sheaths around nerve fibres are characterised by *reduced conduction velocities,* as in the middle set of traces shown in the figure. The space constant is diminished as the myelin is lost and local, contiguous conduction replaces the much faster saltatory conduction. Causes again include chronic, hereditary disorders and acute, acquired disorders such as the demyelinating form of GBS. Since the myelin sheaths appear easier to repair than the axons, demyelinating GBS, while just as acutely devastating, usually has a shorter duration of illness and fuller recovery than its axonal counterpart.

Conduction block (bottom trace) occurs when there is a focal interruption to a group of nerve fibres. If the distance between stimulating and recording electrode is increased there is an increased probability that a conduction block will be interposed between them, hence conduction block is suggested when the signal area and amplitude are reduced as the separation between the stimulating and recording electrodes is increased. Without block, the signal amplitude is maintained irrespective of distance due to the all-or-none, self-regenerating nature of the action potential. Multifocal motor neuropathy with conduction block is an acquired autoimmune condition manifesting with progressive weakness and disability. Appropriate therapy depends on the diagnostic information provided by the NCS.

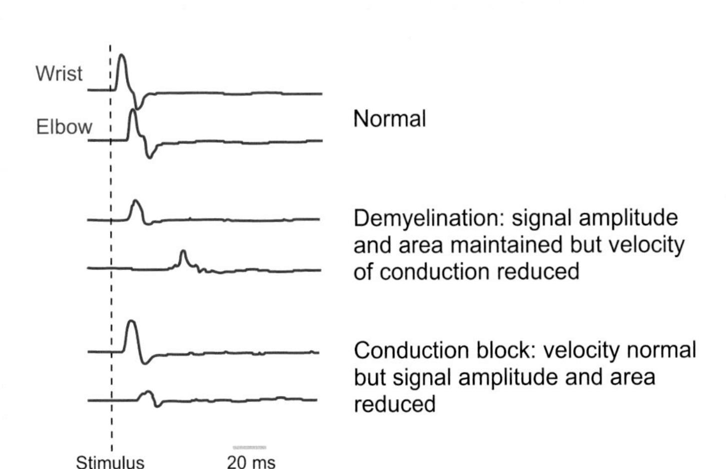

NCS allow distinction between axonal, demyelinating or conduction block pathologies, and identify predominantly sensory or motor patterns of loss. They also distinguish mononeuropathy (one nerve, e.g. following a nerve injury), multifocal polyneuropathy (several nerves in several spots) and uniform polyneuropathy (all nerve segments).

distortion. Noise includes both the hiss that arises inevitably in any electrical system – including neurons – from the random movements of the electrons or ions in its conductors and the disturbances picked up from external sources of interference. Now imagine a thousand hi-fi amplifiers connected end to end, so that the output of one forms the input of the next. The noise generated by each one of them will be amplified all the way down the line and added to those of the others, making the final output very much noisier than if there were only one amplifier. Similarly, distortion – inaccuracy in the linearity of the amplification – also becomes exaggerated if a number of amplifiers are connected in series. If for example the gain of each amplifier is 1 per cent greater than it should be, then the gain of the whole set of a thousand will be too large by a factor of some 2000; and if 1 per cent smaller than it should be, then the overall gain will be 1/2000 of the correct value. Thus accurate transmission of quantitative information becomes almost impossible: the system almost automatically becomes all-or-nothing in character, since signals either vanish or become saturatingly huge.

The only solution is to be less ambitious about *what* one is trying to signal. If for example one limits oneself to only two possible signals – 'yes' or 'no' – then distortion no longer matters: the signal is either there or not there, and no regard need be paid to how large it is. If we also arrange for each amplifier to have a threshold that is higher than the normal noise level but allows through the signal 'yes', then we can get rid of noise as well. In other words, the only kind of system that is capable of transmitting messages reliably over distances that are much bigger than the length constant is precisely what we have found in the nerve axon itself: a series of regenerative amplifiers (the voltage-sensitive sodium channels) exhibiting a threshold that prevents the fibre from producing spurious signals in response to its own noise. There is no advantage in such a system for conduction over shorter distances, and in practice it is found that short neurons (as for example the bipolar cells of the retina) never use action potentials. They rely on the much simpler and more informative method of passively propagated electronic potentials: stimuli of different sizes are converted into potentials of different sizes, and these are conducted passively and faithfully to the other end, where they cause the release of similarly graded amounts of transmitter. But over distances longer than a few length constants, this will not work: propagation has to be active. This is bad news, as it means a large expenditure of energy; it also slows things down, since the regeneration process takes time. Thus there is nothing intrinsically desirable about action potentials: they are a necessity imposed by the need for nerves to be small and severely constrain the way in which information is coded. If – because of the all-or-nothing law – each action potential is exactly the same as any other, how can a nerve communicate how strong the original stimulus was?[7]

Neural codes

In fact there are many ways in which such graded information could be coded despite all-or-nothingness. At one time, in the very early days of digital computers, when people were thrilled by the way in which these tremendously complex devices with their miles of wiring and hundreds of repetitive units seemed so very much like the brain, it was natural to compare the all-or-nothing action potential with the similarly all-or-nothing digital pulses on which these computers were based. (There are still textbooks that say that nerves code information digitally.) But this is completely wrong. The essential difference between the two basic ways of coding information, analogue and digital, is this:

- In *digital* coding we convert it into a pattern using a more or less arbitrary code which is discrete and not continuously variable at all, with only a finite number of possible values: think of the digits on a digital watch.
- In *analogue* coding, we convert the incoming variable into another continuously variable quantity that directly represents its size in a one-to-one sort of way: in other words, it's an *analogue* of it, like the minute hand on an analogue watch (**Fig. 2.32**).

Nowhere in the brain do we ever see what you need to have in digital pulse-coded systems, which is impulses either occurring or not occurring in fixed positions. In computers, there has to be some kind of internal clock within the machine that defines what the timing of the pulses means. It is unlikely that this form of coding is used by the brain, both because the slowness of conduction would make it difficult to maintain accurate timing between events, and also because of the apparent absence of anything equivalent to an internal reference clock. There are other possibilities: a familiar example of a different kind of binary code is the Morse code, where information is carried in the temporal pattern of the only two possible signals – dot and dash. However, coding of such sophistication has never been observed in actual neurons, and we shall see that the mechanism by which neurons are caused to fire repetitively makes it unlikely

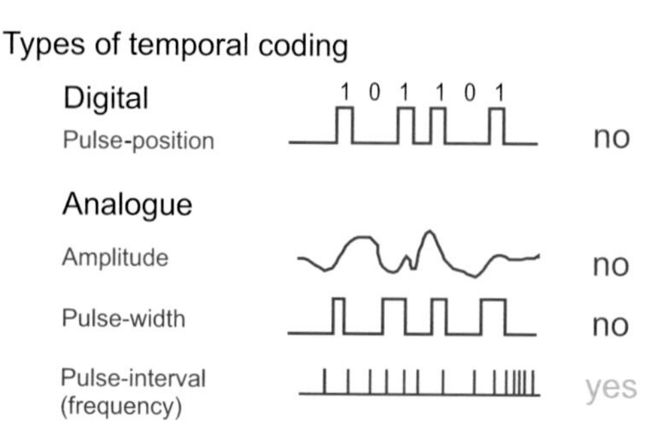

Figure 2.32 Types of neural coding that could be used by nerve fibres. Over long distances, only pulse-interval coding is actually used.

that information could actually be carried by the nervous system in this form.

So in this respect the brain is not in the least like a computer. In nerves, information is mostly coded by the frequency of firing, and as this is something that that can vary continuously, it provides an analogue code. This kind of signalling is technically called frequency modulation. So although we can't control the *size* of the APs or spikes, what we can do is alter *how often they happen*.

Frequency coding

Frequency coding has one very great advantage, the same as the advantage of FM (frequency modulated) as opposed to AM (amplitude modulated) radio. If you add interference to a signal it messes up its amplitude but hardly affects its frequency at all; as a result FM is much less prone to noise and interference than AM. In AM, the amplitude of the radiofrequency carrier wave is a direct copy of the sound wave being transmitted (**Fig. 2.33**); the radio receiver decodes this signal by converting the envelope of the radio wave back into a sound wave. The disadvantage of such a system is that any variations in the amplitude of the wave caused by transmission itself – fading, or noise generated by radio interference – get incorporated in the sound reproduced by the receiver. In FM transmission this is no longer the case: here it is the frequency of the radio wave rather than its amplitude that conveys the sound information, and disturbances that affect its amplitude no longer matter, since it is only the frequency of the received signal that is decoded by the receiver, producing essentially noise-free transmission. If a peripheral nerve is stimulated by a touch receptor in the skin so that 50 impulses are despatched from the periphery, it is virtually certain that exactly 50 impulses will be received by the central nervous system. So action potentials are extremely *reliable*, but very *costly* in terms of the energy consumed by the sodium pumps needed to mop up after the repeated action potentials.

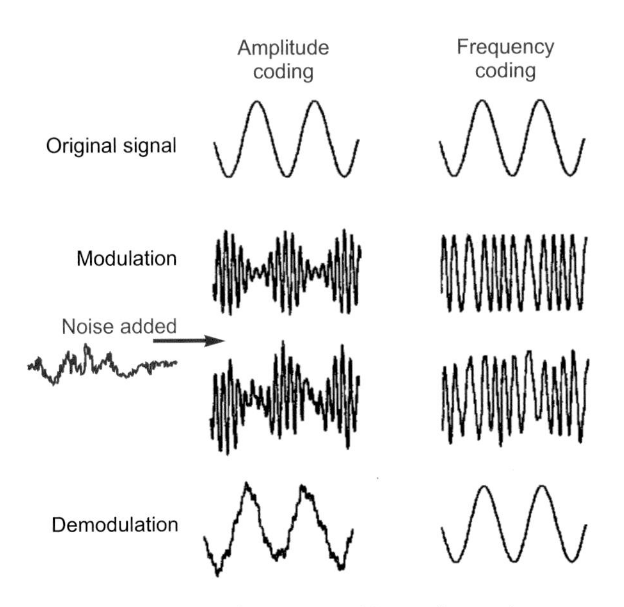

Figure 2.33 Amplitude and frequency modulation, showing how noise added to the modulated radio wave results in more interference in the decoded audio signal in the case of amplitude modulation than for frequency modulation.

Spatial coding

So far we have only considered *temporal* codes; but in addition, nerves can signal information *spatially*. This is related to what at first sight is a strange feature of nerves, that they contain quite so many fibres. Take a muscle like the gastrocnemius, for instance: it can only do one thing – at any moment it can exert a certain force. So one might well think that all you need to control it is just one nerve fibre, whose frequency of firing would tell it what force to generate. Yet the nerve innervating the gastrocnemius has far more than just one fibre – in fact, hundreds and hundreds of them. The reason is that this provides an additional mechanism for providing fine grading of the force of contraction, called *recruitment*, the extra dimension enabling the *total* number of action potentials per second to be varied over a wider range. If you record from any one fibre, you find that its firing frequency increases with the strength of contraction, as you'd expect; but there is a certain threshold below which it doesn't fire at all. If you move to a different fibre, you find the same thing, but typically with a different threshold. Because all the thresholds are different, as the force of contraction increases, it isn't just that any particular fibre fires faster, it is that *more* of them start to fire at all (**Fig. 2.34**). Recruitment enables the muscle to be controlled with a wider range of commands than would be possible if there was only one afferent nerve fibre. It does not just apply to motor nerves: a good example of this, as we shall see, is in the nerve from the vestibular apparatus. Here the fibres are all found to have different stimulus thresholds, so that increasing stimulation leads to more and more of them firing at once.

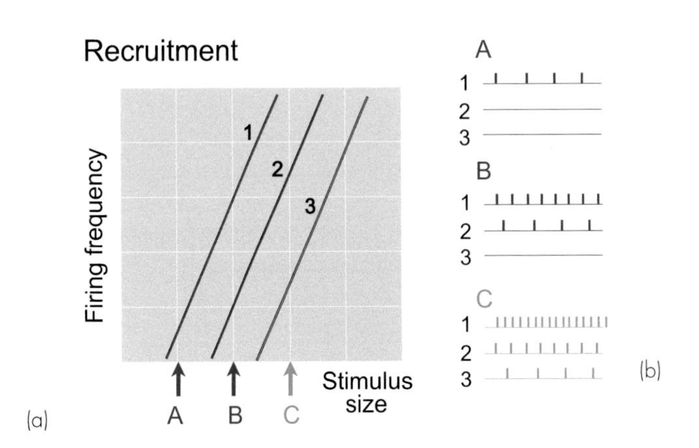

Figure 2.34 Recruitment. The lines show how firing frequency is related to stimulus size in three hypothetical nerve fibres of different thresholds (a). With increasing stimulus size, the number of fibres firing as well as the rate of firing both increase, leading to an acceleration in the total number of action potentials (b).

Coding and decoding the frequency

How are frequency codes generated? And how are they decoded again?

The second question is easier to answer. Recall that the function of nerves is to release controlled amounts of transmitter from their terminals. As we shall see in more detail in the next chapter, when action potentials finally reach the end of the axon they open voltage-gated channels, that let *calcium* in. This then makes the terminal release the transmitter it contains from the vesicles in which it is normally stored. Since each spike is identical, it releases the same quantity of transmitter; consequently, altering the frequency causes the *rate* of transmitter release to change, so that the original information is passed on to the target cell.

Converting stimulus size to firing frequency is a little more complex. Consider a neuron with a set of receptor channels which when open tend to short-circuit the membrane and lead to an equilibrium potential around zero. Somewhere between the resting potential *ER* and zero there will be a threshold potential θ for triggering an impulse. If we suddenly open the receptor channels and keep them open, the potential will move towards zero and must at some point cross the threshold, setting off an action potential. The usual stereotyped sequence of changes in permeability will then ensue, terminating in a recovery phase in which P_K will be elevated and the neuron relatively hyperpolarised as its potential is pulled towards *ER*. As P_K declines to normal after the impulse, the potential will rise again, not just to the original resting potential but past it (since we suppose that the receptor channels are still open) towards zero. What happens next will depend on the rate at which this depolarisation occurs. If it is sufficiently fast (and correspondingly low), the threshold will be crossed once more, and a second action potential will be generated; then a third, a fourth, and so on: impulses will continue to be generated so long as the receptor channels remain open (**Fig. 2.35**). (But if the rate of depolarisation after the first action potential is too slow, θ may rise so much because of accommodation that the membrane potential never reaches it, and the neuron will fail to fire for a second time).

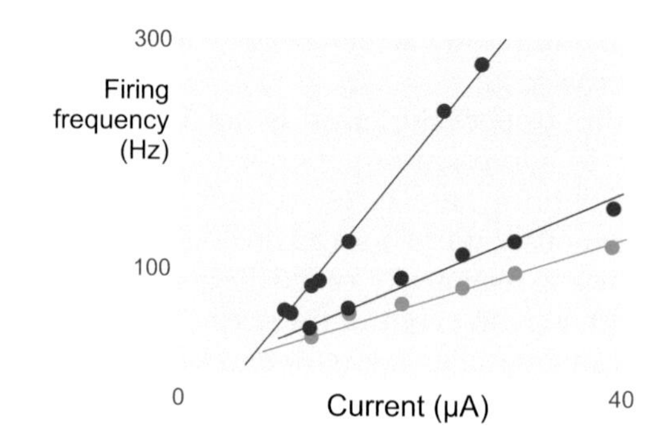

Figure 2.36 Experimental relation between injected current and resultant steady firing frequency for three motor neurons. (Data from Grant *et al.*, 1963; with permission from John Wiley and Sons.)

The greater the short-circuiting current, the faster the rate of depolarisation will be after each impulse, and so the sooner the nerve will fire off again. Thus the frequency of the repetitive firing will depend on the degree of short-circuiting that we have produced: the more receptor channels are open, the higher the frequency. Under suitable conditions, a steady current injected through a microelectrode will imitate the effect of a short-circuiting permeability change and will elicit repetitive firing. In such cases the frequency is often a linear function of the applied current, as in the data from three motor neurons shown in **Figure 2.36**. Even though the current is held constant, one often observes in such a preparation that the frequency of action potentials declines from an initial high value to a lower steady state (**Fig. 2.37**). A decline of this kind in the response to a steady stimulus is called *adaptation*, and is discussed much more fully in the next chapter. This particular example of adaptation seems to be a general property of all kinds of neurons, including receptors, and is called *membrane adaptation*. It is sometimes described as accommodation, but this is very misleading since it is not in fact due to the same mechanism as that underlying the true accommodation described earlier.

One mechanism that appears to contribute to membrane adaptation is the entry of calcium ions during the action potentials. The calcium then acts on a type of potassium channel that is distinct from the voltage-sensitive

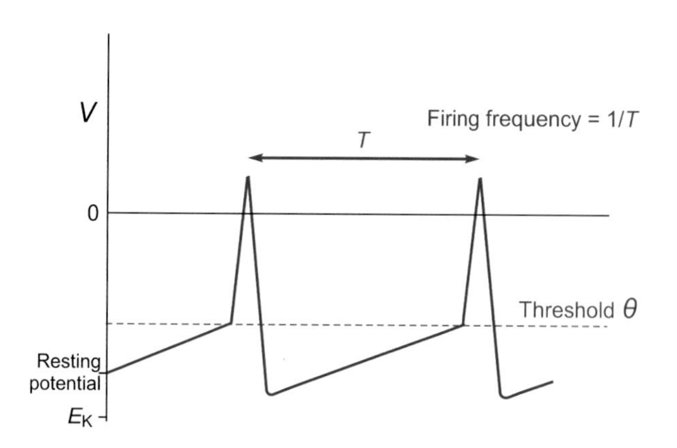

Figure 2.35 Mechanism by which a steady current may initiate repetitive firing.

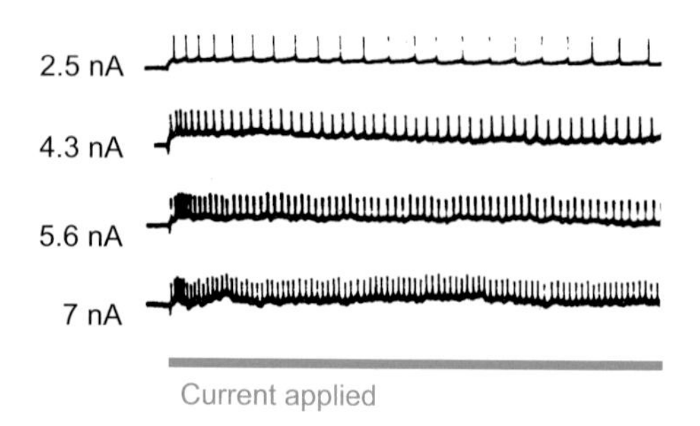

Figure 2.37 Membrane adaptation in motor neurons: spike responses to steadily injected currents of the strengths indicated. (Oshima, 1969; permission pending.)

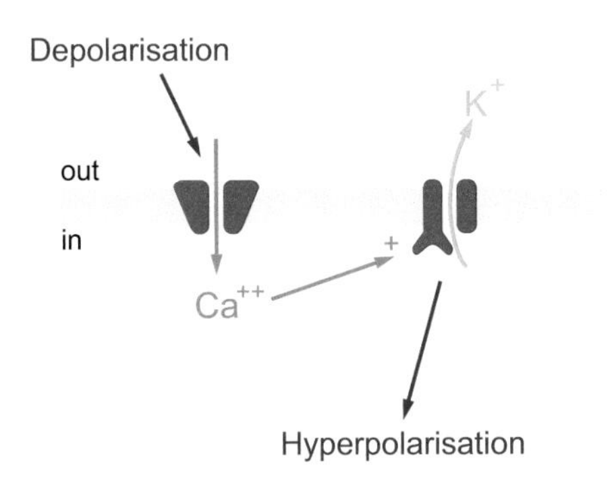

Figure 2.38 One mechanism that contributes to adaptation. Depolarisation increases internal calcium concentration, which in turn activates calcium-dependent potassium channels. The resultant increase in P_K forms a negative feedback loop that tends to restore the original potential.

ones we have come across so far, causing it to open and thus increase P_K, stabilising the membrane and raising the threshold for generating action potentials (**Fig. 2.38**). This probably represents a general mechanism for regulating the resting potential, rather than specifically intended for adaptation.

In addition to adaptation, there are other more complex kinds of temporal patterns that can occur, especially in central neurons, in response even to steady stimulation. An example is the rhythmic occurrence of *bursts*, clusters of high-frequency action potentials separated by periods of quiescence. This more complex behaviour can be due to channels with longer periods of activation or inactivation, or to specialised low-threshold voltage-gated calcium channels, particularly when, as in **Figure 2.38**, calcium entry then interacts with some other channel to alter its properties. A particular instance of unusual discharge patterns with an identifiable functional meaning occurs in the thalamus in relation to sleep and arousal (discussed in Chapter 14).

Conduction velocity

So far, very little has been said about the *speed* at which all these processes occur. We have traced the sequence of events by which one active region of nerve can trigger off a similar pattern of activity in another one at a distance from it by means of local currents: conduction velocity is simply a matter of how *far* and how *quickly* these currents spread, and how long it takes for them to be regenerated.

You might be forgiven for thinking that conduction of electricity down a nerve would simply happen at the speed of light, but that is completely untrue. Currents travel down cables – whether nerve fibres or any other kind of man-made cable – at speeds that are considerably less than this. When the first transatlantic telegraph cable was laid in 1866 from Valentia in Ireland to Newfoundland – over 2000 miles – people were astonished that it took three seconds for current to travel from one end to the other, about 3 per cent of the speed of light. That was a *good* cable: with really *bad* cables like nerve fibres, passive conduction is very slow indeed – sometimes less than 1 m/s. Why is it so slow?

Capacitance

The answer lies in an electrical property of the membrane called *capacitance*. Any two conductors separated by a layer of insulation act as a capacitor. The larger the opposed areas of the conductors, and the thinner the insulating layer between them, the larger the capacitance will be. In the case of nerve fibres, the membrane is both a good insulator and extremely thin, and makes a splendid capacitor: it has a capacitance, C_M, of about 1 μF/cm². So our equivalent circuit should really be redrawn in the form shown in **Figure 2.39**.

Now the effect of having capacitance in a circuit of this sort is to make it more sluggish in its responses. If we suddenly pass a current I through a resistor R_M on its own, the voltage across it immediately reaches the value $V = IR_M$; but with a capacitor as well it now takes *time* for the voltage to reach this value, because part of the current must be used to charge up the capacitor to the new level. What is observed is that on injecting a step of current of this kind, the voltage rises only slowly to its final value of IR_M, with a time-course that is exponential and given by $V = IR_M(1 - e^{-t/\tau})$. τ here is the *time constant* of the circuit (the time taken for the discrepancy ($IR_M - V$) to fall by a factor e), and is equal in this case to $R_M C_M$. For many nerve fibres, this time constant is of the order of a few milliseconds, setting a limit on the rapidity with which the membrane can generate voltages in response to local currents (**Box 2.2**).[8]

The question of how *far* the local currents spread was considered earlier in this chapter. We saw that a voltage generated at a particular point on the membrane declines exponentially as a function of distance, with a length constant, λ. The length constant and time constant together give a measure of the speed with which an electrical disturbance is propagated passively along the axon, regarded as a simple cable. This speed is in fact equal to λ/τ, which has the dimensions of a velocity.

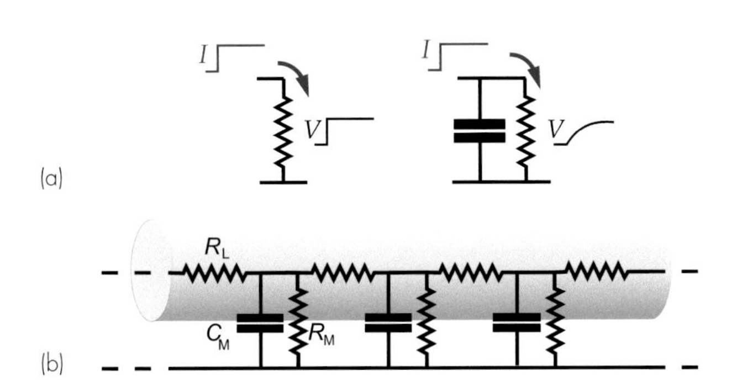

Figure 2.39 (a) Voltage response of a resistor (left) and of a resistor and capacitor in parallel (right) to an applied step of current, showing the slow, exponential rise of voltage in the second case. (b) Modification of the equivalent circuit of Figure 2.5 to include membrane capacitance, C_M, as well as resistance, R_M.

Box 2.2 Capacitance

Capacitance means the ability to store charge. All objects can store charge to a certain extent, but as you add charge the voltage quickly rises and makes it difficult to add more. The ratio of charge Q to voltage V is what is called the capacity or capacitance C, and for most objects C is very small. But if two conductors are separated by a thin insulator, as in the early Leyden jars, capacity is much bigger: if you take electrons from one side and add them to the other, creating a positive charge on one side and a negative on the other, the pluses and minuses attract one another and partially neutralise themselves, and you get a smaller potential difference than would otherwise be the case.

It's called capacity for the very good reason that there is an exact parallel here with, say, water stored in tanks of different capacity: actually many people who feel their grasp of electricity is shaky find it helpful to think in terms of water instead of currents and voltages. Voltage or potential is the same as water *pressure*; charge is equivalent to the quantity or *volume* of water, and current to *rate of flow*. Then you can see that a given quantity in a tank of small capacity is going to fill it up more, to a higher level and therefore a higher pressure, than a large tank. Now if we attach an outlet tap to the tank, water leaks out, at a rate that depends on the pressure and on how much resistance the tap offers to the flow. The flow is equal to the pressure divided by the resistance, just as $I = V/R$. As the level in the container falls, the rate of flow falls too, because the pressure is dropping all the time – so you get an exponential decline.

Exactly the same is true of the capacitor: if you connect a resistor across a charged capacitor, it gradually discharges it, the voltage dropping rapidly at first but then slowing down, again exponentially. And just as we use the space constant as a measure of how far current spreads exponentially, so we can use something called the *time constant*, τ, to express how quickly the capacitor discharges. It's simply the time it takes to fall by a factor e, just as for the space constant. If we call the size of the capacitor C and the resistance R, the time constant is in fact given by the product $R.C$: if the resistance is bigger it will take longer to discharge, and so it will if the capacity is bigger.

Regeneration time

Now we need to consider what difference it makes having *active* rather than passive conduction. Passive conduction involves just λ and τ; active conduction – perhaps paradoxically – is actually slower than passive because of the extra time, T, needed to regenerate the action potential from threshold to full size. We need to modify the formula for the velocity to take this into account: in effect, instead of λ/τ we now need something like $\lambda/(\tau+T)$ (an over-simplification but one which will suffice for the purposes

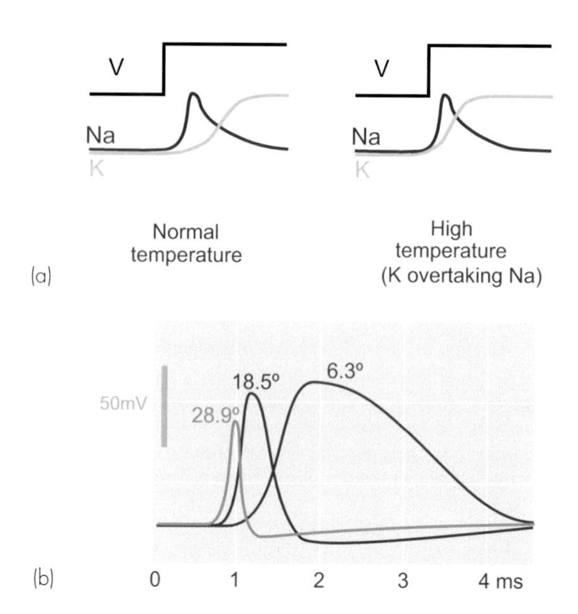

Figure 2.40 Temperature and conduction velocity. (a) Because the opening of potassium channels is a higher-order process than that of sodium channels, its speed is more affected by temperature. Thus, at higher temperatures, the potassium response to a step of depolarisation tends to catch up with the sodium response. (b) As a result, action potentials travel faster at higher temperatures, but also get smaller. (Squid axon, after Huxley, 1959; with permission from John Wiley and Sons.

of this book). T is mostly due to the time it takes for the sodium permeability to respond to the change in potential and normally is very short, so that T is small in comparison with τ.

Although T is not normally a very large factor, there are circumstances when it can alter. Higher temperatures speed the permeability changes up a great deal, because they are both high-order reactions, but they affect the fourth-order potassium more than the third-order sodium (**Fig. 2.40**). As a result, potassium gradually catches up with sodium and the action potential actually gets briefer and smaller as the temperature is raised. In some cold-blooded animals, conduction ceases altogether if the temperature exceeds 37°C, as in the squid axon shown in **Figure 2.40**.[9]

The size of the local currents also depends on the *ionic concentrations* inside and outside the fibre – low external sodium, for instance, reduces the velocity of conduction because it makes the sodium current smaller – and is influenced by local *anaesthetics* and other pharmacological agents acting on the permeability mechanisms. It is also a function of the density of sodium channels in the membrane; the nodes of Ranvier have a much higher density of sodium channels than do ordinary unmyelinated fibres, another factor contributing to the increased conduction velocity of myelinated nerves.

Diameter and myelination

Finally, we need to consider second-order factors that might influence conduction velocity by acting on λ and τ. One such factor is the *diameter* D of the fibre. How will this affect τ? τ is equal to the product of C and R_M, both of which depend on the surface area of the fibre: if D increases, the surface area increases in proportion. This makes the capacitance increase, but it makes the resistance decrease;

Box 2.3 Summary of factors affecting conduction velocity

Factors that affect conduction velocity concern either the time it takes things to happen or how far the effects spread: *temporal* factors or *spatial* ones. Velocity is given approximately by:

$$\frac{\lambda}{\tau + T}$$

where λ is the space constant, τ the time constant for passive propagation down the axon, and T a measure of how long it takes for a threshold depolarisation of the membrane at any point to regenerate itself to full size. These three *primary* factors in turn depend on *secondary* factors.

The space and time constants depend on the longitudinal and transverse resistances of the axon (R_L, R_M) and the membrane capacitance (C_M):

$$\lambda = \sqrt{\frac{R_M}{R_L}}$$

$$\tau = R_M C_M$$

These in turn are influenced by:

1. *Diameter*
   ($R_M \propto D^{-1}$, $R_L \propto D^{-2}$, $C \propto D$; so $(\lambda/\tau) \propto \sqrt{D}$)
2. *Myelination*
   (λ is increased but not τ; the effects on R_M and C cancel out)
3. The external resistance
   (Because it contributes to the effective value of R_L)

The effective regeneration time, T, depends on:

1. Temperature
2. Density of gates
3. Ion concentrations
4. Anaesthetics, anoxia, etc.

oddly enough these two effects cancel out, so that the time constant doesn't vary with diameter at all. What about λ? Earlier we looked at the effect of diameter D on the length constant, and we saw that in fact λ varies with the *square root* of the diameter. So if the velocity is proportional to λ/τ, and τ is constant, then velocity will also vary with the square root of D as well, which is indeed what is found (see **Box 2.3**).

The second factor is *myelination*. Animals rarely have unmyelinated fibres larger than about 1 µm in diameter: the reason for this is that myelination offers a far better way of increasing the conduction velocity of large fibres than simply increasing their size. As we saw in the previous chapter, the effect of myelination is to greatly thicken the layer of insulation round the fibre (except at the nodes of Ranvier); this has the desirable consequence of greatly increasing R_M and reducing C_M. As you might expect, this enormously increases R_M, but once again it has the opposite effect on C_M, which gets smaller. Again, the two factors

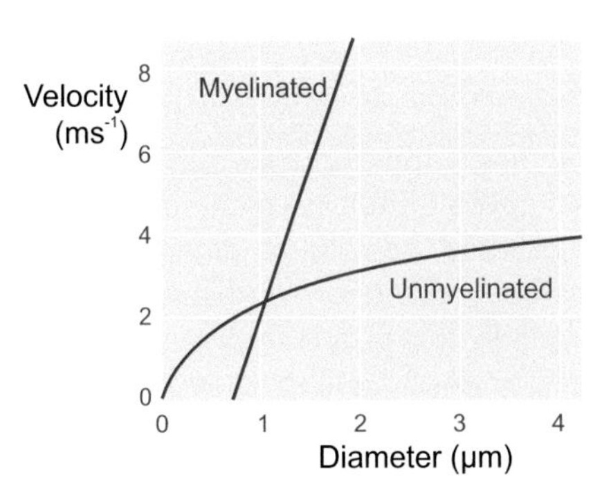

Figure 2.41 Theoretical dependence of conduction velocity, V, on axon diameter, D, for unmyelinated (U) and myelinated (M) axons. M is extrapolated from observations; U is scaled to fit observations on fast C-fibres (considerably idealised). (After Rushton, 1951; with permission from John Wiley and Sons.)

cancel out, so that the time constant is no different. But as we saw before, the extra insulation *does* increase the length constant, so as a result conduction is greatly speeded up: in effect, the myelin forces the external local currents to travel further before they can gain access to the axoplasm through the nodes. A curious thing about myelinated nerve fibres is that they don't show the square root relationship for velocity and diameter, but something nearer a *linear* relation (**Fig. 2.41**). The reason is to do with optimisation; in real life it turns out that the myelin thickness is not constant: there is an optimum thickness for myelin, which varies with the diameter, and the effect of this is to make the curve more or less linear rather than showing the square root relation characteristic of unmyelinated fibres (fast C-fibres in the figure). An important consequence of the linear relation for myelinated fibres as opposed to the square root one for unmyelinated fibres is that the two curves *cross*, at about 1 µm diameter. This answers a question that may already have occurred to you; if myelin's so wonderful, why aren't all fibres myelinated? The reason is that although there is a speed advantage in myelinating larger fibres, it is actually better to leave the smaller ones alone, because for a given *overall* diameter, the myelin takes up space that impinges on the conducting axoplasm. So there is no point in having myelinated fibres smaller than 1 µm in diameter, or unmyelinated ones larger than this (squids don't seem to have heard of myelin).[10]

Since in myelinated fibres the active, voltage-sensitive, sodium and potassium channels are virtually confined to the nodes, the action potential moves rather quickly from one node of Ranvier to the next, but lingers at the node itself while it is being regenerated, like a car on a motorway stopping for petrol: something called *saltatory* conduction, a word that just means 'jumping'.[11] Obviously myelinated nerves would conduct even faster if there were no nodes at all, but you have to have *some* in order to make up for the loss of current that still occurs despite the thick layers of myelin. In fact the nodes are separated by something of the order of a length constant, which provides enough of a safety margin that even if one or even two nodes are poisoned the nerve can still just conduct. The importance of myelination can be seen in *multiple sclerosis*, a condition in which one's own immune system

Clinical box 2.4 Multiple sclerosis (MS)

No clinical discussion about the physiology of nerve conduction would be complete without a comment on MS, the commonest cause of non-traumatic disability in young people. Immune-mediated damage to myelin leads to slowed, and often blocked, conduction along nerves, resulting in symptoms such as weakness, sensory loss, reduced visual acuity, loss of coordination, and speech and swallowing difficulty; reflecting the myriad nerve pathways that can be affected. Initially this occurs in discrete episodes ('relapses') which recover after a period of a few weeks as the myelin is repaired and voltage-gated channels redistribute themselves to bridge the demyelinated segments. However, over years, the picture often develops to one of progressive disability as nerves and axons start to degenerate; it seems that myelin (and its associated oligodendrocytes) serves an additional protective role to the axon, and that the movement of sodium channels to the demyelinated section causes a harmful energy demand in the previously myelinated area, eventually resulting in calcium accumulation and cell death.

Fortunately, in the last twenty years, an extensive therapeutic armamentarium of immunomodulatory treatments has been developed to control the relapsing phase. However, a treatment to prevent neurodegeneration is the greatest unmet need. One particularly exciting approach that is emerging is to stimulate the stem cells (oligodendrocyte progenitors) already present in the brain to reform myelin – this 'remyelination' could be an important treatment for all people with MS.

attacks myelin in the brain and spinal cord, causing such symptoms as weakness and lack of co-ordination.

The compound action potential

A typical peripheral nerve has a mixture of myelinated and unmyelinated fibres all jumbled up together, conducting at a wide range of speeds (Fig. 2.42). The biggest nerves in

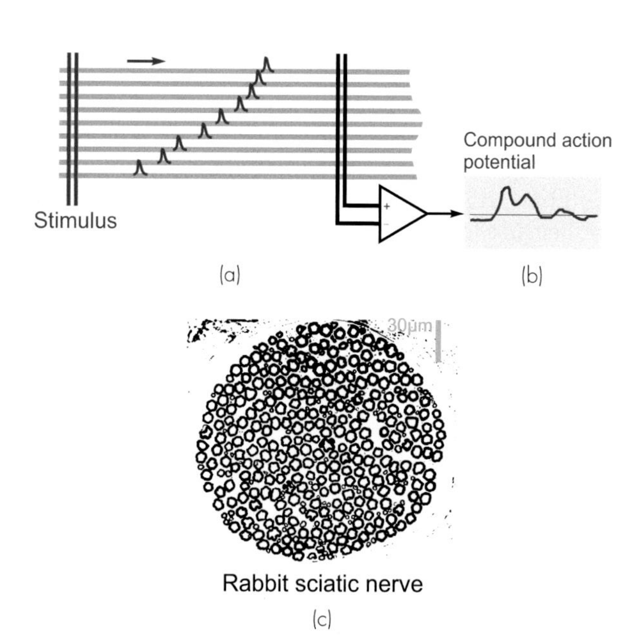

Figure 2.42 The compound action potential. (a) Biphasic recording from whole nerve: the compound action potential is spread out because of 'straggling' by action potentials in smaller fibres. (b) Actual compound action potentials from frog sciatic nerve with the A group shown on an expanded time scale in the inset. (Data from Erlanger and Gasser. Copyright © 1937 University of Pennsylvania Press; permission pending.) (c) Section of part of a rabbit sciatic nerve, showing a mixture of fibre sizes, mostly in groups Aα and Aγ.

Box 2.4 The classification of nerve fibres

Unfortunately, two different systems for classifying nerve fibres according to their s are in use.

Erlanger's system

	Diameter (µm)	Velocity (m/s)
A		
α	8–20	50–120
β	5–12	30–70
γ	2–8	10–50
δ	1–5	3–30
B	1–3	3–15
C	<1	<2 (unmyelinated)

This is used for motor nerves (whose fibres are mostly groups Aα ('alpha fibres') and Aγ ('gamma fibres'), and for skin afferents, mostly groups Aβ, Aδ and C (see Chapter 4).

Lloyd's system

	Diameter (µm)	Velocity (m/s)
I	12–20	70–120
II	4–12	24–70
III	1–4	3–24
IV	<1	<2 (unmyelinated)

This system is used for afferents from receptors in muscle, which fall into classes I and II; consequently classes III and IV are not in practice used.

your body are extremely fast, conducting at about 120 m/s or around 270 mph: this means that the time taken for information to get from say your toe to your brain can be as little as 10 ms; on the other hand, a small fibre conducting at less than 1 m/s would take more than a second for the same journey.

As a result of this mixture of speeds, if you take such a nerve, stimulate one end and record some distance down it, the action potentials behave rather like horses in the Grand National: the further they go, the more the whole pattern is spread out. Recording from a mixed nerve therefore produces rather a complicated electrical response called the *compound action potential*, the sum of many different action potentials all occurring at different times. Under these circumstances, the pattern of peaks in the compound action potential gives a sort of spectrum of the conduction velocities of the fibres in the nerve, though not a very quantitative one, since large peaks may simply be due to large fibres rather than to a large *number* of fibres of a particular velocity. Often the fibres appear to fall into groups based on their diameter and therefore their conduction velocity. A common way of classifying fibres is basically into fast, medium and unmyelinated slow, but with subdivisions of the fast (A) category. As a result, clinical neurophysiologists can only reliably measure conduction velocity in people by recording the latency between stimulating a nerve and the start of the compound AP; they are only assessing the integrity of the fastest conducting (most myelinated) nerve fibres, and diseases confined to smaller nerve fibres will not be seen (see **Clinical Box 2.3**).

Notes

There are many excellent books on electrophysiology. Aidley, D. J. (1989). *The Physiology of Excitable Cells.* (Cambridge: Cambridge University Press) and Keynes, R. D., Aidley, D. J. & Huang, C. L. (2011). *Nerve and Muscle.* (Cambridge: Cambridge University Press) are excellent all round; as is Nicholls, J. G. *et al.* (2012). *From Neuron to Brain.* (Sunderland, MA: Sinauer) (with the added advantage of an appendix on electrical circuits for those who missed out on their physics at school). Matthews, G. (1986). *Cellular Physiology of Nerve and Muscle.* (Oxford: Blackwell) is more general in its scope, as is Levitan, I. B. & Kaczmarek, L. K. (2001). *The Neuron.* (Oxford: Oxford University Press), a book with unusually clear text and illustrations. Aidley, D. J. & Stanfield, P. R. (1996). *Ion Channels.* (Cambridge: Cambridge University Press) provides detailed and comprehensive information from a molecular viewpoint; at the opposite extreme is the outstandingly thoughtful and functional. Rieke, R., Warland, D., Ruyter van Stevenunck, R. & Bialek, W. (1999). *Spikes: Exploring the Neural Code.* (Cambridge: MIT Press). Cotterill, R. M. J. (2002). *Biophysics: An Introduction.* (Chichester: Wiley) is a wonderfully clear and intelligent introduction to the underlying biophysical science. A recent historical account of the discovery of nerve conduction is McComas, A. (2011). *Galvani's Spark: the Story of the Nerve Impulse.* (Oxford: Oxford University Press). Another excellent account is Ashcroft, F. (2012). *The Spark of Life* (Allen Lane).

1. Galvani Galvani's description of his experiments, and his careful reasoning from them, are well worth looking at. His summary of how nerve operates is, in essence, a remarkably percipient description of passive conduction: *'For what pertains to voluntary motions, perhaps the mind, with its marvellous power, might make some impetus either into the cerebrum, as is very easy to believe, or outside the same, into whatever nerve it pleases, wherefrom it will result that neuro-electric fluid will quickly flow from the corresponding muscle to that part of the nerve to which it was recalled by the impetus, and when it has arrived there, the insulating part of the nerve substance being overcome through its then increased strength, as it goes out thence, it will be received either by the extrinsic moisture of the nerve, or by the membranes, or by other contiguous parts, and through them, as through an arc, will be restored to the muscle from which, as we are pleased to think, it previously flowed out, from the positively electric part of the same, through impulse in the nerve'.* Galvani, L. (1791). *De Viribus Electricitatis in Motu Musculari Commentarius*; translated R. M. Green. (Cambridge, MA: Licht).
2. Rocked in the cradle of the deep Some people go even further and suggest that just as ECF represents the composition of seawater in relatively recent evolutionary history, ICF, with its high potassium, represents the composition of the sea much longer ago, when the various fundamental molecules of life first began to be packaged up inside their cells – so ICF is a kind of oceanic memory or ionic fossil. While this is an attractive idea, geologists are not absolutely certain that the sea really has been getting steadily more concentrated in sodium in relation to potassium. But whether or not this is the true evolutionary explanation, presumably this high-potassium micro-environment must represent in some way an ideal ionic environment for all the molecular systems within the cell to function. For a more general introduction to life's origins and an excellent overview of cell biology in general, read Nick Lane's *The Vital Question, Energy, Evolution and the Origins of Complex Life* (2015). (London: Norton).
3. The voltage clamp technique Is well described in Alan Hodgkin's own *Conduction of the Nervous Impulse.* (1964). (Liverpool: Liverpool University Press). Something of the atmosphere in the lab in the exciting time that led up to these findings can be felt in Hodgkin, A. (1992). *Chance and Design: Reminiscences of Science in Peace and War.* (Cambridge: Cambridge University Press).
4. Tetrodotoxin Is a powerful poison from the pufferfish, a high-risk Japanese delicacy that adds a certain frisson of excitement to dining out in Japan, since it regularly poisons a certain percentage of the gourmets who eat it. One shouldn't blame the pufferfish, actually, as it turns out to be due to bacteria that associate with it; reared in isolation, pufferfish are not poisonous at all.
5. Calculations In *Chance and Design* (1992) Alan Hodgkin describes how at this stage, in March 1951, the only computer in Cambridge was out of use for 6 months, and how Andrew Huxley spent 3 weeks of gruelling labour literally cranking out the calculations on a hand-operated mechanical calculating machine: it is rather awful that what took him 3 weeks is done by my computer in a millisecond.
6. Voltage-driven permeability changes One additional factor, however: it turns out that in many cell bodies, terminals and dendrites (though less so in axons), calcium as well as sodium may enter during action potentials. Since calcium concentrations outside cells are normally very much larger than those inside, this calcium entry can also contribute substantially to membrane depolarisation (an important example of this is cardiac muscle). But we shall see in Chapter 3 (p. 50) that calcium entry is also important in another way, for in many situations it also acts as a chemical messenger, triggering other kinds of responses from the cell apart from changes in potential. A good

example of this is in synaptic transmission, discussed in Chapter 3; another is of course muscular contraction. In addition, when calcium enters it may indirectly contribute to membrane potential by altering the permeability to potassium, through more than one type of channel: these channels are distinct from the purely voltage-sensitive ones discussed so far. For the most part, these mechanisms tend to stabilise the resting potential.

7. Better not to use action potentials An antidote to the common misconception that all nervous communication has to be through action potentials is Roberts, A. & Bush, B. M. H. (1981). *Neurons without Impulses*. (Cambridge: Cambridge University Press).

8. Time constant One might wonder why R_L does not contribute to the time constant. The reason is that just as the space constant is defined in terms of what happens when we disregard time (by considering what happens when everything reaches equilibrium), so the time constant is defined in terms of what happens when space is entirely neglected: that is, when a current is applied uniformly along the fibre. Since there is then no spatial variation, no current flows through R_L, so it cannot contribute.

9. 37° causing nerve block Good news for oysters, anaesthetised as they are swallowed. And possibly for lobsters, traditionally brought slowly to the boil while still alive.

10. Conduction velocity and diameter Apart from Rushton's classical paper mentioned above, you may care also to look at Arbuthnott, E. R., Boyd, I. A. & Kalu, K. U. (1980). Ultrastructural dimensions of myelinated peripheral nerve fibres in the cat and their relation to conduction velocity. *Journal of Physiology* **368** 125–157. One might wonder why the myelinated curve doesn't go through the origin: is there really a certain size at which the fibre stops conducting altogether? The answer is that in the model it is assumed that one can alter the thickness of the myelin for optimum conduction velocity; as the diameter goes down, this optimum thickness gets relatively bigger. There comes a point where the model says that one does best with solid myelin and no axoplasm at all! You can investigate this yourself with the NeuroLab conduction velocity exhibit.

11. Saltatory conduction Students sometimes get the impression that saltatory conduction is fast because the action potential jumps in this way. This is really to think of it back-to-front: each node causes the action potential to be delayed while it is regenerated, like pit stops in a motor race. The nerve would conduct faster if there were no nodes, but not very far.

CHAPTER 3

COMMUNICATION BETWEEN NEURONS

In the previous chapter we saw how information is conveyed electrically from one part of a neuron to another: by passive conduction when the distances are short enough to permit it, and otherwise by means of action potentials. However, as we established at the beginning of the previous chapter, the purpose of a neuron is not to generate action potentials – or any other kind of potential – but to *release neurotransmitter in response to stimuli*. Later in this chapter we look at the mechanism by which the output terminals of neurons do this; but first, we need to look at how the whole process is initiated in the first place, either by stimuli in the outside world, or by the action of other neurons.

Common features of all neurons

The schematic neuron

Despite the huge variety of shapes and sizes and functions of neurons, from the tiny sensory hair cells in the ear to the huge neurons that carry commands from the cerebral cortex to the bottom of the spinal cord, there is a basic ground plan for intercellular communication that applies to all of them, and indeed to endocrine cells as well (**Fig. 3.1**).

All neurons have an *output* region (the terminals that release the neuron's transmitter, T) and an *input* region (the dendrites, or the receptive region in the case of sensory receptors). The mechanism at the terminal end is as far as we know absolutely identical in all neurons and receptors: depolarisation opens voltage-sensitive calcium channels, and the resultant rise in intracellular calcium causes exocytosis of *vesicles* about 50 nm across, containing the transmitter substance that is to act on the next cell along.

In the input region, a stimulus S acts on membrane channels that open (or sometimes close). As we have seen, this action may be a direct one, or may be indirect, mediated by intracellular mechanisms. Either way, these channels create changes in ionic permeability that in turn give rise to currents and potentials (ΔV) that in short neurons may spread passively to the terminal, or in longer ones may induce repetitive action potentials as described in Chapter 2 (p. 40). Often – for example in the hair cell receptors of the ear – the receptor and the nerve axon that joins it to the central nervous system are quite separate, the receptor cell acting on the axon by means of a chemical transmitter whose release depends on the amount of depolarisation of the ending. The only fundamental difference between sensory receptors and central neurons is the nature of the stimulus that acts on the channels in the first place.

Once this basic neuronal ground plan is understood, differences between particular neurons and receptors becomes a matter of filling in the blanks in a rather simple table (**Table 3.1**). We need to know only: the *type of stimulus*, the *type of receptor channel*, what *permeability change* occurs, whether *action potentials* are used, and what *transmitter* is finally released. What could be simpler?[1]

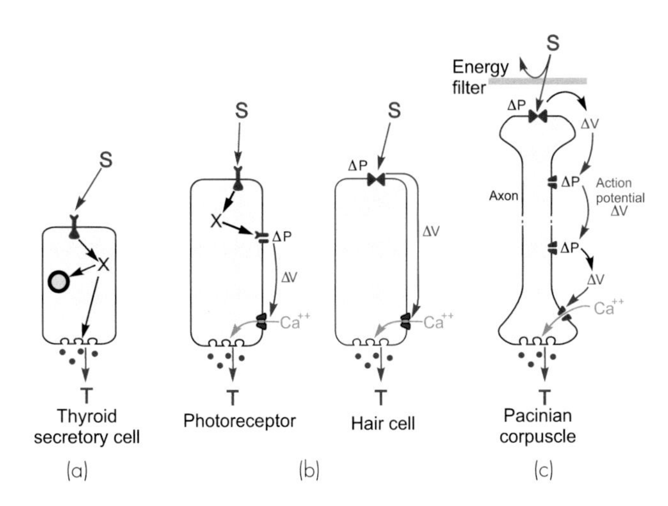

Figure 3.1 Types of intercellular communication. (a) A cell that responds to a stimulus, *S*, by producing a secondary messenger, *X*, which may cause the release of a transmitter (or hormone), *T*, in addition to purely internal effects. (b) Two examples of short neurons without action potentials: in each case, a depolarisation of the terminal opens voltage-sensitive calcium channels that trigger the release of *T*; in the photoreceptor this depolarisation is the result of the opening of membrane channels by *X* (indirect transduction), whereas in the hair cell the membrane receptor that responds to the stimulus also opens the channels (direct transduction). (c) A longer neuron: the initial depolarisation results in the propagation of repetitive action potentials, which again result in calcium entry at the terminal. *S* may be a transmitter or hormone rather than an external stimulus, as in interneurons.

Table 3.1 Theme and variations: some types of neuron

	Nicotinic	M_2 muscarinic	Photo-receptor	Olfactory receptor	Hair cell	Pacinian corpuscle
Stimulus	ACh	ACh	Light	Chemical	Mechanical	Mechanical
Indirect?	No	Yes	Yes: cGMP	Yes: cAMP, and Ca^{2+}	No	No
ΔP_{Na}?	+		–	+	+	+
ΔP_K?	+	+		+	+	+
Action potentials?	Yes	(No)	No	Yes	No	Yes
Result	Mechanical	Mechanical	Glutamate	Glutamate	Glutamate	Glutamate

The initiation of activity

The simplest of all kinds of intercellular communication is when currents pass directly from one cell to another. But rather stringent structural conditions have to be met before this mode of synaptic transmission will work. **Figure 3.2** shows an idealised electrical synapse and its equivalent circuit. It is clear that the current I generated by the presynaptic bouton has two alternative routes: it can either cross the gap and enter the postsynaptic cell, or it can simply leak out sideways through the synaptic cleft. The greater the fraction of current that takes the former route, the greater will be the degree of electrical coupling between the two neurons, since by entering the postsynaptic cell the current will cause potential changes that may, if large enough, trigger a new action potential. More formally, if I is to create a sufficient depolarisation V_{out} of the postsynaptic cell, this sideways leakage through R_L must be small relative to the forward resistance R_F. One way to minimise the sideways leakage is to use *gap junctions* (**Fig. 3.3**). A gap junction is a sealed molecular bridge along which electrolytes can move from one cell to another without escaping to the extracellular fluid, effectively increasing R_L, as well as reducing R_F.[2]

A familiar example is conduction between the muscle fibres of the heart, where the currents pass through the gap junctions that form part of the intercalated discs. Electrical transmission between neurons appears to involve either casual sets of gap junctions as are found between photoreceptors in the retina, or more organised regions of contact called *electrical synapses* which include gap junctions. But even if no leakage at all occurs, and the transmembrane resistance at the junction is reduced to zero, there is *still* no guarantee that an action potential will be able to pass successfully from one cell to another. The size of the local currents that flow during the passage of an action potential along an axon are strongly dependent on the size of the axon itself. The larger it is, the greater the number of sodium channels per unit length and so the larger the active currents that can be generated. But equally, larger axons have greater capacitance and smaller transverse resistance per unit length; consequently the currents have to be that much larger to achieve a particular threshold level of depolarisation. In other words, the currents automatically keep up with the increased requirements as the fibre's diameter is increased. But if we imagine a small axon whose diameter suddenly gets bigger at a particular point, or – as comes to the same thing – a small axon with a low-resistance electrical synapse joining it to a cell body of larger size, this is obviously no longer the case (**Fig. 3.4**). Just as a burning thread may not generate enough heat to ignite a rope to which it is attached, so the current generated by the small axon may well be too weak to trigger an action potential in the larger cell. One can calculate, for example, that electrical transmission across the ordinary neuromuscular junction is in principle impossible even if the junction were a low-resistance one (which it is not): the impedance ratio on the two sides is

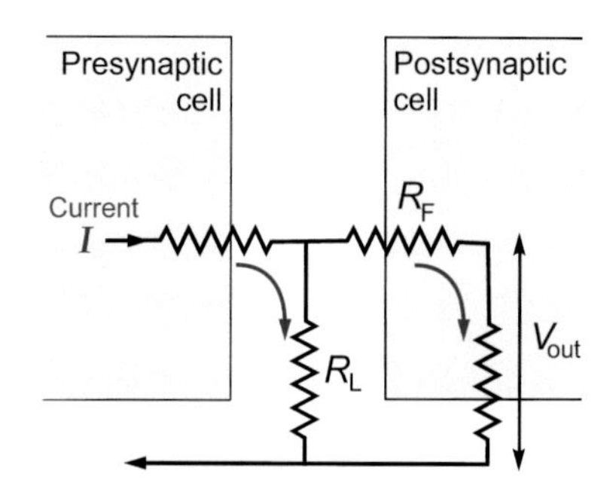

Figure 3.2 The requirements for electrical transmission. If a pre-synaptic current, I, is to create a sufficiently large depolarisation, V_{out}, of the post-synaptic cell, sideways leakage through R_L must be small: so the forward resistance R_F needs to be much less than R_L.

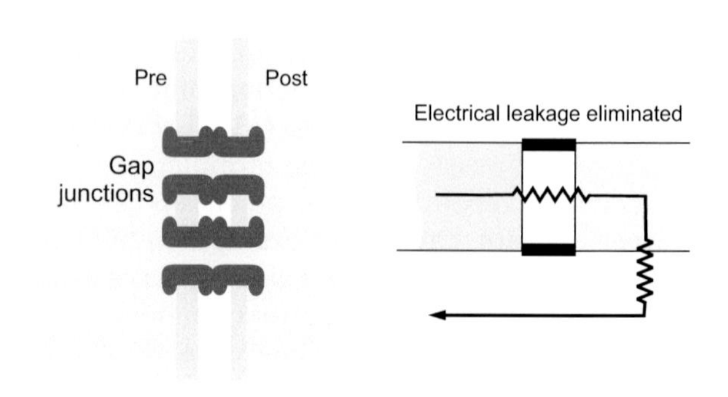

Figure 3.3 A gap junction or tight junction.

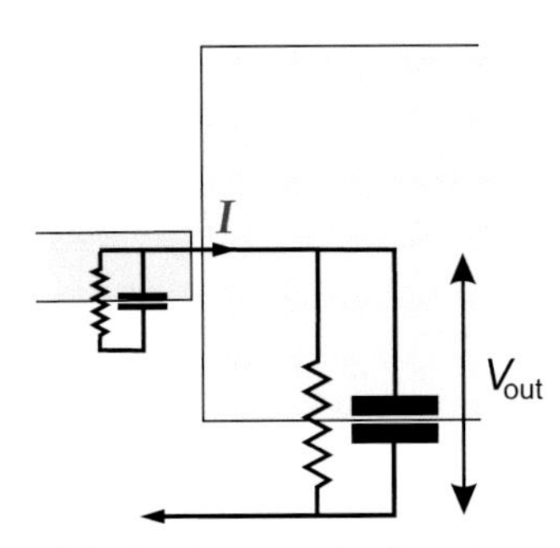

Figure 3.4 The size of the local currents that flow during an action potential are dependent on the size of the cell. Even if loss through R_L is negligible, if the postsynaptic area is large in comparison to the pre-synaptic ending, its large capacitance and small membrane resistance result in a low impedance, and I may still be insufficient to cause a threshold change in V_{out}. This would be the case for an ordinary neuromuscular junction.

much too large for the axonal currents to make any significant impression on the potential of the muscle cell. It is clear, therefore, that in such cases an *extra* source of amplification in addition to that provided by the action potential mechanism is needed, equivalent to soaking the knot between thread and rope in petrol.

Ligand-gated channels

This amplification – the petrol – is provided by membrane channels sensitive not to voltage but to chemical transmitters. In general, membrane channels may be controlled by one of three things: by mechanical stimulation, by depolarisation, or, in the case of *ligand-gated* channels, by the arrival of particular molecules at sites that can either be on the inside or the outside of the membrane (**Fig. 3.5**). There is, in fact, very great similarity between the way channels in sensory receptors respond to stimuli and those in interneurons that respond to transmitters. Whether it is a sensory stimulus acting on a sensory receptor or transmitter released by one neuron acting on the postsynaptic membrane of another, what happens in every case is the opening (sometimes the closing) of particular ionic channels in the cell membrane. This leads to changes in the ionic permeability (nearly always to one or more of sodium, potassium or chloride), which must inevitably cause a change in membrane potential. This change has different names at different sites: in receptors it is called the receptor potential, at the neuromuscular junction the endplate potential, and at central synapses the postsynaptic potential; but they are all the consequence of the same underlying process. However, there are two fundamentally different ways in which a stimulus can open channels, *directly* or *indirectly*. These two modes are mediated by what are officially and inelegantly called ionotropic and metabotropic mechanisms respectively. The basic difference between the two kinds of channel is that the direct ones are looking *outward*, waiting for signals to arrive from the outside world, whereas the indirect ones are looking *inward*, for messages that are generated within the cell itself.

Direct (ionotropic) mechanisms

One example of a *direct* gating mechanism is found at the muscle endplate, or neuromuscular junction (NMJ) (**Fig. 3.6**). This is technically a *cholinergic* synapse, meaning that the transmitter is acetylcholine (ACh), and the receptors belong to the sub-class of cholinergic receptors called nicotinic, because they also respond to the well-known substance nicotine. Recognition of the transmitter causes opening of an unselective, short-circuiting channel, permeable to sodium and potassium. The result is to cause depolarisation and the generation of an action potential in the muscle; it is discussed in more detail later in this chapter (p. 52). In addition, all mechanoreceptors appear to work by a direct mechanism of this kind, with the mechanical stimulus acting immediately to cause opening of the channel. Examples include the hair cells in the cochlea of your inner ear that respond to sound vibration, or the touch receptors in your skin.

Indirect (metabotropic) mechanisms

The other possibility is *indirect* gating; here the channel is inward-looking, responding only to chemical messages from within the cell. The link with the outside world is provided by a second protein that straddles the membrane and responds to transmitters or stimuli in the outside world by triggering off a chemical response on the other side of the membrane, which then results in the required message being sent to the channel. This intracellular communication may involve just one intermediate, or a cascade of several of them; very often the first link in the chain is formed by a *G-protein* – guanosine triphosphate (GTP)-binding protein. An example of short indirect coupling is the M_2-muscarinic acetylcholine receptor, where the G-protein, activated by an acetylcholine receptor, then acts directly on a potassium channel to cause hyperpolarisation (**Fig. 3.7a**). Sometimes the G-protein activates the

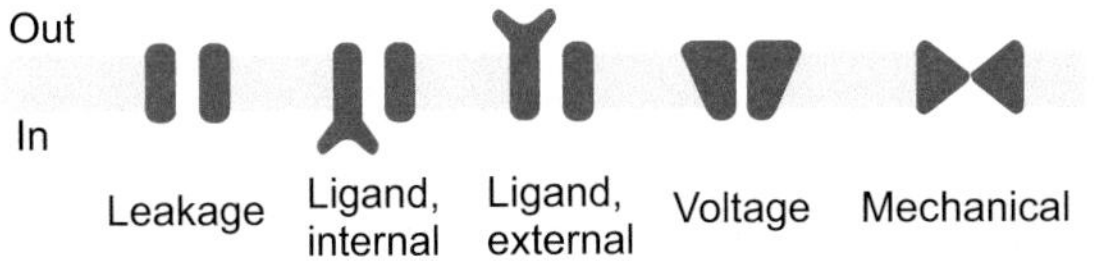

Out
In
Leakage | Ligand, internal | Ligand, external | Voltage | Mechanical

Figure 3.5 Types of membrane channel, shown in the symbolic form used throughout this book.

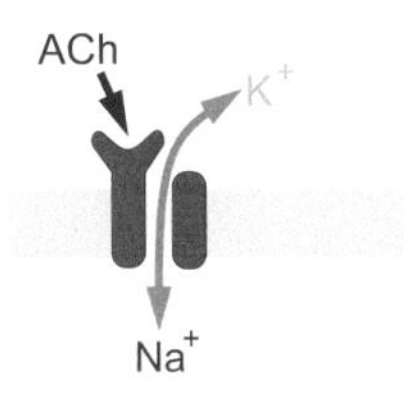

Figure 3.6 A direct, ligand-gated channel.

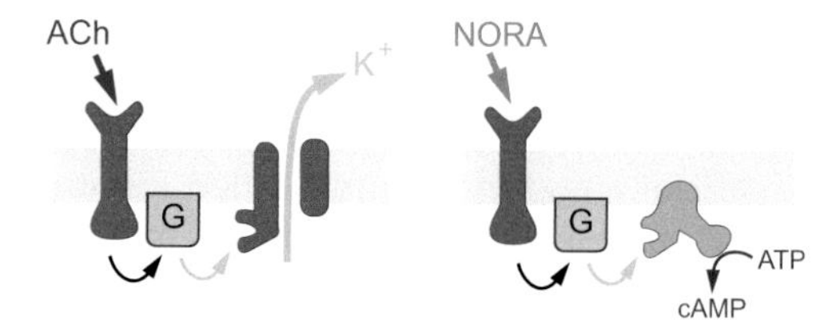

Figure 3.7 Examples of indirect, G-protein–coupled receptors.

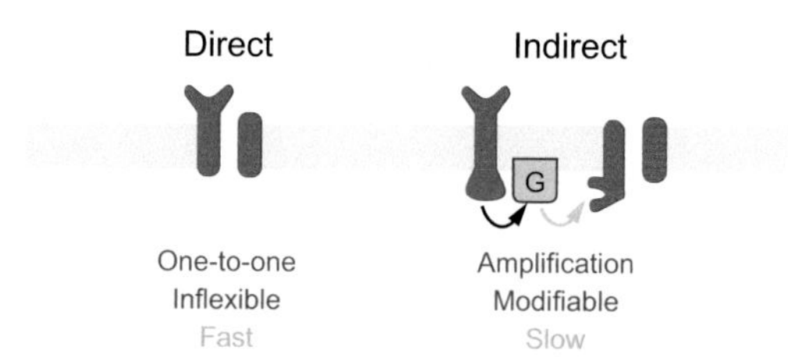

Figure 3.9 Direct vs indirect activation.

production of a second messenger, which may in turn either have intracellular effects (as with the β-adrenergic receptor illustrated in **Figure 3.7b**), or again may have an effect on membrane channels. The transduction mechanism in retinal rod receptors that allows them to respond to light is a good example of a *cascade* of this kind: here a photolabile pigment molecule, rhodopsin, is coupled to a G-protein that activates a phosphodiesterase (PDE), that in turn results in the conversion of cGMP to GMP; since cGMP opens sodium channels on the surface of the receptor cell, the effect of light is to close them, and thus to cause hyperpolarisation (**Fig. 3.8a**). A similar cascade, but with cAMP instead of cGMP, and further amplified by a second stage involving calcium, can be found in olfactory receptors (**Fig. 3.8b**).

These two general methods of generating permeability changes each have their advantages. Direct activation is fast and secure. Indirect activation provides for *amplification* (in rods, one photon can trigger the breakdown of a million or so cGMP molecules), for *prolongation* of effects, for *control* by the cell (which can intervene in the link between receptor and channel), and for *intracellular effects* (**Fig. 3.9**). An example of cellular control is again in photoreceptors, where intracellular calcium modifies the sensitivity and time-course of the cGMP changes, thus effectively altering the receptor's sensitivity to light. An obvious example of intracellular effects is the β-adrenergic response to noradrenaline, where cAMP production is again coupled via a G-protein to the receptor itself, the cAMP then having metabolic effects within the cell.

Calcium

This is a good moment to introduce calcium, an ion which is used as a form of intracellular communication by very many kinds of cells, not just neurons and muscles. The significance of calcium lies in the fact that its concentration inside cells is normally very low indeed, something of the order of 0.1 μM, because of specific calcium pumps, by its sequestration within the cell, and in some cases because of its storage in organelles that act as internal stores. Because cellular calcium concentration is effectively zero, the sudden appearance of even a tiny amount of free calcium inside is a spectacular event. Often the cell uses this as a means of telling the interior that something has happened at the membrane surface, very much like ringing the cell's doorbell (**Fig. 3.10**). There are two ways in which this signalling can occur. Calcium can enter from outside, through channels triggered either by a transmitter or hormone or by voltage, or these signals may operate indirectly, for instance by causing the production of a second messenger such as inositol (1,3,5) triphosphate (IP3), that causes calcium to be released from the internal stores. Generally speaking, calcium entry has significant consequences: transmitter release from synaptic terminals, muscular contraction, enzyme secretion, or – as in many sensory receptors – there can be profound regulatory effects on the behaviour of the entire cell.

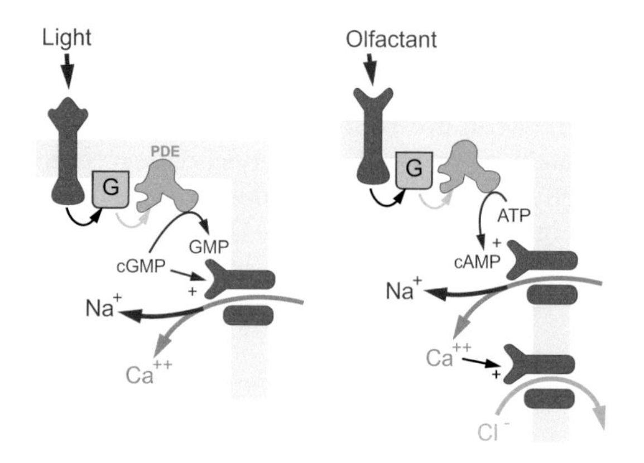

Figure 3.8 Sensory receptor mechanisms, both using a G-protein, in one case (olfaction) to increase membrane permeability, and in the other (light) to decrease it (PDE = phosphodiesterase).

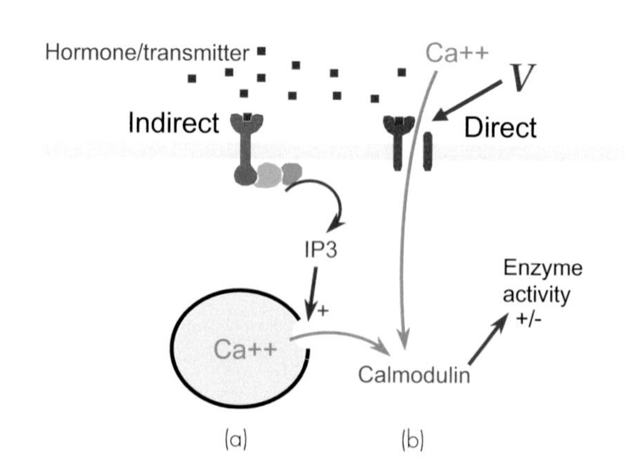

Figure 3.10 Mechanisms increasing internal Ca^{2+} in response to external stimuli. (a) Indirect: hormone or transmitter indirectly promotes the production of an intracellular messenger (in this case IP3), which releases calcium from internal stores. (b) Direct: a transmitter, or depolarisation, opens channels that allow calcium to enter from outside.

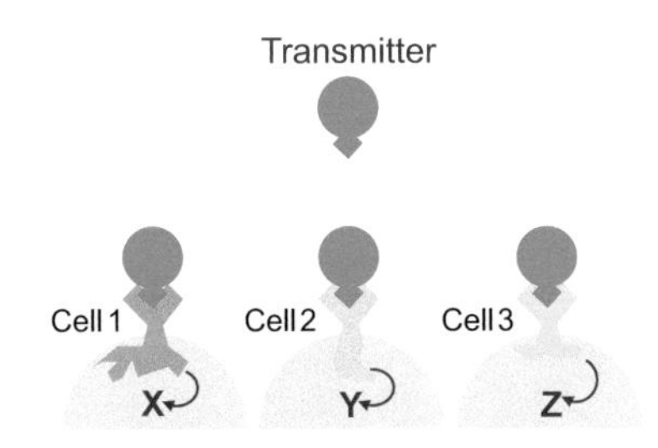

Figure 3.11 One transmitter may have different actions.

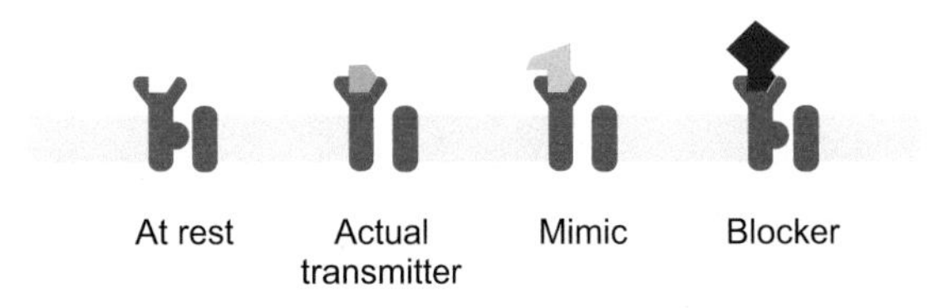

Figure 3.12 One receptor may respond to different substances.

Different transmitters

Bear in mind that the same transmitter may have quite different effects on different cells: there is no logical or necessary connection between the identity of a transmitter and what it does to the target cell – everything depends on what receptors are expressed on the target membrane (**Fig. 3.11**). You are probably already familiar with the different effects that acetylcholine can have in the autonomic system, mediated by nicotinic or muscarinic receptors. A particularly clear example that we come across later, in Chapter 7, is the fact that although retinal photoreceptors release only glutamate as their transmitter, the bipolar neurons on which they act may be depolarised or hyperpolarised, depending on whether they have one kind of glutamate receptor or another.

Finally, in general a receptor 'designed' for a particular transmitter will in general also respond to a range of other substances that may mimic the transmitter, or block it by becoming attached but refusing to budge, or in more complex ways. Many of these substances are natural poisons: many more are the artificial creations of pharmacologists and are as important in trying to elucidate how the receptor works as they are in clinical therapy (**Fig. 3.12**).

Synaptic transmission

Central neurons are driven not by sensory stimuli in the outside world but by the activity of other neurons that make contact with them at specialised regions, the *synapses*. At a typical synapse, a branch of the afferent axon forms a swelling, the terminal *bouton*, the further side of which forms an enlarged area of intimate contact with the postsynaptic cell body: in the case of the neuromuscular synapse, the *muscle endplate*, this area is much increased by the presence of invaginating folds (**Fig. 3.13**).

Clinical box 3.1 Synapses as drug targets

The synapse provides one of modern medicine's most common drug targets. Valium, for example (also known as diazepam), is one of a group of drugs called benzodiazepines used to control anxiety and seizures. The drug binds to a specific benzodiazepine binding site on the γ-aminobutyric acid (GABA)-A receptor causing the chloride channel to favour the open state. As a result, the neuron bearing the receptor is reduced in its activity. This causes a general reduction in activity in the central nervous system, which clinically manifests as relaxation or cessation of seizure activity. In overdose, predictably, patients become sedated, comatose, hypoventilate and can even die.

Indirectly, most antidepressants also act by augmenting synaptic activity. While the organic aetiology of depression is poorly understood and almost certainly multifactorial, one line of thought holds that at least part of the underlying problem is a lack of serotonin activity in the brain. This has stimulated the development of serotonin selective reuptake inhibitors (including *Prozac*) which inhibit the reuptake of serotonin by the presynaptic neuron: serotonin is present in the cleft for longer and activity in the postsynaptic neuron is prolonged.

Even less nuanced, Parkinson's disease is characterised by reduced dopaminergic transmission in the nigrostriatal part of the basal ganglia. While there remains no treatment capable of modifying the disease trajectory, we can achieve a degree of symptomatic control with drugs capable of increasing the activity of the dopaminergic pathway. The most common treatment is to give the precursor of dopamine (L-dopa), which it replaces after its conversion by the enzyme dopamine decarboxylase; giving dopamine would not be effective as it does not cross the blood-brain barrier. Other options include inhibiting enzymes responsible for the breakdown of dopamine in the synaptic cleft (COMT inhibitors are often given alongside L-dopa to prolong its effect) or using dopamine receptor agonists to directly stimulate the receptor.

What is astonishing is not our clever design of pharmacological agents to delicately alter neural physiology, but rather that splashing empirical, undirected drugs around the brain, little more sophisticated than throwing a can of petrol at a car, should prove to be so safe and effective!

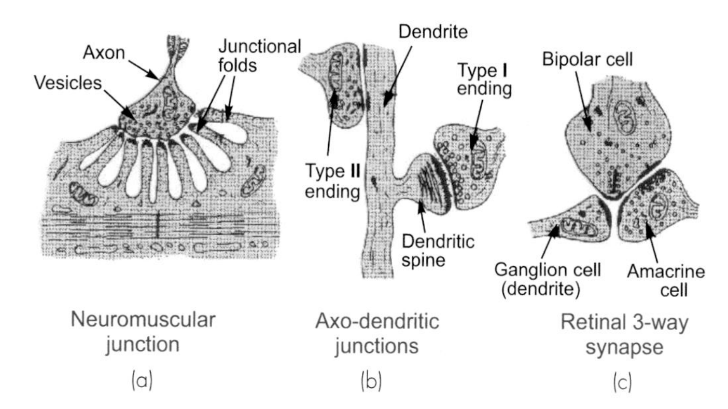

Figure 3.13 Somewhat stylised representations of some synaptic types. (a) Neuromuscular junction. (b) Two types of pre-synaptic axonal endings synapsing on a dendrite; the synapse on the right is with a dendritic spine. (c) A three-way synapse from the retina; the junction between bipolar and amacrine cell probably permits the transfer of information in both directions.

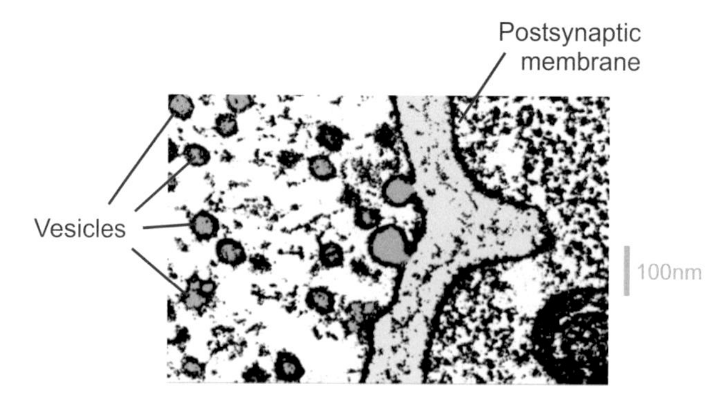

Figure 3.14 Electron micrograph of rat neuromuscular junction, showing vesicles fusing with the pre-synaptic membrane prior to their release. (Heuser, 1977.)

In most cases there is a clear *synaptic cleft* between pre- and postsynaptic membranes, typically some 20 nm wide. Transmitter is released from the presynaptic side and diffuses to the postsynaptic side, where it causes permeability changes through the various mechanisms already outlined. Anatomically, many variations on this basic pattern of *axo-somatic* contact can be found. Most neurons have an elaborately branched dendritic tree which is smothered in *axo-dendritic* synapses; dendrites may also make contact with each other in *dendro-dendritic* contacts; and in *axo-axonic* contacts one axon may terminate on the terminal of another and modify its transmitter release. But it is convenient to begin with the most familiar type of synapse of all, whose working is most thoroughly understood: the NMJ between motor axon and striated muscle fibre. Although the postsynaptic cell is not a neuron, the fundamental mechanism by which it is activated is closely similar to many kinds of synapses within the brain.

Transmission at the neuromuscular junction

The transmitter at the NMJ is ACh. The vesicles in which it is contained are normally created by the pinching off of parts of the transmitter-filled Golgi apparatus in the cell body and are then transported down microtubules in the axon to the terminal (**Fig. 3.14**). Each is about 40–60 nm in diameter and contains some 10^5 molecules of ACh, and when an action potential arrives at the endplate the consequent entry of calcium triggers the release of the contents of some 200–300 of them. Vesicles appear to obey a kind of all-or-nothing law in that they either empty completely into the synaptic cleft or not at all. The transmitter thus released must diffuse across the synaptic cleft – a process that takes a millisecond at most – before it can act on the muscle cell. When it arrives there, it interacts with direct (ionotropic) channels that open to increase the permeability to sodium and potassium, thus depolarising the membrane and initiating an action potential. How do we know all this?

The effect of acetylcholine at the endplate: the reversal potential

Early experimenters had to work backwards from the changes in potential they observed to make deductions about the action of acetylcholine on the receptors. If we put a microelectrode in the muscle fibre close to the endplate, and poison the muscle with tetrodotoxin so that our observations of the primary electrical events are not obscured by any subsequent action potentials that may be generated, we see that a single action potential in the afferent nerve is associated with a characteristic electrical response in the muscle cell, called the *endplate potential* (EPP) (**Fig. 3.15**). Although the original action potential only lasts a millisecond or so, the EPP is relatively prolonged. Most of this prolongation is due to the capacitance of the membrane, which is first rapidly discharged and then slowly recharged to the resting potential E_R. If we know the value of the appropriate time constant, we can estimate the duration of the current that flowed through the ACh-gated channels to produce the potential change. It turns out to have a time-course (blue trace) not very different from that of the original action potential, though delayed in time by the millisecond or so of synaptic delay that is the result of diffusion across the synaptic cleft. This brief current discharges the membrane capacitance, which subsequently must recharge relatively slowly through the resting membrane resistance, giving a long tail to the resultant synaptic potential. The fact that the current hardly lasts longer than the afferent impulse is largely due to the presence at the ending of high concentrations of the enzyme *cholinesterase* that mops up the ACh almost as soon as it arrives. If this enzyme is blocked by an anticholinesterase such as eserine, one finds that the current flow, and hence the EPP, is enormously prolonged, leading to a depolarisation block of the muscle fibre. Patch-clamping demonstrates that the channels operate in a stochiastic, or 'all-or-nothing' manner, being either fully open or fully closed, as can be seen in the recording from rat NMJ in **Figure 3.16** – which also shows the increase in the probability of opening with increasing calcium concentration. In addition, this technique can also be used to demonstrate

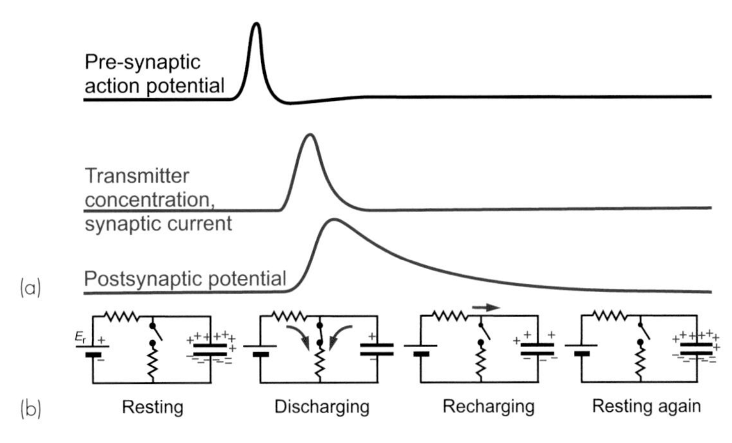

Figure 3.15 (a) Relation between pre-synaptic action potential, endplate potential and synaptic current (somewhat idealised). (b) A schematic equivalent circuit of the postsynaptic membrane showing membrane capacitance charged to resting potential, E_r, rapid discharge through opening of unselected channels, channels closed, capacitance recharging relatively slowly, until equilibrium is finally restored when the capacitor is fully charged.

that two acetylcholine molecules are required to trigger the opening of one channel.

To understand what these channels do, the first thing we need to know is what ions they let through when they are open. One way to find out is to measure something called the *reversal potential*. This is a fundamental technique, that forms a basic way of determining what permeability changes are going on in sensory receptors and in postsynaptic membranes. The principle is a simple one: since every combination of permeabilities results in some corresponding equilibrium potential E_s (from the constant-field equation), then a change in equilibrium potential implies a change in one or more of the permeabilities. The problem is that most sensory and synaptic events are short-lived, so that there is no time for the membrane potential actually to settle down at its new value of E_s. But the *direction* of its movement tells us whether the new equilibrium is above or below the resting potential; and if we have some way of setting the resting potential artificially to different levels, we can see how this influences the direction of the response. As the resting potential is made to approach E_s, the response will get smaller and smaller, then reversing in sign as the resting potential passes through E_s. The reversal potential, defined as the value of the resting potential at which stimulation has no effect, is simply equal to E_s. Once we know E_s we can make an informed guess as to what permeability change must be causing it.

In the case of the NMJ, a steady current is passed into the muscle cell in order to set the resting potential at a new artificial level, and the size of the EPP is then observed. As illustrated schematically in **Figure 3.17**, what is found is that the EPP gets smaller and smaller as the resting potential is reduced to near zero, and that if the membrane is hyperpolarised, the EPP is reversed. The conclusion is therefore that the effect of ACh is to open channels that allow sodium as well as potassium to pass through the postsynaptic membrane, and thus produce something like a short-circuit. Since the number of ACh molecules released by each impulse, and hence the number of channels

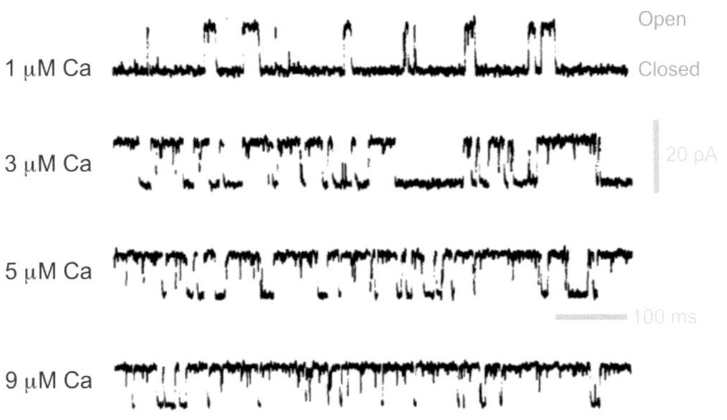

1 μM Ca   Open   Closed
3 μM Ca   20 pA
5 μM Ca   100 ms
9 μM Ca

Figure 3.16 Patch-clamped single acetylcholine channel, from rat neuromuscular junction. (Moczydlowski & Latorre, 1983; with permission.)

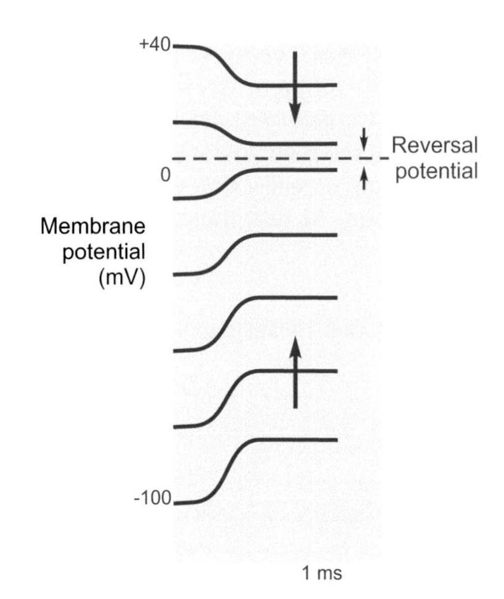

Figure 3.17 The reversal potential. Excitatory postsynaptic potentials recorded at different initial resting potentials produced by passing current steadily in or out of the neuron by means of a double-barrelled microelectrode. The response reverses at the reversal potential, E_{rev}, not far from zero. (Partly after Curtis & Eccles, 1959; and Coombs *et al.*, 1955; with permission from John Wiley and Sons.)

opened, is very large, it is clear that this is the mechanism whereby the relatively small currents in the axon can trigger off the relatively enormous currents needed to initiate an action potential in the muscle cell: the source of these currents is in fact the muscle cell itself.

The release of neurotransmitter

The ultimate action of all neurons, whether interneurons or sensory receptors, is to release a chemical transmitter from their terminals. As far as we know, the mechanism by which this occurs is identical in every case: depolarisation opens voltage-gated calcium channels, calcium enters and causes the neurotransmitter to be released from its vesicles. The process has been most extensively studied at the NMJ because of its relatively greater size and accessibility. In brief, the transient rise in calcium is detected by synaptotagmin, which acts as a sensor and triggers membrane- and vesicle-associated SNARE proteins to fuse and exocytose the neurotransmitter.

Recording from the endplate with very high sensitivity, one finds that even when the afferent fibre is not stimulated there are continual spontaneous potential changes taking the form of a random succession of *miniature endplate potentials* (mEPP) having roughly the same shape as a normal evoked EPP, but about 0.2–0.3 per cent of its size (**Fig. 3.18**). They are due to the fact that the presynaptic ending, even at rest, releases individual vesicles randomly at a very low rate. This rate of spontaneous release is strongly dependent on the resting potential across the presynaptic terminal, and if this is artificially reduced – for example by changing the external potassium concentration – the average rate of vesicle release increases sharply. By extrapolation, one can show that the size of a normal EPP is about what would be expected if the action potential simply had the effect of temporarily increasing the rate of spontaneous release of vesicles. The size of the mEPPs does not alter with different degrees of depolarisation, implying that the transmitter packaged in a vesicle is either released or not released, in packets of fixed size called *quanta*. It is important to realise that vesicle release is a *probabilistic* phenomenon. At rest, the probability of a vesicle being released per unit time is very low but not zero; what happens when the terminal is depolarised and calcium enters is simply that this probability is increased. Thus the *frequency* of the mEPPs goes up, but their individual *size* doesn't. mEPPs provide a convenient way of determining the mode of drug action of synaptic transmission. Some well-known poisons (for instance δ-tubocuarine, found in the arrow-poison curare, or the snake venom α bungarotoxin) work by blocking the action of acetylcholine on the postsynaptic membrane, whereas others (for instance botulinum toxin, found in decaying meat) operate by interfering with release from the terminal; the former affect mEPP size but not frequency, the latter frequency but not size.

Finally, the role of calcium entry into the terminal can be demonstrated by looking at the effect on quantal frequency of altering calcium concentrations outside the terminal, or of adding magnesium, which blocks its entry. The calcium inside the terminal can in addition be directly visualised by using a rather handy substance called aequorin, derived from luminescent jellyfish, that lights up in response to calcium.

Figure 3.19 summarises the sequence of events at the neuromuscular junction, and their relation to the three kinds of membrane channel found there: voltage-gated sodium and potassium (grey), voltage-gated calcium (white) and cholinergic ligand-gated (red). The action potential arrives at the terminal, channels in the presynaptic ending are opened that permit the entry of calcium, and this in turn stimulates the emptying of the vesicles into the synaptic cleft. The ACh that they contain diffuses across, reaching nicotinic receptors in the postsynaptic membrane that respond by increasing the permeability to sodium and potassium. This generates a short-circuiting current tending to pull the membrane potential towards zero, that in turn depolarises the surrounding membrane sufficiently to initiate an action potential. Finally, acetylcholinesterase breaks down the ACh in the synaptic cleft, terminating the response and restoring the status quo.

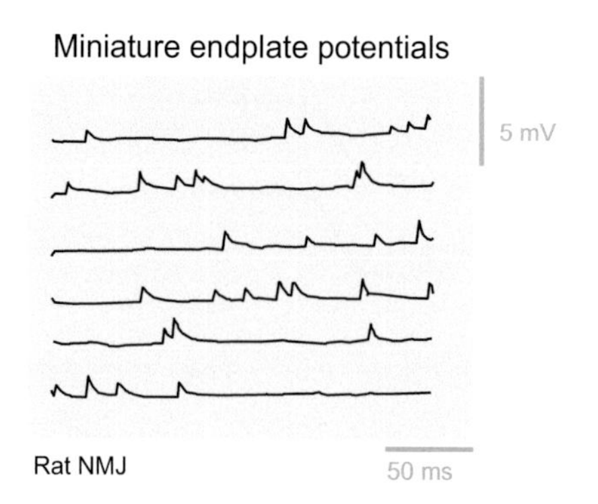

Figure 3.18 Miniature endplate potentials (mEPPs) recorded from rat neuromuscular junction. (Reprinted by permission from *Nature* 166 597–598 (1950). Copyright 1950 Macmillan Publishers Ltd.)

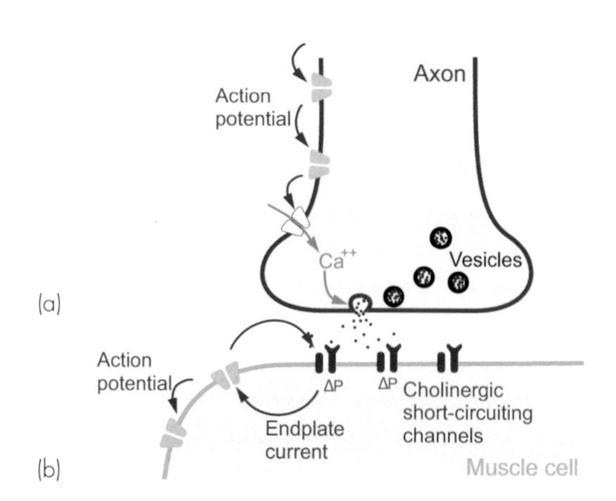

Figure 3.19 Schematic representation of the sequence by which nerve action potentials (a) lead to muscle action potentials (b) at the neuromuscular junction. Voltage-sensitive sodium and potassium channels are shown in grey, calcium in white; channels sensitive to the transmitter acetylcholine, which is represented by the red dots, are shown in red.

Clinical box 3.2 Neuromuscular blockade facilitates certain types of surgery

Even when a patient is deeply unconscious and not experiencing any pain, the muscles can be active. For example, during abdominal surgery contraction of the abdominal muscles in an effort to breathe makes suturing the abdominal wall together difficult; often the surgeon asks the anaesthetist to 'paralyse' the patient and let the ventilator take over the work of breathing while the surgeon completes the suturing. Similarly, during orthopaedic surgery, the trauma and irritation of cutting open a joint, whilst not 'felt' by the anaesthetised patient, does result in reflex contraction of the surrounding muscles. This can make relocating a dislocated joint impossible without paralysis. Even more obviously, an uninvited cough or wriggle during brain or eye surgery may be disastrous and a patient is paralysed prophylactically.

Paralysis is achieved using two different types of agent: depolarising and non-depolarising neuromuscular blockers. The former is an analogue of ACh itself and activates the nicotinic ACh receptor at the NMJ. Unlike ACh however, the associated cation channel is maintained in its open state so that after the initial depolarisation and twitch, the muscle fibre is held in its refractory, relaxed state. These agents have the advantage that they act quickly and then wear off in a few minutes, ideal for a very brief intervention such as endotracheal intubation. However, they have a physiologically interesting danger. ACh receptors are usually tightly clustered on the muscle fibre surface at the motor endplate. But diseases which result in 'denervation' – a reduction in the number or activity of the motor nerves reaching the muscle (such as spinal trauma or Guillain–Barré syndrome) – trigger a compensatory increase in the number of functional receptors which appear all over the entire muscle fibre surface. Administration of a depolarising agent in this context can liberate enough intracellular potassium to cause hyperkalaemia, cardiac arrest and death.

Non-depolarising agents are analogues of the poison curare which is derived from tree frogs and used to hunt in certain parts of the world. They act simply by sitting on the ACh receptor at the motor endplate and preventing the ACh liberated from the presynaptic membrane from reaching the receptor, activating the cation channel and depolarising the fibre. They have no intrinsic activity of their own. It is possible to reverse the action of nondepolarising neuromuscular blocking agents by administering an acetylcholinesterase inhibitor, which allows enough ACh to accumulate in the NMJ to overcome the receptor blockade.

Central excitatory synapses

Once the principles of operation of the neuromuscular junction are understood, the operation of the synapses by which one neuron can excite another presents little extra difficulty. The general sequence of events is identical: vesicles containing neurotransmitter are released from the presynaptic terminal by an influx of calcium triggered by depolarisation; these vesicles were manufactured in the Golgi apparatus of the cell body and transported to the terminal using microtubules (though one difference is that the incorporation of small transmitter molecules into the vesicle may occur in the terminal rather than in the cell body). When released, the transmitter is recognised by receptors located in the postsynaptic membrane, that in turn alter the permeability of ion channels, leading to the generation of currents and voltages, and other effects. This whole sequence of events is summarised in the diagram of the schematic neuron in **Figure 3.1** (p. 47).

The most frequently studied synapse of this kind is one forming part of the *monosynaptic reflex arc* in the spinal cord that generates the tendon jerk response. This reflex pathway, whose functions will be discussed in Chapter 10, consists of exactly two neurons: a primary (Ia) afferent fibre carrying impulses from stretch receptors in a muscle synapses excitatorily with a motor neuron in the ventral horn of the spinal cord, whose axon returns to innervate the same muscle from which the afferent fibre came (**Fig. 3.20**).

Tapping the tendon of the muscle causes a brief stretch of the sensory ending, firing the Ia fibre, which then excites the motor neuron and causes a reflex twitch of the muscle – the familiar knee-jerk response, if we use the patellar tendon.

The advantage of this reflex pathway from the experimenter's point of view is that the afferent fibres are readily accessible in the dorsal root for controlled stimulation,

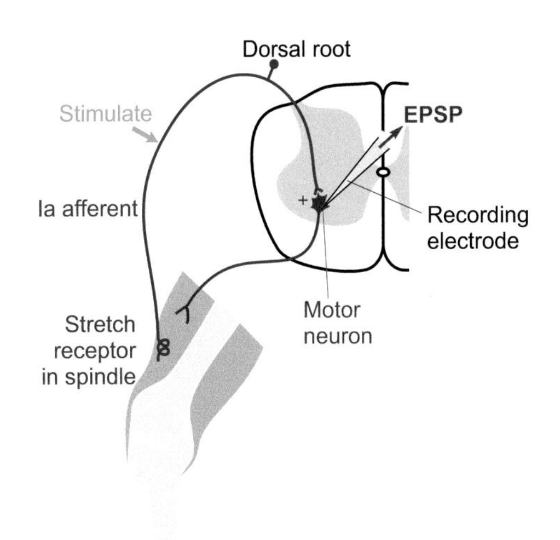

Figure 3.20 Excitatory synaptic action in the monosynaptic stretch reflex. Ia afferents from stretch receptors in a muscle enter the dorsal root and then synapse excitatorily with motor neurons in the ventral horn that innervate the same muscle.

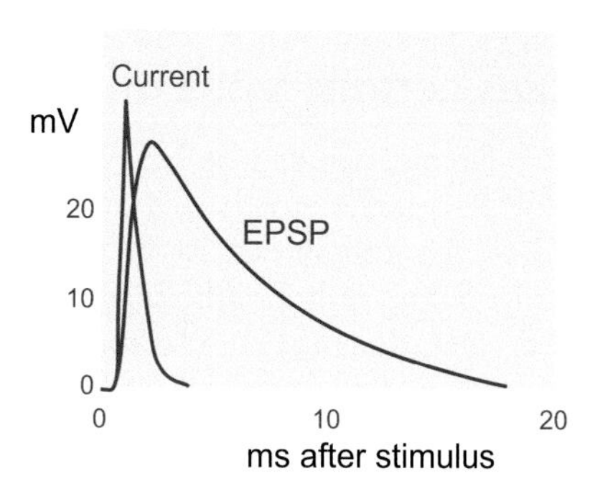

Figure 3.21 Excitatory postsynaptic potential (EPSP; blue) recorded with a microelectrode in a motor neuron after a single stimulating shock to the afferent fibres. The purple line shows the time-course of the synaptic current.

while the postsynaptic cell bodies are large enough to be punctured easily by a microelectrode.[3] If we apply a single brief shock to the Ia fibres whilst recording from the motor neuron, we find that the postsynaptic response consists of a small depolarisation rather similar in shape to the EPP, called the *excitatory postsynaptic potential* (EPSP) (**Fig. 3.21**); if it is large enough, it may trigger off an impulse in the motor neuron. By measuring the reversal potential in exactly the same way as in the case of the NMJ, it is possible to show that the EPSP is the result of a transient increase in permeability to sodium and potassium ions. Although this increase lasts only about as long as the action potential, just as in the case of the endplate, the EPSP itself is relatively prolonged because of the long time constant of the cell membrane. One may therefore assume that the arrival of an impulse releases some transmitter substance from the vesicles visible in the presynaptic endings and that this substance diffuses across the synaptic cleft and causes the opening of short-circuiting channels. Could this transmitter also be ACh?

It would be nice if we could simply extract the vesicles from the ending and see what was in them, but this is technically difficult. However, there are histological stains (and more recently and usefully, *immunohistochemical* stains) that are selective for particular transmitters or their metabolic precursors, or for enzymes that are associated with them, which can help to identify transmitter substances. In the case of ACh, which is widely found as a transmitter within the brain as well as at the NMJ, one may stain for the enzyme cholinesterase, whose presence suggests strongly the use of acetylcholine itself. But the mere presence of a possible transmitter in the presynaptic endings is not sufficient evidence by itself that it is actually being *used* as a transmitter. There are a number of further criteria that have to be met. We need to confirm, for example, that application of the supposed transmitter actually causes the *same effects* as the real transmitter. It is not enough simply to note that both are excitatory: they must both open the same channels, and so have the same reversal potential. Ideally it should also do so in plausibly small concentrations, but this is a difficult criterion to meet because the postsynaptic membrane is very much less accessible from outside than it is to transmitter released in the proper way from the terminal. We must also demonstrate that the real and supposed transmitter have the *same pharmacology:* that

Box 3.1 Examples of types of neurotransmitters and modulators

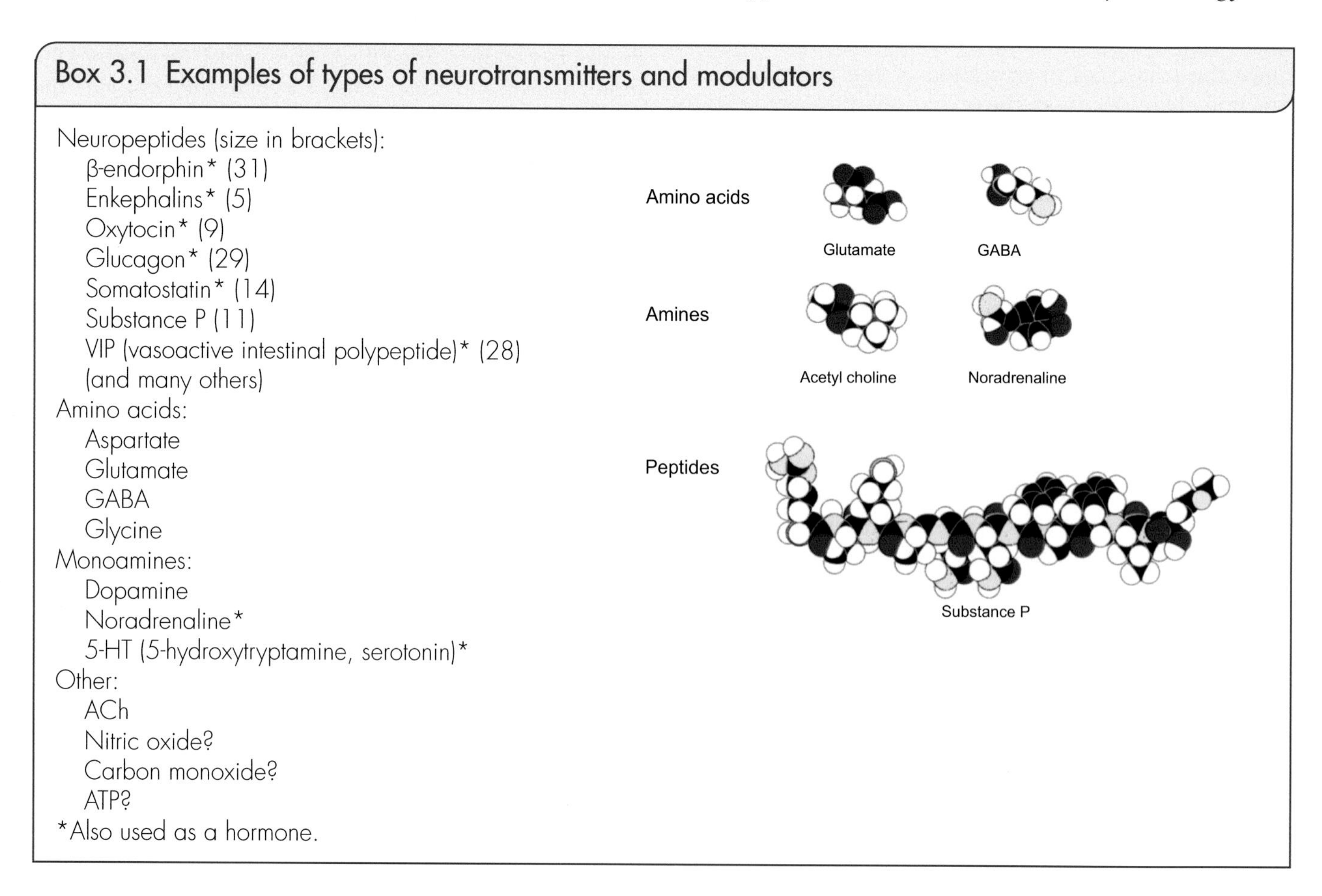

Neuropeptides (size in brackets):
- β-endorphin* (31)
- Enkephalins* (5)
- Oxytocin* (9)
- Glucagon* (29)
- Somatostatin* (14)
- Substance P (11)
- VIP (vasoactive intestinal polypeptide)* (28)
- (and many others)

Amino acids:
- Aspartate
- Glutamate
- GABA
- Glycine

Monoamines:
- Dopamine
- Noradrenaline*
- 5-HT (5-hydroxytryptamine, serotonin)*

Other:
- ACh
- Nitric oxide?
- Carbon monoxide?
- ATP?

*Also used as a hormone.

they are blocked by the same pharmacological agents, and that substances that inhibit the inactivating enzyme for one, and hence prolong its action, do so for the other. Ideally, one should also be able to show that afferent action potentials really do release the supposed transmitter, but this also is often technically extremely difficult to establish adequately.

In fact it is only at a relatively small proportion of the synapses in the central nervous system that we are certain of the identity of the transmitter in the sense that all these criteria have been met. But if one is satisfied with circumstantial evidence, then one can generate quite long lists of putative transmitters (**Box 3.1**) and create maps indicating their distribution throughout the brain.[4] It would be nice to think that such maps would reveal some deep pattern of meaning as regards which transmitter does what, but disappointingly this is not really so. Knowing what the transmitter is at a particular synapse does not generally help us to understand what the synapse *does*, not least because that is a function of what the receptor site is linked to: the same transmitter may do several quite different things that bear no obvious functional relation to each other (**Table 3.2**).

Table 3.2 Some varieties of receptors for neurotransmitters

Neurotransmitter	Receptor type	Direct/ indirect	Action	Agonists	Antagonists	Comments
Glutamate	α-amino-5-hydroxy-3-methyl-4-isoxazole propionic acid(AMPA)	Direct	$\uparrow P_{Na'}$ $\uparrow P_K$	AMPA	CNQX	
Glutamate	B	Indirect, ↓cGMP and other mechanisms	$\downarrow P_{Na}$	trans-ACPD	α-methyl-4-carboxyphenyl-glycine	Several types, some presynaptic
Glutamate	N-methyl-d-aspartate (NMDA)	Direct	$\uparrow P_{Ca}$	NMDA	AP5	Voltage-dependent
Glycine	GlyR	Direct	$\uparrow P_{Cl}$		Strychnine	Mostly spinal cord
ACh	nAChR	Direct	$\uparrow P_{Na'}$ $\uparrow P_K$	Nicotine carbachol	Tubocurarine	
	mAChR	Indirect	$\uparrow P_K$	Muscarine	Atropine	
GABA	GABA-A	Direct	$\uparrow P_{Cl}$	Muscimol	Picrotoxin, bicuculline	Also presynaptic
	GABA-B	Indirect, ↓cAMP	$\uparrow P_K$	Baclofen	CGP 335348	Also presynaptic
Serotonin (5-hydroxytryptamine; 5-HT)	5-HT_1	Indirect, ↓cAMP	$\uparrow P_K$	(Various)	(Various)	Several sub-types, often presynaptic
	5-HT_2	Indirect	$\uparrow P_K$	α-methyl-5-HT	Ketanserin	
	5-HT_3		$\uparrow P_{NA'}$ $\uparrow P_K$?	2-methyl-5-HT	Tropisetron	Periphery only
Dopamine	D_1	Indirect, ↑cAMP	↓PK	Dehydrexidine	SCH 23390	
	D_2	Indirect, ↓cAMP	$\uparrow P_K$	Haloperidol, spiperone	Spiroperidol	Also presynaptic
Noradrenaline	α_1-adrenoceptor	Indirect		Phenylephrine	Prazosin	
	α_2-adrenoceptor	Indirect		Clonidine	Yohimbine	Presynaptic
	β_1-adrenoceptor	Indirect		Dobutamine	Practolol	
	β_2-adrenoceptor	Indirect, cAMP		Procaterol	Butoxamine	

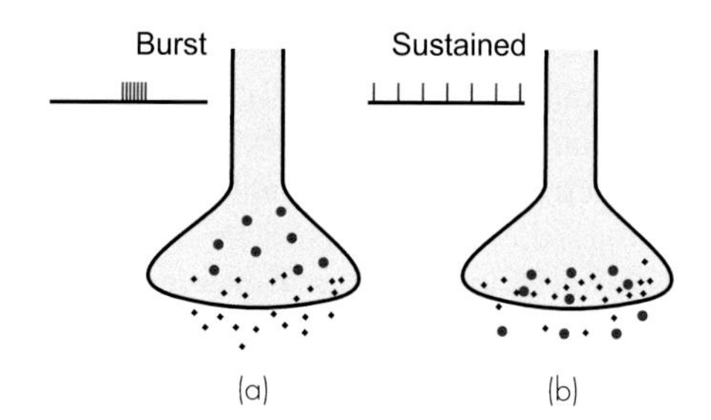

Figure 3.22 Possible mechanisms of differential release of neuropeptide (red) and conventional transmitter (black). With brief, intense stimulation (a), there is insufficient time for the larger neuropeptide vesicles to reach the pre-synaptic membrane, so less is released than in the case of milder but sustained stimulation (b).

In addition, some synaptic terminals are known to release more than one type of molecule, often a conventional transmitter in conjunction with one or more small peptides such as substance P, vasoactive intestinal peptide (VIP), cholecystokinin (CCK), etc., called *co-transmitters*, often acting as *neuromodulators*. Examples of co-transmitters are VIP and Substance P with acetylcholine, and enkephalin and neuropeptide Y with noradrenaline. Because the peptides are often packaged in larger vesicles, further from the synaptic junction than the conventional transmitter, it has been suggested that different patterns of afferent stimulation could result in different proportions of the two kinds of transmitter being released; *sustained* activity would favour the peptide, *bursts* of activity the conventional transmitter (**Fig. 3.22**). Some evidence that this actually happens has been found in peripheral autonomic synapses; if it is true, the functional implications are profound, since it would provide a mechanism potentially capable of discriminating more subtle aspects of the action potential code than simply the mean frequency of firing. We shall see later (p. 59). that many neurons in the CNS have mechanisms by which they can *generate* more complex patterns of firing such as periodic bursts. In addition, there is increasing evidence that neuropeptides often have different kinds of effects from 'conventional' transmitters. As we shall see later, they often appear to modify the action of transmitters, rather than initiating action themselves, and sometimes they are released in a more diffuse way than is typical for conventional transmitters, suggesting a function that is almost intermediate between neurotransmitter and hormone. (Many of them *are* of course hormones in other contexts). Conventional transmitters – ACh and 5-HT for instance – may also act as diffusely-acting neuromodulators.

Synaptic integration

Functionally, a central synapse is very different from an NMJ. Endplates are not meant to be intelligent: we just want them to obey orders. Consequently they generally fire one action potential in the muscle fibre for every action potential that arrives down the nerve fibre. (However, in some muscle fibres – for example, the slow fibres of the frog – no action potential is generated at all: here there is not just one endplate on the cell, but a large number of synapses distributed all over its surface. Activation of the afferent fibres thus causes a widespread passively summated EPP over the whole cell, leading to a slow contraction of the muscle fibre, rather than a rapid action potential and a relatively fast mechanical twitch, a process a little more akin to what happens in central neurons.)

Clinical box 3.3 Synapses as sites of disease

An antibody, also known as an immunoglobulin, is a Y-shaped protein produced by B-cells of the immune system. The tips of these structures are highly variable (up to 10 billion different types), which allows them to recognise, bind to and neutralise most invading pathogens (particularly viruses and bacteria) by acting as a 'signal' to direct the effector arms of the immune system. Unfortunately, there are occasions, especially in the setting of a recent infection or cancer ('post-infectious' and 'paraneoplastic' respectively), where antibodies are produced that mistakenly target components of the body, leading to autoimmune destruction and disease.

At the neuromuscular junction, two illnesses arise in this way. In myasthenia gravis, antibodies are produced against the nicotinic ACh receptors, leading to destruction of receptor and, ultimately, the motor endplate. This classically results in a 'fatigable' weakness, with patients showing symptoms more after exercise or at the end of the day. It is treatable, with immunosuppressants and pyridostigmine – which blocks the acetylcholinesterase to potentiate the effect of ACh on the remaining receptors – but can affect the respiratory muscles in particularly severe 'crises'. A similar illness, Lambert-Eaton myasthenic syndrome (LEMS), sees antibodies produced against presynaptic voltage-gated calcium channels. This also causes weakness but, by contrast, this gets *better* with exercise as increasing activity in the motor nerves leads to intracellular calcium accumulation and resumption of vesicular release at the NMJ.

Of course, receptors in the CNS are similarly vulnerable to autoreactive antibodies and, over the past decade, a host of different syndromes have been recognised. For example, antibodies against the glycine receptor – the predominant inhibitory pathway in the spinal cord – causes muscle spasms, rigidity, and sudden jerks (myoclonus). Meanwhile, anti-NMDA receptor antibodies first manifest through a florid psychosis, probably due to the abundance of these receptors in the limbic system. In each case, an understanding of the underlying neurophysiology enables recognition of the culprit antibody, sometimes many weeks before confirmatory tests are completed.

Table 3.3 Types of potential

Action potential	Endplate potential (EPP)	Excitatory postsynaptic potential (EPSP)
Actively regenerated	Decays passively	Decays passively
All-or-nothing	Graded	Graded
Threshold	No threshold	No threshold
No summation	Summation	Summation
Large (>60 mV)	Moderate (ca. 10 mV)	Small (ca. 10 μV)

But things are very different at a typical central neuron (see **Table 3.3**). In the brain the EPSP is much smaller than the EPP, so that a large number of them need to coincide to generate firing in the postsynaptic cell. This is just as it should be, when you realise that a typical neuron in the brain may have something like 10 000 synapses on it: the whole *point* is that each neuron actually weighs up all this incoming information and decides whether to fire as the result of a complex and essentially intelligent evaluation. Postsynaptic activity is thus a function of the *integrated* discharge of all its afferent terminals, and neurons exhibit what is called *spatial and temporal summation*. The classical experimental set-up, with huge synchronised volleys in the dorsal root, followed by just one action potential in a motor neuron, is a good way to find out the basic mechanism of synaptic action, but a ludicrous travesty of how neurons actually go about their daily business. Bear in mind that in ordinary life, most neurons fire most of the time.

Temporal summation occurs because the potential produced by a brief synaptic current falls off relatively slowly, so it is possible to get summation of the effects of repeated stimulation of a single ending if the frequency of firing of the afferent fibre is high enough. In fact, because of both this smoothing effect of the membrane time constant, and the very large number of endings synapsing with each neuron, most of which will probably be tonically active at any moment, it is probably more helpful to forget about the quantised nature of the action potentials at afferent synapses. Indeed it is misleading to think in terms of individual EPSPs at all – they are essentially artefacts. Rather, one should think of the way in which all of them together provide a combined inward *current* that is continually varying as the result of changing patterns of afferent discharge. Think in terms of flowing water: a trickle in the brooks and streams of the furthest dendrites – which flow together into the rivers of larger and larger dendrites – that finally empty themselves into the lake formed by the cell body (**Fig. 3.23**). This river system receives a continual patter of rain from the afferent synapses, with local cloud-bursts and deluges from time to time; and at the end of the lake is a waterwheel (the axon hillock) which turns repetitively to generate action potentials in response to the total flow of water. When a motor neuron is excited to fire by its afferent connections, the action potential actually starts not in the region near the excitatory synapses themselves but rather at the axon hillock (where the threshold is lower), from which it spreads both forwards down the axon, and also backwards over the cell body and also possibly the dendrites (a phenomenon that fits less easily into the watery metaphor.)

This summation is not in general linear unless the synapses are sufficiently separated from one another for no interaction to occur *between* them. If a particular point on the cell surface is short-circuited by an excitatory synapse, short-circuiting another point very close to the first will have little additional effect; because the membrane is already depolarised, the second synapse will contribute less additional current than it would do if it were acting on its own (**Fig. 3.24**). Thus, although sometimes the effect of stimulating two separate afferents to a motor neuron is the sum of the effect of stimulating each separately, because they are too far apart to interact, more often it is substantially less, a phenomenon known as *occlusion*. In addition, the nearer an excitatory ending is to the axon hillock – other things being equal – the greater will be its influence on the firing of the cell; synapses on distant dendrites will be relatively less effective. Like real riverbeds, the dendrites are leaky, so that much of the current generated at distant sites is lost. (However there is some evidence

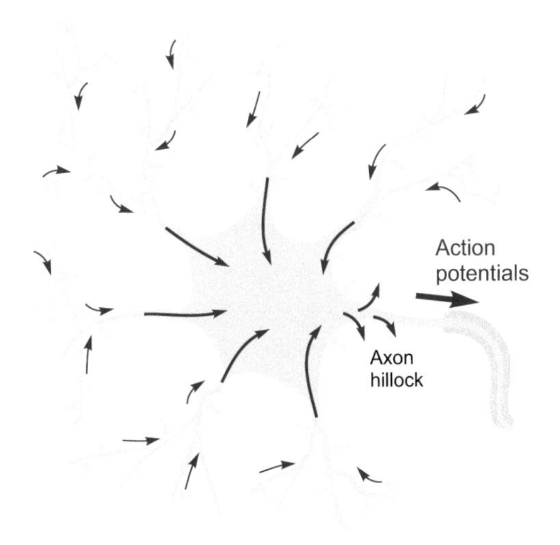

Figure 3.23 Currents from many synapses flow down dendrites and thence into the soma, like rain gathering in rivulets, streams and rivers before entering a lake; the firing of action potentials depends crucially on the axon hillock region.

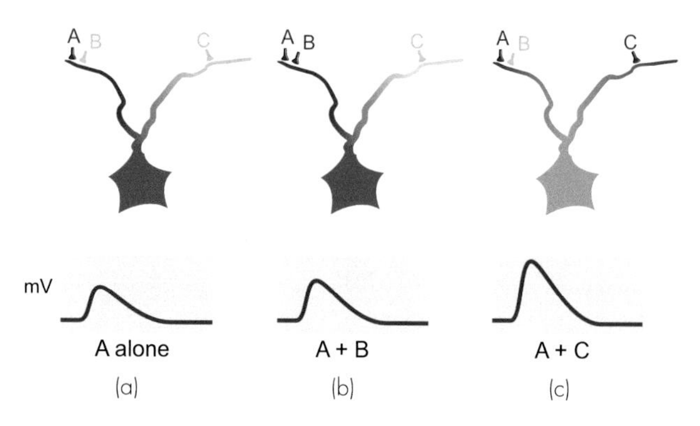

Figure 3.24 Linear and non-linear addition of excitatory postsynaptic potentials (EPSPs) and spatial summation. (a) Excitatory synapse, A, on its own generating and EPSP in the soma of the neuron (below). (b) If a neighbouring synapse, B, is activated as well, there may be a little increase in the EPSP, because the membrane in the vicinity of B is already depolarised. If a distant synapse, C, is activated instead of B (c), the combined effect is greater because the synapses do not interact with one another.

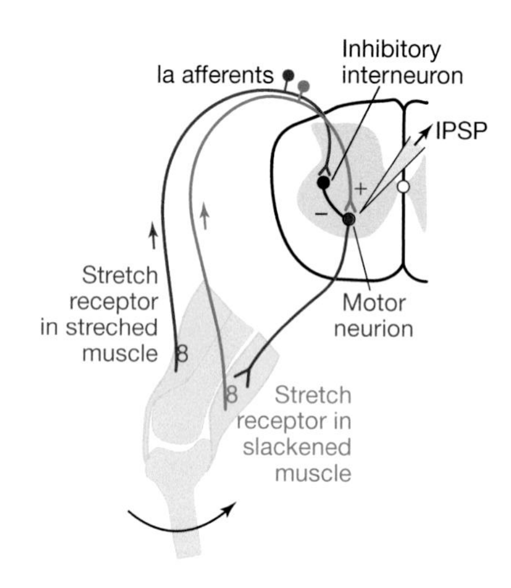

Figure 3.25 Schematic circuit for reciprocal innervation of flexor and extensor muscles by receptor afferents.

that distant synapses may initiate larger currents, to make up for this.)

This is the classic view of summation by neurons: recent studies have shown that this picture must be modified in a number of respects. In addition to simple electrical interactions between short-term effects of synaptic activity, there are also modulatory systems (see later) that can alter transmission by synapses and also the transfer of information from one part of a neuron to another. In addition, it is now clear that many neurons in the brain have significant numbers of voltage-gated sodium, potassium and calcium channels, that can be capable of initiating and propagating action potentials in dendrites. This has considerable theoretical importance, for as we shall see in Chapter 13, many important kinds of memory, mediated by changes in synaptic strength, depend on a synapse recognising that there has been simultaneous activation of both the presynaptic and the postsynaptic cell. Back-propagation of action potentials from the axon hillock and soma could therefore be crucial to ensuring that distant dendritic synapses 'know' that the postsynaptic cell has actually fired. Furthermore, neuromodulators that specifically regulate the activity of these active dendritic processes may therefore regulate not only the extent to which one part of a neuron talks to other parts, but also whether or not learning processes are permitted to take place.

Inhibitory synapses

Now, a nervous system in which the only connections were excitatory would not be a very useful one: clearly there are occasions on which the proper response to a stimulus is inhibition rather than excitation, withdrawal rather than attack, relaxation rather than activation. In particular, the way in which our muscles are generally arranged in pairs that oppose one another implies that excitation of one muscle is usually associated with inhibition of the other, by a process of *reciprocal innervation*. In the tendon jerk, for example, the reflex contraction of the muscle that is stretched is accompanied by a relaxation of its antagonist. In this case, the inhibition of the corresponding motor neurons is brought about by branches of the afferent fibres from the stretch receptors, which after entering the dorsal cord send excitatory branches to interneurons, which in turn form *inhibitory synapses* with the motor neurons in the ventral horn (**Fig. 3.25**).

One might wonder why a seemingly unnecessary interneuron is interpolated in this pathway. The reason may lie in a general rule that seems to be true of the transmitters used by cells in the central nervous system, namely that a neuron always releases the same transmitter, acting in the same way, at all its terminals (*Dale's hypothesis*).[5] Since the stretch receptor fibres are excitatory to the motor neurons of the agonist muscle, they cannot also be inhibitory to the motor neurons of the antagonist: thus they must first excite an interneuron, by opening the same kind of short-circuiting channels that they open in the motor neurons, and this interneuron must then inhibit the antagonist motor neurons.

The existence of reciprocal inhibition in the tendon jerk reflex is convenient from the experiment's point of view, since it is possible first to insert an electrode in a motor neuron, and then stimulate various dorsal root fibres until some can be found that produce either excitation or inhibition of the motor neuron. If we then give a single shock A to the inhibitory fibres, and follow it with an excitatory stimulus delivered after different time delays, we can measure the time-course of the inhibitory effect by measuring the size of the subsequent response to the excitatory stimulus given at different times after A: some results of this kind are shown in **Figure 3.26**. Here, an inhibitory shock clearly leads to a depression of the excitatory response that lasts for many milliseconds (black, dashed line). One may also of course simply see what happens to the motor neuron's potential when the inhibitory shock is delivered in the absence of any excitation (blue line): the stimulus is followed by a potential change in the neuron, of rather

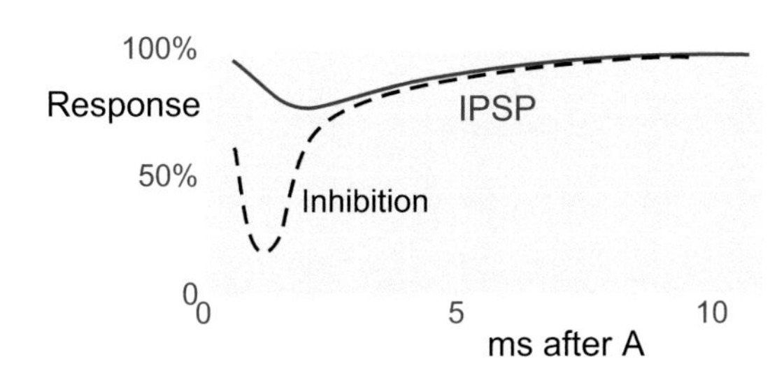

Figure 3.26 Inhibition of the monosynaptic reflex and its electrical correlates.

similar time-course to an EPSP, but of opposite polarity: this hyperpolarisation is called the *inhibitory postsynaptic potential* (IPSP).

To find the origin of the IPSP, we can follow our usual procedure of passing various steady currents in or out of the cell by means of one half of a double-barrelled electrode, and using the other half to measure the resultant size of the IPSP, and hence determine its reversal potential. What we then find is that the IPSP, unlike the EPSP, gets larger and larger instead of smaller as the resting potential is reduced to zero; but if we artificially hyperpolarise the membrane, the potential is reduced in size and eventually reverses at about –80 mV, the reversal potential for the IPSP (**Fig. 3.27**). This voltage lies somewhere between the equilibrium potentials for potassium and chloride ions, which might suggest an increase in permeability of chloride and potassium (**Fig.3.28**). However, it is now clear that – as in most central sites – inhibition occurs through changes in chloride permeability alone; the apparent hyperpolarisation may well have been due to a background depolarisation caused by leakage resulting from the cell's impalement by the microelectrode.

The transmitter in this particular instance is likely to be glycine. Elsewhere in the central nervous system both gamma aminobutyric acid (GABA) and glycine have been confirmed as inhibitory transmitters, working through an increase in chloride permeability, but GABA is very much more common. At this site in the spinal cord, the convulsant poison strychnine blocks this synaptic inhibition, and strychnine is known to block glycine receptors but not GABA. (Where GABA is known to be an inhibitory transmitter it is blocked by another convulsant, picrotoxin, which is ineffective at this site in the spinal cord.) Incidentally, the convulsant effect of these inhibitory blockers illustrates another general function for inhibition in the brain: if we have a large network of cells that are connected together in highly convergent and divergent pathways that are entirely excitatory, we have a situation that is potentially explosive. Stimulation of any one cell is likely to lead to a chain reaction involving the progressive spread of activity over a large area, and this is precisely what is observed with convulsants like strychnine and picrotoxin. At least as much inhibition as excitation is required if this sort of explosive response is to be avoided, and we shall see in Chapter 14 that special systems exist in the brain to regulate the general level of neural activity through diffuse inhibition – very like the damping rods in a nuclear reactor – and thus prevent seizures of this kind from occurring.

There is another site in the spinal cord where inhibition is relatively easy to study. The axons that leave the motor neurons of the ventral horn on their way to the muscles also send off branches that turn back into the cord and innervate – excitatorily – small interneurons called *Renshaw cells* (**Fig. 3.29**). From Dale's hypothesis we would expect the transmitter at this synapse to be ACh, and this is found to be the case; it acts in the usual way, by causing an unselective increase in membrane permeability, though here the response is much more prolonged than at the NMJ, the response to a single shock being a burst of firing at high frequency. But these Renshaw cells themselves send off short axons that in turn synapse with the pool of motor neurons by which they are stimulated, and their synapses are inhibitory. They are relatively easy to study because one can activate the Renshaw cells by stimulating the motor neurons antidromically in the ventral root. This kind of feedback inhibition is a common one in sensory systems as well, providing one of many mechanisms for *adaptation*

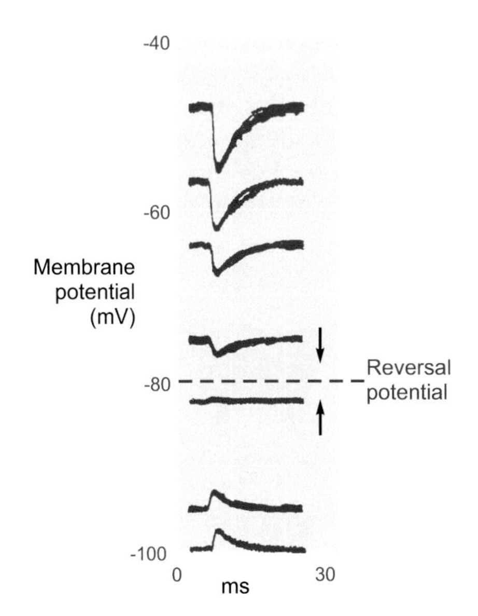

Figure 3.27 Reversal potential of the inhibitory postsynaptic potential (IPSP). IPSPs were recorded at different initial levels of depolarisation, as in **Figure 3.17**. They show a reversal potential of about –80 mV. (After Eccles, 1964; reproduced with permission.)

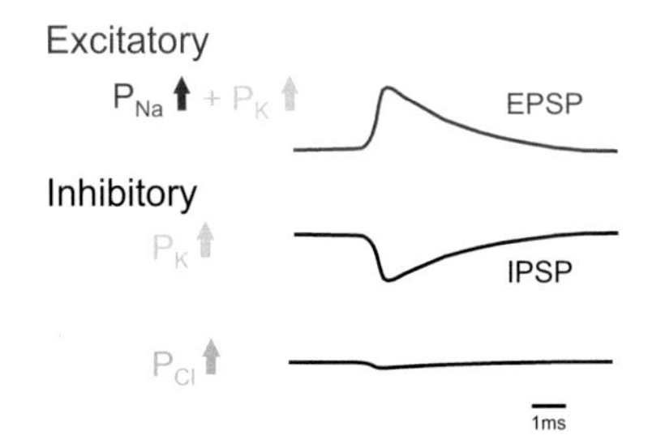

Figure 3.28 Postsynaptic potentials in response to activation of different kinds of channel (idealised). Excitatory postsynaptic potential (EPSP) generated by opening of channels permeable to sodium and potassium. Inhibitory postsynaptic potential (IPSP) generated by opening of potassium channels (P_K) or chloride channels (P_{Cl}).

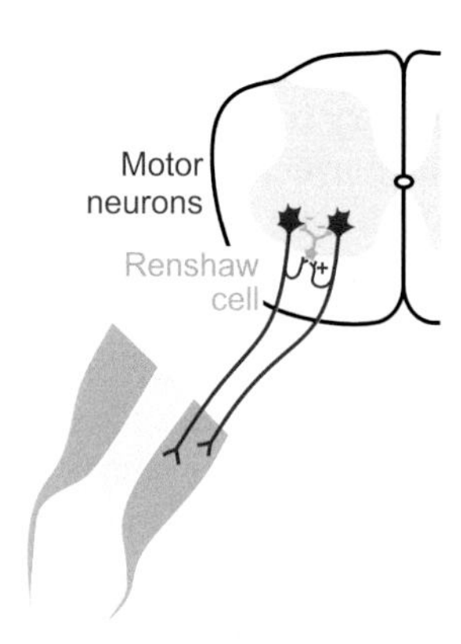

Figure 3.29 Renshaw cell (blue) in the spinal cord, showing feedback inhibition of motor neurons (red).

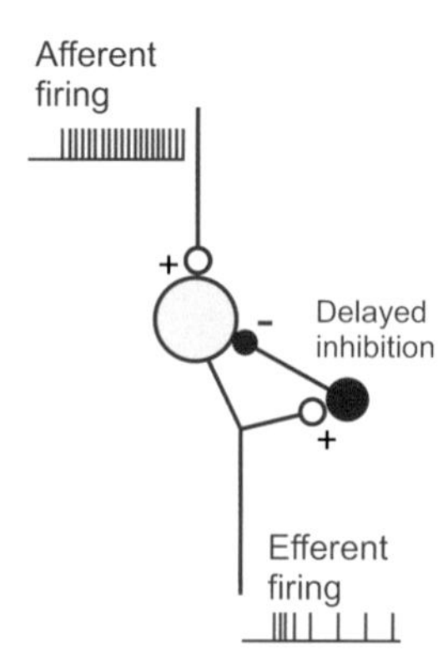

Figure 3.30 Schematic representation of how a sudden maintained onset of afferent activity is converted into an adapting efferent response.

(p. 69) (**Fig. 3.30**). However, it is not clear that this is actually the primary function of Renshaw cells; rather, they probably serve to discourage synchronised firing that would lead to unwanted clonus in the muscle. Indeed, when the Renshaw cells are damaged in tetanus, the result is widespread activation of the muscles ('tetany'), starting in the jaw ('lockjaw') before progressing to abdominal, limb and respiratory muscle groups (see notes in Chapter 10).

Voltage and current inhibition

We have now completed the chain of events by which stimulation of inhibitory afferents leads to release of a transmitter that then opens up hyperpolarising channels in the postsynaptic membrane. But how does this result in actual inhibition? The answer is not quite as simple as might appear at first sight. In some cases, the time-course of the inhibition mirrors quite accurately the time-course of the IPSP itself. Other things being equal, a hyperpolarisation means that the potential has to be driven further than would otherwise be the case in order to reach threshold, and so one might well expect to find a substantial correlation between the degree of hyperpolarisation at any moment and the degree of inhibition. But there is often an extra peak of inhibition at the beginning that cannot be explained in this way, and in some cases one may find this short-term component even in the absence of the IPSP-like slow component (**Fig. 3.31**). In this figure the red dots show the time-course of the inhibition itself, and it is obvious that the shape of this peak is very similar to the time-course of the burst of current that generates the IPSP: this current is shorter in duration than the IPSP, being roughly the same as that of the action potential, because as usual the decline of potential back to its resting level is prolonged by the membrane capacitance. Thus there appear to be two separate components of the inhibition that may be observed: one that is closely related to the *potential* at any moment (voltage inhibition), and is relatively easily explained, and one that seems to be associated with the *current* (current inhibition).

To see how current inhibition arises, consider two synaptic endings, one excitatory and one inhibitory, lying close to one another on the postsynaptic cell membrane. It is clear that if both happen to be active simultaneously, they will to some extent cancel each other out, since one is a current source and the other a current sink (**Fig. 3.32**). The currents that would otherwise be generated by the excitatory ending, and might eventually initiate an impulse at the axon hillock, are mopped up before they have got any distance at all. But this inhibitory effect would clearly only operate while the inhibitory channels are actually open: as soon as they close the excitatory current would be free to exert its effects at a distance as before.

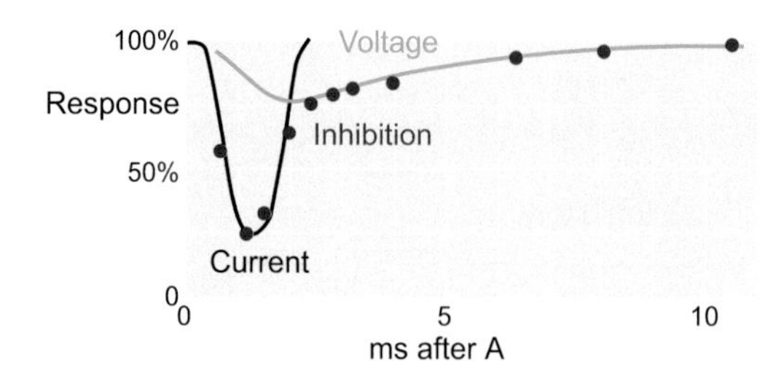

Figure 3.31 Data points (red) show the size of reflex evoked by stimulation of a motor neuron at different times after shock, A, that stimulates inhibitory afferents. The blue line shows the time-course of the inhibitory postsynaptic potential (IPSP) associated with the inhibition and the black line shows the approximate time-course of the inhibitory synaptic current. The degree of inhibition in this case appears to be related to both current and potential. (Data from Araki *et al.*, 1960; with permission from John Wiley and Sons.)

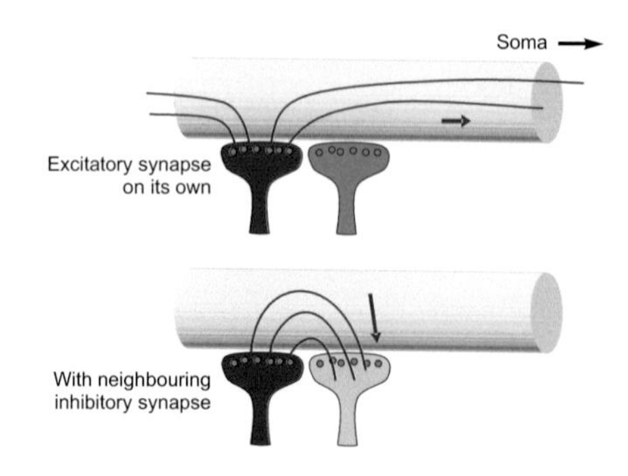

Figure 3.32 Current inhibition. (Top) An excitatory current from an excitatory synapse (red) progresses along the soma unheeded. (Bottom) If an adjacent inhibitory synapse (blue) is simultaneously activated, it acts as a current sink, impeding the flow of excitatory current.

This kind of inhibition is quite distinct from the effect of hyperpolarisation, and may indeed produce inhibition in the absence of an IPSP. Imagine for example a hypothetical channel whose associated reversal potential happened to be exactly the same as the resting potential. Clearly such a mechanism could not generate an IPSP; but it would still be inhibitory because while it was open it would tend to clamp the membrane potential firmly at the resting level, by draining away current generated by any nearby excitatory synapses that happened to be active. As sodium enters at one site, chloride enters at the other and effectively neutralises it. In a sense, an increase in P_{Cl} during the IPSP does just this: because E_{Cl} is normally close to the cell's resting potential, this increase contributes nothing to the hyperpolarisation: in fact it actually makes the amplitude of the IPSP less than it would be if there were an increase in P_K. The importance of Cl^- lies in its current effect, in *clamping* the membrane potential close to its resting level. In terms of the river analogy, inhibitory synapses are sluices that modify the local flow.

We can now perhaps see why it is that some inhibitory effects seem to be of the current type and some of the voltage type, or a mixture of the two. If the excitatory and inhibitory synapses involved happen to be close to one another, the inhibition will be predominantly of the current type, and of short duration. If they are separated, for example on different dendrites, they will not interact directly with one another, but their effects will simply summate at the site of initiation of the action potential, producing the voltage effect. Spatial summation of excitation and inhibition is thus rather complex and – as in the case of summation of EPSPs – not just a matter of simple linear addition. Again, inhibitory endings that are very close to the axon hillock region will be particularly good at preventing the cell from firing, because they will ambush the excitatory currents just before they reach the detonator region. Indeed it is frequently found in the central nervous system that powerful inhibitory synapses are found clustering near the axon hillock and acting as a sort of guard ring around it: a sluice-gate that is right next to the waterwheel.

It is important not to underestimate the complexity of the processes of spatial and temporal summation in central neurons. Each individual neuron in the brain is a little microcomputer that calculates the size of its output on the basis of the whole spatiotemporal *pattern* of excitation and inhibition that it receives from the enormous number of afferent fibres that drive it. When we look at the varying dendritic shapes of different kinds of neurons – sometimes of extraordinary complexity (p. 8) – we are in a sense looking at something like a diagram of the rules that determine the neuron's behaviour. Furthermore, although we have been treating the production of action potentials in the axon as the final output of the cell, it is important also to remember that the intermediate slow changes in potential that occur in remote dendrites as the result of local synaptic activity may, in some cases, generate other outputs as well. There are many sites in the brain (notably the retina and olfactory bulb) where dendrodendritic synapses occur, and transmitter is released locally by dendrites in response to the potential changes caused by nearby synaptic currents, rather than to action potentials. Indeed, neurons of this kind need not possess axons at all.

Presynaptic modulation

A phenomenon that puzzled early investigators was that sometimes they found clear evidence of inhibition of a cell – in the sense that stimulation of certain fibres resulted in a reduction of the usual response to stimulating other afferents – *without* any corresponding change in its potential or permeability. In some cases this could be explained as 'remote inhibition': in other words, an interaction between neighbouring inhibitory and excitatory endings of the type just described, on a dendrite so far from the recording site that although the inhibitory synapse was capable of cancelling the effect of the excitatory one through current inhibition, it could not generate currents large enough to be measurable to the cell body. Then histologists noticed that not all the synaptic endings that could be seen in the cord were between axon and cell body or dendrite: on the contrary, many terminated on *another* ending that in turn synapsed in the conventional way (an *axo-axonic* synapse: **Fig. 3.33**). It was therefore clear that another explanation of inhibition in the absence of any detectable change in the postsynaptic cell was possible: that of *presynaptic inhibition*. The notion here is that the ending on the second terminal, the *modulatory* ending, may somehow hinder the latter's excitatory action, and so cause inhibition of its effects without influencing the postsynaptic cell in any way: in the monosynaptic reflex, the size of the EPSP is reduced, with a time-course similar to that of the depression of the reflex response itself (**Fig. 3.34**)

This inhibition can occur in a number of different ways. The most common is simply through the activation of conventional inhibitory channels at the terminal, resulting in a reduction in the size of the action potential and thus a decreased entry of calcium. GABA-B receptors are often found on presynaptic terminals, implying very strongly that they mediate such a mechanism; but since these receptors are indirect, increasing cAMP, it is conceivable that this operates directly to reduce transmitter release; other receptors found on presynaptic membrane are the α_2-adrenoceptor and 5-HT1 (serotonin) receptor. Paradoxically,

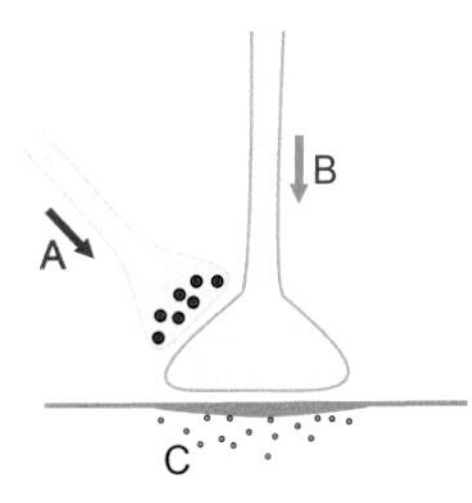

Figure 3.33 Presynaptic inhibition. Schematic representation of a terminal, A, that pre-synaptically inhibits the excitation of C by B, by synapsing with B's terminal.

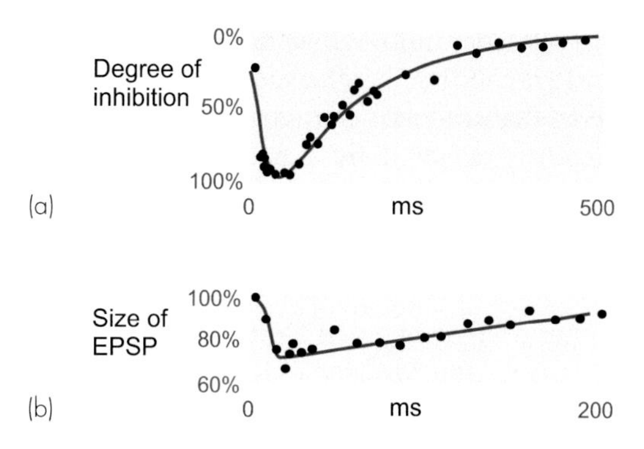

Figure 3.34 (a) The time-course of depression of monosynaptic reflex after brief stimulation of afferents producing presynaptic inhibition. (b) The time-course of the excitatory postsynaptic potential size in the same experiment. (Data from Eccles, 1963, with permission from Elsevier; Eccles et al., 1961; with permission from John Wiley and Sons.)

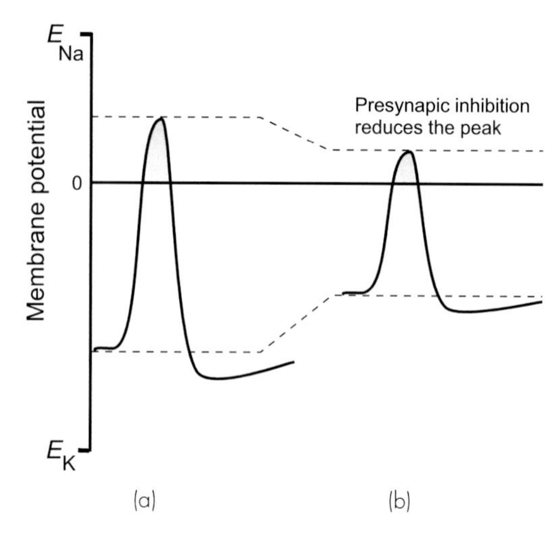

Figure 3.35 Effect of depolarisation of terminal on size of peak action potential (schematic). (a) A normal action potential. Because the rate of transmitter release depends heavily on depolarisation, only the peak (shaded) contributes significantly to the amount of transmitter released. (b) Partial short-circuiting causes a tonic depolarisation, but also reduces the peak of the action potential because the effect is to pull the potential at every moment towards zero (dashed lines). Consequently, there is a disproportionately large reduction in the amount of transmitter released.

a conventional 'excitatory' mechanism, with an E_{rev} around zero, can have a similar effect. Because there is a very sharply rising increase in the rate of release of transmitter as a function of the potential across the ending, during the entry of the action potential it is really only the *peak* of the impulse that contributes significantly to the number of vesicles that are released (**Fig. 3.35**). So if we open up short-circuiting channels in the terminal itself, although admittedly the consequent depolarisation of the resting potential should increase the steady rate of transmitter release, equally it will reduce the peak potential of any afferent impulses, by pulling the membrane potential towards zero, and also through the reduction in excitability caused by steady depolarisation (see Chapter 2, p. 35). Since the peak is what counts, this latter effect will more than compensate for the resting depolarisation, and the consequence will be a *reduction* in the amount of transmitter released by the terminal, and hence in the size of the ensuing EPSP. One site where this may be the mechanism for presynaptic inhibition is on the terminals of primary afferents from mechanoreceptors, in the dorsal horn. Selective stimulation of the small fibres of the dorsal root tends to cause a quite long-lasting inhibition of the monosynaptic response to stimulation of the Ia afferents.

In some invertebrates such as *Aplysia* an analogous process of presynaptic *facilitation* may be seen. Here the mechanism is much better understood: the modulatory ending releases a transmitter, serotonin, that enhances transmission by increasing the level of cAMP within the presynaptic ending. This in turn increases the size of the presynaptic action potential by reducing potassium permeability.

What's the point of it? Functionally, presynaptic inhibition has the advantage of being rather more precise and specific in its actions than postsynaptic inhibition. In the latter case, an inhibitory ending acts on the postsynaptic cell as a whole, regardless of what the source of excitation may be (although it will affect some excitatory afferents more than others because of the spatial effects outlined earlier). But presynaptic inhibition provides a mechanism whereby certain inputs may be disabled while others are left unhindered: a gating function, which implies control over *which* of the many types of input to a cell may or may not be allowed to influence it. In the case of a motor neuron, we shall see that many different neuronal pathways converge to form synapses on it: tendon jerk reflex afferents, inhibitory afferents for reciprocal inhibition, descending fibres from the higher levels of the brain, afferents controlled by pain and other receptors in the skin, Renshaw cells, and many others. It would clearly be an advantage to be able to alter the strength of some of these inputs independently of the others, and presynaptic inhibition provides a mechanism for doing this.

Modulators can act at other sites apart from the presynaptic terminal, for instance altering postsynaptic responses and probably also altering the extent to which dendrites can transmit potentials. The latter could be particularly significant: as was noted earlier, prevention of back-propagation would be expected to impair or prevent the learning that occurs through synaptic strengthening, and there is good evidence that ACh, acting as a modulator, may function in this way.[6]

Long-term changes in excitability

All the phenomena described so far in this chapter are really rather brief in duration, and fall into the category of what has been described as 'millisecond physiology'; but some other synaptic phenomena are of a rather longer timescale.

The first of these concern the effects of *repetitive* stimulation, effects most easily seen at the NMJ or at autonomic ganglia. If we stimulate a frog NMJ with a short train of action potentials, under suitable conditions we may find that size of the evoked EPPs gradually increases throughout the period of stimulation, a phenomenon that is often called *facilitation* (sometimes *potentiation*), though of course it is distinct from the serotonergic facilitation described earlier. When, as here, a change in synaptic strength is induced by its own activity it is called *homosynaptic;* induced by activity in a different synapse it is called *heterosynaptic*. Homosynaptic facilitation is a presynaptic rather than postsynaptic phenomenon, and due to an increase in the amount of transmitter released by each action potential. One fundamental mechanism seems to be that not all the calcium that enters in one action potential is pumped out again before the next, so that its concentration within the ending steadily rises, resulting in increasing numbers of vesicles being released. Confirmation for this explanation comes from the fact that the rate of mEPPs is raised after the end of the pulse-train.

If, however, we stimulate for a long period at a rather high frequency, we find a reduction rather than increase in the size of the synaptic potential: this reduction is called *post-tetanic depression,* and is thought to be due simply to a depletion of the amount of transmitter in the ending that is available to be released. If the vesicle release is blocked by lowering the calcium content of the medium, it is found that instead of declining, the size of the evoked potentials actually increases (*post-tetanic potentiation*); this is observed in cases of Lambert-Eaton myasthenic syndrome (see **Clinical Box 3.3**, p. 58). A possible explanation for this phenomenon is that the increase in calcium following the arrival of an action potential at the nerve ending not only causes the release of transmitter, but also in some way promotes the mobilisation of stored transmitter so that it is more readily available, as well as facilitating the release of vesicles. Normally, this mobilisation is insufficient to keep up with the rate at which transmitter is being used up when the nerve is tetanically stimulated, and post-tetanic depression is observed; but if the rate of release is artificially reduced the transmitter begins to accumulate, resulting in a gradual increase in the amount of transmitter which is released by each action potential.

The mechanism by which neural activity stimulates transmitter mobilisation as well as release is at present unknown. In this context it is worth remembering that in many neurons the transmitter is synthesised in the cell body rather than in the endings. It is transported there by means of a specialised system that is associated with the activity of neurotubules and neurofibrils within the axon, at rates that may be as high as 15 mm/hour, and it is quite possible that this transport mechanism may also be stimulated by the electrical activity of the axon. Another way in which the effectiveness of a synapse may undergo slow modification, which has been demonstrated in some invertebrate neurons, is by gradual inactivation of the calcium channels in the presynaptic ending as a result of its activity: this will reduce the calcium entry and so cause a reduction in transmitter release. Yet another mechanism is the existence of *autoreceptors*, receptors on the presynaptic ending that respond to the transmitter that the ending releases, and then cause longer-term regulation of presynaptic function.

The most interesting long-term alterations in synaptic activity are those which are associated with *learning*. We shall see later that it appears to be possible to explain all of the different kinds of memory and learning that the brain is capable of by postulating changes in synaptic effectiveness that are a function of the patterns of activity that the pre- and postsynaptic cells have experienced, and synaptic receptors with exactly these postulated properties are now well established (*N*-methyl-D-aspartate (NMDA) receptors). These phenomena are called *long-term potentiation* (LTP) and *long-term depression* (LTD), though these terms are misleading in that LTP and LTD are almost always induced heterosynaptically. They are discussed in the context of learning, in Chapter 13 (p. 260).

A rather simpler phenomenon that is more appropriately discussed here because of its obvious relation to adaptation is *habituation*. Like adaptation, habituation is a decline in response to a constant stimulus; but whereas adaptation means a decline during the application of a continuous stimulus, habituation implies a decline in the successive responses to a stimulus that is *repeatedly* applied. It is essentially a high-level phenomenon, seen not in the responses of sensory receptors but rather in behavioural responses to stimuli and in the parts of the brain that control behaviour. As a consequence, it is typically very specific to one particular pattern of stimulation – in a way that cannot be explained by simple adaptation of peripheral receptors. For instance, when one reads for the fourth time in a month of some appalling plane crash one is rather less shocked than the first time; but this is evidently not because of adaptation in the eye. Again, the fact that one is not continually aware of the somatic sensations produced by one's own clothes is often attributed to adaptation by touch receptors in the skin. But this is clearly wrong, for the stimulus here is repeated rather than continuous, and sensitivity to other patterns of stimulation is not affected: it is actually due to habituation.

Finally, an entirely different phenomenon that can contribute to relatively long-term changes in excitability arises because of the way in which the neural elements of the central nervous system are densely packed together with apparently rather little extracellular space. Under such circumstances, ionic concentration changes, of a kind that we can usually neglect when considering transmission in peripheral axons, may cause long-term changes in permeability and excitability. Potassium is the culprit here, for although the concentration changes of Na and K are equal and opposite as a result of the passage of action potentials, because external $[K^+]$ is normally very low, a given change in concentration has a relatively larger effect on E_K than on E_{Na}; furthermore, the E_K has far the greater influence on the resting potential. Thus activity in any one neuron tends to cause neighbouring neurons to depolarise, over a long time-scale.

To an extent this effect is mitigated by the ubiquitous *glial cells,* that occupy most of the space not taken up by the neurons themselves and which they outnumber by two or more to one. One kind of glial cell is the *oligodendrocyte,* responsible for laying down myelin in the CNS, as Schwann cells are in the periphery. Another variety are the *microglia*, which are believed to function very like macrophages. But the third type, the *astrocytes*, are perhaps the most interesting.[7] Like neurons, they have highly branched dendritic structures with which they make intimate contact with neurons, blood vessels, and other astrocytes, but they do not carry conventional action potentials; microelectrode recording shows that they help to buffer potassium changes, and the astrocytes themselves exhibit wave-like changes in potential that reflect the local concentration of potassium in their environment and are spatially integrated through the gap junctions that they make with one another. These potential changes can influence calcium entry, which in turn can trigger release of transmitter substances (*gliotransmitters*), including glutamate. They also have a role in mopping up glutamate, which is converted into glutamine and offered back to neuronal terminals. This is an essential and clinically extremely significant function, as excess glutamate is toxic to neurons: this *excitotoxicity* is liable to cause further neuronal death after brain injury. For all we know, astrocytes may play some more active role than merely providing housekeeping services for central neurons. In a sense they form a communication system within the brain that is related to, but independent of, the actual neurons and it is not wholly impossible that they perform some computing functions over a larger time-scale than neurons do, and perhaps on a wider scale: in particular, they might modulate learning processes.

The development of synaptic connections

Finally, there is the interesting and really rather basic question of how synaptic connections are set up in the first place. The human brain is perhaps the most complex structure known to science, and if we understood the rules that govern the way in which its innumerable and intricate synaptic connections are specified and formed, we would have come a long way in our understanding of its function. Bearing in mind that there are some 10^{12} neurons in the brain, and that each one receives and gives on average some 10^4 synaptic connections, it is quite inconceivable for these patterns to be specified in *detail* by the instructions for building the brain embodied in our DNA. Though the broader structure of the brain – in terms of tracts that connect nuclei and other sub-populations that are relatively homogeneous within themselves – might be genetically specified, one must conclude that the connections between individual neurons are either essentially random, or more probably that they are in some way governed by our own sensory experience. The latter is an attractive hypothesis, since it implies that the structure of the brain may in a sense be capable of *self-organisation*, of adapting itself to the particular tasks and the particular types of sensory stimulation it has to cope with. In this sense the brain may be thought of as rather like a telephone exchange, in which, when first built, only the broad outlines of the connections between its various elements are specified by its designer, and up to a point almost every unit in it is potentially capable of connection with every other. Once in use, the actual pattern of links at any moment is clearly a function of the patterns of impulses that subscribers have sent to it from their telephone dials. It is also clear that something of this sort must be present to explain the modification of synaptic connections as a result of experience that is implied by the existence of memory, a topic that will be pursued further in Chapter 13. This is a very active area of research, and it is becoming clear that in particular cases there do exist mechanisms by which the brain can in effect build its own connections so as to adapt itself to a particular task, and that the instructions given it by the genetic code are essentially rather vague. This is an area rather beyond the scope of this book, and there is not room here to do more than mention a few underlying principles and simple examples; for further information there are several up-to-date and readable accounts that may be consulted.[8]

Consider first a question that may already have occurred to you. Clearly a synapse will only function properly if there are receptors in the postsynaptic membrane that match the transmitter released by the presynaptic terminal. Often one finds nuclei containing a mixture of cell types, and groups of incoming fibres that connect specifically with one cell type or another. Is there then some mechanism that guides the tip of a developing axon towards only those cells that have receptors corresponding to its own transmitter? Or is it rather that when a nerve fibre approaches another neuron, in some way it stimulates the manufacture of the appropriate kind of receptor site? One piece of evidence on this point comes from the study of the formation of the NMJ. If ACh is applied locally to different parts of a muscle cell's surface, it is found that only the region underlying the NMJ will respond with depolarisation: presumably the cholinergic receptors are confined to the postsynaptic membrane. If we now cut the nerve fibre to the muscle we find that progressively more and more of the muscle cell's surface becomes sensitive to the transmitter, a phenomenon called *denervation hypersensitivity*. But if a new axon starts to grow towards the muscle cell, this process is reversed, and once again we find that the response to ACh becomes limited to the region where the new junction is developing. It seems therefore that the presence of the nerve ending, presumably by the release of some substance, either attracts the receptors or at least encourages their formation while suppressing those that are present elsewhere. We saw in Chapter 1 that damage to an axon not only causes degenerative changes in the parent neuron, but may also produce transneuronal degeneration in the neurons with which it is in contact. The implication is that synapses are not merely for communication, but are also *trophic:* they not only transmit messages, but also contribute to the maintenance of the cells they contact.

Furthermore, it appears that the development of hypersensitivity is itself in turn a stimulus that attracts nearby axons and leads them to form new synaptic junctions. A normal frog muscle fibre has only one NMJ, and if a severed motor nerve is placed in its vicinity it will not form additional endplates to it. But if the original innervation is cut, it is found that the resulting hypersensitivity is also accompanied by the acceptance of a synaptic junction from a fibre that previously was ignored. In the same way, transplanting an extra limb at an inappropriate site in many amphibia leads to new fibres growing out from the central nervous system to innervate it. Similar work on the regeneration of neural connections in amphibia (regeneration is not observed – at least not over such large distances – in the mammalian central nervous system, although it does occur in the peripheral nervous system) has shown that this guidance of neurons on to their targets can sometimes be even more specific; not just on to the correct type of cell, as defined by its receptor properties, but even on to the correct part of an extended mass of such cells. In the frog, there is an orderly projection of the fibres of the optic nerve to the frog's 'visual brain' (the tectum), that preserves the topology of the retinal image. If the optic nerve is cut, it is found that the fibres not only regenerate back to the tectum, but do so in such a way as to retain, at least approximately, their correct spatial arrangement.

The details of the mechanism by which this specificity of connection arises is an active area of current research. The basic mechanism appears to be that the tip of a developing axons – the *growth cone* – is not unlike an amoeba: motile and responsive to a wide range of chemical substances, some being attractive and some repulsive, some acting at a distance and others essentially on contact. It is perhaps worth emphasising that in these experiments the guidance is anatomical rather than functional: if, after cutting the optic nerve, the frog's eye is rotated in its orbit through 180, the pattern of the regenerating fibres is not also rotated through 180. As a result, the animal's subsequent visual behaviour is inverted, with upward movements in response to objects in the lower visual field, and so on. In other words, there is no suggestion in these experiments that the pattern of activity in the incoming fibres can influence the pattern of their connections.

Other experiments in mammals have shown that connections can in certain circumstances be altered by the pattern of neural activity, in such a way that only useful connections are formed, or useless ones are lost. For example, the cells of a cat's visual cortex (Chapter 7) are usually found to be driven in almost equal numbers by each eye, and many by both. But if one eye of a kitten about 5 weeks old is kept closed, even if only for a few days, one finds when it has grown up that the number of its cortical cells that are driven by the eye that had been closed is very greatly diminished. There has been no anatomical interference here: the only difference between the two eyes was the degree of their neural activity during the period of closure, so it must follow in this case that the synaptic connections have been influenced by the pattern of neural activity experienced. What seems to happen in the course of development is that initially there is an excess of synaptic contacts, that form randomly in a somewhat promiscuous way. But at a certain critical period in development, perhaps because of a fall in the general level of neurotrophic factors, there is intense competition between endings and neurons, and those which are in some sense less successful than others simply degenerate or die. There is thus a *critical period*, a delimited window of opportunity for laying down the basic pattern of neuronal connections within the brain. Recently, some of these neurotrophic factors have also been shown to have immediate modulatory effects on presynaptic activity, thus blurring still further the distinctions between development, memory, and short-term modulation.

Sensory receptors

Types of transduction

A process by which energy of one form is converted into energy of another is called *transduction*. The energy incident on a receptor cell, whether thermal, electromagnetic, mechanical or chemical, must be turned into electrical energy in the form of potentials across the cell membrane, that eventually cause release of transmitter. As we have seen, in general the effect of the stimulus energy is to alter the *permeability* of the cell membrane to certain ions, resulting in a flow of current and a movement of the potential towards some new equilibrium value.

But not all kinds of energy are able to do this in any particular receptor. It is clearly important, if the brain is to make any sense of the outside world, that receptors respond specifically to certain *kinds* of stimulus, perhaps light, or heat. For this reason, transduction is in effect preceded by a specialised *energy filter* that allows certain types of energy through to cause electrical effects, and not others (**Table 3.4**). This filtering is seldom absolute: the receptors of the eye, for example, though exquisitely sensitive to light, will also respond to mechanical deformation if it is severe enough. If in the dark you shut your eye and press on the side of it with your finger, you will see a faint blue patch of light called a phosphene as the receptors respond to mechanical stimulation by sending the brain exactly the same message that they would have sent if a real blue light had been present. Specificity of receptors is in fact a relative matter, and many receptors are actually surprisingly unspecific in what they respond to. Fine discriminations between one type of stimulus and another are to a large extent the work of the central nervous system rather than of the receptors themselves.

Transduction in the Pacinian corpuscle

One receptor that because of its peculiar anatomical structure happens to be very specialised indeed has been studied in great detail, and that is the *Pacinian corpuscle*, a mechanoreceptor found in the skin and mesentery (**Fig. 3.36**). Here the naked tip of an otherwise myelinated axon

Table 3.4 Classification of sensory receptors

Mechanoreceptors	*Special sense*	Cochlear hair cells
		Vestibular hair cells
	Muscle	Spindles
		Tendon organs
	Skin and visceral	Pacinian corpuscles
		Ruffini endings
		Merkel discs
		Meissner corpuscles
		Lanceolate endings
		Free endings
		Nociceptors
	Vascular	Arterial baroceptors
		Venous and atrial stretch receptors
Thermoreceptors		Skin (warm and cold, nociceptors)
		Hypothalamic
Photoreceptors		Retina
Chemoreceptors		Olfactory
		Gustatory
		Hypothalamic
		Vascular
		Visceral
		Nociceptors

is sheathed in concentric onion-like layers called *lamellae* that shield it from virtually every type of stimulus except that of mechanical deformation. It is a good preparation for studying the transduction process in general, since one can easily isolate individual corpuscles and – as originally done by Loewenstein – record their electrical responses to precise mechanical stimuli applied to the capsule's surface. A convenient means of stimulation is to hold against it a probe mounted on a small piezoelectric crystal. When a voltage is applied across the crystal, it changes its shape slightly and causes a controlled deformation of the capsule, and a change in membrane permeability. Now a complicating factor when trying to measure the relation between a stimulus and the permeability changes that result from it is that if the permeability changes are big enough they will trigger off action potentials that will in turn interfere with the very permeabilities one is trying to measure. Consequently it is helpful to disable the active properties of the axon by poisoning it with a substance like tetrodotoxin that blocks the voltage-dependent sodium channels. If this is done, we find that mechanical stimulation of the ending results in a depolarisation of the axon – the *receptor potential* (sometimes, misleadingly, called a generator potential) – whose magnitude depends on the size of the stimulus.

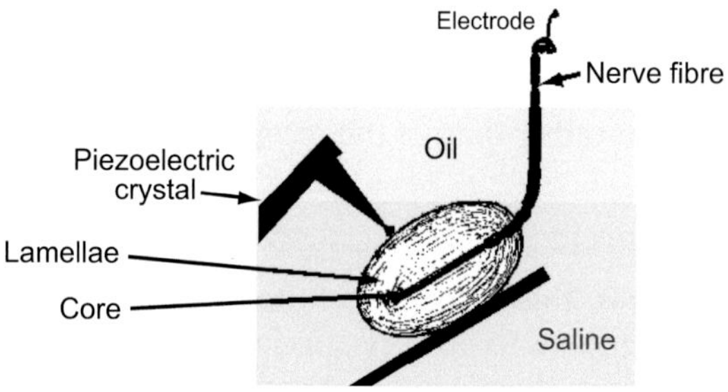

Figure 3.36 A Pacinian corpuscle dissected out and arranged for electrical recording and mechanical stimulation by means of a piezo-electric crystal.

To find out more about the permeability changes giving rise to these potentials we can do what was described in the case of the neuromuscular junction: measurement of the reversal potential – which turns out to be very close to zero. The simplest kind of permeability change that would generate an equilibrium potential around zero would be an increase in both sodium and potassium permeability. So when we deform the cell membrane it seems that we open up ionic channels in it which are permeable to all ions and so act as a sort of short-circuit: perhaps distorting the membrane simply increases its leakiness.

Adaptation

A striking feature of the response that is obvious in **Figure 3.37** is that its time-course is very different from that of the stimulus itself. If we apply a prolonged but constant deformation to the surface, we find that the generator potential rises quite rapidly to a peak, and then spontaneously falls back again to the resting potential; when the stimulus is removed a second peak is generated. The cell seems, in fact, to respond to *changes* in the degree of stimulation rather than to its steady level, a very common type of receptor response that is called *adaptation*. It turns out that much of this adaptation is due to the mechanical properties of the lamellae that surround the ending, that act as an energy filter. If we strip them off, and apply the stimulating probe directly to the surface of the axon, we find that although the generator potential still falls off after its initial rise, it remains depolarised so long as the stimulus is maintained, with no hint of a second peak of depolarisation when the stimulus is removed (**Fig. 3.37**, (b)). This kind of response is known as *incomplete* adaptation, as opposed to the complete adaptation seen when the lamellae are intact, and the response falls to zero.[9]

There is a simple mechanical model of the Pacinian corpuscle that explains how it filters out steady levels of stimulation (**Fig. 3.38**). We can think of the end of the axon itself as behaving in a simple elastic manner, so that any force applied to it results in a corresponding deformation, and hence in a change in permeability. The lamellae on the other hand behave very differently because they are separated from one another by layers of viscous fluid: when a steady deformation is applied to a particular part of the capsule, the lamellae in that region slowly collapse as the fluid between them oozes sideways to neighbouring regions. This slow collapse is characteristic of what engineers call *viscous elements*, as in the oil-filled cylinders or dampers (also known as dashpots) that make up part of the shock absorbers fitted to car suspensions or to the tops of

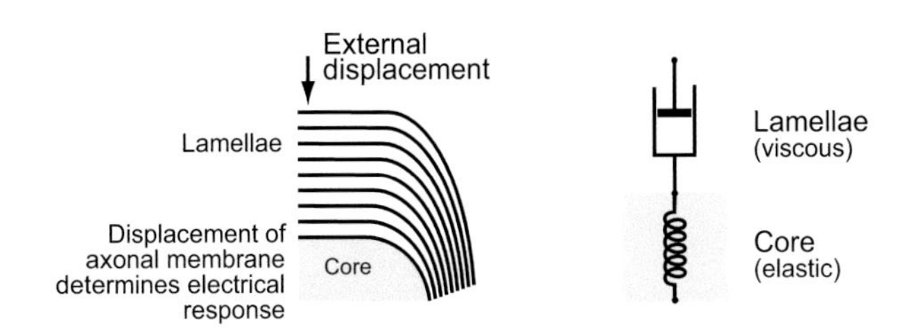

Figure 3.38 Highly schematic representation of the mechanical elements of the Pacinian corpuscle and their mechanical equivalent 'circuit' in the form of a viscous element (dashpot) and elastic element (spring).

swing doors, contrasting with the purely elastic element in the axon itself. (Diagrammatically, elastic elements are shown as spring symbols, viscous elements as schematic dampers.)[10] In the case of the Pacinian corpuscle, both these elements are connected in series: if we apply a sudden steady displacement to the outer end of the viscous element, at first there is no time for it to collapse, and the displacement is taken up by the elastic element, which is thus compressed. But in being compressed (B in **Fig. 3.39**), it exerts a force on the viscous element which then tends to collapse (C and D in **Fig. 3.39**), through the sideways oozing of fluid described above. The elastic element, the axon itself, therefore gradually resumes its original shape, and the potential returns to its resting value. But if the stimulus is removed (E in **Fig. 3.39**), the whole process is reversed: at first the elastic element must stretch to take up the new displacement, but in doing so it pulls on the viscous element, expanding it again (F and G in **Fig. 3.39**), so that in the end all is as it was originally.

It is important to realise that this adaptation occurs only when, like Loewenstein, we apply constant *displacements* rather than constant *forces*. To see why this is, consider the alternative model of the corpuscle shown in **Figure 3.40**: an old-fashioned spring balance with a squashy cushion on top. If we thrust our fist into the cushion, and then hold it in a fixed position, the needle on the dial will initially be deflected, and then return to its original position, as the spring balance pushes back on the cushion and makes it collapse. But if instead we suddenly place a weight on the cushion, the deflection is maintained (though the weight will gradually sink into the cushion), because the force felt by the balance is constant. Whether, in real life, Pacinian corpuscles are relatively non-adapting pressure receptors,

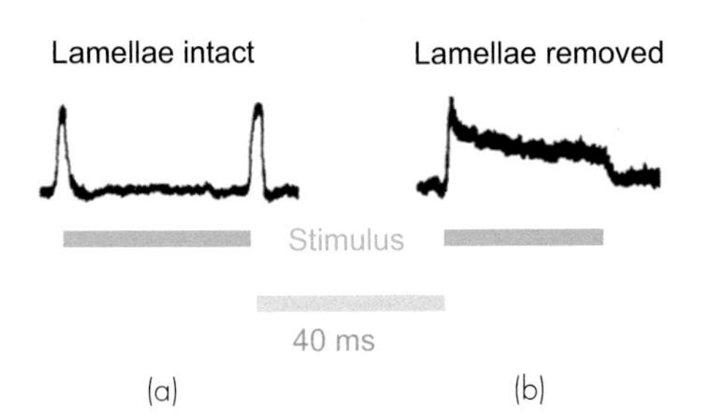

Figure 3.37 Generator potentials recorded from a Pacinian corpuscle before (a) and after (b) removal of the outer lamellae, in response to a brief maintained deformation (green bar). (After Loewenstein & Mendelsohn, 1965; with permission from John Wiley and Sons.)

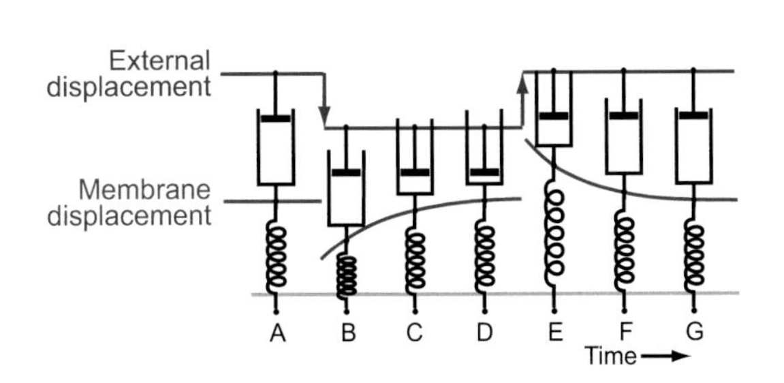

Figure 3.39 Response of a mechanical model of the Pacinian corpuscle to steady displacement applied just before B and removed just before E, showing adaptation of the degree of distortion of the elastic element, and hence of the electrical response of the receptor.

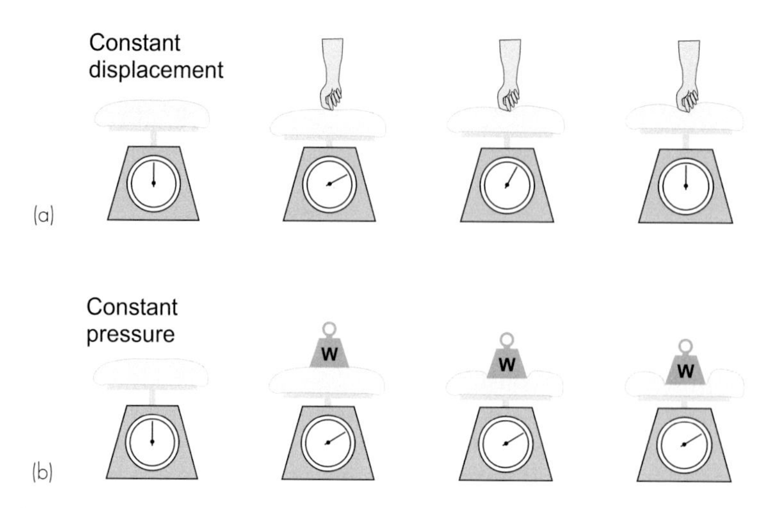

Figure 3.40 An alternative model of the Pacinian corpuscle: a spring balance with a cushion on top. There is a differential response to a fixed, sudden (step) deflection (a) and constant force (b).

rather than rapidly adapting displacement receptors, is hard to establish, as this depends on the mechanical properties of the surrounding tissue.

In one respect the Pacinian corpuscle – although a popular example of adaptation – is not at all typical of receptors in general: essentially the same response is produced at both stimulus onset and stimulus offset. If we suppose that inward and outward deformations of the cell membrane are equally good at causing changes in permeability, because both of them simply result in membrane stretch, then we can see how what is termed the *biphasic receptor potential* comes about. This indifference to the sign of a stimulus is quite exceptional among receptors, however.

Another oddity of the Pacinican corpuscle is that a steady deformation normally produces only one action potential when the stimulus is applied, and another when it is removed: the accommodation of the ending is too great for the rate of depolarisation after the action potential. But this is not at all typical of sensory receptors, which generally respond to a steady stimulus with a continuous train of impulses. Under these conditions, adaptation will manifest itself as a gradual decline in firing frequency after starting the application of the stimulus, as may be seen in the responses from a touch receptor in a cat's paw in **Figure 3.41**. If the receptor is of the completely adapting sort, it will eventually stop firing altogether; otherwise the frequency will decline to some steady level. Adaptation is also manifested when a fibre responds more when the stimulus is increasing than when it is steady. **Figure 3.42** shows a recording from a muscle receptor that responds to stretch: you can see that it fires more briskly while it is being stretched than it does during the period of constant stretch: conversely, when the stretch is reduced, it stops firing altogether. These kinds of responses can be called *phasic* and *tonic*. Receptors vary a great deal in the speed with which the decline during constant stimulation occurs. Some, like the Pacinian corpuscle, are very fast and phasic; others, like the receptors in the semicircular canals that adapt over some 20 seconds, are very slow and tonic. It is important not to confuse *speed* with *degree* of adaptation: fast-adapting fibres may be completely or incompletely adapting, and so may those that adapt only slowly (**Fig. 3.43**). A confusing terminology that you will often come across is to muddle these two aspects by calling a completely adapting receptor 'fast adapting' and one that has a tonic response 'slow adapting'. This kind of mechanical filtering is a common one in the body, and we shall meet it again in discussing the stretch receptors that are found in muscles. It is sometimes called a *high-pass filter* because it passes high-frequency vibratory stimuli much better than low-frequency vibrations of the same amplitude, since for a given amplitude of vibration the *rate* of movement is higher at high than at low frequencies.[11]

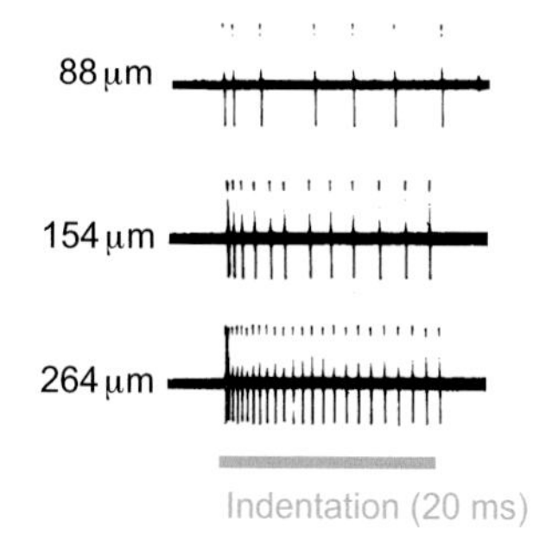

Figure 3.41 Tactile receptor in the skin of a cat's paw, responding to different indentations maintained for the duration shown by the bar below. (After Mountcastle, 1965; with permission from Springer Nature.)

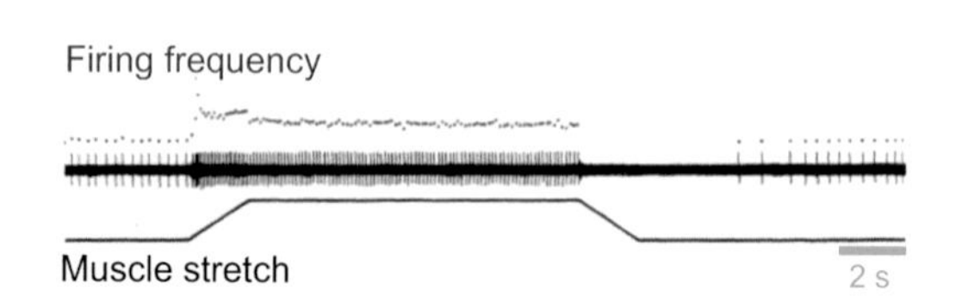

Figure 3.42 Response of a single afferent from a muscle spindle receptor to a gradually applied and released stretch (time-course shown underneath). (After Roll & Vedel, 1982; with permission from Springer Nature.)

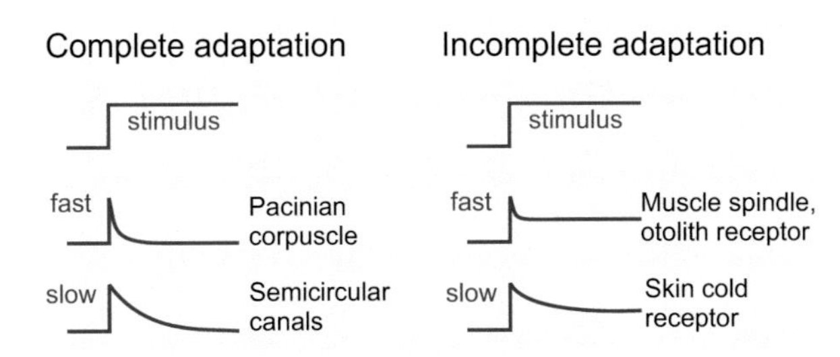

Figure 3.43 Types of adaptation. Adaptation may be complete or incomplete – depending on whether, during a steady stimulus, the response declines to the unstimulated level – and fast or slow – depending on how long it takes to reach an equilibrium response.

There are many receptors in which adaptation is at least partly due to energy filtering, similar to what is done by the lamellae of the Pacinian corpuscle. The muscle spindle stretch receptor (see Chapter 5) for example has a mechanical high-pass filter that behaves in a very similar way. Adaptation in the Pacinian corpuscle cannot be *entirely* due to the lamellae, since we have seen that there is still a decline in response at the beginning of a period of constant stimulation even when they have been stripped away. In most cases at least some adaptation is also the result of the *membrane adaptation* that was described in Chapter 2 (p. 40), which is a universal feature of neurons of all kinds. One way to demonstrate this is to bypass the transduction stage altogether and send currents directly into the cell through a microelectrode. A third way in which adaptation may occur is by *cellular modification* of an indirect transducer mechanism. In the eye, for example, incident light causes intracellular calcium to fall, which lowers the receptor's sensitivity during steady illumination.

Figure 3.44 may help to summarise the whole sequence of events in sensory receptors, and the stages at which adaptation may occur. The energy impinging on a receptor is first *filtered*, possibly by virtue of surrounding structures as in the Pacinian corpuscle, or in the visual receptors of some birds where little coloured oil droplets contribute to their colour selectivity. This filtering may incorporate an element of adaptation. It then brings about – by mechanisms that are yet unknown – *a change in the permeability* of the neuronal membrane, which if indirect may also be subject to adaptation. This in turn produces a *generator current* causing depolarisation. In 'long' receptors (those with their own axons), if conditions are right, this may be followed by *repetitive* firing at a frequency dependent on the degree of stimulation, although this frequency will in general decline because of membrane adaptation. Otherwise, if the degree of accommodation is too great in relation to the rate of depolarisation after the first impulse, only a single action potential will occur.

Efferent control

Another way in which this chain of events may be modified is through *efferent control*. Here, signals from the central nervous system are sent back to the receptor, and may

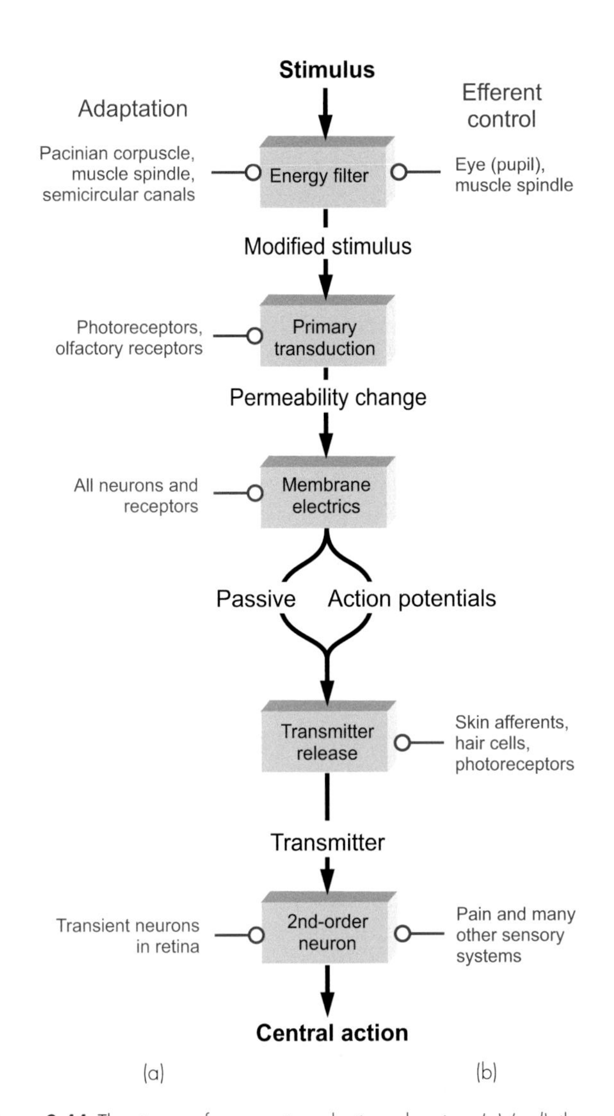

Figure 3.44 The stages of sensory transduction, showing, (a) *(red)*, the various ways in which adaptation may occur and (b) *(blue)*, points at which efferent control of sense organs may occur.

act either on the energy filter or on the coupling between terminal depolarisation and transmitter release, as shown in **Figure 3.44** (b). If these efferent signals are driven by the signals coming from the receptor, and they are such as to reduce the receptor's sensitivity, then they will effectively function as *negative feedback*, a further source of adaptation. A well-known example of central control over energy-filtering is the iris of the eye, which shrinks as the ambient light level increases; this reduces the stimulus and thus the degree of neural stimulation. (An alternative way of considering this 'pupil light reflex' is that the iris is actually *relaxing* in response to the *dark*, to admit more light to the retina when the ambient light is poor but at the expense of a shallower depth of field, as a consequence of widening the aperture of the box camera that is the eye.) But very often the control is exerted quite independently of the afferent signals. As we shall see in Chapter 5 (p. 100), muscle spindles receive efferent fibres that affect their sensitivity to stretch, that essentially specify the range over which the receptor should be working. Efferent control over transmitter release, typically through inhibitory efferent synaptic terminals, is often found on mechanoreceptors in particular, for instance on hair cells in the inner ear and on terminals of afferents from the skin, in the dorsal horn of the spinal cord (see below, p. 94), but they do not seem to be essential for adaptation.

Functions of adaptation

It may perhaps seem strange that the pattern of impulses generated by what is supposed to be a pressure receptor should be so very different from the time-course of the stimulus that it actually experiences, as we saw in the case of the Pacinian corpuscle: it looks as though the receptor is throwing away useful information. Is there any advantage in signalling changes rather than steady levels?

In the first place, it is not quite true to say that information has been thrown away. So long as the brain 'knows' how an adapting receptor responds to different patterns of pressure, it can in principle reconstruct the time-course of the original stimulus from the coded signals that it receives from the receptors, so that no information is really lost. But we have certainly lost a lot of action potentials, and one advantage of adaptation may be that it makes for economy of energy expenditure in generating fewer nervous impulses. If a stimulus tends to remain constant for long periods of time, with only occasional shifts to some new value – for example the pressure sensed by Pacinian corpuscles in one's buttocks during a long lecture – then there is little point in sending a stream of information to the brain which only tells it, in effect, that nothing has happened.[12] Since the same information could have been carried by many fewer impulses, by sending a message only when something new happens that might call for a response, one general function of adaptation could be said to be to get rid of unnecessary action potentials – it reduces the *redundancy* of the messages that are conveyed. Similar mechanisms are used in computers when storing data on disk: since computer data often contains long strings of repeated bytes, by storing only information about *changes,* files can be considerably compressed.

A second possible reason for the widespread existence of adaptation in sensory systems is that it may improve sensitivity by increasing the *signal-to-noise ratio* of the receptor.[13] This is a concept that is fundamental to understanding the coding of sensory information, and is well worth the little investment of intellectual effort needed to master it. Any signal, whether it consists of frequencies of action potentials or simply of varying voltages as in a telephone wire, is inevitably subject to a certain degree of uncertainty on account of the all-pervasive random *noise* that is an inescapable feature of the physical world. For example, if we measure the frequency of firing of a sensory fibre under conditions that are as constant as we can make them, we shall find nevertheless that the frequency we observe is not fixed, but undergoes continual random perturbations. This *noise* may be due to small changes in the temperature or chemical environment of the receptor, to slowly acting properties such as fatigue that are not under our control, or, ultimately, to the fact that the ions whose movement generates the potentials we measure are themselves in continuous random thermal motion, so that the currents they carry must equally be subject to a certain degree of unpredictability. Thus we can never assert that a nerve is firing exactly 70 times a second: the best we can do is to estimate with more or less confidence that its frequency lies somewhere between 69 and 71 Hz. As a result, if we plot firing frequency as a function of stimulus size S for a sensory receptor on a large number of occasions, we will find a band of scatter of width N associated with any particular value of S. This *noise* inevitably puts a limit on the amount of information that a fibre can carry (**Fig. 3.45**).

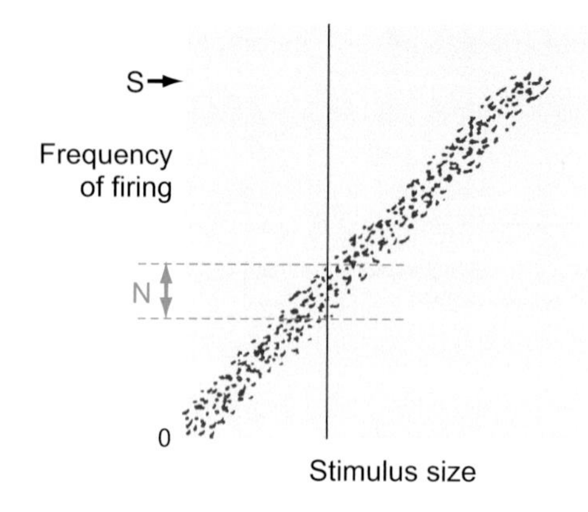

Figure 3.45 Signal-to-noise ratio. Schematic relationship between stimulus size and firing frequency for a hypothetical sensory fibre, measured on a large number of occasions (single data points), showing the band of frequency scatter of width N associated with any particular value of the stimulus. As a result, frequencies must be separated by N before they can be discriminated reliably from one another. Thus the number of significantly different frequencies, and hence the number of discriminable stimuli, is given by the *signal-to-noise ratio:* S:N, where S is the maximum firing frequency of the fibre. In this case, S:N is only about 6.

Now the function of a nerve axon is to convey messages from one place to another: in the jargon of information theory, it is a *communication channel.* The most fundamental thing we need to know about any communication channel, whether it is a fibreoptic broadband connection or the cable that joins a computer to its printer, is what its *capacity* (*bandwidth*) is. How many different messages can it convey in a given time? In the case of nerves, this amounts to asking how many distinguishably different frequencies it can fire at. Clearly, the refractory period sets an upper limit to firing frequency, perhaps around 500 Hz or so. One might think that since it could fire at any frequency below this limit, the number of possible frequencies (and thus the number of different messages it could send) must be infinite. But this is not so, for the existence of the noise that has just been described means that the brain may not be able to *discriminate* between frequencies that lie close together, because of the impossibility of determining exactly what the frequency actually is at any moment. More specifically, if we call the size of the largest signal that a nerve fibre can carry S, and the amplitude of the ever-present noise N, then the number of different frequencies that can be discriminated reliably from each other is only of the order of S/N. For example, if the noise in a fibre leads to an uncertainty of about 1 Hz in determining its frequency, the ratio S/N – the signal-to-noise ratio – will be 500: in other words, at any moment the nerve can only convey one of 500 distinguishably different messages.

If one thinks of this as being equivalent to an accuracy of one part in 500, or 0.2 per cent, this may not seem too bad a performance. But the problem is that the dynamic *range* – the ratio of the largest stimulus normally encountered to the smallest – over which most receptors have to operate is exceedingly large. In the case of the eye, for example, the dynamic range corresponds to the ratio between the brightness of the sun and the visual threshold in the dark, and is of the order of 10^{15}. If there were no adaptation in the eye, and each receptor coded a particular level of light intensity directly as a particular steady frequency of firing, its 500 possible output levels would have to be spread – pretty thinly – over the entire 10^{15} range of possible inputs – just as a thermometer measuring from 0 to 500 degrees would not be much use for monitoring body temperature. Clearly, a just-discriminable difference in receptor firing would correspond in general to a very large difference in light intensity, and our power of perceiving small differences in intensity would be very much worse than it actually is. But in practice the whole of this 10^{15} range is never present in our field of view at the same time, and – for reasons that are explained in Chapter 7 – the ratio between the darkest and lightest parts of our field of view at any particular instant is typically only about 1:100. It is true that in the course of the day the *absolute* level on which this range of brightness is centred may fluctuate very widely indeed, as the sun rises and sets, and night follows day (**Fig. 3.46**). As a result, the rapid fluctuations that convey visual information are dwarfed by large but uninformative changes due to ambient lighting; a receptor capable of responding to the entire range of steady intensities would necessarily be insensitive to the smaller but more important variation. But these shifts of absolute level, as well as being of little interest to us, are comparatively slow. Adaptation, acting as a high-pass filter, will tend to get rid of them, leaving behind the significant and relatively rapid changes of intensity that are generated by our eye movements as we look around, and which lie in a dynamic range that is not so very different from that of the nerve fibres themselves. In other words, adaptation provides a kind of automatic sliding scale by which the limited signal-to-noise ratio of our neurons can be shifted to match the range of inputs we are interested in, ignoring slower and larger changes of the baseline that could only be accommodated by sacrificing overall sensitivity. The receptor now need only cope with some two log units of intensity rather than 10, and can thus respond better to the important details. This argument assumes, of course, that the slow changes really are unwanted: one man's noise is another's signal. In fact: in the eye, there are a very few cells that do not adapt; they are relatively insensitive but do provide information about static levels, acting in parallel with adapting channels, and are used to signal time of day and to control the pupil. But as a general rule, things that don't change don't demand a response and need not be coded.

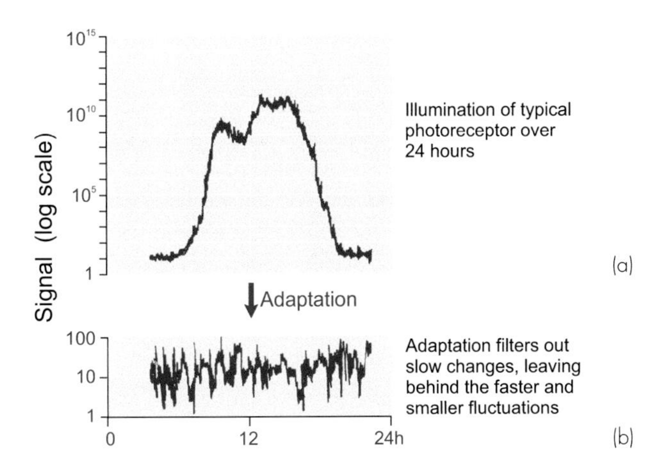

Figure 3.46 Adaptation reducing low-frequency noise. (a) A plot of the intensity of light falling on a typical retinal photoreceptor throughout one day, showing that the rapid fluctuations that convey visual information are dwarfed by large but uninformative changes due to ambient lighting that are much slower. A receptor capable of responding to the entire range of steady intensities would necessarily be insensitive to the smaller but more important variation. (b) Adaptation has the effect of filtering out the slow changes of intensity level: the receptor now need only cope with some 2 log units of intensity rather than 10, and can thus respond better to the important details.

Adaptation thus acts like the automatic gain control sometimes fitted to recording equipment, which automatically adjusts the amplification to compensate for the average level of sound that is being recorded, but lets through the more rapid fluctuations that constitute the sound itself. In the case of the Pacinian corpuscle, for example, it means that we can be made aware of very small changes even when superimposed on large steady background pressures – for example in the soles of the feet when standing – because adaptation has again shifted the scale of the receptor to allow for the steady background.

Finally, for proprioceptors such as joint receptors and muscle spindles that essentially give information about position of the limbs, the fact that adaptation implies response to *rate of change* rather than to steady levels means that such receptors, if completely adapting, will essentially signal *velocity* of the limbs rather than position, which may be more relevant in the control of certain kinds of movement such as throwing. Rate of change is a particularly important thing for the brain to measure, because it enables it to predict the future; apart from anything else, this enables feedback systems to compensate for the inevitable time delays caused by the slowness of nervous conduction.

Notes

Neurons in general Excellent general coverage of this area is provided by Bradford, H. F. (1986). *Chemical Neurobiology.* (New York, NY: W. H. Freeman); Hall, Z. W. (1992). *An Introduction to Molecular Neurobiology.* (Sunderland, MA: Sinauer); and Shepherd, G. (1994). *Neurobiology.* (Oxford: Oxford University Press); a particularly clear account of the basics is given by Levitan, I. B. & Kaczmarek, L. K. (1997). *The Neuron.* (Oxford: Oxford University Press). Other highly recommended books include: Hille, B. (1992). *Ionic Channels of Excitable Membranes.* (Sunderland, MA: Sinauer); Revest, P. & Longstaff, A. (1998). *Molecular Neuroscience.* (Oxford: Bios); Hammond, C. (2001). *Molecular and Cellular Neurobiology.* (San Diego, CA: Academic Press); Brady, S. T., Siegel, G. J., Albers, R. W. & Price, D. L. (2011). *Basic Neurochemistry.* (San Diego, CA: Academic Press); Shepherd, G. M. (2004). *The Synaptic Organization of the Brain.* (Oxford: Oxford University Press); Davies, R. W. & Morris, B. J. (eds) (2006). *The Molecular Biology of the Neuron.* (Oxford: Oxford University Press); and the elegant and thoughtful Fain, G. L. (1999). *Molecular and Cellular Biology of Neurons.* (Cambridge MA: Harvard University Press). An engaging historical account of part of this area is Rapport, R. (2005). *The Discovery of the Synapse: the Quest to Find how Cells Communicate.* (London: Norton).

1. Similarities between channels The evolutionary relationships between these various types of channel have been discussed in Hille, B. (1989). Ionic channels: evolutionary origins and modern roles. *Quarterly Journal of Experimental Physiology* **74** 785–804.
2. Gap junctions . . . and thus remarkably like the famous *Ponte dei Sospiri*, or Bridge of Sighs, in Venice, though which prisoners could be taken from the ducal prisons to trial in the Doge's Palace without the risk that they might prefer to jump into the canal below. A similar bridge, serving an analogous purpose, can be found at St John's College, Cambridge.
3. Spinal synaptology The classic account of the pioneering experiments in this area is Eccles, J. C. (1964). *The Physiology of Synapses.* (Cambridge: Cambridge University Press).
4. Diversity of transmitters Nieuwenhuys, R. (1985). *Chemoarchitecture of the Brain.* (Berlin: Springer Verlag) provides a comprehensive survey of the location of transmitters in different parts of the brain, but is becoming out of date. A more recent book in the same area is Bloom, E., Björklund, A. & Hökfelt, T. (1997). *The Primate Nervous System Part 1: Handbook of Chemical Neuroanatomy.* (Amsterdam: Elsevier).
5. Dale's hypothesis It is not at all clear that Dale's hypothesis is actually true, at least as far as 'acting in the same way' is concerned. The glutamate released by retinal receptors, for instance, causes inhibition of some bipolar cells and excitation of others. What is probably true is that a cell of a certain class, A, cannot have two different effects on cells of another class B (or any effect at all on cells of class A). Such a rule would prevent certain difficulties in the way the brain wires itself up.
6. Modulation This area is thoughtfully discussed in Katz, P. S. (ed.) (1999). *Beyond Neurotransmission: Neuromodulation and its Importance for Information Processing.* (Oxford: Oxford University Press). A more recent account is Krames, E. S., Peckham, P. H. & Rezai, A. R. (2009). *Neuromodulation: A Comprehensive Handbook.* (London: Academic Press).
7. Astrocytes See for example Scemes, E. & Spray, D. C. (2011). *Astrocytes: Wiring the Brain.* (London: CRC Press).
8. Neurons wiring themselves up A useful recent account is Kolodkin, A. L. & Tessier-Lavigne, M. (2010). *Neuronal Guidance: the Biology of Brain Wiring.* (New York, NY: Cold Spring Harbour); Price, D., Jarman, A. P., Mason, J. O. O. & Kind, P. C. (2011). *Building Brains: an Introduction to Neural Development.* (Chichester: Wiley-Blackwell) is an approachable introduction.
9. Loewenstein The paper is Loewenstein, W. R. & Mendelsohn, M. (1965). Components of adaptation in a Pacinian corpuscle. *Journal of Physiology* **177** 377–397.
10. Dashpots Dashpots are used to introduce damping into mechanical systems. Very often they consist of an oil-filled cylinder with a leaky piston: the rate of movement of the piston is proportional to the force applied. You can see them in car suspensions, or sometimes in old-fashioned door-closing mechanisms, where they stop the door slamming.
11. High-pass filters Less common are *low-pass* filters, but we shall come across an example in the photoreceptors of the eye, where in the dark there is a mechanism that accumulates information over a significant time, improving sensitivity at the expense of rapidity of response (see p. 152).
12. Information The information carried by a message can be regarded as a measure of the change in perceived probability that the message produces. If I get a telephone call saying I've won the national lottery, my perception of the probability that I have won it suddenly changes from about 3×10^{-8} to exactly 1. A second, identical, telephone call carries no information at all, since the probability remains 1. So identical signals do not always carry identical amounts of information.
13. Signal-to-noise ratio An excellent account of this area for the general reader, recently reissued, is Pierce, J. R. (1980). *An Introduction to Information Theory, Symbols, Signals and Noise.* (New York, NY: Dover Publications); a more technically demanding book is Ash, R. B. (1991). *Information Theory.* (New York, NY: Dover Publications). Another useful account is Barlow, H. B. (1982). General principles: the senses considered as physical instruments. In Barlow, H. B. & Mollon, J. D. (eds) *The Senses.* (Cambridge: Cambridge University Press).

PART 2

SENSORY FUNCTIONS

CHAPTER 4
SKIN SENSE

This chapter is concerned with the information that comes from *cutaneous receptors*, sensory receptors in the skin, and from the very similar ones that can be found in the gut and other visceral organs. The whole system is often loosely termed the *somatosensory system*, but this strictly also includes receptors from muscles and joints, which as proprioceptors are considered in Chapter 5.

Preamble: sensory processing in general

The conventional thing at this point would be to launch immediately into a description of the cutaneous sensory receptors, their afferent fibres, the spinal cord, the central pathways, and so on and so on. But in physiology in general, and above all in neurophysiology, if there are difficulties to be faced they lie not so much in assimilating these undeniably essential but nevertheless not-always-very-enlightening details, but in understanding the system as a whole, and above all what it is *for*. We make no apology then, since this is the first sensory system you have encountered, for beginning at a much deeper level, inviting you to think a little about the purpose not just of the cutaneous system, but sensory systems in general. What *use* is sensory information? How does its use determine how it is processed?

It is not easy to think clearly about our own sensory systems, as we all have this overpowering feeling of sitting inside our head as in a cinema, with information from these sensory systems being projected in front of us and providing us with 'conscious sensation'. Being an egocentric species, we naturally imagine that the *purpose* of sensory systems must be to deliver as accurate a picture of the outside world as possible to this little man in the head. Leaving aside the fact that consciousness is itself a difficult subject to think about scientifically – it is left to the very end of the very last chapter – intellectually, this is scarcely a very satisfying concept. For a start, it raises the natural question whether the little man in the head has a little man in *his* head. Nor is it a very *biological* way of regarding sensory processing. Systems do not evolve unless they are useful to their owners, ultimately in enhancing their ability to reproduce. The question we should be asking therefore is what we *do* with our sensory information, and what sensory systems need to do to make this information as biologically *useful* as possible. An analogy may make this clearer.

In **Figure 4.1** we see two representations of the same area of Dorset countryside. Which is better? In one sense, the aerial photograph: it tells us exactly what we would see if we were actually there in the sky. But for an afternoon's walking, we take the map, for it tells all the things we need to know if we want to interact with the landscape. More specifically, it provides information about *what, where* and *how*. To plan our walk we need to know *what* there is out there, the goals that motivate the walk itself – perhaps the village pub, perhaps a tourist attraction of some kind. To get to the goal, we need to know *where* it is in relation to our own location so that we can orientate ourselves and set off in the right direction; and finally we need hints about *how* to achieve our goal with the least effort and discomfort – avoiding the steepest inclines, crossing rivers without getting wet. What the helpful people at Ordnance Survey (OS) have done in making the map is to separate off these different kinds of information with different colour codes, so that we can convert them directly into actual directed movements: the map is the first stage in a process of *translation*, from sensory stimuli to motor responses.

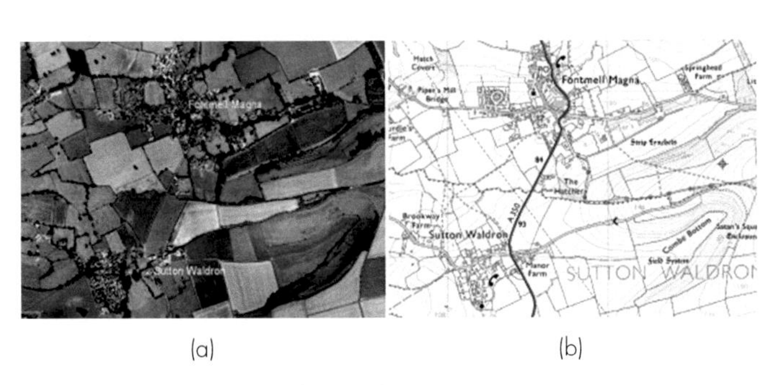

Figure 4.1 Two representations of an area of Dorset; a satellite picture (a) and an OS map (b).

Again and again we shall find when looking at sensory systems that the information from sensory receptors is almost immediately segregated into different pathways reflecting different kinds of use that is going to be made of the information, in terms of generating movement. In vision, for instance, we shall see that one stream of information is devoted to localisation of objects – and ourselves – in relation to the visual world so that we can go towards them, or reach out and grasp them. Another stream is devoted to recognising visual objects so that we can make decisions about our reactions to them. In the somatosensory system, we shall see an even greater degree of specialisation, which starts in the receptors themselves. One whole class of receptors in the skin is concerned exclusively with recognising stimuli that demand immediate responses of one kind or another – withdrawing from pain, brushing off an insect – while others provide information that improves the precise control of manipulation. A particularly clear example is temperature sense. One might think the logical thing would be to have a single receptor whose firing frequency was simply proportional to temperature over some useful range. But in fact what we have are *two* kinds of receptor, conveying two very different types of information: one signals warmth, the other cold. They have evolved this way because each corresponds to a very specific kind of homeostatic response: it is the need to make particular responses that has driven the evolution of different receptors and sensory systems, rather than the other way round.

If the only purpose of sensory systems was to relay as exact an image as possible of the outside world to the little man in the head, there would be no need for 'sensory processing' at all. As soon as you realise that sensory systems are simply mapping from stimuli to responses, much that is otherwise incomprehensible in the brain suddenly begins to make sense. So we urge you, as you read the following chapters, to ask yourself all the time: *What is this for?* Only function makes sense of structure.

Structures and pathways

Types of cutaneous receptor

The afferent fibres from cutaneous receptors are pseudo-unipolar: their bodies lie in the *dorsal root ganglia* near the spinal cord, while their axons run all the way from the sensory endings in the skin to their terminals within the central nervous system (**Fig. 4.2**). The dorsal roots are connected in an orderly way to different areas of the skin, and one may draw maps of the body surface showing the *dermatomes* or regions projecting to each dorsal root (**Fig. 4.3**).[1] The demarcation of the different zones is not actually as sharp as such idealised representations imply, and because of overlap between adjacent dermatomes, each point on the body surface is connected to at least two dorsal roots. Overlap is more marked for touch than it is for pain or temperature, and thus pain sensation can more accurately define a segmental nerve injury than assessments of touch.

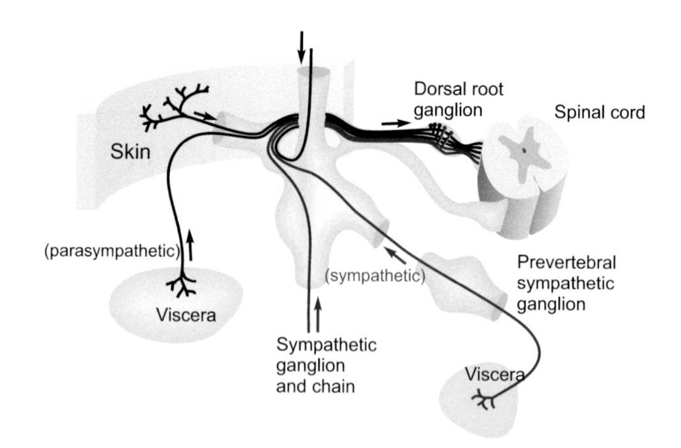

Figure 4.2 Afferent pathways from the skin and viscera (schematic). Cell bodies of the fibres reside in the dorsal root ganglia, lying alongside the spinal cord, while those from the head reside in the trigeminal ganglia.

Sensory fibres from the *viscera* are found in both the sympathetic and parasympathetic divisions of the autonomic nervous system.[2] The former pass in peripheral sympathetic nerves to the sympathetic chain, and thence via the dorsal root ganglia (where their cell bodies are) to the dorsal root; parasympathetic afferents of the sacral region travel with the corresponding efferents and again have their cell bodies in the dorsal root ganglion, while the cell bodies of the cranial afferents (in the vagus) are in the inferior (nodose) ganglion. 'Vagus' means 'wandering', and despite its localised connection to the central nervous system, it has an extremely widespread distribution in the thorax and abdomen.

At the peripheral end, the fibres branch and terminate either as naked endings or in terminal *encapsulations*, of which many varieties have been described.[3] One such is the *Pacinian corpuscle*, whose responses to deformation were discussed in Chapter 3. Others include the *Meissner*

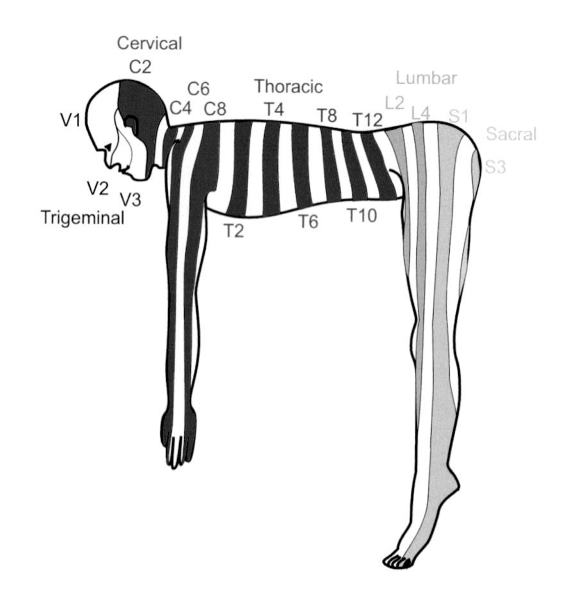

Figure 4.3 Pattern of dermatomes in humans. The boundaries are not actually as sharply defined as this schematic representation implies, resulting in discrepencies between the maps produced by different authorities.

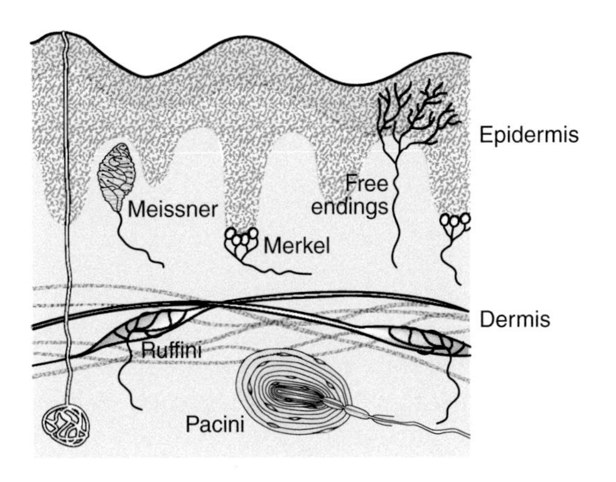

Figure 4.4 Representative types of endings in glabrous skin (schematic).

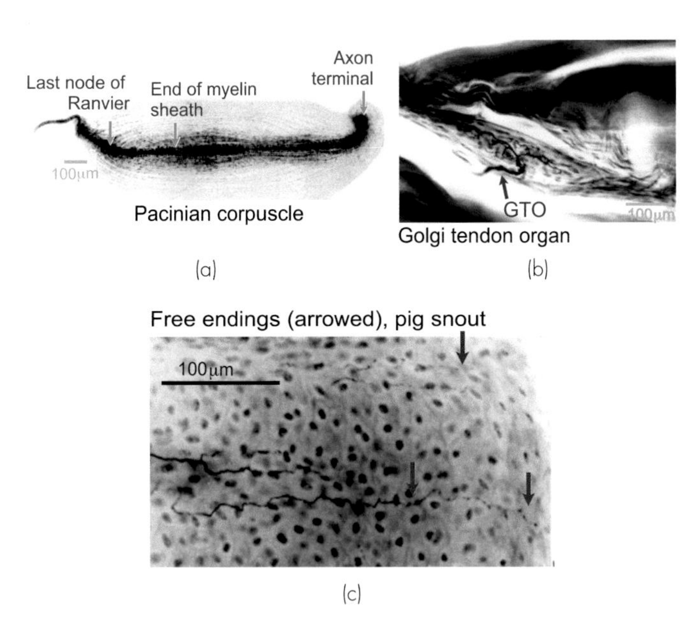

Figure 4.5 Some somatosensory receptors. (a) Pacinian corpuscle (human plantar). (b) Golgi tendon organ, similar in structure to a Ruffini organ, silver stain. (c) Free nerve endings in dermis, penetrating the stratum corneum (pig snout).

corpuscle, the *Merkel* disc, and the *Ruffini* ending (the latter is very similar to the Golgi tendon organ found in muscle) (**Figs. 4.4** and **4.5**). Encapsulated endings are found mainly in hairless or *glabrous* skin: the palms of the hands and soles of the feet, the lips, eyelids, mucosal surfaces and parts of the external genitalia. Some, notably the Pacinian corpuscles, are also distributed in visceral structures, joints, ligaments and deep connective tissue. Free or naked endings are abundant in both glabrous and hairy skin, as well as in deep fascia and visceral organs. Others innervate the hair follicles themselves and sense hair movement. Their afferent fibres are small and sometimes unmyelinated, falling into group C and group Aδ (or III and IV, with a few in II): the fibres from encapsulated endings are mainly of group Aβ (or II). (See **Table 4.1**; cf. also **Box 2.1**.) In addition, there are specialised sensory structures associated with *sinus hairs* (the eyelashes, for example, or the whiskers of a cat, **Fig. 4.6**). The hair is surrounded at its base by pressure-sensitive Pacinian corpuscles; in the middle it is encircled by a ring of Merkel discs, and along part of its length the hair is linked by thin filaments to a palisade of fast-adapting lanceolate endings. The result of this battery of mechanoreceptors is to provide the brain with exquisitely sensitive and highly directional information about displacements of the hair brought about by contact with external objects.

It is only recently that we have begun to form a clear picture of what this great variety of receptor types in the skin is actually *for*. Broadly speaking, there is a clear division between the large afferents (Aβ) coming entirely from encapsulated mechanoreceptors, and the small afferents (Aδ and C) from free endings and other fibres that respond to a wide range of stimuli, thermal and chemical as well as mechanical. Obviously these receptors give rise to what we *feel* when the skin is stimulated. But the question of the correspondence between the categories of feelings from the skin (sensory modalities, classically warmth, cold, pain, various *mechanical* categories, 'deep pressure', 'light touch' and so on) and categories of *nerve fibre* is a complex and controversial one, and is discussed more fully later in this chapter (p. 91).

Table 4.1 Types of skin receptor

Name	Type of adaptation	Fibre size	Stimulus
Meissner	Complete	Aβ	Shear
Merkel	Incomplete	Aβ	Contact
Pacinian	Complete	Aβ	Deep pressure
Ruffini	Incomplete	Aβ	Tension, folding
Krause	Complete	Aβ	Sexual??
Free endings	None	C	Nociceptive, mechanical, chemical
(Warm)	Incomplete	C	Warmth
(Cold)	Incomplete	Aδ, C	Cold, hot
(Nociceptive)	Probably none	Aδ, C	Nociceptive, mechanical

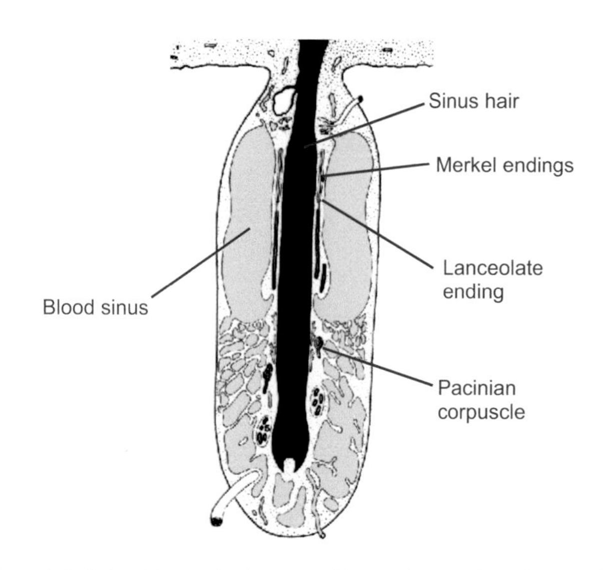

Figure 4.6 Stylised longitudinal section of base of a sinus hair showing the mechanoreceptors with which it is associated. (After Halata, 1975; with permission from Springer Nature.)

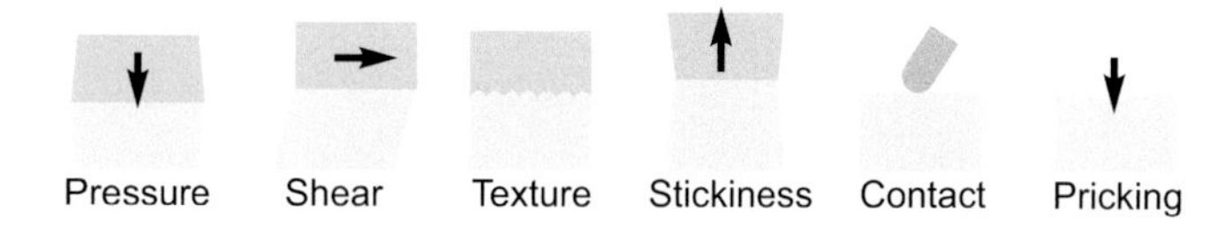

Figure 4.7 Types of mechanical stimuli.

One might perhaps have expected each type of ending to correspond to one of the classical modalities, but this turns out not to be the case. The naked or free endings seem to serve the modalities both of warmth and pain, and are probably sensitive to mechanical stimuli as well. Specific encapsulated endings for cold have recently been described, but do not yet have a name: unusually for encapsulated endings, their axons fall into the Aδ category. All the other endings that are known are mechanoreceptors of one kind or another, and it is clear that the classical descriptions of 'light touch' or 'deep pressure' are completely inadequate for classifying the various kinds of mechanical stimuli to which they respond. A stimulus may generate uniform pressure over an area of skin, or it may cause shear if it exerts a sideways force, or tension if it is sticky, or stretch as a result of folding (**Fig. 4.7**); there may be localised indentation, or more complex spatial textures; and all these stimuli may vary in time.

In many cases the structure of the endings makes it fairly clear what they do. The concentric layers of the Pacinian corpuscle imply a rather non-directional sensitivity to local *deformation*. As we saw in Chapter 3, it is often used as an example of complete adaptation; but it is important to realise that this is only true of applied displacements or deformations, and not of *pressure*. We saw that a suddenly applied, maintained increase in pressure will not show adaptation, because collapse of the lamellae will not alter the force that is felt by the central receptive structure. Ruffini organs consist of branched naked nerve endings twisted in between collagen fibres anchored to nearby muscle cells and other structures. *Tension* in these fibres appears to distort the nerve endings and thus stimulates them; they adapt incompletely. (Very similar endings, responding to tension, are found in the tendons of muscles: as proprioceptors, these *Golgi tendon organs* are considered in Chapter 5.) Merkel discs lie much closer to the outside world, at the bottom of the epidermis, to which they are attached by desmosomes: they are extremely sensitive to deformation of the skin, and show incomplete adaptation. In classical terms, they might be regarded as light touch receptors; more functionally, they may serve to indicate local skin deformation and *contact*. Finally, Meissner corpuscles, like Ruffini endings, have nerve endings that are associated with collagen fibres, and are completely adapting. The corpuscles are found in the dermal folds beneath the epidermal ridges, and the collagen fibres are connected sideways with the epidermal cells on each side. Thus they are ideally placed to register sideways *shearing* of the skin, of the kind that is experienced, for example, when holding an object in the fingers and then lifting it (they are in fact most commonly found in the fingertips). A curious feature is that their density declines dramatically with age, from some 50/mm^2 at 10 years to about 10/mm^2 at 50.

The fact that many of these receptors show adaptation means that it is only when the pattern of stimulation to the skin is *changing* that we perceive very much. Shut your eyes, and run your hand over some nearby object, perhaps the table-top: your hand gives you a vivid impression of its texture, of its cracks and dents and all its other surface properties. Now keep your hand still: at once this perception vanishes, and it is hard to be sure that one is sensing anything at all except its temperature. It is evidently the *temporal patterns* of firing of mechanoreceptors from the skin, combined with knowledge of the movements we make, that determines what we feel when we touch something.

Afferent pathways

The broad division of afferent fibres into two groups (*small*, mostly from free endings; *large*, from encapsulated endings) is reflected in their mode of termination within the central nervous system. In the spinal cord, they both terminate in the dorsal horn, but in different parts of it. The dorsal horn is conventionally divided into a set of six roughly parallel layers, *Rexed's laminae* (**Fig. 4.8**). Smaller fibres enter directly from the dorsal root and terminate in layers I and II; the largest fibres sweep round dorsally on their way up to their main destinations higher up, but some branches terminate in layers III–V where, among other cells, they make contact with short interneurons conveying mechanical information back to layer II. This is a fact of some significance in the processing of pain signals. The neurons of the dorsal horn are also under firm control from the brain; if the descending pathways that enter layer II are experimentally blocked the receptive fields of

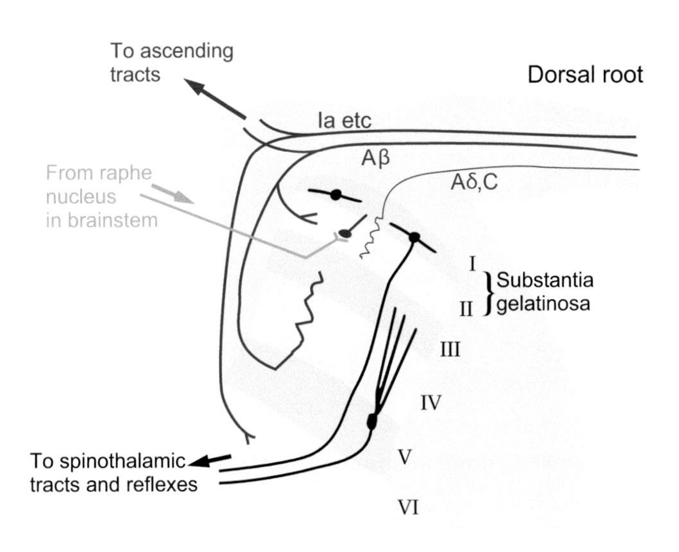

Figure 4.8 Schematic representation of laminae in the dorsal horn of the spinal cord, showing the approximate sites of termination of different kinds of sensory afferent, and the cells of origin of the spinothalamic tracts. Also shown are lateral inhibitory neurons found in layers I and II, and interneurons relaying information from large (mechanical) afferents from layer IV to layer II. There are also descending fibres that enter layer II and inhibit the transmission of pain signals.

dorsal horn cells undergo radical alteration. We shall see that this descending control is also of huge importance in the case of pain.

The mode of projection to higher levels of these two divisions is also distinctive. Branches of the larger fibres, from encapsulated mechanoreceptors, essentially turn their back on their spinal cord, turning upwards soon after entering the dorsal horn, to form a pair of large ascending tracts called the posterior or *dorsal columns* (**Figs. 4.8** and **4.9**). These continue ipsilaterally up to the level of the medulla and terminate in the *dorsal column nuclei* (*gracile* and *cuneate*); the gracile receives afferents from sacral, lumbar, and lower thoracic segments, and the cuneate from higher regions.[4] The second-order neurons decussate and then ascend through the brainstem in the medial lemniscus before, after synapsing again in the thalamus, reaching the primary somatosensory cortex. The smaller fibres, in contrast, project only to the cord itself, and ascending fibres are derived from second-order neurons. These decussate just anterior to the central canal of the spinal cord before ascending in the anterolateral tracts. As will become apparent, this difference between large and small fibres reflects a profound difference in their function. The large fibres are not much concerned with the initiation of responses: their main job is to provide feedback that is used by the brain, particularly by the primary motor cortex, to improve the way in which it manipulates objects. This phylogenctically new system, sometimes described as *epicritic* (implying the provision of precise, objective information), is the opposite of the ancient system of small fibres, sometimes called *protopathic* ('primitive feeling'), whose function is to cause appropriate responses to quite specific types of stimuli (**Table 4.2**), and incidentally to give rise to corresponding feelings that can be extremely marked. Such responses as withdrawal from a noxious stimulus, brushing off an insect, paroxysms induced by tickling, thermoregulatory responses to warm and cold, and sexual reactions to caresses, all fall in this category, and are characterised by instinctive and often powerful behavioural responses that are strongly resistant to conscious control. Of these, pain is of such fundamental clinical importance that it is discussed separately later in this chapter.

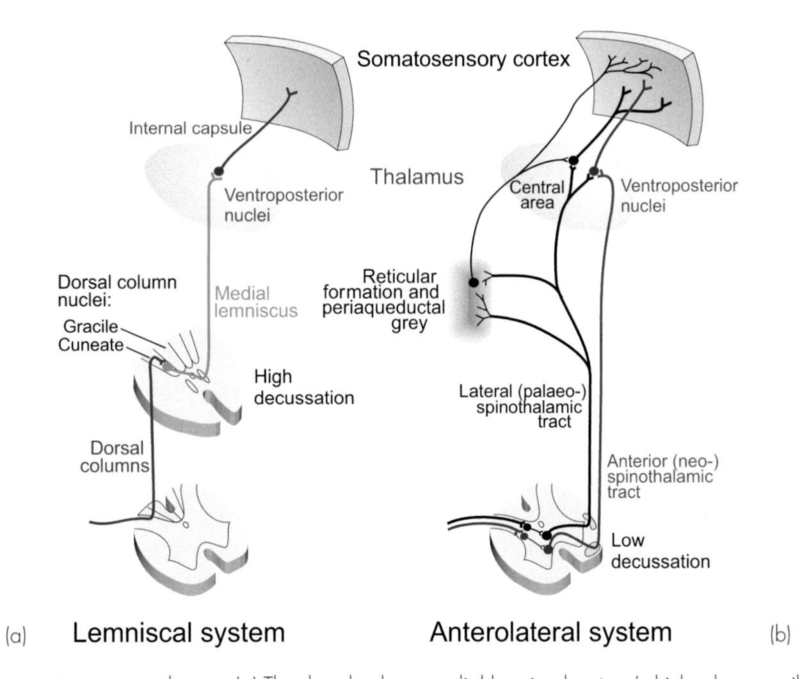

Figure 4.9 The main ascending somatosensory pathways. (a) The dorsal column medial lemniscal system (which subserves vibration and proprioception). (b) The neo-spinothalamic and palaeo-spinothalamic divisions of the anterolateral system (which communicate information about pain and temperature).

Table 4.2 Protopathic versus epicritic

Protopathic	Epicritic
• Evolutionarily old • Synapse in cord • Contralateral course in spinal cord • Initiate actions (demands a response) • Small fibres • Pain, temperature, tickle, itch	• Evolutionarily more recent • Synapse above cord • Ipsilateral course in spinal cord • Feedback and modification of actions • Large fibres • Encapsulated mechanoreceptors

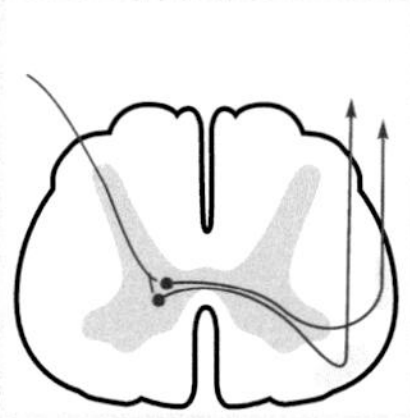

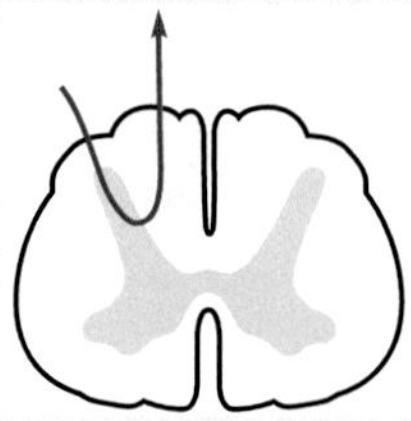

The final sensory destination of these ascending pathways is in the thalamus and then to cerebral cortex, and this is a convenient point to have a preliminary look at these areas, structures that we will be coming across again and again in subsequent chapters, since they dominate all the neural systems of the brain.

Clinical box 4.1 Peripheral neuropathy

Diseases that affect peripheral nerves, such as diabetes or vitamin deficiencies, tend to affect the longest nerves first. Typically, this results in a 'glove and stocking' distribution of sensory loss; numbness and paraesthesias usually start in both feet and gradually ascend to the level of the knees before starting to involve the hands. While this sounds inconvenient, leading to a tendency for spilling one's tea or struggling to put on gloves, the consequences can be far more catastrophic than this. At every moment, sensory signals from our skin and joints are informing us whether we have a stone in our shoe, or if we are twisting our ankle, and we make rapid adjustments to prevent tissue injury. Patients with long-standing sensory loss develop terrible non-healing ulcers from continual abrasion and unrecognised infection, or grossly deformed and unstable joints (termed *Charcot joints,* originally coined by Charcot as he observed patients with sensory loss complicating syphilis). By far the most common causes in modern times are diabetic neuropathy (as shown in **Figure 4.10**) and alcohol excess.

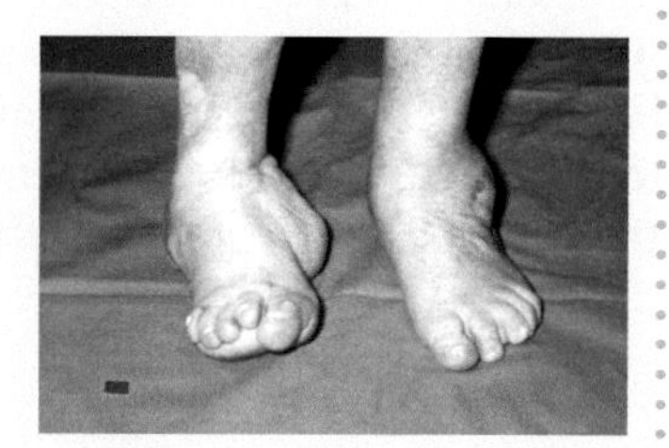

Figure 4.10 Diabetic feet.

Other peripheral neuropathies, meanwhile, may be much more acute and rapidly progressive; Guillain-Barré syndrome, first mentioned in **Clinical box 2.3**, is one that should not be missed by doctors. It (usually) starts with an ascending paralysis and can rapidly affect vital functions including swallowing, breathing (through loss of diaphragmatic innervation) and even the heart rhythm (due to involvement of the autonomic nerves). While this usually is a self-limiting illness (though recovery may be more prompt if given intravenous immunoglobulin), these patients must be monitored very closely and often require a prolonged stay on an intensive care unit.

Spinal cord lesions

The anatomical differences between the dorsal column-medial lemniscal and anterolateral/spinothalamic systems can be of diagnostic value in localising lesions within the spinal cord. For example, a hemisection of the cord that interrupts all ascending fibres on one side, will result in a loss of deep pressure and vibration sense below the level of the section on the same side, and loss of pain, temperature and light touch on the other (the *Brown-Séquard syndrome* (**Fig. 4.11**)). Alternatively, a *central cord syndrome* arises when the central, fluid-filled, spinal canal becomes distended as a result of congenital abnormalities of cerebrospinal fluid (CSF) flow, trauma or inflammation (termed *syringomyelia*). The expanding canal damages the nearby fibres, notably those crossing to form the antero-lateral system. Consequently there is a dermatomal band, corresponding to the site of the lesion, in which pain and temperature sense is bilaterally impaired; as this typically occurs in the cervicothoracic cord it often manifests in a 'cape-like' distribution. Another example of the anatomical segregation of sensory modalities is the posterior column syndrome in which the afferents ascending in the dorsal columns are injured, leading to a bilateral loss of proprioception and vibration sense below the area of damage; this can manifest with a characteristic 'stamping gait' due to the difficulty co-ordinating walking when lacking proprioceptive input.

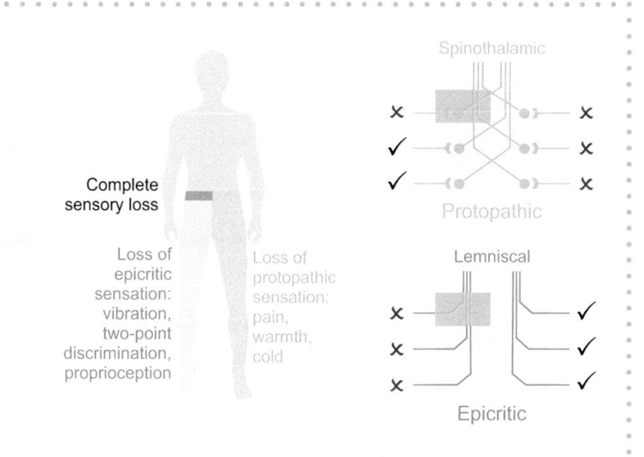

Figure 4.11 *Brown-Séquard syndrome.*

Cerebral cortex

Cortex is the rind of the brain. We perhaps tend to over-estimate it, partly because it is on the outside and therefore relatively easy to investigate, and partly because for us humans it is the bit we're most proud of, in which – because of its huge expansion – we differ most from other species. But we need to remind ourselves not to be totally infatuated with humans. When we look at sheep or gerbils or sparrows, we see that despite their cortical deficiency they lead happy and useful lives, and do all the biologically important things in life no worse than us, and in some ways better. At the risk of stating the obvious, cortex is for *refinement,* a bolt-on extra whose function is to add flexibility and programmability to what otherwise would be rather robotic. Cortex is the icing on the cerebral cake.

Apart from expansion, another evolutionary trend has been the superseding of a simpler three-layered sheet of neurons called *archicortex* by the more elaborate six-layered form, *neocortex* (**Fig. 4.12**). Neocortex first appears in reptiles, but in marsupials and mammals it starts to expand dramatically in area. In most mammals it forms by far the largest part of the cortex, elaborately wrinkled into the complex pattern of sulci and gyri that enables a large surface area to be stuffed into a relatively small volume. By the time we get to humans, about 90 per cent of the cortex is neocortex – its total area is about the same as an unopened broadsheet newspaper – with some 10^{10} neurons. Under each square millimetre of cortical surface lie nearly a quarter of a million neurons.

The structure of neocortex is essentially quite uniform, with only two basic kinds of neurons (**Fig. 4.13**). The main output comes from the large *pyramidal cells* of layer V, forming what are known as the projection efferents, which go to subcortical destinations. However, in a sense the whole point of the cortex is to bring many diverse types of input and output into functional association, and to a large extent this is brought about by smaller pyramidal cells having cell bodies in layer III, that send off axons (association fibres) terminating in vertical columns in some other area. Pyramidal cells have a long apical dendrite, and a group of basal dendrites at their base, where the axon exits downwards, so that the cells are capable of being influenced by all layers of the cortex, within a radius of half a millimetre or so. They are excitatory, releasing glutamate or possibly aspartate. The other main type of neuron, the *stellate cell,* is entirely confined to the cortex itself. It comes in two main kinds: *smooth* stellates are inhibitory and release γ-aminobutyric acid (GABA); *spiny* stellates are excitatory and are covered in a profusion of dendritic spines (**Fig. 4.14**). These neurons are organised horizontally into layers and vertically into columns some 0.5 mm in diameter. Corresponding to the projection efferents, on the input

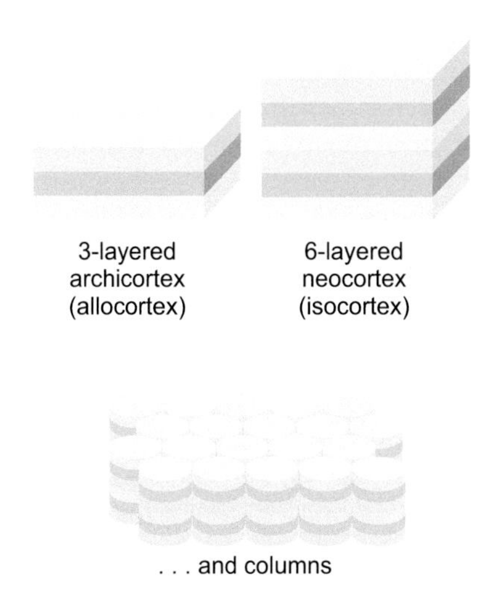

Figure 4.12 The cortex is arranged as discrete layers and columns.

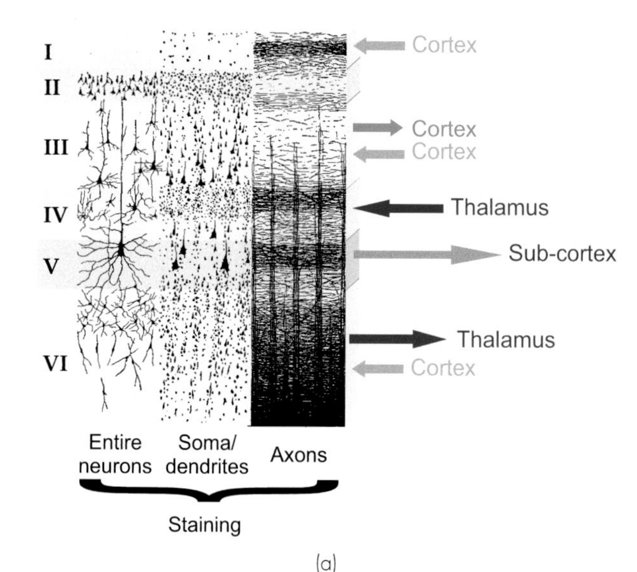

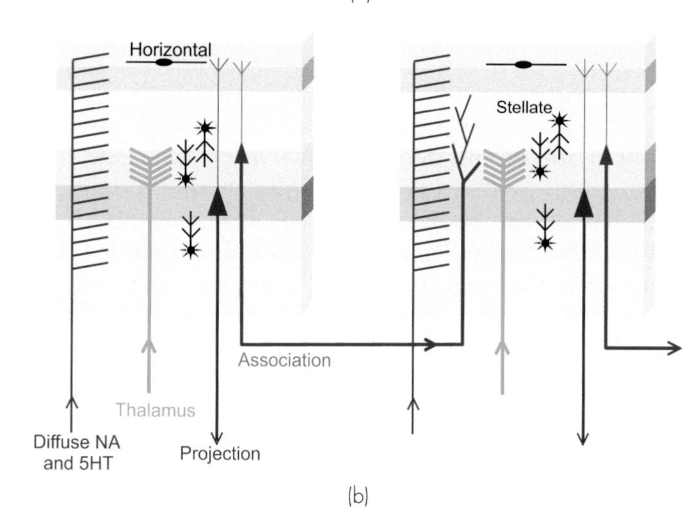

Figure 4.13 (a) Representation of the distribution of some types of neuron in the cerebral cortex, and their principal connections. (b) Equivalent schematic representation of the corresponding neural circuits.

side there are the specific projection afferents of subcortical origin (mostly in fact from the thalamus) that ramify into large terminal trees around layers III and IV and tend to end on stellates rather than on pyramidal cells. There is also a diffuse input mostly from reticular formation, influencing the activity of cortex in a rather global way: this is discussed in Chapter 14 (p. 288). In addition to these primary types of neuron, other varieties of interneuron communicate from layer to layer as well as horizontally; many of them are inhibitory in nature and probably carry out functions analogous to lateral inhibition (see p. 88).

Of the six neocortical layers, III, V and VI contain most of the pyramidal cell bodies, while the stellates are mostly in layers II and IV, and there are horizontal interneurons in layer I. Because different regions have different proportions of input and output, the relative size and appearance of these layers varies from place to place. Because the larger projection pyramidals are mostly in layer V, it tends to be more prominent in motor areas;

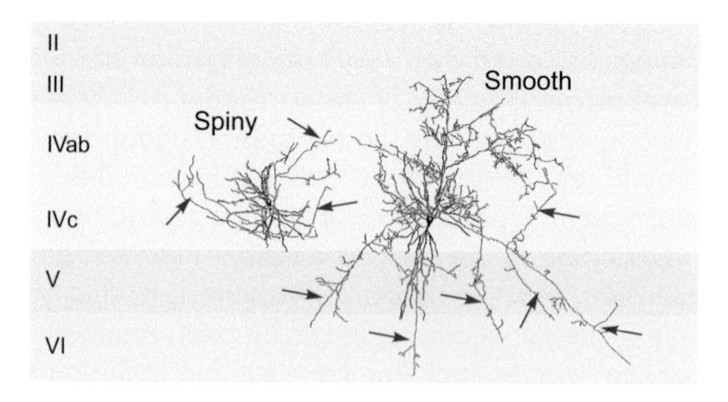

Figure 4.14 Spiny and smooth stellate cells, Golgi stained.

conversely, because layer IV is where the bushy afferents from the thalamus terminate, it tends to be more prominent in sensory cortex. A striking example is at the edge of a visual region of cortex, Area 17, whose massive inflow of fibres from the eye produces the dense *stripe of Gennari* (**Fig. 4.15**). There is a broad division into *granular* cortex (sensory, input layers bigger) and *agranular* (motor, output), with other intermediate types. Finer differences in sizes of layers and density of fibres and so on form the basis of the division of cerebral cortex into the Brodmann areas, whose numbers are a convenient shorthand for referring to different cortical regions (**Fig 4.16**). Increasingly, however, it is found that the classic Brodmann areas need to be further subdivided because of obvious *functional* differences between one sub-region and another.

Thalamus

The thalamus is strategically placed, right in the centre of the cerebral hemispheres like jam in a doughnut. Its neurons project entirely to the cerebral cortex, and much of its input comes from the various sensory systems, so it is natural to think of it as a *relay* for afferent fibres to the cortex. But if that were all it was it would be perfectly pointless, and in fact it is clear from its other connections that it is far more than that. In the first place, fibres don't just go from the thalamus to the cortex: they go the other way round as well (**Fig 4.17**); there is a massive, quite topographically precise, projection from layers V and VI of the cortex back down to the thalamic nuclei with which they are

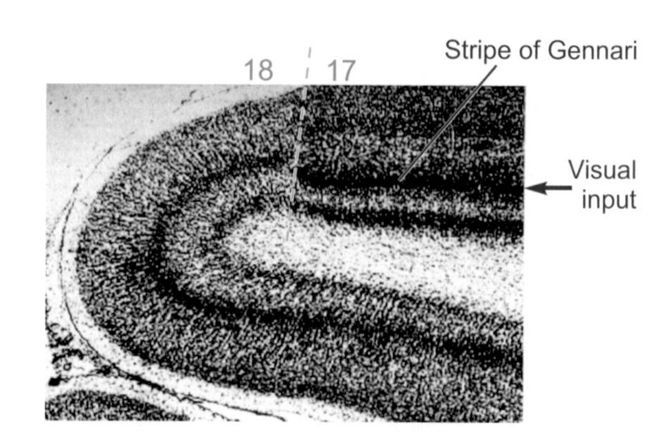

Figure 4.15 The transition from area 17 to area 18 of the macaque, showing the stripe of Gennari.

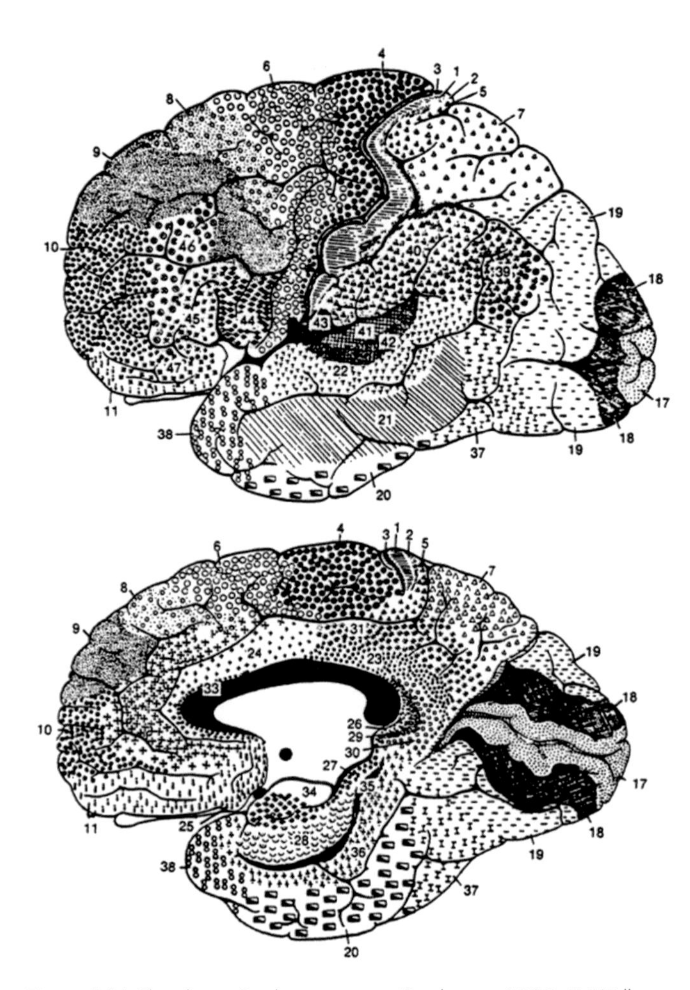

Figure 4.16 The classic Brodmann areas. (Brodmann, 1909. © Wellcome Collection. Source: https://wellcomecollection.org/works/vrnkkxtj)

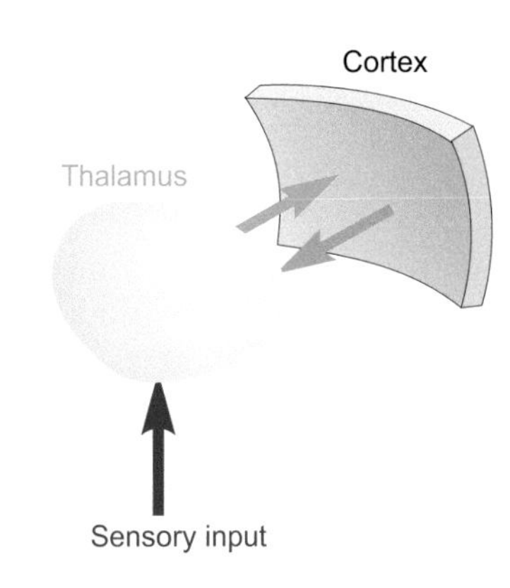

Figure 4.17 The thalamus and cortex project to each other.

subdivisions of these regions are illustrated diagrammatically in **Figure 4.19**, and listed in **Table 4.3**. For the moment, in the context of the somatosensory pathways, we should note two nuclei, the *ventroposterolateral* (VPL), which receives somatosensory information from the entire body apart from the head and face, which are dealt with by the *ventroposteromedial* (VPM) nucleus. Second-order afferents from the dorsal column nuclei first cross the midline, and then continue through the brainstem as the *medial lemniscus* to VPL, from which third-order fibres travel first through the *thalamic reticular nucleus* (TRN: blue) and then through the *internal capsule* to a region of cerebral cortex called the *somatosensory cortical area or* S1 (Brodmann areas 3, 2 and 1).[6]

Throughout the projection system linking encapsulated mechanoreceptors to the cortex via the thalamus – often associated. Second, if it were just a relay, then you would only expect sensory areas to have thalamic connections. But in fact nearly all cortical areas have projections from the thalamus, including areas that are very definitely motor. It is important to appreciate that cortex and thalamus work together very much as a single unit.[5]

In cross-section the thalamus is divided by the internal medullary lamina (shown in purple in **Figure 4.18**) into three major areas: anterior, medial and lateral. The

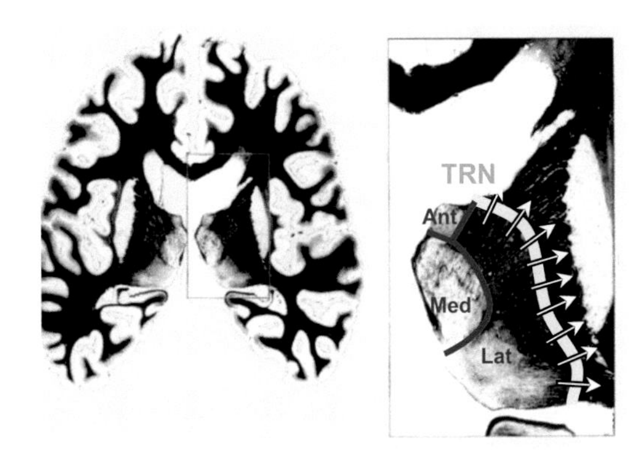

Figure 4.18 Horizontal section of the human brain, showing the position of the thalamus and its major divisions. The arrows show the thalamocortical projections traversing the thalamic reticular nucleus (TRN).

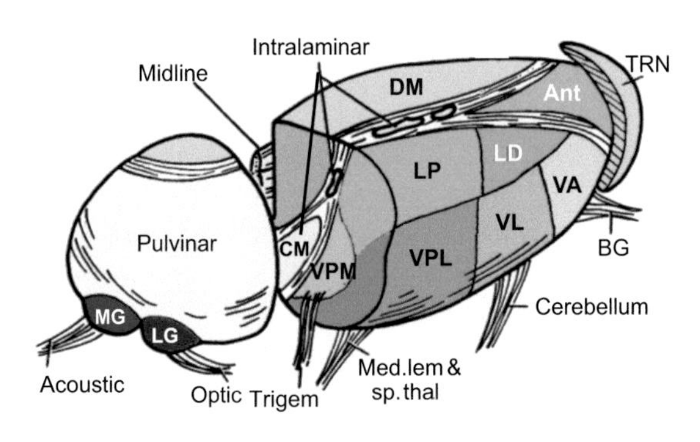

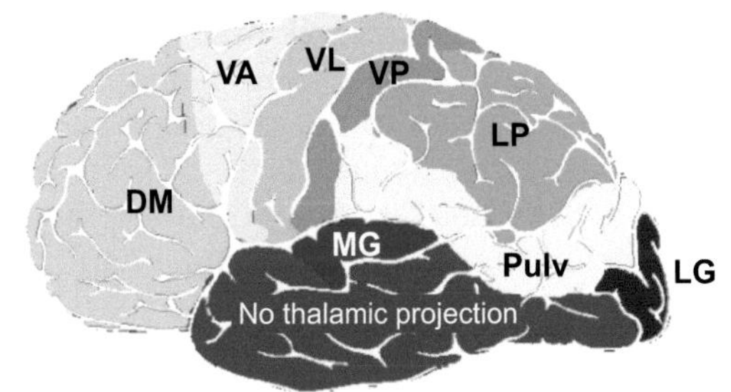

Figure 4.19 The subdivisions of the thalamus, and the corresponding areas of cerebral cortex. MG, LG: medial and lateral geniculate; VPM, VPL, VL, VA: ventroposteromedial, ventroposterolateral, ventrolateral and ventroanterior nuclei; LP, LD: lateral posterior and lateral dorsal; DM: dorsomedial; CM: centro median; Ant: anterior; TRN: thalamic reticular nucleus; BG: basal ganglia. (Partly after Brodal, 1998; permission pending.)

Table 4.3 The principal thalamic nuclei

	Input	Cortical projection
Anterior (Ant)	Mammillary bodies	Cingulate gyrus (limbic paleocortex)
Dorsomedial (DM)	Limbic, including amygdala, dorsal striatum, hypothalamus	Frontal associational cortex
Lateral dorsal (LD)	Uncertain	Cingulate gyrus
Lateral posterior (LP)	Associational visual and somatosensory and visual cortex	Parietal-temporal-occipital (PTO) associational cortex
Pulvinar	LGN, MGN, associational and visual cortex	PTO associational cortex
Ventroanterior (VA)	Basal ganglia	Motor cortex, especially the supplementary motor area (SMA)
Ventrolateral (VL)	Mostly from cerebellum	Motor cortex, especially primary and premotor area (PMA)
Ventroposterolateral (VPL)	Somatosensory, from body	Somatosensory cortex (areas 1, 2, 3)
Ventroposteromedial (VPM)	Somatosensory, from head	Somatosensory cortex
Lateral geniculate (LGN)	Visual	Primary visual cortex (area 17)
Medial geniculate (MGN)	Auditory	Primary auditory cortex (area 41)

called the *lemniscal system* – the general topological relationship between the representations of different areas of the skin is preserved, so that the somatosensory cortex itself embodies a map of the opposite side of the skin surface (**Fig. 4.20**). This map is topologically correct in the sense that neighbouring parts of the body surface are on the whole represented by neighbouring regions of cortex, but very much distorted in shape; those areas with the greatest cutaneous sensitivity and acuity such as the hands and lips have a much larger area devoted to them than regions such as the trunk and back. These distortions can be represented by a correspondingly distorted *sensory homunculus*. A second somatosensory area, S2, is found in primates and differs from S1 in receiving somatosensory information from both sides, and to some extent in the modalities to which it responds.

We noted earlier that the smaller afferents, derived from free and also some encapsulated endings, and concerned with temperature, pain and light touch, do not immediately ascend on entering the cord: instead, either directly or via an interneuron, they activate neurons in layers I and V whose axons cross to the other side and proceed upwards as part of the *spinothalamic* projection (cf. the diagram on p. 81). There are two of these spinothalamic pathways, the *anterior spinothalamic* tract and the *lateral spinothalamic* (the whole system is called the *anterolateral system*). Though they are both evolutionarily older than the more recent lemniscal system, the anterior tract is more highly developed in higher animals than the lateral, and for this reason they are also known as the neo- and palaeo-spinothalamic pathways; however, more recent work suggests that one should not exaggerate the difference between them. The former projects only to the border of VPL, and to nearby regions that are not wholly somatosensory, terminating in large bushy arborisations (in contrast to the more compact lemniscal endings). The palaeo-spinothalamic afferents project to central and intralaminar regions of the thalamus, and also rather diffusely to the reticular formation of the medulla and pons and indirectly to the anterior cingulate and insula: they

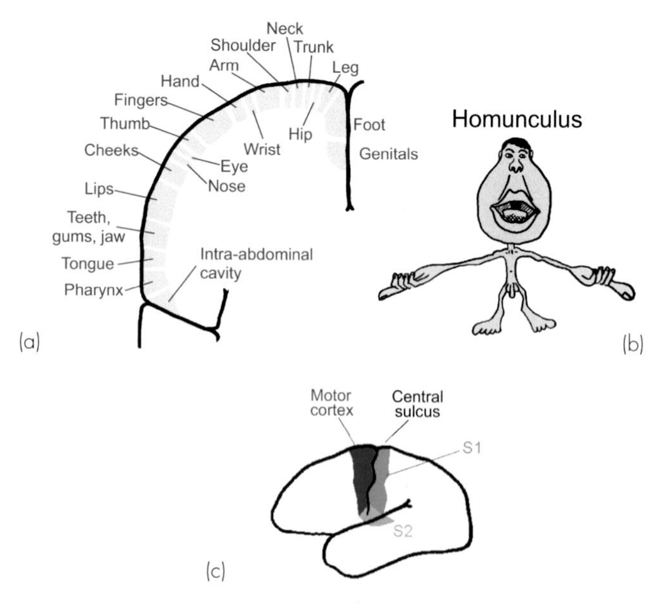

Figure 4.20 (a) Frontal section through the primary somatosensory cortex in a human, showing the approximate areas associated with different parts of the body. (b) Sensory homunculus, distorted so as to indicate the relative size of different parts of the body and the relative areas devoted to each in somatosensory cortex. (c) Lateral view of human cerebral cortex, showing approximate positions of somatosensory areas S1 and S2. (After Penfield & Rasmussen, 1950; permission pending.)

are concerned more with pain and temperature than with touch.

These differences reflect the epicritic/protopathic distinction mentioned earlier. On the one hand we have the new, fast lemniscal system with its precise and orderly projection of accurate mechanical information directly up to the cortex; and on the other, the older, slower, and more diffuse projection of less precise, but in a sense more immediately important information – often with an affective or emotional quality to it – by the anterolateral system, which scarcely projects to the cortex at all. One provides objective information which we can take or leave as we please; the other demands a response. One goes straight up the brain with hardly a nod to the spinal cord; the other terminates in the spinal cord, which may or may not bother to inform the brain about it. Finally, some cutaneous fibres ascend to the cerebellum in the *posterior spinocerebellar* tract (see p. 101).

Neural responses

Larger mechanoreceptive afferents

The larger, Aβ or group II, fibres from the skin all respond specifically to mechanical stimuli. As we have seen, some are completely adapting, and thus only respond to changes in the deformation of the skin: they originate from the Pacinian and Meissner corpuscles and from some of the endings in hair follicles. Because of their adapting properties, they are particularly sensitive to vibration, but that does not of course mean they are 'for' detecting vibration, an artificial stimulus found essentially only in the laboratory: nevertheless, it is true that they are well suited to contribute to the sense of roughness when the hand is passed over a textured surface. They are also extraordinarily sensitive: the threshold of a Pacinian corpuscle is of the order of 10 μm of skin displacement (about 1.5 times the width of a red blood cell), provided it is rapidly applied. They may well help one to sense when an object being lifted between the fingers begins to slip, and thus assist in regulating the pressure with which such an object is grasped. The importance of this kind of information can be seen by comparing the strength used to grip and lift an object covered with surfaces of different slipperiness: normally grip force is modified with short latency to take account of the nature of the surface, but if the skin of the fingers is anaesthetised these rapid responses to slip do not occur, and objects get dropped.[7]

Other fibres show only incomplete adaptation and can therefore signal static deformation as well: they come mostly from Merkel discs and Ruffini endings. The former are present in large numbers, and being close to the epidermis are ideally suited to providing information about light touch; the Ruffini endings, very similar to Golgi tendon organs, supply information about underlying tensions and stretch, and thus indirectly about joint position. Useful information has come from micro-electrode recording from afferents from the hand running in the medial nerve, in conscious human subjects, the particular advantage being that one can also perform micro-stimulation and see what the subject feels. One can demonstrate, for instance, that a single action potential in some of the fastest-adapting fibres is sufficient to evoke a sensation (see **Table 4.1**).[8]

Responses from smaller afferents

The cutaneous fibres of groups A and C that are associated with light touch, pain and temperature, show response patterns that are a little more complex than those of the larger fibres. In thermoregulation we have already seen that, instead of having a group of receptors signalling absolute temperature, there are two separate populations (*warm* and *cold*), as each demands a different behavioural response. These fibres, of Aδ size, fire tonically at a rate that is a function of temperature, with a peak for warm fibres around 45°C and for cold receptors around 30°C (**Fig. 4.21**). Both also show incomplete adaptation: sudden warming of the skin results in a transient increased discharge of warm fibres, whose activity then settles down to a new level, while sudden cooling has the same effect on cold fibres. However, it appears that the cold receptors also respond transiently to warming above some 45°C, giving rise to the familiar sensation of *paradoxical cold:* a hot object, when briefly touched, may often give the immediate impression of being intensely cold.

These adapting properties of the thermoreceptors dominate one's sense of skin temperature. Thus a swimming pool that seems appallingly cold when one first dives in soon seems quite comfortable, and a bowl of warm water may feel simultaneously cold to one hand and hot to the other, if previously the two hands have been held, respectively, in hot and cold water. In fact, it might be more accurate to think about thermoreceptors in the skin as signalling information about temperature loss or temperature gain. The receptive fields of thermoreceptors are very small and – far from overlapping – are actually separated from one another by large areas of skin that do not respond at all. Thus one may find warm and cold *spots* on the skin; on the hand the cold spots are about 5–10 mm apart and the warm spots some 15 mm; on the arm this difference is even more marked. (Temperature is not of

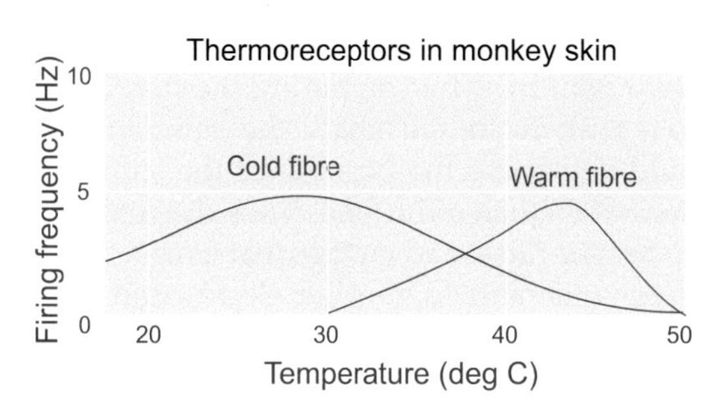

Figure 4.21 Tonic firing frequencies of cold and warm fibres from monkey skin in response to different temperatures. (Data from Kenshalo, 1976 © Pergamon 1976; with permission from Elsevier.)

course a cutaneous sense for which accurate localisation is particularly important, and as a consequence the spatial resolution of these small-fibre afferent systems is poor; this is reflected in the very small number of fibres that ascend in the spinothalamic tract – in humans perhaps as few as a thousand.) Thermoreceptive properties similar to those described above are also found among C fibres, more of them responding to cooling than to warming; other C fibres respond to light touch in the same general way as A fibres. The remaining Aδ and C fibres serve the sense of pain, produced by noxious stimuli, and other protopathic sensations: they are discussed later in this chapter.

Receptive fields

In order to be effective in stimulating a fibre, a stimulus must lie within a particular area of the skin called the *receptive field*.[9] The size of this field is partly a consequence of the unavoidable spread of the stimulus itself – any indentation of the skin, however localised, will cause deformation of the layers of the skin over a much wider area – but is also the result of the branching of the afferent fibre and consequent distribution of its endings over an extended region. In the case of the fibres innervating hair follicles, for example, one finds that each fibre may innervate as many as a hundred follicles, and each follicle in turn receives branches from several fibres. Thus the receptive fields of the individual fibres are quite large, and also show a considerable degree of *overlap* (**Fig. 4.22**). Such a pattern of overlapping receptive fields is a common one in all kinds of sensory systems. What is the point of it? Surely, one would think, it would be better to avoid this duplication by making the receptive fields smaller, resulting in an improvement of the precision with which a stimulus can be localised. Yet it turns out, on closer analysis, that a system in which the receptive fields overlap is not only just as good as one in which they are discrete, but in many ways actually much *better.*

Consider first the question of *accuracy of localisation* of a stimulus. If the fields were discrete, then all the brain could tell about the position of a stimulus is that it must lie somewhere inside a particular receptive field; there is no way in which it could find out *where* it lay within that field. But consider the case of two overlapping fields: if a stimulus lies within the area of overlap, then it will stimulate the two fibres in different proportions, depending on its exact position. So by analysing the pattern of discharge of the two neurons – by comparing the frequency of firing of one with that of the other – the central nervous system could determine the position of the stimulus much more accurately than if the fields were discrete (**Fig. 4.23**). Overlap has the further advantage that it makes the system much less vulnerable to damage: destruction of any one fibre will still leave each area of skin with the innervation of its neighbours.[10]

Now it is perfectly true that if one thinks of the receptors as converting the spatial pattern of the stimulus into a kind of 'neural image' – a corresponding pattern of firing among the array of afferent fibres – then it must follow that the effect of having large receptive fields will be to blur this neural image. Sharp stimulus boundaries will be converted into a rather fuzzy gradation between those fibres that are firing maximally and those that are not firing at all. However, there is a simple way in which the brain can mitigate the effects of this kind of neural blur on the precision of localisation, called *lateral inhibition.*

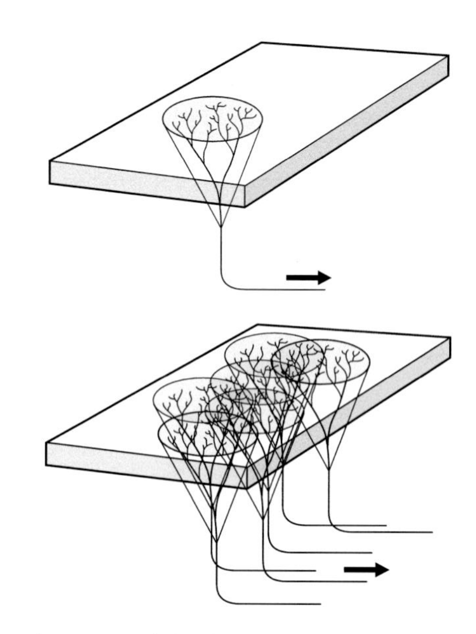

Figure 4.22 The receptive field of a single idealised cutaneous afferent fibre and the overlapping receptive fields between neighbouring afferents.

Lateral inhibition

Imagine that at the level at which the incoming fibres first relay their signals on to ascending, second-order, neurons – in this case, in the gracile and cuneate nuclei – they excite local interneurons as well; and suppose that the interneurons in turn send inhibitory connections to neighbouring second-order cells (**Fig. 4.24**). What will happen now is that each incoming fibre will stimulate its own second-order cell, but inhibit the ones that surround it: as a result the receptive field of a neuron such as S will have a central region of excitation, with a surrounding region that is inhibitory. Functionally, this introduces *competition:* the cells will, in effect, be pushing on each others' shoulders. Thus a cell that is stimulated more than average will have a larger effect on its neighbours than they will in turn have on it. The result will be to exaggerate changes in intensity, compensating to some extent for the blurring effect of the original overlap of the receptive fields. This mechanism of lateral inhibition is of fundamental importance in understanding the processing of neural information, and occurs in every kind of sensory system. It is also found universally within the central nervous system itself, since blurring can occur not only through having large receptive fields, but also whenever there is convergence and divergence in the projection from one neuronal level to another.

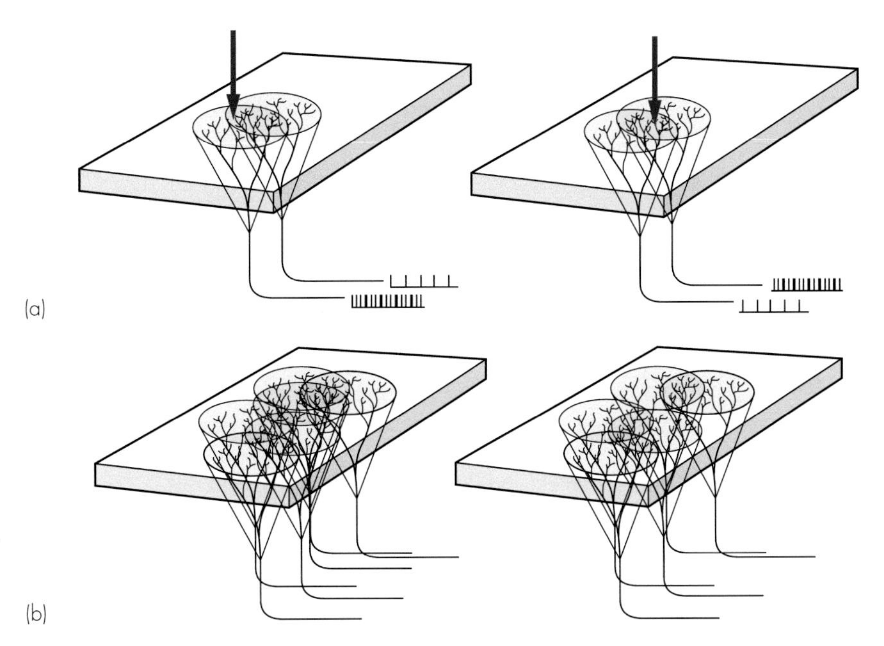

Figure 4.23 Advantages of overlap. (a) The brain can determine the exact position of a stimulus within an area of overlap by analysing the relative activities of the corresponding fibres. (b) Overlap means that damage to any one fibre does not necessarily produce an area of anaesthesia.

Lateral inhibition also tends to enhance *edges*. A cell lying just within the edge of an extended uniform stimulus will receive less lateral inhibition than one further in (**Fig. 4.25**), resulting in a pattern of neural activity that is maximal around the border of the stimulus. In many ways, lateral inhibition is analogous to adaptation, but in the spatial rather than in the temporal domain: it makes neurons sensitive to a change in activity across a *pattern* of neural activity, as opposed to a change in any one neuron as a function of time. Just as adaptation causes a burst of activity at the onset of a steady stimulus, so lateral inhibition emphasises spatial boundaries. The illusion in **Figure 4.26** demonstrates the same phenomenon in vision: because of lateral inhibition, the series of grey areas do not appear uniform, but lighter on the left where they are next to a darker region, and darker on the right.

One can sense this quite easily in one's own skin: stepping into a bath when the water is almost too hot to bear, the maximum discomfort is localised not so much in the foot, but rather at the line formed by the surface of the water around the leg, where the spatial rate of change of temperature is greatest. Or if you put your finger into a beaker of mercury (since mercury is poisonous you need to wear a light rubber glove), what you feel is a tight constriction round the meniscus, even though the pressure is of course greatest at the fingertip. And like adaptation, lateral inhibition helps reduce the *redundancy* of neural signals. An analogy may help to make clear why this is.

Imagine a central weather bureau whose function is to gather information from a network of weather stations to

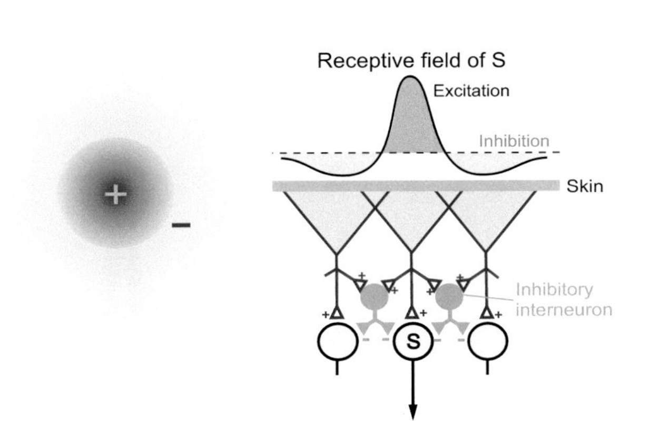

Figure 4.24 Lateral inhibition. If a second-order neuron, S, is excited by one receptor but inhibited by interneurons driven by its neighbours, the result will be to reduce the size of the excitatory receptive field, and to surround it with an area in which stimulation will give rise to inhibition.

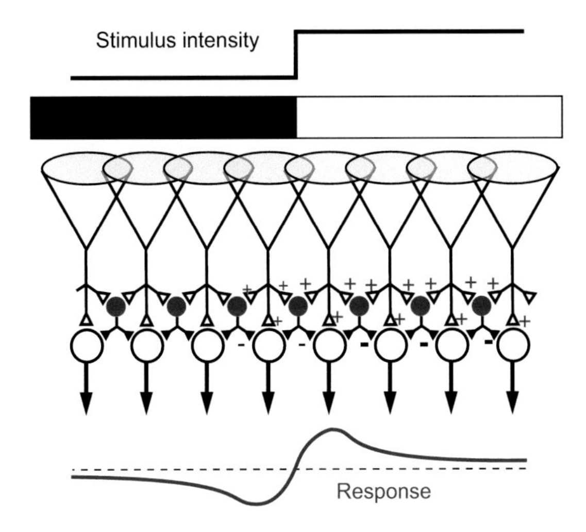

Figure 4.25 Lateral inhibition will exaggerate the response of second-order neurons to an edge, compared to the uniform areas on each side.

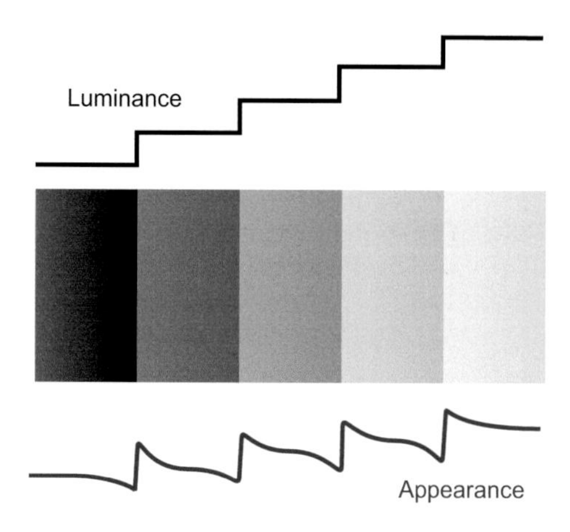

Figure 4.26 The effect of lateral inhibition can be perceived direcly as an illusion when viewing a series of steps in intensity.

compile up-to-the-minute charts of the changing patterns of weather in the region as a whole. One way of obtaining the necessary information would be to get the local weather stations to ring up every 5 minutes and describe local conditions. But this would not be a very economical arrangement; quite apart from the cost of the enormous number of telephone calls that would be required, the central office would have to employ a very large number of staff simply in order to receive them. Clearly the weather at any one place does not in practice vary much from one 5-minute period to the next, so that most of the phone calls in such a system would be redundant, consisting simply of the message 'same as before'. The first rationalising step would be to instruct the local stations to ring only when a *change* in the weather has occurred; to act, in fact, like completely adapting sense organs. The next improvement would be to recognise the existence of *spatial* redundancy in their reports; in general, the weather experienced by any one station is likely to be much the same as that experienced by its neighbours. So by telling the local stations to call only when they are aware that they are on the *edge* of a particular condition (a cold front, for instance), the number of calls, and the staff required to process them, could be reduced still further. We shall see later that this process is particularly prominent in the visual system (see p. 153). It helps to explain why outline drawings are as effective as they are in evoking the appearance of what they represent, even though topologically they are so different: since the visual system in effect converts everything it sees into an outline drawing anyway, it doesn't mind if what it normally throws away isn't there.

However, the one thing that lateral inhibition cannot do – contrary to a popular misconception – is to improve *acuity*. Acuity is a measure of how well a sensory system can transmit in its neural images the spatial detail that is presented to it. A common test of cutaneous acuity is the *two-point discrimination test:* the skin is stimulated at two points simultaneously – a pair of dividers does this very well – and the distance between the points is gradually increased until the subject has the sensation that there are two points and not one. Cutaneous acuity varies greatly from one part of the body to another, almost in proportion to the size of its representation in the somatosensory cortex, from a few millimetres on the fingers to nearly 50 on the calves. This variation essentially reflects the size of the cutaneous receptive fields, for if two points of stimulation are separated by less than the receptive field size, no amount of subsequent neural processing in the form of lateral inhibition or anything else will enable one to distinguish the neural image from that produced by a single point. Localisation on the other hand is in general more accurate than acuity, because of the extra information provided by the overlap of receptive fields. Consequently, one finds the apparently paradoxical situation that with the dividers set to a distance less than the local acuity, so that we cannot tell whether there is one point or two, one can nevertheless still tell, when only one of the points is applied, which one it was.

An analogy may help to make this surprising conclusion more intuitive (**Fig. 4.27**). Imagine a man holding a tray above his head, on which weights can be put. He has just two channels of information about the weights, namely the force felt by each of his arms. He will still be able to sense the position of a single weight on the tray quite accurately, from the relative forces felt in each arm. Yet he will be completely unable to tell the difference between a single central weight and two smaller weights arranged symmetrically on each side. Remove one of the smaller weights, however, and he will be able to tell you which has been taken away.

Finally, it is worth mentioning that the neural circuit for lateral inhibition presented on p. 89 is only one of several possible arrangements that have broadly similar effects. The one that is shown there is a *feedforward* system – the source of the inhibition is the incoming fibres, and the inhibition itself is of the postsynaptic type. In fact, it appears that the lateral inhibition actually observed in the dorsal column nuclei is primarily presynaptic. At higher levels of the somatosensory system, in the thalamus and cortex, lateral inhibition appears to be mainly postsynaptic. A further possibility is that it may be not the incoming fibres but collaterals of the *outgoing* ones that excite the inhibitory interneurons: this is called *feedback* lateral inhibition

Figure 4.27 Localisation is more accurate than acuity because of the extra information provided by the overlap of receptive fields. The man holding a tray above his head can sense the position of a single weight on the tray from the relative forces felt in each arm. He will be unable to tell the difference between a single central weight and two smaller weights arranged symmetrically on each side. Remove one of the smaller weights, however, and he will be able to tell you which has been taken away.

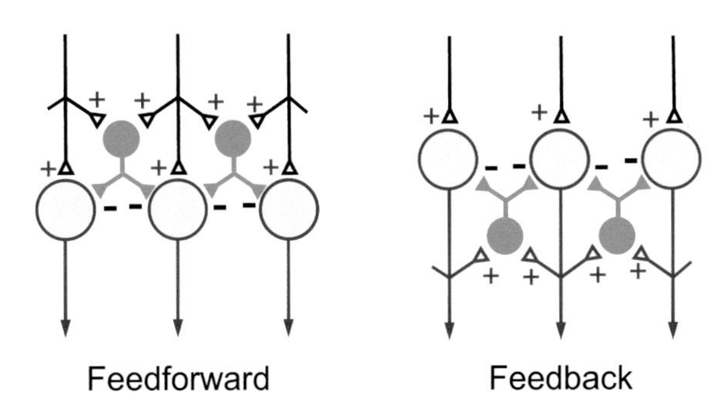

Figure 4.28 Two varieties of lateral inhibition

(Fig. 4.28), and is found for instance at the level of the thalamic relay of ascending somatosensory pathways; its functional properties are slightly different. Feedback inhibition may originate not from the outgoing fibres themselves but from the higher levels to which they project; the cortex can be shown to inhibit both thalamic and dorsal column relays in this way, as well as the spinothalamic pathways at the level of the cord itself.

We shall see later (Chapter 13, p. 256) that the concept of lateral inhibition can be extended considerably, particularly to cases in which the 'laterality' is not literally spatial, and inhibition acts on neurons which are adjacent in a more abstract sense (for example, responding to stimuli that are similar in terms of modality).[11]

Central responses

Thalamic responses to lemniscal and anterolateral afferents are not particularly interesting: they show the modality specificity that would be expected from the fibres that project to them, in contralateral receptive fields that may sometimes be larger than those of neurons at lower levels in the somatosensory system. There is some segregation of modality; for instance, neurons in the posterior ventromedial region respond quite specifically to painful or thermal stimuli. In somatosensory area S1 of the cortex (Brodmann areas 1, 2 and 3), responses are again not qualitatively very different from those in the thalamus. One striking feature of the distribution of responses over the cortical surface, apart from its large-scale organisation in the form of the sensory homunculus already described, is the fact that the cortical organisation appears to be in the form of a mosaic of *columns* a few hundred micrometers in diameter, such that responses from cells at any depth within a particular column are confined both to a particular modality and also to a localised region of the skin. In general, each column is surrounded by neighbours of different modality but similar location, and there are mutually inhibitory connections between columns that presumably accentuate differences in their activity, by a kind of lateral inhibition, that helps to refine sensory discrimination. Sideways inhibition within columns can also give rise to the directional responses often seen in S1, to cutaneous stimuli moving in specific directions. There also appear to be differences between areas 1, 2 and 3 in the predominance of the different classic modalities, at least as regards deep (2 and 3a) versus superficial (1 and 3b) receptors, and the more completely adapting responses of area 1. What is also striking is the extent to which these maps are labile and dynamic: simply bandaging a monkey's hand is sufficient to cause a reduction within a matter of hours in the hand's representation in the cortical map, and after amputation or severing of peripheral nerves the cortical representation is lost altogether, taken over by other parts that are more functional.

In the second sensory area, S2, we begin to find evidence of more complex kinds of analysis of afferent information. Units here are on the whole bilaterally activated, with a receptive field on one side of the body that is approximately the mirror image of that on the other; these relationships are partly but not entirely the result of connections crossing in the corpus callosum. Many of the cells show specific responses to stimuli that move across the skin in particular directions. As one examines more and more outlying regions of the somatosensory cortex, one begins to find cells that may respond to more than one stimulus modality, and also to painful stimuli, a property not usually reported in the main part of S1 and S2. Electrical stimulation of the somatosensory cortex in conscious human subjects tends to produce tingling, 'electrical' sensations rather than the illusion of actual tactile stimulation; pain is only rarely reported. Lesions here result in raised tactile thresholds, in reduced two-point discrimination, and a general impairment in the finer somatosensory judgements such as estimating weights. In humans, lesions affecting the further, posterior parietal, regions of the somatosensory cortex are sometimes associated with *astereognosis,* an inability to 'put together' somatosensory information in judging such things as the shape of an object held in the hand, even though primary somatic sensibility – as measured by tests such as the two-point threshold – may be relatively unimpaired. Posterior parietal cortex is further considered in Chapter 13.

Sensory modalities

Many types of stimulus can produce sensations from the skin, and one of the functions of the somatosensory system is to distinguish between them. Thus to a large extent, as we have seen, the receptors and pathways of cutaneous sensation are modality-specific, responding preferentially to such specialised categories as pressure, cold, warmth and so on. But the concept of 'modality specificity' is not quite as straightforward as might be thought at first sight, as will become particularly obvious when we come to consider pain. Some preliminary reflection on what exactly is *meant* by a 'sensory modality' is needed.

The concept of modalities comes about through our natural urge to classify the objects around us; the reason that difficulties arise in its use is that there are many *criteria* by which objects may be classified, and unless one is clear about which type of classification is referred to,

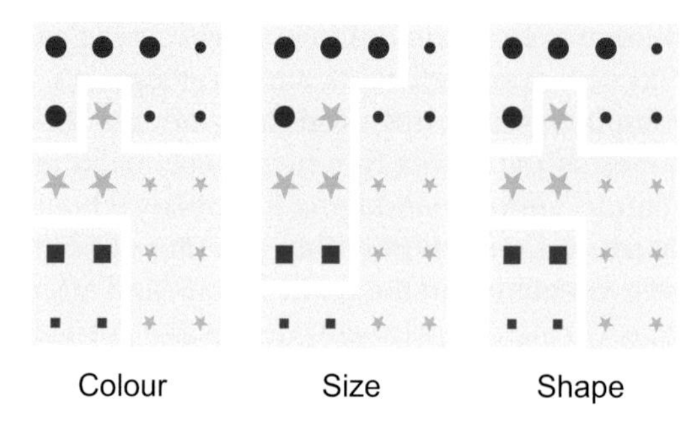

Figure 4.29 A set of miscellaneous objects classified according to colour, size and shape.

misunderstandings become inevitable. If we consider all the kinds of things that may come in contact with the skin, we might group them according to their *physical effects* (as mechanical, thermal, etc.), or according to the *sensations* they produce (pain, tickle, softness), or even in terms of the types of peripheral *nerve fibres* they stimulate. Each of these classifications will in general divide the whole set of stimuli into different patterns of subsets which may or may not correspond with one another, like a set of symbols that could be classified by colour, by size or by shape (**Fig. 4.29**). Now if it happened that in each system of classification the boundaries were identical, as is true for colour and shape in this example, then no difficulties would arise, and we could say with certainty that the fibres were modality-specific. For example, if we found a particular type of fibre that responded only to heating of the skin, and that this in turn was also a clear and distinct class of sensation, then one could say that the fibres in question were specific for that particular stimulus or sensory modality. But in practice, things are seldom so simple, and there is no uniquely valid way of classifying either the physical attributes of objects or the sensations they evoke. In particular, there is a danger of introducing a degree of tautology: one may be influenced by one's knowledge of one of three levels of classification when drawing up the boundaries for the others.

Arbitrariness of classification

If, just for a moment, you forget all you have been taught and ask yourself what you really *feel* to be the categories of cutaneous sensation, your list is likely to include not just the familiar stereotypes of pain, warmth, pressure and so forth, but also other sensations that are just as immediate and apparently 'primary': tickle, itch, softness, roughness, hardness, stickiness, wetness, sharpness and many others. It is doubtful whether someone who had never read a physiology book would naturally consider the classic modalities to be more 'primary' than the others. Our classification of the physical classes of stimuli is almost equally biased. Some physical attributes are left out: for instance, the all-important factor of local curvature of the skin, that gives rise to the sense of sharpness and roughness, is usually wholly ignored. Other, mythical, physical stimuli are simply invented. In an effort to try to produce some physical quality that could be said to correspond with the obvious sensation of pain, it is customary to invent a special class of physical stimulus, whether mechanical, thermal or even chemical, that causes tissue damage of some kind and may therefore be called 'noxious', and sensed by 'nociceptors': yet many kinds of pain are not associated with tissue damage at all.[12]

In other words, there is a danger of unconsciously falsifying what might be called 'natural' classifications of sensations or physical types of stimulus; and if our modalities are thus defined by what is observed in sensory fibres, then it must follow tautologically that one fibre responds only to one modality. A discussion about whether a particular system is modality-specific, or whether on the contrary a particular mode of stimulation gives rise to a characteristic *pattern of activity* among a set of afferent fibres that is indicative of that class of stimulus (as for example seeing the letter 'A' does to our retinal fibres), amounts in the end simply to an argument about how we happen to name what we perceive.

A further, insidious, bias that may creep into investigations of sensory systems – this applies with equal force to other special senses such as vision and audition – is that by having preconceived notions as to what the categories of stimulus are, based on categories of primary fibres rather than on what might be important to the organism in controlling its behaviour, in experiments one may tend to limit oneself to those categories when trying to evoke responses from higher levels of the brain. If one explores the neurons of the somatosensory cortex using only stimuli of light touch, warmth, cold, or one of the other traditional modalities, then naturally all the cells that respond at all must fall into one of these categories. If there were cells responding to more useful things like stickiness or wetness, one would never discover them; and so the myth would be perpetuated. So it is very important to bear in mind these reservations about over-simple categorisations of stimuli into modalities when considering the specificity of receptors and of central neurons to cutaneous and other kinds of stimulation. Our cutaneous sensory world is vastly richer than the pathetic number of 'modalities' derived from studies of skin fibres.[13]

Pain

The International Association for the Study of Pain defines pain as 'An unpleasant sensory and emotional experience associated with real or potential tissue damage, or described in terms of tissue damage'. Pain is not simply the physiological response to a physically applied stimulus, but a perception deeply coloured by context, experience and expectation. Furthermore, pain is dynamic: exposure to a painful stimulus influences the future response to the same stimulus. Pain research is effervescent with new concepts and discoveries, links to the immune system and endocrine pathways as well as the genetics underlying susceptibility to pain disorders. Climbing from receptor to the centres where pain is 'experienced' we shall outline the physiology of the system that underlies this most central human condition, and also exemplifies the complexities of the relation between stimuli and sensations.

Fast and slow pain

It is a common experience that there are two qualities of pain sensation, often called pricking pain or first pain, and burning pain or second pain.[14] If one stubs one's toe against something, the feeling is a sort of immediate 'Ow!' followed by a more drawn-out 'Ooooh!', and these two kinds of pain are thought to be the result of stimulating the Aδ and C fibres respectively. The main evidence for this comes from experiments in people in which the conduction of peripheral nerves is partially blocked either by anoxia or by local anaesthetics. Anoxia, which can most easily be produced by inflating a cuff round the arm, affects the largest fibres first and the C fibres only after a considerable delay. The subject loses pressure and position sense first; then, as the Aδ fibres begin to be affected, temperature sense and pricking pain; and lastly burning pain and itch. The sequence of block for local anaesthetics is different: the C fibres are the first to suffer, and the largest A fibres the last; as a result it is then burning pain and itch that are the first to go, followed by temperature and pricking pain, and lastly pressure. Recordings from single afferents have shown that the Aδ pain fibres are specifically sensitive to mechanical deformation of the skin and that although their receptive fields are quite large, *within* each field the sensitivity is limited to specific 'pain spots' similar to those found in the case of thermoreception. The C fibres on the other hand are of various types: some show a response to mechanical stimulation, while others respond specifically to noxious chemical stimuli or extremes of temperature. Both these and the Aδ fibres responding to noxious stimuli are thought to originate in the free endings of the skin, muscle and connective tissue. An important point is that with prolonged anaesthesia, knocking out everything except the Aβ fibres, no pain is felt at all even with the very greatest degree of mechanical stimulation: pain is certainly not just a matter of overstimulation.

One might wonder what purpose is served by having both fast and slow fibres signalling pain. As was emphasised at the start of this chapter, when faced with peculiarities in sensory coding, it is enlightening to think not so much of the sensations they produce, but rather what use they are to the body – what *behaviour* they are meant to control. Pain elicits two very different kinds of response. One is reflex *withdrawal*, as when touching a hot object; the other is *immobilisation*, protecting the affected part from being further injured by movement (particularly obvious after a back injury). Withdrawal demands a rapid response and fast-conducting fibres; immobilisation is a long-term response where slow fibres will do perfectly well. (The question of why we have warm and cold endings rather than a single temperature receptor may be answered in an analogous way.) It is significant that visceral pain is mediated by C fibres only, since withdrawal here is not an option.

Pain may also be experienced by certain kinds of stimulation of the viscera, particularly severe distension or constriction; yet the digestive tract is quite insensitive to some stimuli – notably cutting and burning, and some chemical stimuli – that are painful when applied to the skin. Visceral pain is often poorly localised and diffuse, evoking autonomic effects such as high blood pressure or sweating. Visceral and cutaneous afferents converge on second-order dorsal horn neurons. For this reason visceral pain may be 'referred' or felt in the region of body surface that shares the same dorsal root: thus pain is felt in the groin in response to a stone in the ureter, and in the left arm in angina pectoris.[15] A corresponding observation is that all cells in the spinal cord that respond to stimulation of visceral afferents also have a somatic receptive field. Cell bodies of visceral afferents are located in the dorsal root, and fibres travel with sympathetic and parasympathetic axons.

Some unmyelinated primary afferents can be 'sensitised' to discharge vigorously to stimuli to which they would normally be unresponsive. Such sensitisation often accompanies inflammation, the warmth, redness and swelling that accompanies infection or tissue damage. The inflamed tissues release chemical mediators such as histamine, serotonin and prostaglandins which lower the threshold for signal generation in otherwise silent pain receptors. This concept will be familiar to anyone who has received a comradely clap on the shoulder following an intramuscular injection, or the lightest application of pressure on a twisted ankle. Conversely, localised tissue damage can lead to the expression of opioid receptors on peripheral nerve terminals. Opioids are peptides usually associated with suppression of pain throughout the nervous system. It is possible that endogenously synthesised opioids, released by immune cells, have a local analgesic, or pain-killing, effect.

Sensitisation is not limited to the peripheral terminals. The dorsal horn neuronal response to a painful stimulus of given intensity progressively increases with time. This is termed wind-up, and is mediated by excitatory neurotransmitter release and activity at the N-methyl-D-aspartate (NMDA) receptor. The receptive field of the spinal neuron enlarges, the threshold for activation falls, and the response to stimulus becomes greater and longer-lasting. This phenomenon manifests as difficult-to-treat acute post-operative pain and may also play a key role alongside long-term potentiation (LTP; discussed elsewhere) in the development of pathological chronic pain states. Strategies to prevent wind-up include the administration of pre-emptive analgesia before surgical incision or the use of ketamine, an NMDA receptor antagonist that seems to reduce the opioid requirement for postoperative pain control.

Itch is not well understood. The blocking experiments described above indicate that the information generating the sense of itching is carried in C fibres; but specific itch

Clinical box 4.2

Allodynia

Allodynia is defined as 'pain in response to a non-noxious stimulus'. Clinicians use a specific stimulus, gentle brushing of the skin, and diagnose allodynia if the brushing elicits the sensation of pain. The mechanism is uncertain but a popular model proposes that an inciting injury activates nociceptors and a signal is transmitted from the periphery to the spinal cord. The input from the peripheral afferent to the spinal cord reinforces the connection, leading to a heightened response in dorsal horn neurons. The sensitised dorsal horn wide dynamic range (WDR) neurons can now be activated by non-noxious mechanoreceptive afferents. Unlike the sensitisation that is associated with inflammation at the injury site, allodynia can persist after the injury has resolved, as part of a complex regional pain syndrome (CRPS). As well as allodynia, CRPS also features local swelling, vasodilation and sweating, even leading to trophic changes of surrounding skin and bone. Why does the sensitisation persist? CRPS requires ongoing input from peripheral nerve to dorsal horn neuron. One putative mechanism by which a persistent signal is established is through a reflex loop involving the sympathetic efferents liberating noradrenaline near the nociceptive nerve terminals. These nerve terminals develop a hypersensitivity to noradrenaline, possibly by upregulating α-adrenergic receptors such that they are stimulated by the sympathetic outflow triggered by the afferent signals from the nociceptive fibres themselves! This poorly understood condition is common and difficult to treat – indeed difficult to diagnose and even define – demonstrating why pain management is one of the most challenging areas in modern medicine.

Synaesthesia

This describes a condition whereby a sensory experience in one modality is involuntarily experienced in another modality. For example, hearing particular numbers evokes different colour sensations; different days of the week are attributed specific personalities or different flavours. Theories to explain synaesthesia include a failure to regulate feedback from higher sensory centres, responding to polymodal stimuli, to early unimodal pathways, such that touch could evoke activity in a pathway dedicated to carrying auditory information. Synaesthesia can be inherited or acquired; in the latter case it is associated with brain injuries such as tumours, trauma, haemorrhage or infection.

fibres have never been found. It may, like tickle, simply represent the sensation produced by a particular pattern of stimulation of the C fibres, perhaps as the result of the release of histamine from damaged tissue, an extremely powerful stimulus for itch when injected locally. Both stimuli, as always with C fibres, *demand* a response.

Central pathways

The central pathways for pain start with the anterolateral system, section of which causes complete peripheral analgesia for both pricking and burning pain. While the majority of afferents ascend at most two to three spinal segments before crossing to the other side of the cord, a significant minority of fibres ascend ipsilaterally; fibres arising caudally ascend posterior to those arising more cranially. The exception to this is the recent observation that some primary visceral afferents do not travel in the anterolateral tracts; rather their second-order neurons ascend in the dorsal columns, adjacent to the midline.

At higher levels the two types of pain show slightly differing distributions; ascending fibres concerned with pricking pain go to the somatosensory thalamus and from thence to the cortex, particularly area S2, while the pathways for burning pain appear to be both older and more diffuse, involving the more central thalamic regions, with their rather general projections to the cortex, the ascending reticular formation, periaqueductal grey and hypothalamus. In humans, interference with the thalamus generally has more effect on pain than with the cortex. Electrical stimulation of the ventrobasal region may produce sensations of pricking pain, and of the central regions a general sense of intense unpleasantness. Lesions of the thalamus can have widely varying effects ranging from relief from pre-existing chronic pain to the production of unendurable spontaneous pain, while the sensation of pain is usually unaffected, or at most slightly reduced, by cortical damage. Correspondingly, stimulation of cortex has never been reported as producing pain sensations. Interestingly, while acute pain results in increased thalamic activity, chronic pain is associated with reduced thalamic activity.

The central pathways for pain are in fact rather complex and poorly understood, partly because the sensation of pain is itself very complex. The relation between the type or intensity of a stimulus and the degree of pain that is felt is a highly variable one and depends to a large extent on the emotional state of the subject and on any implications or meaning that the pain may have. We have all had the experience of injuring ourselves inadvertently, and of not feeling pain until we actually *see* what we have done. In states of excitement, as frequently reported by soldiers severely wounded in battle, there may be a general insensitivity to injuries that would certainly be painful under normal circumstances. In one study, more than a third of patients admitted to an emergency clinic said that they felt no pain when they were injured.[16] Conversely, some patients, perhaps when particularly apprehensive, may show exaggerated responses to quite mild stimuli: in the dentist's chair, we may respond violently to almost any unexpected dental stimulus. One might almost go so far as to call pain an *emotion* that is simply triggered off by certain patterns of cutaneous stimulation in certain behavioural circumstances but not others, in much the same way that, for example, an identical pattern of skin stimulation may produce erotic sensations when done by one person but not by another. One of the clearest pieces of evidence that there is a degree of separation between peripheral neural discharges and the objective sense of the existence of a noxious stimulus on the one hand, and actually *feeling* the pain on the other, comes from patients who have undergone frontal leucotomy, or lesions of the anterior cingulate cortex, to relieve intractable pain. As described later in Chapter 12, on questioning they may indicate to the doctor that they sense the pain, yet from their attitude and mood it is evident that they do not, in any normal sense, 'feel' it. Neuroimaging studies suggest that painful stimuli 'light up' diffuse areas of subcortex as well as sensory, motor, limbic and association cortex. It may be that the objective identification of pain in time and space is mediated by traditional 'sensory' cortical areas in the parietal regions while the emotional 'affective' experience relates to frontal cortex.

Pain is also influenced to a larger extent than other sensory modalities by other modes of skin stimulation, being reduced for example by warmth and by mechanical stimulation such as rubbing, or for that matter by acupuncture; self-stimulation with implanted electrodes has been used successfully for many years for the relief of certain kinds of intractable pain. Conversely, in certain circumstances specific damage to the larger cutaneous afferents may result in an increased sensitivity to painful stimuli. A plausible model of the neural mechanism for this antagonism between the large cutaneous afferents and the small pain fibres was originally proposed by Melzack and Wall. Interneurons in the substantia gelatinosa, in layers I and II of the dorsal horn, receive excitatory information relayed from incoming large mechanical fibres, and inhibit the neurons of the ascending anterolateral system; thus the size of central response to a nociceptive stimulus will depend in general on the balance between the degree of stimulation of the large and small fibres. Although the precise details of this gating mechanism are unclear and to some extent controversial,[17] recordings from the anterolateral neurons show that the majority of them, called *wide dynamic range* or WDR cells, do indeed have the kinds of properties that would be expected from such an arrangement. Typically, they have a concentric receptive field arrangement, in which the centre responds to light touch as well as noxious stimuli but the surround is inhibited by mechanical stimulation (**Fig. 4.30**).

One can think of this as a mechanism for making the signals that are sent to the brain more specifically nociceptive

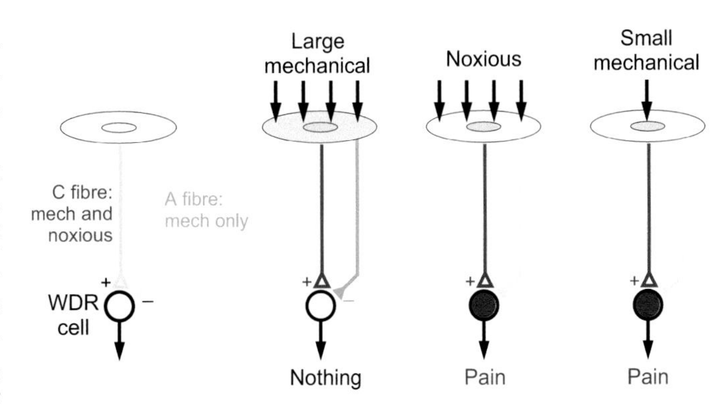

Figure 4.30 Lateral inhibition in WDR cells between large and nociceptive afferents increases specificity to noxious stimuli, as well as causing pain from mild, small mechanical stimuli.

than the Aδ fibres themselves that respond to mechanical stimulation as well as noxious stimuli. With purely mechanical stimuli, provided they are large enough, the effects of centre and surround will cancel each other out, and an erroneous pain message will not be transmitted. This makes good sense, since application of a force through the tip of a knife has very different consequences from the same force being applied through its handle; correspondingly the former stimulates only one or two fibres and generates a pain signal, while the latter stimulates surrounding mechanoreceptors which extinguish the pain signal. So what we have is a kind of lateral inhibition, used not to reduce the effects of spatial overlap, but rather to reduce *overlap between modalities*. In addition, it also means that a mechanical stimulus that is sufficiently small in extent will stimulate the central area but not the periphery. As a result it will also cause pain long before it actually causes tissue damage. It is a matter of common experience that such stimuli – a thorn, the point of a drawing pin – do indeed cause pain without damaging the skin in the slightest; the advantage of responding in this way when walking barefoot is obvious enough.

Direct evidence that the feeling of pain is not linked in any simple, direct way to 'pain' fibres has come from recording from C fibres in conscious human subjects. Heat is applied to the skin until pain is just felt, and the frequency of discharge is noted; then pressure is applied instead, and increased until the nerve is firing at the same frequency: yet no pain is felt until the pressure is increased much further and the frequency is some four or five times higher than the original threshold.[18]

Some of the descending control exerted by the brain on the transmission of pain messages seems to be related to the release of the natural opioids, the *endorphins* and *enkephalins*. The functions of these neuropeptides, which are widely distributed as transmitters throughout the nervous system, and to some extent as hormones as well, are not yet fully understood. Opioid receptors are found on first- and second-order afferents where they synapse in the dorsal horn; activation of these receptors suppresses the transmission of pain signals. Pathological changes at this level are thought to underlie some pain disorders and

Clinical box 4.3 Pharmacology of pain

There now exists an extensive analgesic armamentarium to allow doctors to treat both acute and chronic forms of pain. While a comprehensive account of these is beyond the range of this book, the main classes are:

Non-steroidal anti-inflammatory drugs (NSAIDs): these act to inhibit the action of the cyclooxygenase (COX) enzymes, which catalyse the metabolism of arachidonic acid to prostaglandins – a potent sensitising agent. In this way, they are particularly effective in managing inflammatory conditions such as arthritis. Examples include ibuprofen, naproxen and diclofenac; paracetamol is tentatively included in this group also. However, their use has to be balanced against their potential side effects: antagonism of the COX-1 isoform can lead to excessive acid secretion in the stomach (and thence ulceration and bleeding), diminished autoregulation in the kidney (necessitating caution in those with renal insufficiency), and can also cause bronchospasm in asthmatics. As the COX-2 enzyme appears to be the one induced in reponse to inflammation, more selective agents for this isoform have been developed (such as *celecoxib*).

Opioids: this group, including codeine, tramadol and morphine, act at opioid receptors (particularly the μ subtype). They reduce pain through actions at peripheral (counteracting the sensitising effects of prostaglandins), spinal (reducing neurotransmitter release at the first synapse) and supraspinal (activation of the peri-aqueductal grey) targets. However, long-term use can lead to tolerance – larger doses required to have the same effect – and side effects including constipation, nausea and, in some cases, respiratory depression.

Local anaesthetics: these agents act at voltage-gated sodium channels to reduce the probability of action potential firing. They have a preferential effect on smaller, unmyelinated, C fibres and so when being used for procedures, such as suturing a wound, one should remind patients that, while they will not feel pain, they are still likely to feel pressure (as this is mediated by larger, Aβ, fibres).

Neuropathic agents: finally there are medicines which act at a diverse range of targets to control pain; often they were not developed with their analgesic properties in mind. Examples include carbamazepine (an anticonvulsant), which can be particularly effective in trigeminal neuralgia, or amitriptyline (a tricyclic antidepressant), which can be used for prophylaxis against migraine.

tolerance to exogenously administered opioids such as morphine. Some opioids produce marked analgesia when injected either intravenously or into particular regions of the brain. One such area is the *periaqueductal grey* (PAG); it is thought that an excitatory pathway exists from this region to the *raphe nucleus* (*magnus*) of the medullary reticular formation, which in turn sends descending fibres down into the superficial laminae of the spinal cord which ultimately inhibit the transmission of afferent pain impulses through a spinal interneuron that itself releases enkephalin (ENK) (**Fig. 4.31**). Stimulation of the raphe or of PAG in humans can produce profound anaesthesia and abolish behavioural responses to noxious stimuli. PAG is in turn probably stimulated indirectly by nociceptive afferents, providing a negative feedback system by which pain might modify its own transmission. It is also stimulated by limbic areas – especially during preoccupying behaviour such as copulation – and by opioids, through a mechanism of disinhibition. It also seems to be stimulated by quite meaningless stimuli, if they are strong enough.[19] Descending modulation uses not only opioid transmitters, but also an array of peptides and amines, including serotonin and noradrenaline. This might explain the breadth of pharmacological agents that have analgesic properties, including serotonin and noradrenaline reuptake inhibitors and α2 adrenergic receptor agonists.

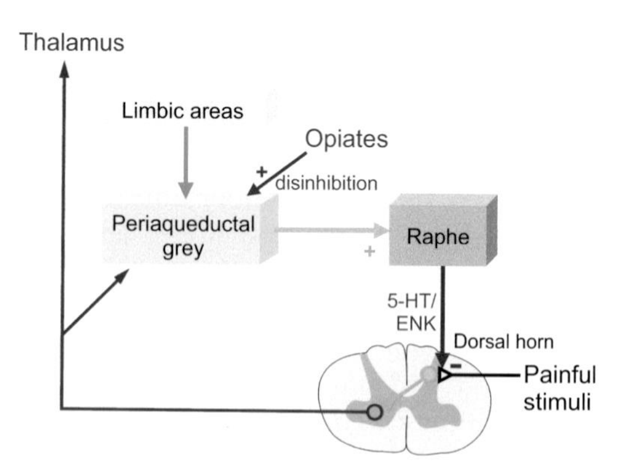

Figure 4.31 A descending system from the raphe nucleus that may serve to modify the transmission of afferent impulses generating the feeling of pain and associated responses.

Notes

The skin Useful general accounts include Sinclair, D. (1981). *Mechanisms of Cutaneous Sensation*. (Oxford: Oxford University Press); Willis, W. D. & Coggeshall, R. E. (1978). *Sensory Mechanisms of the Spinal Cord*. (New York, NY: Wiley); Zotterman, Y. (1976). *Sensory Functions of the Skin in Primates*. (Oxford: Oxford University Press); Hertenstein, M. & Weiss, S. (eds.) (2011). *The Handbook of Touch*. (New York, NY: Springer).

1. Dermatomes The boundaries between dermatomes are nothing like as sharp as the figure implies, and there is a certain amount of disagreement between authors concerning some of the details. An excellent source, with comparisons with the earlier maps of Head, Elze, Richter and others is Hansen, K. & Schliak, H. (1962). *Segmentale Innervation. Ihre Bedeutung für Klinik und Praxis.* (Stuttgart: Georg Thieme Verlag); another is Keegan, J. J. & Garrett, F. D. (1948). The segmental distribution of the cutaneous nerves in the limbs of man. *Anatomical Record* **102** 409–437. Corresponding to the dermatomes on the motor side are the *myotomes* – they also overlap one another considerably.

2. Visceral afferents They – and probably small somatic afferents as well – contain an extraordinary range of peptide transmitters, including VIP, somatostatin, angiotensin, substance P, CCK-like peptides, and so on. What are they all for? A useful account of 'interoception' is Cameron, O. C. (2002). *Visceral Sensory Neuroscience: Interoception.* (Oxford: Oxford University Press).

3. Morphology of endings See for instance Iggo, A. & Andres, K. H. (1982). Morphology of cutaneous receptors. *Annual Review of Neuroscience* 5 1–31; Halata, Z. (1975). The mechanoreceptors of mammalian skin. Ultrastructure and morphological classification. *Advances in Anatomy, Embryology and Cell Biology* **50** 1–77; and Hsiao, S. S. (2011). Biomechanical and neurophysiological basis of the processing of tactile stimuli. In Hertenstein, M. & Weiss, S. (eds) *The Handbook of Touch.* (New York, NY: Springer).

4. Cuneate and gracile Cuneate means 'wedge-shaped', as in cuneiform, the Babylonian script created by pressing a stick into wet clay to create wedge-shaped indentations. Gracile means 'slender'.

5. Thalamus and cortex Indeed, far from the thalamus being subservient to the cortex, perhaps it is really the other way round. Experimenters tend to think that the cortex is important because it is easy to work on!

6. Internal capsule Because of its blood supply, it is peculiarly vulnerable to the damaging effects of stroke.

7. Slip and grip See, for example, Johansson, R. S. & Westling, G. (1987). Signals in tactile afferents from the fingers eliciting adaptive motor responses during precision grip. *Experimental Brain Research* **66** 141–154. Robot hands designed for grasping and lifting objects are sometimes provided with a similar sense, in the form of microphones built into the gripping surfaces, whose output is used in a feedback loop to increase the pressure when the object is slipping.

8. Single units of the human hand See, for example, Johansson, R. S. (1978). Tactile sensibility in the human hand: receptive field characteristics of mechanoreceptive units in the glabrous skin area. *Journal of Physiology* **281** 101–127; Johansson, R. S. & Vallbo, Â. (1979). Tactile sensibility in the human hand: relative and absolute densities of four types of mechanoreceptive units in glabrous skin. *Journal of Physiology* **286** 283–300; Vallbo, Â., Olsson, K. Â., Westberg, K. G. & Clark, F. J. (1984). Microstimulation of single tactile afferents from the human hand. *Brain* **107** 727–749.

9. Receptive field The concept of the receptive field is one we shall be coming across repeatedly, and applies most obviously to spatial systems like the somatosensory and visual. At slightly higher levels, parts of the fields may be excitatory and some inhibitory. It can also apply to neurons at much higher levels still, which may be responsive to a particular kind of stimulus (in vision, a line of a particular orientation, for example) but still only within a defined area. Finally, the concept can usefully be extended to cover attributes which are not literally spatial, but can be thought of as arranged along an axis, for instance in the case of visual neurons responding to a range of wavelengths, auditory cells tuned to certain frequencies, or olfactory neurons responding to some odorants and not others.

10. Overlap providing redundancy This principle seems to extend right up to higher levels of the sensory pathways. Recordings from sensory cortex in conscious cats have shown that local anaesthesia in the periphery can cause an almost immediate restructuring of cortical receptive fields, with the appearance of new areas that previously had no effective contribution. See Metzler, J. & Marks, P. S. (1979). Functional changes in cat somatosensory and motor cortex during short-term reversible epidural blocks. *Brain Research* **177** 379–383.

11. Universality of lateral inhibition Looking at spatial differences is one of the most basic and primitive biological functions. Chemotactic bacteria swim towards sugar and away from acid. Yet because their bodies are only a couple of micrometres in length, this means that they have to be able to sense concentration differences of the order of 1 part in 10 000.

12. Modalities are just names An example may make this clearer. Imagine a simple-minded creature – perhaps some kind of slug – whose cutaneous sensations fall into only three categories: 'wet', 'earth', and 'nice' – the last being the result of contact with a slug of the opposite sex. A slug who studies physiology and investigates the responses of his own somatosensory neurons would find that some fibres – what we call 'light touch' receptors – fire during both 'earth' and 'nice', while others ('cold') fire during 'wet' and sometimes during 'earth', and so on. He would deduce in fact that his fibres were not modality-specific, but that 'earth', 'wet' and 'nice' were coded in the form of particular patterns of activity. A human physiologist would completely disagree. What he would report would be highly specific fibres responding to the traditional categories of 'warm', 'cold', 'light touch' and so on, but the argument would clearly be about the naming of sensory categories, not about the observations themselves. Furthermore, many feelings have no name at all. Think of the feeling that you need to urinate: extremely clear, specific and presumably universal, yet we have no name for it, and cannot begin to *describe* the sensation itself.

    A thoughtful discussion of this whole area (one that many students find difficult) can be found in Melzack, R. & Wall, P. D. (1962). On the nature of cutaneous sensory mechanisms. *Brain* **85** 331–356.

13. Complexity of skin sensation Three accounts that do justice to this area: Katz, D. (1989). *The World of Touch.* (Hillsdale, NJ: Lawrence Erlbaum); Sathian, K. (1989). Tactile sensing of surface features. *Trends in Neuroscience* **12** 513–519; and Grunwald, M. (ed.) (2008). *Human blapfic Perception: Basics and Applications.* (Berlin: Birkhauser). An imaginatively free-ranging discussion is Paterson, M. (2007). *The Senses of Touch: Haptics, Affects and Technologies* (Oxford: Berg).

14. Pain Intelligent accounts may be found in Melzack, R. & Wall, P. D. (1982). *The Challenge of Pain.* (Harmondsworth: Penguin); Hunt, S. & Koltzenburg, M. (eds) (2005). *The Neurobiology of Pain.* (Oxford: Oxford University Press); and Zhuo, M. (ed.) (2008). *Molecular Pain.* (New York: Springer). A popular account is Wall, P. D. (2002). *The Science of Pain and Suffering.* (London: Weidenfeld and Nicolson).

15. Referred pain It is interesting to speculate on whether this convergence of function is perhaps due to a common embryological origin, although the evidence on this point is uncertain.
16. Injury without pain See Wall, P. D. (1985). Pain and no pain. In C. W. Coen (ed.) *Functions of the Brain*. (Oxford: Clarendon).
17. Gating theory This theory generated a quite astonishing degree of hostility when it was first proposed, partly, one suspects, because people did not like to have their comfortable, simplistic ideas of 'one fibre, one modality' unsettled. See Melzack, R. & Wall, P. D. (1982). *The Challenge of Pain*. (Harmondsworth: Penguin), p. 233.
18. Human single C fibres See van Hees, J. & Gybels, J. M. (1972). Pain related to single afferent C fibers from human skin. *Brain Research* 48 397–400.
19. Pain reduced by meaningless stimuli In the seventeenth and eighteenth centuries large numbers of itinerant mountebanks would travel to fairs round the country drawing people's teeth, a popular public spectacle. Often they would have a drummer with them whose job was to perform a prolonged drum-roll close to the victim's head during the operation, which apparently induced a certain degree of anaesthesia – and of course helped drown their screams, which might have put off other customers.

CHAPTER 5

PROPRIOCEPTION

This chapter is concerned with those mechanoreceptors that provide us with information about ourselves: about the positions and movements of our limbs, the forces generated by our muscles, and our attitude and motion relative to the earth. The brain mostly uses this information to help control movement; consequently a discussion of how proprioception is used to control muscle length is left until Chapter 10, and its use in maintaining an upright posture to Chapter 11.

Muscle proprioceptors

Two distinct kinds of proprioceptors are found in voluntary muscles, specialised for providing information about two quite different things: muscle spindles that respond to muscle length and rate of change of length, and Golgi tendon organs that signal muscle tension or force. Both are essentially stretch receptors: their difference in function comes about because of their different situation in the muscle as a whole (**Fig. 5.1**). Whereas the spindles are in parallel with the main contractile elements in the muscle, so that their stretching is simply a measure of the degree of stretch x of the muscle itself, the tendon organs are situated in the muscle tendons, in series with the contractile elements and the load, so that their stretch is proportional to the tension T exerted by the muscle.

Consequently, when a muscle contracts the tendon organs are stimulated by the extra tension, but the activity of the sensory elements in the spindles is reduced as they are less stretched.

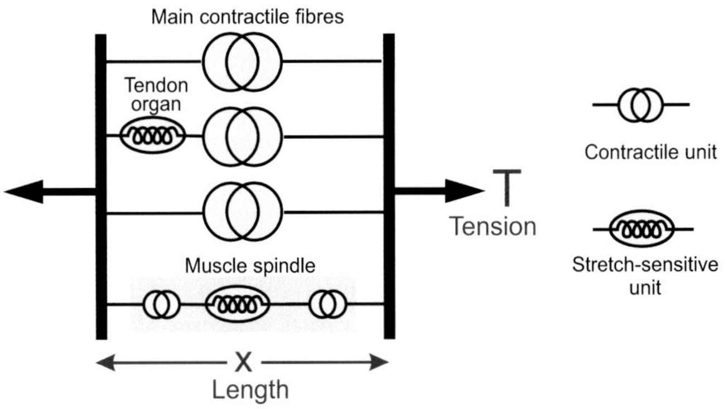

Figure 5.1 Schematic representation of contractile and stretch-sensitive elements in muscle. The contractile elements within the spindle (intrafusal) are innervated separately from the main (extrafusal) muscle fibres, and make only a negligible contribution to overall muscle tension, *T*. Thus, whereas tendon organs respond essentially to muscle tension, spindles respond to length, *X*, but in a manner modified by the activity of their own contractile elements.

Spindles

Muscle spindles are found in practically all the striated muscles of the body, but are greatly outnumbered by the striated muscle fibres themselves: in cat soleus, there is only one spindle for every 500 or so ordinary fibres. Each consists of a fluid-filled spindle-shaped capsule some 2–4 mm long and a few hundred micrometres in diameter, whose ends are attached to the exterior sheaths of neighbouring muscle fibres (**Fig. 5.2**). Inside is a small number of modified muscle fibres called intrafusal fibres (the 'fus-' root means 'spindle'), each having contractile ends, and a region in the middle that is not contractile but contains the nuclei. Though contractile, their direct contribution to muscle tension is negligible: this is done by the ordinary extrafusal fibres, with a completely separate motor innervation.

Two main types of intrafusal fibre are found, differing in the way in which these nuclei are distributed. Nuclear chain fibres are thinner, and their nuclei are lined up in a

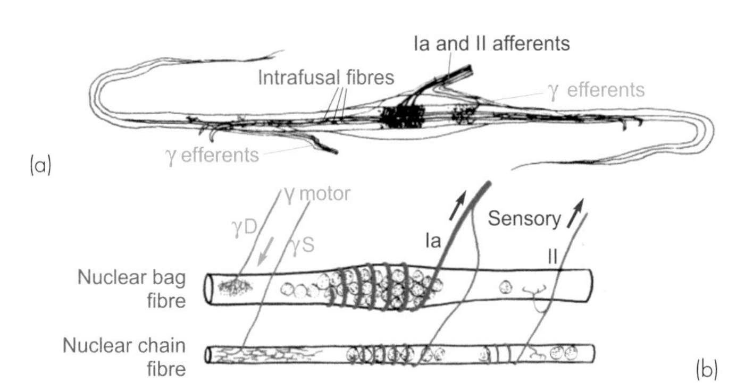

Figure 5.2 (a) A typical mammalian muscle spindle (simplified from Barker, 1948). (b) Schematic representation of the central region of a nuclear bag and nuclear chain fibre, showing the afferent Ia and II innervation, and the two kinds of γ fibre. (Copyright © 1948 by the Company of Biologists Ltd.)

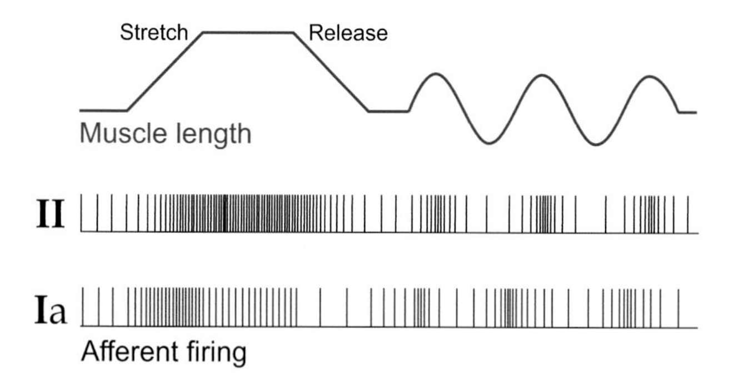

Figure 5.3 Idealised responses of primary (Ia) and secondary (II) fibres to various patterns of muscular stretch. (After Matthews, 1964; ©1964, The American Physiological Society; with permission.)

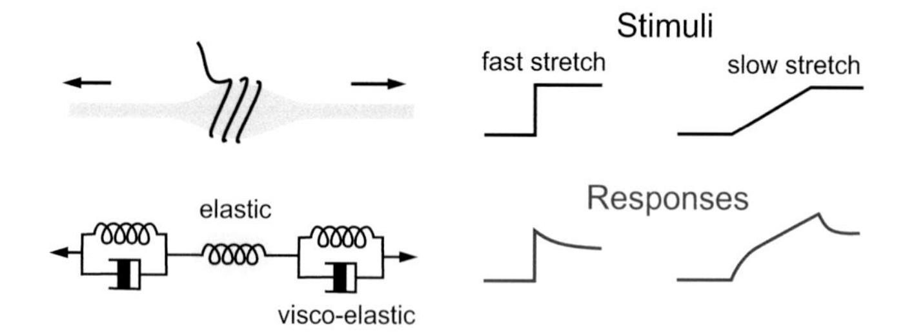

Figure 5.4 A simple model of the viscoelastic properties of spindle fibres in response to both sudden and slow stretch.

row along the central portion like peas in a pod; nuclear bag fibres have a pronounced bulge in the middle in which the nuclei are bunched together. A typical spindle has some half-dozen intrafusal fibres, the nuclear chain fibres generally being in the majority. Two kinds of afferent or sensory fibre innervate the spindle: the larger, primary fibres, belonging to group Ia, send branches to the central portions of both types of fibre and have annulospiral endings; the smaller secondary fibres are of group II and terminate partly as annulospiral and partly as flower-spray endings mainly on the nuclear chain fibres, more peripherally than the Ia endings.

These two kinds of nerve fibre respond very differently to muscle stretch. The secondary fibres are in a sense simpler: their signals are more or less directly proportional to the degree of stretch of the spindle at any moment. Whether the muscle is suddenly stretched to a new length, stretched more slowly, made to shorten, or alternately stretched and relaxed in a sinusoidal manner, their frequency of firing mirrors quite accurately the instantaneous value of the muscle length (**Fig. 5.3**). They are therefore essentially non-adapting or static. The Ia fibres, by contrast, are dynamic and show very pronounced adaptation. During a sudden stretch they fire maximally during the period of stretching and only at a reduced rate when the muscle is held at its new length; during a slow stretch they again respond most during the movement, at a frequency nearly proportional to the rate of stretch; and during sinusoidal stretching their maximum firing is not at the moment of maximum stretch but near the point of maximum rate of change of stretch. In other words, they respond partly in proportion to muscle length but mostly in proportion to its rate of change, or velocity: this means that they are in effect predicting the future length of the muscle, an important feature in providing feedback to the motor control system (see p. 193 and p. 208).

Now we saw in Chapter 3 that adaptation in sensory receptors can in general be due to two distinct processes: there may be energy filtering, in which static information is wholly or partly thrown away before it even reaches the transducer element itself; or there may be membrane adaptation, in which even a steady conductance at the ending results in a fall-off in firing frequency. Whereas in the Pacinian corpuscle both of these mechanisms contribute almost equally to the adaptation that is observed, in the case of the muscle spindle it appears that nearly all is of the energy-filtering kind, and not very different from what is produced by the concentric lamellae of the Pacinian corpuscle. The contractile portions of the intrafusal fibres behave as if they were very much more viscous than the central portion (this is particularly true of the nuclear-bag fibres), so that the mechanical properties of the fibres as a whole may be represented by a mechanical model like that of **Figure 5.4**. In a brief stretch, there is no time for the viscous elements to lengthen, and the stretch is entirely taken up by the central region, where the annulospiral endings are. But if the stretch is maintained, the viscous elements gradually yield, releasing the strain on the middle portion, and causing the frequency of firing to drop. The difference between the non-adapting properties of the secondary fibres and the marked adaptation of the primaries seems to be due partly to the fact that the former go only to chain fibres, which are less viscous, and also to the fact that they innervate more peripheral parts of them.

Besides the two kinds of afferent fibre, spindles also receive motor innervation from the γ fibres, belonging to group Aγ and around 6 μm in diameter. There are two types of fusimotor fibre, called γs and γd – static and dynamic – and they innervate, respectively, the nuclear chain (mostly) and nuclear bag fibres, causing contraction of the peripheral regions.[1] For any given muscle length, such a contraction must of course stretch the sensory elements and thus increase the firing of the afferent fibres, and in fact the effect of γ stimulation is in general much the same as if an extra stretch had been applied to the muscle as a whole, though they may also increase the sensitivity of the endings to stretch, by altering the elasticity of the intrafusal fibres and thus changing the proportion of the stretch that is experienced by the stretch receptors themselves. **Figure 5.5** shows how the firing frequency of stretch receptor afferents (from the eye muscles of a goat) respond to different degrees of stretch, when the corresponding γ fibres are also stimulated at different rates: the interaction between external stretch and the internal stretch produced by the γ activation can be clearly seen. The static and dynamic γ fibres produce slightly different effects on primary and secondary afferent responses, as would be expected from their differing distribution to the energy-filtering bag and non-adapting chain fibres. Dynamic γ fibres increase the sensitivity of group Ia

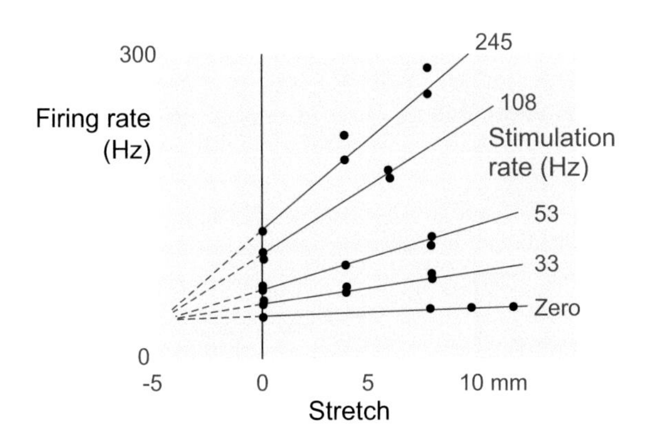

Figure 5.5 Effect of γ-fibre stimulation on afferent response to stretch, in an eye muscle from the goat; the rates of stimulation are shown on the right. (Data from Whitteridge, 1959; with permission from John Wiley and Sons.)

fibres but have no effect on group II fibres, whereas the static ones increase the sensitivity of both the secondaries and primaries to static stretch, and actually decrease the primary sensitivity to rate of stretch (the mechanism is unclear). Thus the central nervous system can, through the γ efferents, control not just the sensitivity of the spindle afferents, but also in a sense their adaptational properties. Consideration of the way in which γ control is actually used by the motor system is postponed to Chapter 10.

Golgi tendon organs

The Golgi tendon organs have received much less attention from experimenters. In appearance they are very similar to the Ruffini organs in the skin (Chapter 4) and, like them, they appear to respond to tension in the fibres with which they are associated in the tendon; they are innervated by afferents of group Ib. At one time their importance was underestimated because they seemed to have such high thresholds: large tensions had to be applied to the tendon as a whole before they could be induced to fire. But it is now clear that this was because in these circumstances the total tension applied is in effect shared out among the tendinous fascicles, so that each tendon organ feels only a small fraction of it. They respond in fact very briskly to the modest, active tensions generated by the actual muscle fibres to which they are joined: the contraction of less than a dozen motor units can be enough to generate significant activity in a tendon organ. Because they register tension rather than muscle length, during active movements their discharge generally has a reciprocal relationship to that of the muscle spindles: extrafusal activity simultaneously increases the tension in the tendons and decreases muscle length; but during passive movements, both kinds of response are normally in step with one another. Like many other mechanoreceptors, they respond partly to change, in this case changing load or tension.

The function of the tendon organs is not entirely clear. Information about the forces being generated during movement is certainly vitally important to the central controlling mechanisms, particularly in sensing load and in learning to generate appropriate commands to achieve particular movements: this is discussed in Chapter 10 and Chapter 12. The clasp-knife reflex has an obvious protective function, in helping to ensure that the tension in a tendon does not reach the point of rupture; it has also been suggested that Golgi tendon organs ensure that each group of motor units contributes equally to the development of total force by the muscle as a whole.

Central projections

The central pathways of afferents from both muscle spindles and tendon organs are quite similar. Fibres enter the dorsal roots in the usual way, and branches of the majority synapse in a spinal nucleus extending roughly from L3 to T1 called Clarke's column (or dorsal nucleus) with neurons whose fibres ascend in the homolateral posterior spinocerebellar tract to an extremely important region in the control of movement, the cerebellum (**Fig. 5.6**). However, since Clarke's column fades out above T1 or so, more rostral afferents turn upwards and ascend to the *accessory cuneate nucleus* of the medulla, from which fibres run in the *cuneocerebellar tract*; they carry information from the forelimbs, which are not represented in the dorsal pathway. Another route (not shown in the figure) by which both cutaneous and proprioceptive information may reach the cerebellum is via the *spino-olivary tract* to the inferior olive, which in turn projects through climbing fibres to the cerebellar cortex (see Chapter 12). Apart from ascending to

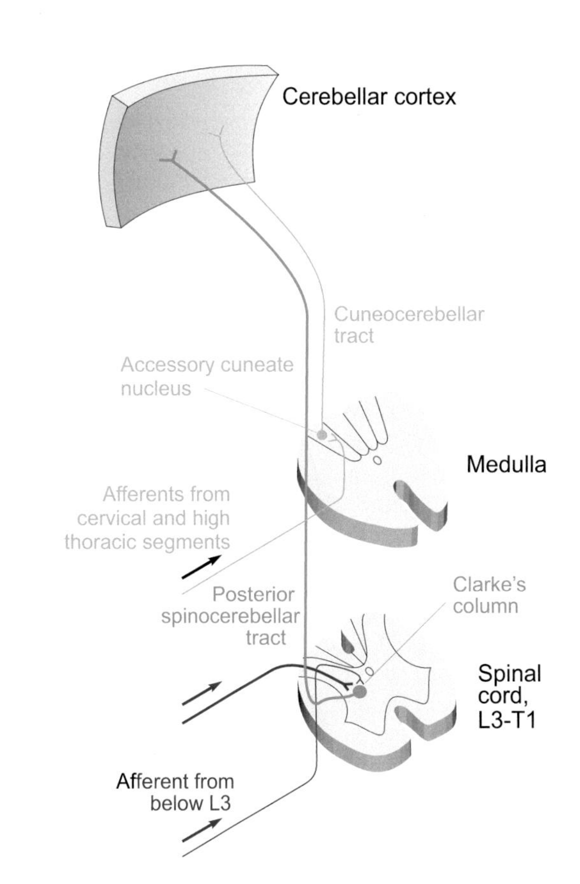

Figure 5.6 Schematic representation of the principal spinocerebellar pathways. An additional route, not shown, is via the inferior olive and climbing fibres.

the cerebellum, branches of fibres from muscle proprioceptors are involved in various reflex mechanisms within the spinal cord, notably the stretch reflex (which in its simplest form consists of a monosynaptic excitation of a motor neuron by a Ia afferent) and the clasp-knife reflex: these are discussed in Chapter 10. Muscle proprioceptors also project to the cerebral cortex, via the ventral posterior thalamus.

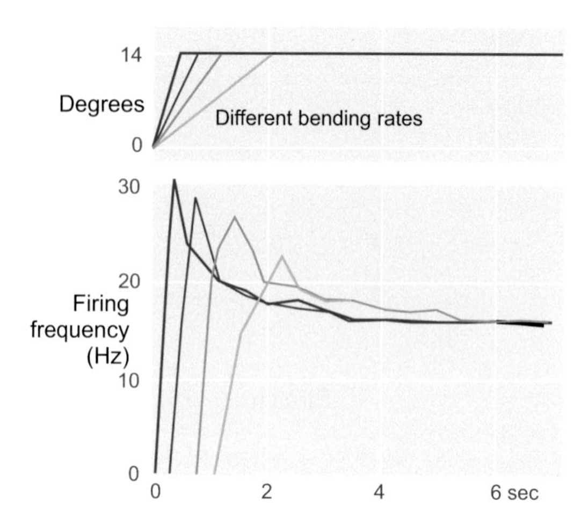

Figure 5.7 Firing frequency of an afferent from a cat's knee joint in response to flexion through a fixed angle at the different rates shown at the top, demonstrating incomplete adaptation. (Data from Boyd & Roberts, 1953; with permission from John Wiley and Sons.)

Joint receptors

Another important source of information about limb position and movement comes from the mechanoreceptors that are found in the ligaments and capsules of joints. They are of a variety of morphological types, some very similar or identical to those found in the skin. Thus Pacinian corpuscles and Golgi-like endings are found with large axons (group I), Ruffini endings (II), and also small nerve fibres with unencapsulated endings. Some show complete adaptation and are thus more sensitive to rate of change, but most show incomplete adaptation and thus signal limb position as well (Fig. 5.7). The patterns of response are complicated to a certain extent by the fact that few receptors are able to respond over the whole range of movement that a joint is capable of: their 'excitatory angle' is typically less than half the entire possible range, thus increasing sensitivity to changes in position within that range. This means that information about the position of a limb is partly coded by frequency of firing, but also by which neurons are firing. Although in some joints, most of the afferent fibres fire preferentially at extremes of joint position and are presumably intended to give warning that the joint is about to become dislocated, nevertheless there appear to be sufficient fibres responding at mid-range positions to provide adequate proprioceptive information; in other joints the majority are mid-range anyway. Afferent information from joints follows the same route as that from Pacinian and other corpuscles in the skin (see p. 81): fibres ascend in the ipsilateral posterior columns, relay in the cuneate and gracile nuclei, cross and proceed via the medial lemniscus to the ventral posterolateral thalamus

Clinical box 5.1 The Romberg test

The term '*ataxia*' describes a gross failure of motor coordination. Commonly, ataxia arises after injury to the cerebellum (the part of the brain that integrates sensory input and motor planning to generate coordinated movement; see Chapter 12) resulting in lack of co-ordination, difficulty with rapidly alternating movements (dysdiadochokinesis), and an unsteady, broad-based, gait. However, ataxia can also occur following injury to those parts of the sensory system yielding proprioceptive information. This is termed *sensory* ataxia, and classically causes a 'stomping' gait.

In the early nineteenth century, neurologists noted that patients with *tabes dorsalis*, a complication of neurosyphilis causing dysfunction of the dorsal columns, exhibited ataxia that was markedly worse at night. The 'sensory ataxia' arose because the information from pressure receptors in the feet, and joint and muscle receptors, relating the limbs to each other and to the world, was not reaching the brain. The deterioration at night, they reasoned, was because the sufferers could no longer compensate for their lack of proprioception by visual feedback.

Moritz von Romberg made use of this observation to generate a new diagnostic test, to distinguish sensory from cerebellar ataxia: making the patient stand still with his or her arms to their sides, they would then observe the patient's posture both with their eyes open and shut. If, on closing their eyes, the patient promptly fell over, a defect in proprioception, such as tabes or severe vitamin B_{12} deficiency was suggested. In contrast, ataxia due to cerebellar disease should be unaffected whether the eyes are open or not, since the failure is not with sensory information but its integration and analysis.

Today, the causes of sensory ataxia are different: tumours compressing the posterior aspect of the spinal cord, multiple sclerosis lesions affecting the dorsal columns, or hereditary disorders such as *Friedrich's ataxia*. A slight ambiguity occurs in the case of vestibular disease, which, in theory at least, should also be associated with a positive Romberg test. However, this can usually be distinguished by the presence of associated vertigo, nystagmus and sometimes hearing loss. Furthermore, the loss of balance seems to take longer to develop in the case of vestibular pathology.

and thence to the somatosensory cortex; some also contribute to the spinocerebellar pathways.

Conscious proprioception

We shall see in later chapters that the information these proprioceptors provide is essential in regulating movement and posture. A slightly different question to ask is to what extent they contribute to kinaesthesia, our conscious sense of the position and movement of our limbs. For a long time this question was the subject of bitter controversy between those who believed that such sensations came only from the muscles and those who favoured joint receptors, each side seeking to show that the other mechanism contributed nothing at all. It seems strange in retrospect that the modern view that both contribute, and indeed other sources of information such as the skin as well, was not grasped sooner.[2]

The main evidence that mechanoreceptors in joints contribute to conscious proprioception comes from anaesthetising the receptors either by direct injection into the synovial fluid or by putting an inflatable cuff around the limb in such a way that it stops the blood flow to the joint but not to the muscles that move it. The sense of limb position is then greatly impaired, though the subject will generally still be able to sense movement: static proprioception is more affected than dynamic. It is clear therefore that the joints contribute to proprioception, but that they are not the sole source. Indeed, after operations in which the hip joint is replaced with a prosthesis and the mechanoreceptors are entirely lost, the patient can still sense joint position (though with reduced sensitivity), presumably by using information from muscles and from the skin. Conversely, it is not difficult to demonstrate the muscle contribution directly. Conscious patients have often reported a sense of limb movement when surgeons pull on exposed tendons without moving the joint. A less traumatic way of stimulating the stretch receptors is to apply a vibrator to a muscle or its tendon (**Fig. 5.8**), which acts preferentially on the Ia endings because of the high rates of stretch it generates. Such a stimulus produces an illusion that the muscle is shortening even though it is in fact held stationary, as can be seen if the subject is asked to match the felt position of the vibrated arm with the other one: this feeling is greatly enhanced if at the same time a cuff has been applied in such a way that the corresponding joint is anaesthetised. Under these conditions it can be shown that the illusion is essentially one of a roughly constant rate of movement rather than of static position.

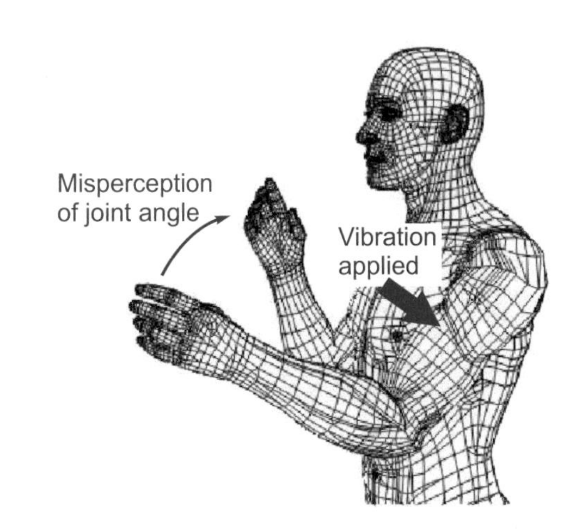

Figure 5.8 A vibrating stimulus is applied to the left biceps and the subject tries to match the felt position of the left arm with the right.

Table 5.1 Contributions to sense of limb position and muscle loading	
Muscle	Predominantly change of position
Joints	Predominantly static position
Skin	Predominantly load; may also contribute to position sense
Efference copy	Predominantly load

Experiments like these leave little room for doubt: both joints and muscles play a part in providing proprioceptive information, with limb position being sensed more by joint receptors, and velocity by spindles; it also seems highly probable that skin receptors round joints play a part as well, though this is less well established. The other sensation that we seem to get from our muscles is that of the forces we exert with them, and the sense of weight. In this case it is clear from experiments that as well as mechanical information from the skin, an important factor is the sense of effort, the size of the commands that we send to them. In circumstances where extra effort is needed for the same load, with muscle fatigue, or because the usual reflex contribution from sensory receptors has been blocked by local anaesthesia (see Chapter 10), subjects feel that loads are heavier and that they are generating greater tensions. It may be that Golgi tendon organs also contribute to the sense of load, but this has not been demonstrated unequivocally.

The vestibular apparatus

The vestibular apparatus forms part of the labyrinth of the inner ear, which – as its name suggests – is a complex structure. It is, in a sense, a specialised version of the closely related lateral line organ found in many aquatic species. This is a system of tubes or grooves lying just under the skin cells and in communication with the surrounding water, and provided with a series of epithelial sensory organs called neuromasts. Their ciliated sensory cells are stimulated by the flow of water through the tubes, as a result either of something moving in the outside world and setting up fluid currents, or of the fish's own motion through the water. In the labyrinth, these tubes are sealed off from the outside world, and its two functions – one

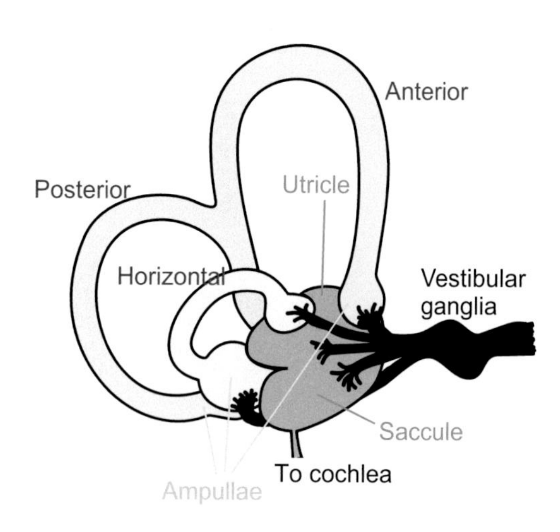

Figure 5.9 Gross structure of the vestibular part of the labyrinth, viewed laterally (somewhat schematic).

exteroceptive, one proprioceptive – have come to be carried out by two separate organs: the cochlea, signalling movement of the surrounding air in the form of sound waves (discussed in Chapter 6), and the vestibular apparatus, signalling movement of the head itself.[3]

The vestibular part of the labyrinth is divided functionally into two components: the semicircular canals, of which there are three on each side of the head, and the otolith organs, of which there are two on each side, the saccule and utricle (**Fig. 5.9**). The ciliated sensory cells are very similar in all parts of the vestibular apparatus, and will be described first; it is the accessory structures that enclose them that make them specifically responsive to different types of stimuli.

The sensory cells

The sensory epithelium of the vestibular apparatus is made up of a mosaic of sensory cells and supporting cells, the former being divided into two morphological types: a flask-shaped type I cell, and a roughly cylindrical type II cell. Each sensory cell has a characteristic pattern of cilia projecting from its upper surface, consisting of a single flexible kinocilium, whose root is near the edge of the receptor, and between 60 and 100 stereocilia, relatively thin and stiff, and arranged in a regular array in the more central area. The latter are graded in size rather like a set of organ pipes, the longest ones being nearest to the kinocilium (**Fig. 5.10**). The kinocilium is a much more elaborate structure than the stereocilia, having the '9 + 2' arrangement of longitudinal filaments characteristic of motile cilia, and basal bodies. The asymmetrical arrangement of kinocilium and stereocilia defines a direction of polarisation for each cell, and it is found that bending in the direction of the kinocilium leads to excitation, while bending in the opposite direction gives inhibition. The deformation caused by bending alters the ionic permeability of the stereocilia, through mechanisms that are presumed to be essentially the same as in the hair cells in the cochlea, discussed in more detail in the next chapter, and probably involve direct

Table 5.2 Divisions of the vestibular apparatus

Vestibular apparatus	
Semicircular canals (horizontal, anterior, posterior)	**Otolith organs (utricle, saccule)**
Angular velocity	Linear acceleration, and hence angular position relative to gravity

attachments from mechanically opening channels at the tip of one stereocilium that are linked by filaments to its longer neighbour. This generates currents that alter the membrane potential of the far end of the cell, causing calcium entry and release of transmitter, which is probably glutamate.

The sensory cells are innervated by branches of the vestibular nerve: type I cells are almost completely enclosed in a nerve calyx or chalice, and there are often regions of close apposition, which, together with the large synaptic area, suggest that the transmission process here may be partly electrical as well as chemical in nature. Type II cells generally receive more than one ending; the terminals are smaller in size, and some of them are efferent rather than afferent. There seems to be much more convergence from type II cells on to their afferent fibres than in the case of the type I cells, and more still in the case of the efferent fibres: there are only some 200 fibres in a cat's vestibular nerve going to the receptors, in contrast with the 12 000 or so afferents. Sensory and efferent fibres travel together in the vestibular division of the VIIIth nerve to the region of the vestibular nuclei, of which the lateral nucleus is the

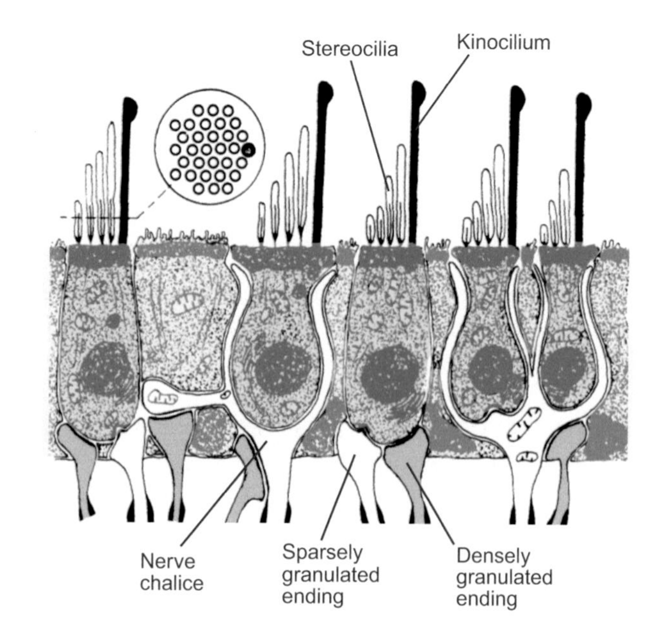

Figure 5.10 Diagrammatic cross section of sensory epithelium of the vestibular system, showing type I and type II cells with their innervation. (Inset) A typical arrangement of the cilia seen in horizontal section.

origin of the efferents. Most of the afferent fibres terminate within the nuclei, but some carry on through and project ultimately to the cerebellum. Most afferent fibres fire spontaneously, and one finds that a stimulus that bends the kinocilium in one direction accelerates the rate of firing, and in the other decreases it: thus each cell has a specific direction of polarisation, which is generally similar to its neighbours'. Different cells tend to have different spontaneous firing frequencies so that they function over different parts of the total possible range of stimulation, in a manner reminiscent of joint receptor afferents.

Otolith organs (utricle and saccule)

The utricle and saccule are a pair of hollow sacs containing endolymph, a fluid whose ionic composition is of the intracellular type (with a high K+/Na+ ratio), and is continuous throughout the whole of the labyrinth. Surrounding the endolymphatic sac is a second sac containing perilymph, whose composition is low in K+ and more like that of a typical extracellular fluid. The receptor cells of the otolith organs are confined to a special region of the sac called the macula, and their projecting cilia are embedded in a jelly-like mass, the otolith, whose density is increased by the incorporation of large quantities of calcite crystals called otoconia. If the head is tilted, this mass moves relative to the macula, bending the cilia and thus resulting in stimulation of the afferent nerve fibres. The cells also respond to linear accelerations of the head, since in this case the otoliths tend to get left behind and hence cause the same sort of bending. Thus the afferent signals from the otolith organs are dependent simply on the vector sum of the acceleration due to gravity and any linear acceleration that may be occurring at the same time: in other words, on the effective direction of gravity (**Fig. 5.11**). There is necessarily no way in which the brain can distinguish between head tilt and linear acceleration – since 'gravity' is itself of course simply a kind of linear acceleration – nor is it desirable that it should. From the point of view of controlling posture (as for example in trying to stand upright in a bus that suddenly accelerates), it is the effective direction of gravity that matters (see Chapter 11, p. 217).

Information about the direction of the acceleration vector is available because the maculae of utricle and saccule lie in different planes – in the utricle roughly horizontal, in the saccule roughly vertical – and also because in each case the direction of polarisation of the hair cells varies in a systematic way over the macular surface, each with two opposing polarities represented on either side of a curved ridge: the striola (**Fig. 5.12**). Recordings from individual otolith fibres show that each one fires maximally at a particular orientation of the head, and that there is a fall-off in frequency as the head is tilted away from that position. Thus the spatial pattern of discharge from the whole population of receptors provides information about the direction of acceleration. As might be expected, most units show little adaptation, and for the most part faithfully signal head position without decrement over indefinitely long periods of time. Some do, however, show adaptation, firing most rapidly during changes of head position, but their adaptation is not complete and their tonic discharge still gives a measure of head position: it is the semicircular canals that primarily signal changes in the attitude of the head.

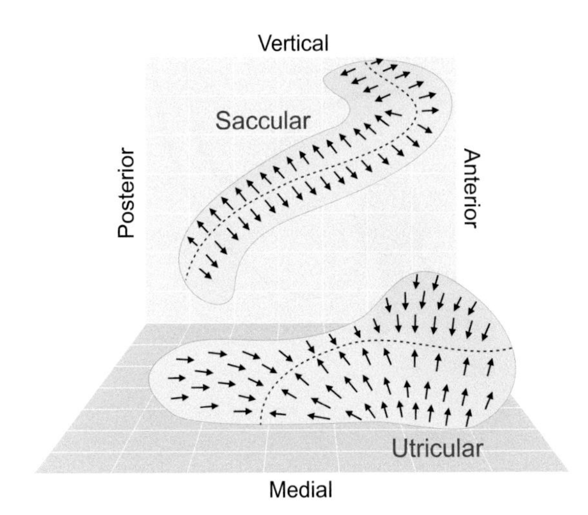

Figure 5.12 Stylised representation of the orientation of the utricular and saccular maculae in the head. The small arrows indicate the approximate direction of polarisation of receptors at different points on the surface. Consequently displacements in all possible directions can be represented.

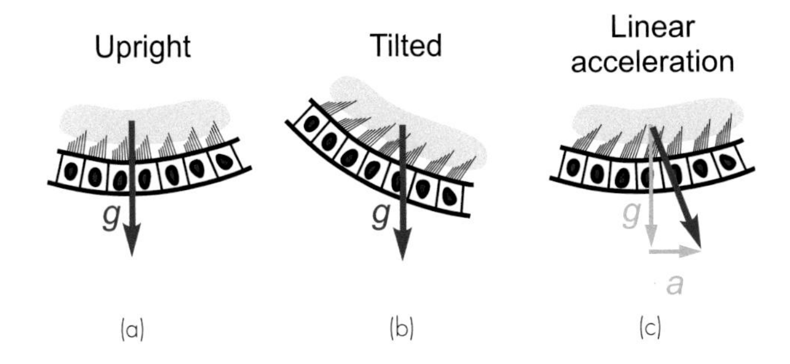

Figure 5.11 Equivalence of head tilt and linear acceleration. Schematic representation of action of macular receptors (a) at rest, (b) with head tilt, and (c) under horizontal linear acceleration, *a*. The last two conditions are indistinguishable as far as stimulation of the hair cells is concerned.

The semicircular canals

Each canal consists of a looped tube containing endolymph and having a swelling at one point along its length, the ampulla, into which projects a crest, the crista, which is covered with sensory hair cells (**Fig. 5.13**). Their cilia are embedded in a jelly-like structure called the cupula, forming a kind of flap which can swing backwards and forwards in response to movement of fluid along the canal, thus bending the cilia and causing neural excitation. Unlike the otolith, this jelly does not contain calcareous granules, and is in fact of exactly the same density as the endolymph surrounding it: it is very important that this should be so, as otherwise the hair cells would respond to gravity in the same way as the macular receptors. What

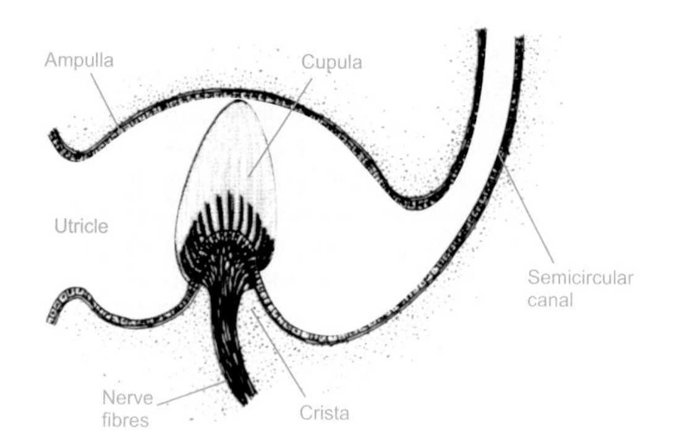

Figure 5.13 Diagrammatic section through a single canal in the region of the ampulla showing the cupula, the hair cells of the crista and their innervation.

they do respond to, in fact, is rotation of the head. When the head is turned, the fluid in the canals tends to get left behind and pushes on the trap-door–like cupula, thus bending the sensory cilia.

Because the three canals on each side of the head are arranged in more or less mutually perpendicular planes (**Fig. 5.14**), they are able to signal rotations about any axis in space. In most animals, with the head in the normal upright position, the horizontal canals are parallel with the ground, and the superior and posterior canals lie at about 45° to the sagittal plane. The cells in each crista are all oriented in the same direction: thus the rate of firing of fibres from the left horizontal canal is increased by turning the head to the left, of those from the right is reduced (**Fig. 5.15**), and similar mutual antagonisms exist between the superior canal of one side and the posterior canal of the other. Corresponding to this arrangement, one finds cells in the vestibular nuclei that are excited by a particular canal but inhibited by its opposite number on the other side, thus effectively combining the signals from the two sides in a 'push–pull' manner (**Fig. 5.15**). For this reason, rotation is not a good stimulus to use if we want to test each of a subject's vestibular organs separately; instead, we need to use a stimulus that only acts on one side at a time. In clinical practice, caloric stimulation of the canals is sometimes used; the outer ear is irrigated with warm water, which appears to set up convection currents in the endolymph of the canals and thus produces unilateral vestibular stimulation.[4] There is an important consequence of this push–pull arrangement: with unilateral vestibular damage, at first there is a compelling illusory sense of rotation because of the unopposed spontaneous activity of the functioning vestibular apparatus, often leading to debilitating vertigo.

Clinical box 5.2 Vertigo

This is best described as an illusion of movement of either the patient or their surroundings; one should always establish this is truly the symptom in question as often those feeling 'faint' or 'dizzy' might mistakenly describe their problem as 'vertigo'. The causes can then be divided into those occurring centrally (such as in multiple sclerosis, brainstem stroke or tumour) or peripherally (due to dysfunction of the vestibular apparatus – below).

Benign Paroxysmal Positional Vertigo

This can occur at any age, but is more common in the elderly, and has a characteristic history: patients report disabling attacks of vertigo when the head is rotated, usually to one particular side, for example when turning over in bed. The pathology is thought to be the presence of calcium debris in the posterior semicircular canals (probably loose fragments of otoconia), which generate currents in the endolymph and give the sensation of rotational movement even when the head is stationary. Amazingly, this can be treated by a simple manoeuvre of the head, bearing the name of John Epley whom first described it, to use gravity to encourage the debris back into the utricle.

Vestibular neuronitis and labyrinthitis

This is believed to be caused by a viral infection and, as it results in a more persistent imbalance in signals from the left and right, it leads to continuous and severe vertigo even when not moving (sometimes causing vomiting). An attack can persist for 7 days before gradually subsiding. Labyrinthitis is often used interchangeably with vestibular neuronitis but, as the former can lead to hearing loss while vestibular neuronitis does not, they are probably best diagnosed in isolation.

Ménière's disease

This is an illness of older age groups and causes a triad of vertigo, hearing loss and tinnitus. Attacks of vertigo last from 20 minutes to several hours, but the hearing loss can be more persistent with (unusually for this age group) a tendency to affect lower frequencies.

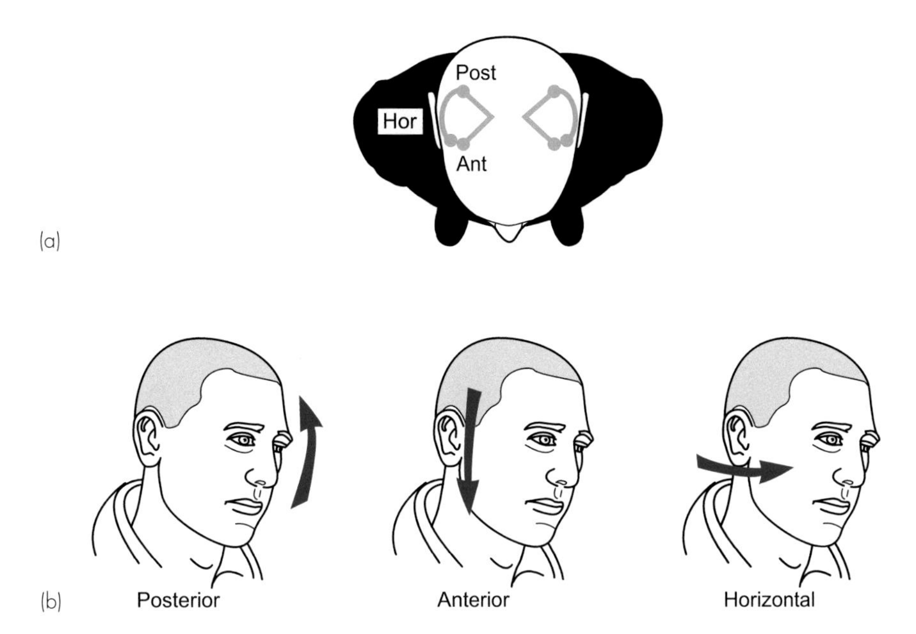

Figure 5.14 (a) Approximate orientation of the semicircular canals in humans. The arrows show the direction of fluid movement that is stimulatory in each case. (b) Directions of head movement that stimulate each of the canals on the left side of the head.

Adaptation in the semicircular canals

The adaptational properties of the canals are very important, and are almost entirely of the 'energy-filtering' type. If the cupula were completely unrestrained, and moved in unison with the fluid of the canal, then the bending of the cilia of the sensory cells would simply signal rotational position of the head; the endolymph would be acting rather like the gyroscope in an inertial guidance system. A highly stylised mechanical model of such a system is illustrated in **Figure 5.16a**. A large truck, representing the

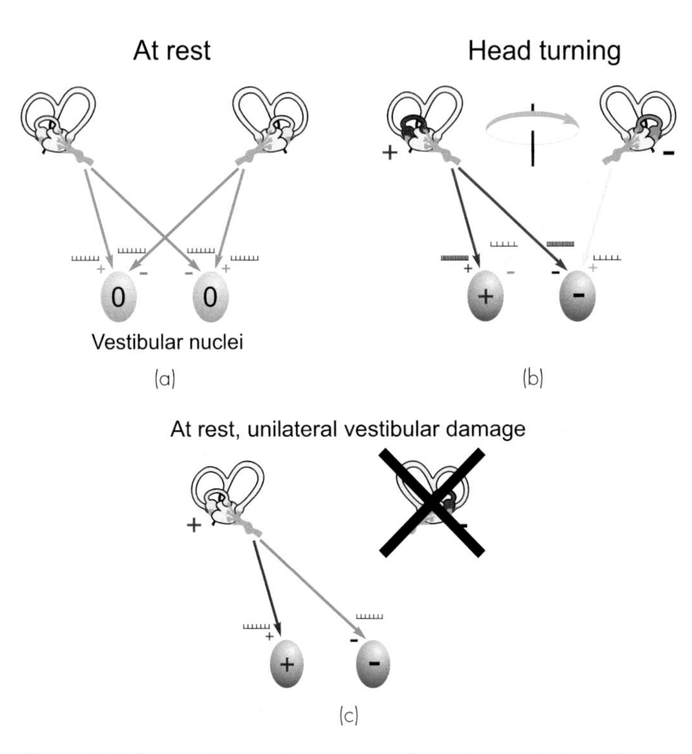

Figure 5.15 Balanced bilateral projections of canals to vestibular nuclei. At rest (a), the spontaneous activity from each canal cancels out, through reciprocal inhibitory innervation. When the head turns (b), there is excitation of one side, and inhibition of the other, leading to a differential signal. With unilateral vestibular damage (c), at first there is an illusionary sense of rotation because of the unopposed spontaneous activity of the functioning vestibular apparatus.

Clinical box 5.3 Diagnosing brain death

In many countries, the diagnosis of brain death requires that all evidence of brainstem reflexes is lost (see Chapter 14, p. 293). One part of the examination to establish brain death involves caloric stimulation of the vestibular apparatus. This test is based on the fact that heating or cooling the semicircular canals can set up convection currents capable of deflecting the cupula. Cooling tends to cause flow away from the utricle and the cupula is inhibited, while heating causes flow in the opposite direction and stimulates the associated afferents. One semicircular canal being more active than its counterpart simulates the condition of head rotation toward that side. As a result, this drives the eyes in the opposite direction as part of the vestibulo-ocular reflex (see Chapters 9 and 11). If the instillation of warm or cold water into the ear canal triggers nystagmus, then the diagnosis of brain death cannot be made as it demonstrates that the VIIIth cranial nerve, the vestibular nucleus, the oculomotor nuclei and the oculomotor nerves are still functional. This has enormous implications in the field of organ transplantation and it cannot be overstated how important clinically, and ethically, the performance and understanding of this test is.

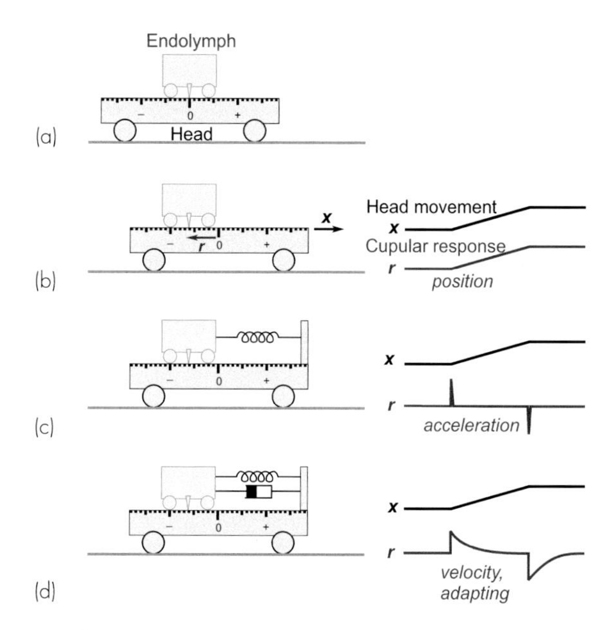

Figure 5.16 Mechanical model of the action of the cupula in semicircular canals. (a) and (b) A small truck of mass M, representing endolymph inertia, rests freely on a larger truck representing the head. The reading, r, of the pointer provides a measure of the position, x, of the truck, because the little one remains stationary. (c) If the two trucks are coupled elastically, r indicates not position but acceleration, a. (d) Actual cupular movement is as if the coupling were partly elastic but predominantly viscous: this produces an adapting velocity response.

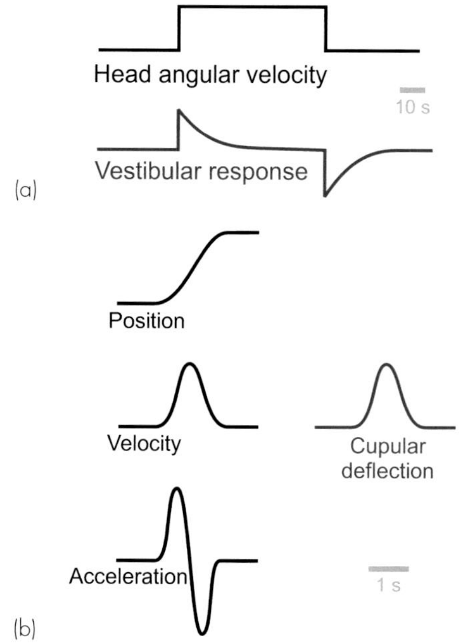

Figure 5.17 Responses of the semicircular canals. (a) During a head rotation at constant angular velocity, canal response declines exponentially over some 20 s; on stopping the movement, there is an opposite response that again declines in the same way. (b) Time-course of head angular position, velocity, and acceleration during a natural turn of the head (left). Cupular deflection (right) behaves more like the velocity curve than either of the others.

head, has a smaller truck that is free to move on top of it, representing the cupula and the inertial mass M of the endolymph. If the big truck moves a distance x, the small truck will remain (in absolute terms) exactly where it was before, resulting in a relative displacement between the two – in effect, what is signalled by the hair cells – of x, the new position of the big truck (**Fig. 5.16b**).

But consider now what would happen if the little truck, instead of being completely free to move, were coupled to the big one by an elastic element, a spring (**Fig. 5.16c**). The system is now no longer a position detector; what we have done is to make it into an acceleration detector, or accelerometer. For if the truck moves off with acceleration a the spring will experience a force equal to Ma, and will therefore stretch by an amount that is simply proportional to the acceleration. Does the cupula in fact behave as a rotational accelerometer? It is perfectly true that it is indeed elastic, and if pushed to one side exerts a restoring force that tends to bring it back to the middle. But it turns out that the cupula and the endolymph are both very viscous. The effect of this viscosity is to slow down the mechanical response and produce an adapting velocity response (**Fig 5.16d**).

Consider for instance a subject on a swivel chair that is suddenly set into rotation at constant angular velocity. A true accelerometer would give a brief response only at the instant at which the rotation started, since this is the only time at which acceleration occurs; but recordings of the firing frequency of vestibular units from the canals show that in these circumstances the period of response is very much drawn out by the damping effect of the viscosity, lasting for some 20 seconds or so after the acceleration has stopped. Under natural circumstances – continual rotation of the head in one direction is not after all very common in real life – the deflection of the cupula, and thus the frequency of firing, mirrors much more closely the instantaneous rotational velocity of the head than its acceleration (**Fig. 5.17**).

In other words, the semicircular canals are not really rotational acceleration detectors at all, but rather velocity detectors. It is only under very peculiar laboratory conditions that their acceleration sensitivity (which can be thought of as an adapting velocity response, just as velocity sensitivity is equivalent to an adapting positional response) manifests itself. A consequence of this adaptation is that a subject who has been set spinning at constant

Clinical box 5.4 Alcohol

It is quite possible that one or two of you *may* be aware that alcohol intake can be associated with vertigo. The mechanism for this appears to reside in the change in relative density between endolymph and cupula that occurs when alcohol first diffuses into the endolymph before equilibrating with the cupula, then diffuses out of the endolymph before leaving the cupula. At each point where the relative densities differ, the abnormal floating or sinking of the cupula in the endolymph gives rise to the sensation of the room spinning.

velocity has a gradually decreasing sense that he is actually rotating; even with his eyes open, after 20 seconds or so he has the strong impression that he is actually sitting still, and that the world is spinning round him, as anyone who has had a ride in a fairground 'rotor' will know. Furthermore, if the chair is then suddenly stopped, the cupula, which had previously resumed its resting position, will now be pushed in the opposite direction by the tendency of the endolymph to retain the previous motion of the head; the subject will then have the very strong impression of rotating in the opposite direction, although he is in fact stationary – what is commonly called vertigo. Some postural consequences of these adaptational properties are discussed in Chapter 11.

Notes

1. Spindle motor innervation Chain fibres also receive some innervation from branches of α fibres, sometimes (unhelpfully) called β fibres. Their function is unclear: see note 2, below.

2. Conscious proprioception Some useful accounts: Brodie, E. E. & Ross, H. E. (1984). Sensorimotor mechanisms in weight discrimination. Perception and Psychophysics **36** 477–481; Burgess, P. R., Wei, J. Y., Clark, F. J. & Simon, J. (1982). Signalling of kinaesthetic information by peripheral sensory receptors. Annual Review of Neuroscience 5 171–187; Matthews, P C. B. (1982). Where does Sherrington's 'muscular sense' originate? Muscles, joints, corollary discharges? Annual Review of Neuroscience 5 189–218.

3. The vestibular receptors A very thorough account of the vestibular periphery is Wilson, V. J. & Melvill Jones, G. (1979). Mammalian Vestibular Physiology. (New York, NY: Plenum). A more recent book is Goldberg, J. M., Wilson, V. J., Cullen, K. E., Angelaki, D. E., Broussard, D. M., Buttner-Ennever, J., Fukushima, K. & Minor, L.B. (2011). The Vestibular System: A Sixth Sense. (New York, NY: Oxford University Press). A comprehensive account of clinical aspects is Baloh, R. & Kerber, K. (2011). Clinical Neurophysiology of the Vestibular System. (New York, NY: Oxford University Press).

4. Caloric nystagmus Not entirely through convection currents, for it still occurs under zero-gravity conditions: see Schere, H., Brandt, U., Clarke, A. H., Merbold, U. & Parker, R. (1986). European vestibular experiments on the Spacelab-1 mission. 3. Caloric nystagmus in microgravity. Experimental Brain Research **64** 255–263.

CHAPTER 6

HEARING

To hear is to feel the touch of vibration in the air, so that its transduction mechanisms are not so very different from those of the somatosensory system. But to obtain and analyse information about stimuli as rapid as sound waves, very specialised neural mechanisms have had to evolve inside the inner ear. To appreciate them you need first to understand the sound waves themselves, a topic that regrettably often seems to get squeezed out of the school physics curriculum.

The nature of sound

Sound is generated in a medium such as air whenever there is a sufficiently rapid movement of part of its boundary – perhaps a moving loudspeaker cone, or the collapsing skin of a pricked balloon. What happens is that the air next to the moving boundary is rapidly compressed or rarefied, resulting in a local movement of molecules that tends to make the pressure differences propagate away from the original site of disturbance, at a rate that depends on the density and elastic properties of the medium. This velocity v is around 340 m/s in air, and about four times as great in water. If the original sound source is undergoing regular oscillation at a frequency f – like the prongs of a tuning fork – the sound is propagated in the form of regular waves of pressure. Their *wavelength*, λ – the distance from one point of maximum compression to the next – is given by their velocity divided by their frequency, or v/f. *Pitch* is the sensory perception that corresponds with frequency, just as colour corresponds with the wavelength of a light, but there is not always an absolutely direct relationship between the two.

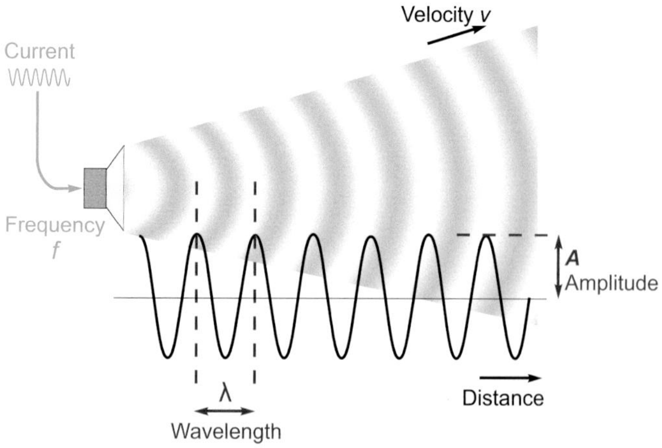

Strictly speaking, not all vibrations of this kind are sound; to be audible, the frequency must lie somewhere between about 20 and 20 000 Hz. The simplest of all vibrations are those in which the variations of pressure along the wave are sinusoidal as a function of time, in other words proportional to sin $(2\pi ft)$, where t is the lapsed time. In such a case we may describe the wave completely by means of its frequency and its *amplitude*, A, the value of the additional pressure at the peak of compression. We can also talk about the *period* of the wave, the time taken for one whole cycle ($= 1/f$) (**Fig. 6.1**).

Figure 6.1 A loudspeaker driven by a sinusoidal current of period T, or frequency $f = 1/T$. The resultant waves of compression and rarefaction of the air travel with a velocity v and are of wavelength λ, where $v = \lambda f$. The amplitude, A, is the difference between the mean pressure and the peak pressure.

The intensity of sounds

Sound waves also carry energy: the rate at which energy is delivered per unit area, the *intensity* of the sound, is proportional to the square of its amplitude, and also to the density of the medium. In addition it increases with the square of the frequency: it obviously takes more energy to move something backwards and forwards very fast than if it is done more slowly. Because the range of intensities to which the ear can respond without damage is a very large one indeed – a factor of some 10^{14} – it is convenient to use a logarithmic scale to describe sound intensities. For this purpose, a standard reference level is used (its value being near the threshold of hearing under ideal conditions, 10^{-12} Wm^{-2}), and the log of the ratio of the actual intensity to this standard gives the intensity of the sound in Bels (named after Alexander Graham Bell, an inventor

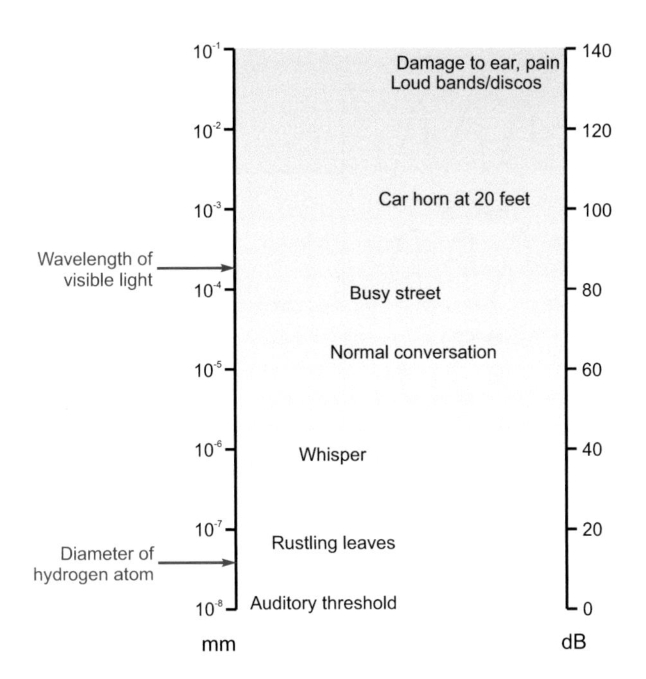

Figure 6.2 Approximate intensities of various sounds, measured in absolute decibels (dB) and also in terms of the amplitude of the corresponding movement of the molecules in the air (0 dB = 10^{-12} W/m^2)

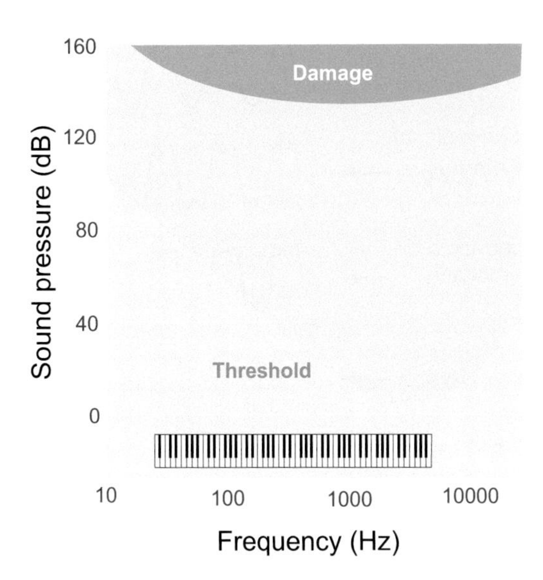

Figure 6.3 Audiometry curve: the shaded region shows the intensities of sounds that may be heard at different frequencies, averaged over many subjects. (Data from Dadson & King, 1952; Copyright © JLO (1984) Limited 1952.) The piano keyboard may help to relate the frequency scale to ordinary musical pitch.

of the telephone). In practice, the unfortunate custom has grown up of dealing in tenths of a Bel, or *decibels* (dB), so that the formula becomes:

$$\text{Intensity in decibels} = 10 \log_{10} \frac{\text{Intensity of unknown}}{\text{Intensity of standard}}$$

Figure 6.2 shows the intensity of some common kinds of sound, expressed in decibels and also in terms of the approximate distance moved by the air molecules. The standard threshold of 10^{-12} Wm^{-2} corresponds to a movement of the air molecules of some 10^{-11} m – considerably less than the diameter of a hydrogen atom.[1]

A tricky point is that because intensity is proportional to the *square* of the amplitude, multiplying the latter by a factor of 10 results in a 20 dB increase in intensity.

Finally, although the measure defined above is an absolute scale of intensity, decibels can also be used to compare two intensities: thus if one sound has one-tenth the intensity of another, it can be described as having a relative intensity of –10 dB, or as being *attenuated* by 10 dB.

The smallest intensity that can just be perceived, the auditory threshold, depends markedly on frequency; a graph of measurements of threshold for a sine wave as a function of frequency is called an *audiometric curve* (**Fig. 6.3**) and often has clinical diagnostic value: the shaded area in the figure represents an average over many subjects.

Phase

One other parameter that is necessary to describe a sinusoidal vibration in certain circumstances is its *phase*. A sinusoidal wave of constant amplitude and frequency that is sampled simultaneously at two fixed points in space – as for example by the two ears – will by definition have the same amplitude and frequency at each point, but the peak pressure at one point will not occur at the same moment as at another point because of the time delay. In general, one wave will appear to be displaced in time with respect to the other, and the phase difference between the two is a measure of the fraction of a whole cycle by which it appears to be shifted (**Fig. 6.4**). It is conveniently expressed as an angle, so that a phase shift of 180° brings the waves into antiphase, the peaks of one then corresponding to the troughs of the other, and a further 180° shift, making 360° (or 0°) in all, brings them back into coincidence.

Sound spectra

Pure sinusoidal waves are actually rather uncommon in real life: musical instruments, for example, produce waves whose profile, though repetitive, is not sinusoidal. However, over 200 years ago the French mathematician

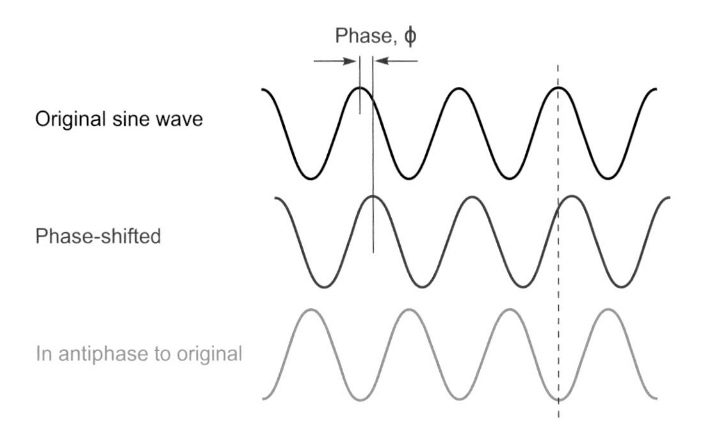

Figure 6.4 Phase. These sine waves have the same frequency and amplitude but are shifted in phase by an angle φ (in the first case a phase advance by 60°). A sine wave shifted by 180° relative to another (dashed line) is described as being in antiphase to it.

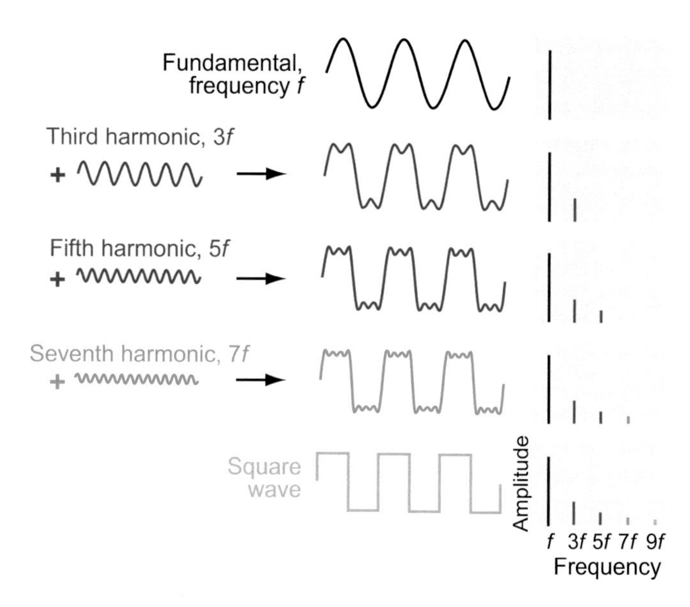

Figure 6.5 Fourier synthesis: in this case, the gradual approximation to a square wave is achieved by adding together successive sine waves of frequency *f*, 3*f*, 5*f*, etc.; their amplitudes are proportional to 1, 1/3, 1/5 etc., as shown in the amplitude spectra on the right.

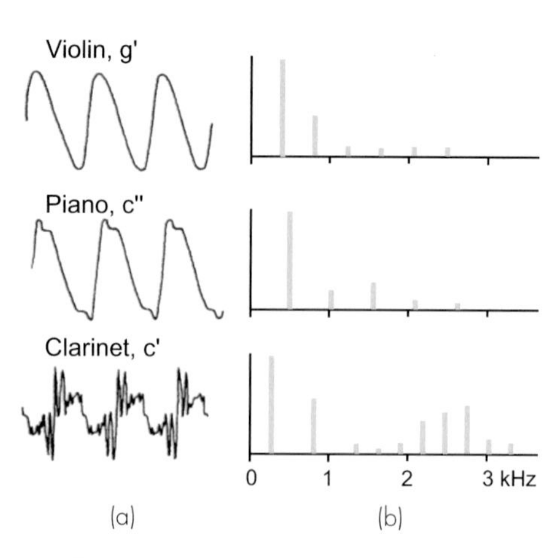

Figure 6.6 Waveforms (a) and amplitude spectra (b) for different notes played by different musical instruments. (After Wood, 1930.)

Fourier proved that every repetitive waveform can be decomposed into a set of simple sinusoidal components, whose frequencies are integral multiples of the frequency (*fundamental* frequency) of the original wave, that will recreate the original waveform if added together. For instance, a square wave of frequency *f* (**Fig. 6.5**) can be *synthesised* by adding together sine waves of frequency *f*, 3*f*, 5*f*, 7*f* and so on, with amplitudes in proportion to 1, 1/3, 1/5, 1/7, etc. A recipe of this kind can be represented graphically in the form of a Fourier *spectrum* (as on the right of the figure) that shows, as a function of frequency, the amplitude of each of the components *(harmonics)* that make it up. (A complete spectrum would show the phase of each harmonic as well, but in practice the phase information is often omitted, for reasons that will become apparent later.) **Figure 6.6** shows the sound spectra of different kinds of musical instrument: in each case, the spectral line of lowest frequency shows the amplitude of the fundamental, and those of higher frequency show the amplitudes of the harmonics. The Fourier spectrum and the shape of the waveform itself are thus in a sense interchangeable: each contains the same information as the other. If we know the spectrum, we can add all the components together and recreate the original waveform, and conversely it is possible by a process of Fourier analysis to translate any given waveform into its equivalent spectrum.[2]

Furthermore, it turns out that Fourier analysis can even be applied to waveforms that are *not* repetitive: unpitched continuous sounds like that of a boiling kettle, or transients of the kind produced by dropping a teapot. To see why this is so, consider what happens if you continuously lower the frequency of a repetitive waveform of given shape. Since the individual components of the spectrum are spaced out on the frequency axis by *f*, the fundamental frequency, as you lower the frequency, *f* gets smaller, and the lines of the spectrum get closer and closer. In the limit, when the frequency of the wave becomes zero – in other words when its wavelength is infinite, and it never repeats itself at all – the spectral components are infinitely close together: so the spectrum, instead of being a sequence of discontinuous spikes, is now a smooth curve. Thus repetitive waveforms give rise to spectra with discrete harmonics, whereas unrepetitive waveforms produce continuous spectra. In either case, it turns out that the quality of a sound – for example the timbre of a musical instrument – is much more closely related to the overall shape of its spectrum than to the shape of its waveform. Musical instruments with lots of high harmonics sound bright and strident (trumpets, clarinets); those with most of their energy in the fundamental sound smoother and more rounded. Sharp sounds, hisses, clicks, all have continuous spectra with prominent high frequencies (the only difference between a hiss and a click is in fact in the phase relationships of its components), whereas thumps, roars and rumbles have their energy concentrated at the low-frequency end.[3]

To summarise, we have two different and largely independent psychological sensations that arise from a sound, that are closely related to two different aspects of the sound's spectrum. On the one hand, *pitch* depends on the fundamental frequency: two instruments may have spectra of entirely different shapes but will be recognised as playing the same note if their fundamental frequencies are identical (and sounds like bangs and hisses with continuous spectra have no pitch at all). *Timbre* on the other hand is governed entirely by the overall shape of the spectrum, regardless of the fundamental frequency; an oboe playing a succession of different notes is still recognisably the same instrument, because the general shape of its spectrum remains essentially unchanged. Together, these two characteristics provide important information about the object that gave rise to the sound in the first place. The fundamental frequency tells us about its size: large objects generate lower frequencies. Timbre on the other hand tells us something about its shape: smooth, round objects produce fewer harmonics, generating a smoother, rounder sound. These factors are very obvious if you compare the appearance of different musical instruments with how they sound, for example a tuba and a triangle.

The human voice

One of the most striking examples of the use of sound spectra comes from studying the human voice. The larynx, isolated from the rest of the voice-producing apparatus of head and throat, is essentially not very different from the double reed of an instrument like the oboe. As the air passes between the vocal cords, alternately they are forced apart and spring back, producing a repetitive series of compressions and rarefactions whose frequency can be modified by altering the shape and tension of the cords themselves; the resultant spectrum is roughly of the quite flat form shown in **Figure 6.7**. But in real life, the sounds it produces have to pass through a number of hollow cavities – the throat, nose and mouth – before they reach the outside world. These cavities tend to resonate, absorbing certain frequencies and reinforcing others, so that the original sound spectrum becomes distorted. The two or three main resonance peaks in this spectrum simply reflect the fact that the tongue effectively divides the mouth cavity into separate compartments, and each compartment acts as an independent resonator, at a frequency that depends mainly on the shape of the tongue and the degree of jaw opening. What is striking is that different vowel sounds are associated in a closely reproducible way with particular positions of these resonant peaks (called *formants*), and these characteristics are largely independent of the speaker, the pitch of their voice, and whether the vowel is spoken or sung: yet the *waveforms* produced by different speakers pronouncing the same vowel are generally very different. In other words, it seems that vowels are recognised, independently of the quality of the voice or its pitch, by the overall shape of the spectrum and not by the shape of the wave itself. It is the frequency of the fundamental that produces the sense of the pitch of the voice and very often conveys information in its own right (as in the different emphasis in 'I *love* you' and 'I love *you*'), particularly of an emotional nature. The spectra in **Figure 6.7c** show the vowel sound in 'EAT' sung at different pitches: the resonance pattern remains identical, but the spacing between the harmonics increases as the fundamental frequency rises.

Finally, it seems to be the fine structure of the spectrum, the little bumps and hollows on it that are the results of idiosyncrasies of the way our own particular mouths and throats are constructed, that enable us to differentiate one speaker from another. It turns out, as we shall see, that almost the first thing the ear does to the sound waves it receives is to Fourier-analyse them, and the pattern of activity of the fibres of the auditory nerve is – at least for medium and high frequencies – in effect a representation of the spectrum of the sound that is heard: only at low

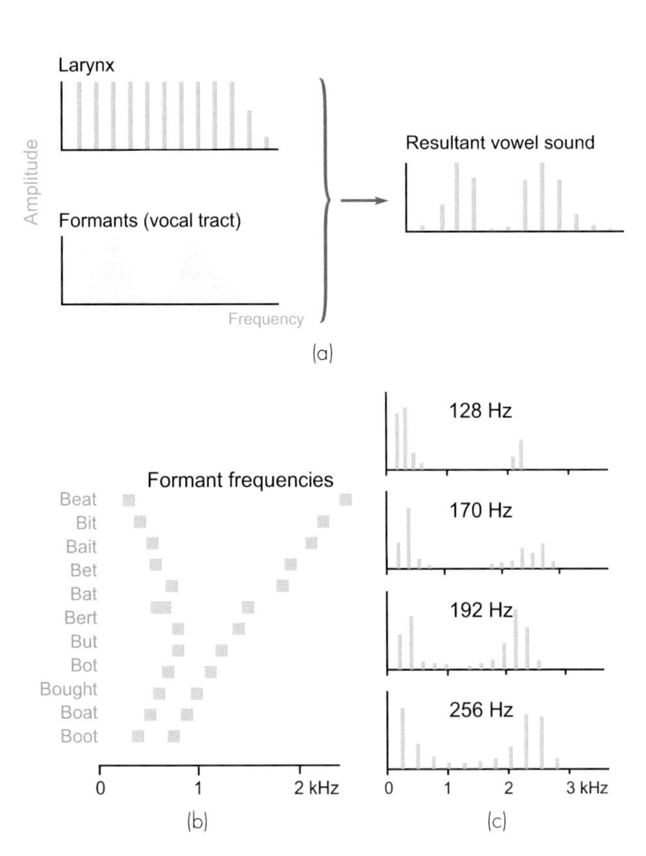

Figure 6.7 The human voice. (a) The spectrum of vibrations of the larynx is modified by the resonances of the vocal tract (shaded) to produce the spectrum of the sound finally emitted (right). (b) The approximate positions of the two principal formants for a number of different English vowel sounds. (c) The vowel sound 'EAT' sung at the various frequencies indicated. Although the spacing of the spectral lines increases as the frequency rises, the overall shape, defining the vowel quality, remains largely unchanged. (After Wood, 1930.)

frequencies is information about the actual shape of the waveform available to the brain (**Table 6.1**). Uniquely among all the senses, most of this analysis has already been done before the stimulus has reached the receptors.

The structure of the ear

External and middle ear

The visible external ear, or *pinna,* has little effect on incoming sound except for colouring it by superimposing little idiosyncratic resonances on it in the high-frequency region that are dependent on the direction from which the sound is coming and can, as will be described later, provide quite accurate information about the position of a sound source. The auditory canal, the *external meatus,* similarly does little in humans except to add its own rather broad resonance peak around 3000 Hz. In other animals with less vestigial pinnae, they may have a more significant part to play in gathering sound, and if movable can provide a great deal more information about where a sound is coming from.

At the end of the meatus, the sound impinges on the eardrum (*tympanic membrane*) that separates the outer ear from the middle ear. On the inner side, the tympanic

Table 6.1 Information carried by sound waves

Physical quantity	Sensory correlate	Coding
Amplitude	Loudness (interaural differences contribute to localisation: higher frequencies)	Total neural activity (including recruitment)
Frequency	Pitch (lower frequencies)	Predominantly temporal periodicity
Spectrum	Timbre (higher frequencies); also contributes to monaural localisation	Predominantly spatial pattern
Phase	Interaural differences contribute to localisation: (lower frequencies)	Temporal

membrane is attached to the *malleus* ('hammer'), the first of a set of three tiny bones, the *ossicles,* whose function is to transform vibrations of the eardrum into vibrations of the fluids that fill the inner ear (**Fig. 6.8a**).

The malleus is joined to the *incus* ('anvil') which in turn bears on the *stapes* ('stirrup'), whose footplate rests on the *oval window,* a membrane separating the middle and inner ears (**Fig. 6.8b**). This chain of ossicles acts as a kind of lever system, converting the movements of the eardrum, which are of comparatively large amplitude but small force, into the smaller but more powerful movements of the oval window; this increase in the pressure of the vibrations is further enhanced by the fact that the tympanic membrane is much larger in area than the oval window. Thus the amplitude is reduced by a factor of around 18, and force is increased by the same amount.

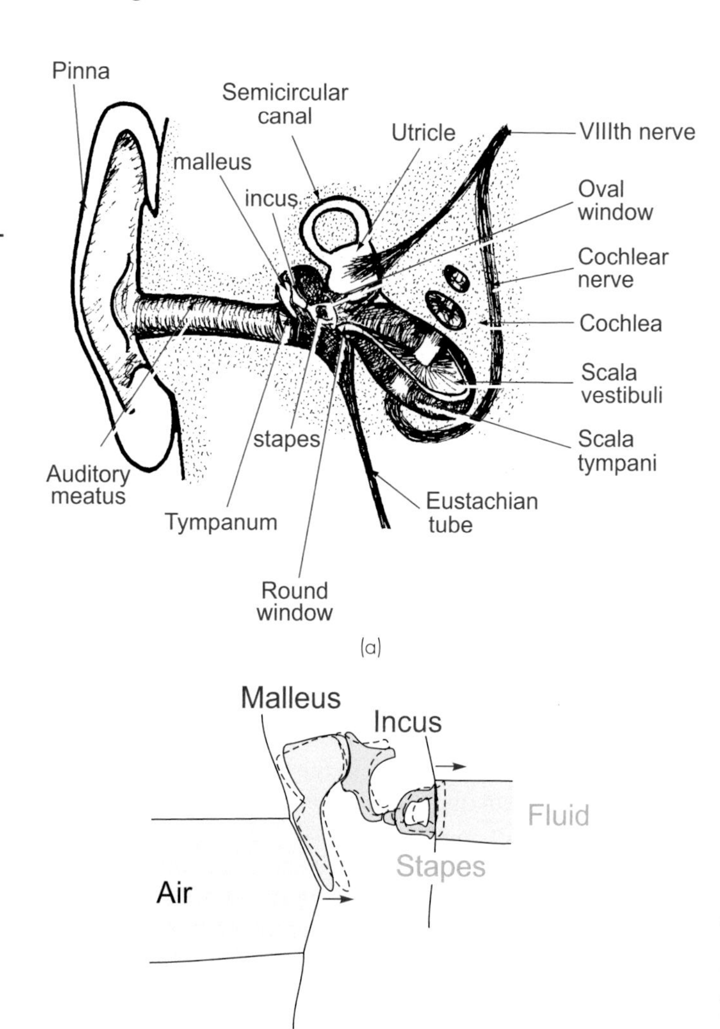

Figure 6.8 (a) Diagrammatic section through the ear. (b) How the ossicles convert low-pressure waves in the air into high-pressure, small-displacement waves in the cochlear perilymph (impedance matching).

The reason why this transformation is necessary – it is *not* amplification, for like all passive systems it cannot increase the *energy* of the waves that are transmitted – is because the fluid of the inner ear is very much denser than air. If a sound wave in air strikes a dense medium like water, the pressure changes of the air are too small to make more than a slight impression on the fluid, and most of the sound is reflected back. To ensure the most efficient transfer of energy from air to fluid, we need some way of increasing the pressure changes in the sound wave to match the characteristics of the new medium, and this *impedance matching* appears to be the primary function of the middle ear; without it, only some 0.1 per cent of the sound energy reaching the eardrum would reach the inner ear.

A second function of the middle ear is that it is capable of acting as a kind of censor: it contains muscles – the *tensor tympani (innervated by cranial nerve V)* and *stapedius (innervated by cranial nerve VII)* – that effectively disable the transmission system when they contract, protecting the inner ear from damagingly powerful sounds. Weakness of these muscles, such as with the CN VII dysfunction seen in Bell's palsy syndrome, can lead to hyperacusis – a painful oversensitivity to sounds.[4] In addition to all this, the middle ear also further shapes the audiometric spectrum, mostly by reducing low-frequency sensitivity.

Inner ear

The inner ear is a component of the labyrinth, described in Chapter 5; the part of it that is concerned with sensing sounds is the *cochlea,* in effect an elongated sac of endolymph some 35 mm long, shaped as if it had been pushed sideways into a corresponding tube of perilymph, rather like the sausage in a hot dog (**Fig. 6.9**). The upper half of the perilymph is called the *scala vestibuli*, the lower is the *scala tympani*, and the endolymphatic sausage is the *scala*

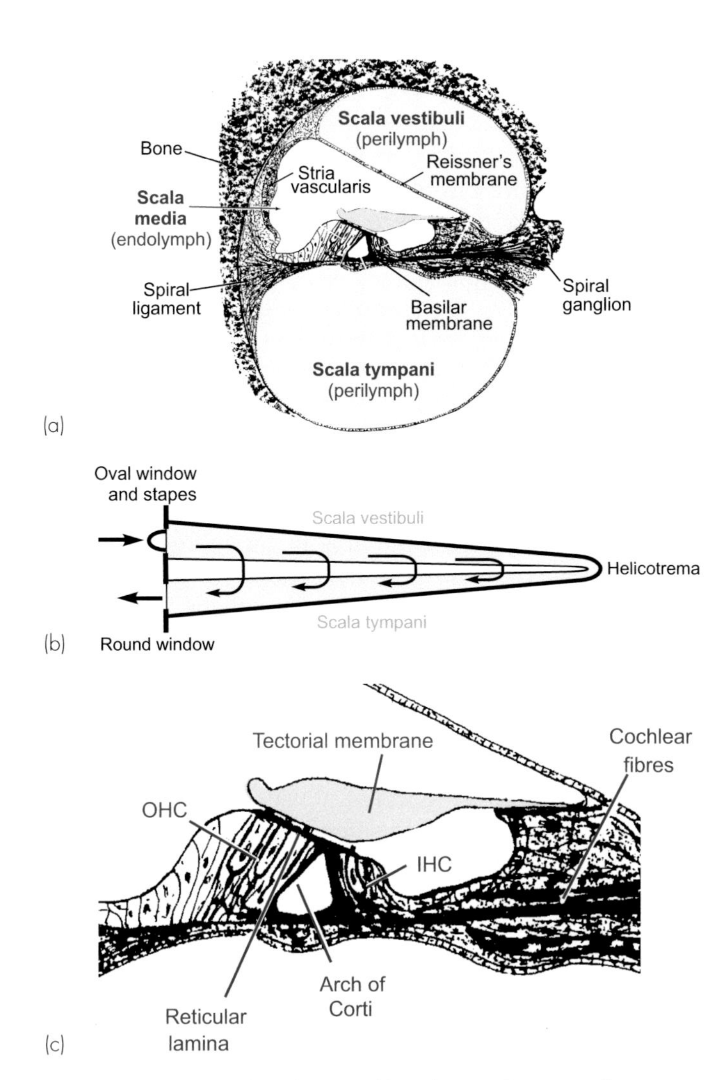

Figure 6.9 (a) Section through the cochlea, showing the organ of Corti – shown in (c) in more detail. (b) A stylised representation of the relationship of the scala vestibuli, scala media and scala tympani, and the path of sound between them. OHC, outer hair cell; IHC, inner hair cell.

media. At the far end of this structure, the two perilymphatic regions join up through an opening called the helicotrema; and finally the whole thing is rolled up into a conical spiral, giving it the shape of a snail shell. The scala vestibuli and scala media are separated by only a very thin membrane, *Reissner's membrane,* while the boundary between scala media and scala tympani is much more complicated and contains several layers of cells, including the receptors themselves, resting on the *basilar membrane.*

The oval window faces onto the scala vestibuli, while a similar structure, the *round window,* separates the scala tympani from the air of the middle ear. This air, incidentally, is in communication with the outside atmosphere through the Eustachian tube which joins it to the pharynx; this tube is normally closed, but opens briefly during swallowing and yawning, causing a characteristic modification of one's hearing: one may get relief in this way from the ear-drum pain sometimes experienced in air travel as a result of pressure differences between the atmosphere and the middle ear. Occasionally, such pressure gradients across the tympanic membrane (or very rarely the oval window), if significant or unrelieved (for example if the Eustachian tube is blocked during hay fever or a cold), can cause the membrane to rupture.

Now endolymph and perilymph are virtually incompressible fluids, so any movements of the oval window must result in corresponding movements of the round window. In other words, sound energy has no option but to pass from the scala vestibuli to the scala tympani, either by crossing through the endolymph or by travelling right to the end of the cochlea and traversing the helicotrema (**Fig. 6.9b**). In the former case, it will cause the basilar membrane, with all its elaborate superstructure, to vibrate as well. Now the cochlear receptors – the hair cells – are peculiarly well adapted to respond to exceedingly small up-and-down movements of the basilar membrane. Their cell bodies are attached at the bottom to the membrane itself, and at the top to a rigid plate called the *reticular lamina,* which is in turn attached to a strong and inflexible support called the *arch of Corti,* that also rests on the basilar membrane. Consequently, any movement of the basilar membrane results in an exactly similar displacement of the hair cells. These hair cells are very like the vestibular hair cells described in the previous chapter; they possess a number of stereocilia, arranged this time in a characteristic V- or W-formation and graded in size, but the auditory hair cells of adults do not appear to have kinocilia. The cilia project through holes in the reticular lamina, and in the case of the *outer hair cells* (OHCs) the tips of the cilia are lightly embedded in the lower surface of the *tectorial membrane,* a flap of extracellular material analogous to the cupula or otolith, that extends from the inner boundary of the scala media and lies along the top of the inner and outer hair cells. The cilia of the *inner hair cells* (IHCs) do not appear to make contact with the tectorial membrane, but lie just short of it.

A consequence of the geometry of this arrangement is that any up-and-down movement of the basilar membrane will result in horizontal sliding, or shear, between the reticular membrane and the tectorial membrane. This in turn will bend the cilia of the hair cells through an angle which will be enormously greater than the original deflection of the basilar membrane (**Fig. 6.10**). There is then a further stage of mechanical amplification: as the cilia bend, protein filaments running from the tip of one cilium to the tip of its shorter neighbour appear to be directly linked to the opening of membrane channels (**Fig. 6.11**). These tip

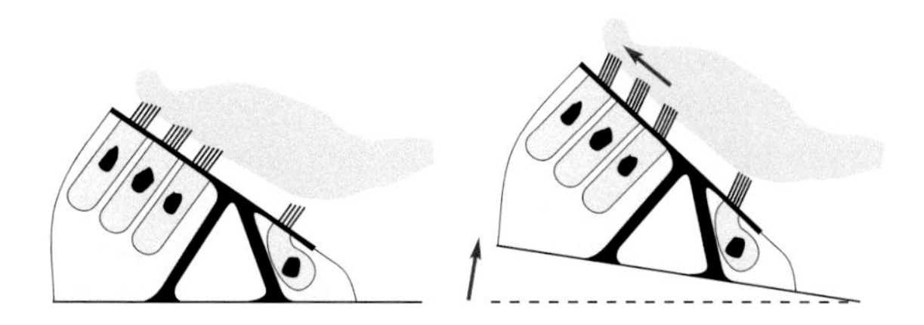

Figure 6.10 Shear amplification. Vertical deflection of the basilar membrane causes shear between the reticular lamina and the tectorial membrane, thus bending the cilia of the hair cells. (For clarity, the movement of the basilar membrane is, of course, greatly exaggerated.)

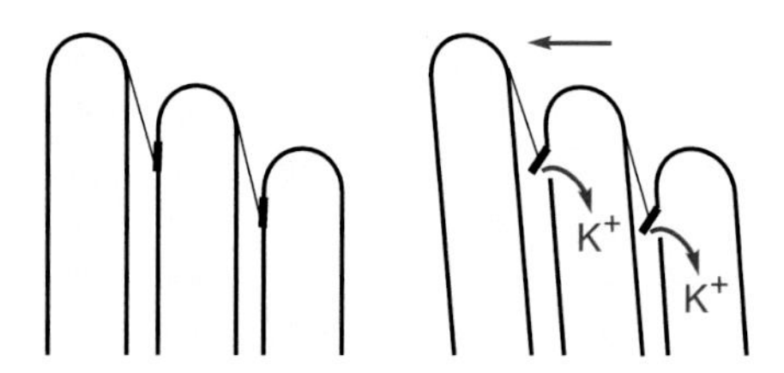

Figure 6.11 Mechanoelectrical transduction at hair cells. When the stereocilia are displaced towards the tallest, cation-selective hair cell, mechanoelectrical transduction (hcMET) channels open, allowing K^+ to depolarise the cell. The ensuing generator current leads to opening of voltage-gated calcium channels and release of glutamate to stimulate the afferent auditory nerve.

links mean a small angle of ciliary bending will generate a disproportionate mechanical effect. The resulting permeability change gives rise to generator currents that eventually lead to depolarisation of the terminal, calcium entry, release of transmitter (glutamate) and stimulation of the auditory fibres. This process is remarkably sensitive: as little as 0.3 nm of movement of the stereocilia is required at the threshold of hearing. It is also incredibly fast – displacement of the stereocilia can yield a generator potential in 10 µs – this is necessary for accurate localisation, as will be discussed below.

But although it is fairly clear how the cochlea may act as a transducer of sounds into nervous energy, nothing has yet been said about whether it also carries out any analysis of the sound at the same time, in the way that for instance the cones of the retina begin the analysis of colour by responding preferentially to light of different wavelengths. In fact, it turns out that different regions along the length of the cochlea are especially responsive to different sound frequencies, providing a rough kind of Fourier analysis of the type described earlier in this chapter. Before going on to discuss the electrophysiological responses of auditory nerve fibres, it is useful first to consider the mechanical properties of the cochlea that enable it to do this.

Cochlear function

Fourier analysis by the cochlea

The physicist and physiologist Hermann von Helmholtz (1821–1894) was the first to appreciate the general way in which sound quality was related to its frequency spectrum. Observing that the basilar membrane gets wider as one approaches the helicotrema, he suggested that it might be the cochlea itself that carried out this Fourier analysis. His notion was that individual transverse fibres of the basilar membrane might act rather like the strings of a piano, tuned to different frequencies and resonating in sympathy whenever their own particular frequency was present in a sound. (If you open the lid of a piano and sing loudly into it with the sustaining pedal depressed so that the strings are undamped, you will hear it sing back to you as particular strings are set into sympathetic vibration: if there were some device that signalled which strings were vibrating and which weren't, you would have a kind of Fourier analyser.)

However, direct measurements of the mechanical properties of the basilar membrane show that it cannot in fact behave in this simple mechanical way. There is too much longitudinal coupling of the membrane, much as if neighbouring strings in the piano were glued together; and in any case it is easy to show, simply by cutting it and observing that it does not twang back, that there is hardly any tension in the basilar membrane at all, and certainly not enough to make it resonate in the way Helmholtz suggested.

Nevertheless, it is clear from a number of pieces of evidence – for instance, that lesions of the basal end of the cochlea are associated with the specific loss of high-frequency auditory sensitivity – that different regions of the cochlea really are sensitive to different frequencies, in a systematic way. The foundations for our knowledge of the mechanism by which this comes about were laid when Georg von Békésy, some 50 years before, applied his

Clinical box 6.1 Weber and Rinne tests

The first task of the clinician in assessing a patient complaining of hearing difficulty is to distinguish between *conductive* or *sensorineural* hearing loss. The former is the term given to lesions preventing transmission of external sound to the inner ear (such as middle ear infection, ossicular chain fixation, ruptured tympanum, tumour or foreign body in the auditory canal), the latter refers to a lesion of the inner ear, cochlea or auditory nerve (such as chronic noise–induced hearing loss, or a cerebellopontine angle tumour pressing on the VIIIth cranial nerve).

Weber test: a (usually 512 Hz) tuning fork is held to the forehead and the subject asked to identify in which ear the sound is loudest. If it is the same in both ears, then there is either normal hearing or symmetrical hearing loss. If the sound lateralises to one side, then there is either a sensorineural deficit on the contralateral side or a conductive deficit on the ipsilateral side (as sound reaching the ear on the affected side is not attenuated by environmental noise).

Rinne test: The tuning fork is alternately placed on the mastoid or half an inch from the external auditory canal until it is no longer heard in one of these positions. This compares bone and air conduction; the latter should be louder in normal subjects (confusingly referred to as being Rinne positive). A Rinne negative test is when the tuning fork is heard longer when placed on the mastoid than when placed in front of the ear canal, and points to a conductive deafness as bone conduction bypasses the impaired ossicular chain.

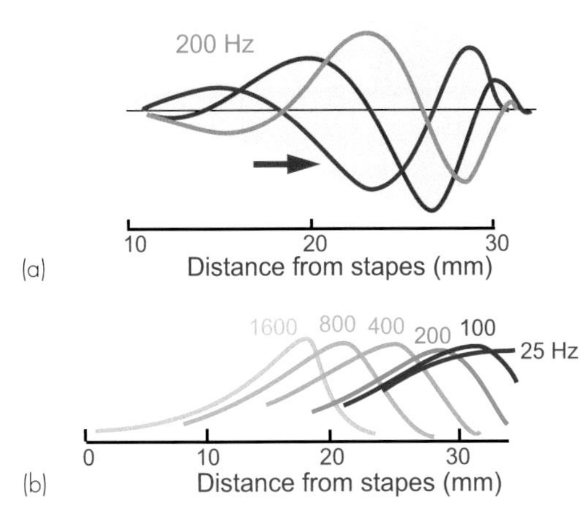

Figure 6.12 (a) Three consecutive 'snapshots' of the displacement of the basilar membrane in response to a sine wave (vastly exaggerated in amplitude, as usual). The whole waves move from stapes to helicotrema, with an envelope (shaded) that depends on the frequency. (b) Peaks of the envelopes associated with waves of the frequencies indicated. (After von Békésy, 1960; permission pending.) Had the displacements been measured using modern techniques, the peaks would be seen to be much sharper fibres.

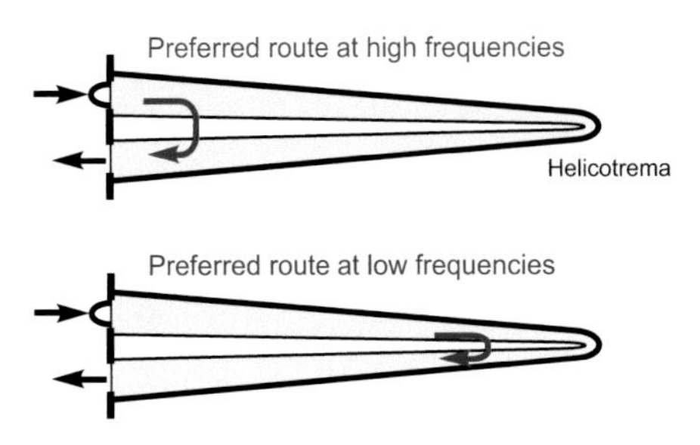

Figure 6.13 The basilar membrane as a Fourier analyser. The peak position varies with distance from the helicotrema.

superlative experimental skill to an investigation of the way the basilar membrane actually responded during stimulation by sounds of different frequencies.[5] He found that at any particular frequency, as one explored from the base towards the apex, the amplitude of the vibration of the membrane increased relatively slowly up to a maximum and then fell off again more sharply. The position of this maximum along the cochlea was dependent on the frequency, and the lower the frequency, the nearer it was to the helicotrema: by 50 Hz or so the maximum had more or less reached the end of the cochlea. There were also phase differences at different points along the membrane: the greater the distance from the basal end, the greater the phase lag between the movement of the membrane and that of the oval window.

Consequently, when one looked at the basilar membrane as a whole, the appearance was of a *travelling wave* progressing towards the apex, growing larger as it went until the point of maximum response was reached, and then abruptly decaying to zero (**Fig. 6.12**). More recent work using preparations in a more physiological condition than von Békésy's cadavers has shown that the peak of amplitude is actually much sharper than this, for reasons that will become apparent. But this remains a good description of the passive contribution of the basilar membrane, considered in isolation.

The mechanical properties that give rise to this behaviour are rather complex, and an elementary account must necessarily be over-simplified. In essence what happens is something like this: the sound waves, entering the scala vestibuli at the oval window, can only get out again through the round window, but there is a choice of routes by which they can get there. They might, for example, cross straight through the basilar membrane at the basal end: the membrane is stiffer here, but on the other hand this would be a short route involving less movement of perilymph. Alternatively, the sound waves might prefer to travel further along the cochlear duct before crossing into the scala tympani, involving a longer mass of fluid to have to move, but an easier crossing because of the decrease in stiffness of the basilar membrane at increasing distances from the base.

Now the relative hindrance offered by the stiffness of the membrane and the inertia of the perilymph depends very much on what the frequency of the sound is; the energy required to move a mass backwards and forwards increases enormously with increasing frequency, so that at high frequencies the path of least resistance is for the sound to cross from scala vestibuli to scala tympani near the base. At very low frequencies, the opposite is true: it is not much trouble to move the entire mass of perilymph backwards and forwards, and by doing so the sound can make the easier crossing at the apical end, where the stiffness is least (**Fig. 6.13**). In other words, the preferred route will represent a compromise between the relative disadvantages of membrane stiffness and perilymph inertia, and as a result the peak of the wave moves further and further from the stapes as the frequency falls (again, under more natural conditions than in von Békésy's experiment, shown above, the peaks would be sharper). Put differently, the structure as a whole acts as an *auditory prism*, sorting out vibrations of different frequencies into different positions along the membrane: behaving, in fact, like a Fourier analyser. One can find out how well it carries out this task by looking at the responses of the auditory fibres themselves.

Responses from auditory fibres

The synaptic connections between primary auditory fibres and the hair cells are broadly similar to those found in the vestibular system; corresponding to the distinction there between type I and type II cells, there are clear differences in innervation between the inner and outer hair cells of the cochlea; both probably release glutamate. Each IHC has afferent connections from some 20 or so radial fibres (**Fig. 6.14**), each of which appears to terminate on a single hair cell. By contrast, the spiral afferents that innervate the outer cells run along the cochlea for a millimetre or so, and send afferent terminals to large numbers of hair cells. Consequently, although OHCs greatly outnumber IHCs, about 5 per cent of the auditory nerve fibres come from the outer hair cells and 95 per cent from inner. Thus there

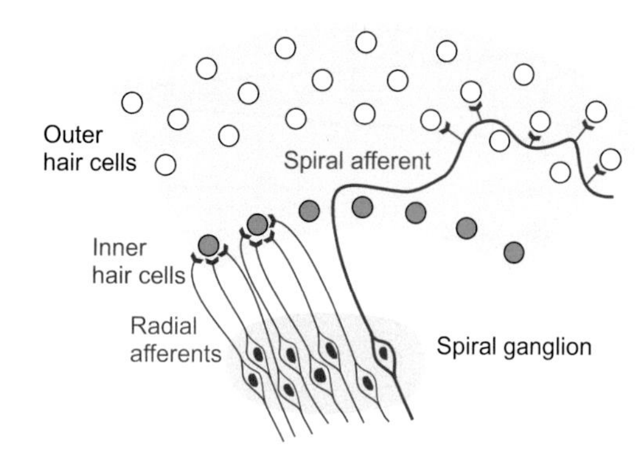

Figure 6.14 Afferent innervation of outer hair cells and inner hair cells, showing convergence in the former case and divergence in the latter. (After Spoendlin, 1968, Copyright © 1968 Ciba Foundation; with permission from John Wiley and Sons.)

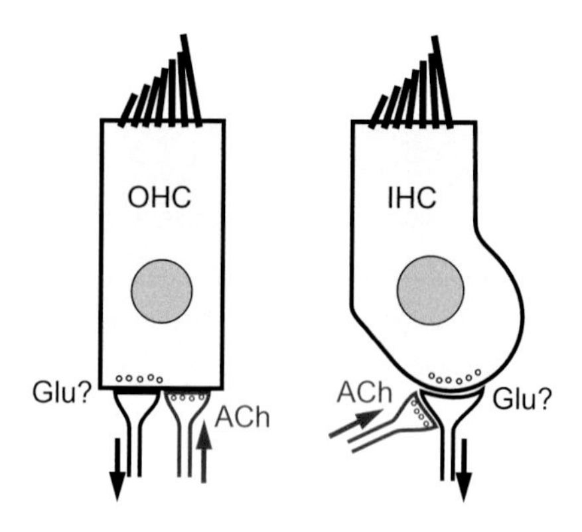

Figure 6.15 Differences in the innervation of individual outer and inner hair cells. Outer hair cells (OHCs) receive afferent and direct efferent innervation; inner hair cells (IHCs) receive direct afferents only, but the terminals themselves are pre-synaptically influenced by the efferent innervation.

is a great deal of convergence from outer hair cells on to the afferent fibres, but overall a divergence from the IHCs: as we shall see, there is a rough analogy here with the rods and cones of the retina. The OHCs have (like rods) a low threshold for stimulation – partly because their cilia actually stick into the tectorial membrane – and are grouped together in large receptive fields; whereas the IHCs, like cones, have reduced sensitivity but a more discrete connection with the brain, suggesting that here too the inner hair cells may have better 'acuity', in this case to small differences of frequency; there are some 400 IHCs per octave, or more than 30 per semitone[6]; but the OHCs might be better at registering differences in amplitude. However, we shall see later that the main function of the OHCs may not be transduction at all, but to do with modifying the response of the basilar membrane. The hair cells also receive an *efferent* innervation, originating from a nucleus in the brainstem called the superior olive. These fibres, which are cholinergic and divergent, terminate presynaptically on afferents to inner hair cells but directly on OHCs (**Fig. 6.15**), suggesting a more direct control of their function (see p. 120).

There are two different sorts of electrical response that can be measured in the cochlea; with microelectrodes one can record action potentials from the auditory nerve fibres, and with larger electrodes in various areas within and around the cochlea one may in addition record several kinds of slower potential, some being essentially static and some related to stimulation by sound. The hair cells have *resting potentials* that are some 45 mV negative to the perilymph in the scala tympani, while an electrode in the scala media records a standing potential of some 80–90 mV positive with respect to perilymph. This endocochlear potential appears to come about through an electrogenic Na^+/K^+ pump in the stria vascularis[7] and has the desirable consequence that the voltage difference across the top end of the hair cells is half as big again as what is usually found across neural membranes, so that any given conductance change will give much more than the usual generator current, thus presumably increasing sensitivity (**Fig. 6.16**). Because the equilibrium potential for potassium is around zero, they depolarise when their (non-specific) channels open, mostly due to an inward potassium current.

One may also record *cochlear microphonic* potentials with large electrodes almost anywhere in the vicinity of the cochlea; these are thought simply to be the summed generator potentials of large numbers of hair cells, and appear to be more or less proportional to the local displacement of the basilar membrane. Thus they follow the shape of the sound wave itself quite accurately, though with different degrees of frequency-filtering at different distances from the oval window. If amplified and used to drive a loudspeaker, the result is a quite faithful reproduction of the sound entering the animal's ear.

The auditory nerve fibres show similar properties to primary vestibular fibres, most having a spontaneous resting discharge, whose frequency is increased when the basilar membrane moves towards the scala vestibuli and decreased when it moves in the opposite direction. Outer hair cells seem to respond to the actual deflection of the basilar membrane at any moment, whereas IHCs seem to

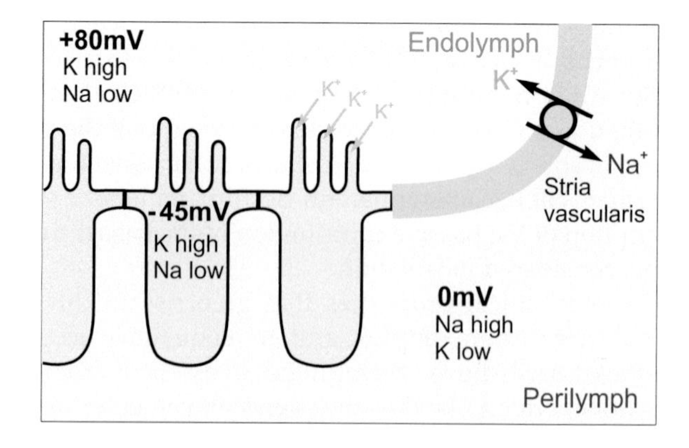

Figure 6.16 Electrical and ionic aspects of hair cells. Ionic pumps in the stria vascularis make the endolymph rich in potassium, low in sodium and about 80 mV positive to the perilymph. The hair cells also have high potassium and low sodium, with a resting potential around −45 mV. Consequently, the potential difference across the stereocilia is unusually large (some 125 mV), but the equilibrium potential for potassium is around zero. They therefore depolarise when their (non-specific) channels open, mostly due to inward potassium current.

respond to its velocity, probably because their cilia are not directly linked to the membrane but only viscously coupled, providing a mechanism of filtering adaptation. However, we know very much less about the contribution of outer hair cells to the afferent signals in the auditory nerve than we do about that from the inner, since their afferent fibres are very small and vastly outnumbered by the others. Different fibres have response curves lying on different positions along the stimulus axis, and the effect of stimulation of the efferent fibres seems to be to reduce their sensitivity. This effect is rather small under experimental conditions, being some 15 dB, and it is difficult to believe that this is all they are for. We shall see later that there is good reason to think they play a much more fundamental part in receptor tuning.

The second filter

When tested with sine waves of different frequencies, individual auditory fibres show a marked frequency selectivity (**Fig. 6.17**). Each has a 'best' frequency for which the threshold is lowest, and as the frequency is shifted from that optimum, the threshold rises. In general the curves are asymmetrical, rising more steeply on the high-frequency side than the low-frequency one. This is exactly what would be expected from the form of the travelling wave envelope, which, as we saw, tends to rise gradually to its peak, then fall off more abruptly. As a result, if we start with the best frequency for a particular spot on the membrane, a small increase in frequency will make the amplitude at that point fall off more than a small decrease.

Broadly speaking, then, the shapes of these tuning curves are roughly what might be expected from von Békésy's measurements of the response of the membrane itself to pure tones. However, subsequent work has shown that the system is actually much more sharply tuned than was originally thought. When the preparation is in good condition, careful measurements show that there is an extra, very narrow, component at the foot of the tuning curve. It seems as though there must be some extra mechanism – a *second filter* – that makes the responses more selective than they would otherwise be. It appears to be an active, energy-requiring process rather than the kind of passive filtering provided by the basilar membrane's mechanics, for, under the influence of anoxia or cyanide, the tuning curves revert to something more like what von Békésy originally observed for the basilar membrane. What is the mechanism?

In some species, individual hair cells respond in a frequency-selective manner even to *electrical* stimulation at auditory frequencies (which of course bypasses the mechanical filter provided by the basilar membrane), suggesting that the second filter is an intrinsic property of the receptors themselves. This suggests that there might be some kind of resonant circuit, perhaps generated in the outer hair cells by mutual interactions between mechanical displacement and electrical depolarisation. It is not hard to imagine a process in which not only displacement of cilia causes a change in potential, but changes in potential, with consequent calcium entry, in turn generate mechanical forces on the cilia (in most ciliated cells the cilia are, after all, motile; **Fig. 6.18**). Such a mechanism would tend to resonate, emphasising a particular narrow range of frequencies that would depend on the delay round

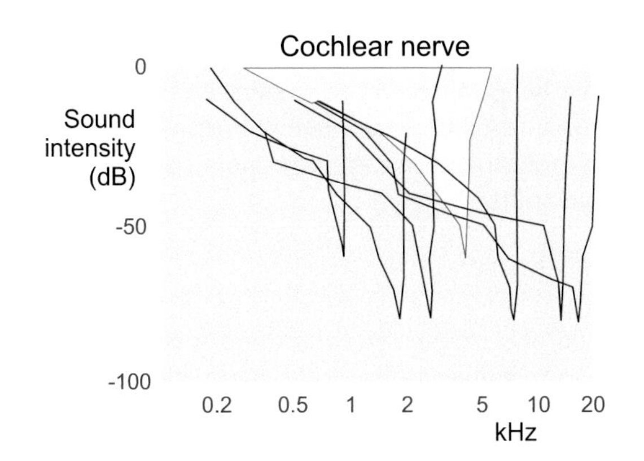

Figure 6.17 Threshold response curves for individual units in cochlear nerve and inferior colliculus, as a function of frequency. (After Katsuki, 1961; by permission of The MIT Press.)

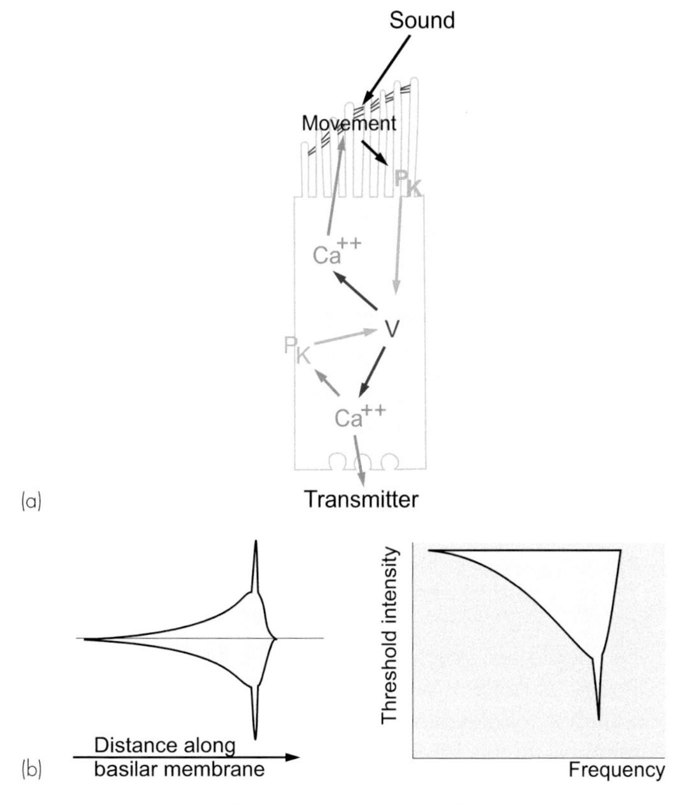

Figure 6.18 The second filter: sharpening of hair cell tuning by a mechanism within the outer hair cells. (a) Two ways in which such intrinsic tuning might arise. Changes in potential cause calcium entry, which in turn increases P_K: if the delay around this feedback loop were large enough, it could cause resonance. In addition, the calcium can cause the cilia to move, creating another feedback loop through the transducer mechanism itself. (b) The effect of the second filter on cochlear function. A typical basilar membrane envelope for one frequency is shown alongside a tuning curve for a cochlear afferent, both in highly schematic form. The second filter sharpens the response, which disappears, for example, after poisoning or with poor experimental techniques.

the loop. One would then expect the second filter to be observable not only in the electrical responses of the cells, but also (since their cilia are coupled to the basilar membrane) in the sharpness of the peak in the travelling wave. Observations of the pattern of movement of the basilar membrane, using less disruptive techniques than were available to von Békésy, appear to support this idea. If great care is taken to maintain the animal in reasonable condition, and to cause minimal interference with the membrane itself, the envelope of the travelling wave appears to be much more sharply tuned than was originally observed and corresponds more closely to the tuning curves of the hair cells themselves.

Many other observations point in the same direction. One of the distressing symptoms of the high-frequency hearing loss associated with progressive disorder of the cochlea is *tinnitus*, imaginary sounds taking the form of continuous high-pitched whistling (as in the well-known case of Ludwig van Beethoven) or hissing noises. Occasionally, there have been reports of 'objective' tinnitus in which the whistling that is complained of by the subject can actually be heard by another person listening at his ear. This too can be readily explained on the assumption that the spontaneous oscillation of the hair cells causes rhythmical movements of their cilia and thus movement of the basilar membrane. A system like the ear that is designed to transfer vibration with the minimum loss of energy from air to fluid is of course equally efficient at transferring vibration in the opposite direction. Similarly, with a sensitive microphone in the auditory meatus one may record cochlear echoes to very brief sound pulses, resulting from a short-lived ringing of the resonant mechanism: the phenomenon can be used to check auditory transduction in very young babies. In man-made systems, instability of this kind is a very common problem with highly resonant feedback systems: if the feedback is too great it can easily turn into spontaneous oscillation. It may well be that a function of the efferent fibres to OHCs in particular is to provide some general kind of central control of the feedback, incidentally detuning their selectivity.[8]

A further mechanism that helps to sharpen up the spatial patterns of neural activity in response to auditory stimulation is the existence, as in all sensory systems, of lateral inhibition. If one determines the tuning curve for a single auditory unit using a single tone, and then adds to this a second tone of different frequency, one finds that in the regions immediately neighbouring on the original tuning curve the response of the cell is actually reduced by the extra sound. In other words, each fibre has – in terms of frequency – a central excitatory area and an inhibitory fringe, which serves further to sharpen its selectivity. For various reasons, however, it is clear that this *two-tone suppression* is not, as elsewhere, due to inhibitory synaptic connections, but rather to an intrinsic non-linearity in the initial part of the transduction process that is not fully understood. But at subsequent levels in the auditory

Clinical box 6.2 Noise-induced hearing loss

Chronic hearing loss

Nearly 300 years ago it was noticed that chronic exposure to loud noise was associated with hearing loss, termed 'boilermaker's deafness', particularly affecting the detection of high-frequency tones. Although both sensory and supporting cells of the inner ear are believed to be injured by loud noise, the OHCs are the most vulnerable. They lie against the part of the basilar membrane with the greatest range of movement and, unlike the IHCs, their cilia actually stick into the tectorial membrane. Histologically, the cilia are fractured or dysmorphic and the cells become swollen, a sign that cell membrane injury has caused loss of ionic regulation, leaving the cell susceptible to osmotic injury. In addition, there is evidence that metabolic overload with excessive free radicals may compound the damage.

The stapedius and tensor tympani muscles are the effector limb of the acoustic reflex: sudden intense sound activates the Vth (trigeminal) and VIIth (facial) cranial nerves, which stimulate the muscles to contract and stiffen the middle ear structures, limiting the amplitude of displacement of the tympanic membrane and the oval window. Studies of patients with Bell's palsy (injury to the facial nerve, often as a result of viral infection) have demonstrated a temporary elevation of stimulatory threshold following exposure to moderate noise. Whilst unethical to perform in humans, studies in animals have suggested more permanent threshold shifts.

Acute hearing loss

Unfortunately, the initiation of the acoustic reflex requires the detection of a loud noise, and there is a finite delay in the generation of the reflex. For this reason, the protection is limited to sustained noise; loud intermittent noises, such as the fall of a hammer or crack of a gunshot, exert their full force upon an unshielded hair cell. Sometimes a sudden noise may be so loud, usually with pressure levels around 130 dB or more, that it causes acoustic trauma potentially accompanied by permanent hearing loss. Macroscopically, there may be evidence of ruptured tympanum or fractured ossicles.

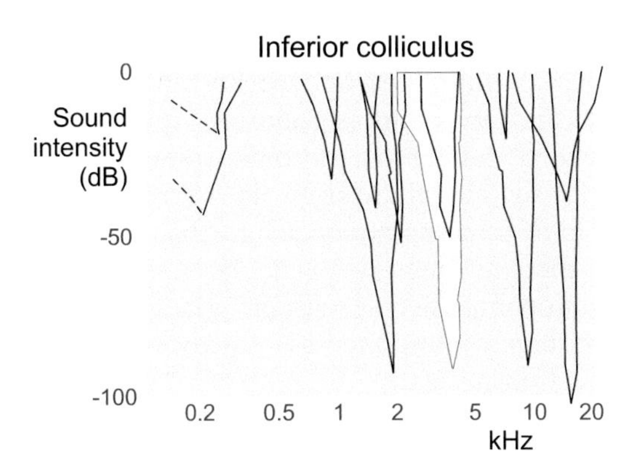

Figure 6.19 Threshold response curves for individual units in the inferior colliculus, as a function of frequency, showing sharpened responses due to lateral inhibition.

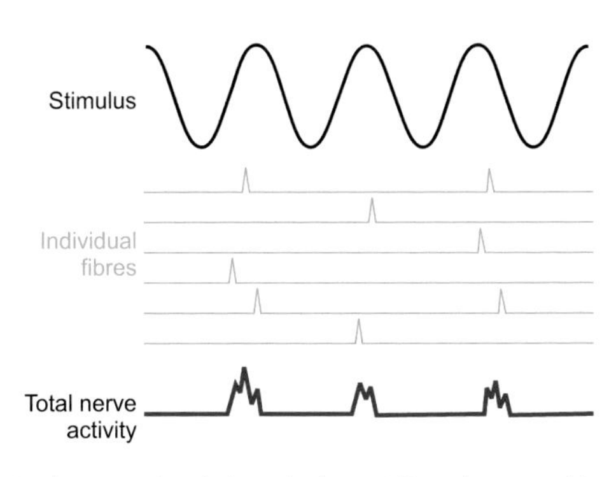

Figure 6.20 The principle of phase locking. Although no one fibre out of the whole ensemble fires in every cycle of the sound wave, nevertheless the modulation of the total activity reflects the frequency of the original stimulus. With a larger-amplitude wave, the probability of firing per cycle would increase.

pathway – for instance at the inferior colliculus – the tuning curves are further sharpened, by a mechanism that does appear to be conventional synaptic lateral inhibition (**Fig. 6.19**).

Temporal coding of low frequencies

The mechanisms of frequency analysis described so far provide a means for coding the spectrum of a sound into a spatial neural pattern, giving rise to the sense of timbre or tone quality. They operate essentially at medium and high frequencies, and indeed at frequencies above a kilohertz or so there is no other way that information about frequency *could* be transmitted to the brain except by peripheral analysis and recoding, since individual nerve fibres are incapable of firing more frequently than about a thousand times a second at the very best, and hence cannot reproduce the pattern of the sound waves reaching the ear. But this is not the case at low frequencies, and in any case we saw earlier that the frequency analysis produced by the basilar membrane begins to become ineffective at low frequencies because the peak of activity has nearly reached the helicotrema. Recordings from single auditory units show that as the frequency of a stimulating tone is decreased, there is an increasing tendency for firing to be *phase locked* to the stimulating frequency; even if the frequency is too high for any one fibre to be able to fire once in every cycle, it may do so every two cycles, or every three or more (**Fig. 6.20**). So although no single fibre will be firing at the frequency of the stimulus, the average activity over the whole set may nevertheless be modulated at this frequency.

In practice, phase locking is not quite as rigid as this: even at low frequencies where the fibres would be perfectly capable of following the imposed frequency, unless the stimulus intensity is very great what one observes is simply that there is an increased probability of firing during one part of the cycle rather than another. At all events, phase-locking provides a method of conveying auditory information to the brain without peripheral frequency analysis and has the advantage that it retains information about the phase of incoming sound, information which is thrown away at high frequencies, above some 5 kHz. (Though phase information contributes very little to sound quality, we shall see later that it is extremely important in localisation.)

This mechanism of phase-locking, working at low frequencies, is of course much better suited to transmitting information about the fundamental frequency of a sound than its harmonics, and a number of kinds of observations suggest very strongly that it is the frequency at which the activity in the auditory nerve repeats itself that generally determines the *pitch* of a sound. Frequencies higher than those that can be coded by phase-locking are not heard as pitches at all, which is why the piano keyboard stops where it does rather than at our high-frequency hearing limit: the top notes on a piano (above 3 kHz) sound more like clicks than tuned notes, just as the top notes on a violin have a hissing rather than a truly musical quality. Sounds that are modulated in amplitude at an auditory frequency generally appear to have the corresponding pitch, even though there is no energy at that frequency and no corresponding peak of neural activity along the cochlea. Similarly, if the fundamental frequency is removed from a complex sound, the pitch remains unchanged, because the remaining harmonics generate a temporal pattern whose repetition frequency is the same as before. Finally, human subjects who have been provided with auditory prostheses in the form of an implanted electrode that stimulates the cochlea or auditory nerve report that although they cannot perceive speech very well (a task that requires analysis of the shape of the spectrum, which their prosthesis cannot provide since it can generate only temporal patterns), they can still perceive pitch, though not very accurately. It is clear that much of the sense of pitch must be due essentially to central analysis by the brain rather than to something done by the cochlea, since in this case it has been bypassed altogether.

To summarise, it seems that the perceived *pitch* of a sound depends essentially on the periodicity of the afferent neural activity – i.e. on its *temporal* pattern – whereas the sensation of *timbre* or quality, which requires the

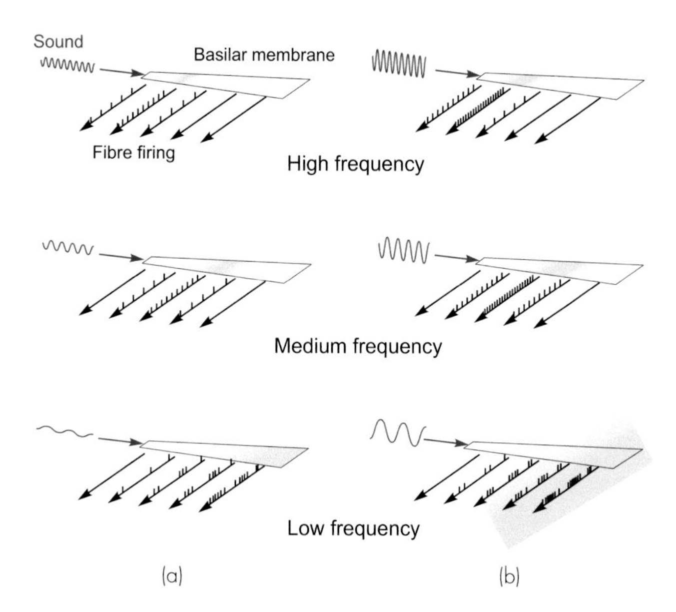

Figure 6.21 Summary of spatiotemporal coding, showing sounds of different frequencies, of small (a) or large amplitude (b), and a schematic representation of the resultant firing patterns of neurons originating from different points along the cochlea.

detailed perception of the high-frequency power spectrum of the sound, is coded by the relative activity of fibres from different parts of the cochlea, i.e. by their *spatial* pattem of activity. *Loudness* is presumably simply a matter of the total amount of auditory activity; as auditory intensity is increased, there is both an increase in the firing of any one fibre and an increase in the total number of fibres that are firing at all, through recruitment (Chapter 2, p. 39): in effect, the patterns on the basilar membrane become broader. These different aspects of auditory coding are summarised in **Figure 6.21**, which shows the spatial pattern of activity in the basilar membrane and the firing patterns of different fibres, at different frequencies, with high or low intensities. It may help to emphasise the way in which sounds are represented in the auditory nerve as an extremely complex spatial and temporal pattern of activity.

Spatial localisation of sound

Distance

There are two components to localisation – distance and direction – and these are analysed in very different ways by the auditory system. Since the energy of a sound wave decreases with the square of the distance it has travelled, one could, in principle, judge *distance* if one knew in advance the power of the sound source and the degree of absorption of the intervening structures; but in practice intensity can give only very approximate information. Rather more useful is the fact that not all frequencies suffer equal attenuation with distance: in an ordinary sort of environment, shorter wavelengths tend to be reflected or absorbed by physical objects, whereas long wavelengths simply ignore them. For this reason, the further one is from a sound source, the more of its high frequencies are lost: as a marching band approaches, it is the bass drum and then the tubas and euphoniums one hears first. Or again, when listening to a radio play one has a clear sense of how far the actors are from the microphone from the ratio of high to low frequencies in their speech sounds: close-to, the consonants – particularly sibilants like 'S' – are predominant, whereas with increasing distance it is the lower-frequency components – mostly vowels – that are heard most prominently.

Direction

Locating the *direction* of a sound source is a much more precise business, and is carried out by at least three separate mechanisms. Contrary to popular belief, one can localise sounds quite accurately using one ear alone. As was mentioned earlier, the peculiar pattern of bumps and whorls that decorate our pinnae add colouring to all the sounds one hears, a pattern of small peaks and troughs in one's frequency sensitivity curve that is dependent on the angle at which the sound waves impinge on the ear. In the course of growing up one presumably learns that particular kinds of coloration of familiar sounds are associated with particular directions, and in the adult this mechanism has been shown to provide localisation of sound accurate to a few degrees. Using both ears produces only a slight improvement, to perhaps 1–2°. The extra information provided by binaural listening is of two distinct kinds: differences in interaural *intensity* and in interaural *phase*.

Intensity differences come about because the head casts a 'sound shadow' that screens the ear to certain extent from sounds coming from the opposite side (**Fig. 6.22a**). For the head to cast a shadow of this kind, it needs to be at least of the order of magnitude of the wavelength of the sound itself; thus screening of this type can only cause significant effects at frequencies higher than some 2–3 kHz. You can explore this effect for yourself by using a transistor radio as a source of sounds of different frequencies, covering one ear and listening to the changes in the intensities of low- and high-frequency components as you move the radio around your head. Intensity differences alone, even at high frequencies, do not permit sounds to be localised very accurately unless one is also allowed to move one's head to find the direction for which the intensity is most nearly equal in the two ears.

Phase differences arise because a sound coming from one side takes slightly longer to reach one ear than the other (**Fig. 6.22b**). Since by using phase information alone a subject may detect movement of a sound source of only 1–2°, one can calculate that the brain must be sensitive to interaural time differences of the order of 10 μs, or about 1/100 part of the duration of an action potential! However, although phase differences can give accurate information about the direction of sounds, pure tones cannot be localised in this way if their frequencies are higher than some 1–2 kHz. The reason for this is that once the wavelength

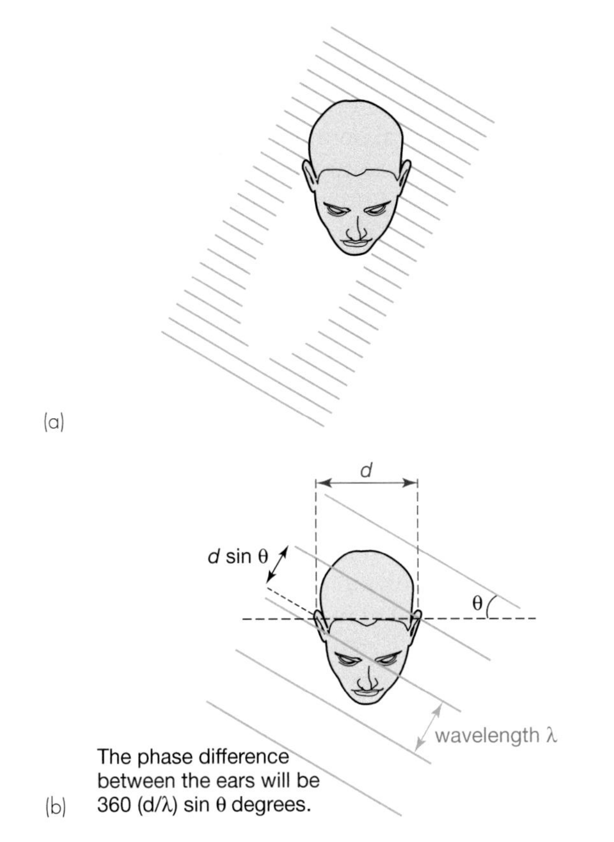

Figure 6.22 Two binaural methods of localising sounds. (a) Sounds of sufficiently high frequency cast a sound shadow on the far side of the head: the wavelength, λ, must be less than the order of magnitude of the head diameter, *d*. (b) A sound coming from a direction at a bearing θ is associated with a phase difference between the ears of 360 (*d*/λ) sin θ degrees, equivalent to a time difference of (*d*/*v*) sin θ milliseconds, where *v* is the velocity of sound in km/s and *d* is expressed in metres.

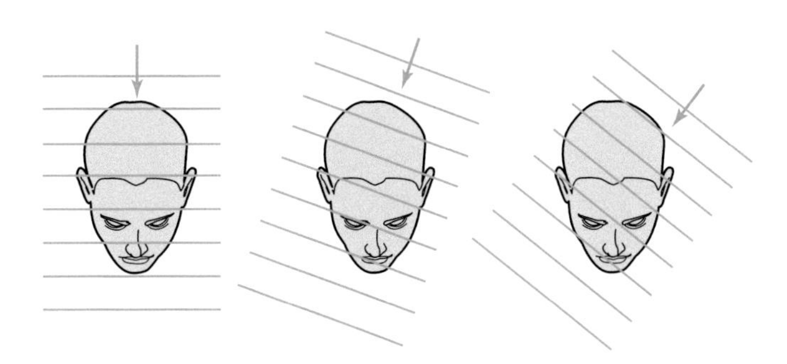

Figure 6.23 At high frequencies, a given phase difference (in this case, zero) could be due to more than one sound direction.

of the sound is less than the distance between the ears, ambiguities can occur, in the sense that a given phase relationship could be the result of more than one possible source direction (**Fig. 6.23**). For instance, a sound straight ahead will be exactly in phase at both ears; but the same will equally be true if the sound is coming at such an angle that the distance between the ears is exactly one wavelength. In any case, we have already seen that phase is simply not transmitted by auditory nerve fibres at high frequencies.

Thus the two fundamental binaural mechanisms of location are – rather conveniently – exactly complementary to one another (see **Table 6.2**). At low frequencies, only phase can be used, and at high frequencies, only intensity: the crossover point is a function of the size of the head. With such a system one might expect to be able to cancel out a time delay on one side by increasing the corresponding intensity; this kind of *time–intensity trade* can in fact be demonstrated quite easily in the laboratory by arranging for a subject to hear a click delayed in one ear, and asking him to adjust the relative loudness in each ear until they sound as if coming from straight ahead – a sort of titration.

Neither of these binaural mechanisms can do more than tell you the angle between the direction of a sound source and an imaginary line joining the two ears (**Fig. 6.24**); in particular, they cannot distinguish between a sound lying immediately behind the head, immediately in front, or somewhere overhead in the sagittal plane; this results in a 'cone of confusion', the locus of all points lying at a given angle to the axis between the ears. For a complex sound of known frequency composition, this extra information can be provided by the monaural mechanism of directionally selective coloration by the pinna, described earlier. When this is impossible, then moving the head can provide an extra 'fix' on the sound that will enable its exact three-dimensional direction to be established, as when a dog cocks his head on one side when trying to locate the source of a sound.[9]

Table 6.2 Contributions to auditory localisation

Distance	Spectral pattern: high versus low frequencies
Horizontal direction	
Monaural	Spectral pattern: high-frequency detail
Binaural	Phase difference (low frequencies) Intensity difference (high frequencies)
Vertical direction	
Monaural	Spectral pattern: high-frequency detail
Head movement	Resolves ambiguity

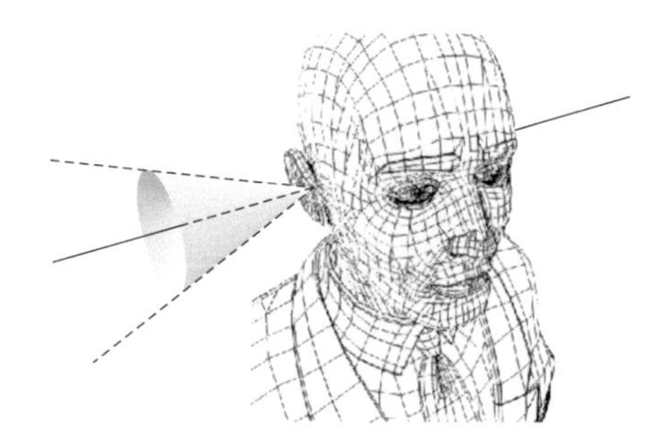

Figure 6.24 The cone of confusion. A sound producing a given phase delay, even at low frequencies, may lie anywhere on the cone, which is the locus of all points lying at a given angle from the axis formed by the two ears. (This is not strictly the case if the source is close to the head.)

Central pathways and responses

After leaving the cochlear ganglion, the primary auditory fibres synapse first in the *cochlear nuclear* complex, a group of three nuclei, each of which has a systematic tonotopic representation of the basilar membrane, so that neighbouring areas correspond to neighbouring frequencies. Functionally, the cells of the dorsal and anteroventral parts (DCN and AVCN) have very different properties and project to different areas, with the intermediate posteroventral nucleus (PVCN) showing a mixture of the two types. The AVCN cells behave very like auditory nerve fibres, showing relatively simple responses to particular frequency bands and an incompletely adapting response to tone bursts; at low frequencies they show phase-locked responses. In the dorsal region one finds cells with entirely novel and complex specialisations. Some show only a brief burst of activity at the start of a sustained tone, while others respond with a slow increase in activity or with repetitive bursts of spikes. **Figure 6.25** shows some of these categories: the shaded areas represent probability of firing at different times after presentation of a 40 ms tone burst. Many show tuning curves in which the main excitatory peak is flanked by prominent areas of inhibition that narrows the range of frequencies to which they respond.

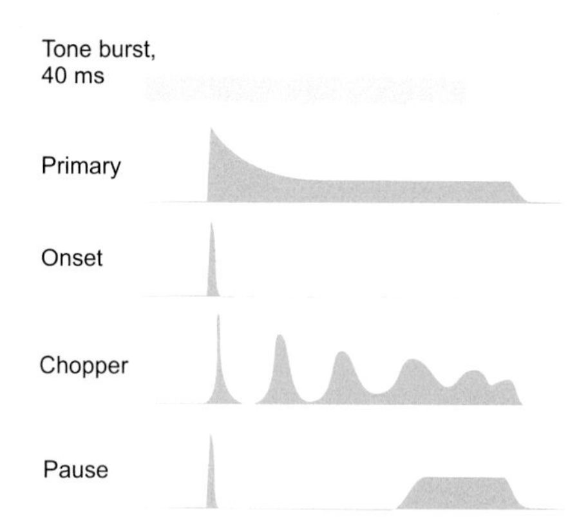

Figure 6.25 Responses from cochlear nucleus. Typical, simplified profiles of summed responses from some categories of neuron found in the cochlear nucleus are shown, in response to a 40 ms tone burst. The heights of the curves effectively represent the instantaneous probability of a spike occurring.

The central auditory pathways are complex (**Fig. 6.26**) and not fully understood. Cells in this dorsal region project straight up to the next highest level in the ascending pathway, the (contralateral) inferior colliculus, whereas those in the simpler, ventral, region first have to undergo an additional stage of processing in the *superior olive* (medial superior olive (MSO) and lateral superior olive (LSO)) of the brainstem. This represents the lowest level at which information from one ear meets information from the other, and seems to be concerned with auditory localisation rather than recognition. Cells in the lateral part of the superior olive are typically excited by the ipsilateral ear and inhibited by the contralateral one (through a relay in the nucleus of the *trapezoid body*, TB), and are concerned mostly with high frequencies: it seems very likely therefore that they form the neural basis for the use of interaural intensity differences in judging the direction of a sound. In the MSO the cells are predominantly low frequency, respond in the same way to both ears, and appear to be interested in *time differences:* many of the cells respond best when there is a particular time interval between the arrival of sound at each ear, so that sounds from different directions preferentially stimulate different neurons according to the interaural phase difference (**Fig. 6.27**).[10] When a sound arrives sooner at one ear than the other, the resultant neural response will have travelled further than

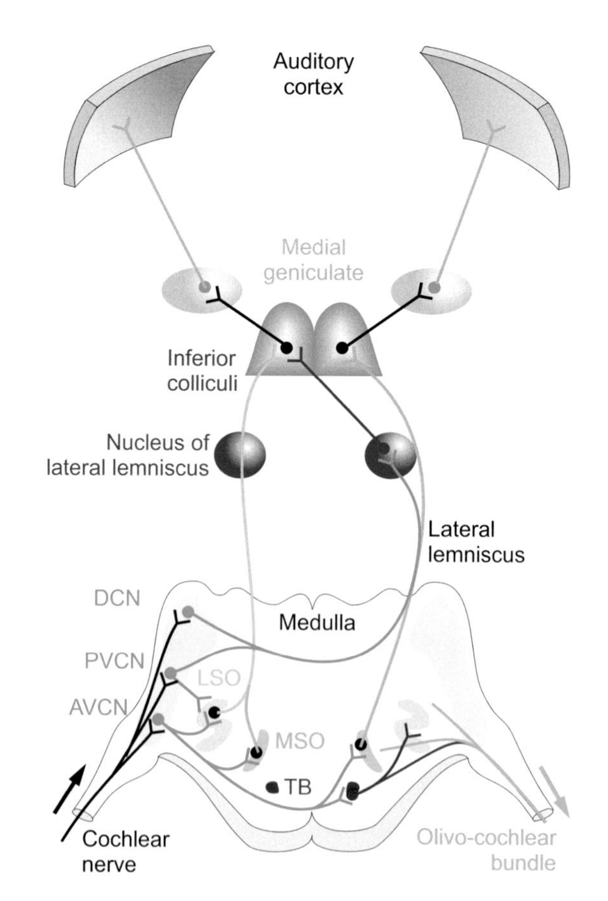

Figure 6.26 Schematic diagram of ascending auditory pathways. DCN, PVCN, AVCN: dorsal, posteroventral and anteroventral cochlear nuclei; LSO, MSO: lateral and medial superior olive; TB: trapezoid body.

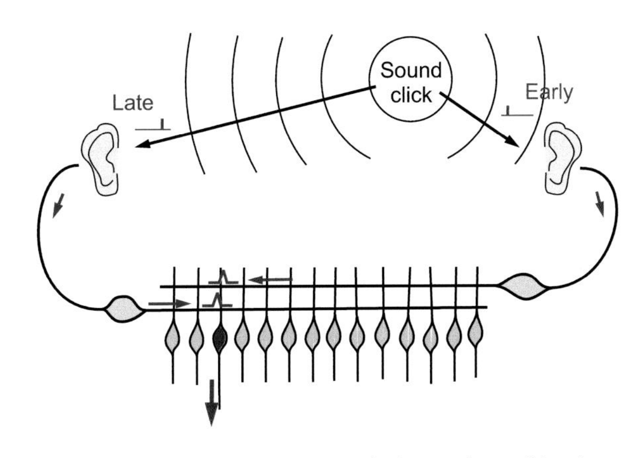

Figure 6.27 A possible neural mechanism for binaural sound localisation. Each of two neurons with axons pointing in opposite directions is driven by one ear. When a sound arrives sooner at one ear than the other, the resultant neural response will have travelled further than its opposite number when they meet. An array of cells responding only when excited simultaneously by both axons would then code spatially for different interaural time differences.

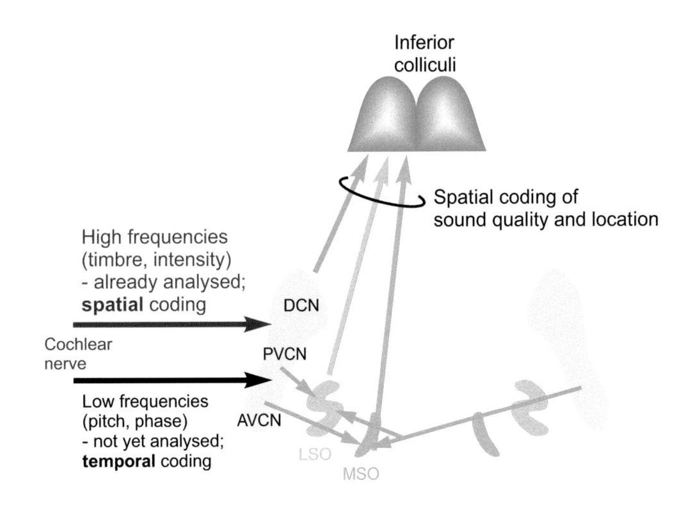

Figure 6.28 A functional interpretation of the pathways shown in **Figure 6.26**. (See **Fig. 6.26** for explanation of abbreviations.)

its opposite number when they meet. An array of cells responding only when excited simultaneously by both axons would then code spatially for different interaural time differences.

At the *inferior colliculus* the two pathways recombine, bringing together information about the kind of sound and about where it is. Cells here may show the same types of complexity in their response as can be seen in the dorsal cochlear nucleus, as well as coding for localisation; in some species a systematic topological mapping of the directional responses has been described, with some somatosensory responses as well. Crossed and uncrossed projections ascend to the next highest level of the auditory system, the medial geniculate nucleus, which in turn projects to auditory cortex. Cells in the ventral part of the medial geniculate show much the same properties as those in the inferior colliculus, and go to the primary *auditory cortex* (Al). But in the medial part, and in the secondary cortex (A2) to which it projects, some neurons seem to respond specifically to more complex sounds. A1 is essentially tonotopically arranged, and in addition there is evidence for 'columns' (actually stripes) running perpendicular to the frequency axis in which neurons respond alternately to the sum of the signals from both ears, and to their difference. We shall see in Chapter 7 that a rather similar arrangement is found in primary visual cortex, in relation to information coming from each eye. Tonotopicity is much less obvious in the adjoining areas of auditory cortex. Many units can be found that are not simply tuned to one particular frequency: some respond best to two different frequencies (the kind of response needed to recognise speech sounds), to changing frequencies, or react preferentially to such specialised, 'real' stimuli as clicks, whistles, hisses and voices. When in Chapter 7 we look at the way the visual cortex processes information from the eye, we shall see that there are cortical cells that respond very specifically to fragments of the retinal image such as lines and edges and thus provide detailed information from which one's recognition of visual objects could easily be derived. It would be nice to demonstrate a similar process unequivocally in the case of the auditory cortex, but it has been a less fashionable area of study and we simply do not have enough data, from enough species, to be able to make the same kinds of generalisations, except that the cortex adds to the analysis that the lower auditory system has already performed, by responding to the temporal pattern of sounds whose frequencies are changing, an

Clinical box 6.3 Cochlear implants

Cochlear implants are distinct from hearing aids in that they do not simply amplify sound. Instead, an array of electrodes are surgically implanted in the scala tympani, bypassing the hair cells by electrically stimulating the afferents of the spiral ganglion. Most implants consist of an external microphone which communicates wirelessly with a receiver under the skin; this processes sounds into discrete frequency bands in order to stimulate the residual neural elements in a tonotopic manner. In this way, they can restore a degree of sound perception, even to those with profound deafness, providing the central auditory processing stream is intact. However, the scala tympani arrays typically do not extend right to the apical turn of the cochlea – where lower frequencies are represented – and there is a fundamental challenge of replicating the highly specific functioning of the hair cells with an electrode placed in a conducting fluid, relatively remote from the target afferents. Consequently, most implant users still report poor pitch recognition which can impact upon their understanding of speech and enjoyment of music.

important component in recognising speech, for example. It is striking that ablation of auditory cortex in both cats and rats leads to very little impairment of frequency discrimination per se, but does cause difficulties in discriminating temporal patterns. On the other hand, destruction of the auditory cortex in rats does not seem to impair their localisation of sounds.

While the multiple auditory pathways may seem complex, some aspects are not difficult to explain by thinking about what kinds of information are analysed in the *cochlea* and what has to be analysed by the *brain* (**Fig. 6.28**). In the first place, timbre has already been sorted out by the cochlea and is essentially represented as a spatial code in the auditory nerve. Therefore, relatively little processing remains to be done, and this information passes almost directly from the cochlear nuclei up to the inferior colliculus, coded spatially. Pitch, on the other hand, derives from the repetition frequency of the temporal pattern of auditory nerve activity. While this is *transduced* quite faithfully at low frequencies by the cochlea, it is not *analysed*. Some additional mechanism is required to convert periodicity into the kind of spatial code that more central areas understand: we have already seen how this could be done by an extension of the processing carried out in the MSO. At all events, it is clear that this kind of information must undergo brainstem modification before being allowed to proceed to the inferior colliculus. In the same way, information from the two ears needs to be compared in the brainstem to compute location – and again encode it as a spatial code – before it can be allowed access to higher levels. This dichotomy in the auditory system – between low and high frequencies, pitch and timbre, central and peripheral analysis – is in many ways the key to making sense of the system's complexities.

Notes

Hearing Excellent general accounts of the physiology and psychophysics of hearing include: Gelfand, S. A. (2017). *Hearing: an Introduction to Psychological and Physiological Acoustics.* (New York, NY: Marcel Dekker); Moore, B. C. J. (2014). *Introduction to the Psychology of Hearing.* (Cambridge: Cambridge University Press); Pickles, J. O. (2013). *An Introduction to the Physiology of Hearing.* (London: Academic Press); Stebbins, W. C. (1983). *The Acoustic Sense of Animals.* (Cambridge, MA: Harvard University Press); Yost, W. A. (2013). *Fundamentals of Hearing: An Introduction.* (London: Academic Press). A more elementary account is Plack, C. J. (2014). *The Sense of Hearing.* (London: Psychology Press).

1. Auditory sensitivity To appreciate what this astonishing sensitivity means in practical terms, in theory a 10 W loudspeaker of moderate efficiency situated in London and sending out a 1 kHz tone ought to be audible in Cambridge, 50 miles (80 km) away! In fact, of course, not only would much of the sound be absorbed by intervening structures, but prevailing background noise will tend to drown the incoming signal: maximum sensitivity can only be obtained when all other sound sources are silenced. Nevertheless, there are well-authenticated reports from World War I of heavy shelling at a particular location being heard over wide areas in Europe.
2. Fourier analysis The best general book on Fourier analysis and synthesis is probably still Bracewell, R. (1999). *The Fourier Transform and its Applications.* (New York, NY: McGraw-Hill).
3. Musical instruments There are many good accounts of specifically musical aspects of hearing: see for instance Pierce, J. R. (1983). *The Science of Musical Sound.* (New York, NY: Scientific American Books); Roederer, J. G. (2008). *The Physics and Psychophysics of Music.* (New York, NY: Springer); and Sethares, W. A. (2005). *Tuning, Timbre, Spectrum, Scale.* (London: Springer). An approachable book, based on the Royal Institution 1990 Christmas Lectures, is Taylor, C. (1992). *Exploring Music: Science of Tones and Tunes.* (London: Taylor & Francis).
4. Protective function of middle ear The reaction time of such a response to intense sounds is such that it cannot in fact provide much protection against things such as loud bangs, since by the time the muscles contract the damage has been done. But in discos and other hostile environments they may help by acting as automatic earplugs.
5. von Békésy His *Experiments in Hearing.* (1960). (New York, NY: McGraw-Hill) still makes stunning reading.
6. Inner hair cells per octave This happens to be of the same order of magnitude as what a musician can discriminate; but since pitch discrimination is more likely to be a central phenomenon, based on periodicity, this appears to be simply a coincidence. Note also that this is not strictly a test of 'acuity' in the sense it is used in vision, but of 'localisation' along a frequency axis. Acuity in the sense of hearing whether there are two waves present or just one is meaningless in these terms, because of beats.
7. Enhanced potential gradients See for instance Anselm, A. Z., Wangemann, P. & Jentsch, T. J. (2009). Potassium ion movement in the inner ear: insights from genetic disease and mouse models *Physiology* **24** 307–316.
8. Hair cell mechanisms See for instance Ashmore, J. F. (1991). The electrophysiology of hair cells. *Annual Review of Physiology* **53** 465–476; and Dallos, P & Corey, M. E. (1991). The role of the hair cells in cochlear tuning. *Current Opinion in Neurobiology* **1** 215–220.
9. Binaural localisation These mechanisms are obviously of very great importance in the design of stereo audio systems. Ordinary stereo heard through a pair of loudspeakers is not very realistic for a number of reasons. First of all, it can provide an impression only of right–left localisation and not of vertical localisation; but more important, it messes up the normal time delays between the ears that are vital in low-frequency localisation: each ear hears the sound from both speakers, so that a single recorded click reaches the brain as four separate clicks, two to each ear. This problem can obviously be got round by listening through headphones; and although this can certainly give a greatly improved sense of localisation, one is then up against a different problem. In order to achieve good balance between different instruments in an orchestra or band, sound engineers like to use a vast array of microphones scattered about in different locations, and then mix them all together to form the two stereo channels. As a result, the phase relations between the same sound on the two channels are more confused than ever, and a single click is likely to end up as many dozens of clicks by the time it reaches the

listener. Since most people listen through loudspeakers which muddle up the phase relationships anyway, this is not thought to matter much; but it means that recordings of this type do not work very well even through headphones.

A very great improvement is the use of dummy head microphones. Here each channel is recorded through its own single microphone, which is placed in a dummy head, carefully designed with detailed modelling of the external ears in such a way that it produces the same kind of directional coloration that a real head would. If one now listens to the recording through headphones, the effect is extraordinarily realistic, partly because both amplitude and phase information is preserved, and also because for the first time it is possible to perceive the vertical localisation of a sound. The one snag, which is true of all headphone systems, is that moving the head moves the sound image with it, reducing the illusion.

10. Binaural coincidence One might speculate on whether a similar mechanism, with signals from the same ear sent in from *both* directions, might not serve to convert the periodicity of lower-frequency auditory signals into a spatial pattern suitable for processing by higher levels. The same sine wave sent in on both sides will obviously be in phase – and therefore reinforce itself – in the centre; but it will also be in phase one wavelength on each side, producing 'sidebands' whose spatial position can tell the brain what the frequency is.

CHAPTER 7
VISION

Like hearing, vision depends on our being able to analyse vibrations: not of mechanical waves, but of electromagnetic ones. This implies an entirely different kind of transduction mechanism, and – as always – some understanding of the underlying physics.

Light and dark

Light is a form of energy propagated by electromagnetic waves travelling at an immense velocity – some 300 km/ms – and carried in discrete packets called quanta or photons. Only a very small range of all the wavelengths of electromagnetic radiation is *visible* (**Fig. 7.1**). The longest waves that we can just see, forming the red end of the spectrum, are some 0.7 μm in length, slightly less than twice as long as the shortest waves at the blue end. In nature, most electromagnetic radiation is generated by hot objects: the hotter they are, the more of this energy is radiated at shorter wavelengths. The peak of the spectrum of light from the sun – an exceedingly hot object – corresponds quite closely with the range of wavelengths seen by the cone receptors in the eye. Of man-made sources of light, many, like the ordinary incandescent electric lamp, radiate as hot bodies and have a smooth and broad emission spectrum; others are quite different, and emit light only at a few discrete wavelengths. The older sodium lights used for street lighting, for example, are effectively monochromatic, their energy being concentrated in a very narrow band in the yellow region. Domestic fluorescent lamps have a spectrum consisting of a number of emission lines superimposed on a continuous background.

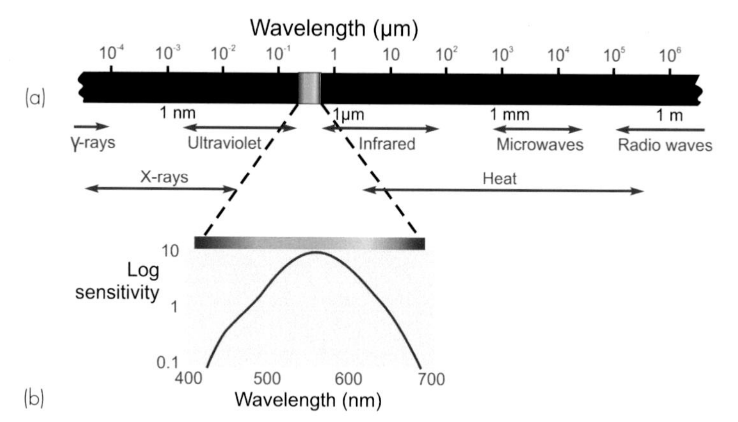

Figure 7.1 (a) The electromagnetic spectrum. (b) The visible portion of the spectrum expanded, showing the relative sensitivity of the human eye to different wavelengths in the light-adapted state.

The spectrum in a sense defines the *quality* of a light; determining its *quantity* is called photometry, and is complicated by the fact that there are two kinds of photometric measurements: first, how much light is *emitted* by a source of radiation (which may be self-luminous, like the sun, or illuminated, like the moon), and second, how much light is *received* by an illuminated object. The *candela* (cd) is a measure of the rate of emission of light by an object: an ordinary 60 W bulb emits about 100 cd. The amount of light received by an object per unit area is its *illuminance*, and is measured in lux. This unit is defined as the degree of illumination of a surface 1 m from a source of one candela radiating uniformly in all directions. Full sunlight may provide about 100 000 lux.

Objects in the real world scatter back some of the light that falls on them, so that in general an illuminated surface is also a luminous one, emitting a certain amount of light per unit area: this is described by its *luminance*, measured in candelas per square metre (cd/m^2). Finally, the ratio of luminance to illuminance in these conditions is a measure of the surface's whiteness or *albedo* (**Fig. 7.2**). If we shine one lux on a perfectly white object that is also a perfect diffuser, it will have a luminance of about 0.32 cd/m^2, and such a surface is said to have an albedo of unity. Ordinary white paper has an albedo of about 0.95; paper printed with black ink, about 0.05. The photometry of coloured objects, which scatter back light of a different spectral composition from that which illuminates them (so that their albedo is a function of wavelength), is much more complex and requires special definitions and methods of measurement.[1]

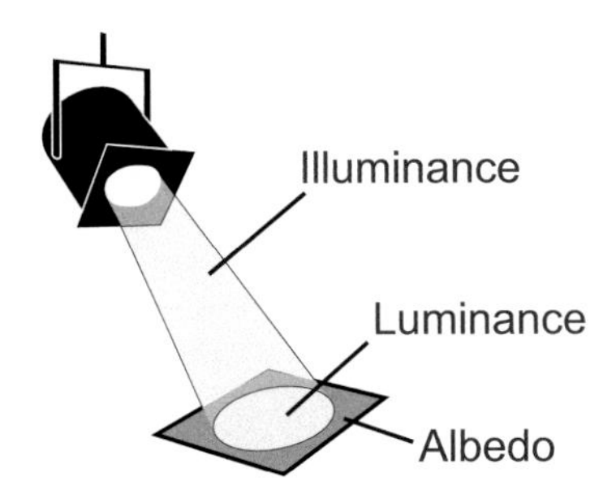

Figure 7.2 Illuminance measures how much light falls on a surface (lux); luminance measures how much it emits (cd/m²). Albedo is a measure of the extent to which the surface scatters back the light that falls on it: a perfect diffuser has an albedo of 1.

Figure 7.3 gives some idea of the range of luminances found in nature. At the bottom end, the eye can function at light levels measured in terms of single photons; at the upper end, the brightest lights we can tolerate without retinal damage are an amazing 10^{15} times – 15 log units – stronger than this. These are extremes: the receptors in the eye, the rods and cones, provide useful vision over the middle 10^{12} or so of all this. This is an extraordinary performance that no man-made device can begin to challenge: in television studios, for instance, absurdly high levels of lighting have to be used to produce decent pictures. So here is something very special about vision – a function whose technical name is *light and dark adaptation* or simply adaptation for short.

Adaptation: a sliding scale

In practice, however, *at any one moment* the actual range of luminances to which the eye is exposed is very much smaller than this. The albedos of natural objects vary only from about 0.05 to 0.95. As you look round a uniformly illuminated room, the range of luminances that you see is therefore only about 20:1. This is true whatever the illumination: whether a room is bathed in sunlight, or dimly lit by light bulbs, black objects still look black and white objects, white. Black and white are thus *relative* terms: the eye operates on a sliding scale of brightness that can be moved up and down the whole range in such a way as to match the prevailing level of luminance (**Fig. 7.4**); this property is the result of various mechanisms of *adaptation*.

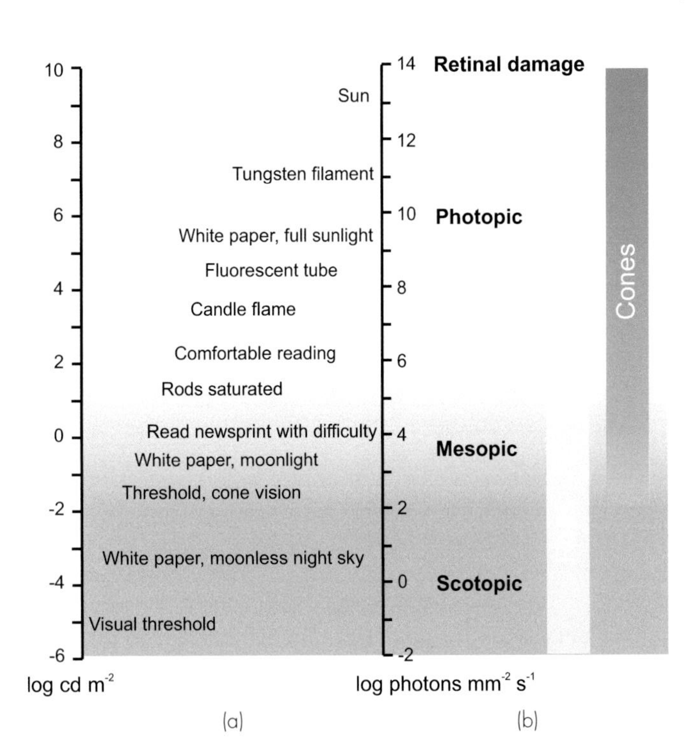

Figure 7.3 The range of luminances (a) and retinal illumination (b, approximate) found in the natural world.

As a result, the range of the visual system is dynamically matched to the range of luminance in your visual surroundings. If the general illumination changes, the visual system quickly follows suit. It follows that the eye responds not so much to the luminance of natural objects as to their *albedo*: a much more useful sensory quality, since albedo is an intrinsic property of objects that helps us to identify them, whereas their luminance depends on how much they happen to be illuminated. A brightly illuminated patch of black on a white background still looks black, even though its luminance may be more than that of neighbouring dimly illuminated patch of white (**Fig. 7.5**). So adaptation has two functions: it enables you to cope

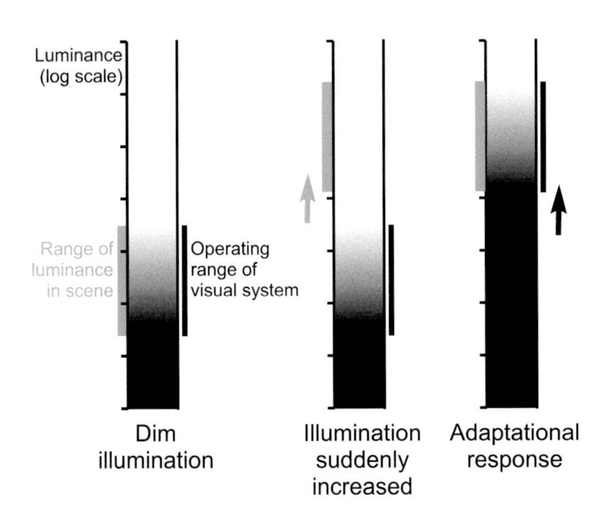

Figure 7.4 The eye's sliding scale of brightness. The range of the visual system is usually matched to the range of luminance in the visual scene. If the latter changes, the visual system quickly follows suit. As a result, perceptions correspond to albedo more than to luminance.

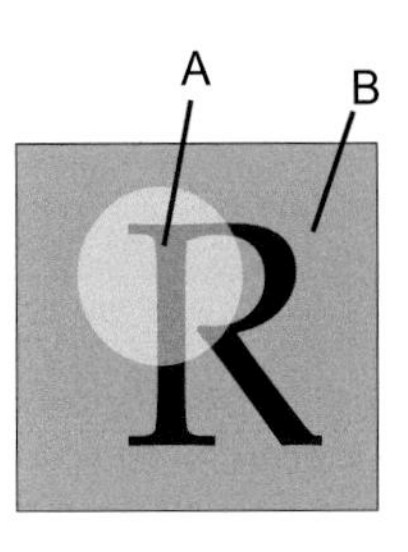

Figure 7.5 Brightly illuminated black (A) may have a higher luminance than dimly lit white (B), yet it still looks black.

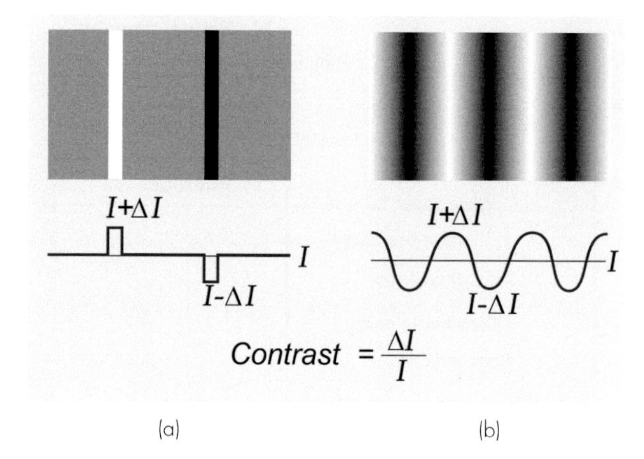

Figure 7.6 Contrast is defined as $\Delta I/I$ for both light and dark lines on a grey background (a) and for a sinusoidal grating (b).

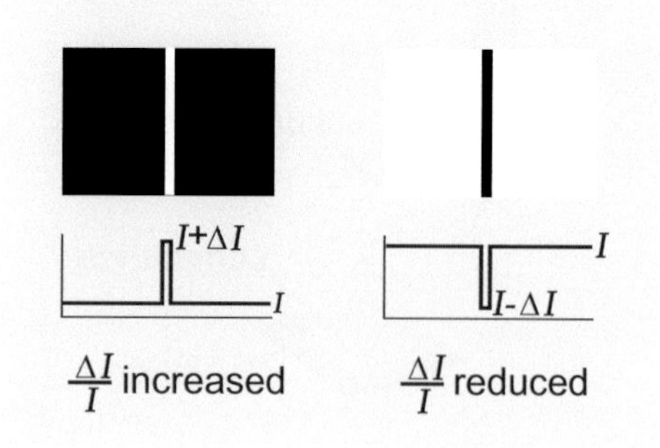

Figure 7.7 Even though the width of the lines and the intensities of the black and white parts of the targets are matched, the visibility of the white line will always be greater than that of the black, because its contrast, $\Delta I/I$, is higher.

with a very wide range of light levels, but it also provides information about albedo, the first step towards *recognition.*

Contrast

The existence of adaptation has profound effects on what we are able to see and how things look. It is easy to show that in a very wide variety of situations our perceptions are *scaled* to the prevailing luminance level. Much of the time we are trying to see small differences of luminance on a relatively uniform background, and in many situations it turns out that their perception depends crucially on something called *contrast*. Contrast is the ratio $\Delta I/I$ between some difference in luminance ΔI and the background luminance I. It can be defined equally well for positive or negative increments on a steady background (e.g. a star in the evening sky, a fly on a sheet of paper), or for the repetitive stimuli often used by visual scientists called *gratings* in which the luminance varies – often sinusoidally – around a mean level I (**Fig. 7.6**). In general, we detect an object if its contrast exceeds a certain threshold value, otherwise it will be invisible. This explains, for instance, why we may be able to see a very thin white line on a black background, yet not see a black line of identical width on a white background (**Fig. 7.7**). Though ΔI is the same in each case (but of opposite sign), I is not, so the contrast is considerably bigger for the dark background. Or consider stars, which have a fixed ΔI: as the sun rises, I increases, their contrast drops, and one by one they disappear (**Fig. 7.8**). In good conditions, threshold contrast is typically around 1 per cent.

Photopic and scotopic vision

Several distinct mechanisms contribute to this ability of the eye to adapt to the prevailing level of illumination (discussed in more detail on p. 150). Some respond quickly to a sudden change in the ambient level, others more slowly. If we go from daylight to a dark room we find that it takes nearly 40 minutes for the eye to adjust its sensitivity fully to the reduced level of illumination. The simplest way to demonstrate this process of dark adaptation is to plot a graph of a subject's *absolute threshold* – the luminance of the dimmest light he or she can just perceive – at regular intervals during this adapting period. Such curves normally show two distinct components (**Fig. 7.9**): an initial one that levels off after some 8 minutes, and a further, slower increase in sensitivity that takes another 30 minutes or so to reach completion.[2]

Apart from demonstrating the slowness of dark adaptation, the two-stage recovery also demonstrates the existence of two populations of receptors in the retina.

- *Cones* are found particularly in the middle of the visual field and provide very detailed information about the retinal image, being a little more than 2 µm in diameter, and are also responsive to colour. But cones have a high threshold, and can only function when the light is above some 10^{-2} cd/m^2 (the *photopic* region). Below this, in the *scotopic* region, we're forced to use the rods. At in-between levels we have an intermediate kind of vision called *mesopic*. The top of this mesopic region is at about 100 cd/m^2, when the rods stop functioning because they're completely saturated.
- *Rods* are much more sensitive – in fact as sensitive as they could possibly be, since one individual rod can respond to a single photon of light – but the bad news is that in order to achieve this sensitivity they have to group themselves together into functional teams – numbered in the thousands – by means of their neural connections in the retina. By pooling their information

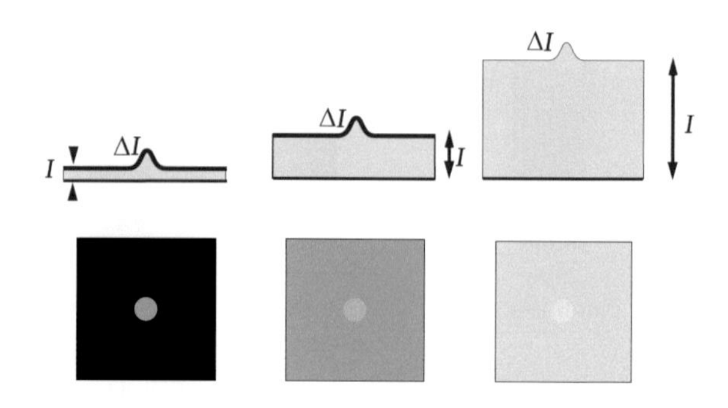

Figure 7.8 On not seeing stars. As the sun rises, the contrast of the image of a star will fall as the background intensity, I, against which it is seen gradually rises, even though ΔI is constant.

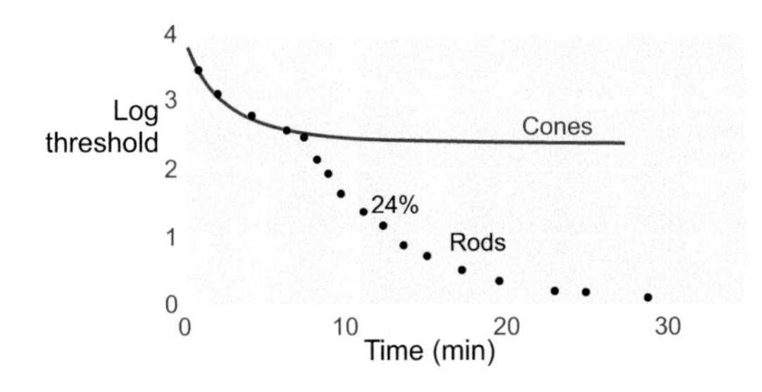

Figure 7.9 Dark adaptation curves. The points show measurements of absolute threshold at different times after a strong light has bleached 24 per cent of the pigment in the rods. The purple line indicates the approximate time course of recovery of sensitivity of the cones alone. It is clear that the recovery occurs in two stages: the first is due to cones, and the second, delayed, stage to rods. (Data from Rushton & Powell, 1972; with permission from Elsevier.)

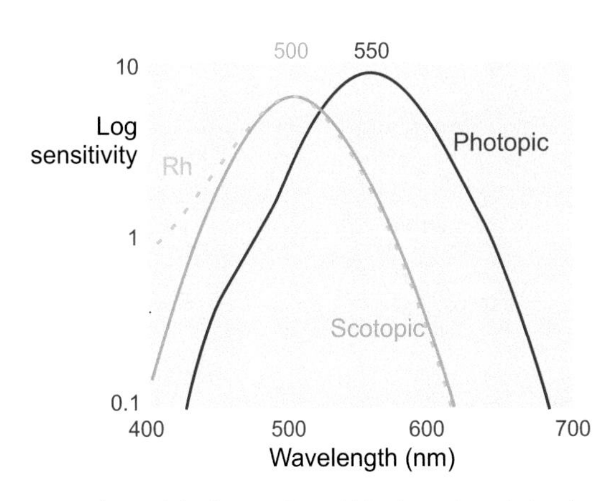

Figure 7.10 Purkinje shift. The purple and blue lines show light-adapted and dark-adapted spectral sensitivity curves; for comparison, the dashed line shows the approximate absorption spectrum for rhodopsin (Rh).

they enormously increase their sensitivity (whereas the random background noise tends to cancel out by being pooled, the signal one is trying to detect does not). But this is at the cost of throwing away a lot of information about the spatial detail of the retinal image, and also sacrificing the ability to distinguish wavelengths or colours; they are also slow, and they respond to a range of wavelengths that is slightly shifted in the blue direction – something called the *Purkinje shift* (**Fig 7.10**).[3] The difference in the *spectral sensitivity curves* relates to the degree to which rods and cones absorb light of different wavelengths.

So we have a sort of dichotomy between these two types of vision. Some of these differences between photopic and scotopic vision are summarised in **Table 7.1**, and further explanations of the terms used are given in the sections that follow.

Image-forming by the eye

Optics

The eye is not just a device for sensing light and dark, it forms an *image* of the outside world and encodes it as a spatial pattern of neural messages for the brain. Image formation is a matter of basic physics: when light enters a region of higher refractive index (or leaves one of lower) it is bent towards the normal, by an amount that depends on the refractive index, μ. So if the surface is *curved,* the further out bits bend parallel rays more and the inner ones less, and if you're lucky they come to a *point.* To a first approximation, the way to get them to come to a point is to use a *spherical* surface. In the eye, there are three surfaces of this sort that act together to bring the images of distant objects to a focus on the retina: they are the *cornea* and the front and back surfaces of the *lens* (**Fig. 7.11**). The refractive index of the aqueous humour that separates the cornea and lens is much the same as that of the vitreous humour that fills the rest of the eye and is about 1.34; that of the crystalline lens is only slightly greater than this, about 1.41.

Opticians describe the power of refractive surfaces by the reciprocal of their focal length in metres, and these units are called *dioptres* (D); for a concave lens, the power is negative. An advantage of dioptres is that they add up: if you have several lenses in series, the total power is simply the sum of the powers of each individual component.

Table 7.1 Vision under photopic and scotopic conditions

	Photopic	**Scotopic**
Sensitivity	Low; best vision in fovea	High; best vision outside fovea
	Light entering periphery of pupil less effective than centre (Stiles–Crawford effect)	No Stiles–Crawford effect
Spatial properties	High acuity; contrast sensitivity reduced at low spatial frequencies (lateral inhibition)	Low acuity; less lateral inhibition
Temporal properties	High flicker fusion frequency; reduced sensitivity at low frequencies (fast adaptation)	Low flicker fusion frequency; less fast adaptation. Increased latency
Wavelength	Most sensitive at around 550 nm	Most sensitive at around 500 nm (Purkinje shift)
	Trichromatic colour discrimination	Monochromatic

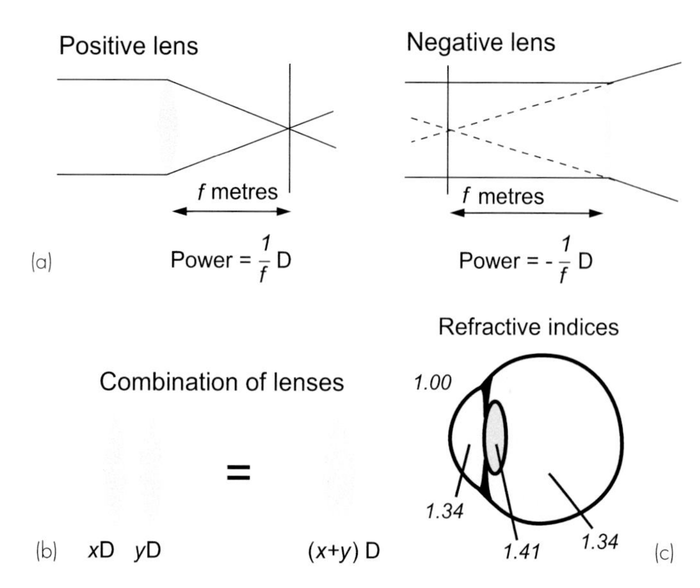

Figure 7.11 (a) A positive (convex) lens with focal length f metres has a power of $1/f$ diopters; for a concave lens, the power is negative. (b) For thin lenses, close together, diopters simply add linearly. (c) Images are formed in the eye by refraction at several interfaces between media of different refractive indices; the lens itself contributes less than might be expected, because its refractive index is not very different from that of what surrounds it.

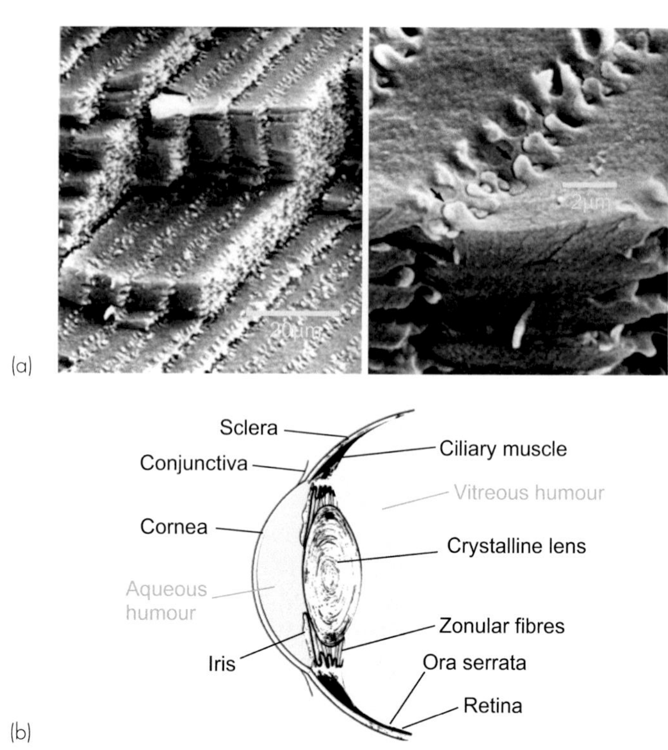

Figure 7.12 (a) Scanning electron micrograph of cells of human lens, showing interdigitating processes. (b) The anterior part of the human eye, as shown in parasagittal section.

Since the distance from the cornea to the retina in humans is about 24 mm, you might think the total refractive power of the eye when focused on a distant object would be some 43 D, but because the focal length is being measured within a refractive medium, this figure must be multiplied by the refractive index (1.34), giving about 57 D. Of this, about 48 D are due to the cornea alone, and only 19 D or less to the lens. So, contrary to popular belief, the focusing power of the eye is *not* mainly due to the lens: the reason is that its refractive index is not very different from that of the aqueous humour and vitreous humour on each side, so it doesn't contribute much. The cornea is powerful because its interface is with air, whose refractive index is much smaller (1.0008 to be exact).

Accommodation and the lens

Though the lens contributes little to the total refractive power, it still has a very important function: it can alter its shape and hence fine-tune the eye's effective focal length; this function is called *accommodation*. It's able to do this because it is elastic and flexible, made of remarkably long (12 mm) thread-like cells, rectangular in cross section and – for transparency – lacking nuclei, knitted together in a series of concentric layers by zip-fastener–like ball-and-socket joints that provide flexibility (**Fig. 7.12**). You can think of the lens as a kind of jelly held in place by guy-ropes all round the edge, the *suspensory ligaments*, that are normally under tension from elastic elements attached to the wall of the eye and tend to flatten the lens out. Encircling the lens is a ring of muscle called the *ciliary muscle*. When it contracts, it relieves the tension on the filaments, and the lens relaxes into a more bulging, rounded shape with a shorter focal length (**Fig. 7.13**).

The lens presents something of a design problem for nature, since – being transparent – it obviously cannot have a blood supply. It obtains its nutrients and oxygen from the *aqueous humour* that bathes it on both sides, and it is able to penetrate the lens because of its fibrous nature. It is a fluid similar to plasma but with only some 1 per cent protein concentration and a peculiarly large amount of ascorbic acid. The aqueous humour is continuously secreted by the *ciliary body*, and passes through the iris into the anterior chamber where it eventually filters its way out into the *canal of Schlemm*, whence it ends up in the bloodstream. The resistance to its outflow generates an intraocular pressure of some 10–20 mmHg. Blockage may raise this pressure to the point where the flow of blood into the back of the eye is hindered, a serious condition called *glaucoma*, which is a common cause of blindness.

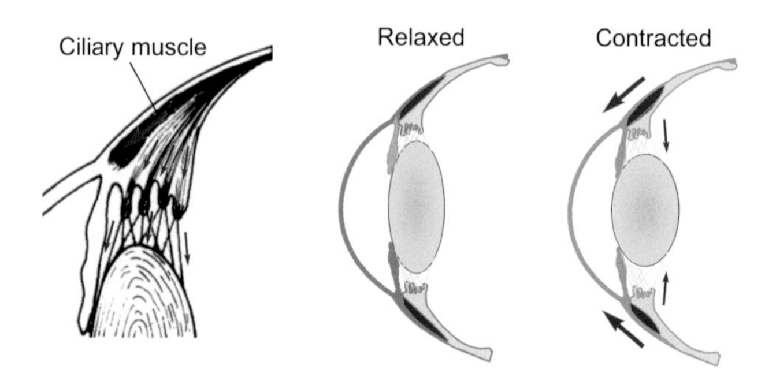

Figure 7.13 When the ciliary muscle contracts, the lens can adopt a more rounded shape and brings closer objects into focus.

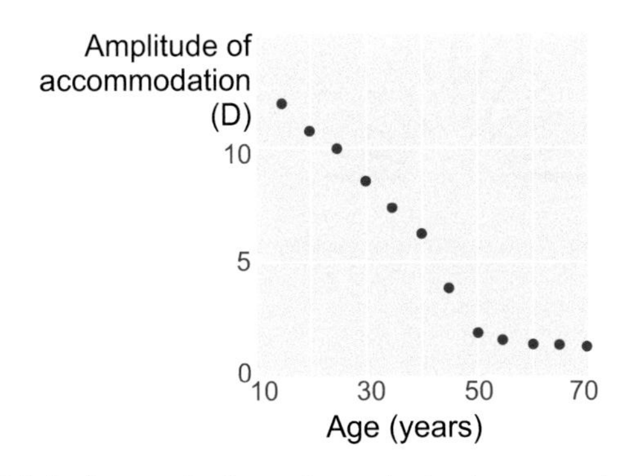

Figure 7.14 Presbyopia. Decline in the amplitude of accommodation as a function of age: average from three large groups of subjects. (Data from Fisher, 1973; with permission from John Wiley and Sons.)

The *range* of accommodation is the difference between the maximum and minimum power of the optics, expressed in dioptres. It can easily be measured by finding the positions of the *near and far points* of the eye, the nearest and furthest distances at which objects can just be brought into focus. For a normal or *emmetropic* eye with accommodation fully relaxed, the far point will be at infinity (0 D), and the range of accommodation will be given by the reciprocal of the distance of the near point in metres. For instance, a young emmetrope's near point will generally lie at around 80 mm, so that the range of accommodation is 12 D. As one gets older, however, the lens begins to seize up, so that it no longer bulges so much when the ciliary muscle contracts, and the range of accommodation falls. By the age of 60 the possible amplitude of accommodation may have fallen to 1 D or so, a condition known as *presbyopia* (**Fig. 7.14**). It is the *near* point that moves further away, and when you find you can no longer read the newspaper even at arm's length, then you have to start wearing either *half-glasses*, or *bifocals*, in which the bottom half of the lens has a stronger power than the top.[4]

Errors of image formation

That's the basic optics. How *good* are they?[5] Few people are exactly emmetropic; for most, one finds that when the accommodation is fully relaxed the total refractive power is either too strong or too weak in relation to the distance from the cornea to the retina: they suffer from *refractive error.* If it is too *strong*, the image of a distant object lies inside the vitreous instead of on the retina and we have the condition called *myopia.*[6] An optician corrects myopia by using a *negative* or concave spherical lens. For example, if your far point is 1 m away, then that means your eye has 1 D too much power, and you need a –1 D lens to make it up. If you're long-sighted or *hypermetropic,* the eye is too weak and needs an additional positive lens to correct it (**Fig. 7.15**). In either case, you can describe the degree of disability by the power and sign of the lens needed to bring the eye back to emmetropia: thus a mildly short-sighted patient might require a correction of –1.75 D. This is called the *spherical* correction, part of an optician's prescription, and in general is not the same in both eyes.

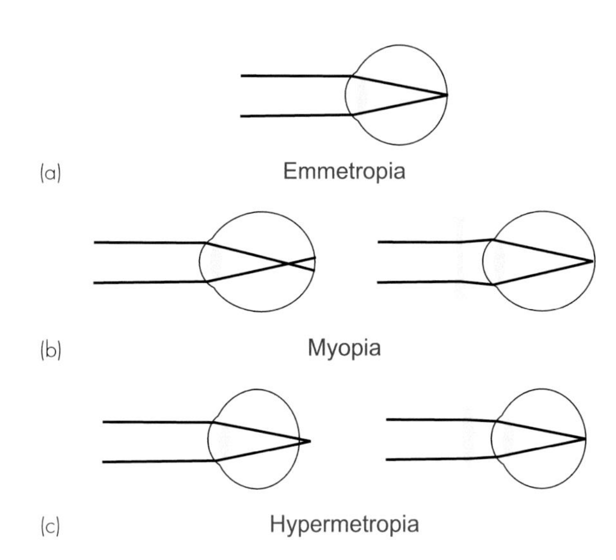

Figure 7.15 Errors of focusing. (a) An emmetropic eye with relaxed accommodation that focuses parallel rays exactly on the retina. (b) A myopic eye brings parallel rays to a focus that is too close to the lens. The defect may be corrected with a negative (concave) lens (right). (c) A hypermetropic eye cannot bring parallel rays to a focus at all; a positive lens is needed for the correction.

Astigmatism

A more subtle but very common type of focusing error occurs when the curvature of the cornea is different in different meridians. It will then focus a horizontal line at a different focal length from a vertical one, a condition called *astigmatism* (**Fig. 7.16**). If, for example, it has a smaller radius of curvature in the horizontal plane than in the vertical, the far point when measured with a vertical line as test object will be closer than when a horizontal line is used. It is corrected with a cylindrical lens – in effect a section cut from a cylinder, just as a spherical lens is from a sphere: it focuses only in one meridian (**Fig. 7.17**).

Opticians test for astigmatism by means of a target like that in **Figure 7.18**, called an *astigmatic fan*; an astigmatic subject will see some of the lines more sharply than others, and this will tell the optician the angle at which a cylindrical lens should be placed in front of the eye to make the refractive power as nearly as possible equal in all meridians. The power of the cylindrical lens that is needed to do this, together with its meridional angle, make up the

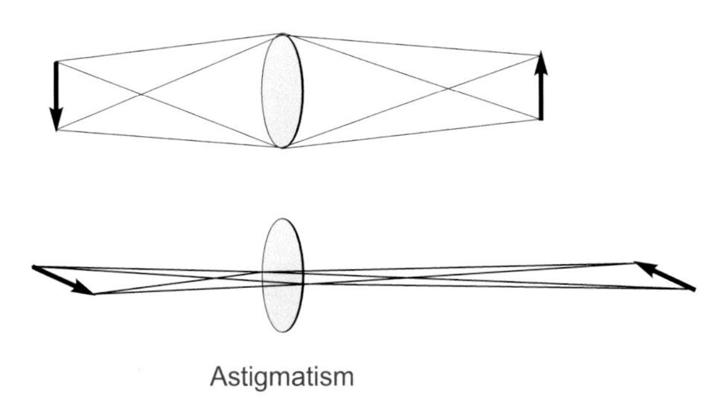

Figure 7.16 An astigmatic lens has different focal lengths for line targets in perpendicular directions.

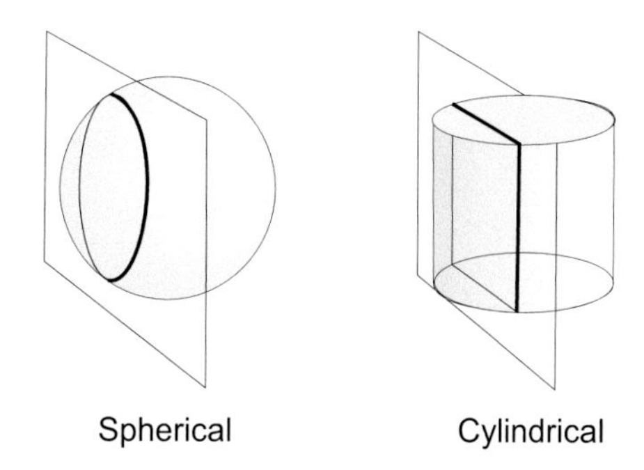

Figure 7.17 The origin of spherical and cylindrical lenses.

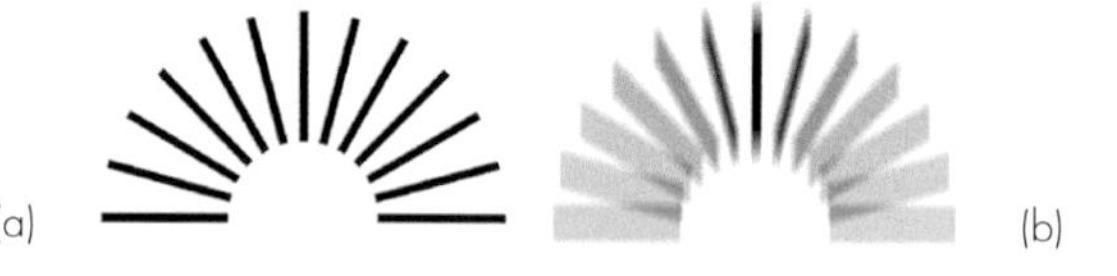

Figure 7.18 (a) An astigmatic fan, a target used for testing for astigmatism, and (b) its appearance to a subject with marked astigmatism in the horizontal/vertical directions.

cylindrical correction that is the second part of a prescription for spectacles.[7]

Astigmatism and incorrect refractive power are not the only faults that may be found in the eye's optics, and – as in many man-made optical systems – the cornea and lens together produce a number of different types of optical defects in addition to refractive error.

Chromatic aberration

The first of these is due to the fact that the refractive indices of the various optical media of the eye depend on the wavelength of the incident light. In general, the refractive index increases with decreasing wavelength, so that blue light is refracted more than red. This phenomenon, called dispersion, means that the focal length of a lens depends on the wavelength, with blue being shorter than red, amounting to some 2 D over the whole visible spectrum. This gives rise to defects in the resultant image, in the form of coloured fringes, called *chromatic aberration* (**Fig. 7.19**). So if you look at a blue object and a red object lying side by side at the same distance from the eye, they cannot both be in focus simultaneously, and a subject who is emmetropic when his far point is measured in red light will be short-sighted if it is measured in blue: his far point will then be only a metre or so away. This forms the basis of a simple clinical test for errors of refraction, consisting of an illuminated screen divided into three portions that are red, green and white: identical test figures are superimposed on each field, and the subject is simply asked which figure he sees most clearly. A myope, whose lens is too strong anyway, sees red targets more clearly than green, and vice versa for a hypermetrope; an emmetrope will see the one on the white background best (**Fig. 7.20**).

The importance of chromatic aberration can be seen by the fact that visual acuity is improved by some 25 per cent in monochromatic yellow light. The eye mitigates the effects of chromatic aberration in two ways: by a yellow pigment over the fovea that reduces the blue component, and by the fact that very few blue cones are found in the centre of the fovea, where the finest spatial vision is found.

Spherical aberration

The other aberration is *spherical aberration*. We saw earlier that to make a surface bring parallel rays to a point it needs to be roughly spherical. Although ordinary man-made lenses are nearly all spherical, simply because they're easier to make, a spherical lens does *not* in fact bring rays to a point focus: they get bent too much as you go further out, and it is the resultant blur that is spherical aberration (**Fig. 7.21**). The shape you really need is not a sphere but an ellipsoid. For surfaces that are small in comparison with their radii of curvature, the difference is slight, and spherical aberrations are often negligible. But in the case of the eye, the aperture is of the same order of magnitude as the radius of curvature of the cornea, and the result is that rays entering near the periphery of the cornea are bent too much, and form a closer focus than those entering near the centre. To some extent Nature has compensated for spherical aberration, first of all by making a cornea that is not exactly spherical but tends towards the desired ellipsoid; and second, in that the refractive index of the lens is not constant throughout, but graded from a maximum of some 1.42 at its centre to about 1.39 at the edge, thus cancelling out, to some extent, the extra bending of peripheral light rays. The degrading effects of both spherical and chromatic aberration, and of other defects due to irregularities of the refracting surfaces, get worse as the *pupil* or aperture of the eye increases, and this in turn depends on the state of the *iris*: this is discussed on p. 138, in the context of visual acuity.[8]

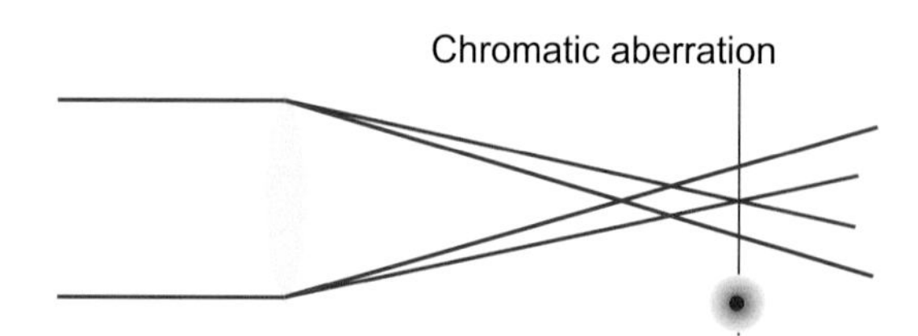

Figure 7.19 Chromatic aberration is the result of focal length depending on wavelength.

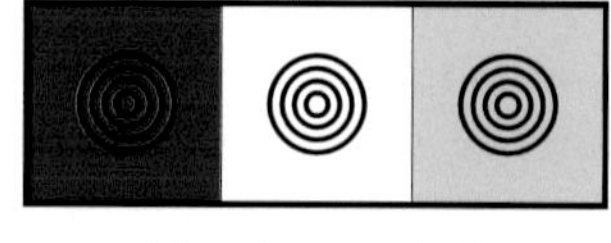

Figure 7.20 The duochrome test. A myope sees the object on the red background most clearly, while a hypermetrope sees that on the green background best.

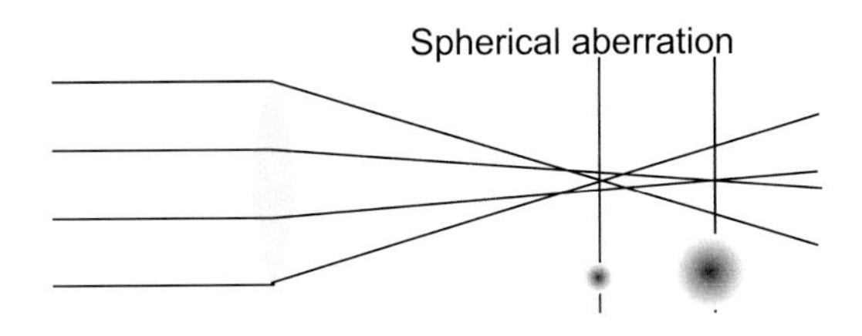

Figure 7.21 Spherical aberration arises because the focal lengths of different regions of the lens are not the same.

Scatter and diffraction

There are two other problems with the eye's optics that are not exactly errors of focus, but do degrade the quality of the retinal image. When light enters the eye it tends to get scattered, particularly by the cornea and lens but also to some extent by bouncing off the back of the retina. This effectively superimposes on the image a more or less uniform background, perceived as *glare*, whose illuminance is of the order of 10 per cent of the mean illuminance of the retina. This means that if we look at a target whose actual contrast is 100 per cent, the effect of this scatter is to reduce the contrast of the retinal image to something nearer 90 per cent. Because of the progressive opacity of the lens, glare gets worse as you get older; it is particularly obvious when driving at night in the face of opposing headlights. The second problem is something that is caused by the pupil, called *diffraction*. Whenever light passes through a restricted aperture it tends to spread out and therefore degrades the retinal image: the smaller the aperture, the worse this gets. More precisely, the width of the resultant pointspread function is of the order of λ/d radians, where λ is the wavelength and d the aperture of the system. In practice, so long as the pupil is bigger than some 3 mm, diffraction contributes rather little in comparison with the other optical problems.

The pupil

Thus the ideal size for the pupil is a compromise. An important factor is the ambient light level: under bright photopic conditions the eye can take advantage of the excess light by constricting the pupil and improving the quality of the retinal image. In scotopic conditions, however, the eye needs all the light it can get and the quality of the retinal image is of secondary importance: in any case, we shall see later that the rods are not capable of passing on accurate information about the detailed structure of the retinal image. It is important to emphasise that pupil dilatation contributes very little to the enormous changes in sensitivity that accompany dark adaptation, since it can only vary the incoming light by a factor of 16 at most, or 1.2 log units. The effect of pupil diameter on visual acuity is discussed in more detail on p. 138.

Another factor is that the control of the pupil is closely linked to accommodation: when the ciliary muscle contracts in order to focus on a near object, there is normally an associated constriction of the pupil (the *near reflex*), which may help to increase depth of focus (see p. 139). As these responses are usually also combined with binocular convergence movements of the two eyes, the whole pattern of response (constriction, accommodation, convergence) is also known as the *triple response*. Under certain clinical conditions, notably in neurosyphilis, one may find that the pupillary response to near objects remains despite loss of the response to bright lights: this condition is known as the Argyll Robertson pupil, and is an important diagnostic neurological sign. The fact that pupil dilation is also a measure of general sympathetic activity and of emotional or sexual excitement also has it uses.[9]

Visual acuity

Measurement

Visual acuity is a measure of the fidelity with which the visual system can transmit fine details of the visual world: it is the equivalent of the ability of a camera to produce sharp pictures. In a camera there are essentially two stages at which sharpness may be lost: either through optical defects that blur the patterns of light in the image, or subsequent factors – a lack of pixels – that limit the density of detail. These correspond in the eye to the quality of the *optics* and to the density of the *retinal receptors*. But in the case of the eye there is a third factor: the possible degradation of the image that may occur in the course of the *neural processing* that takes place in the retina.

The pointspread function

The effect of optical blur is relatively straightforward. Consider for instance the simplest of all visual objects, a star. Stars are so far away that they can in effect be regarded as infinitely small point sources. But the retinal image of the star will certainly *not* be a point, because the optics will spread the light out into a sort of heap on the retina. This distribution of light is called the *pointspread function*, and its size is a useful measure of how good or bad the optics are. In the human eye, under the best possible conditions, the pointspread function has a diameter of about 1.5 arc min (measured half-way up); the worse your optics, the bigger this becomes (**Fig. 7.22**).[10]

This pointspread function is of fundamental importance, since it absolutely determines what sort of patterns we can see and what we can't. For instance, if we have two stars rather than one, each with its own pointspread, then if they are far apart, they will be seen correctly as two separate stars; if closer, eventually there will be no little dip in between them, and the brain will have no way of knowing that there are two stars and not one (**Fig. 7.23**). For normal observers, the angle of separation for which this kind of resolution can just be performed provides a quantitative measure of visual acuity. Its value – around 30–45 arc sec – is an order of magnitude greater than the width of a black line that can just be seen. This figure, that is found to apply in many similar tasks of *resolution*, can

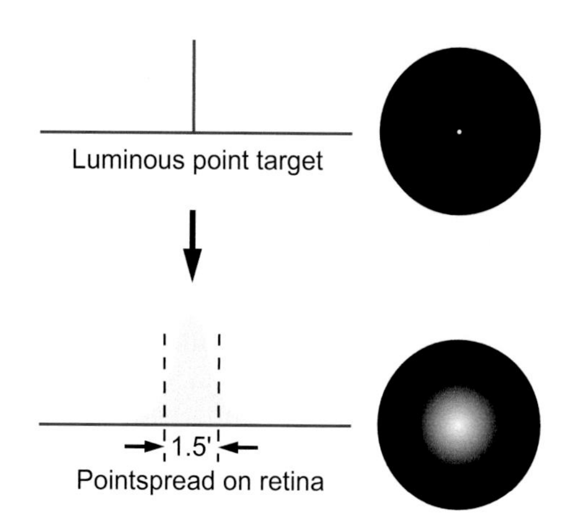

Figure 7.22 The pointspread function. An infinitely small point of light generates an image of finite width on the retina, the pointspread function. Its size determines the spatial quality of retinal images.

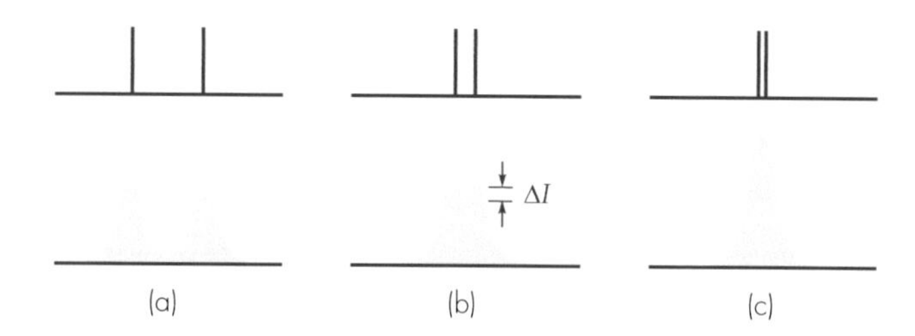

Figure 7.23 Resolution and contrast. As a pair of point sources are gradually brought together, their retinal images begin to overlap: (a) is easily resolved, as long as the points can be seen at all; (b) will be resolved but only if the contrast is sufficiently high; whereas (c) can never be resolved, whatever the contrast.

be taken as a measure of the *visual acuity* or *resolving power* of the eye.

Resolution

We need to pause at this point to consider exactly what is meant by 'seeing' something. Seeing can mean *detection*, or *resolution*, or *recognition*. An ornithological friend points up in the sky and says '*Can you see the crested willow-warbler?*' If you can't, it may be either because it's so far away you can't *detect* it, or because although you can see a formless speck you can't *resolve* the pattern of its markings, or finally because – although you can perceive every aspect of it – you haven't the least idea what a crested willow-warbler is *meant* to look like. In each case, there is a failure to 'see', but for completely different reasons.

The existence of spatial spread has implications for detection as well as resolution. Since the incident energy from the point source is spread out over a larger area, the maximum intensity at the central peak is necessarily reduced, leading to a decrease in the *contrast*, $\Delta I/I$, which determines whether it will be seen against its background (see p. 130). For objects of intrinsically high contrast, such as stars seen against the void of space, this will not matter much, and subjects with poor visual acuity as measured conventionally (see later) are not as bad at seeing stars as one might expect, bearing in mind the fact that such objects subtend an almost infinitely small angle. In the dark, whether one sees a star or not is almost entirely a matter of whether a sufficient number of photons from it fall upon a rod summation pool; as the sun rises its visibility depends on whether $\Delta I/I$ exceeds the threshold contrast. Thus – rather as in the case of the skin, discussed on p. 90 – we may find that we can localise a visual object to a much higher degree than our ability to tell whether there is one object or two.

While the main effect of contrast is on detection, it also has some effect on resolution. With the two stars, in a case like that of **Figure 7.23c**, it is clear that we cannot improve resolution simply by increasing the contrast, and such a stimulus may be described as absolutely unresolvable. But in an intermediate case like **Figure 7.23b**, whether resolution is possible or not will depend on the contrast of the original object as well as on the width of the pointspread function. This interaction between resolution and contrast can best be investigated by using *grating* patterns as test targets. A grating is simply a regular pattern of stripes; if the stripes are uniformly black and white, it is called a square-wave grating, because a plot of intensity as a function of distance across the grating would have a square-wave profile. In the same way, sinusoidal gratings have an intensity profile that is sinusoidal (for an example, see p. 130). In each case, one can describe the grating in terms of its *spatial frequency* (i.e. the number of cycles per degree) and its *contrast* (defined as the difference in intensity between peak and mean intensity divided by the mean intensity). Thus a pattern of alternate pure black and pure white strips, each 1° across, could be described as a square-wave grating of 100 per cent contrast and spatial frequency 0.5 cycles per degree. A simple experiment is to ask a subject to view a sinusoidal grating of a particular spatial frequency, and then reduce its contrast until they report they can no longer see it. If we plot this threshold contrast as a function of spatial frequency, we typically obtain a curve such as that in **Figure 7.24**. Because a blurred pointspread function affects high spatial frequencies much more than low ones, the contrast required to see the grating increases sharply as its frequency is increased, until at about 40–50 cycles per degree (the cut-off frequency), the subject cannot even see a grating of 100 per cent contrast. Because of the steepness of this cut-off, a small amount of extra blur causes a large increase in the contrast needed, and so the method provides a sensitive measure of acuity. The reason for the fall-off in contrast sensitivity at *low* frequencies is discussed later (p. 154).

Of course, dispensing opticians don't bother with all that. A rough-and-ready measure of visual acuity is to make up charts of letters of standardised shape and graded in size, and see at what point the patient is unable to read them. A common chart of this kind is the Snellen chart (**Fig. 7.25**), in which rows of letters of diminishing size are to be read. By discovering the row at which the subject finally stops and knowing the size of letters and the subject's viewing distance, one can estimate his minimum resolvable

Clinical box 7.1 The pupils in clinical practice

While the eyes might not be the window to the soul, the pupils certainly provide a window to the brain. If pupils are normal and equal in size, exhibiting intact direct and consensual light reflexes, then we can be reassured of the integrity of the retina, optic nerves and tracts, midbrain, oculomotor nuclei, oculomotor nerves and intraocular muscles. *Abnormal* pupils, however, provide clues to the nature of brain injuries.

Box Table 7.1

	Mechanism	Causes (not exhaustive)
Mydriasis (>5 mm)		
Unilateral	Direct trauma	Local injury to the sphincter pupillae or its innervation
	Raised intracranial pressure squashes the oculomotor nerve* between the innermost aspect of the temporal lobe and the rigid tentorial membrane (uncal herniation)	Intracranial tumour, bleed or trauma
	Injury to the oculomotor nucleus in the midbrain	Ischaemia, bleed, multiple sclerosis
Bilateral	Bilateral oculomotor nucleus damage	Midbrain infarction or haemorrhage
		Severe hypoxic brain injury, brain death
	Anticholinergic action, catecholamine reuptake inhibition at synapse	Tricyclic antidepressant drugs
	Interruption to parasympathetic pathway	Atropine
	Overwhelming sympathetic outflow	Amphetamines, cocaine, other sympathomimetic drugs
Miosis (<2 mm)		
Unilateral	Interruption to sympathetic* pathway from the pons, via the sympathetic trunk of the thoracic spinal cord and sympathetic ganglion, over the apex of the lung, along the sheath of the carotid artery to the sphincter pupillae	Ischaemia or trauma to pons or medulla, injury to the upper thoracic spine and root of the neck, tumours in the lung apex, aneurysm or dissection of the carotid artery
Bilateral	Interruption to sympathetic outflow	Large pontine or thalamic injury (usually infarct or haemorrhage)
	Excessive cholinergic activity	Acetylcholinesterase inhibitors (such as organophosphate pesticides)
	Centrally mediated effects	Opioids (e.g. morphine), metabolic encephalopathy (e.g. liver failure)

* The oculomotor nerve carries the parasympathetic (pupil constricting) fibres from the brainstem to the sphincter pupillae. Sympathetic (pupil dilating) fibres arise from the pons and follow a circuitous route to the iris.

angle. You can see that a letter E, for instance, is a bit like a miniature grating, and you need to be able to resolve it in order to read it. Each line of text is marked with distance at which you should just be able to read it. If at 6 m you can only read the line for 12 m, then your acuity can be described as 6/12; a normal person is in theory therefore 6/6.[11] But Snellen charts are very conservative, since they are calculated not on the basis of 45 but of 60 arc sec, or one minute. So at 6 m away, with very good vision you should in fact be able to read the line marked 5, in which case your vision is described as 6/5. The idea is no doubt to make the glasses the optician has just sold you seem better than they are.

The difficulty of the Snellen chart for scientific work is that the test is only partly one of resolution. Apart from the assumption that you are literate, it is also clear that some letters are recognised more easily than others because of their overall shape; and for some purposes the Landolt C

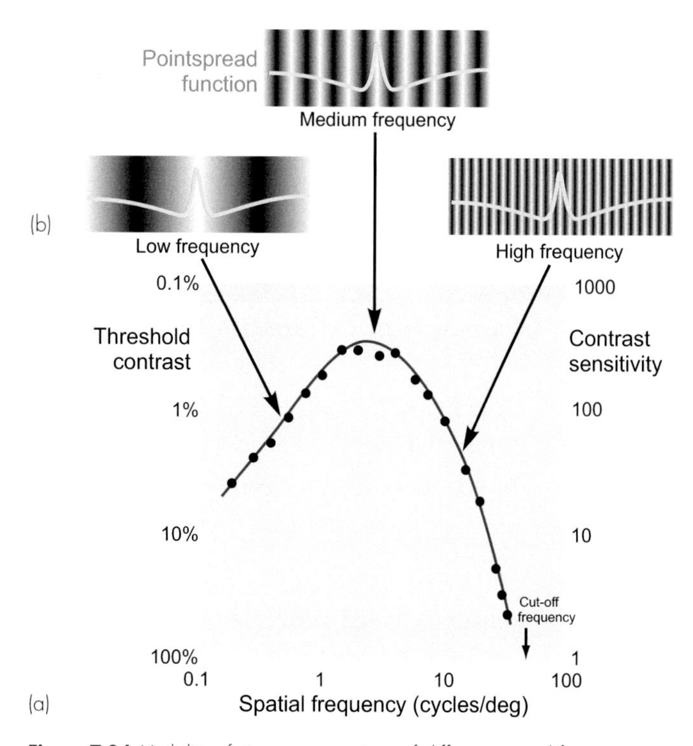

Figure 7.24 Visibility of sine-wave gratings of different spatial frequency. (a) Frequency sensitivity function for sine-wave gratings, showing a peak around 2–3 cycles/degree and a fall-off in sensitivity both at higher and lower frequencies. (b) Schematic representation of a sinusoidally modulated image falling on a receptive field having an excitatory centre and an inhibitory surround. When the centre is roughly matched in size to the peak of the sine wave, the response is maximum. With lower or higher frequencies there is a relative increase in the degree of stimulation of the inhibitory surround. (Data from Campbell & Robson, 1968.)

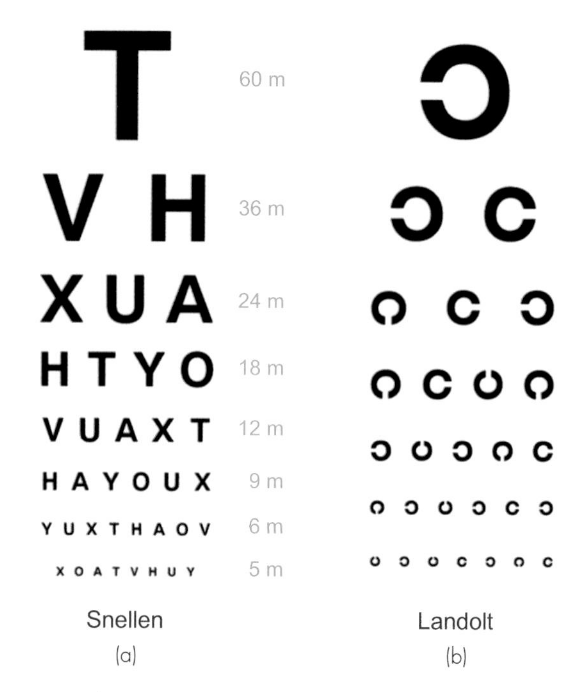

Figure 7.25 Charts commonly used for routine testing of visual acuity. (a) Snellen chart. (b) A similar chart using Landolt Cs: the subject must name the position of the gap in the circle.

chart (**Fig. 7.25**), used in the same way, is preferable because it provides no extraneous clues to the subject. Even this is not ideal, since it is still possible to detect the overall orientation of the C even though it is not really resolved: for this purpose, simple barred patterns are better (**Fig. 7.26**), since when blurred it is completely impossible to guess the original pattern, which is not true either of Snellen letters or even Landolt Cs. Another kind of chart in which all the letters are the same size but are graded in contrast can be used to test for certain kinds of defects in the visual system in which sensitivity to contrast is specifically impaired (the Pelli-Robson chart).

The tests described so far are all genuine tests of acuity in that they require that detail of some kind be resolved. Other tests, that at first sight might also appear to be acuity tests, are really tests of detection or localisation, and give apparent acuities far better than 30–45 arc sec. A well-known example is *vernier acuity* (**Fig. 7.27**), where a subject is required to move two lines into alignment, as for instance in the scale of vernier callipers. Here one does incredibly well, typically of the order of a few seconds of arc. But the task is not resolution but *localisation:* even if the retinal image is blurred, one can still estimate where the peak is quite accurately. One can show that the *longer* the line, the better one is, showing that accuracy is also being improved by averaging information over the whole line. Another pseudo-acuity task is the detection of stars, which may subtend extremely small angles at the eye. For instance, the bright star in Orion called Betelgeuse subtends only some 1/20 arc sec. But in a sense that figure is quite irrelevant: because of the pointspread, its image *still* has a width of 45 arc sec, and whether you detect it or not depends simply on whether its luminance exceeds the absolute threshold, ΔI_0.

The influence of the pupil

Unlike the lens, the size of the pupil is under the control of two different muscles: one, the *sphincter pupillae*, lies circumferentially round the iris, and the other, the *dilator*, lies radially. So the two muscles have opposed effects, the first causing contraction of the pupil and the second dilatation,

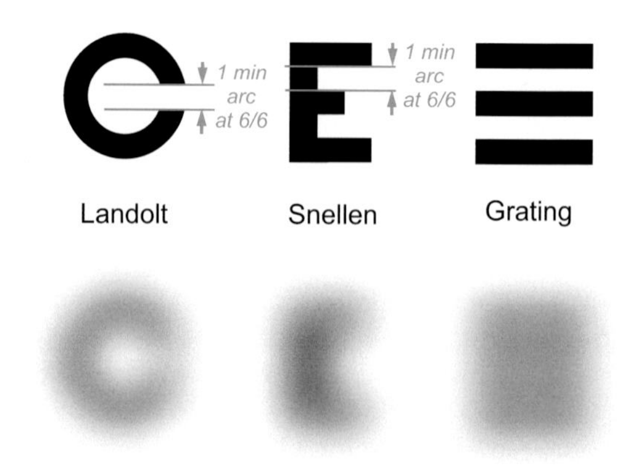

Figure 7.26 (a) Criteria for the Landolt and Snellen charts. At the standard viewing distance, critical parts of letters subtend 1 minute of arc. (b) When equally blurred, Snellen and Landolt figures may still be recognised, even though not resolved; this is not the case for a grating target (right).

Vernier acuity

Aligned

Misaligned

Figure 7.27 Vernier acuity (look closely).

and they are respectively under the control of the parasympathetic and sympathetic systems (**Fig. 7.28**).

It is not entirely clear which of the two branches of the autonomic nervous system is responsible for normal tonic control of pupil size, and one may cause *mydriasis* (enlargement of the pupil) by drugs that either block the action of acetylcholine on the sphincter (e.g. atropine) or simulate the effect of noradrenaline on the dilator (e.g. phenylephrine). Light causes constriction through the parasympathetic route from the ciliary ganglion, in turn activated by the *Edinger–Westphal* nucleus high up in the brainstem, close to the oculomotor nucleus, which receives sensory information about overall light level from neurons in the pretectum, which themselves receive fibres from the optic nerve.

The advantages of a large pupil size are first that the eye receives more light (over the normal range of pupil diameters, about 2–8 mm, the amount of light caught by the eye varies by a factor of 16), and second that the diffraction effects that always occur when light passes through a small aperture are minimised. The advantages of a small pupil, on the other hand, are an increased depth of field (a greater tolerance of errors of focus) and a reduction in the magnitude of the optical aberrations and of glare (**Fig. 7.29**). The extent of this effect can be seen for oneself by looking through a pinhole, which acts as a very small artificial pupil. Aberrations are reduced because the smaller the pupil, the more nearly the optical surfaces will approximate to their ideal forms, and the less noticeable the aberrations will be. The effects of diffraction can be calculated without much difficulty. For a pupil of diameter 2.5 mm, and with green light, diffraction alone creates a pointspread of a little less than one minute of arc. In other words, under these conditions acuity is effectively limited by diffraction at the pupil. In the dark, with a pupil of some 8 mm diameter, the corresponding figure for diffraction alone is about 17 arc sec, but the actual pointspread is very much wider than this because of the increased contribution of the aberrations when the lens is widely exposed: in fact the pointspread function actually gets wider with increasing pupil diameter beyond 3 mm or so. As a result, a graph of visual acuity as a function of pupil size is U-shaped, with a distinct optimum around 3 mm (**Fig. 7.30**). Thus if acuity were the sole consideration, we might expect to find the pupil always fixed at that value. But as we shall see, under conditions of dark adaptation the intrinsic acuity of the neural processing of the retinal image is so low that the poor optics contribute little to the overall blur, and the advantage of being able to increase retinal sensitivity by catching more light with a dilated pupil outweighs the disadvantage of slightly decreased acuity.

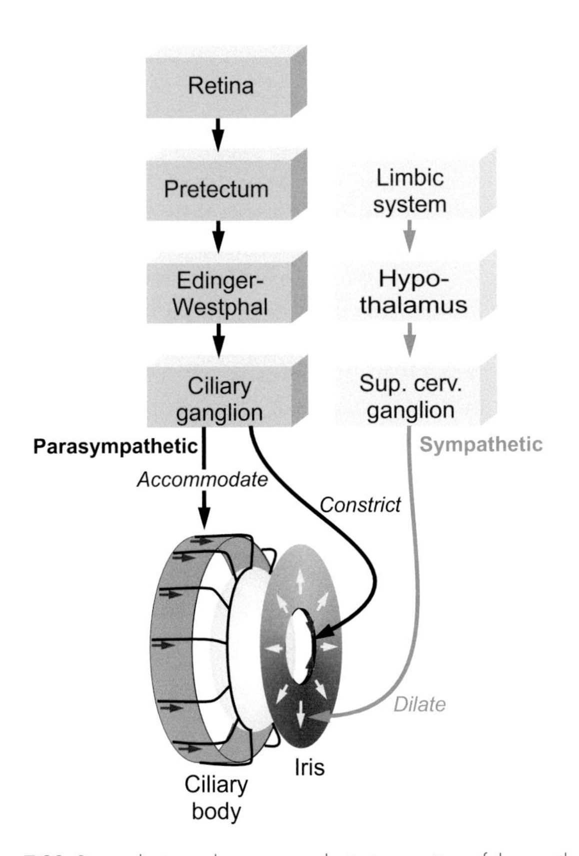

Figure 7.28 Sympathetic and parasympathetic innervation of the pupil and ciliary muscle.

The influence of the retina

The previous section may have given the impression that acuity is purely a matter of the *optics*, and that what the retina and brain do doesn't matter very much. Under photopic conditions, for most of us, in practice this is probably true. But even when there is an abundance of light, and the optics are free of any defects, then one finds that the visual acuity that one measures is not *quite* as good as one would expect from the quality of the retinal image; what is happening is that the retina and brain are introducing some extra deterioration of their own.

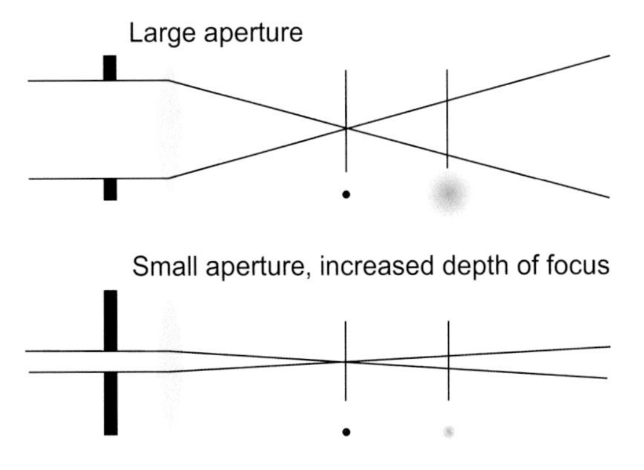

Figure 7.29 A small pupil minimises the effects of bad focus on the size of the image, so it increases the depth of focus as well as reducing the effects of aberrations.

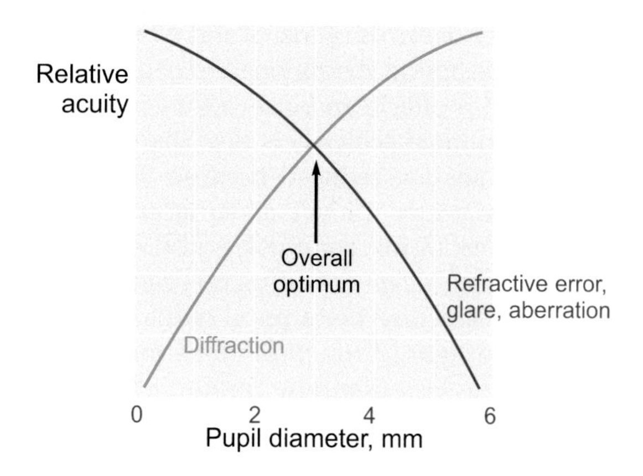

Figure 7.30 Pupil and visual acuity. Highly schematic representation of the effects of pupil diameter on different optical factors affecting acuity and the overall effect.

Table 7.2 Advantages and disadvantages of small and large pupils

Smaller pupil	Larger pupil
Increases depth of field	Receives more light
Minimises optical aberrations	Minimises diffraction effects
Minimises glare	

The most obvious way in which the retina influences visual acuity is simply that the receptors are a finite distance apart from each other, which in itself inevitably limits the detail that can be seen. In the very centre of the fovea, receptors are at their closest, with spacing of a little over 2.3 µm, or half a minute of arc, which happens to be about the same as the half-width of the pointspread. Actually this is no coincidence, since obviously it wouldn't make much sense to have wonderful optics with a beautifully detailed image combined with huge receptors incapable of transmitting the information; nor would it be helpful to have finely packed receptors able to respond to details that in real life they could never possibly experience, because of optical blur (**Fig. 7.31**).

The relative importance of optical as opposed to retinal and neural factors determining acuity can be determined directly by arranging to project a grating on the retina in such a way that its contrast is unaffected by the quality of the optics. One way of doing this is to generate interference fringes on the retina by means of two point sources of coherent light from a laser: the resulting interference pattern is in effect a sinusoidal grating, whose frequency depends on the separation of the two sources, and whose contrast is substantially unaffected by the quality of the optics. One can then measure the subject's threshold contrast as a function of frequency, as already described, and compare the result with what is found when viewing a 'real' sinusoidal grating. Although there is some improvement when the optics are bypassed in this way, even when the eye is fully corrected it is not a very great one. This suggests that the retina is in a sense *matched* to the eye's optical properties.

But it's not just the receptor spacing that counts; we need to consider what happens to the neural signal as it goes through the retina to the brain. In many ways one can think of the spatial pattern of activity in the receptors and subsequent neurons as another kind of image – a neural one – that may be subject to the same kinds of degradation as an optical image.

Neural blur

If the receptors simply had a one-to-one connection to bipolars, and bipolars to ganglion cells, then one would not expect any deterioration to occur as the neural image is passed along. We shall see later that this is perfectly true in the centre of the fovea, where acuity is highest, but is certainly not the case further out. Here one finds many receptors pooling their information, funnelling onto one bipolar, and many bipolars funnelling onto one ganglion cell. A telling statistic is that each eye has about 130 million receptors but only about one million ganglion cells. In fact the degree of convergence in the periphery is actually even bigger than that implies, because of huge overlap of receptive fields. In the periphery, a typical alpha ganglion cell may receive input from a staggering 75 000 rods, via 5000 rod bipolars (**Fig. 7.32**). This is good news from the point of view of trying to *detect* things, because it increases the difference between the size of the signal you are trying to detect and the background noise which would otherwise obscure it: in fact it accounts for nearly all the difference in sensitivity between rods and cones (since the difference in threshold between one isolated rod and one cone is only about a log unit). But it is not so good for acuity.

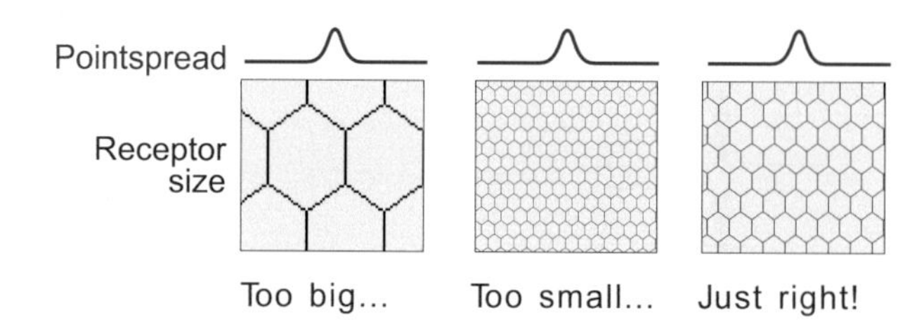

Figure 7.31 Receptor spacing in the retina is tailored to the optical properties of the eye.

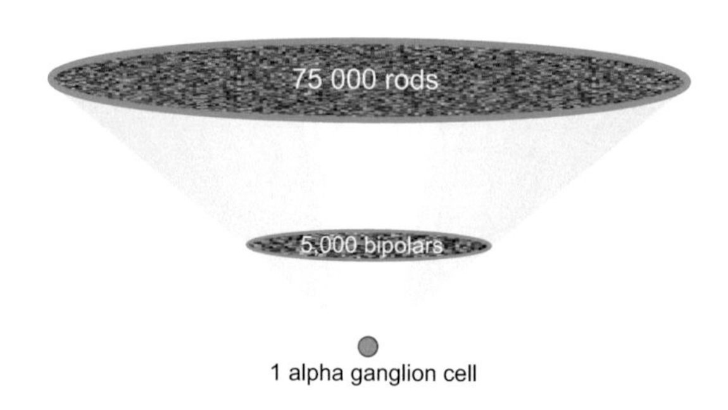

Figure 7.32 Receptor pooling, which is more marked further from the fovea and particularly limits visual acuity in the dark-adapted state.

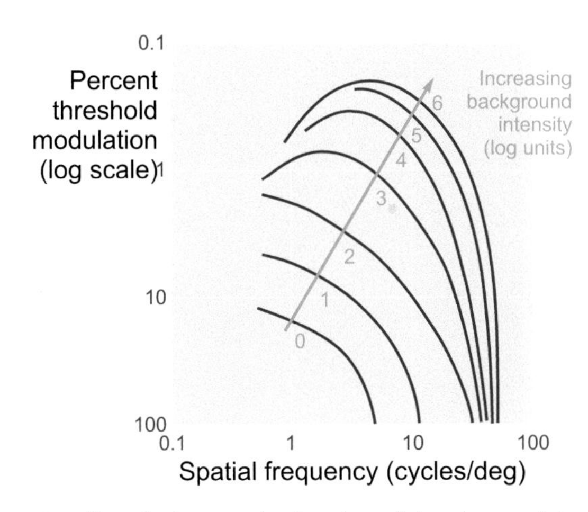

Figure 7.33 Effect of adaptation level on the visibility of sinusoidal gratings. Each curve is a contrast sensitivity function of the kind shown in **Figure 7.24**, measured at a different background intensity level ranging over 6 log units. (After van Nes & Bouman, 1967; permission pending.)

Because this *neural* pointspread is broader the further from the fovea, acuity is very much worse in the periphery than in the centre. And most important of all, as you dark-adapt, changing from central cone vision to rod vision with their enormous pooling of information, acuity drops off dramatically. If the contrast threshold as a function of spatial frequency is measured with a fixed pupil during progressive stages of dark adaptation, one finds a steady decrease in the cut-off frequency (**Fig. 7.33**), the result of changes in the neural organisation of the retina. One of the adaptational responses to reduced light levels, as we shall see, is an increase in the effective size of the ganglion cells' summation pools, so that they can catch more light. But this obviously has the effect of increasing the degree of neural blur, and hence of reducing the overall acuity.

And this brings us back to the pupil once again. The fact that as you dark-adapt, the neural blur introduced by the retina increases and begins to dwarf the real optical blur caused by bad optics, means that the pupil can now afford to get bigger. Whereas a large pupil in bright light is a very bad thing because of the effect it has in increasing the aberrations and defects of focusing, in dim light these are not so important, so that one can enjoy the benefits of having more light for the purposes of detection. Thus there is a kind of necessary reciprocal relation between sensitivity and acuity; the *better you are at one, the worse you are at the other.* A big pupil provides more sensitivity but worse acuity; pooling of receptors onto ganglion cells also provides more sensitivity but less acuity. We shall see other examples of this kind of trade-off later on.

To summarise, there are many factors that contribute to visual acuity, and their relative contributions depend on the state of adaptation of the eye: they are summarised in **Table 7.3**. In good light, an emmetrope's acuity is limited by diffraction, and thus about as good as could be expected from an eye of the size that we actually have. But most people are not emmetropes, so that without spectacles it is refractive error that limits their acuity.

The retina

One might hope to be able to see another person's retina directly by eyeball-to-eyeball confrontation: if both eyes are emmetropic and relaxed, each retina should be clearly in focus on the other (**Fig. 7.34**). The reason why this doesn't

Table 7.3 Factors affecting visual acuity

Target	Optical	Receptor/neural
Contrast	Aberrations Chromatic Spherical	Receptor spacing: matched to best optics
Colour Wavelength – decreased diffraction if shorter; Monochromatic light – decreased chromatic aberration	Diffraction	Neural convergence and divergence: decreased spatial resolution with dark adaptation
Luminance Very low luminance: increased quantum fluctuation Dark adaptation: large pupil degrades optics; increased neural convergence	Refractive error Myopia Hypermetropia Astigmatism	
	Glare, scatter Pupil size Decrease: increases diffraction Increase: all other optical factors worse	

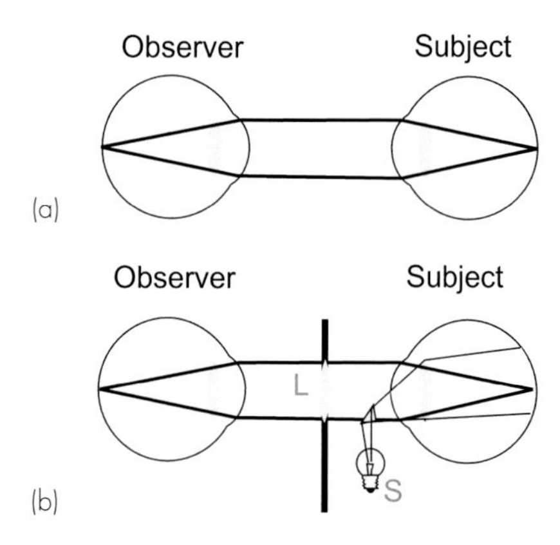

Figure 7.34 The principle of the ophthalmoscope. (a) If an observer looks into a subject's eye and both are emmetropic, the retina of one will be focused on that of the other. But the observer's eye prevents light from reaching the subject's retina, so that nothing can be seen. (b) However, the ophthalmoscope introduces an extra source of light (S) to illuminate the subject's retina, and is fitted with a set of lenses (L) that correct for errors of refraction.

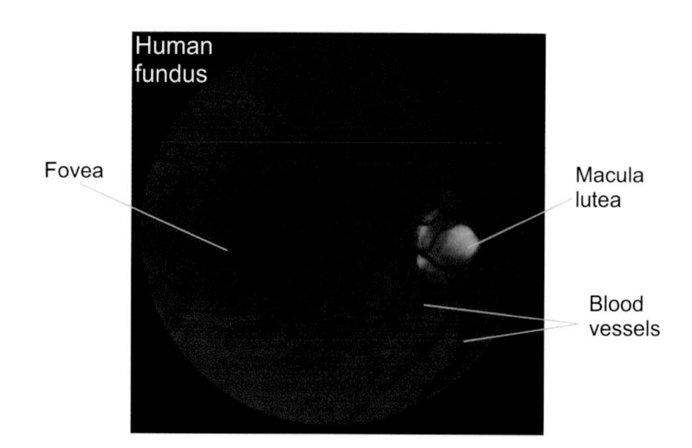

Figure 7.35 Photograph of a living human retina.

in fact work is that the presence of the observer's eye also prevents light falling on the other's retina, so that nothing can be seen: under normal conditions the pupil of the eye is always dark. The *ophthalmoscope* is a device that gets round this problem by projecting a small beam of light into the subject's pupil while the examiner is peering into it. It also has an arrangement whereby one of a set of negative and positive lenses can be introduced into the optical pathway: the power of the lens that exactly brings the subject's retina into sharp focus is equal and opposite to the combined refractive errors of observer and subject. Thus so long as an oculist knows his own correction, the ophthalmoscope provides an objective method for determining what spectacles the subject requires, as well as permitting the examination of the retina for signs of disease.

Two features of the retina are immediately obvious when seen through the ophthalmoscope (**Fig. 7.35**), both of them the consequence of what seems like a massive error of judgement on the part of Nature, namely the decision to have the retina inside out. The tips of the photoreceptors face outward, and they pass their information backwards through the retina: as a consequence, the nerve fibres from the retina find themselves inside the eye when they want to be outside. What they do is to come together to form the optic nerve, and crash their way to the exterior, together with the central retinal artery and vein, through a region called the *optic disc*, at about 15° to the nasal side of the optical axis. Since this area is consequently incapable of responding to light, subjectively it forms the *blind spot*. Although it is some 5° across, one is usually unaware of its existence because the brain tends to fill it in with whatever background colour or pattern immediately surrounds it (**Fig. 7.36**).

The other gross feature of the retina that is visible with the ophthalmoscope is that very close to the centre of the retina is an area about 15° across that is free of large blood vessels – they arch around on each side to supply it from the edge – and is also distinctly yellower than the rest of the field. This is the *macula lutea* (yellow spot), and at its centre is a very small dot – actually a depression or pit – called the *fovea centralis*. When we look at a small object in the outside world, it is the fovea that is directed to the corresponding part of the retinal image. This region is specialised for high-quality, photopic vision: the central fovea is quite without rods, and the cones themselves are tightly packed to give the maximum information about image detail (**Fig. 7.37**): the angular size of the rod-free region is about that of one's fingernail with the hand fully extended. Cones in this region are about 2.3 μm across, corresponding to a visual angle of some half minute of arc. The depression arises because the retinal structures that elsewhere in the retina lie between the receptors and the lens – remembering again that the retina is inside-out in its layered structure – are here displaced to one side so as to cause the minimum scattering of incoming light. The supply of oxygen and nutrients for this region must derive almost entirely from the blood vessels that richly supply the *choroid*, the layer immediately superficial to the

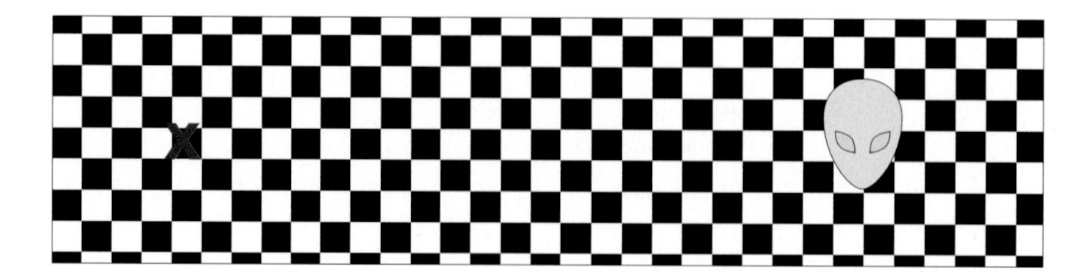

Figure 7.36 Demonstration of the blind spot. Close your left eye and fixate the red cross from a distance of about 30 cm: the alien will disappear, yet no discontinuity in the background will be apparent.

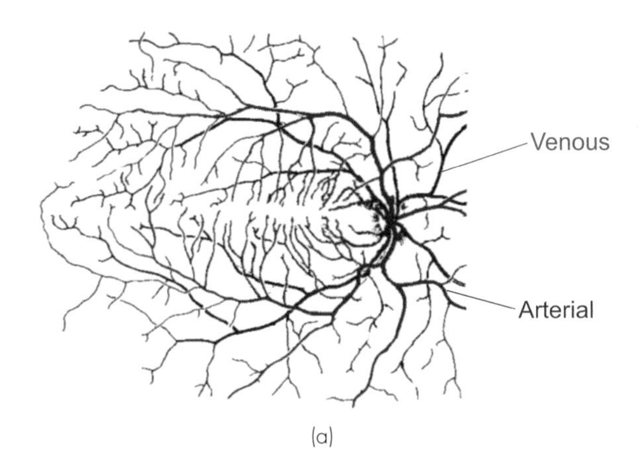

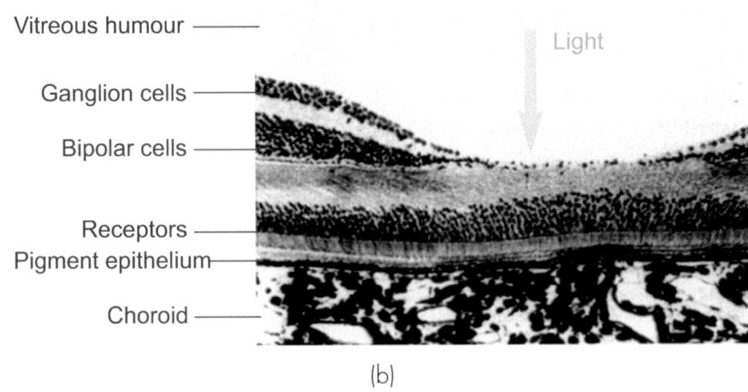

Figure 7.37 (a) The blood supply to the retina. Note you can see this on yourself by rubbing a lit pen-torch up and down against a closed eyelid. (b) Section through a monkey fovea. The direction of incident light is from above, passing through the layers of neural elements (which in the central fovea are pushed to one side) before reaching the receptors.

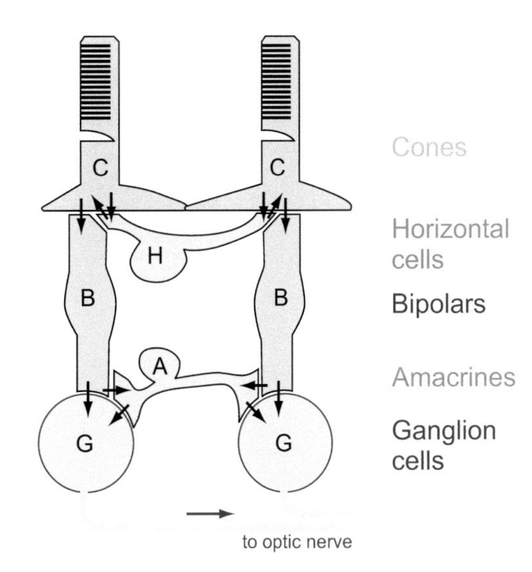

Figure 7.38 Retinal neurons and their connections. Rods and cones (C) form synapses with horizontal cells (H) and bipolars (B). The bipolars connect with amacrine cells (A) and ganglion cells (G), whose axons form the optic nerve.

receptors and separated from them by the thin *pigment epithelium*, which helps to reduce scatter by absorbing the incident light.

The retina is quite different from any of the sense organs we have met so far, in that a good deal of the neural processing of the afferent information has already occurred before it reaches the fibres of the optic nerve; it is in effect part of the brain. No doubt the reason for this is that the eye is a highly mobile organ, and if each of the 130 million or so receptors sent its own individual fibre into the optic nerve, the latter would have to be some 11 times thicker than at present and would be a considerable hindrance to rapid movement of the eye; and of course the blind spot would be correspondingly larger as well. Thus some degree of compression of the afferent visual information is needed, and overall there is indeed a 100-fold convergence of information from large groups of receptors, particularly from rods in the periphery. The fibres of the optic nerve are in fact at two synapses removed from the cones: receptors synapse with *bipolar cells,* and these in turn synapse with the million or so *ganglion cells* whose axons form the optic nerve. These two types of neuron form consecutive layers on top of the receptor layer – except in the fovea, where we have seen that they are pushed to one side – and are mingled with two other types of interneuron that make predominantly sideways connections. These are the *horizontal cells* at the bipolar/receptor level, and, at the ganglion cell/bipolar level, the *amacrine cells* (some of which also provide another stage of convergence for the rod signals; see p. 147). The arrangement of the connections of all these types of interneuron is shown schematically in **Figure 7.38**; the general arrangement is quite constant across species, though details vary. We shall see that there are marked differences in the electrical behaviour of these different neurons: although ganglion cells and amacrines show action potentials in response to retinal stimulation, the bipolars, horizontal cells and the receptors themselves do not. They are small enough to be able to interact electronically without the need for active propagation.[12]

The photoreceptors

Rods and cones both consist of two distinct parts: an *outer segment*, apparently a grossly modified cilium, and an *inner segment* containing the nucleus. The outer segment possesses a high concentration of photopigment, associated with a richly folded set of invaginations of the outer surface, which are formed at the bottom and gradually move up to the tip over the course of a month or so, then breaking off and being destroyed. In the case of rods, they seal themselves off near the bottom, to form a stack of flattened saccules or discs (**Fig. 7.39**); in the cones they remain partially open: it is not obvious why. At the base of the outer segment the remains of the ciliary filaments and centrioles can be seen. The inner segment has mitochondria as well as the nucleus, and its inner end forms the synaptic junction with bipolar and horizontal cells. There is no doubt that the *photopigment* straddling the membranes of the outer segment discs plays a key role in transforming incident light into electrical changes, for if the pigment is isolated from the receptor, it is found that its absorption of light of different wavelengths corresponds closely with the spectral sensitivity of the receptors themselves.

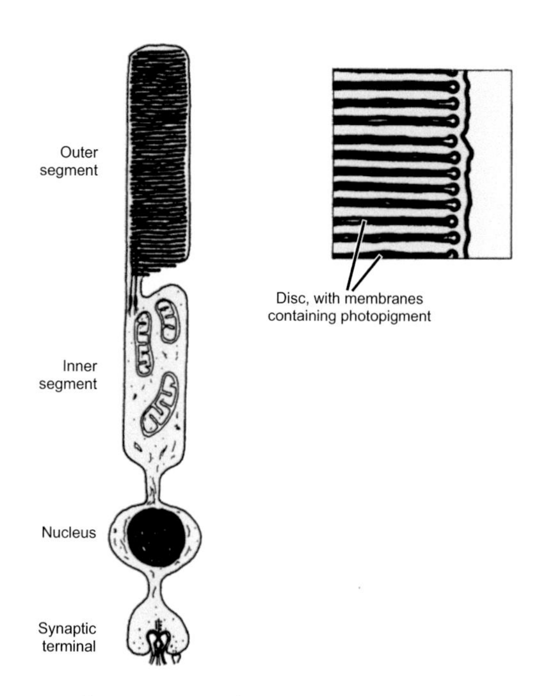

Figure 7.39 Schematic section of a monkey rod showing inner and outer segments.

Retinal photopigment consists of two portions: a chromophore called *retinal* or retinene (a derivative of retinol, better known as vitamin A), in association with a protein/oligosaccharide complex with a molecular weight of around 40 000, called an *opsin.* It is slight differences in the composition of the opsin part that give rise to the different spectral sensitivities of rods and cones. In the case of the rods, the combination of rod opsin with retinal is called rhodopsin, sometimes also known as visual purple. In frog rods, there are about 1.5 million rhodopsin molecules in each disc, and about 1700 discs per rod.[13] Much more is known about rhodopsin than other pigments, but there is no reason to believe they are essentially different. Rhodopsin absorbs over nearly all the visible spectrum, peaking in the green region, at around 500 nm (See **Fig. 7.10**). We can compare this with the spectral sensitivity of vision itself, by measuring the absolute threshold for lights of different wavelength, in the dark-adapted state when only the rods are operating. We saw earlier (p. 131) that this curve, the *scotopic sensitivity curve*, corresponds very closely to the absorption spectrum for rhodopsin, implying that absorption by the pigment is indeed the first step in the transduction process.

The first effect of light on rhodopsin is to cause an isomerism of the retinal from the normal, bent, 11-*cis* form to the straightened *all-trans* configuration (**Fig. 7.40**). This in turn leads to a series of changes in the configuration of the rhodopsin, producing a number of more or less short-lived intermediates, at the end of which opsin and retinal part company, and the rhodopsin is said to be *bleached*. In the test tube, this is the end of the matter; but in the retina, the ingredients are all recycled by enzymes present in the receptors and in the pigment epithelium that lies behind them. The first stage of this process consists of the

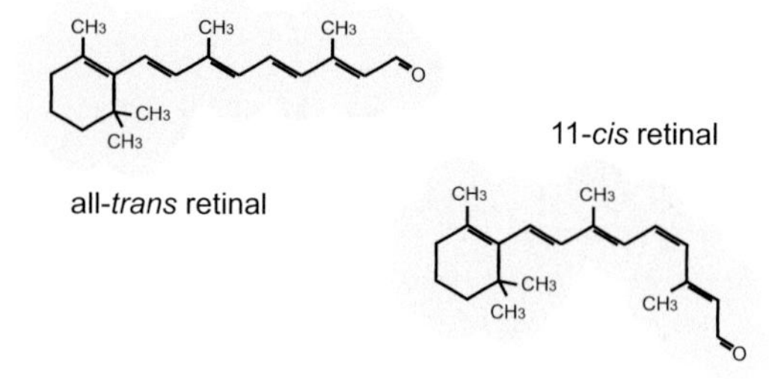

Figure 7.40 All-*trans* retinal and 11-*cis* retinal.

reconversion of the free all-*trans* retinal back to the 11-*cis* form, a relatively slow process (**Fig. 7.41**). The significance of these wanderings of pigment back and forth between receptor and pigment epithelium is unclear.[14]

We shall see later that it is this slow regeneration of pigment that determines the long time-course of recovery of rod sensitivity during dark adaptation that has already been mentioned (p. 130). In bright light, most of the rhodopsin is in the bleached form: an equilibrium is reached in which the rate of bleaching equals the rate of regeneration. Estimates of the amount of pigment in the receptors of a living eye during particular stimulus conditions may be made by the technique of *retinal reflection densitometry*, in which one measures the amount and spectral composition of the light scattered back from the retina when a light is shone into the eye. In this way it is possible to track continuously the amount of rod or cone pigment in bleached form under relatively natural visual conditions. Alternatively, in microdensitometry, the spectral absorptions of individual receptors may be measured in a preparation on a microscope slide. As far as we know, the reactions that occur in rods and cones are fundamentally similar, though the

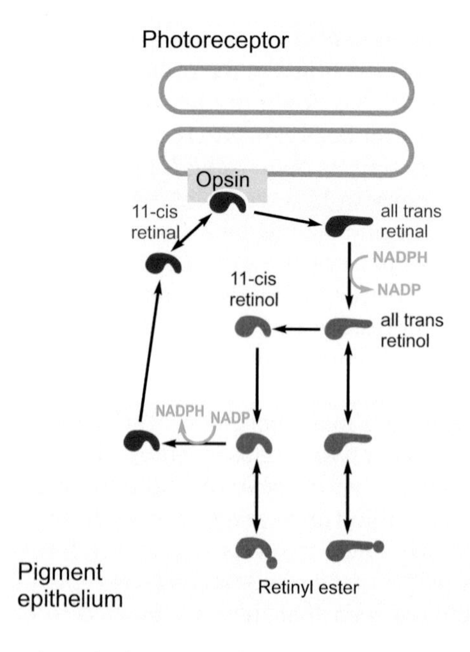

Figure 7.41 The cyclical sequence of events by which light leads to isomerisation of retinal and its dissociation from opsin, followed by the relatively slow processes that lead to the final regeneration of rhodopsin.

regeneration of cone pigment is substantially quicker than in rods, so that under photopic conditions a smaller fraction of the cone pigment is in the bleached state than is the case for rods: this is one of the reasons why the cones are able to function at much higher light levels.

Electrical responses to light

The nature of the basic transduction process was outlined in Chapter 3 (p. 50). One of the stages in the sequence of photopigment bleaching – it is not certain which – is coupled by a G-protein called *transducin* to the activation of *phosphodiesterase* (PDE), that converts cyclic guanosine monophosphate (cGMP) to GMP. Since cGMP tonically promotes the opening of sodium channels in the plasma membrane, the effect of light on the outer segment is to *reduce* sodium permeability by reducing the level of cGMP (**Fig. 7.42**). Because the effect of light is to reduce the amount of cGMP, you can see that – perhaps paradoxically – light will cause a *reduction* in sodium permeability instead of an increase. As a result, the receptor hyperpolarises from a resting value of some –30 mV to a maximum of –60 mV, when the response saturates because all the channels are closed. As in many such cascades, there is a huge amplification of effects along the way: in rods, each quantum absorbed appears to cause the breakdown of about a million cGMP molecules, although the next stage is a bit of an anticlimax, since it takes three cGMPs to open a channel. Measurements of the absolute threshold for seeing dim flashes of light when the eye is fully dark-adapted show that a single rod is capable of responding to a single absorbed photon. Individually, cones are an order of magnitude less sensitive (photopic vision as a whole is *several* orders of magnitude less sensitive, because it enjoys less convergence and pooling of neural signals).[15]

A disadvantage of cascades is that they tend to be slow, and the time-course of the hyperpolarisation generated by a brief flash of light is very prolonged – a characteristic of indirect transduction with a lengthy cascade – and shows a pronounced plateau with very large stimuli, corresponding to closure of all the sodium channels, as can be seen in the cone hyperpolarisations shown in **Figure 7.43**, in

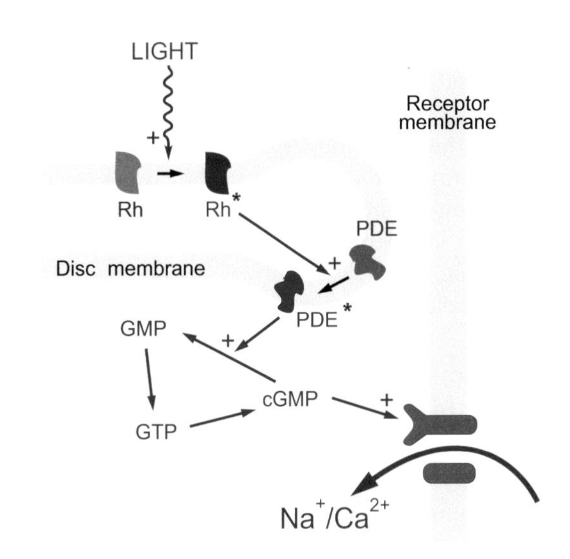

Figure 7.42 The cascade linking bleaching of rhodopsin (Rh to Rh*) to sodium permeability, as believed to occur in vertebrates. The entry of calcium at the same time as sodium is believed to be a mechanism contributing to receptor adaptation, in part by slowing the regeneration of cGMP. PDE, phosphodiesterase.

response to flashes of intensities ranging from 0 to 4.2 log units.

As a result, if one plots the size of this receptor potential as a function of the intensity of the flash (**Fig. 7.44**) one finds a characteristic S-shaped, or saturating, relationship. The effect of different levels of light adaptation is to shift this curve along the intensity axis, providing one of the mechanisms by which the sensitivity of the retina is adjusted to suit the prevailing luminance. Bright backgrounds shorten the response as well as reducing its size; the significance and mechanism of this are discussed later (p. 149).

Students are sometimes upset to discover that photoreceptors respond to a positive stimulus, light, by what might be regarded as a negative response (hyperpolarisation and consequent reduction in the rate of transmitter release at the synaptic ending). It is understandable enough that a physicist should regard turning on a light as a positive signal; but in the natural world, *dark* is just as often

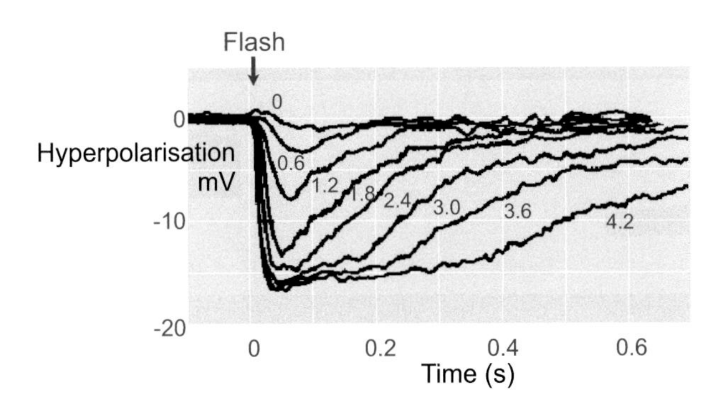

Figure 7.43 Electrical responses from cones. Hyperpolarisations generated by a very brief flash of various intensities ranging from 0 to 4.2 relative log units in steps of 0.6 log unit.

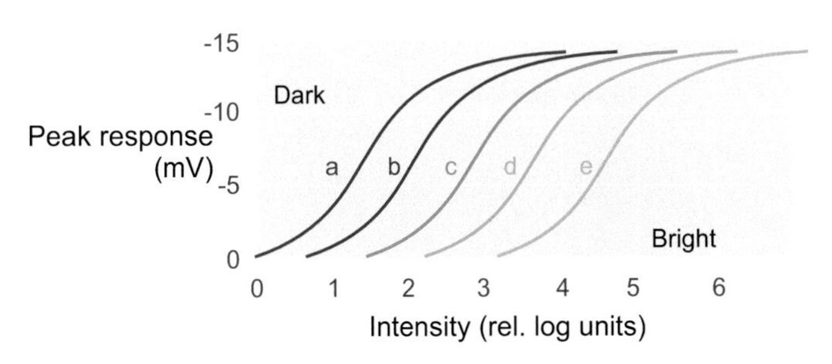

Figure 7.44 The effect of pre-adaptation to the hyperpolarising response of cones in the context of different backgrounds. Each curve is the result of an experiment like the one in **Figure 7.43**, where peak electrical response is plotted as a function of flash intensity. Curve a was measured in the dark, curves b–e under different increasing background adaptation levels over a range of some 3.5 log units.

something one may want to react to as light; indeed more so, since objects are probably more often dark against a light background than vice versa – think of a frog catching flies, or a student reading. And for many small creatures sudden dimming of the whole visual field is a terrible sign that something is either about swoop down on them or step on them. So it is not surprising that darkness is something one specifically wants to recognise, coded for at a very early level. Photoreceptors are really scotoreceptors.

Horizontal cells and bipolars

The *bipolar* cells represent the next stage in the transmission of visual information from receptors to brain. Photoreceptors tonically release their transmitter, glutamate, from their pedicles or feet, and because they hyperpolarise in the light they release less of it. Their connections with the bipolars are quite complex (**Fig. 7.45**). Rods synapse with a particular kind called the *rod bipolar,* which depolarises in response to light. There is tremendous scope for getting muddled here. One way to keep a clear head is to avoid talking about excitatory and inhibitory synapses, and instead describe them as *sign-preserving* and *sign-reversing*. A *sign-preserving* synapse is in effect one with a conventional 'excitatory' transmitter: if the presynaptic cell depolarises, it releases more transmitter, so the postsynaptic cell depolarises as well; conversely, hyperpolarisation of the presynaptic cell leads to hyperpolarisation of the postsynaptic one. At a sign-reversing synapse, depolarisation of the presynaptic cell causes postsynaptic hyperpolarisation, and vice versa (**Fig. 7.46**).

Whereas the bipolars driven by cones in the *central fovea* have a one-to-one relation to individual cones, those driven by rods in the *periphery* show an enormous amount of convergence (**Fig. 7.47**), contributing to the rod pooling mentioned earlier. Cones make contact with two kinds of bipolars. With one class, the *flat bipolar,* the sign-preserving synapse is at the base of the receptor and is relatively conventional in appearance; the bipolar hyperpolarises to light. The other kind is of an unusual, sign-reversing, type in which invaginations of the foot of the receptors receive processes from both *invaginating bipolars* (depolarising to light – very unusually, the glutamate is inhibitory, reducing sodium permeability indirectly via cGMP-mediated channels) and also from *horizontal cells* (hyperpolarising) in a kind of three-way junction. Here, the receptor affects both horizontal cells and bipolars, and in addition transmission to the bipolar is modulated by the horizontal cells acting on the receptors (they also make conventional GABAergic sign-reversing synapses with both types of bipolars). Since each horizontal cell is stimulated by a large number of receptors, what the horizontal cell is doing in fact is summing the activity of the receptors over a wide area, forming an estimate of something like the average background luminance. In many species there are gap junctions between horizontal cells, so that they respond to light over an even wider area than one would expect from their anatomical size. In fish their effective fields are well over a millimetre in diameter, even though the horizontal cell itself extends over some 150 μm at most; these gap junctions are disabled in the dark-adapted state, under the control of dopamine released from interplexiform cells. All of this means that – under light-adapted conditions – the horizontal cells can provide a mechanism of lateral inhibition (**Fig. 7.48**); in the dark, it is turned off to maximise sensitivity.

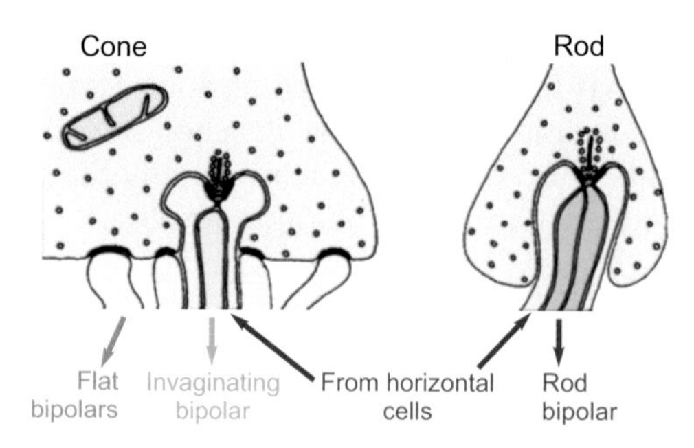

Figure 7.45 Interactions between photoreceptors, bipolars and horizontal cells.

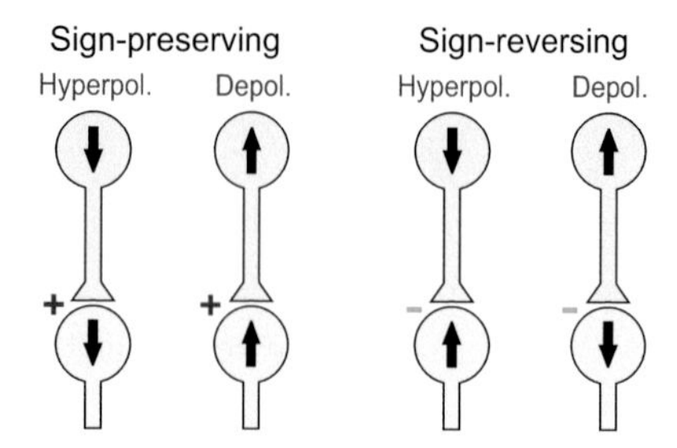

Figure 7.46 Sign-preserving and sign-reversing synapses. While the cone bipolars come in sign-preserving and sign-reversing forms, rod bipolars are restricted to the sign-reversing variety.

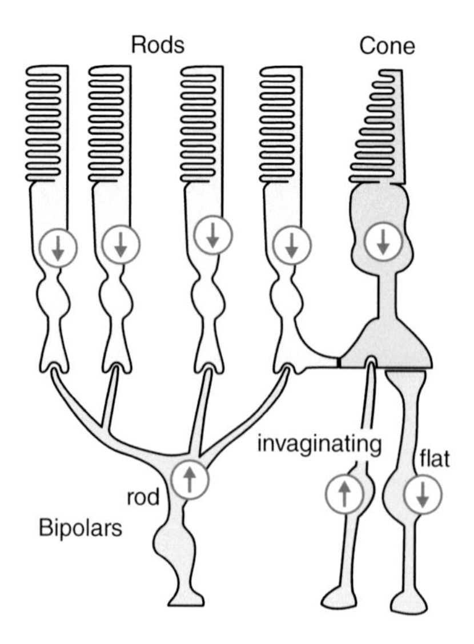

Figure 7.47 Diagrammatic representation of the functional relationship between primate rods and cones, with the different types of bipolar cell. The arrows show the sign of the electrical response: upward means depolarisation.

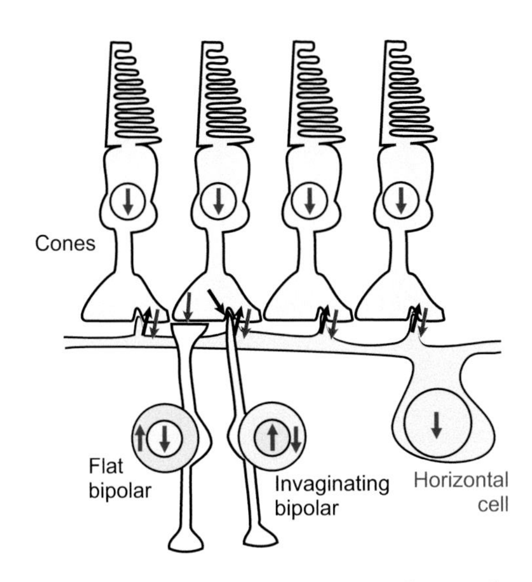

Figure 7.48 Lateral inhibition in bipolar cells. A set of cones influences a horizontal cell; bipolars may receive a hyperpolarising input from a receptor and a depolarising one from the horizontal cell, or vice versa, resulting in a receptive field with opposed centre and surround. Arrow conventions as in **Figure 7.47**.

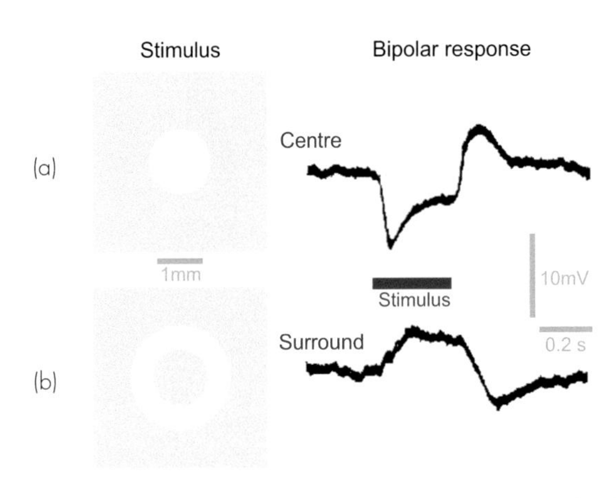

Figure 7.49 Responses of a bipolar cell to a disc (a) and annulus (b) of light, showing antagonism between centre and surround.

This can be demonstrated directly by electrical recording from bipolars. Although their electrical responses are generally similar in size and timescale to the slow potentials that can be recorded from the receptors themselves, they show receptive field properties that are quite different from those of rods and cones. Whereas the receptive field of a receptor is very simple (a small area over which light causes hyperpolarisation), in the case of a bipolar the field is much larger, and is non-uniform: light falling in its centre has the opposite effect to light falling in the periphery. Thus a cell that hyperpolarises when a bright spot shines on the centre of its receptive field will depolarise if light is shone only on the surround, and vice versa (**Fig. 7.49**). The effect of this antagonism between centre and surround is to make the bipolar respond more vigorously to small stimuli in the centre of its field than to large areas that cover both centre and surround; the existence of the two populations of bipolar cell, depolarising and hyperpolarising, no doubt corresponds to the need to be able to detect objects that are lighter than their backgrounds as well as those that are darker. It is similarly intuitive that the rod bipolar is restricted to a depolarising response given the need to detect light stimuli on a dark background in scotopic circumstances. More generally, this centre-surround antagonism is a mechanism for lateral inhibition, whose desirability was discussed in Chapter 4. In this case, it helps to detect the edges that separate one visual object from another. In a more abstract sense, because a bipolar shows a response in one direction to a light in the centre, and the opposite in its surround, it is in effect calculating the difference between the two, or in general ΔI. So you can see that it is beginning the computation of *contrast* that dominates visual perception. As always, it is *change* – either in time as in adaptation, or in space as here, that the brain wants to know about.

In addition to pooling of information through horizontal cells, direct electrical synapses between the receptors themselves, and also between horizontal cells, can sometimes also be seen. Pooling of this kind is extremely desirable under scotopic conditions, when it is helpful to average responses over large areas of the retina in order to increase sensitivity and thus distinguish feeble stimuli from background noise (but it doesn't, of course, do much for one's visual acuity; see p. 140). In the periphery of the retina, pooling of this kind is also mediated by the bipolars, some of which receive synapses from large numbers of receptors and hence provide the first stage of convergence, the funnelling of information that enables the number of optic nerve fibres to be so much smaller than the number of rods and cones, and supplements the pooling effects of the electrical synapses. *Rod* bipolars do not synapse with ganglion cells directly, but only through a particular class of amacrine (AII) that in turn passes its information on indirectly through the feet of *cone* bipolars, providing a second stage of convergence (**Fig. 7.50**). As a result, one (alpha) ganglion cell may receive an input from perhaps 200 AII amacrines, in turn driven by a few thousand rod bipolars, which ultimately gather information from over 70000 rods. In the fovea this convergence is much less evident, and most of the bipolars contact only a single cone: acuity is thus preserved, at the expense of sensitivity.

There is a curiously complex changeover between these different systems as we go from the light-adapted state to scotopic vision (**Fig. 7.51**). When photopically adapted, rods are completely saturated and the detailed cone pathway is the only functional one. In the mesopic region, rods add their signals to those of cones through mutual electrical synapses in the receptor pedicles, but the AII route is switched off by another kind of amacrine cell (A18). In dark adaptation, the cones are below their threshold, but the A18 (which is dopaminergic) switches the AII route on, creating the high-convergence, high-sensitivity visual system needed to cope with low light levels (we saw earlier that dopamine also decouples horizontal cells, reducing lateral inhibition).

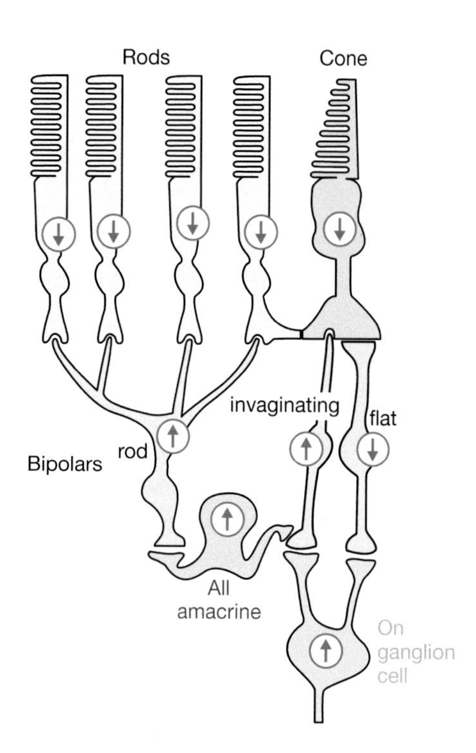

Figure 7.50 Schematic representation of the relationship between various types of bipolar cells, a typical 'on' ganglion cell, and an AII or rod amacrine cell. Rod bipolars receive information from many rods and depolarise in response to light; their signals reach ganglion cells via AII amacrines, which are deactivated except under scotopic conditions, under the control of another kind of amacrine (A18, not shown), which releases dopamine.

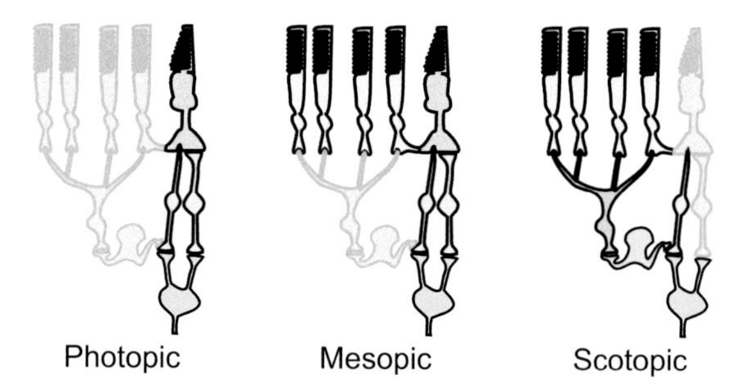

Figure 7.51 Diagrammatic representation of the functional pathways at different light levels. Under photopic conditions the rods are saturated and cannot contribute. In mesopic conditions, their signals are added to those of cones through electrical synapses. In the scotopic range, the cones cannot function, and the rods operate through the highly convergent AII system previously described.

Ganglion cells and amacrines

We noted earlier that amacrines and ganglion cells differ from all the other types of retinal neuron in that they respond to light with repetitive spike discharges. Of the ganglion cells, the simplest are W or wide-field cells that show sustained responses to steady light level over a large field, with very slow conduction velocities: clearly some such source of information must be projected into the optic nerve to explain such tonic responses to constant illumination as the tonic pupil light reflex, or the various hormonal responses to day length and time of day. But most show more complex responses, and fall into more or less distinct functional classes, often related to their size or the general shape of their dendritic tree. One broad classification is between large neurons with high conduction velocities that respond *transiently to* movement or changes of light intensity, and small neurons showing *sustained* responses with simple linear summation when different parts of the receptive field are simultaneously illuminated. The former are called magnocellular (M) or parasol cells in the primate (Y cells in cat); the latter, which form some 80 per cent of the ganglion cell population, are parvocellular (P) or midget, or X in the cat. Some of these differences seem to be related to whether their input comes primarily from bipolars (sustained) or from amacrines (transient); in mammals, all signals in the dark-adapted state must be transmitted via the amacrines.

Like bipolars, both M and P ganglion cells generally have receptive fields consisting of a centre region with an antagonistic surround. M cells have larger receptive fields than P, and respond only transiently when retinal illumination is suddenly changed from one level to another. Sometimes a transient burst of firing is found in response

Clinical box 7.2 Inherited retinal dystrophies

Defects can occur in genes that encode proteins involved at various stages of phototransduction: the retinal dystrophies. Retinitis pigmentosa is the most common of these, affecting 1 in 3000 people, but rather than being a single disease entity, it represents a final common pathway that arises from any number of insults yielding rod photoreceptor degeneration. Clinically this manifests with night blindness, followed by an insidious loss of peripheral vision and, eventually, loss of visual acuity (as the cones eventually degenerate). Pathologically retinal pigment epithelial cells disintegrate, photoreceptors die and strange reticular patterns can be traced across the back of the eye as abnormal pigment accumulations thread alongside blood vessels.

Advances in genetic medicine are successfully identifying the causative mutations for inherited retinal diseases; for example, over 60 can lead to retinitis pigmentosa. This field is particularly exciting as treatments, such as ocular gene therapy, appear increasingly within reach. Indeed, a gene therapy for a mutation affecting the enzyme RPE65 (a component of the retinal pigment epithelium) is now licenced, although this represents a major challenge for health economics (at the time of writing, the price stands at $425 000 per eye!).

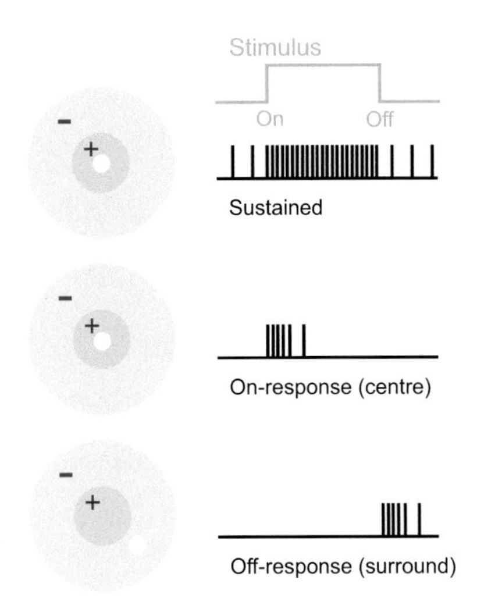

Figure 7.52 Schematic representation of types of ganglion cell response.

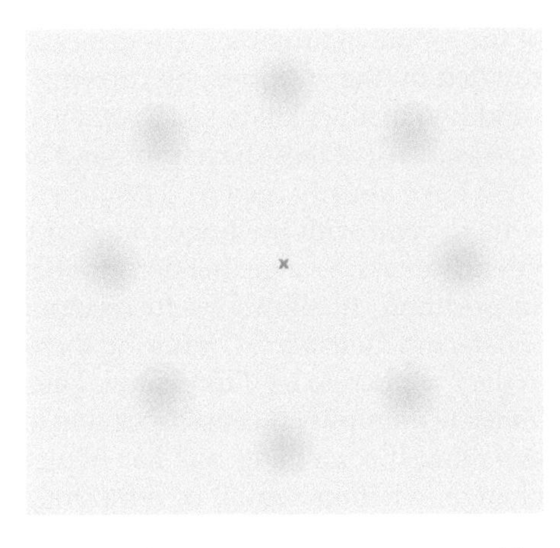

Figure 7.53 Perceptual fading. Fixate the centre cross carefully: after some 10–20 seconds you will find the peripheral spots disappear, but come back as soon as you make an eye movement.

to an increase of illumination (an 'on-response'), sometimes to a decrease ('off-response'); sometimes one may find a burst response both at the beginning and end of a period of steady illumination (an 'on–off response'). A cell with an on-response at its centre will normally show an off-response in its surround, and *vice versa* (**Fig. 7.52**), and they show on–off responses in intermediate regions. The P cells normally show sustained responses, and many code specifically for colour, showing colour-opponency (see p. 165).

More specialised ganglion cells can often be found, responding to movement in particular directions, and a host of other things; 23 different classes of ganglion cell have been described in the cat retina, but not all these kinds of response are found in all species. It has also been shown that a few retinal ganglion cells are intrinsically photosensitive, containing a pigment called melanopsin; they appear to project particularly to the hypothalamus and suprachiasmatic nucleus, presumably mediating such functions as triggering circadian rhythms and monitoring day length. It is noteworthy that diseases centred on photoreceptors lead to loss of vision but do not affect circadian rhythms.

The complete adaptation shown by the M cells may correspond with the fact that images that are stabilised on the retina, for example by projection through a device attached to the cornea, disappear from view in a matter of seconds; the function of this kind of fast adaptation was discussed earlier (p. 71). Even without strict stabilisation, you can see the effect for yourself with low-contrast, low-spatial-frequency stimuli such as the one in **Figure 7.53**. At all events, it is clear that the pattern of neural activity that is sent from the retina to the brain is not just a simple map of retinal illumination, but that various kinds of information have already been computed and extracted from the retinal image, in preparation for the still more specific analysis that is performed by the brain itself.

It seems very likely that these new functional features seen at the ganglion cell level but not in bipolars – especially transient behaviour and movement sensitivity – are due to the actions of the amacrine cells. It is easy to see how in principle they can generate transience by delayed inhibition (**Fig. 7.54**), and the same circuit could also provide the lateral inhibition of the centre/surround arrangements.

Amacrine cells come in a large number of distinct types, each with a characteristic morphology and often with its own particular synaptic transmitter, including some fancy peptides, and it is likely that each type has a different specialised function. In some cases, such as the AII and A18 neurons mentioned earlier, particular types are known to carry out very specific functions, for example mediation of the transient and field properties of ganglion cells, but in general we currently have rather little idea what different roles amacrines may play in rendering the coding of information in the optic nerve more intelligent.

Early processing

Having completed our tour of the retina, we are now in a position to consider the more difficult and perhaps more interesting problems that arise when we try to correlate what we actually perceive (the psychophysics of vision) with what we know about neural networks like those in the retina (the neurophysiology). The most important

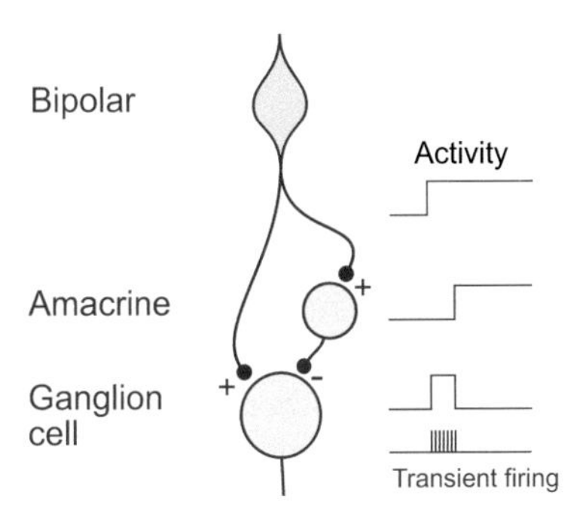

Figure 7.54 Schematic wiring diagram showing how amacrine cells may mediate the transient properties of ganglion cells.

function of the retinal neurons is to pre-process the information provided by the receptors, by carrying out *lateral inhibition,* and two distinct kinds of *adaptation:* changes in *gain* and *transience.* All of these have profound implication for vision. We have already seen (p. 129) how changes in gain allow us to cope with the huge range of luminance that the eye encounters, a range that defeats all man-made devices. In addition, it allows us to recognise objects through registering their *albedos,* ignoring the accident of how much they happen to be illuminated. Lateral inhibition and transience simplify the messages sent to the brain by only signalling if something *new* has happened: they enhance change, whether spatial or temporal, and hide from the brain things it doesn't need to know.[16]

Gain adaptation

There are several different mechanisms in the eye that contribute to adaptation. The first is a rather obvious one, but not actually very important: the diameter of the *pupil,* controlled by the iris of the eye. The pupil of course expands in dim light and contracts when it's bright, which must help to reduce the range of light levels reaching the retina. The reason why it's not all that important is that the range over which it can vary is quite small. In very bright light it may be 2 mm in diameter, in the dark about 7–8 mm. This means the *area* changes by a factor of only a little more than a log unit, which is only a small fraction of the 12 log units over which we have good vision.

The other mechanisms are of two distinct kinds. When we change the overall illumination of the visual scene, part of the resultant change in sensitivity occurs almost immediately, and is simply a function of how intensely the visual field is illuminated at any moment: this kind of adaptation is called *field adaptation.* But in addition, we find that having adapted to bright visual surroundings, when the brightness is subsequently reduced it takes an appreciable time for sensitivity to return to its original value; the time-course of this slower component of adaptation, which persists after the adapting stimulus has been removed, turns out to be closely related to how much of the retinal pigment is in the bleached form: it is called *bleaching adaptation.*

Field adaptation

The simplest way to demonstrate the changes in sensitivity that accompany field adaptation is by means of an *increment threshold* experiment (**Fig. 7.55**). Typically, what we do is set up a screen with a background I on it, and then add an extra test light, ΔI, and see how big it has to be just to be seen against the background; the sensitivity is then the reciprocal of this threshold value, or $1/\Delta I$. It turns out that over a moderate range of background intensities the ratio $\Delta I/I$ is constant: in other words, the sensitivity is inversely proportional to the light level to which one is adapted. This is known as the Weber – Fechner relationship, and the quantity $\Delta I/I$ as the Weber fraction, k.[17] We saw earlier that it also represents *contrast,* which can be

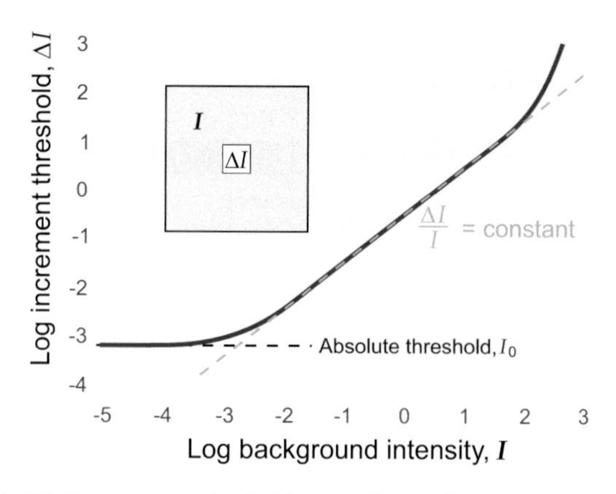

Figure 7.55 The increment threshold curve. The smallest increment, ΔI, that can just be seen against a background I is plotted; for very small values of I, the curve levels out at the absolute threshold, ΔI_0; whereas in the mid-range, ΔI is proportional to I. At very large values of I, saturation occurs, and ΔI begins to rise more steeply. (Data from Aguilar & Stiles, 1954.)

defined for something like a star or bright line on a dark background, for sinusoidal grating patterns (see p. 130) or indeed for any spatial pattern, and is absolutely fundamental to vision. Whether we perceive something or not nearly always depends simply on whether the contrast of the retinal image is greater or less than a relative fixed threshold, typically around 1 per cent.

This proportionality in the increment threshold experiment breaks down both at very high and very low luminances. At the high end, the size of flash needed increases out of proportion to the background: this can be explained very well in terms of the kind of saturation of receptor response shown in **Figure 7.44**. At low luminances, the value of ΔI levels off to a fixed quantity, ΔI_0, which is the absolute threshold (i.e. the 'increment' threshold for a flash when there is no background present at all). In fact you may already have spotted that in the formula $\Delta I/I = k$ something rather *silly* is going to happen when I tends to zero. In that case ΔI will also tend to zero, which means we ought to be able to see *anything,* no matter how dim it is; whereas we know perfectly well that even in pitch darkness there are always *some* things so dim you can't see them.

In fact, if you go on turning the intensity of the background down, the value of ΔI gradually levels off to a constant value, called the *absolute threshold,* ΔI_0. So the formula $\Delta I/I = k$ is not quite right: but it turns out that we can describe this low-intensity behaviour very economically with a simple alteration to the formula:

$$\frac{\Delta I}{I + I_0} = k$$

where I_0 is a constant; so at zero I, $\Delta I_0/I_0 = k$.

That is a neat mathematical trick, but what does it mean? It is as almost as if there were some source of light I_0 within the receptors that is somehow being added to any true light being received, even in pitch darkness. That's nonsense, of course, but what *is* clear is that even in the dark, the pigment in receptors tends to break down spontaneously in exactly the same way as if it *had* really received light. Even when no light falls on them, there is always a

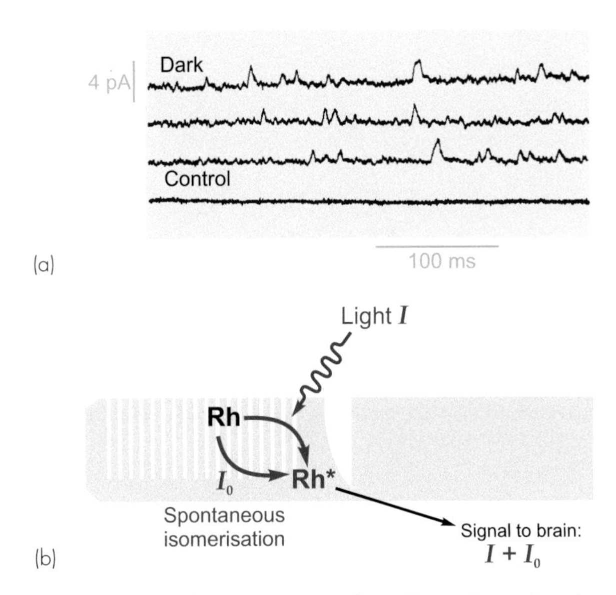

Figure 7.56 Noise in photoreceptors. (a) Three electrical records in the dark, with a control record below showing the level of instrumental noise. (b) Schematic diagram highlighting the signal to the brain includes the degree of spontaneous isomerisation of photopigment.

certain probability that the rhodopsin molecules will isomerise spontaneously through thermal activation, initiating the same train of events that would normally be triggered by the arrival of a photon. This rate of breakdown is actually quite small – in rods, of the order of one molecule per receptor every 2 minutes or so, but because there is such an enormous number of rods – of the order of 100 million – taken altogether it is rather significant. So even in the dark the receptors are all the time sending out these false messages about non-existent light – called receptor noise (**Fig. 7.56**); it is sometimes also known as the *dark light*. You can see the dark light yourself when you're fully dark-adapted – not black, exactly, but a sort of changing fluctuating grey: in German it is called the *Eigengrau* – 'intrinsic grey'. Consequently, even if no background I is present, the receptors will still generate a neural signal that will look to the rest of the visual system exactly like a 'real' background. The reason that there is an absolute threshold at all is that visual targets have to be detected against this virtual background of dark light.

The Weber–Fechner relationship implies that the sensitivity is not constant, but decreases as the background intensity gets bigger. How might this be done? One possibility is to arrange for the sensitivity to be turned down automatically by the prevailing level of the input ($I + I_0$). Such a device is called an *automatic gain control* (AGC),[18] and is used in radio receivers, for example, to maintain a roughly constant average level of output to the loudspeaker despite fluctuations in the strength of the received signal: the analogy with visual adaptation is obvious. In particular, if we arrange for the gain to be equal to $1/(I + I_0)$, then the output in response to an increment ΔI will be $\Delta I/(I + I_0)$, and on the assumption that there is some constant threshold at which this is just detectable, the Weber–Fechner law is explained.

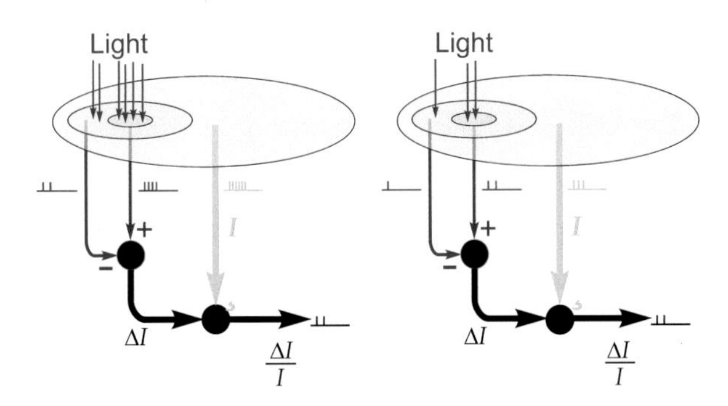

Figure 7.57 Two aspects of adaptation. Sensitivity to change, ΔI, comes about primarily through centre-surround arrangements of receptive fields. Changes in gain or sensitivity are logically distinct and may come about in part through the monitoring of larger areas of the visual field. Together, they create a final signal proportional to $\Delta I/I$ that is essentially independent of changes in overall illumination (compare left and right).

It is perhaps worth pointing out that the Weber-Fechner formula contains *two* kinds of adaptation built into it, which are logically entirely distinct (**Fig. 7.57**). On the one hand we have the change in gain implied by the division by ($I + I_0$): we have examined at length the reasons why this kind of adaptation is necessary if we are to perceive objects independently of their illumination. But as well as this *divisive* adaptation, which scales the visual signals according to the prevailing light level, there is another kind of adaptation, implied by the ΔI, which is quite distinct, and is *subtractive*, making us sensitive to differences or change, whether in time or space. The function of this *transience* is discussed later (p. 155).

Where in the retina is this automatic gain control operating? There are almost too many places where it *might* occur, and at different times various proposals have been made. Any kind of negative feedback could do the trick (see Chapter 3, p. 62), and suitable feedback circuits are evident at the ganglion/amacrine cell level and at the horizontal/bipolar level; but on the whole they are probably more concerned with adaptation of a slightly different kind, that generates the transient properties obvious in many ganglion cell responses. An attractive candidate is a feedback mechanism that can be demonstrated within the receptors themselves, involving calcium. When light falls on receptors, as we have seen, sodium channels in the outer segment close; but these channels are also permeable to Ca^{2+}, so that a consequence of a raised level of illumination is that calcium concentration within the receptor starts to fall. Calcium has several intracellular effects, including one that is relevant to the gain of the transduction process: it inhibits the recycling of GTP to form cGMP, that opens the sodium channels (**Fig 7.58**). So in the light, when calcium levels are low, cGMP is quickly replaced and the responses to light are relatively small and brief; in the dark, the abundant calcium slows cGMP regeneration, so that the response to light is increased in *duration* as well as size (**Fig. 7.59**).

This mechanism appears to be supplemented by a further mechanism: *recoverin*, a protein resembling calmodulin in certain respects, responds to calcium by inhibiting

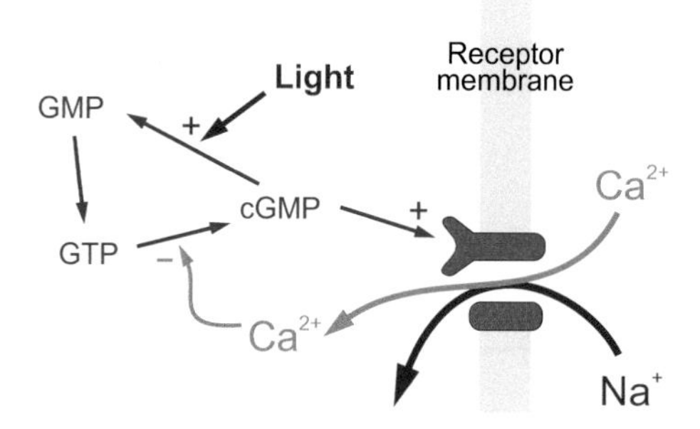

Figure 7.58 A mechanism for adaptation. Calcium acts to inhibit the conversion of GTP to cGMP such that, in the light, cGMP is more rapidly regenerated.

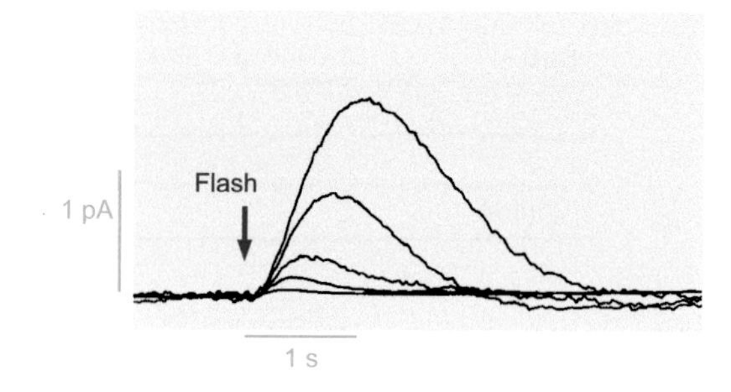

Figure 7.59 Effect of adaptation on time-course of toad rod response. Stimuli of constant intensity were delivered in the presence of backgrounds of increasing intensity, starting from zero (top trace). It is evident that brighter backgrounds make the response shorter as well as smaller. (After Lamb, 1984; with permission from John Wiley and Sons.)

rhodopsin kinase. Since rhodopsin kinase arrests the phototransduction cascade by phosphorylating, and thereby inhibiting, the activated form of rhodopsin, its reduced activity at low light levels prolongs the response to light. If you record from rods and compare the electrical responses to the same flash at different background levels, you can see not only how the size increases as background level reduces, but also how dark adaptation dramatically draws out the timescale of the whole thing. This slowing down also indirectly increases sensitivity, because it means that at low light levels the cell is integrating or summing the signals over a longer period of time.

It is very easy to see this summation effect directly for yourself by looking at flickering lights. The slower the receptors, the less good they are at following rapid changes in luminance, so as you turn the frequency up there comes a point where you can no longer see that the light is not simply steady. This *flicker fusion frequency* is directly related to summation time, and the more dark-adapted you are, the lower the flicker fusion frequency. Incidentally, flicker is in *time* what a grating is in *space*, and there is a clear parallel between the effects of dark adaptation in increasing sensitivity on the one hand by increasing summation time, and on the other by increasing the effective sizes of receptive fields, through spatial pooling.

Bleaching adaptation

Bleaching adaptation is quite different in its properties. It can be demonstrated by means of the same apparatus as for measuring increment thresholds, exposing the subject to an adapting field I which is then turned off before the eye's sensitivity is tested with the test flash ΔI. The results are very different from the previous case. Whereas in field adaptation ΔI depends directly on I, now it is found that ΔI is a function not of I on its own, but rather of *how much pigment was bleached* during the adapting period: for fairly short adapting periods this is proportional to the product of I and the time of exposure. The second difference is that the adaptation lasts a considerable period after the adapting field has been switched off (see **Fig. 7.9**): the time-course of the recovery also goes hand-in-hand with the time-course of regeneration of pigment, and the extent to which the threshold is elevated depends in a quantitative way on how much pigment was bleached.

From our knowledge of the photochemistry of the pigment we might well have expected some such effect: obviously sensitivity must depend in part on the amount of active pigment present, and if say 20 per cent of it is in the bleached state, then we would expect the eye to be 20 per cent less sensitive. But it turns out that the changes in sensitivity that result from pigment bleaching are vastly greater than the simple proportionality that would be expected by this argument. **Figure 7.60** shows the time-course of recovery of a rod monochromat's log threshold (relative to

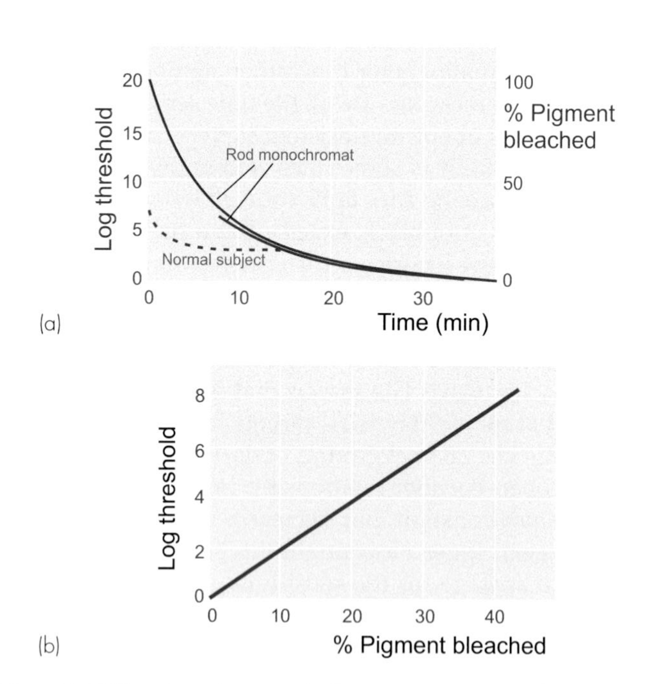

Figure 7.60 The relation between rhodopsin concentration and sensitivity during dark adaptation. (a) Continuous red line shows percentage of rhodopsin in the bleached state at various times during recovery after a full bleach. The curves are closely similar in a normal subject and in a rod monochromat. Also shown are the simultaneous measurements of absolute threshold by a rod monochromat (solid blue line) and normal subject (dashed line), plotted on a logarithmic scale as shown on the left. (b) The relation between percentage pigment bleached and log threshold obtained from this experiment in the case of the rod monochromat. This relationship is a linear one, and threshold is proportional to 10^{aB}, where B is the percentage of pigment bleached and a is a constant.

the absolute threshold) after a very strong bleaching light: by means of the technique of retinal densitometry (p. 144) it is also possible to measure the proportion of pigment that is bleached in this experiment at the same time (red curve). The two curves agree rather well: for a normal subject (dashed blue line) the threshold curve shows a clear cone component as well. In the case of rods, it turns out that it is not ΔI, but $\log(\Delta I)$, that is proportional to B, the fraction of pigment bleached, and that very small bleaches produce very large changes in sensitivity: a 20 per cent bleach produces not a 20 per cent reduction in sensitivity, but a reduction by a factor of 10 000!

Another feature of bleaching adaptation that shows that it is not just the consequence of a simple lack of pigment is that if we bleach a patch of retina with a pattern that affects some receptors and not others, we find that not only is the sensitivity of the bleached receptors reduced, that of their neighbours is as well. It is difficult to avoid the conclusion that bleached receptors are sending some kind of message into a pool – perhaps to horizontal cells – that turns down the gain over a relatively wide area.

We can get a clearer idea of how this comes about by measuring the increment threshold curve after bleaches of different sizes. What one then finds is that, although the values of ΔI are unchanged at high background luminances, as I is reduced, the curve flattens off sooner to a higher value of ΔI_0 the more pigment is in the bleached state (**Fig. 7.61**). It is as if the presence of free opsin as a result of bleaching caused an increase in the level of retinal noise or 'dark light', and indeed such curves can be very well explained by supposing that the dark light, I_0, is increased by a factor of $10^{\alpha}B$, where α has a value of about 20. In other words, bleaching an area of retina has much the same effect on sensitivity as shining a light of luminance $I_0 \times 10^{\alpha}B$ on it; this imaginary light is called the *equivalent background*. What seems to be happening is simply that bleaching the pigment greatly increases the rate of spontaneous isomerisation and hence the background of retinal noise (**Fig. 7.62**).

But why do we not see this light? In fact we often do, but ignore it. After viewing a light that is bright enough to bleach a significant amount of pigment – an ordinary light bulb does very well – at first we see a bright after-image (*positive after-image*), which is in effect $I_0 \times 10^{\alpha}B$; but being stabilised on the retina, it undergoes complete adaptation, as already described, and after some seconds it fades from view. If we now look at an illuminated surface, the after-image may reappear, but in its negative form: the areas that have been bleached are less sensitive than the rest of the retina, because their steady dark light turns down than AGC, so that the bleached areas look darker than their surroundings. When we go from a brightly lit room into a dark one, the reason we cannot see clearly at first is that everything we look at is superimposed on an invisible background consisting of all the after-images that we have accumulated over the past 20 minutes or so. Although short-term field adaptation prevents us from seeing the dark light that is generated by bleaching, the pathways that

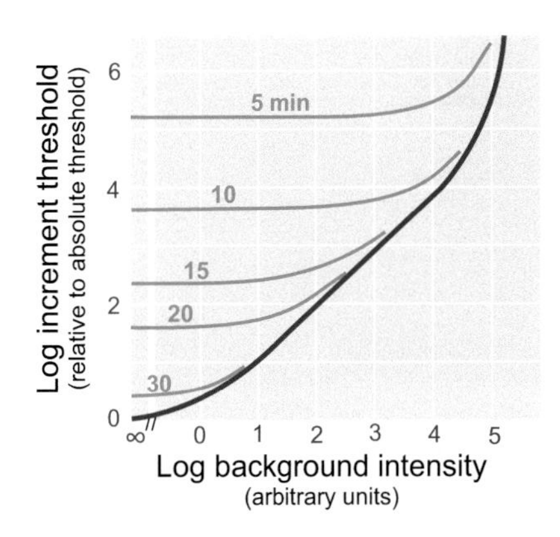

Figure 7.61 Effect of bleaching on increment thresholds. Increment threshold curves are plotted as in **Figure 7.55**, at various times (shown in minutes above each curve) after a very intense bleach of the rods. (Data from Blakemore & Rushton, 1965; with permission from John Wiley and Sons.)

control the size of the pupil do not show this complete adaptation, and during the course of dark adaptation the pupil responds to the dark light in exactly the same way that it would to real lights. Incidentally, the lack of visual sensitivity in the dark that is caused by vitamin A deficiency can be explained very easily in these terms, since it will result in a pile-up of opsin that cannot be reconverted to rhodopsin, which will increase B and turn the sensitivity down.

What is not clear in this system is how the function $10^{\alpha}B$ comes about, and whether the dark light signal is conveyed in the same neural pathways as those from real lights. It is difficult to think of plausible mechanisms by which a receptor's output would be the sum of a signal corresponding to the amount of light falling on it, i.e. proportional to the rate of bleaching, and a signal so nonlinearly dependent on the amount of bleach. The answer almost certainly lies in the complex interactions that occur between the many intermediate products produced by the action of light on the retinal pigments.

Lateral inhibition

The general properties of lateral inhibition, and its desirability in sensory systems, were discussed in Chapter 4. It

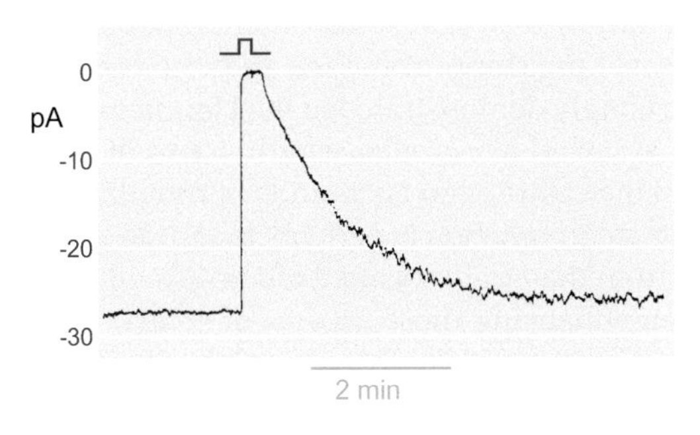

Figure 7.62 The effect of bleaching a receptor followed by recovery in the dark. You can see in this record spontaneous depolarisation and noise after exposure of toad rods to a brief but intense light, that has bleached a small proportion (0.7%) of their pigment: the subsequent electrical noise has the same properties as the random responses that would occur during actual illumination.

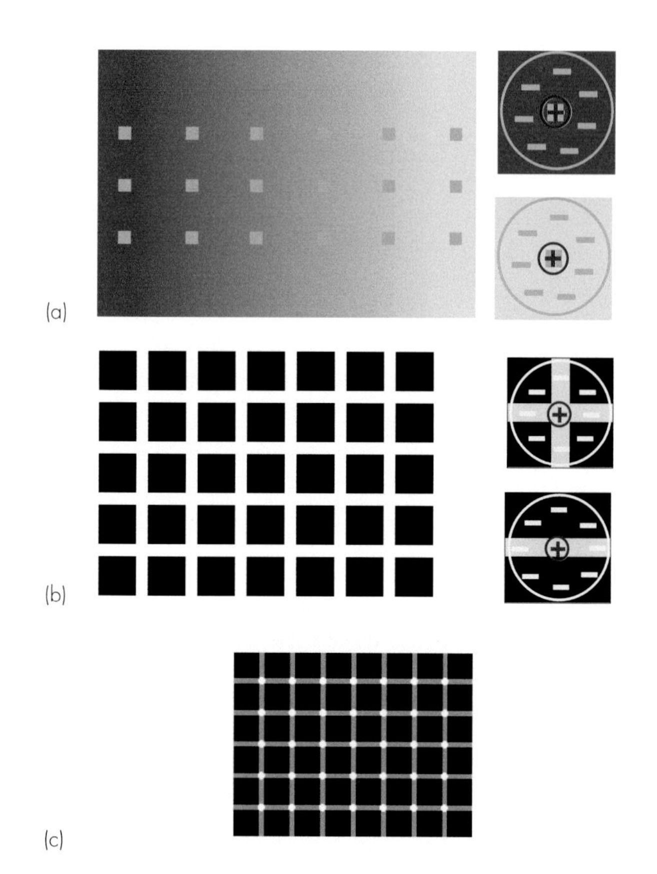

Figure 7.63 Illusions caused by lateral inhibition. (a) Simultaneous contrast: the small grey squares are of equal luminance, but appear different in brightness because of the backgrounds they are seen against. (b) The Hermann grid: illusionary dark spots are seen at the intersections of the white bars. Both illusions can be explained in terms of units with centre-surround organisation. The effect is much less striking at the fovea, perhaps because of a difference in the size of the inhibitory surrounds. (c) An enhanced version of the Hermann grid, not easy to explain.

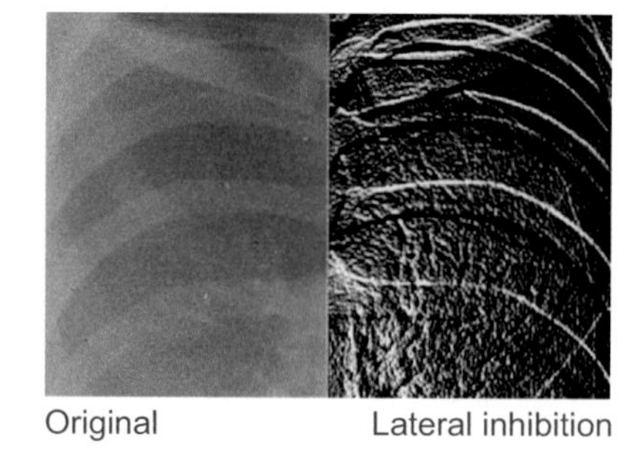

Figure 7.64 The effect of lateral inhibition on a plain chest X-ray.

is a very important feature of visual processing, and is indeed one of the first things that the retina does – at the horizontal cell level – to the signals that come from the receptors. It accounts for some of the phenomena of simultaneous contrast (**Fig. 7.63**), in which the appearance of a grey patch of constant luminance is so enormously influenced by the background. Like adaptation, it helps to ensure that the subjective sensation of brightness is on a sliding scale and thus more closely related to albedo than to luminance: a bright surround reduces the response at the centre, thus making it appear darker, and vice versa. It is even more strikingly demonstrated in the Hermann grid, which can be explained in a similar way, in terms of interactions between the centres and surrounds of receptive fields.

But its main function is perhaps to recode spatial information from the receptors more efficiently, while retaining and even enhancing those aspects of spatial information which are important in recognising objects. To recognise something against a background, the really important thing is the *edge*, because the shape of the edge defines the shape of the object itself. So you don't need a vast number of neurons in the middle all saying exactly the same thing, in effect 'This bit's grey!' 'Yes this bit's grey too!' 'So's this bit!' and so on: lateral inhibition helps reduce *redundancy*. Lateral inhibition can also be thought of as enhancing aspects of an image that are of some use, and suppressing what we don't really need to know. Exactly similar processes are used for example to enhance the pictures from security cameras, space probes and clinical X-rays (**Fig. 7.64**), and to compress graphics files in computers (.JPG files, for instance): it is interesting that the compression factors of around 100 that are commonly achieved without significant loss of quality are similar to the compression ratio that is achieved in going from receptors to optic nerve fibres.

A good way to study lateral inhibition quantitatively is to measure contrast threshold as a function of spatial frequency, as already described in **Figure 7.24**. You will recall that as the frequency increases, you need more and more contrast in order to be able to see it, and that this can be explained in terms of the width of the pointspread function, influenced by optical blur and also by neural convergence. Lateral inhibition has in a sense the opposite effect: it reduces the contrast of low-frequency gratings, producing what is called the *low-frequency cut*. To see why this is so, consider the response of a unit with a receptive field consisting of an excitatory centre and inhibitory surround, as it views sinusoidal gratings of various spatial frequencies. At lower and lower frequencies, the intensity is changing more and more *slowly* (in spatial terms), so that the excitatory centres and inhibitory surrounds are about equally excited and tend to cancel out. You can see that a cell like that is going to be *most* excited by a grating that roughly matches it, in the sense that if a peak in the grating corresponds to the peak of the excitatory centre, and the troughs on each side correspond to the inhibitory troughs of the field, then the response of the unit will be a maximum. Consequently the shape of the contrast-threshold curve as a function of frequency for such a unit will be as shown in **Figure 7.24**; and just as the high-frequency cut-off tells us about the size and shape of the excitatory centre, so the low-frequency part of the curve tells us about the inhibitory surround. You can demonstrate the low-frequency cut quite easily for yourself by means of **Figure 7.65**, which is a sine-wave grating of low contrast and low spatial frequency. At a distance of a metre or so, you will easily perceive the grating. But, paradoxically, the more closely it is examined, the more difficult it is to see: eventually, when you get really close, it disappears altogether because its spatial frequency is reduced to below the cut-off for the contrast concerned. Lateral inhibition means *sensitivity to*

Figure 7.65 Demonstration of the low-frequency cut from **Figure 7.24**. This sine-wave grating, of low spatial frequency and contrast, is more easily visible at distances of 1 metre or so; when viewed more closely its spatial frequency is too small.

Figure 7.66 'Caricature di Cardinale' by Bernini (1598–1610).

change, and at low spatial frequencies the rate of change is too small to be perceived.

By measuring a number of contrast-threshold curves at different levels of light adaptation, one can follow the changes in effective field configuration as a result of changes in the retina: under bright conditions, the excitatory centre is small and the surround prominent, so that there is a marked low-frequency cut but a good high-frequency response. As the illumination is reduced, the increase in size of the excitatory area brings the high-frequency cut down to lower frequencies, as already described (p. 141), and the simultaneous reduction in lateral inhibition gradually flattens the low-frequency response (see **Figure 7.33**). At the lowest light levels the low-frequency cut cannot be seen at all, corresponding to the fact that under dark adaptation, inhibitory surrounds of the ganglion cells in experimental animals become progressively less and less prominent, and finally disappear altogether. The fundamental process at work here is one we have met several times: that of a basic dichotomy between *sensitivity* and *resolution*. If we try to maximise sensitivity by pooling information over wide areas, this can only be at the expense of acuity.

Lateral inhibition helps to explain what is perhaps otherwise rather puzzling, how it is possible for *drawings* of objects to be accepted by the visual system as representations of the real thing. As a stimulus, nothing could be more different from a real face than the sketch by Bernini in **Figure 7.66** – there are no areas of colour or light and shade, just a thin black line where in real life there would be merely a transition from one colour to another. Yet we accept it instantly because our retina is sending essentially the same kind of signals to the brain that the real things would.

Transience

Just as lateral inhibition makes you respond to *spatial* changes, so *transience* makes you more responsive to *temporal* change, and less so to steady levels. *Lateral inhibition is in space what adaptation is in time*. Consequently, there are many tests of temporal function that are exactly analogous to tests for spatial processing. Just as the spatial characteristics of vision can be determined with a sinusoidal grating, an area whose luminance is modulated sinusoidally as a function of distance, but is constant in time, so its temporal properties can be determined with a stimulus that is spatially uniform, but whose luminance is altered sinusoidally as a function of time. With a sinusoidally flickering stimulus of this kind, we can perform an experiment that is analogous to determining the threshold contrast of a sinusoidal grating as a function of its frequency, previously described; this time we ask the subject to reduce the contrast of the flicker until he can only just see it, and determine how this threshold contrast varies with the temporal frequency of the flicker. The resultant curve shows many points of similarity with the spatial one. At the high-frequency end, there is a cut-off frequency (the *flicker fusion frequency*: see p. 131) at which the flicker cannot quite be seen even with 100 per cent contrast; and at the low-frequency end, one again finds that sensitivity begins to fall off as the rate of change of luminance declines, reflecting the inability of the eye to perceive slow changes because of its adaptational mechanisms. A further striking parallel between the two experiments is that if the flicker sensitivity curve is measured at progressively lower light levels, it undergoes very similar changes to those observed with sinusoidal gratings. The increasing sluggishness of the system when luminance is low (see for instance the long-drawn out electrical responses on p. 145) is reflected in a progressive lowering of the flicker fusion frequency, while the gradual loss of the fast adaptational component causes a flattening of the low-frequency end. The close parallelism of the two effects, spatial and temporal, tempts one to speculate that the same fundamental mechanism might be responsible for both. Once again, sensitivity is achieved only at the expense of resolution: in time, rather than in space (**Table 7.4**).

The central analysis of vision

We have seen something of the wide variety of signals being sent down the optic nerve from ganglion cells, reflecting the fact that vision is used for a variety of different tasks. Boundary or edge detectors for recognition; movement detectors for proprioception; whole-field tonic units for working out the time of day or controlling the pupil; and so on. Much of this information is packed off to different destinations within the brain, that correspond with the three main uses to which visual information is put (**Fig. 7.67, Table 7.5**).

We use our eyes:

- To *recognise* objects in the outside world (primarily a cortical function);
- To *locate* them (superior colliculus);
- For *proprioception*, a source of information about our position and movement relative to the outside world (largely pretectum and associated structures in the brainstem).

Table 7.4 Integrating and differentiating the visual signal trades-off absolute sensitivity and sensitivity to change

	Space	Time
Test stimulus	Grating	Flickering light
Integration	Spatial summation	Temporal summation
	Increases sensitivity	*Increases sensitivity*
	Reduces acuity: blur	*Reduces flicker fusion frequency*
Differentiation	Lateral inhibition	Adaptation
	Enhances sensitivity to edges	*Enhances sensitivity to sudden change*
	Reduces sensitivity to uniform stimuli	*Reduces sensitivity to constant stimuli*

(In addition, a small number of fibres also project, through pathways that are largely unknown, to the areas of the brain concerned with the control of accommodation and pupil size, and hormonal responses to light. Oddly enough, recent evidence suggests that such responses may be driven by retinal ganglion cells containing photopigment that are directly light-sensitive.) Each of these kinds of analysis requires the incoming visual information to be processed in quite distinct ways, and we shall see that this is reflected in the behaviour of the neurons of which they are composed. We shall deal first with the areas that are concerned with recognition.

Recognition

Much of the brain is divided pretty strictly down the middle, and you are probably aware that at higher levels, the left half of the brain is concerned with the right half of the outside world, and vice versa. This has rather important implications for the topology of the incoming fibres from the optic nerve. In animals like the rabbit, whose eyes point almost entirely sideways, it means that the fibres from one eye have to cross over or *decussate* to reach the other, the crossing point being called the chiasm (named from its shape, after the Greek letter chi, χ), and the part lying beyond the chiasm being called the *optic tract* rather than optic nerve.

But in animals like us whose eyes point straight forward this clearly won't do, because each eye is capable of seeing both right *and* left sides of the visual world. So what has to happen is that the fibres that are interested in the right half of the world (which lie on the left side of each retina, because of the inversion caused by the lens) must go to the left half of the brain, and vice versa, turning what used to be a nice simple chiasm into something more like a motorway fly-over (**Fig. 7.68**).

A consequence of their rearrangement is that, whereas lesions of the optic nerve naturally enough cause blindness in one eye (*unilateral anopia*), a unilateral lesion of the optic tract (the continuation of the fibres after the chiasm) results in blindness of the same half field of each eye (*contralateral homonymous hemianopia*); damage to the chiasm itself may give a bitemporal *heteronymous hemianopia*. Not all the fibres decussate completely, however, and the fovea of each eye is to some extent represented in both cerebral hemispheres. This gives rise to the phenomenon of *macular sparing*: lesions that would be expected to produce an exact homonymous hemianopia often show no loss of

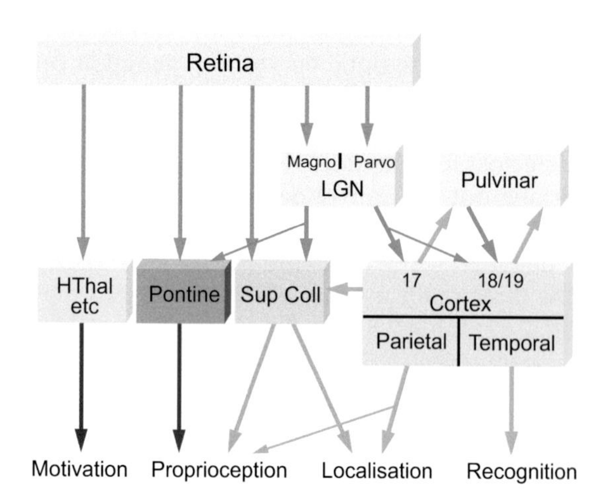

Figure 7.67 The functional destinations of fibres in the optic nerve. LGN, lateral geniculate nucleus; HThal, hypothalamus; Sup Coll, superior colliculus.

Table 7.5 The three types of visual information

Recognition	LGN, cortex	Cells respond to specialised features of a stimulus but are relatively uninterested in where it is
Localisation	Superior colliculus, cortex	Cells respond to the existence of a stimulus at a certain place, regardless of what it is
Proprioception	Pretectum, pons	Cells respond to movement of the visual field as a whole in a particular direction

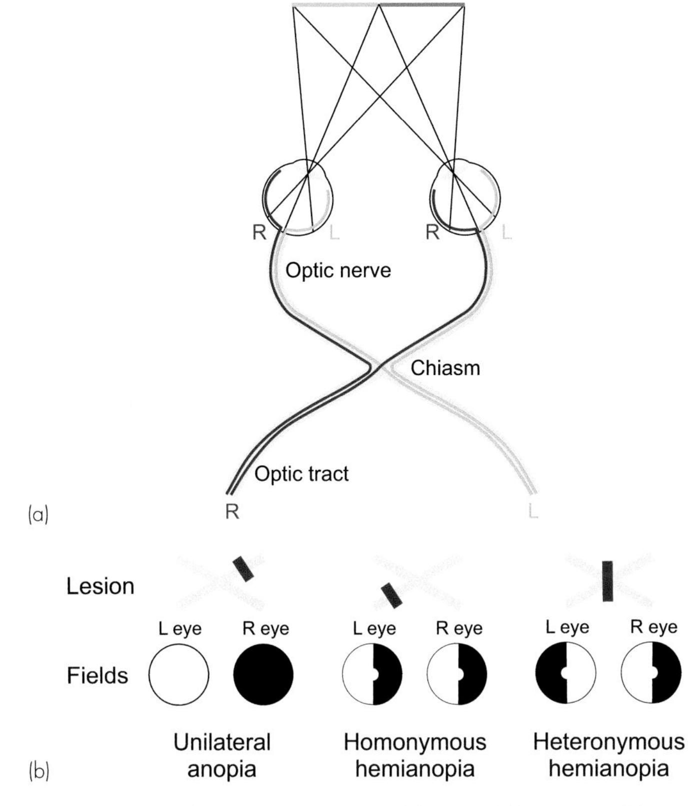

Figure 7.68 (a) Schematic representation of the partial decussation of the optic nerve fibres in the chiasm. (b) The effect on the visual fields of each eye of lesions at three different points on the peripheral visual pathway.

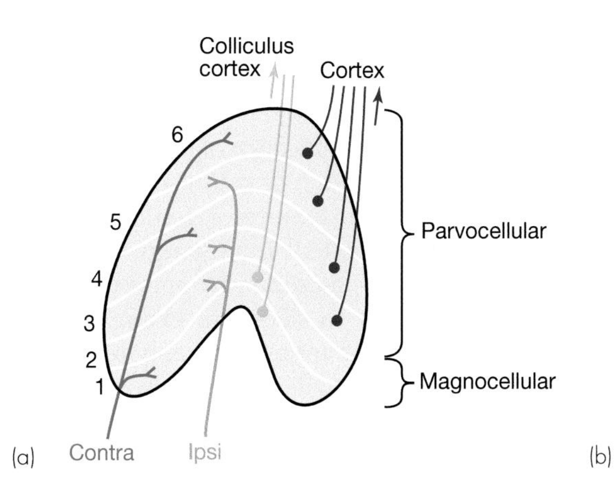

Figure 7.69 Lateral geniculate nucleus: schematic diagram of the layers, showing the termination of optic nerve fibres from each eye (a) and the cells of origin of the optic tract, projecting to different destinations (b).

vision on the affected side near the point of fixation (**Clinical box 7.3**).

Lateral geniculate nucleus

After the chiasm, the majority of optic tract fibres go on to the lateral geniculate nucleus (LGN), which is in effect part of the thalamus and essentially doing the same sort of thing that you have already come across in the somatosensory system, in part a relay up to the cortex. Oddly enough the fibres from each eye are still strictly segregated from one another, in a series of six layers in all on each side, with layers II, III and V receiving their input from the eye on the same side, and layers I, IV and VI from the other (**Fig. 7.69**). Layers I and II have larger cell bodies: they form part of the magnocellular (M) pathway that begins with the larger ganglion cells and project more to the superior colliculus and other sub-cortical regions, though many go to the cortex as well, particularly to those regions concerned with the processing of movement. The other, smaller, parvocellular (P) neurons send their axons up to the cortex, their smaller conduction velocities perhaps reflecting the fact that recognition is a relatively leisurely process whereas information about localisation and movement is often urgently required by the motor system.

The receptive fields and responses of geniculate neurons are not markedly different from those of retinal ganglion cells, showing the same concentric organisation of on- and off-responses, and may be classified into M- and P-types. Many LGN cells also show centre-surround antagonisms that are wavelength-dependent. A cell of this type might, for example, show an on-response in the centre to red light, and an off-response in the surround with green light; or for that matter, vice versa. Yellow-versus-blue cells may

Clinical box 7.3 Pituitary tumours

The pituitary gland nestles just behind and beneath the optic chiasm and, despite being only the size of a pea, pumps out the hormones upon which we depend for growth, metabolism and reproduction. It is also not an infrequent site of (usually 'benign') tumours. Often these come to medical attention when they are still tiny as they can be 'functional' and secrete excessive amounts of hormone; for example, too much adrenocorticotrophic hormone (ACTH) can be released and instruct the adrenal gland to make large quantities of cortisol, resulting in marked weight gain and hypertension, among other things (often referred to as Cushing's disease). However, when these are 'non-functional', or when the hormone in question does not cause an obvious effect – a prolactin-secreting tumour would do very little in a human – the growth can become much larger and, with little room except to expand forwards against the optic chiasm, it compresses the decussating fibres. As a result, the sufferer literally develops tunnel vision. It is hypothesised that Goliath suffered from a growth hormone–secreting pituitary tumour which accounted for his size and the ability of a nimble David to dance out of his field of vision before letting fly with the pebble which killed the giant.

be found as well as red-versus-green ones: in the interlaminar zones are cells of the *koniocellular*, or K-cell, pathway which preferentially respond to information from short wavelength cones. The significance of these spectrally opponent cells in the perception of colour is discussed later (p. 165). Another general feature of geniculate neurons is that, to a far greater extent than ganglion cells, their response to light may be modified by nervous activity in other parts of the brain: in sleep, for example, their responses to flashes of light are considerably reduced. This 'gating' of the geniculate cells, allowing some control of what information reaches the cerebral cortex (see p. 287), seems to be a general feature of thalamic relays, and it is important to remember that the thalamus receives many descending fibres from other parts of the brain in addition to its primary sensory projections.

The various layers of the LGN are all strictly in register in the sense that cells driven by the same area of retina form a single radial column, and thus form a series of superimposed maps of the half-retina. However, there is a certain amount of distortion, in the sense that central regions have a relatively greater representation than more peripheral ones: this is partly a consequence of the greater degree of retinal convergence found in the periphery. The general features of this map are preserved in the projection of neurons of the lateral geniculate through the *optic radiation* to Brodmann's area 17, the primary visual cortex, in the occipital lobe of the cerebral hemisphere, with the centre of the visual field projecting to the most medial part.[19]

Cerebral cortex

Area 17 is also known as the *striate* cortex on account of the prominent stripe of Gennari (see **Fig. 4.15**)[20] that runs through it – a consequence of the massive inflow of afferent fibres to this region (**Figs. 7.70** and **7.71**). Area 17 is surrounded by other areas (18, prestriate, and 19, middle temporal or MT) that are wholly visual; further out are associational areas which show many visual features. An alternative system of nomenclature denotes 17 as V1, divides 18 into areas V2–V4 and calls 19 V5. The interconnections between these various areas are partly direct and partly through another thalamic nucleus, the *pulvinar.*

It is a relatively simple matter to record from cells in the visual cortex, and this has been a happy hunting ground for neurophysiologists over the past 50 years or so. As in the retina, there is an enormous range of morphological types of cell within the cortex. Two of the six layers of the visual cortex contain pyramidal cells whose axons form the output to other parts of the brain, including other cortical areas: in between are interneurons of different kinds, including stellate cells. Although some cells, particularly in layer IV close to the layer of afferent fibres from the geniculate, show receptive field properties that are very similar to the roughly circular and concentric fields seen in retinal ganglion cells and in the LGN, most (and all of them outside area 17) show receptive field properties that are quite novel. Many have fields that are not circular, but when mapped with single spots of light consist of a central strip with antagonistic strips flanking it (**Fig. 7.72**): the centre may be excitatory or inhibitory, and the orientation of the entire field is different for different cells. As would be expected, what these cells respond to best is a line of a particular position and orientation: they can be called *line-detectors*. Moving or flashed stimuli are in general more effective as stimuli than steady ones, and diffuse illumination is generally completely ineffective: this is of course only what would be expected in a system designed for recognition.

Cells whose behaviour in response to different kinds of stimuli can be easily explained from the receptive fields mapped with small spots of light, whether centre-surround or linear, are called *simple* cells. *Complex* cells are not like this: it is not possible to map out excitatory and inhibitory areas of their fields as it is with simple cells, because they do not respond to single spots of light. What they respond to best is a bar or edge of a specific orientation, but in this case they will respond to such a target placed anywhere within their field of view (**Fig. 7.73**).

Moving stimuli are again more effective, and the neurons often show responses of opposite sign to movement in the opposite direction. One can think of a complex cell as extracting information about what kind of object is present, while throwing away information about exactly

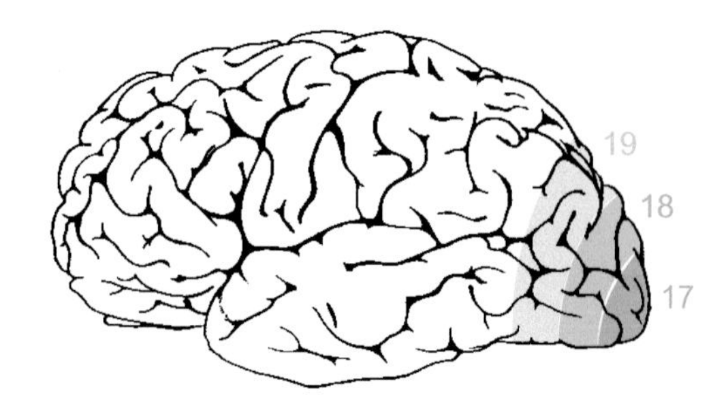

Figure 7.70 Cerebral cortex, showing the general location of areas 17, 18 and 19.

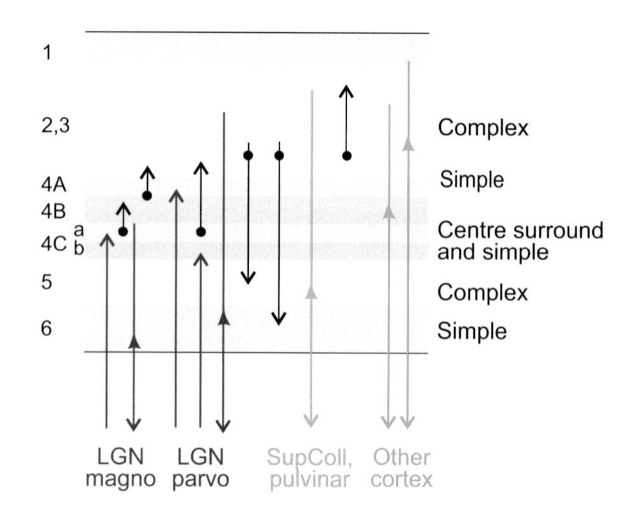

Figure 7.71 Schematic representation of the segregation of input and output in visual cortex. LGN, lateral geniculate nucleus.

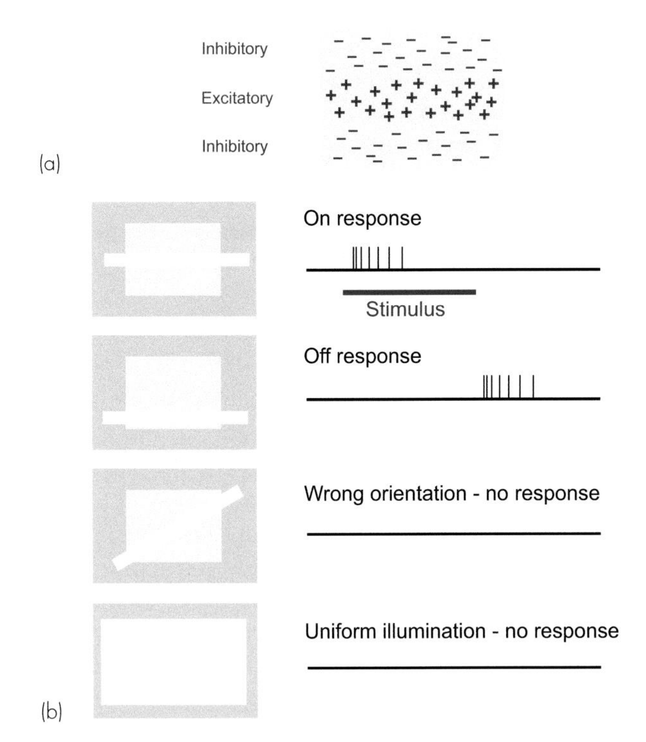

Figure 7.72 (a) Receptive field of a typical simple cell, showing regions responding to a localised spot of light being turned on (+) or off (–).
(b) Visual responses of simple cells. Such a unit responds when presented with a white line at particular positions and orientations. Its preference for horizontal stimuli is reflected in the shape organisation of its receptive field; with uniform illumination it does not respond at all, because the excitatory and inhibitory components cancel out.

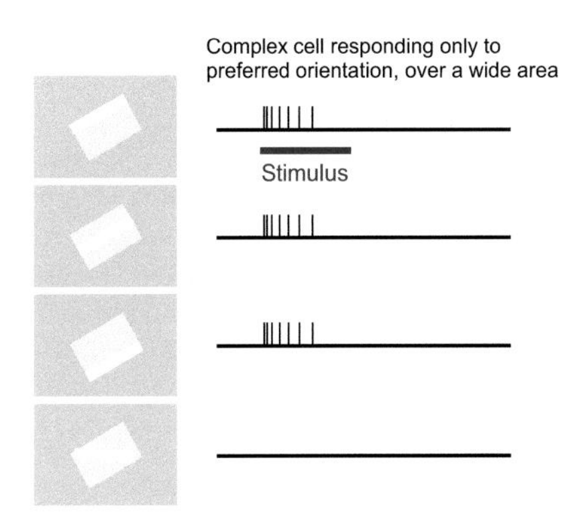

Figure 7.73 Responses of a complex cell to a light bar within its receptive field. It responds best at one particular orientation, but not if the bar is tilted from this.

where it is: recognition without localisation. Some cells, originally called *hypercomplex* cells but now properly called *end-stopped* complex cells, are even more fussy: not only has the orientation to be correct in order to obtain a response, but the length of the stimulating bar or line must also lie within certain limits (**Fig. 7.74**). It is easy to imagine how a simple cell might derive its receptive field from the summation of the outputs of two or more geniculate cells arranged in a straight line; and although it is tempting to extrapolate this notion by supposing that the complex cell response might similarly be the result of the summation of the outputs of simple cells, it appears that in fact this is not so, and that both types of response are actually the result of appropriate connections from geniculate afferents. Responses to colour are less evident, at least in area 17, than in the geniculate, and of the cells that show colour-opponent responses, most are of the simple or concentric type, and gathered together in 'blobs' (**Fig. 7.75**).

It is important to appreciate that these various classifications of specificities amongst cortical cells are to a large extent arbitrary and artificial; as was emphasised in Chapter 4, if an experimenter finds any response at all from a cell, it must necessarily be to a stimulus that he or she has chosen beforehand. Recent examinations of cortical visual cells suggest a richness of variety that is not well conveyed by conventional descriptions in terms of 'simple cell', 'complex cell' and so forth.

Orientation-specific responses appear to be functionally grouped in *columns* perpendicular to the cortical surface, the cells in a particular column sharing the same preferred

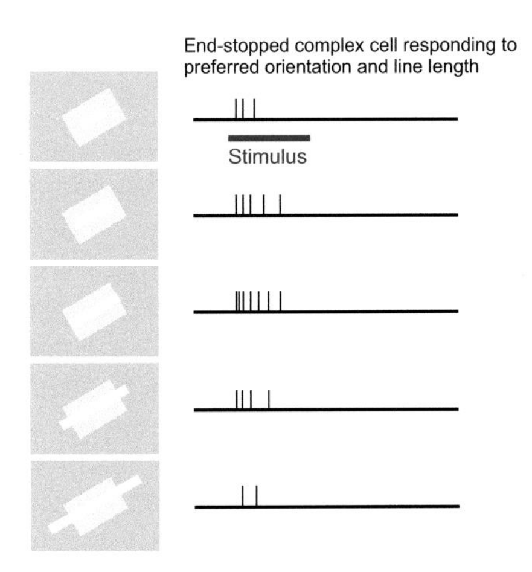

Figure 7.74 Responses of an end-stopped ('hypercomplex') cell to a moving bar of optimum orientation but different lengths. There is clearly an optimum length for evoking maximum activity. (After Hubel & Wiesel, 1965; with permission.)

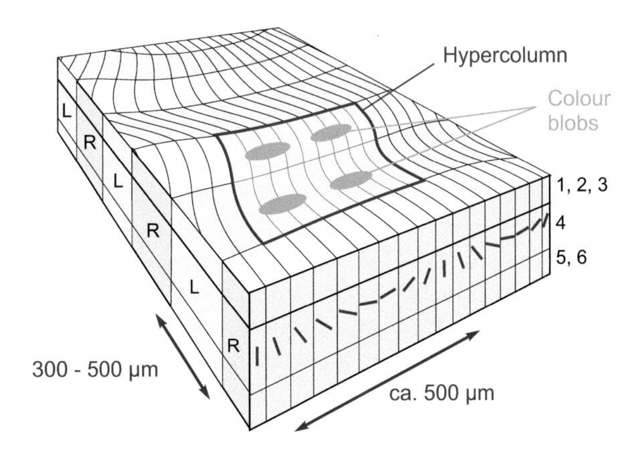

Figure 7.75 Highly stylised representation of a slab of visual cortex, showing its organisation into 'columns', narrow strips in which cells share a common preferred orientation, and roughly perpendicular dominance bands, preferentially driven by one or the other eye. Preferred orientation usually changes systematically (as shown) in passing along a set of orientation columns. An analogous 'columnar' organisation is found in other neocortical regions.

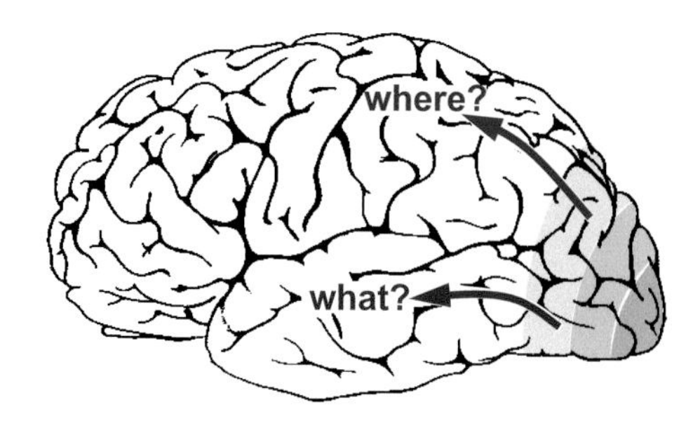

Figure 7.76 Subsequent processing after V1 can be divided roughly into two separate strands: of localisation and movement, to parietal areas ('where?'), and of recognition, including colour and shape, to temporal regions ('what?').

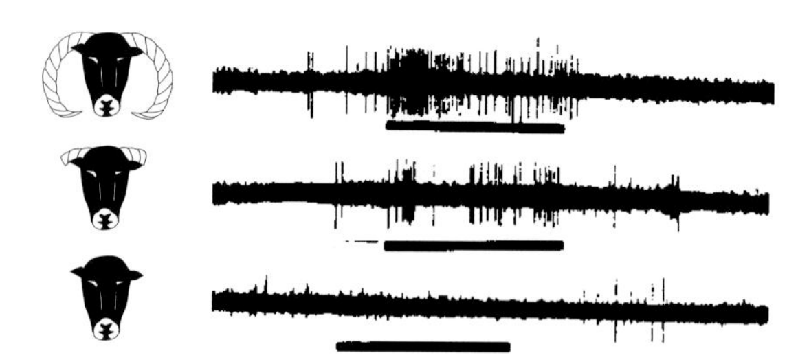

Figure 7.77 Responses of a single cell in the temporal cortex of a Dalesbred (horned) sheep when presented with the three sheep faces shown.

orientation, this orientation changes in a systematic way as one moves across the cortical surface, such that after half a millimetre or so we are back to the first orientation (**Fig. 7.75**). Thus the visual cortex is traversed by a series of bands, within each of which every possible orientation is represented. But columns can differ in another way too, related to the fact that we have two eyes. We saw that in the geniculate the inputs from each eye are strictly segregated. This segregation is maintained in the projections up to the cortex, with each column receiving fibres associated either with one eye or the other, but not both. A column receiving right eye fibres is called right eye *dominant*, and columns having the same dominance form a second series of bands, at right angles to the first. Together they create a sort of chequerboard, with both eyes and all orientations being represented in a patch about one millimetre by a half, called a *hypercolumn*; each hypercolumn typically also contains four colour *blobs*, regions occupying layers II and III staining for cytochrome oxidase and containing neurons that are monocularly driven and respond to colour but not to orientation (see p. 165). The total number of hypercolumns in human striate cortex is around 13 000.

Although the simple cells in layer IVc of a column will normally be driven by one eye only, the complex cells on each side are normally found to be binocular, driven by both eyes. Some of these neurons have receptive fields that are identically situated with respect to the fovea of each eye, others have pairs of fields that do not exactly correspond in the two eyes: this retinal *disparity* is undoubtedly an important source of information about the distances of visual objects from the plane of fixation, and is later discussed in the context of depth perception (see p. 170).

So here are cells coding a wealth of information about the visual world, looking for spots and edges and lines of a certain orientation, of a particular length and moving in a particular direction. Where do we go from here? From area 17 onwards, the two main kinds of use made of vision – localisation and recognition, where and what – seem to be associated with two distinct streams of processing that pass outward into areas 18 and 19 and beyond (**Fig. 7.76**). The *recognition* stream heads downhill, through the inferotemporal regions, areas 20 and 21, to its ultimate destination, the structures in the limbic system, buried deep in the temporal lobe and discussed in Chapter 14, which are concerned with triggering appropriate actions in response to things in the outside world. In the temporal lobe, cells have for instance been described in the monkey that respond to objects as closely defined as particular faces, or the appearance of the monkey's own hand. In sheep, similar cells seem to identify other sheep, and classify them as to the degree of threat they pose, and their place in the social hierarchy (based on the length of their horns; **Fig. 7.77**). Lesions of these areas in the monkey can give rise to visual defects that are more subtle than those associated with lesions in the occipital lobe, such as difficulty in recognising or appreciating the significance of visual objects. In humans, lesions in such areas can similarly result in an inability to recognise objects and people. A famous example of this is the case of 'the man who mistook his wife for a hat', reported by the neurologist Oliver Sacks, who witnessed one of his patients reach out, grab his wife's head, and try to put it on – this, no doubt, caused by a tumour or similar disease process in the temporal lobe of the subject. We will consider these 'visual agnosias' and their implications for sensory processing in Chapter 13. The *localisation* stream, on the other hand, heads forward towards the cortical areas that control the directing of movement, fusing with similar localising information from the somatosensory and auditory systems. These mechanisms are also discussed in Chapter 13.

Colour vision

So far we've completely ignored rather an important property of light, that not all photons are the same. It was Sir Isaac Newton who laid the foundations of our understanding of colour by showing that white light was in fact composed of an infinite number of different components – what we now know as wavelengths – and could be split up into a spectrum. The psychological perception we call 'colour' is a function of the relative energy in different parts of the spectrum.[21]

Now different objects reflect different amounts of light of different wavelengths – a red apple, for instance, is one that tends to absorb short wavelengths and reflect long ones. So the simple concept of albedo presented at the start of this chapter needs to be extended to include different wavelengths: you can see a red apple against green foliage

Table 7.6 Responses of visual neurons

Receptors	Circular uniform fields; tonic
Bipolars	Circular concentric +/– fields, essentially tonic
Ganglion cells	Many types, all with circular fields:
	Mostly phasic
	X-cells: sustained, simple
	Y-cells: transient, complex
	W-cells: slow, may be complex
	Some cells show colour-coding (R/G, Y/B), and in some species there can be responses to movement in specific directions. A small number of cells show tonic responses to light level over a wide dynamic range
Lateral geniculate	Essentially similar to ganglion cells, with double-opponent colour coding (e.g. R+G– centre, R-G+ surround) as an additional feature in some cells
	Little response to diffuse light
Visual cortex	The first level at which binocular cells are found; colour less common than in the geniculate. Many types, some circular, mostly linear. No response in absence of pattern
	Simple: linear +/–, localised
	Complex: linear, often directionally selective; larger response area
	Hypercomplex: linear, generally directionally selective, end-stopped
	In inferotemporal cortex, more specificity and more colour (including responses showing colour constancy)
Superior colliculus	Small uniform fields in centre, larger in periphery; often directionally selective
	Uninterested in shape or colour
Pretectum, pons	Very large fields, directional selectivity when detail in field moves as a whole; firing rate reflects the velocity

even if its *overall* albedo is the same, because in the red parts of the spectrum its albedo is higher than the leaves, but at shorter wavelengths lower. If the light falling on them is monochromatic, so that only the albedo at one wavelength can be compared, then they may happen to be indistinguishable.

So when we say something is a different colour from something else, what we mean is that we perceive that the overall shape of its spectrum is different. But we are not in fact terribly good at determining the overall shape, in fact we do it extremely crudely. What happens is simply that we have three types of cone, long-, medium-, and

Clinical box 7.4 Blindsight

This is a curious consequence of isolated injury to V1 in the occipital cortex. The phenomenon was initially described by Larry Weiskrantz and Elizabeth Warrington reporting the case of patient 'DB', who had undergone resection of the right-sided visual cortex to rid him of a tumour. Following the surgery DB, not surprisingly, described being blind in his left visual field, and when asked to perform visual tasks such as selecting gratings of a certain orientation or objects of a certain colour he would simply state that it was impossible. However, when subjected to a 'forced discrimination' test DB showed remarkable accuracy identifying and localising objects in his 'blind' field – he could discriminate even basic shapes and movements. Knowing the myriad connections the visual tracts send back to the midbrain, the synaptic promiscuity found in the thalamus through which visual information must pass and the considerable area of cortex outside V1 dedicated to visual processing, the retention of visual skills despite the ablation of V1 is not surprising; what *is* exciting about DB's success, however, was that this performance came without any awareness of the visual feat he had accomplished. Despite many years of intensive research, DB remained convinced that he was simply guessing. The demonstration that visual information could be processed whilst the subject experienced 'blindness', accurately plucking the correct object from the correct spot while denying vehemently any awareness of it, generated a thrill of excitement among scientists making the long-suffering pilgrimage towards a neurobiological explanation for consciousness (see Chapter 14).

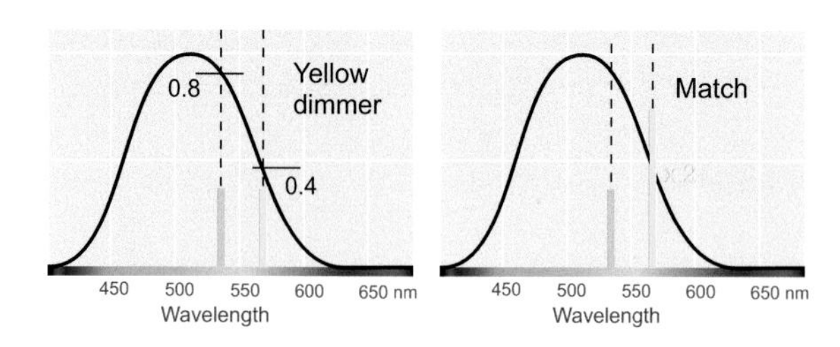

Figure 7.78 A monochromatic system is unable to distinguish two separate wavelengths independently of their intensity. In this case, the output of the receptor would be identical if presenting either a green stimulus or a yellow stimulus of twice the intensity.

short-wavelength, roughly in the red, green and blue bits of the spectrum; every time we look at some coloured surface they send us just three signals – how much red, how much green, how much blue. So what we mean by saying one colour matches another is simply this: that the red signal is the same whether we look at one or the other, the green signal is also the same, whichever we look at, and so is the blue.

Now having three receptors makes things get a little complex and mind-boggling by the end: better to work up to it gradually. So before considering what happens with three receptors we'll consider what would happen if there were only two, and before doing *that* – paradoxically – it is enlightening to consider what colour vision is like when we have only *one* receptor type, and can't discriminate colours at all. This is what happens to us all, of course, in the scotopic state, when we have only rods. For a start, rods are perfectly capable of *detecting* different wavelengths, because their absorption spectrum extends nearly over nearly all the visible range. But what they cannot do is *discriminate* between them. Suppose we have two wavelengths of equal intensity, a green one lying somewhere quite near the peak of the scotopic sensitivity curve, and a yellow one further out, so that the green provides say 0.8 units of stimulation and the yellow only 0.4 (**Fig. 7.78**). Clearly they won't look the same, since the yellow produces a smaller response; but it is equally true that one could make them look the same, just by making the yellow *twice as bright*. So we would then have a situation in which the lights looked identical, even though one is yellow and the other green. You can see that this is going to be true of any pair of colours: *any* two wavelengths can be made to look identical – match – provided you are allowed to make them brighter or dimmer. More exactly, a person with only rods would be *unable to distinguish wavelengths independently of their intensities*, which is what colour is all about. It doesn't matter what shape this curve is: any system that has only one kind of receptor is bound to lack colour vision, that is, it is *monochromatic*.

Colour mixing

Suppose now we allow ourselves to have two receptors, each with a different absorption spectrum. To make things easier we're going to suppose a pair of very improbable absorption curves (**Fig. 7.79**). In general, every wavelength will stimulate the red by a certain amount and the blue by a certain amount, producing a pair of signals (R,B). So, for example, a pure red might give (100,0); a deep blue (0,100); yellow might stimulate each equally – (50,50). A useful way to think about this is to plot a chart whose axes tell you how much each of the two types of receptor is stimulated. All possible wavelengths will produce a point somewhere in this graph. So if we keep intensity fixed but sweep through the spectrum, we get a line called the *spectral locus*. If, on the other hand, we keep the wavelength fixed but alter the intensity, the point moves in and out, keeping the *direction* the same, because the ratio of red to blue is identical. Thus the direction of lines like A and B in **Figure 7.79** represent particular *colours* (technically hue or chroma), and different distances along them represent different *intensities*. Now one of the things that is very obvious about colour is that things do not in general change colour if they're more or less brightly illuminated. If you shine a bright light on an orange it still looks orange. In fact, in a sense what is *meant* by colour is that aspect of a light that remains constant even when we change the intensity.

We can also use the chart to work out what will happen if we mix two colours together, using a 'parallelogram of forces'. For example, red (100,0) plus blue (0,100) = (100,100), which is the same as we would get with a bright yellow. And, in general, it is clear that we can match *any* colour by choosing suitable proportions of just red and blue. These may then be called *primary* colours, stimuli

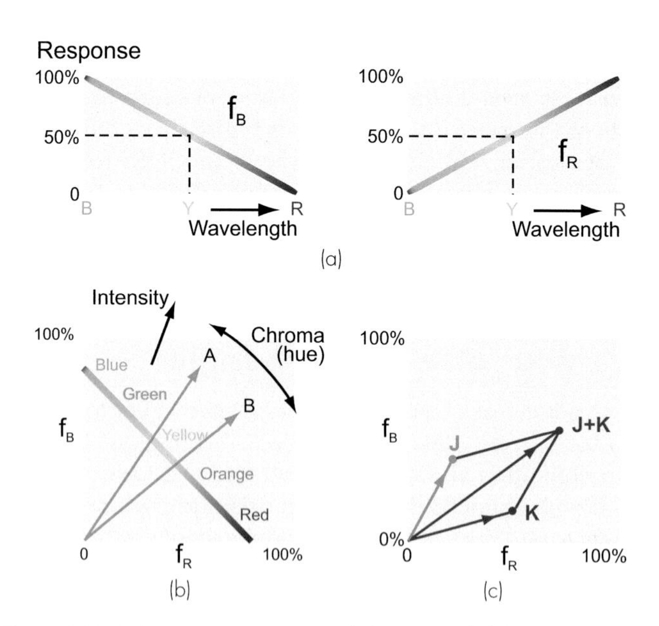

Figure 7.79 Colour vision in an extremely hypothetical dichromat. (a) The spectral sensitivities of the two types of receptor, B (blue) and R (red). (b) The colour response space: the axes show the degree of activity of each of the two channels, and any particular colour will result in a particular point in the diagram. The coloured line indicates how the B and R responses vary as a light of constant intensity is varied in wavelength, while lines such as OA and OB represent stimuli of constant wavelength but variable intensity. In each case, the ratio of B and R activity is constant, and they are thus of constant hue or chroma. (c) J and K represent two stimuli of different intensity and wavelength. If they are added together, the resultant response is given by the vector sum (J + K) of each separate response.

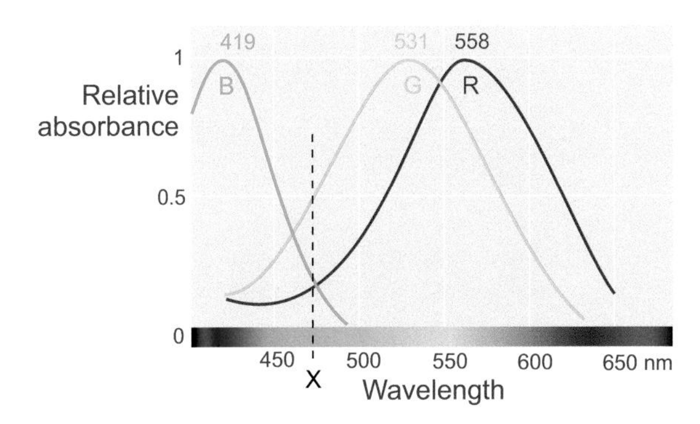

Figure 7.80 Approximate spectral sensitivity curves of the three human cone pigments.

Figure 7.81 If the load can be sensed through three channels, the head as well as the arms, it is then possible to distinguish patterns of weights that would have appeared identical to the person with only two points of contact.

from which all other hues can be made by mixture. There is an exact analogy with the man on p. 90 holding a plank above his head, who can match the sensation generated by *any* distribution of any number of weights by just two 'primaries' of variable weight.

Obviously that isn't true for normal people; so although having two channels certainly provides colour vision, it's not as good as ours: we have to get the blue *and* the red *and* the green right. This kind of two-channel colour vision is called *dichromatic*, because any colour can be matched by mixing *two* primaries. Now, although there are many dichromatic animals, and some colour-blind humans, for whom the above provides an adequate description of their colour sense (except that the shapes of the sensitivity curves in **Figure 7.79** are highly improbable), human vision is *trichromatic*; we have not two but three types of receptors. Their approximate spectral sensitivities are shown in **Figure 7.80**. It is clear that we cannot any more match some given wavelength (such as X in the figure) by using different amounts of just two other wavelengths: we may for instance be able to get the stimulation of the red and blue signals right, but the value of the green will in general be wrong; if we try to correct it we will mess up one of the other channels. In fact we now need *three* primary colours to match any other, to take care of the three degrees of freedom involved: a pair of colours will only match if the blue signal for one is equal to the blue signal for the other, the green to the green and the red to the red. To extend the analogy of the man with the plank (p. 90; **Fig 7.81**), it is as if his tray now rests on his hands but also on his head, providing three channels of information about any weights that are placed on it.

Because of this, the diagram of the response space corresponding to that in **Figure 7.79**, now has to be *three-dimensional* rather than flat; and because the individual sensitivity curves do not have the simple forms that we assumed, the locus of a light of fixed intensity whose wavelength is altered is no longer a nice straight line but one having a twisted three-dimensional shape (**Fig. 7.82**).

It is awkward to have to use three dimensions. One way to get round this is to forget about the brightness part of this and just show the direction or colour part, which only has two dimensions. The trick for doing this was invented by Newton some 300 years ago, and is called the *colour triangle*. Recalling that the distance from the origin to a particular point represents brightness, whereas its direction represents colour (**Fig. 7.83**), if it is only the latter quantity that interests us, we can reduce the whole diagram to a two-dimensional form by the following simple expedient. We set up an equilateral triangle whose vertices are at equal distances along the three axes: clearly every direction – and thus every colour – will be associated with a particular point on the triangle, and the point will be the same however bright the light is. So points on the triangle represent all possible colours, disregarding luminance.

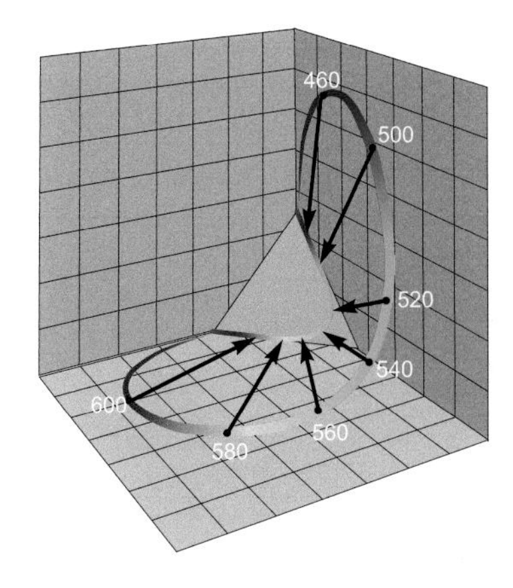

Figure 7.82 Three-dimensional representation of colour space, showing the spectral locus and its projection on the colour triangle.

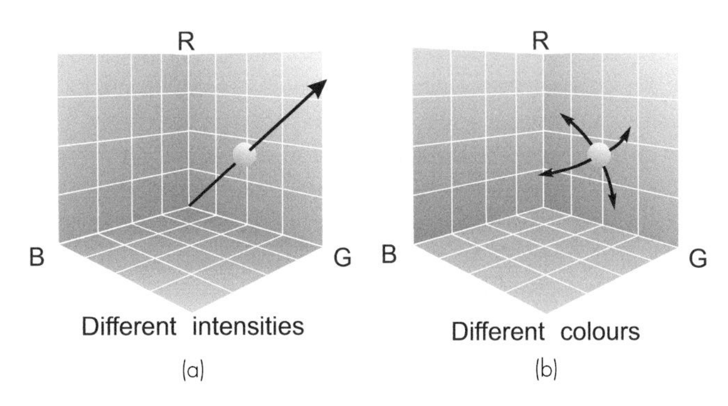

Figure 7.83 Trichromatic colour space. (a) The effect of varying just the intensity of a given colour is to move it along the line joining it to the origin, without affecting its direction. (b) Changing just the colour, on the other hand, affects the direction but not the distance.

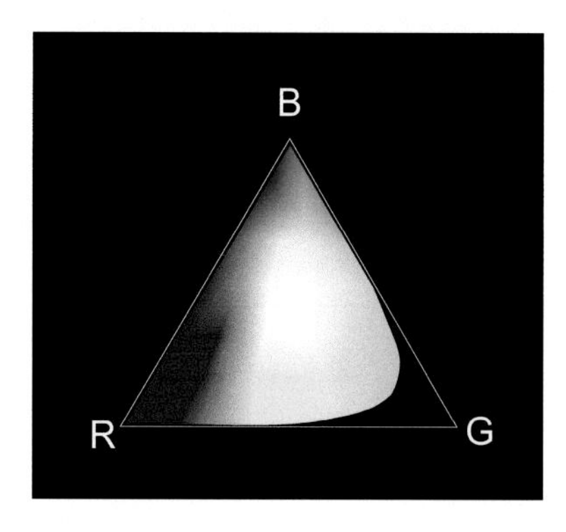

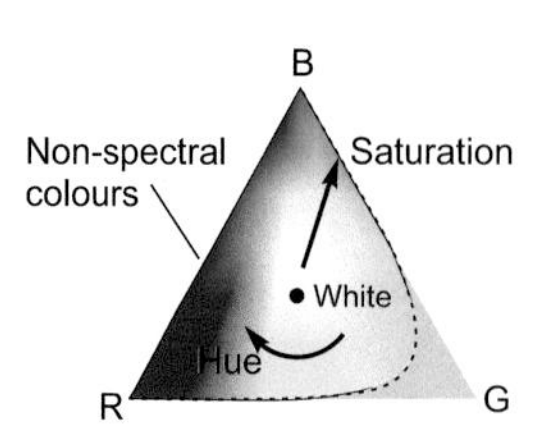

Figure 7.84 The colour triangle, showing the spectral locus. Note that it cannot reach the green vertex, and that along the BR edge are a range of colours not in the spectrum.

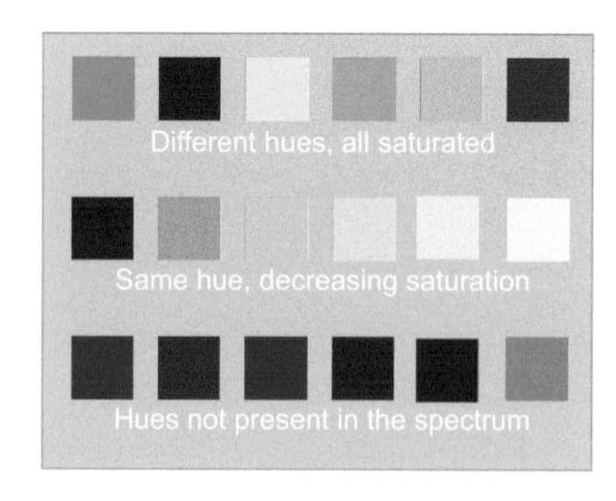

Figure 7.85 Different hues, degrees of saturation and hues absent from the spectrum.

On this colour triangle, each point corresponds to a different colour (**Fig. 7.84**). Its centre corresponds to white light, that stimulates each type of receptor equally, and lines radiating from this point are lines of equal *chroma* or *hue*. Along such a line, the nearer a point is to W the more *unsaturated* it is, i.e. the more it is diluted with white. Thus pink and red lie on the same line of chroma, but pink is nearer the centre. The dashed line in **Figure 7.84** represents the locus of a light whose wavelength is varied over the visible range: it therefore represents colours that are of maximum saturation. The fact that this locus does not reach the G vertex reflects the fact that there is no wavelength that can stimulate G alone without the same time stimulating R or B or both, as can be verified in the spectral sensitivity curves on p. 163. Incidentally, see that some hues *don't appear* in the spectrum, for instance purple. Such hues are not in the spectrum as no *single* wavelength can stimulate R and B without stimulating G as well (**Fig. 7.85**). The edge RB of the triangle represents colours of this sort, saturated violets and purples that are not in the spectrum, but can be formed by mixtures of deep red and deep blue.

The rules for colour mixture can now be stated simply, and were described in this form by Newton in 1704. To find the result of mixing two colours J and K (**Fig. 7.86a**), join JK with a straight line: all the colours on that line can be made by mixing J and K in different proportions, and the more J is used, the nearer the resultant will lie to J. In general it can be seen that the effect of mixing two colours is to produce a new one of intermediate chroma, but less *saturated*, i.e. nearer to W: such mixtures are therefore paler than pure spectral colours. If JK happens to pass through W, J and K are said to be *complementary*: by mixing them in suitable proportions, pure white can be made. By mixing *three* colours (J, K, L) we can produce any colour lying within the triangle JKL; so long as this includes W, this means that any hue can be mixed from any three colours (**Fig. 7.86b**). Obviously, the bigger the triangle JKL is, the more saturated are the colours that can be mixed: but even if we use pure spectral wavelengths as our primaries, there will be certain colours (notably saturated blue – greens or yellow) that cannot be matched. In practical colour mixing, as with colour film or colour television, one tries to choose primaries that make the triangle JKL as large as possible: but one's choice is limited by the dyes or phosphors actually available. The reason why colour television in particular is often unsatisfactory can be seen by looking at the position of the primaries used – they are in fact close to the J, K, L of **Figure 7.86b**. The result is that blue-greens and purples are rather desaturated, though grass-greens and flesh tints are relatively good. The only practical solution to this sort of problem is to use more than three primaries, and high-quality colour printing – for example of reproductions of paintings – may use six or more to get as close as possible to fully saturated mixtures.

The essential point, then, about colour vision is that although there are three degrees of freedom, the sensation of colour itself is two-dimensional – it can be described by the two variables of chroma and saturation – and the third quality, intensity, is not essentially a colour attribute at all. That is not to say that it doesn't contribute to the popular idea of 'colour': the colour brown, for example, is only an orange of low intensity. So everything we see can be described as a combination of colour and brightness. If we ignore brightness, any *colour* can be found somewhere on the colour triangle. And since the colour triangle is two-dimensional, this means that to code for colour we need just two channels of colour information to specify a point on it. It turns out the coordinate system the brain uses is very simple: if we orientate the colour triangle so blue is at the top, the coordinates used by the central areas to code colour are simply horizontal and vertical. You can think of the horizontal one as being a red-versus-green axis, and the vertical as blue-versus-yellow (**Fig. 7.87**).[22]

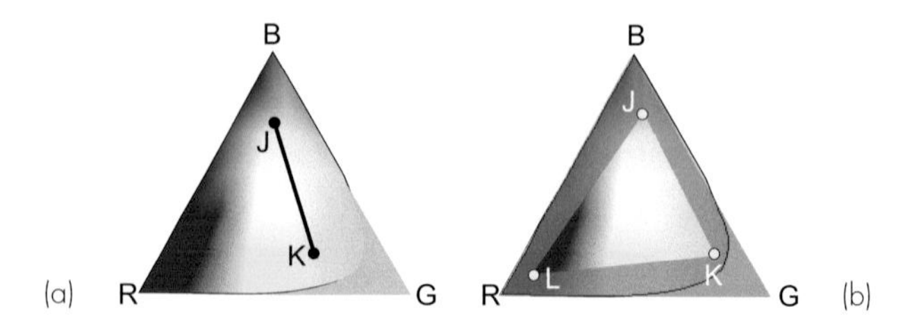

Figure 7.86 The laws of colour mixing. (a) By mixing two colours (J, K) in different proportions, we can form any of the colours lying on the line JK. (b) With three colours (J, K, L), we can form any colour lying within the triangle JKL. If, as here, this triangle encloses the white point, any hue can be formed by mixing them in suitable proportions, though not with full saturation.

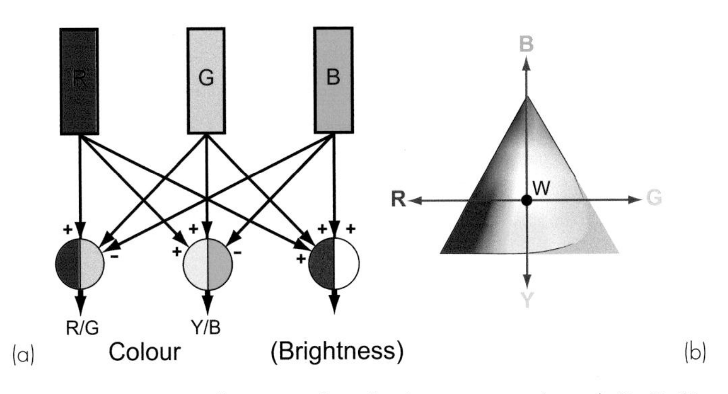

Figure 7.87 (a) How illumination from the three receptor channels (R, G, B) is neurally recoded into a red-versus-green channel (R/G), a yellow-versus-blue channel (B/Y) and a brightness channel. (b) The resultant form of colour space is based on R/G and B/Y axes, with the colour triangle superimposed.

Neural processing

So somewhere in the visual system, something must be recoding the three receptor channels, R, G and B, into two colour ones and a brightness one. It turns out that this recoding is essentially all done in the retina. If you look at retinal ganglion cells, you find that most of them have very wide spectral responses, responding essentially as if they were driven by all three types of cone; they provide the *brightness* channel, which has high acuity and gets further processed in the primary visual cortex for line detection and so on. But some of the parvo cells behave differently, showing excitation for some wavelengths and inhibition for others. Some, for instance, are excited by red and inhibited by green or vice versa, while others behave as if they were similarly wired for yellow versus blue: so in these *colour-opponent* cells we have the neural counterpart of the two axes for the colour triangle: R/G, Y/B (**Fig. 7.87**). Because the cells coding for colour are in a small minority, the true colour channels have poor acuity, as can be shown by using isoluminant sinusoidal chromatic gratings, in which the stripes vary in colour but have the same luminance.[23]

Psychologically there is no doubt that that is how people think about colour. Unless they've been told about vision being trichromatic and three primaries and so on, the natural feeling is that the *basic* colours are not three but *four*: red, yellow, green and blue. Yellow looks fundamental, primary, and certainly not a mixture of R+G. We talk about a yellowish green, or bluish red, but not a reddish green, or a bluish yellow. This primacy of yellowness is odd, given that it is evidently a composite signal derived from two different sets of receptors. Evolutionarily, it seems that the blue/yellow axis is the more fundamental one: it is essentially a division of the spectrum into short wavelengths versus long. The red/green axis seems to have evolved later, probably by duplication of the original opsin gene (though the evolutionary history of colour vision is complex and shows many anomalies), and it has been suggested that it was the need for primates to distinguish between the reds and greens of ripe and unripe fruit that led to the divergence of the original yellow cone into red and green varieties.

Because in a sense the basic analysis of colour has already been done in the retina, there is relatively little representation of colour in primary visual cortex, which is more concerned with *form* perception. The line and edge detectors are not normally colour-sensitive at all, with colour responses being limited to the small patches of cells called blobs (see p. 159), of which there are exactly four per hypercolumn. Colour is also absent from area 19 or V5, which is not concerned with recognition. Colour responses do appear in the *recognition* stream through area 18, especially in areas V2 and V4 – as if we make a detailed black/white drawing first in the primary cortex, then rather roughly colour it in through V2 and V4.

Chromatic adaptation

In the same way that adaptation is important in helping the visual system to register the albedo of an object independently of the intensity of the light that falls upon it, so *chromatic* adaptation – the independent adaptation of the individual colour channels – can help the visual system make allowance for the *colour* of the illumination. Just as we want to recognise a grey object as grey whether it's very brightly illuminated or very dimly, so we want to recognise a yellow object as yellow even when it is bathed in pink light (**Fig. 7.88**). Suppose we have a painting whose full range of colours falls, under white light, within the triangle JKL of **Figure 7.89**. If we now illuminate it with blue light, the effect will be to reduce the signals in the red and green channels relative to those in the blue, and the result will be a distorted triangle (jkl) and hence misperception of hues. But the red and green channels will respond

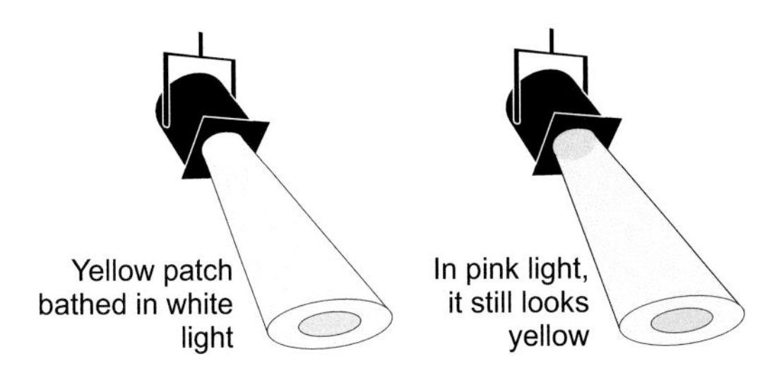

Figure 7.88 Chromatic adaptation.

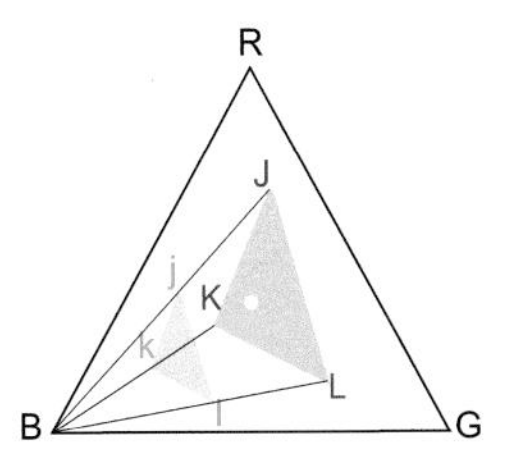

Figure 7.89 Effect of coloured illumination. A picture whose range of colours falls within the triangle JKL under white lights is illuminated instead with blue. The effect is to shift the triangle as shown, to jkl. However, the perceived changes of colour will generally be very much smaller than this, because of the effect of chromatic adaptation: the sensitivities of the R and G channels increase, and that of B decreases.

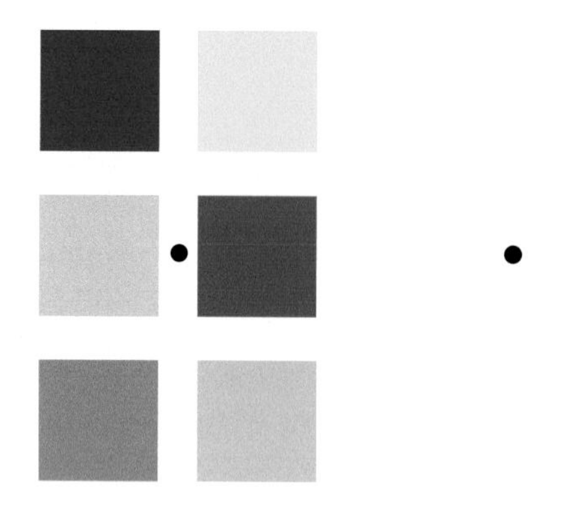

Figure 7.90 The effects of successive contrast (see text).

Figure 7.91 The effects of simultaneous contrast (see text).

to their reduced stimulation by increasing their sensitivities, which will have the effect of restoring the total range of colours perceived to something like its original extent. In fact the eye is surprisingly tolerant of changes in the spectral composition of the illuminating light, and the sensation of colour is much more closely related to an object's albedo for short, long and medium wavelengths than it is to the actual spectral composition of the light reaching the retina. This kind of behaviour – *colour constancy* – is demonstrable in the responses of many colour-sensitive cells in the more temporal regions of visual cortex, which can respond to an object of a particular colour even when the spectrum of its retinal image is quite different, because of changes in the colour of its illumination – a function that Helmholtz called 'discounting the illuminant'.[24]

These adaptational changes can easily be demonstrated by means of *successive contrast*, or coloured after-images. If the black dot in the middle of the coloured squares in **Figure 7.90** is fixated for about 20 seconds in a good light, and the gaze then transferred to the black dot on the right, striking after-images of the complementary colour will be seen that are due to distortion of one's colour space because of adaptation: what was originally a white that fell in the centre of the colour triangle now falls to one side, in a direction opposite to the adapting colour. The same explanation probably underlies *simultaneous contrast*: an area of pale colour lying next to a strong one tends to take on the complementary tinge (**Fig. 7.91**). Like lateral inhibition in general (p. 88), this reduces the redundancy of signals sent to the brain concerning the colours of adjacent area. In the same way, it can give rise to illusions when there is a colour change at a border that does not, in fact, represent the colours of the adjoining areas themselves, as in the *water-colour illusion*, in **Figure 7.92**, where the interdigitated areas take on slightly different, illusory, tints. Simultaneous contrast is probably related to the existence of double-opponent cells, for instance in the blob regions. 'Double' here refers to the fact that they are not merely opponent in the sense of being, for instance, excited by red and inhibited by green, but that they have a colour-based centre-surround organisation as well (**Fig. 7.93**).

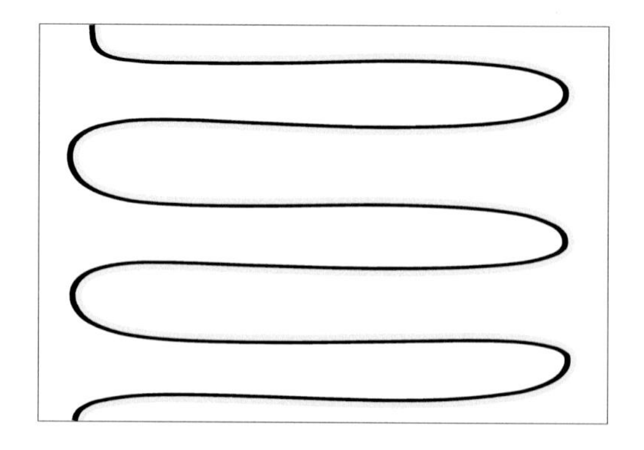

Figure 7.92 Water-colour effect. The coloured edges of the line make the interdigitated areas take on slightly different illusory colours.

Disorders of colour vision

If one of the three channels were inoperative, vision would become dichromatic, and the colour triangle would collapse into a single dimension (**Fig. 7.94**). One would then be able to match any colour in the spectrum by mixing blue and red in suitable proportions. Defects of this kind are often seen: loss of the red mechanism is called *protanopia*, of the green, *deuteranopia*, and of the blue, *tritanopia*. Probably because the division of a single yellow mechanism into red versus green came relatively late in evolution,[25] the first two deficiencies are much commoner than the last and give rise to what is commonly called red-green blindness: the resulting confusion of red with green forms the basis of clinical tests such as simple matching or the Isihara test in which coloured dots are made up into figures that are read differently with different deficiencies (**Fig. 7.95**). A deuteranope, for example, will accept any pair of colours as a match so long as same they have the same R and B components, even if the G components are different: the pairs of colours in **Figure 7.96**, for instance, were accepted as matches by a deuteranopic student.

Some 2 per cent of males are protanopic or deuteranopic, but a much smaller percentage of females: the genes in

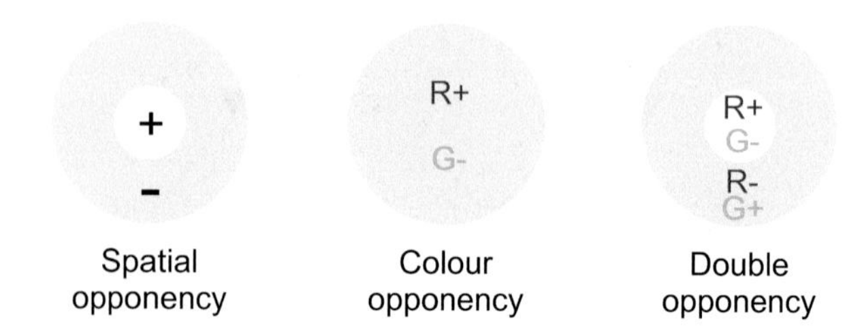

Figure 7.93 Three types of opponency.

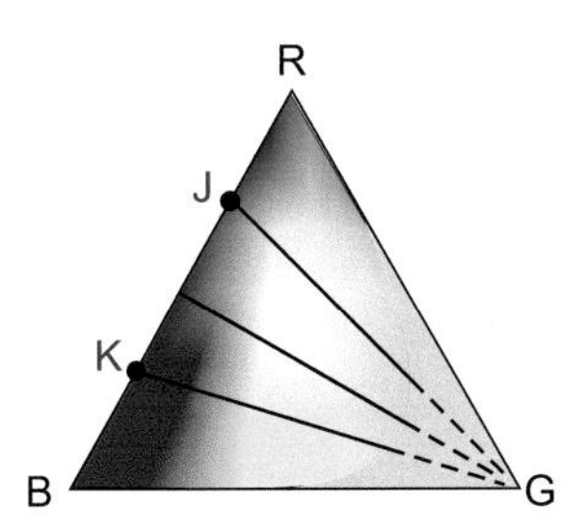

Figure 7.94 Effect of lack of one channel on colour perception. Colours lying on a line such as GJ or GK differ only in the amount by which they stimulate the G channel. If, in a colour-blind subject, this channel were inoperative, such colours would all look identical: lines like GJ and GK could then be called lines of confusion. Consequently, such a subject could match any colour by using only mixtures of deep red and deep blue light, in appropriate proportions.

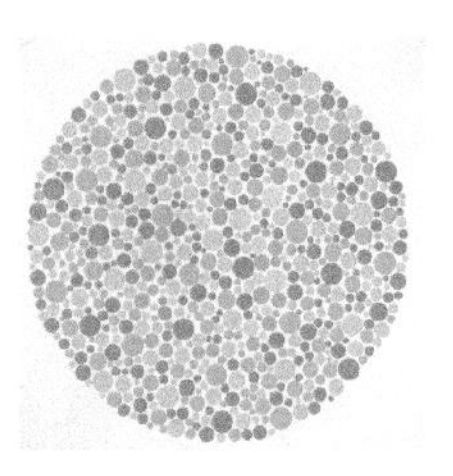

Figure 7.95 Isihara testing. A protonope would see a 3, a normal subject an 8.

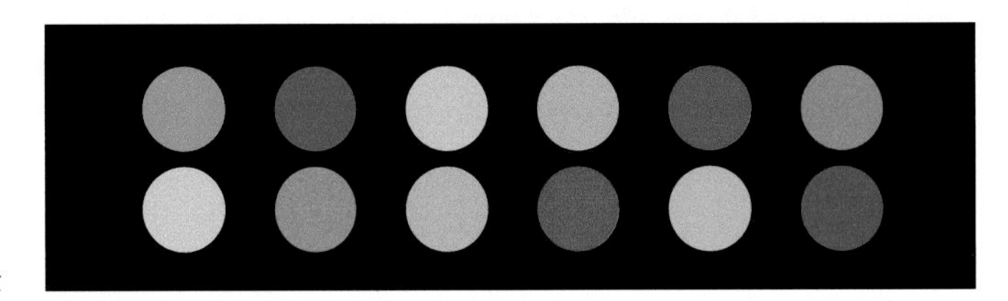

Figure 7.96 Examples of pairs of colours accepted as matches by a deuteranopic student.

question are on the X chromosome, and the impairment is recessive. Loss of the blue mechanism is much less common (and the relevant gene is not on the X chromosome), except that *everyone* is tritanopic in the very centre of the fovea. Over the whole retina, the blue cones are relatively infrequent compared with the red and green ones, and their acuity correspondingly depressed. But in the very *central* fovea there is an area in which there are no blue cones at all – possibly a mechanism for reducing the effects of chromatic aberration. Consequently if you look at very small or distant targets, you find that colours that differ only in the amount of blue, like the four pairs of dots in **Figure 7.97**, cannot be distinguished.[26] More severe defects, involving the functional loss of more than one channel – for example, the rod monochromat, whose retina contains only rods – are also found. Colour deficiencies might be due to a loss of one or more types of pigment, to lack of development of adequate neural connection or possibly to a mixing of pigments or connections, so that discrimination is lost. Studies using retinal densitometry (see p. 144) have shown that in some dichromatic subjects, at least, one of the cone pigments appears to be missing.

One might wonder if it was possible to know what the world actually looks like to a colour-blind person. Thanks to a very rare condition indeed, in which only one eye is colour-blind, the answer is probably yes: such people seem to see everything in terms of blue and yellow with the affected eye, so that the spectrum appears deep blue at one end, fading to white in the middle, and then passing through progressively deeper shades of yellow to the long wavelength end. This is exactly what would be expected if colour space has simply been collapsed down to the yellow/blue axis. **Figure 7.98** shows the old centre of Warsaw as it would then be expected to look with and without a functioning green mechanism.[27]

Less severe than actual colour blindness are the various colour *anomalies*. A protanomalous subject, for example, is trichromatic, but if asked to match a particular yellow by means of red and green will tend to use more red than a normal subject; deuteranomolous subjects use more green. It is likely that such defects are due to imbalance in the amounts or spectral sensitivities of the cone pigments. Finally, disorders of colour vision can arise with damage to specific cortical areas, and can take many forms; in some cases colour constancy may be impaired, so that colours appear to change under coloured illumination more than would be experienced by a normal observer. In humans,

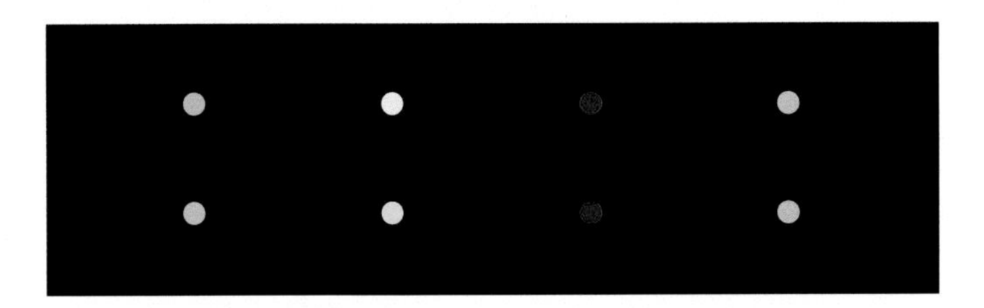

Figure 7.97 If you look at the row of miniature dot pairs at the bottom, at a suitable distance (3 m (9 ft) or more away), they should become indistinguishable.

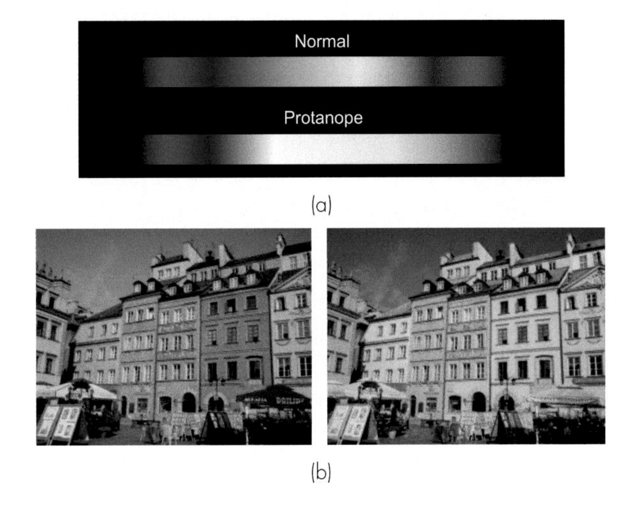

Figure 7.98 (a) A representation of the protanope's spectrum. (b) A picture of Warsaw, showing how colours appear in a deuteranope's world.

lesions in area V4 may cause *achromatopsia*, a complete inability to perceive colour.[28]

Localisation

There are two components to localisation, *direction* and *distance*. Of them, distance is the more difficult feature to extract from visual information; clearly the spatial organisation of the retina provides immediate information about the relative visual angles between different visual objects, since the retinal separation between two receptors that are stimulated is directly proportional to the angle between the corresponding stimuli in the outside world. It is convenient to deal with direction first.

Direction: the colliculus

High at the back of the brainstem, a primitive visual area called the *superior colliculus* (*tectum* in lower animals) integrates the visual information required to compute two things: *Where is this object located?* and *How can I get a better look at it?* Accordingly, it is basically organised in two layers, an upper layer receiving visual information from the retina, the LGN and visual cortex, and a lower layer that is essentially motor, projecting down to areas of the brainstem and upper spinal cord concerned with the generation of eye movements and head movements (**Fig. 7.99**). It also has extensive connections with the cerebellum, where it presumably provides information that helps in making directed limb movements to visual targets. Neurons in the upper layer are interested only in *where* an object is, and care not a bit *what* it is: they show neither orientation nor colour specificity, and have large, overlapping and roughly circular fields. The cells are arranged in an orderly way on the surface of the colliculus, forming a map of visual space. An interesting feature of this map is that electrical stimulation of deeper layers of a region corresponding with a particular part of the visual field often initiates an eye movement or head movement which is of the right size and direction to bring that part of the visual world onto the fovea. This suggests that one of the functions of the colliculus might be to control visually guided *saccades*. These are the very fast step-like eye movements that shift the gaze to objects of interest in the visual field, or when a subject tries to look at a target that is moving: during

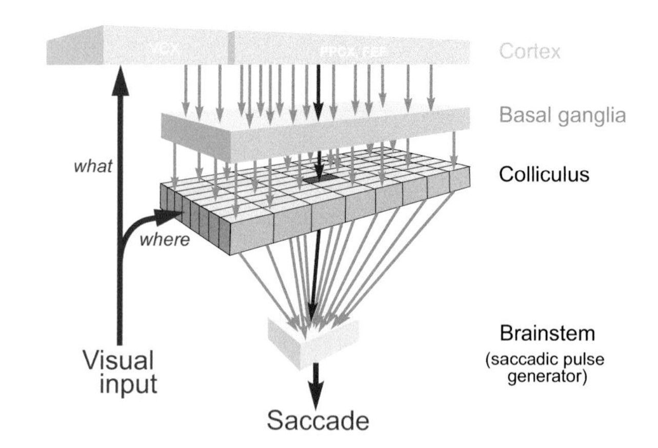

Figure 7.99 Schematic representation of the role of the colliculus and its afferent and efferent connections on the generation of saccades. Although the colliculus has sufficient information to convert information about the location of one saccadic target into a command to the saccade-generating circuits of the brainstem, only the higher levels (cortex and basal ganglia) can decide which of various competing targets should be selected as a saccadic goal.

Clinical box 7.5 Colour 'blindness'

The absence of two or three cone pigments leaving the sufferer unable to discriminate between visual stimuli of different wavelengths, also known as monochromacy, is extremely rare. Loss of one cone pigment leads to dichromacy; for such subjects any colour can be mimicked by mixing just two other spectral colours. Far more common is anomalous trichromacy: subjects are still trichromats, requiring three spectral colours to match any other colour, but a mutation in one of the photopigments alters its spectral sensitivity, changing the mixture of spectral wavelengths used by the sufferer to match a given colour.

In the case of one of the authors, all three normal pigments exist but there is also a fused red/green hybrid gene with a sensitivity peak very close to that of the red (559 versus 563 nm). This fused gene is expressed instead of the green (532), leaving two pigments of very similar sensitivity, reducing the ability to discriminate red, orange, yellow and green hues. To match a yellow colour by mixing red and green the subject will typically add in more green than a subject with normal colour vision. Although the commonest and least troublesome, trichromacy is in some ways the most interesting form of colour anomaly, since the subject still enjoys a three-dimensional colour space but a colour space of different shape. Since the concept of colour, as opposed to wavelength, exists only in the brain and hence can only be conveyed by our own description, necessarily derived from our own experience of the colour, it is impossible to know how each individual's perception of colour differs. Objectively, anomalous trichromats are often unaware of their different perceptual world, having little discernible handicap, and it is interesting to observe that such people may be able to better spot camouflaged objects in certain visual scenes than people expressing the more common photopigments, suggesting some survival advantage that has maintained the anomaly in the genome.

visual tracking of this kind the oculomotor system moves the eye smoothly at a rate that ideally matches that of the target (smooth pursuit), with occasional saccades to correct any errors of position that still remain (the various types of eye movements are listed on p. 194).

Saccades are generated by special neural circuits in a region of the brainstem that lies near the motor neurons for the eye muscles, called the *prepontine reticular formation* (see p. 203). These circuits are normally held in check by inhibitory *pause cells* in the brainstem that are tonically active but stop firing briefly during a saccade. The superior colliculus sends inhibitory fibres to these pause cells, and a burst of activity in such a fibre appears to trigger a saccade to a particular target. Now, of course, there are many objects in the outside world which we might want to look at, and although the colliculus is well adapted to determining where they are and to initiating appropriate eye movements, what it cannot do is decide whether any particular object is *worth* looking at: this requires recognition. Consequently, we would expect to find that the colliculus is itself under tonic inhibition from higher levels of the visual system, and only permitted to act when a possible target is recognised as being interesting. Such a descending inhibitory system to the deeper layer has been demonstrated in the form of a group of neurons in the substantia nigra (part of the basal ganglia: see p. 240). They fire continuously, keeping the colliculus in check, but pause briefly in response to a visual stimulus, well in advance of the subsequent saccade that is made to look at it. Thus, through a curiously bureaucratic cascade of inhibitory neurons, the substantia nigra (which is ultimately cortically driven) allows the colliculus to give its permission to the brainstem to do its work of moving the eyes to a certain position. They in turn appear to be controlled by areas of the cerebral cortex – notably the posterior parietal cortex – that are concerned with the identification and localisation of objects that might be of interest.

The motor system needs to know about the position of objects relative not to the eye but to the body as a whole. We shall see in Chapter 11 how knowledge of the position of objects relative to the eye is combined with information about eye position derived from the commands that are sent to the eye muscles (*efference copy*: see p. 193), in order to compute the position of objects relative to the head; and how in turn this information, combined with signals from the vestibular system and from the neck, enables the motor system to calculate the position of objects both relative to the body as a whole, and also to absolute frames of reference such as the direction of gravity. The deeper layers of the colliculus appear to receive copies of the efferent commands from the oculomotor system, and use this information to work out where targets are in space as well as relative to the eye.

Distance

The sense of distance is a little more complex, relying more heavily than the sense of direction on what might be

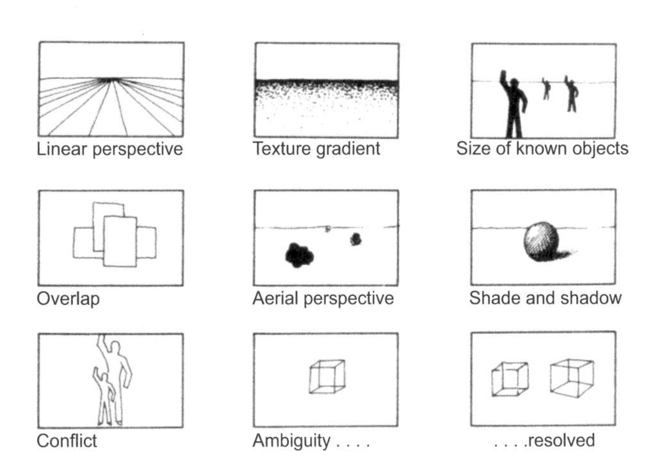

Figure 7.100 Schematic illustration of some of the monocular depth cues. The third row illustrates, first, how conflict may arise (in this case between size and overlap) and, second, how it is possible to construct figures that may be interpreted in more than one way (in this case as a cube viewed either from above or below). Addition of extra depth information resolves the ambiguity.

called 'high-level' cues. Some of this information derives from information about differences in the retinal images of the two eyes (*binocular* cues), while some is essentially *monocular*. The use of one eye rather than two substantially reduces the accuracy with which judgements of distance can be made, but does not abolish it altogether. It is convenient to consider the monocular cues first.

The simplest, though probably the least important, is *accommodation*. In order to focus on objects at different distances, you need to alter the power of the lens, and the effort of accommodation certainly contributes to how far away you think the object is, using efference copy. In isolation, however, this source of information is rather inaccurate.

A much more powerful and accurate physiological cue is something called *movement parallax*. When you move your head, the retinal image of objects that are close to you move much more rapidly than those that are far away, and in fact the relative velocities are linearly related to exactly how far away they are; and if you know the speed of your head, they can give an *absolute* measure of their distance. This is quite important, since the majority of the distance cues tell you only about relative distances. If you look at a cat preparing to leap onto something you can often see it moving its head up and down in order to get precise information about how far it has to jump.

The remaining monocular cues are all the ones that artists have used for centuries to give an impression of depth in their paintings – a hard task, since when you look at a painting, your accommodation, movement parallax and all the binocular cues tell you very firmly that everything is in fact at the same distance. Some of these higher-level cues are illustrated in **Figure 7.100**.[29] They include:

- *Overlap* (nearer objects tend to obscure further ones: crude and unquantitative);
- *Size of known objects* (if we know the actual size of an object and the angle it subtends at the retina we can deduce its distance);

- *Linear perspective* (for example, the apparent convergence of parallel lines);
- *Texture gradient* (that the spatial frequencies of a pattern or texture get higher the further away it is);
- *Position in field* (objects higher in the field are likely to be more distant);
- *Aerial perspective* (distant objects are fuzzier and bluer than near ones);
- *Shadows* (giving information about three-dimensional shape rather than actual distance).

It is possible to create artificial situations in the laboratory in which these cues are contradictory, and work out from subjects' response to them which cues are given more weight by the visual system than others. Many well-known illusions occur because assumptions about the distance of an object affect one's estimate of its size (**Fig. 7.101**).

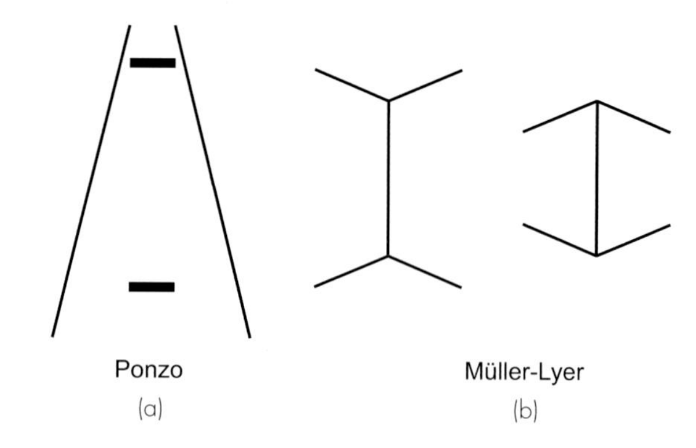

Figure 7.101 Illusions that probably result from false depth interpretations. (a) The Ponzo illusion: the lower bar looks smaller than the upper because it is perceived as nearer (the sloping lines being seen as parallel but receding into the distance). (b) The Müller–Lyer illusion: one upright looks bigger than the other because it is perceived as being the far corner rather than the near corner of a room or box.

What extra information is available if we use two eyes rather than one? Because each eye is a little distant from the other, each has a slightly different view of the world, and these differences are interpreted by the brain as differences in distance. In **Figure 7.102**, the two images are intended each to be seen by one eye: by allowing the eyes to diverge slightly you should be able to bring the two Rs into visual superimposition. The R will then appear to lie in front of the background scene. More precisely, if we imagine the two eyes fixating a point A in the middle distance (**Fig. 7.103**), the image of A will fall on the fovea of each eye (F), and points like B that are as far away as A will be at the same angular distance from A as seen by each eye, and their images will therefore fall on points that are at the same distance and direction from the fovea in each retina: such points are called *corresponding points*.[30] However, a point like C that is at a different distance from the eye will form images in the two eyes that are at different positions relative to the fovea: these are called disparate images. Now it turns out that if we examine the visual fields of cortical cells that have a binocular input (p. 160), although many of them are connected to corresponding areas in the two eyes, others are connected to disparate regions of the retina, and may be called disparity detectors (**Fig. 7.103**). So with the eyes both fixating A, a unit like α will be responding to the image of A, a unit like β with zero disparity will be responding to the image of B, and a unit like γ with marked disparity will be looking at the image of C. Thus even with the gaze fixed on one point in space, some cortical units will be 'looking' at points lying in planes either in front of or behind the target, and will thus provide immediate information about depth.[31]

These mechanisms of disparity detection work automatically and subconsciously even when there seem to be no corresponding objects in the outside world for them to operate on. If one takes pairs of identical patterns of random dots, and then shifts selected portions of them horizontally to create areas of disparity, when they are fused a three-dimensional structure may be perceived of which the individual patterns themselves contain no hint (**Fig. 7.104**).

It is important to emphasise that, though this mechanism of disparity detection is a very sensitive one (under ideal conditions, the threshold may be as little as 2 sec arc), it can only provide information about the distance of an object *relative* to the plane of fixation. As in the case of direction perception, we need further information about the positions of the eyes, about their angle of convergence, before it can be used to compute *absolute* depth. Since it

Figure 7.102 Stereoscopic stimuli. By allowing the eyes to diverge slightly, bring the two red letter Rs into visual superimposition. The R will then appear to lie in front of the background scene.

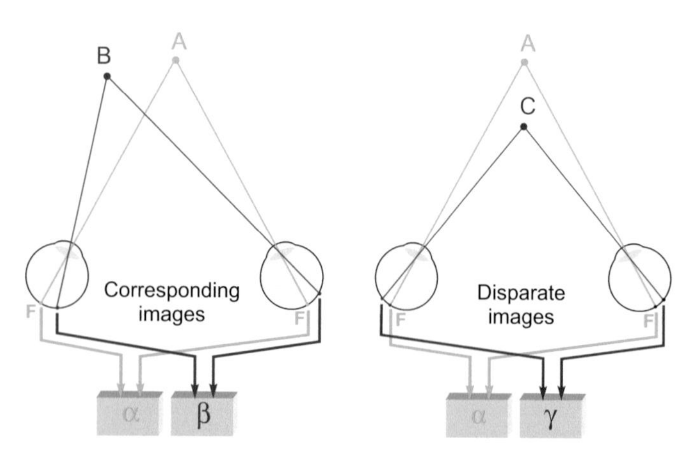

Figure 7.103 Retinal disparity and disparity detectors. When binocularly fixating A, targets such as C that are at a different distance give rise to disparate images on the two retinae. Of the neurons in the visual cortex that respond to signals from both eyes, many are stimulated by corresponding points in each retina (α, β) but some are connected to disparate points (γ).

Figure 7.104 Random dot stereogram. Neither of these on their own appears to contain an image. However, if fused as for the Rs in the earlier figure, a letter appears in depth that is defined by its difference in disparity compared with the background. (The effect may take some seconds to build up.)

turns out that knowledge of the convergence of one's eyes – derived through efference copy – is rather imprecise, it follows that disparity, though highly accurate for determining relative distances, is not of great use for absolute estimates: its main function is probably to provide information about the three-dimensional shape of objects (*stereopsis*), and particularly the detection of camouflaged objects against backgrounds.[32] It is not altogether clear what is in fact used to sense absolute depth: it may well be that the visual size of very well-known objects, particularly of parts of the body such as the hand whose distance can be checked by direct proprioception, provides the ultimate measure by which the rest of visual space – beyond one's own reach – is calibrated.

Proprioception

The use of visual and other kinds of proprioception in improving the control of posture is discussed at length in Chapter 11. A simple way to demonstrate the importance of the visual part of this is to stand on one leg and compare how much you sway about when your eyes are shut and when they are open. The most important contribution of vision to proprioception is information about *retinal slip velocity*, which in the natural world is nearly always caused by one's own movements. Several different areas in the brainstem seem to be involved in sensing retinal image movement of this kind, areas that are not very well defined and show considerable species variation. They include the pretectal nuclei and other nuclei at the very top of the brainstem, and groups of cells in the pons, but are best referred to collectively as the visual proprioceptive system (VPS). Neurons in the VPS typically have very large receptive fields indeed, and respond maximally when the field is filled with a lot of detail moving as a whole in a specific direction (**Fig. 7.105**). This is precisely, of course, what happens if we move our head in natural surroundings: the rate at which such units fire will code for head velocity. Their information is sent to the vestibular nuclei and to the cerebellum, where it is added to information about head position derived from the vestibular apparatus, and generates compensatory postural reactions and eye movement. In the cerebellum it also appears to be used to calibrate the vestibular signals, bringing both sources of information about head movement into correspondence

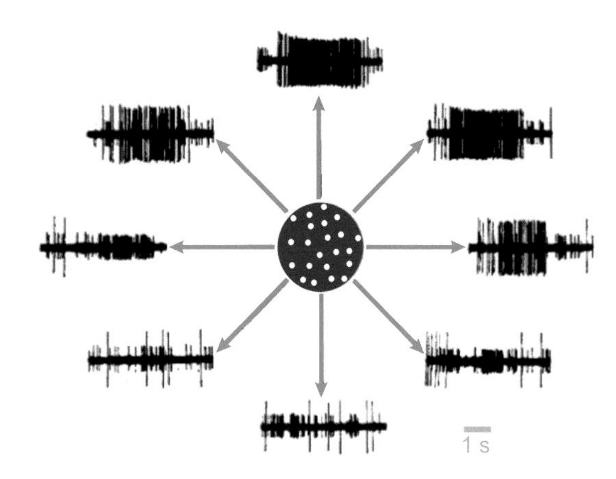

Figure 7.105 Pontine visual proprioceptive cell (cat), showing marked preference for movement of an extended, textured, field in a particular direction. (Baker et al., 1976; with permission from John Wiley and Sons.)

with one another; this function is described in more detail in Chapter 11.

The fact that many of these units adapt under prolonged stimulation gives rise to the striking *waterfall illusion*.[33] If one stares for a minute or so at a surface that fills a substantial part of the field and is in continuous motion in one direction – like a waterfall – and then turns away to look at a stationary scene, one has the strong and persistent notion that it is moving in the opposite direction. Presumably the sense of visual motion depends on the balance between the rates of firing of movement detectors having opposite preferred directions: after adaptation an imbalance is caused by the depression of one set of these, leading to the illusion of movement in the opposite direction.

In addition to assisting the control of posture, retinal slip also provides information that tells us about our locomotion through the outside world, and particularly our heading. As we move, the direction and magnitude of retinal slip changes in a characteristic way across the retina, producing what are known as *optic flow* patterns. In the MT region of parietal cortex, some way along the 'where' processing stream (see p. 158), units have been described whose activity appears to be related to optic flow. Lesions in this region also give rise to disorders of visual motion processing, in which patients can perceive the form and colour essentially normally but show a specific inability to detect or follow object movement. This is termed akinetopsia and leads to peculiar problems such as when pouring drinks: as the motion cannot be perceived, the subject does not know when to stop.[34]

As with the sense of visual direction, the use of retinal slip information relies on knowledge of any movements of the eyes, provided by efference copy. But other assumptions are necessary as well. When two areas of the field move relative to one another, it is usually the larger area that is assumed to be stationary: on a cloudy night, the moon appears to sail through the clouds. Of course it is the moon that is stationary and the clouds that move, but the latter occupy so much more of the visual field that the visual system assumes that they are stationary.

Notes

Vision There are many excellent books on vision in general, intended for readers with different interests. A selection: Cornsweet, T. N. (1970). *Visual Perception*. (New York, NY: Academic); Davson, H. (1990). *Physiology of the Eye*. (London: MacMillan); Hubel, D. H. (1995). *Eye, Brain and Vision*. (New York, NY: Freeman); Oyster C. W. (2006). *The Human Eye: Structure and Function* (Philadelphia, PA: Lippincott Williams & Wilkins), a thoughtful and beautiful book; Rodieck, R. W. (1998). *First Steps in Seeing*. (Sunderland, MA: Sinauer), mostly about the periphery, but intelligent and imaginatively set out; Walls, G. L. (1942). *The Vertebrate Eye*. (New York, NY: Hafner), a classic reference source for those with an interest in comparative aspects; Wandell, B. A. (1995). *Foundations of Vision*. (Sunderland, MA: Sinauer), quantitative and conceptual; Zeki, S. (1993). A *Vision of the Brain*. (Oxford: Blackwell), mostly on cortex and colour: clear and well illustrated. A book with a more clinical slant is Schwartz, S. H. (2017). *Visual Perception: A Clinical Orientation*. (New York, NY: McGraw-Hill); perceptual aspects are intelligently discussed in Snowden, R., Thompson, P. & Troscianko, T. (2012). *Basic Vision: An Introduction to Visual Perception*. (Oxford: Oxford University Press).

1. Units These are all *photometric* units: they take into account the degree to which different wavelengths are actually absorbed by human photoreceptors and contribute to vision. A different system of units, *radiometric*, is purely physical, and is based in effect either on the number of quanta emitted or absorbed, regardless of their wavelength, or on energy (not quite the same thing, since high-frequency photons are more energetic).
2. Dark-adapted vision Hess, R. F., Sharpe, L. T. & Nordby, K. (eds) (1990). *Night Vision*. (Cambridge: Cambridge University Press) has excellent reviews of adaptation and scotopic vision in general.
3. Purkinje shift It is said that Purkinje first became aware of the relatively enhanced sensitivity of the rods at short wavelengths when he noticed that the blue flowers in his garden seemed brighter than the others at dusk.
4. Getting older The lens also starts to lose its transparency: if the tendency to cloudiness and yellowness goes too far, we have *cataract*, with an inability to form proper retinal images. These changes are probably through a gradual accumulation of damage to cell protein and membrane disruption and are accelerated by both short- and long-wave radiation.
5. How good are the optics? Not terribly good. The great nineteenth-century physiologist Hermann von Helmholtz allegedly once said if an optician made me a lens as bad as the one nature gave me, I would send it back.
6. Myopia The incidence of myopia in the USA is about 25 per cent of the adult population. Some of this may be induced by reading: there is increasing evidence that the axial growth of the eye is linked to retinal blur by some kind of internal mechanism, possibly involving dopamine, to create what has been called 'emmetropisation'. See for instance Hung, G. K. & Ciuffreda, K. J. (2000). A unifying theory of refractive error development. *Bulletin of Mathematical Biology* **62** 1087–1108.
7. Astigmatism Astigmatism also results in different magnification in different meridians, and there has sometimes been speculation whether this might explain the distortions that some well-known artists seem to introduce into their paintings – El Greco is a good example. At first sight, there is an obvious logical flaw in this, for the artist would suffer the same distortion when looking at his canvas as well as at the model; nevertheless, there are optical defects that could account for it. This and many other ocular disabilities among artists are discussed in Trevor-Roper, P. (1970). *The World Through Blunted Sight*. (London: Thames & Hudson).
8. Large pupils reveal bad optics Presbyopes often complain that their eyes are 'weaker', in that they need more light in order to read. The reason is not that their sensitivity is low, but that with very bright illumination the pupil shrinks down and increases their depth of focus, enabling a tolerable image to be formed of an object considerably closer than the real near point.
9. Emotional pupils Photographs retouched to make the pupils larger are generally judged more attractive and stimulating than the original. Curiously, the viewer's own pupils then dilate, suggesting a covert system of sexual signalling that would result in positive feedback in certain circumstances – possibly an explanation for 'love at first sight'.
10. Angles You need to be familiar with the units used to specify visual angles. The fundamental unit is the degree (deg), of which there are 360 in a full circle. A degree is divided into 60 minutes of arc (arc min), and these are in turn divided into 60 seconds of arc (arc sec). A radian is 57.3 deg, so a centimetre at arm's length is roughly 1 deg; your thumb will subtend somewhere around 2–2.5 deg when you hold it out (depending on how wide it is).
11. 6/6 vision In the USA, where (oddly enough) imperial measurements still prevail, 6/6 translates as 20/20 (feet).
12. The retina Three classic accounts of retinal structure, of various vintages: Dowling, J. E. (2011). *The Retina: An Approachable Part of the Brain*. (Cambridge, MA: Belknap Press of Harvard University Press); Polyak, S. L. (1941). *The Retina*. (Chicago, IL: University of Chicago Press); Rodieck, R. W. (1974). *The Vertebrate Retina*. (San Francisco, CA: Freeman).
13. So much rhodopsin Thus each rod has over two billion molecules of photopigment, which might seem odd given that – as we see later on – one photon (and therefore just one molecule) is sufficient to initiate a significant electrical response. To a large extent this ratio of 9 log units must reflect the huge dynamic range of vision.
14. Pigment regeneration 11-*cis* retinol is more familiar as vitamin A, and one of the consequences of vitamin A deficiency is incomplete regeneration and hence a condition known as night blindness, cured by eating more carrots.
15. Electrical responses A clear account of this area is Lamb, T. D. & Pugh, E. N. (1990). Physiology of transduction and adaptation in rod and cone photoreceptors. *Neurosciences* **2** 3–13.
16. Adaptation Useful articles in this area may be found in Hess, R. F. (1990). *Night Vision*. (Cambridge: Cambridge University Press). An interestingly historical account of the functional division of rods and cones is Stabell, B. & Stabell, U. (2009). *Duplicity Theory of Vision: from Newton to the Present*. (Cambridge: Cambridge University Press).

17. Weber Ernst Weber (1795–1878) was a pioneer of psychophysics – the application of quantitative analysis to perceptual and behavioural phenomena – and invented, amongst other things, the two-point cutaneous discrimination test. The observation that $\Delta I/I$ is often constant was his; what Gustav Fechner (1801–1887) added was the idea that this relationship could be explained by sensations somehow being scaled logarithmically. If the sensation $S = k \log I$, then $dS/dI = k/I$, so $dI/I = dS/k$. In other words, if one assumes that dS is constant at threshold, then Weber's law is explained.

18. Gain The gain of a system is simply the magnitude of the output divided by that of the input; so, for example, an amplifier that generates 100 mV for a 1 mV input has a gain of 100. In a linear system the gain is always constant, regardless of the size of the input.

19. Optic radiation It is worth emphasising that as we go from geniculate to cortex there is an enormous expansion in the neural *width* of the visual system. Whereas in the retina there are something of the order of 130 million receptors, the LGN has only 1.5 million cells – net convergence; yet in primary visual cortex we are back to something like 200 million. It is clear that cortical cells are elaborating rather than condensing the information that they receive.

20. Stripe of Gennari Discovered in 1776 by Francesco Gennari while he was a medical student, studying his histology: which goes to show that diligence in this area can bring you fame as well as – it goes without saying – immense intellectual satisfaction.

21. Colour vision Mollon, J. D. & Sharpe, L. T. (eds) (1983). *Colour Vision.* (London: Academic) has many useful articles; other useful sources of information on colour are Gegenfurther, K. R. & Sharpe, L. T. (eds) (1999). *Colour Vision.* (Cambridge: Cambridge University Press) and Mollon, J., Pokorny, J. & Knoblauch, K. (2003). *Normal and Defective Colour Vision.* (Oxford: Oxford University Press).

22. Are three dimensions enough? Not for some species. The mantis shrimp has receptors with eight colour pigments, as well as four for ultraviolet (and two specialised for polarised light). See Marshall, J. & Oberwinkler, J. (1999). Ultraviolet vision: the colourful world of the mantis shrimp. *Nature* **401** 873–874.

23. Opponent processing Curiously enough, exactly the same system is used in colour television. The camera provides a red channel, a green channel, and a blue one: but these three signals are then recoded to provide a luminance signal plus two colour or chroma signals that are formed, as in the retina, by subtraction. Black-and-white sets simply ignore the chroma part of the signal. For measurement of colour visual acuity, see Mullen, K. (1985). The contrast sensitivity of human colour vision to red-green and blue-yellow chromatic gratings. *Journal of Physiology* **359** 381–400.

24. Colour constancy This topic (among many others) is intelligently discussed in Zeki, S. (1993). A *Vision of the Brain.* (Oxford: Blackwell).

25. Red versus green More recent sequencing has shown that the red and green cone opsins are more than 90 per cent homologous, much more than in relation to blue.

26. Foveal tritanopia This seems to have been realised empirically for a long time. Tritanopia causes yellow and white to be indistinguishable at a distance, and blue and black. The flags used for signalling at sea are designed such that these confusions can never make one confuse one signal for another; similarly, the strict laws of heraldry forbid a yellow charge on a white field, or vice versa.

27. Monocular colour blindness See for instance Hayes, S. P. (1911). The color sensation of the partially colour-blind. *American Journal of Psychology* **22** 389–407.

28. Achromatopsia Not, apparently, in monkeys: see Heywood, C. A., Gadotti, A. & Cowey, A. (1992). Cortical area V4 and its role in the perception of color. *Journal of Neuroscience* **12** 4056–4065.

29. Monocular distance cues These are the ones that artists have to use to create a sense of depth – a hard task, since both the binocular cues, accommodation and motion parallax, work against them. One trick is the use of the *peep-show*, in which the painted interior of a box is viewed through a small hole; this rather cleverly limits the viewer to monocular vision, eliminates parallax and reduces the effect of accommodation because of the small aperture. Some accounts of psychological and physiological aspects of artistic representation include Gage, J. (1993). *Colour and Culture.* (London: Thames and Hudson); Kemp, M. (1990). *The Science of Art: Optical Themes in Western Art from Brunelleschi to Seurat.* (New Haven, CT: Yale University Press); and Wright, L. (1983). *Perspective in Perspective.* (London: Routledge & Kegan Paul).

30. Corresponding points Strictly speaking, as you can probably prove for yourself geometrically, points like B must lie on a circle that passes through the centres of the two eyes and also through A. Such a circle is called a *horopter*.

31. Fusing disparate points Subjectively, there is indeed an area around the plane of fixation, called *Panums fusional area,* in which one is not aware of the double image normally perceived when an object is out of the plane of fixation; the disparity units seem in some way to have fused the two images back together again in one's perception.

32. Stereopsis It is interesting to note also that in military reconnaissance it is common to use a plane with two cameras some distance apart rather than one, for exactly the same reason. However much you fling nets and leaves and branches over a tank, the one thing you cannot disguise is that it hasn't got zero thickness.

33. The waterfall illusion First described by R. Addams, who wrote in 1834: 'During a recent tour through the Highlands of Scotland, I visited the celebrated Falls of Foyer on the border of Loch Ness, and there noticed the following phaenomenon. Having steadfastly looked for a few seconds at a particular part of the cascade, admiring the confluence and decussation of the currents forming the liquid drapery of waters, and then suddenly directed my eyes to the left, to observe the vertical face of the sombre age-worn rocks immediately contiguous to the waterfall, I saw the rocky surface as if in motion upwards, and with an apparent velocity equal to that of the descending water, which the moment before had prepared my eyes for this singular deception'. *Philosophical Magazine* Series 3, **5** 373–374.

34. Disorders of motion perception A useful review is Nawrot, M. (2003). Disorders of motion and depth. *Neurologic Clinics* **21** 609–629.

CHAPTER 8
SMELL AND TASTE

All neurons in the brain respond to chemical transmitters, so chemosensitivity is hardly a specialisation of function at all. Here we shall be considering only chemical stimuli that originate outside the body, with *olfaction* (smell) and *gustation* (taste). The functions of the chemoreceptors that monitor the composition of the blood, and are used in the regulation of autonomic and hormonal functions, are discussed in Chapter 14.

Olfaction

Receptors

If you have had the privilege of dissecting a human head you will be aware that the nose has an internal complexity that is startling in comparison with its rather drab exterior.[1] The surface area of the nasal cavity is enormously inflated by the presence of three *conchae* on each side, highly vascular organs covered with erectile tissue whose function is primarily to moisten and warm the incoming air, and conversely to limit the loss of heat and water in the air that is expired (**Fig. 8.1**). The olfactory receptors form part of the *olfactory epithelium,* tucked away in the olfactory cleft right at the top of the cavity, and in normal quiet breathing only a very small proportion of the air actually reaches the olfactory epithelium.[2] But when we sniff, turbulences are set up round the conchae, and an appreciable fraction of the air gets to the olfactory receptors. This fraction is critically dependent on the state of the conchae: if you have a cold, they tend to become engorged with blood, hindering the passage of air to the higher regions and causing the familiar partial loss of smell.

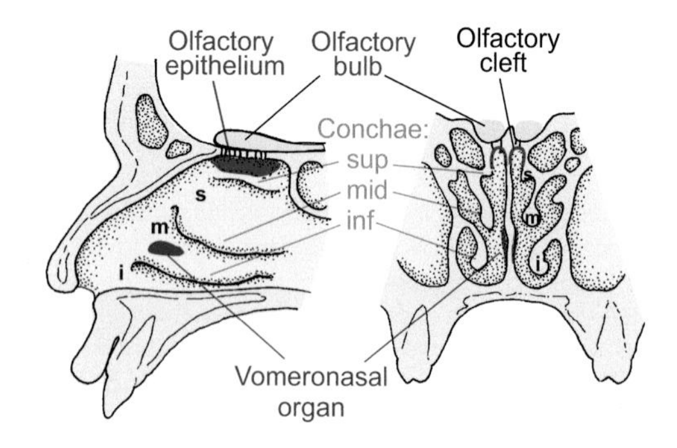

Figure 8.1 The nasal cavity in humans: sagittal and transverse section, showing the superior, middle and inferior conchae, the olfactory cleft, and the approximate position of the vomeronasal organ.

Humans are *microsmatic* animals: smell plays a far smaller part in their sensory world and in the regulation of their actions than in the case of macrosmatic animals such as the dog, and their olfactory sensitivity is in general correspondingly reduced. This is reflected in the small area of our olfactory epithelium: some 10 cm^2 in all, about half that of the cat, despite its very much smaller head. This epithelium has a number of easily recognisable features. It contains *Bowman's glands,* producing a lipid-rich secretion that bathes the surface of the receptors; this means that a substance must to some extent be lipid-soluble to have an odour. Another characteristic is that one of the types of epithelial cells contains granules of *pigment:* the depth of colour in different species is often correlated with olfactory sensitivity, being light yellow in humans and dark yellow or brown in dogs. It has sometimes been suggested that this pigment may play some part in the mechanism of olfactory transduction, perhaps being involved in the absorption of some kind of radiation such as infrared, an attractive but controversial idea discussed later (p. 181).

Finally there are the receptors themselves, of which there are some 10 million in humans. Each plays host to one of nearly 400 different types of odorant G-protein–coupled receptors; the perception of a particular odour ultimately reflecting the pattern of activation across this heterogeneous population of nerves. They are distinguished by a terminal enlargement above the surface of the epithelium, from which project some 8–20 *olfactory cilia* (**Fig. 8.2**). These cilia show the usual 9+2 fibril arrangement at the base, but are not believed to be actively motile: they form a dense and tangled mat that covers the olfactory area, markedly expanding the surface area to which odorants can bind. Vacuoles can also be seen in the terminal enlargement, and experiments with exogenous protein markers have shown that they are actively pinocytotic: fluid is being continually drunk in by the receptors and passed down the olfactory nerves into the brain. The significance of this surprising feature is unclear. A further curiosity of the olfactory receptors

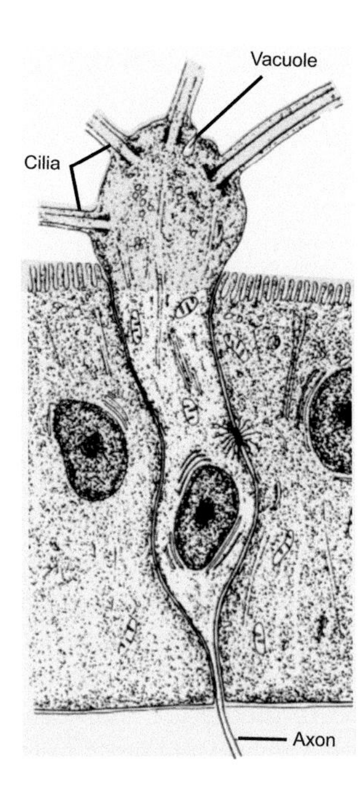

Figure 8.2 Typical olfactory receptor, showing terminal enlargement with cilia and vacuoles projecting above the level of the surrounding epithelium. Note that the cilia are truncated in this picture: in reality, they vary considerably in length, some being shorter than the receptor cell body, and some several times its length.

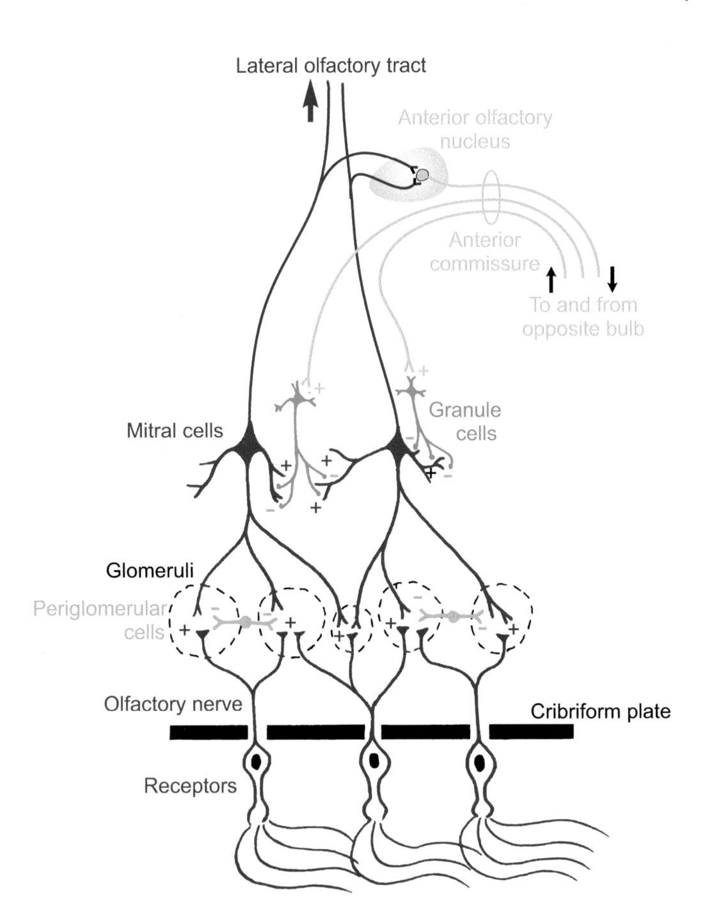

Figure 8.3 Simplified representation of the cell types of the olfactory bulb and their connections.

is their remarkably short life: after a month or two at most they degenerate and are replaced by new ones from the basal cells of the olfactory epithelium.

In addition to the conventional olfactory epithelium, there is another region, much less studied, that appears to contribute to olfaction. This is the *vomeronasal organ*, or Jacobson's organ, connected by a duct to the nasal airways and probably more receptive than the olfactory epithelium to the *pheromones,* which are air- or water-borne signals that can have profound effects on behaviour, but in humans it exists only in the embryo (see p. 178).[3]

Olfactory bulb

Unlike many receptor cells, the olfactory receptors send their own axons to the central nervous system (CNS) without an intervening synapse. These fibres, the *fila olfactaria,* make up the first cranial nerve: they are exceedingly fine and difficult to see with the light microscope: they are also vulnerable to injury from blows to the head. They pass through the *cribriform plate* at the base of the cranial cavity in individual holes ('cribriform' = 'sieve-like') and enter the *olfactory bulb,* which lies just above the olfactory epithelium (**Fig. 8.3**). Here they synapse with dendrites of the large *mitral cells* (which are supposed to look like bishops' mitres) and *tufted cells* in specialised nexuses called *glomeruli*. In rabbits, which have been particularly well studied, each glomerulus receives information from some 26 000 receptors, all expressing the same odorant receptor, and has an output to about 24 mitral cells: there are probably only some 2000 glomeruli in all. The fibres entering any one glomerulus come from a wide area of the epithelium, so that detailed information about any spatial patterns of activity must be largely thrown away. But as is the case for rods in the eye, this enormous degree of convergence must certainly enhance the nose's sensitivity by providing a mechanism by which the contributions of very large numbers of receptors can be added together. The final output of the bulb consists of the axons of the mitral and tufted cells, forming the *olfactory tract*: this has two divisions, lateral and medial, of which only the lateral one

Clinical box 8.1 Anosmia after head injury

Following severe traumatic brain injury, up to 30 per cent of survivors will have anosmia. The underlying mechanisms are multiple, but shearing of the primary olfactory neurons, as they pass through the cribriform plate on their way from the nasal cavity into the cranial vault, is probably the most common cause.

The sense of smell may recover, usually at 6–18 months; in some cases the anosmia is permanent, perhaps as a result of the formation of scar tissue at the cribriform plate, which prevents regenerating primary axons from reaching the secondary neurons of the olfactory bulb.

appears to be important in Man. In those species where it exists, the vomeronasal organ has its own nerve to an outlier of the olfactory bulb, the *accessory olfactory bulb*, whose projections are broadly similar to those of the main bulb itself, except that it has more prominent projections that go indirectly to the hypothalamus, as might be expected from its role in responding to pheromones.

The olfactory bulb is not just a simple relay. It has two other properties that we have already seen to be common to all the sensory systems examined so far, namely lateral inhibition and control of afferent information through negative feedback. The most prominent feedback path is formed by projections from the second-order mitral and tufted cells, which excite *granule cells* and which in turn inhibit neighbouring second-order cells. An interesting feature of these synapses is that they are dendro-dendritic (the granule cells have no axons) and reciprocal: excitation and inhibition occur simultaneously – in opposite directions – at the same synapse. Other collaterals of the second-order cells' axons synapse in the anterior olfactory nucleus with interneurons that also return and synapse with granule cells; but in this case some of the interneurons project contralaterally in the anterior commissure to influence the opposite bulb in the same way. The significance of this mutual inhibition between the two bulbs is not clear: one possibility is that it may serve to enhance differences between the activities of the two bulbs – another kind of lateral inhibition – in a way that might perhaps be useful for localising smells. Careful experiments have shown that even humans are capable of localising odorous objects in a rather approximate way, presumably through slight differences in the timing or intensity of the stimuli in each nostril.[4]

Finally there are the *periglomerular cells*, which appear to carry out local lateral inhibition at the level of the glomeruli, through reciprocal synapses with the second-order cells, and direct connections from the fila olfactaria. All this is strikingly reminiscent of the retina, where the horizontal cells appear to carry out similar functions to periglomerular cells, and amacrine cells to granule cells, using the similar kinds of synaptic mechanism. A similar arrangement is seen in the early stages of both somatosensory and auditory processing, suggesting a basic mechanism carrying out a fundamentally important function (**Fig. 8.4**).

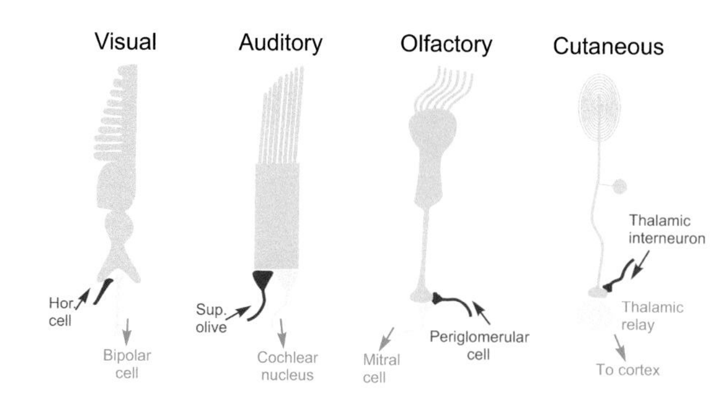

Figure 8.4 Similarities between visual, auditory, olfactory and cutaneous sensory processing.

Central olfactory projections

The central projections of the olfactory system provide something of an *embarras de richesses*, very different, for example, from the visual system, with its orderly projection of the optic tract on to the lateral geniculate nucleus and thence to the cerebral cortex. Indeed olfaction seems to be unique in projecting straight on to cortical areas without relaying in the thalamus or any equivalent structure. You need to bear in mind that the olfactory system is very much older than such senses as vision and hearing, and in the course of evolution the relative size of the olfactory bulb has declined (**Fig. 8.5**). Correspondingly, in more primitive animals a very much larger proportion of the brain is directly or indirectly concerned with olfaction. The reason for this is not hard to see: simple animals depend much more immediately than we do on knowing directly from their senses whether food is in the vicinity, and their motor systems are likely to be more pressingly governed by the need to move towards nutrients and avoid poisons, and to seek out mates by recognising the chemical attractants they release. Their *motivation* – the drive that tells their motor systems what to do – is essentially olfactory.

Even in humans, the remains of this very basic system for chemical motivation and emotion (emotion being the sensory correlate of motivation) can still be seen in the central olfactory projections (**Fig. 8.6**). Many of these structures form part of the *limbic system*, a group of nuclei, cortical regions and connecting tracts of great evolutionary antiquity that appear to be concerned with precisely those kinds of function that one would expect in a primitive animal to be closely related to chemical stimulation: motivation, emotion and certain kinds of memory. The limbic system and its functions are discussed more fully in

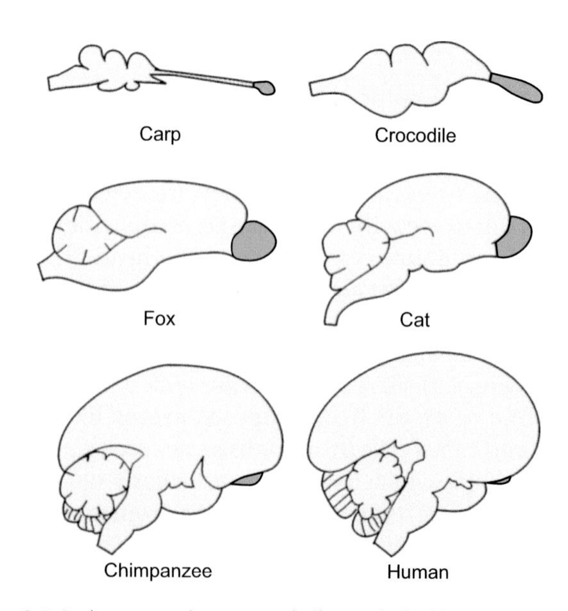

Figure 8.5 Reduction in relative size of olfactory bulb (blue) with phylogenetic development of the brain. (Partly after Carpenter & Sutin, 1983.)

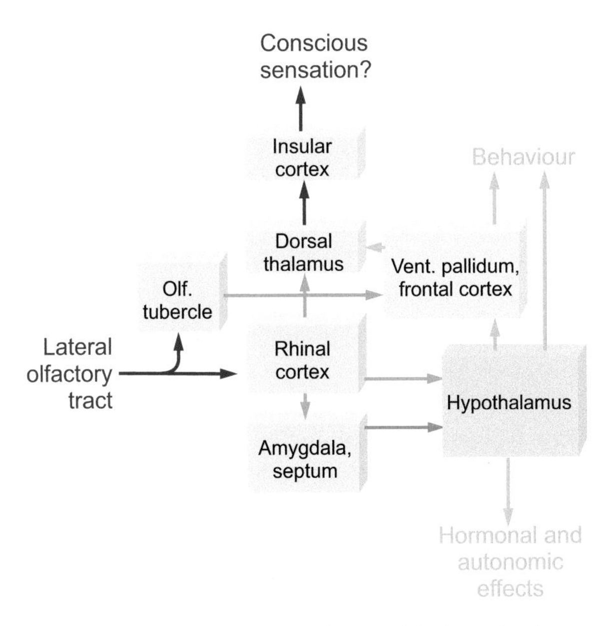

Figure 8.6 Highly schematic diagram of areas of the brain directly or indirectly driven by olfactory stimuli: 'rhinal cortex' includes periamygdaloid, prepyriform and entorhinal cortex, hippocampus and subiculum.

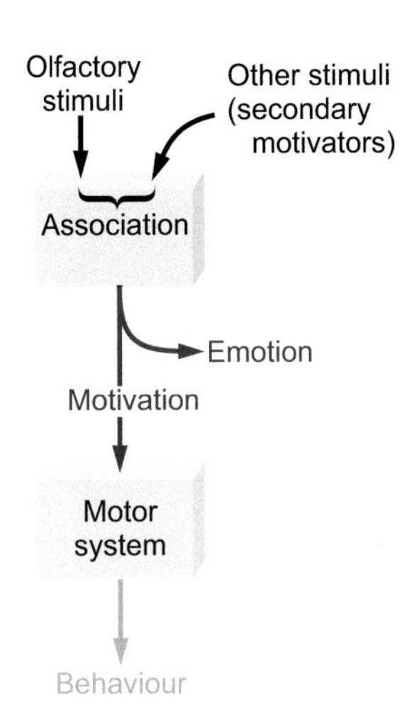

Figure 8.7 Schematic representation of motivation through primary olfactory stimuli and, in a more developed animal, other stimuli (secondary motivators), that become more potent through association.

Chapters 13 and 14; for the moment we may note, for example, that the septal nuclei and amygdala contain regions known as 'pleasure centres', in the sense that when electrically stimulated they may provide a kind of direct positive motivation. 'Rhinal cortex' here includes periamygdaloid, prepyriform and entorhinal cortex, hippocampus and subiculum, areas whose functions are partly olfactory, but mostly to do with associational memory. The hippocampus in particular seems to be concerned with associating different places in the environment with the promise of food or pleasure signalled more directly by olfactory stimulation, and recognising such a stimulus in the future as a source of motivation in its own right (see Chapter 14, p. 277).

What seems to have happened in the course of evolution is that this kind of *secondary motivation* by stimuli that only acquire their meaning through experience and learning has steadily grown in importance relative to that of primary olfactory motivation (**Fig. 8.7**). For humans, *money* is perhaps the most obvious secondary motivator, and the most powerful of all: given the choice between a plate of fish and chips and a plate of £10 notes, unless one was exceptionally hungry there is no doubt which would cause the greater motivational drive.[5] For this reason, limbic structures that were originally subservient to olfaction are now not primarily olfactory at all, so the name *rhinencephalon* ('nose-brain') which is sometimes given to the limbic system is inappropriate in higher animals.

Finally, there are important connections between the *limbic system* and the hypothalamus (see Chapter 14, p. 276), providing routes by which olfactory stimuli can cause such obvious autonomic effects as salivation and other secretory responses to food smells, as well as influencing the choice of food: a single instance of particular odorant, associated with nausea even some hours later, can condition an animal to avoid the substance for the rest of its life.[6] Smell may also have much more widespread hormonal and behavioural effects, of which the best known are perhaps sexual arousal and modification of reproductive cycles, even abortion.

A curious feature of the pheromones deployed by many creatures as long-distance sexual attractants is that they often resemble, in species as different as the civet and the moth, the steroid reproductive hormones themselves (at least in overall shape: **Fig. 8.8**). Furthermore, some of them – such as civet oil and musk – are used in perfumery and so presumably act as lures for human males as well (and they are curiously abundant also in church incense). There are said to be marked sex differences in the olfactory thresholds for some of these macrocyclic compounds, which may also vary with the phase of the menstrual cycle. But it is not just sex life that is dominated by smell. Predators sniff out their prey; fish find their way back to their native streams – sometimes from hundreds of miles away – through smell. Pheromones in the urine or from facial glands may be used as territorial markers; herd animals use them to warn others of the approach of a predator, and personal smells are an essential means by which many animals recognise their own social groups and mothers their young.[7]

It is natural to speculate whether we humans are influenced subconsciously in the same kind of way: if so, it would have the most profound psychological and sociological implications. Firm evidence is lacking, though

Testosterone Civetone Gypsy moth lure

Figure 8.8 The structure of testosterone and two olfactory sexual attractants.

Table 8.1 Examples of behaviour controlled by olfaction

Feeding	Hunting and finding food, eating and aversion, stimulation of digestive secretions
Sexual	Recognition of receptive females, triggering or suppression of ovulation, sexual attraction and arousal
Maternal	Recognition of offspring, triggering of maternal behaviour
Avoidance	Scenting of predators, response to alarm pheromones released by other herd members
Territorial	Recognition of urinal territorial markers, homing and recognition of breeding-grounds
Social	Recognition of individuals and members of species or group, social dominance

pheromone sprays for men apparently sell well.[8] The vomeronasal organ, when it exists, appears to be particularly involved in detecting olfactory stimuli with strong behavioural implications: via the accessory olfactory bulb, it projects extensively to various parts of the limbic system, particularly the medial nucleus of the amygdala (**Table 8.1**).

One striking indication of the psychological links between olfactory and limbic functions in humans is the very vivid way in which odours may call up – often with surprising intensity – recollections of past experience; and it is significant that so often such evocations are not just of the objective circumstances of a particular event, but also of the mood or emotion that was felt at the time, in a way that is seldom experienced with purely auditory or visual stimulation. The direct penetration of the emotional areas of the brain by olfactory fibres seems exactly reflected in what we feel.[9]

Recordings from olfactory cells

Perhaps because smell does not seem as useful or important to us as say vision or hearing, and also because of certain difficulties of experimental technique, our knowledge of the electrophysiology of olfaction is still somewhat rudimentary. In those species where it has been possible to measure the time-course of the response, the transduction process is found to be extremely prolonged, extending over a second or more (**Fig. 8.9**). No doubt there has been less evolutionary pressure to make things happen fast than in other senses, and of course there is no particular benefit in good temporal acuity, unlike in vision or hearing. This, and the extraordinary olfactory sensitivity to be discussed later (with a very steep dose–response curve), suggests some kind of amplificatory cascade similar to what is found in photoreceptors, and – as was mentioned in Chapter 3 – this is now known to be the case. The initial transduction process is of a highly unconventional indirect form, with cyclic adenosine monophosphate (cAMP) acting as an intermediary and causing an increase in both sodium and calcium permeability, the calcium in turn triggering the third-order opening of chloride channels. Normally this would be inhibitory, but because olfactory receptors have unusually high [Cl^-], chloride leaves rather than enters, and further depolarisation results. In addition, there is evidence for a second mechanism in parallel, in which the receptor triggers the opening of calcium channels (and thence chloride) through G-protein activation of inositol phosphate 3 (IP3), and that some odorants vary in the extent to which they affect one pathway rather than the other.

The conclusion of the response sees calcium extruded by sodium-calcium exchange mechanisms (though Ca-ATPase may also contribute). However before doing so, not unlike as was seen in the visual system, the calcium also

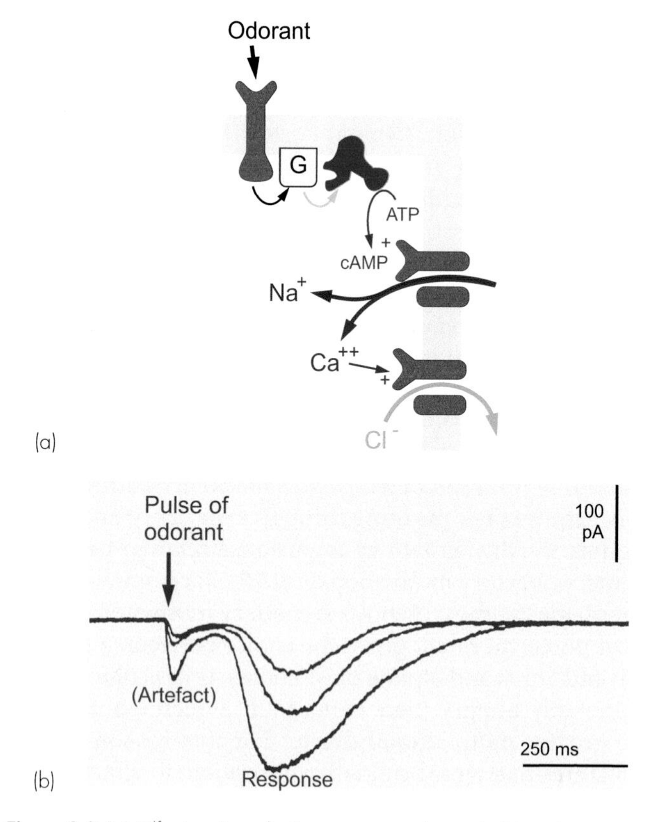

Figure 8.9 (a) Olfactory transduction appears to be a doubly indirect process, in which olfactant triggers a G-protein–linked increase in cAMP, which opens sodium and potassium channels and thus depolarises the cell. But calcium enters as well, and this indirectly causes chloride channels to open. Normally this would be inhibitory, but because olfactory receptors have unusually high chloride concentration, chloride leaves rather than enters, and further depolarisation results. (b) Patch-clamped salamander olfactory receptor responding to brief puffs of odorant of different concentrations. The odorant was mixed with potassium to act as a marker to give the effective duration of the stimulus: the potassium generated the first, brief phase of the response. The later, extended part of the response is due to the odorant itself. (Firestein & Werblin, 1989; with permission.)

mediates the cellular mechanism of olfactory adaptation: in activating calmodulin it reduces the sensitivity of the cation channels to cAMP. Thus people working in uncommonly smelly environments such as sewers or gas works soon become quite insensitive to the smells around them, and people are (in general) unaware of their own body odours.

The details of stimulus encoding – how different odorants give rise to different smells – in the olfactory nerve fibres are not very clear. One of the difficulties is that, whereas it is comparatively easy to stick an electrode into the optic nerve or auditory nerve and record the way in which single fibres respond to particular stimuli, it is rather hard to do the same thing in the case of smell: the olfactory fibres are exceedingly fine, rather short, and buried for much of their length in the cribriform plate. An electrode in the olfactory epithelium tends to pick up not spike responses from individual cells, but an averaged slow potential from many of them together, the *electro-olfactogram* (EOG). The size and shape of the EOG often shows rather little obvious correlation with the kind of substance that is applied, and it is perhaps not a very useful technique.

With care, one may be lucky enough to record spikes from individual fibres, usually with the EOG superimposed on top, but again there is generally no simple relation between firing frequency and the kind of stimulus applied. In fact with a single-unit preparation of this kind one can draw up a list in two columns, showing for a particular cell which substances excite it and which inhibit it. Such lists turn out to be quite chaotic, with apparently similar substances like menthol and menthone (which smell identical to us) often on opposite sides. Even more perplexingly, if one moves the electrode to record from a different cell, one finds in general an entirely different list, with substances that were excitatory for one cell being now inhibitory for the other, and substances that were on the same side of one list being on opposite sides of the other. In fact there seems no system whatever in the way in which chemical stimuli are coded into patterns of firing of the olfactory nerve: each unit has its own private idiosyncratic view of the olfactory world, like a spoilt child who likes baked beans but not bananas, fudge but not fish fingers, in a wholly arbitrary manner. The situation could hardly be more different from a sensory organ like the retina, with each of its units closely specified in terms of position, intensity and colour.

Randomness of this kind is not necessarily a weakness in a sensory system; no information need be lost, since by looking at the pattern of response over the fibres as a whole the nature of the original stimulus can still be reconstructed. Imagine a nursery of spoilt children seated at a dinner table and provided with push buttons with which they can register approval or disapproval of what is set in front of them: if these buttons were connected to an array of lights on a screen, it is clear that although any individual child's preferences may be quite idiosyncratic and quite unlike any other's, nevertheless any particular

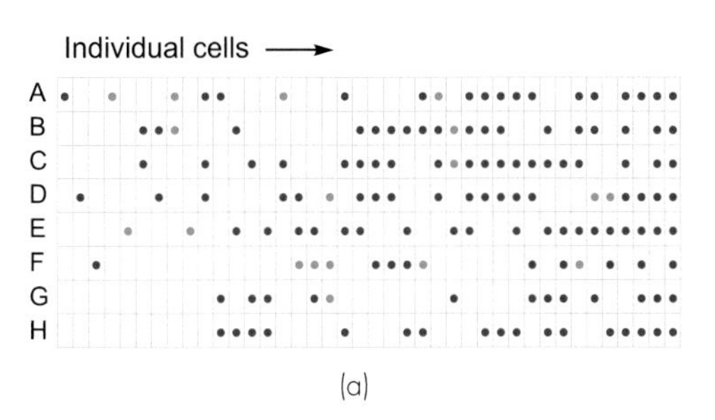

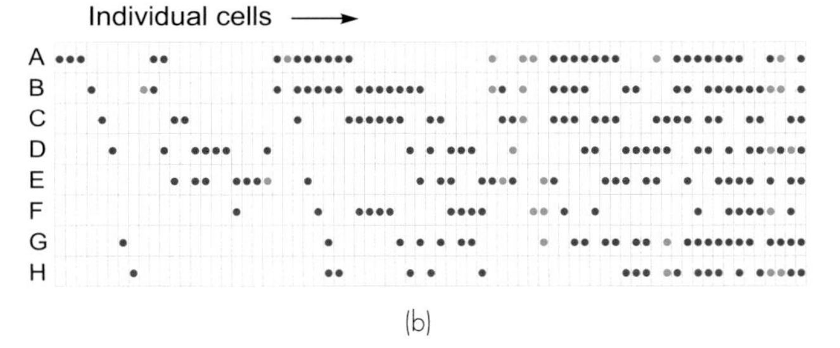

Figure 8.10 Graphical representation of responses of neurons in the olfactory system of a monkey to each of eight different odours (A–H). Pink = excitation, blue = inhibition. (a) Forty different units in the olfactory bulb. (b) Seventy-three units in the prepyriform cortex. A larger proportion of the more central units responds to a smaller set of the stimuli applied.

dish will result in a perfectly characteristic and reproducible pattern of lights by which it may be recognised. The ultimate reason for this randomness is that there is a large repertoire of receptor proteins, each specific for particular groups of odorants and expressed by a particular receptor cell. In the bulb, there is a tendency for fibres from cells of a particular type to converge on the same glomerulus, leading to a degree of spatial organisation.

Recordings from the olfactory bulb show a reduced degree of chaos: **Figure 8.10a** summarises the specific sensitivities of 40 different units in the olfactory bulb of a monkey to a set of eight different chemicals (A–H), and it can be seen that a small proportion of the cells responded to just one of the eight, and most showed excitation (pink) or inhibition (blue) to at least three. This increased specificity is undoubtedly the result of the *lateral inhibition* via periglomerular and granule cells, that occurs early on in the bulb. Just as in the skin, where lateral inhibition reduces functional overlap between receptors by emphasising differences in spatial firing patterns, here too – in a more abstract sense – it is reducing overlap between receptors not literally in space but in 'odorant space'. Of more central areas little is certain: responses to olfactory stimulation can be recorded from wide areas of the brain, not just in the limbic system but as far afield as the basal ganglia as well. One region that has been studied in more detail is the *prepyriform cortex:* here, and in the amygdala, units show properties that suggest that the chaos characteristic of preceding levels of the olfactory system is beginning to be sorted out, and a significantly larger proportion of units respond to just one of a series of chemicals (**Fig 8.10b**). On the whole there is a tendency for a larger proportion of units to respond to 'meaningful' smells like food or urine than to such things as the smell of mothballs.

Psychophysics of smell

The sense of smell shows a number of interesting and unusual features: as well as shedding light on one's understanding of sensory processes in general, they also make experiments on olfaction unusually difficult.

The *sensitivity* of olfaction in many species is astonishing: just as the rods in the eye respond to single photons, and the ear to vibrations of the air of subatomic dimensions, so the olfactory receptors are very near the theoretical limit of their sensitivity and can apparently respond to the absorption of one or two molecules. Tests on tracker dogs have shown that they can respond to 1 mL of butyric acid (an ingredient of stale sweat) in some 1011 L of air. This means that each sniff contains only about 200 000 molecules, and since the dog has roughly 200 million receptors, it follows that the absorption of *two molecules at the most* must be enough to excite an individual receptor.[10] Sensitivity of this kind makes for considerable difficulty in experimentation, for when one measures an apparent response to one particular substance A, unless A is quite exceptionally pure, one can never be quite certain that what one is measuring is not the response to some other substance B present in exceedingly small amounts as a contaminant.

Another characteristic of olfactory sensitivity is that adaptation, though not particularly rapid, is usually absolutely complete. Professional food evaluators have to take special precautions to avoid adaptation of this kind. Wine-tasters, aware of this danger, may nibble a piece of cheese between sips to restore the keenness of their palates (and with intriguing symmetry, Scottish cheese-tasters take a nip of whisky after each sample, ostensibly for the same reason). This may perhaps be why the olfactory epithelium does not lie on the path of the incoming air in normal breathing: in sniffing, a sufficient *change* in odour concentration may be set up that overcomes unwanted adaptation.

The most fundamental way in which smell differs from the other special senses is in the lack of a systematic method of classifying and analysing different types of odour. To some extent this is because we usually pay little conscious attention to smell, having enough to do in coping with the flood of more compelling information pouring in from our eyes and ears. Helen Keller, the writer who was blind and deaf from an early age, was able to develop her olfactory discrimination to an extraordinary degree through not being distracted by her other senses; she could for example recognise people she met and places she visited solely through their characteristic odours.[11]

Part of the trouble is undoubtedly our lack of a proper *vocabulary* for describing smells: if we want to convey to someone what a eucalyptus smells like, we are literally at a loss for words. The difficulty is that there exists no objective, physical, way of *classifying* smells systematically in the way that we can, for example, order colours into a spectrum or tones into a scale. In the case of vision our system of classification leads us to formulate simple rules using the colour triangle that enable us to predict the result of mixing colours in certain proportions to produce other colours: but in the case of smell this is quite impossible. We can never predict in advance what the result of mixing two odours together will be, and the results of doing so are frequently quite paradoxical. For example, the smells of iodoform and of coffee, individually strong and characteristic, are said to cancel each other out if appropriately mixed; other examples of cancellation of this kind have sometimes been exploited commercially to produce specific deodorants. Worse still, many odours smell quite different at different concentrations. Indole, which is a major component of dog excreta, and smells like it when concentrated, has a pleasant floral smell when very dilute and has actually been used in cheap perfumes![12] Many of the organic sulphides smell appalling at close range, but in small quantities turn out to be mainly responsible for the appetising smell of foods such as roast beef and onion. Conversely, the nose is usually unaware without special training whether a particular smell is pure, in the sense that only one kind of molecule is present or a mixture. Many natural odours that seem perfectly unitary and pure, such as that of raspberries, are in fact composed of dozens of components, many of which taken by themselves are rather unpleasant and cannot be detected for what they are in the whole ensemble.

For all these reasons, although in the past strenuous efforts have been made to try to classify smells into primary odour classes, like the nineteenth-century sixfold classification – Henning's 'olfactory prism' (**Fig. 8.11**) – no classification is ever really satisfactory because it does not enable one to predict the results of making mixtures, in the way that the colour triangle does for colours. There is no such thing, in fact, as a *primary* odour. The explanation for this unsatisfying state of affairs lies in the chaotic way in which individual receptor cells respond to particular chemicals, which in turn is a function of the apparently idiosyncratic selectivities of different molecular receptors. If we have two substances A and B which each produce characteristic patterns of activity in the olfactory nerve as a whole, then the response to A and B together will not be simply the sum of the responses to each separately: it will be a new pattern altogether. Thus the number of 'primary odours' will be of the order of the number of types of molecular receptor, which is around 400 in humans: more

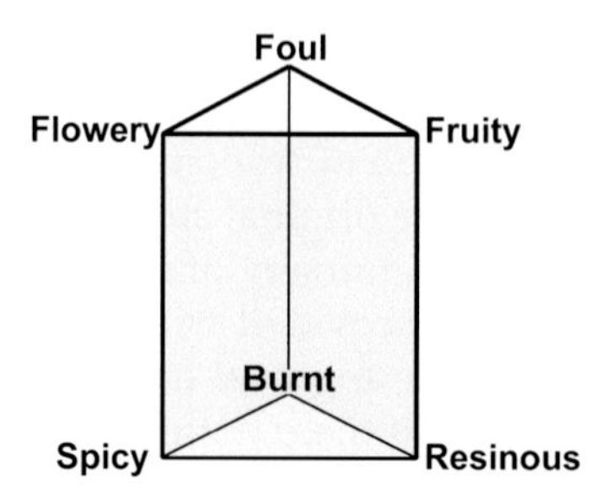

Figure 8.11 Henning's prism: an attempt to divide olfactory stimuli into six 'primary' classes.

than 1 per cent of the human genome is devoted to coding for them.

Transduction mechanisms

Though we know in a general way that the firing of olfactory nerve fibres is a consequence of a non-specific increase in ionic permeability, mediated by cAMP, how odorant molecules actually trigger this change remains something of a mystery. The odorant receptor sites recognise particular molecules or classes of molecule, but the basis of this recognition is not quite as straightforward a matter as one might imagine.

A very simple concept that has been around for some time was proposed by Amoore.[13] He suggested that either on the receptor membrane, or possibly inside, there are hollow receptacles of molecular proportions, which accept or reject odorant molecules according to how well they fit the site (**Fig. 8.12**), like the molecule of geraniol, a constituent of the smell of roses. In his theory he chose just seven types of site, each with its corresponding odour quality – floral, minty, and so on – and two of the sites were for electrophilic and electrophobic molecules. In general terms the theory is plausible enough, except for the very small number of primary classes that is envisaged: we have already seen that a classification with only a few primary odours is quite insufficient to describe the richness and complexity of the real olfactory world. Another problem is that in practice there is often a striking lack of the expected correlation between a molecule's overall shape and what it smells like. Camphor and hexachlorethane smell practically identical to us, yet one could hardly imagine two molecules more different in size and structure (**Fig. 8.13**). Again, optical isomers – molecules having the same structure but mirror images of one another – generally have identical smells: it is difficult to see how the same site could fit both forms.

A second theory that was devised partly to get round this problem of lack of consistency between molecular shape and odour quality is the infrared vibrational theory developed by Wright. The basic notion here is that whereas odour must ultimately be determined by chemical structure – the structure after all defines what the chemical is – overall shape is by no means the only property of a molecule that is determined by its structure. All molecules undergo mechanical vibrations, and the frequencies of these vibrations lie mostly in the infrared region, and depend in a rather complex way on the molecule's structure. In principle one could certainly imagine a sensory system that analysed the spectrum of these frequencies of vibration, perhaps through mutual resonance at particular sites on receptors. In such a case one might well find two substances with similar shape but different smells, because their frequencies of vibration were different; or conversely, substances sharing particular vibration frequencies and thus smelling similar, but of very different overall shape. There are a number of examples of the latter phenomenon that lend some support to the theory:

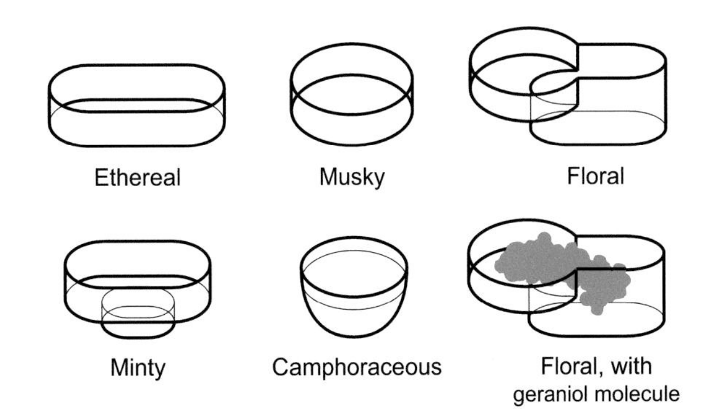

Figure 8.12 Five of the receptor sites proposed by Amoore, and (blue) the floral site occupied by a molecule of geranial, a constituent of the smell of roses.

nitrobenzene, benzonitrile and alphanitrothiophen, all of which smell of bitter almonds, happen to have many of their vibrational frequencies in common, but have widely differing shapes. Optical isomers necessarily have identical vibrational frequencies, and usually smell identical as well. But while the theory has a certain plausibility in the more primitive parts of the animal kingdom,[14] in the case of warm-blooded animals there is a grave physical objection: it is very difficult to see how an olfactory system working with infrared radiation could function with such exquisite sensitivity – responding to single odorant molecules – in the presence of the inevitable background 'noise' generated by the body's own heat. It might conceivably be that the pigmentation of the olfactory epithelium could play some role in the absorption of radiant energy, and its presence is otherwise somewhat puzzling. But few physiologists would care to accept the infrared theory as the explanation of olfaction in higher animals.

It seems most probable that Amoore's theory is basically correct, but with many hundreds of different receptors rather than just seven: the number of odorant receptor genes are now known to be in the region of 1000 in the human and, although not all of these are transcribed, presumably this heterogeneity allows different odorants to elicit a response in the olfactory receptor. However, the affinity for different molecules might still be determined by something rather more sophisticated than simply their overall shape (perhaps vibrational frequency might come into it as well). As mentioned earlier, these receptor proteins seem to be expressed a few at a time in any one

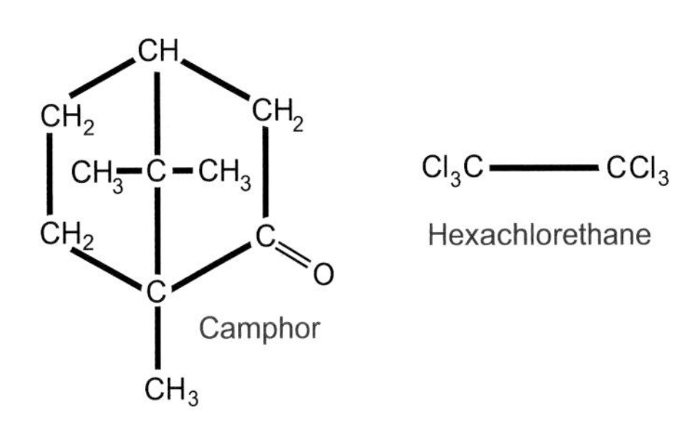

Figure 8.13 Two substances whose smell is extremely similar though their shapes and chemical properties are entirely different.

receptor cell, it what seems a random way. It is interesting in this context that many otherwise normal people show specific anosmias – 'blindness' to particular smells (that of freesias being a common example: more than 60 such specific anosmias are known) – that are inherited as single recessive genes, suggesting perhaps the loss of a single receptor protein.[15]

Taste

The receptors and central pathways

What is popularly called 'taste' is actually very largely *smell*, with purely somatosensory contributions such as texture, temperature and even pain (as in chilli) playing a part as well. People with anosmia, perhaps as a result of a cold, find their sense of 'taste' profoundly disturbed: apples taste like onions, vintage port like blackcurrant syrup.[16] Classically, the human tongue is regarded as having only four modalities of taste apart from ordinary cutaneous sensation: *salt*, *sour*, *bitter* and *sweet*. These four qualities have obvious physiological significance: sweet things are on the whole sources of metabolic energy; bitterness is usually associated with poisons; sourness is simply a measure of acidity, and salt essentially of sodium chloride concentration, two fundamental physiological variables. Some amino acids and their derivatives also stimulate taste receptors, and this is now accepted to constitute a fifth sub-modality *(umami)*; in some species *water* is an effective stimulus as well. To some extent taste thresholds and preferences are under the influence of the body's state of physiological need: salt-deprived rats show a preference for drinking salt solutions instead of water, and will tolerate strong saline solutions that other rats will refuse to drink. Coal miners, who sweat a lot, often used to put salt in their beer, though to others it tasted revolting. Whether the body's acid–base balance similarly affects perceived sourness is unclear.

Stimulating the tongue with solutions applied locally through small pipettes, certain areas seem to be more sensitive than others to particular stimulus modalities (though the zones are not nearly as clearly demarcated as figures such as **Fig. 8.14** inevitably imply). In addition, the sense of taste is confined to special structures on the tongue called the *papillae*. In humans, four main types of papilla have been described: but one of these, the *filiform* papilla, is not concerned with taste at all but is specialised for rasping and is particularly well developed in meat-eaters such as the cat. *Circumvallate* papillae, associated with sour and bitter taste, are found at the back of the tongue and consist of a sort of dome surrounded by a moat (**Fig 8.15**), and in the walls of this moat one can find sensory cells with microvilli arranged in pitlike invaginations (shown by the arrows in the figure) together with accessory cells, 30–50 in all, forming *taste buds*. Similar buds are found in the *fungiform* papillae, which respond to salt and sweet and lie more towards the front of the tongue, and the *foliate* papillae (sour), more around the edge.

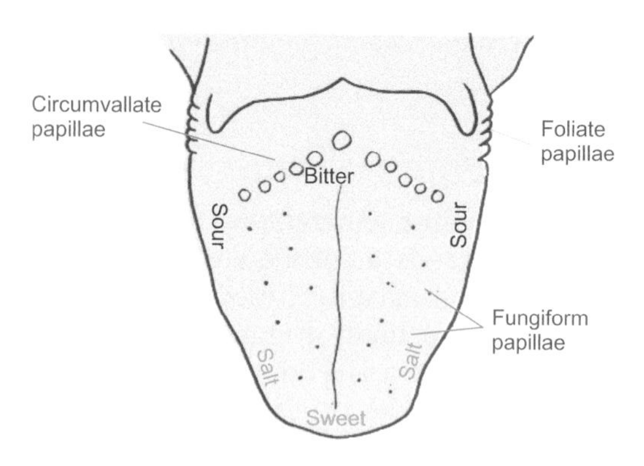

Figure 8.14 Human tongue, showing regions where the four components of taste are most readily evoked, and the approximate distribution of the papillae. The distribution of the different sensations actually overlaps more than this figure implies.

Unlike the olfactory receptors, these do not send their own axons to the CNS, but are innervated by fibres of cranial nerves VII (fungiform papillae) and IX (circumvallate, foliate), whose cell bodies are in the geniculate ganglion and glossopharyngeal ganglion, respectively. The tongue is also innervated by the trigeminal nerve (V), providing ordinary somatic sensibility, and both cold (e.g. peppermint) and pain (e.g. pepper, chilli) are culinary components that are sensed in this way. The afferent taste fibres go to the rostral part of the *nucleus solitarius*, which projects to the medial part of the ventral posterolateral area of the thalamus, whence fibres ascend to a small area of the insular cerebral cortex (**Fig. 8.16**), where they are joined by a part of the olfactory projection. On the whole the

(a)

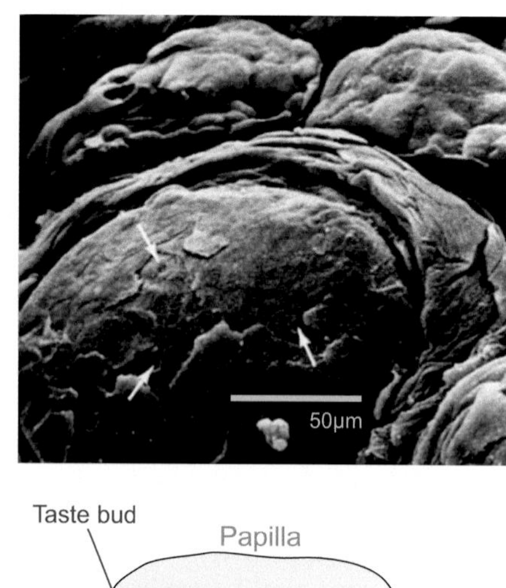

(b)

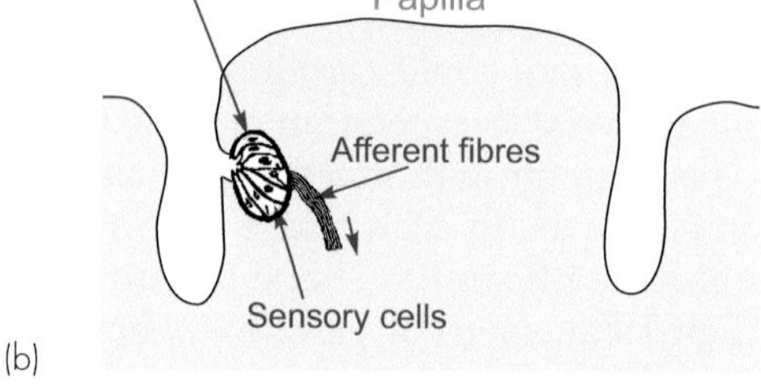

Figure 8.15 (a) Scanning electron micrograph of circumvallate papilla; the arrows show the openings of taste pits or buds on the surface of the papilla. (b) Schematic section of such a papilla, showing the sensory cells lying within the taste bud.

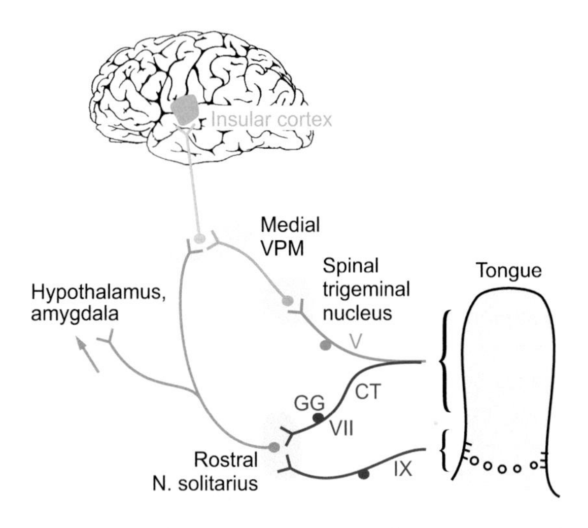

Figure 8.16 Simplified scheme of the main afferent gustatory pathways. Key: V, VII, IX, cranial nerves; CT, chorda tympani; GG, geniculate ganglion; VPM, ventral posterior thalamic nucleus.

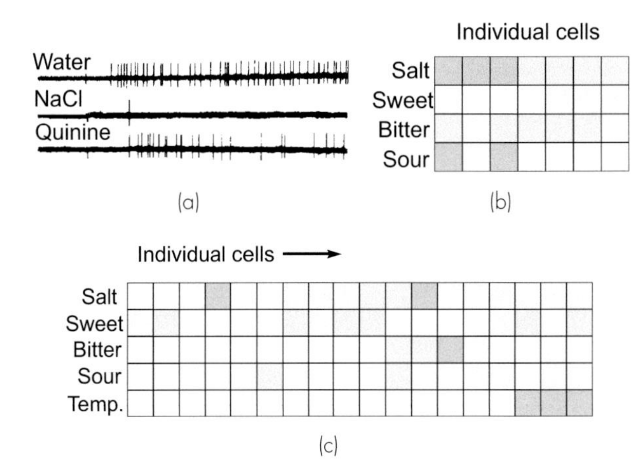

Figure 8.17 Responses of gustatory neurons. (a) Records of activity of a single fibre in the chorda tympani of a cat, showing discharges in response to water and quinine (bitter), but not to salt. (b) Diagrammatic representation of the responses of seven individual taste receptors in the rat to four different stimuli, showing the varieties of stimulus preference. (c) Similar representation for 18 taste units in monkey thalamus. Shading indicates size of response.

anatomy of the gustatory system is much more like that of ordinary cutaneous sense than is the case for olfaction, though gustatory projections to the periamygdaloid cortex, hypothalamus and other limbic areas certainly exist as well.

One branch of the lingual nerve lies conveniently in the chorda tympani, where it is relatively accessible for recording. Single units show almost the same chaotic properties that are seen in the olfactory fibres (**Fig. 8.17**). Instead of responding strictly to just one of the four modalities, they tend to respond to a random assortment of them (including water). Once again, each unit seems to have its own viewpoint of the gustatory world, and stimuli are encoded in the spatial pattern with which the ensemble of afferent fibres discharges. At the thalamic level (**Fig. 8.17**) the situation seems hardly less chaotic, although there may be more of a tendency for units to be selective for one type of stimulus.

Transduction mechanisms

Two of the modalities of taste, salt and sour, seem relatively straightforward. Saltiness is a function mostly of the sodium ion concentrations, though to some extent of lithium as well (**Fig. 8.18**). The response to sodium seems simply to be due to its direct entry through passive sodium channels, altering the Nernst potential for sodium and hence the resting potential. One can also show that the neural response to salt solutions is modified by salt deprivation in the way that would be expected from behavioural observations. Sourness depends on pH, but not in a simple way (not all solutions of equal pH are equally sour, and the anion contributes): while it used to be thought that hydrogen ions reduced the permeability to potassium, it now appears that there is a specific acid-sensing channel.

But in the case of sweet and bitter, we find complexities similar to those seen in olfaction, for instance a lack of correlation between overall molecular shape and taste. Artificial sweeteners are a good example: they are chemically diverse, and their thresholds are vastly lower than for the actual sugars for which the receptors were presumably intended. But the fact that we are dealing here with only two classes instead of indefinitely many simplifies things considerably, and it is now clear that there are specific receptor proteins which in some cases have been extracted and found to bind to sweet or bitter substances, and are 'fooled' by false stimuli like saccharine in the same way that we are ourselves. The sweet receptors – which respond to some amino acids as well as sugars and sweeteners – appear to increase cAMP via activation of G-protein and generate depolarisation through a decrease

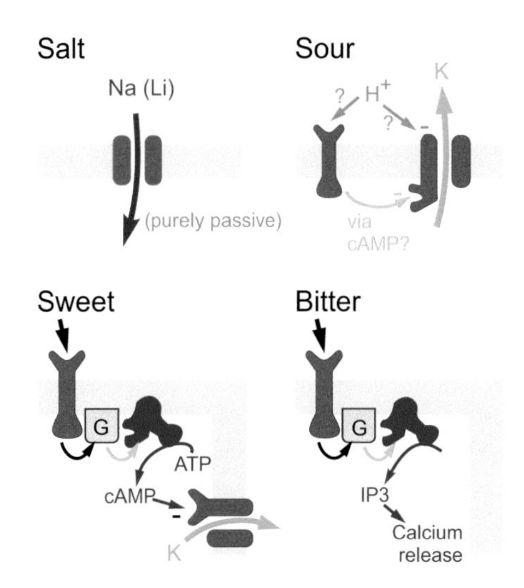

Figure 8.18 The transduction mechanisms believed to be responsible for the four main types of taste receptor. In the case of salt taste, the mechanism appears to be simply one of modification of the Nernst potential for sodium; sweetness appears to be the result of a decrease in potassium permeability, mediated by a G-protein and cAMP; bitter receptors work by activation of phospholipase C to produce IP3, which then triggers release of internal calcium stores, which in turn triggers transmitter release.

Clinical box 8.2 The chorda tympani

Taste sensation in the anterior two-thirds of the tongue passes to the brain through afferent fibres in the VIIth cranial nerve, the facial nerve. These fibres branch from the bulk of the facial nerve as the *chorda tympani.* This branch exits the skull and takes a scenic route, passing over the inner surface of the tympanic membrane on its way to the tongue. At this point the nerve is conveniently exposed for recording, stimulating or anaesthetising.

Anaesthetising the chorda tympani enhances the sensory perception of quinine and diminishes the taste of saline when these chemicals are placed in the region of the tongue supplied by the glossopharyngeal nerve, suggesting the release of some form of inhibition between the nerves. Even more striking, anaesthesia of the facial nerve can lead to phantom taste sensation from the posterior regions of the tongue. Phantom tastes such as this may be the source of unpleasant taste sensations *(dysgeusia)* that accompany the loss of taste *(hypogeusia)* experienced when the chorda tympani is injured, for example in a middle ear infection or following ear surgery.

in P_K. Bitter substances, again via a G-protein, activate the production of IP_3, which then causes calcium to be released from internal stores, and which in turn causes the release of transmitter from the vesicles. The umami taste appears to be due to the activation of specific receptors for glutamate.[17] These findings give some confidence to the idea that olfactory transduction might also be mediated by specific receptor proteins, although necessarily with an enormously greater repertoire of types of binding site. Here and in the case of olfaction, there are also certain parallels with immunological mechanisms, which may suggest a common origin.

Notes

The chemical senses are under-represented in the literature. Three useful sources are: Cagan, R. H. (1989). *Neural Mechanisms in Taste.* (Boca Raton, FL: CRC Press); Davis, J. L. & Eichenbaum, H. (1991). *Olfaction.* (Cambridge, MA: MIT Press); Doherty, P., Brewer, W. J., Castle, D. & Pantelis, C. (2006). *Olfaction and the Brain.* (Cambridge: Cambridge University Press). Moncrieff, R. W. (1967). *The Chemical Senses.* (London: Leonard Hill) has a great deal of detailed information relating to the psychophysics, especially in relation to perfumery, but is otherwise out of date. An entertaining journalistic foray into the field is Avery Gilbert's (2008). *What the Nose Knows.* (New York, NY: Crown). Neurological aspects have recently been covered in Hawkes, C. H. & Doty, R. L. (2009). *The Neurology of Olfaction.* (Cambridge: Cambridge University Press).

1. The outer nose is not without neurological interest: see Critchley, M. (1979). *Man's Attitude to His Nose,* in *The Divine Banquet of the Brain.* (New York, NY: Raven Press).
2. Air not normally reaching the olfactory cleft In 1882, the Viennese physiologist E. Paulsen performed an elegant but somewhat macabre experiment to establish this point. Having cut a human head down the middle, he placed tiny squares of red litmus all over the nasal cavities; then, sticking the two halves together again, he drew air laden with ammonia from a bottle held under the nose by appropriate manipulation of a pair of bellows attached to the trachea. On opening the head he could see by which pieces of litmus had turned blue what course the air had taken: very little reached the receptor region. This and other similar experiments are described in Finger, S. (1994). *Origins of Neuroscience: a History of Explorations into Brain Function.* (Oxford: Oxford University Press).
3. Vomeronasal organ Watson, L. (1999). *Jacobson's Organ: the Remarkable Nature of Smell.* (London: Allen Lane) is an engaging account of the vomeronasal organ and its links with behaviour.
4. Olfactory localisation The experiments were conducted by von Békésy, of auditory fame: see von Békésy, G. (1964). Olfactory analogue to directional hearing. *Journal of Applied Physiology* **19** 369–373.
5. Humans relatively indifferent to smell Aldous Huxley: *'Man's sense of smell is relatively poor and this apparent handicap has proved to be an actual advantage to him. Instead of running round like a dog, sniffing at lamp-posts and becoming deeply agitated by what he smells on them, Man is able to stand away from the world and use his eyes and his wits relatively unmoved'.* Quoted in McCartney, W. (1968). *Olfaction and Odours.* (New York, NY: Springer Verlag).
6. Permanent olfactory aversions See for instance Garcia, J. & Ervin, F. R. (1986). Gustatory-visceral and teloreceptor-cutaneous conditioning: adaptation in internal and external milieus. *Communications in Biology* **A1** 389.
7. Olfaction and behaviour in general Two excellent books by Stoddart review this field very well: Stoddart, D. M. (1980). *The Ecology of Vertebrate Olfaction.* (London: Chapman & Hall), and Stoddart, D. M. (1990). *The Scented Ape: the Biology and Culture of Human Odour.* (Cambridge: Cambridge University Press); Doty, R. L. (1976). *Mammalian Olfaction, Reproductive Processes and Behaviour.* (New York, NY: Academic) is another useful book in this area. Social aspects of smell are also discussed in Bedichek, R. (1960). *The Sense of Smell.* (London: Michael Joseph) – with many entertaining anecdotes – and in Burton, R. (1976). *The Language of Smell.* (London: Routledge & Kegan Paul).
8. Olfaction and sexual attraction Recall Nelson's apocryphal message to Lady Hamilton: *'The fleet's in: don't wash!'.* Conversely, in one experiment men in photographs were rated friendlier when the air was scented with an extract from male armpits: Kirk-Smith, M., Booth, D. A., Carroll, D. & Davies, P. (1978). Human social attitudes affected by androstenol. *Research Communications in Psychological Psychiatry and Behaviour* **3** 379–384.

9. Emotional effects Dementia patients who are confused, restless or aggressive can be calmed if they are exposed to external stimuli which they recognise as having been associated in the past with calm and safety. Since demented patients lose recent memory before early memories, childhood smells can be particularly effective: for instance, one elderly patient who was disorientated and confused (and as a result aggressive) was calmed by exposing her to the smell of lavender, abundant in her childhood home.

10. Sensitivity of olfactory receptors Direct measurements in the laboratory have confirmed that receptors can have very low thresholds: in the frog, 35 molecules may be enough to trigger an action potential, but with many receptors acting in parallel, the threshold concentration may be close to 0.1 pM. Bhandawat, V., Reisert, J. & Yau, K. W. (2010). Signaling by olfactory receptor neurons near threshold. *Proceedings of the National Academy of Sciences of the United States of America* **107** 18682–18687.

11. Helen Keller *'Smell'*, she wrote, *'is a potent wizard that transports us across thousands of miles and all the years we have lived.*

    The odours of fruits waft me to my southern home, to my childhood frolics in the peach orchard. . . . The sense of smell has told me of a coming storm hours before there was any sign of it visible. I notice first . . . a slight quiver, a concentration in my nostrils. As the storm draws near my nostrils dilate, the better to receive the flood of earth odours which seem to multiply and extend, until I feel the splash of rain against my cheek. . . . I know the kind of house we enter. I have recognised an old-fashioned country house because it has several layers of odours, left by a succession of families, of plants, perfumes and draperies.' Keller, H. (1903). *My Life.* (London: Hodder & Staughton).

12. Changes of character on dilution Robert Boyle (1627–1691) describing an odd encounter of this kind: *'An eminent professor of mathematics affirmed to me, that, chancing one day in the heat of summer with another mathematician to pass by a large dunghill that was then in Lincoln' s-Inn Fields, when they came to a certain distance from it, they were both of them surprised to meet with a very strong smell of musk, which each was for a while shy of taking notice of, for fear his companion should have laughed at him for it; but when they came much nearer the dunghill that pleasing smell was succeeded by a stink proper to such a heap of excrements'.* Quoted in McCartney, W. (1968). *Olfaction and Odours.* (New York, NY: Springer Verlag).

13. Amoore's shape theory See for instance Amoore, J. E. (1964). Current status of the steric theory of odour. *Annals of the New York Academy of Sciences* **116** 456.

14. The infrared vibrational theory Support for this idea has also come from a study of the curious way in which male moths are attracted by candles. It turns out that candles have characteristically spiky infrared spectra, and that a number of the emission lines coincide with the vibrational frequencies of the female moth's pheromone: other sources such as hurricane lamps that emit as much infrared but lack the spikes are much less powerful attractants. So it seems that the suicidal fascination of the candle for the moth is a sexual one: the candle is the flamme fatale of mothdom! See Callahan, P. S. (1977). Moth and candle: the candle flame as sexual mimic of the coded infrared wavelengths from a moth sex scent (pheromone). *Applied Optics* **12** 3089–3097. The infrared theory was forcefully argued in Wright, R. H. (1982). *The Sense of Smell.* (London: CRC Press). Recently it has been shown that flies can differentiate between two odour molecules which only differ in a hydrogen isotope (which will drastically change vibrational energy levels of the molecule): Franco, M. I., Turin, L., Mershin, A. & Skoulakis, E. M. (2011). Molecular vibration-sensing component in *Drosophila melanogaster* olfaction. *Proceedings of the National Academy of Sciences of the United States of America* **108** 3797–3802.

15. Molecular recognition See for instance Reed, R. R. (1990). How does the nose know? *Cell* **60** 1–2.

16. Olfaction and 'taste' A recent book explores gourmet neurophysiology in some detail: Shepherd, G. M. (2011). *Neurogastronomy.* (New York, NY: Columbia University Press).

17. Umami receptors See for instance Lindemann, B., Ogiwara, Y. & Ninomiya, Y. (2002). The discovery of Umami. *Chemical Senses* **27** 843–844.

PART 3

MOTOR FUNCTIONS

CHAPTER 9
MOTOR SYSTEMS

Motor systems are intrinsically rather more complex than sensory ones, which is one reason why we know rather less about them. Where do these complexities come from?

Why studying motor systems is difficult

An obvious way to study the motor system, by analogy with recording from sensory systems, is to stimulate various bits and see what happens. But – as we noted in Chapter 1 – there is a difficulty about doing this: how do we know what is an appropriate pattern of stimulation to apply in order to get a response?

A further difficulty in using stimulation to study the motor system is that, whereas sensory systems by and large progress in a straightforward way from level to level, a characteristic of motor control is that every action necessarily results in sensory *feedback*. Lift your hand: immediately, a torrent of sensory activity floods back into the brain, from the skin, from muscle and joint receptors, from vision and the other special senses. This makes the effects of stimulating a particular region of the motor system additionally complex: any movement that may result from it also generates new patterns of afferent activity. By acting on centres whose descending activity then influences that region, this messes things up by altering the pattern of stimulation we are trying to apply.

For both these reasons, electrical stimulation has not been as helpful a way of studying the motor system as one might have thought.[1] A more fruitful approach has been an extension of the methods that have been successful in probing sensory systems. We apply 'real' stimuli to the senses, and trace the resultant patterns of activity as they penetrate deeper and deeper through the levels of the nervous system and emerge triumphantly again at the motor end as movements. In systems as complex as those controlling the human hand, this is not yet technically feasible. But where the number of levels is much smaller, as in more primitive brains like those of insects, or in simpler sub-systems of the mammalian brain (like those controlling eye movements, which may have as few as three neuronal levels between input and output), this approach has taught us a great deal.

One general lesson we have learnt is how much more can be discovered about the brain by studying complete systems with an *output*, compared with purely sensory systems. The visual system is a good example. When people first started exploring it with microelectrodes, at first everything went swimmingly: going systematically from layer to layer into the cortex, investigators such as Hubel and Wiesel found a clear logical progression from ganglion cells to simple cells and to complex cells. People confidently expected this trend to continue, and that after line detectors we would discover square detectors and circle detectors and A detectors and teacup detectors . . . but they didn't; visual neurophysiology began increasingly to lose its way. Why? Because investigating a sensory system without a clear sense of what it is *for* is like trying to understand a television without knowing that it is meant to show pictures. We may think it is 'for' perception; but since we have no idea what perception *is* (or any other aspect of consciousness) it is hardly surprising if we end up bewildered and lost. But once we start looking at systems that actually do something tangible, we make progress.

Motor control and feedback

Feedback means using information about results to improve performance, and feedback from the effects of motor responses is fundamentally important in the control of movement. A good way to begin a study of the motor system is to consider the ways in which this sensory information might in principle be used.

No feedback: ballistic control

There are many circumstances when our motor systems are *forced* to act blindly because for one reason or another they are deprived of normal sensory feedback. When I nonchalantly toss some orange peel into a waste bin, once it has left my hand, no amount of sensory feedback about its trajectory is going to enable me to modify its flight. I have clearly had to work out beforehand the precise sequence of motor commands necessary in order to produce the correct pattern of muscular contractions that I need to achieve my goal. Sometimes this kind of blind behaviour can extend to the overall result of the movement as well as the details of its execution. A classic example is the nest-building behaviour of the brown rat, described by Konrad Lorenz. Having decided to build a nest, it performs a stereotyped series of actions: it runs out to get nesting material, drags it back to the centre of the nest, sits down and forms it into a sort of circular rampart, pats it down and smoothes it, then runs out to get more material; and so on until the nest is finished.[2] This certainly *looks* like the intelligent behaviour of an animal that is aware of the consequence of its actions: yet a simple experiment shows it to be nothing of the kind. If it is not given enough to make a nest, the rat still runs out to grab the (non-existent) material, and goes through the motions of dragging it back, forming it into a rampart and patting and smoothing it, even though in reality there is nothing there. Feedback, in other words, was not in fact being used. And of course we ourselves have all had the experience of carrying out some equally skilled and complex series of actions – making tea, for example – and have embarrassingly revealed its stereotyped, robotic nature by absent-mindedly putting coffee instead of tea into the teapot.

Motor acts of this general type are called *ballistic* – a word meaning 'thrown'– and their control can be represented schematically by a diagram like that of **Figure 9.1**. These kinds of diagrams are central to understanding control systems, and to appreciate them fully we need to get to grips with a certain amount of rather unattractive jargon.[3]

Here we start with the *desired result* or goal, in this case that we want the orange peel in the bin. This is translated by a *controller* into an appropriate pattern of *commands*. These commands produce the *actual result* through their effect on what engineers call the *plant* – the thing that is being controlled, in this case, the body's muscles. If the controller is functioning properly, then the actual result will equal the desired result: the peel will end up in the bin. How well all this works depends on how good the controller is: the more it knows about how the plant will behave in response to any particular command, the better it will perform. So it needs something like a library of motor programs suitable for different acts, where it can look up the rules needed to translate a particular desire into an appropriate command.

A well-known example of such a system is in the control of ballistic missiles. Here the desired result is the destruction of some displeasing portion of the globe; computations are then made, based on knowledge of the missile's characteristics, to determine such parameters as what direction to point it in and how large a thrust is required at take-off: but once it is launched no further action can be taken beyond hoping that the calculations were in fact correct.

Ballistic control is conceptually simple, but it has a fatal defect: it is helplessly vulnerable to what systems engineers call *noise*. Noise is any kind of unpredictable disturbance that makes the actual result different from what the controller expects. If the wind happens to be blowing the wrong way, our ballistic missile may arrive somewhere embarrassing and cause a diplomatic incident. Because the world we live in is *never* entirely predictable – nothing is certain – a given set of motor commands will never produce quite the same result twice running. A particular pattern of activity in motor nerves will produce different movements of a limb on different occasions, depending on a host of internal factors: body temperature, fatigue, the amount of energy available, and so on. Even more important in motor systems is the effect of *load*. When we use our limbs to shift things around, carrying or throwing, a given degree of muscle activity will generate quite different movements, depending whether we are dealing with a lump of rock or a feather. As we shall see, most of the lower levels that control the limbs are devoted to solving this problem of achieving the movements we want despite the noise introduced by unpredictable loads.

Parametric adjustment: feed-forward and feedback

One way of dealing with noise is to have some kind of sensor that monitors the noise before it affects the system, and to use this information to adjust the parameters of the controller to allow for it. This kind of modification of the controller parameters to anticipate the effects of noise is called *parametric feed-forward* (**Fig. 9.2**). If we measure the speed of the wind before we launch our missile, we can allow for its disturbing effects, just as marksmen adjust their sights to allow for deviation by the wind. Much sensory information is used by the brain in this way, especially in making allowance for the effects of different loads. We shall see later that the neural circuits controlling muscle length use information from force detectors in the skin and tendons that monitor load in order to make appropriate adjustments of motor commands.

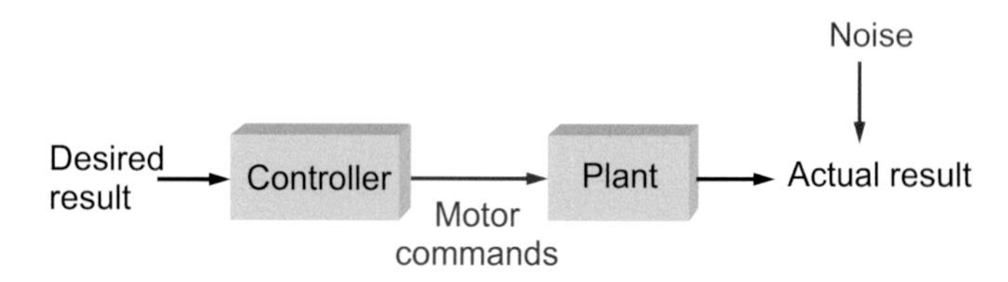

Figure 9.1 A ballistic control system.

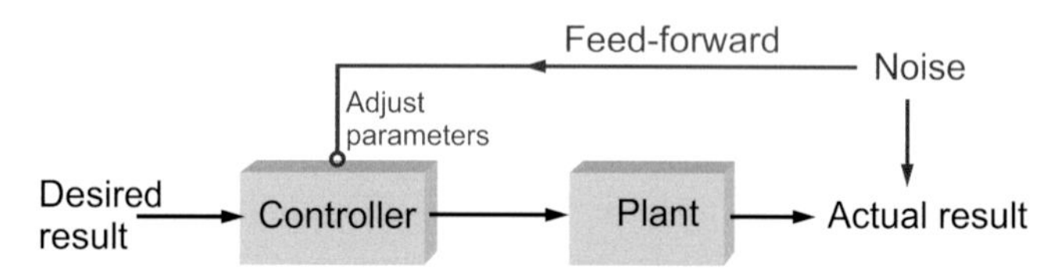

Figure 9.2 Parametric feed-forward. A ballistic system with feed-forward, altering the parameters of the controller in response to noise.

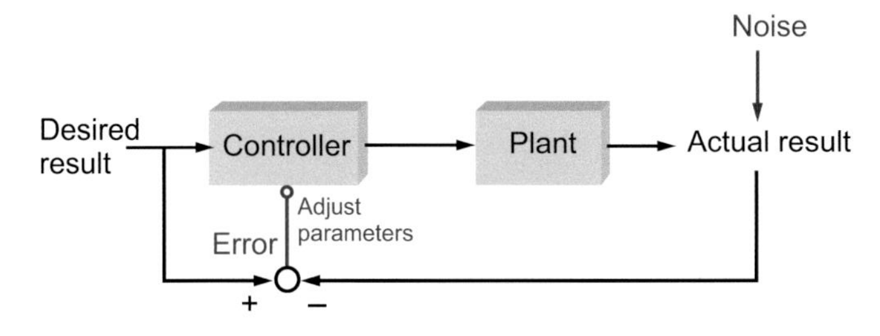

Figure 9.3 Parametric feedback. The controller's parameters are modified in response to errors in performance.

But even this approach is doomed to failure. In general there are infinitely many things that *might* cause perturbations, and the brain clearly cannot have a plan for dealing with every one of them in advance. Instead of trying to anticipate absolutely everything that might possibly occur, one solution is to take a more pragmatic approach, with a system that learns from its own mistakes, using not feed-forward but *parametric feedback* (**Fig. 9.3**). This introduces two exciting new pieces of jargon: a *comparator* (the circle with two converging arrows) compares the actual result with the desired result by subtracting one from the other, and it generates an *error signal* that is used to modify the controller's parameters. In general, the error signal is a measure of how well the system is coping: if it is zero, the controller is doing a good job, and there is no reason to change its parameters. If the system keeps on making a mess of things, generating persistent error signals, then the commands are gradually adjusted until it gets it right. The advantage of this approach is its flexibility. Rather than requiring stored programs ready in advance for any conceivable kind of action, by starting with rather simple, all-purpose programs, one may refine, through trial and error, what is needed for the tasks that are actually encountered.

It goes without saying that this kind of behavior – using error information from one attempt to improve performance on the next – is highly characteristic of the way in which our motor systems learn to execute complex actions. In playing darts, a novice may at first use pre-existing programs developed perhaps from his experience of throwing other objects such as cricket balls, no doubt ultimately from throwing rattles out of his pram. But as he practises, though the feedback from each throw obviously arrives too late to use immediately, it is used to reduce *future* errors, and in the end he may gradually evolve very accurate programs specifically for dart-throwing. A great deal of the learning of motor skills can usefully be thought of as a parametric feedback of this kind, in which errors are used to modify our stored motor programs.

A specific example, discussed in more detail in Chapter 11, is the *vestibulo-ocular reflex*. When we move our head, signals from the semicircular canals are used to drive the eyes by an equal amount in the opposite direction. As a result, they maintain the direction of gaze in space and the retinal image of the outside world remains relatively fixed. This is clearly a ballistic system. Equally clearly, it will go off the rails if the performance of the muscles is degraded through fatigue or disease, or if there is some malfunctioning of the canals. In fact it turns out that the reflex is continually adjusted to ensure that the eye movements really are equal and opposite to the head movement. The error signal in this case comes from neurons that respond to movement of the visual image across the retina; for this retinal slip only occurs when head and eye movement are not matched.

Though parametric feedback and feed-forward can vastly improve the performance of ballistic systems, they are still not ideal. In the first place, the calculations that are needed before the action takes place are in general extremely complex – even throwing orange peel into a bin requires, in effect, the solution of a set of partial differential equations with countless variables – and it is not altogether plausible that the brain could actually have at its disposal a library of such routines so vast as to be able to deal with all the possible motor tasks it might ever encounter during its lifetime. The controller needs to have acquired knowledge about how the plant will behave in response to any kind of command sent to it, and keep this information up to date. So it needs memory as well as intelligence. In addition, parametric feedback only corrects *after the event*, by which time it *may* be too late. But there is another approach, much simpler and often better: *direct feedback*.

Direct Feedback: Guided Control

Here we start, as before, with a desired result (**Fig. 9.4**), which we compare at every moment with the actual result. But now the error signal, instead of being used to tweak the parameters, is used directly as the input to the controller.

Thus errors *immediately* generate motor commands that reduce the difference between the desired and actual result; this is the same kind of negative feedback system that underlies so many of the homeostatic mechanisms of ordinary physiology. Guided missiles are controlled by systems of this type: here the error signal might be something like the angle between the direction in which the missile is pointing and the direction of its target. Whereas a ballistic missile malfunctions disastrously when the wind blows, a guided missile notes the effect of the wind on its relation to the target, and automatically corrects itself. When things go wrong, ballistic systems stay wrong,

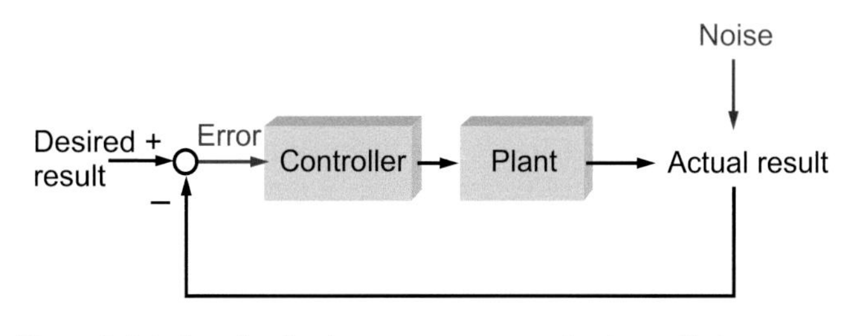

Figure 9.4 A direct feedback system: errors immediately modify the output.

parametric ones gradually get better, and guided systems readjust immediately.

In a guided system, instead of having to calculate what to *do,* you need only specify what you *want.* Like a well-trained servant, the system does all the rest by itself: such control systems are often called *servo systems.* Another advantage is that when errors arise, the computation of the correcting commands is in general very much simpler than the calculations needed in a ballistic system. A familiar example is a domestic central heating system, where the thermostat is the comparator and generates an error signal which consists very simply of one of just two possible messages: either that the actual temperature is above what is wanted, or alternatively that it is below it – too hot or too cold. The subsequent computations of the motor commands could hardly be simpler: in the former case the boiler is switched off, in the latter case it is switched on. Another, physiological, example also illustrates this essential simplicity of guided systems. If we instruct a subject to look at a small light such as A in **Figure 9.5**, and then suddenly move it closer to his nose as at B, we find that his eyes converge smoothly and quite quickly in such a way that in the end the image of the light still falls exactly on the fovea of each retina. The velocity of the convergence movement is high at first, but declines exponentially as the eye gets closer and closer to its target. This is just what would be expected of a guided system, in which the eyes are essentially driven by an error signal.

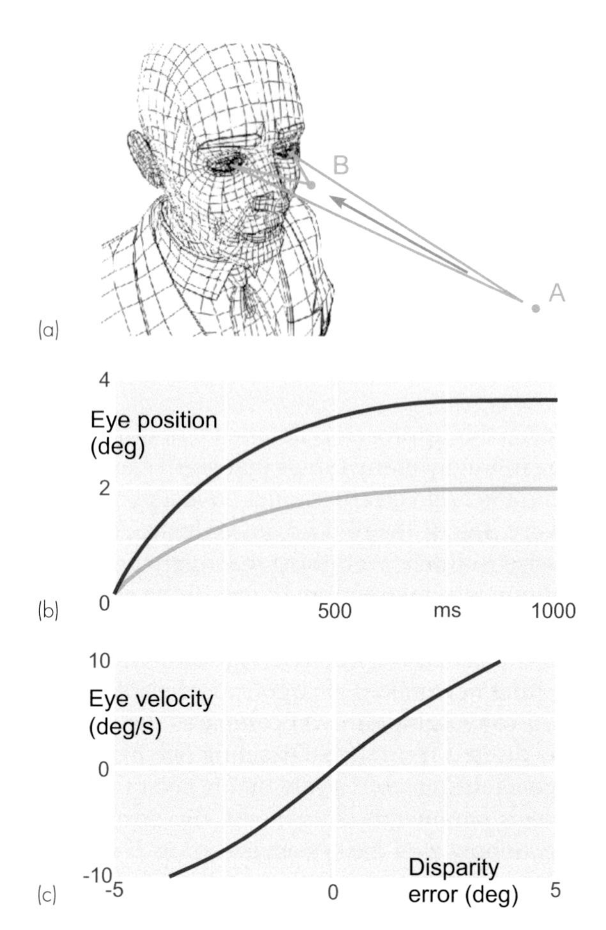

Figure 9.5 (a) Disparity vergence occurs when a subject looks between two objects at different distances. Its time-course (b) shows a slowing towards the end, as the disparity error approaches zero. (c) The vergence velocity is roughly proportional to the disparity error at any moment.

It turns out that this is indeed what is happening, and disparity between the two retinal images, sensed by cortical detectors of the kind described in Chapter 7, provides the error signal which generates the convergence. Experiments show that velocity of the eyes has a very simple relation to the size of this error: it is simply proportional (**Fig. 9.5**). So as the eyes approach their goal, this error gets smaller, and the rate of movement correspondingly declines to zero: hence the time-course of the movement is roughly exponential. Direct feedback is intrinsically a simple process.[4]

The overwhelming advantage of guided systems, however, is not so much their simplicity, but rather the fact that they are almost immune to the effects of noise. For if something unexpected happens that upsets the normal relationship between command and performance, this will be noticed at once (because it will generate error) and the system will instantly generate appropriate commands to achieve the desired result despite the existence of the disturbance. If you leave all the windows in your house open, the thermostat will at once sense the sudden drop in temperature, and the boiler will automatically be switched on until the temperature reaches the desired level once more. Thus the power and elegance of the system are that it guarantees to achieve what it has been designed to achieve, even when upset by types of interference that could not have been anticipated by its creator – including faults in the system itself (like the boiler furring up); it is capable of producing results that *look* intelligent even though – in sharp contrast to the ballistic system with its library of programs for different occasions – it knows very little (only the size of the error) and remembers nothing.

So why are all control systems not of this type? The reason is that direct feedback has a weakness. Its proper functioning depends critically on communicating information about the progress of the action rapidly to the comparator. In neural systems, both the sensory transduction and the consequent transmission may be rather slow, clearly a serious problem when trying to control fast movements. Delay of this kind 'round the loop' means that instead of responding to the error as it actually *is,* the system will be responding to the error as it *was,* however many milliseconds it takes for the information to find its way back to the brain. As a result, direct feedback systems have a tendency to show constant oscillations: if the feedback is turned at off at the moment the sensed error reaches zero, there will be an overshoot in the opposite direction, requiring further correction, and so on: this is called *hunting,* and is obvious for example in ordinary domestic central heating systems. One way to get round it is to monitor not just the actual error x at any moment, but also the rate dx/dt at which it is changing, rather as the captain of a supertanker might stop the engines long before it reaches its berth, using the rate of slowing to predict where it will

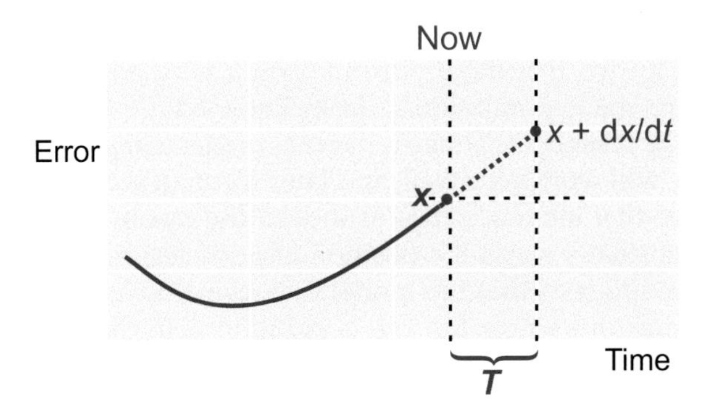

Figure 9.6 Mitigating against oscillation of a system exhibiting direct feedback. Due to the long delays in sensory feedback, another approach is to *predict* future error based on its current rate of change.

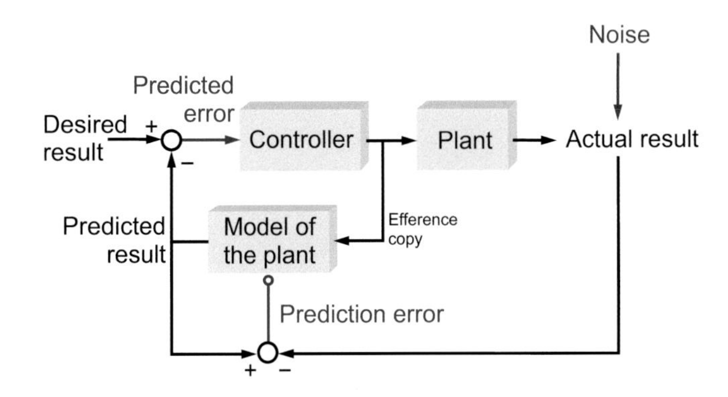

Figure 9.7 A system using internal feedback. The model of the plant predicts errors, and is adjusted if its predictions are wrong.

come to a halt. As can be seen in **Figure 9.6**, $(x + T\,dx/dt)$ is a good estimate of what the error *will* be at a time T in the future, so that if we make T equal to the time taken for observed corrections to have an effect, we can – to a first approximation – avoid these oscillatory responses. A good example of this is (see p. 100) in the Ia fibres from muscle spindles, which do indeed respond partly to error and partly to its rate of change, and for which T is roughly the time taken for information to be sensed and acted on.

Now the visual system happens to be particularly slow, with reaction times of around 200 ms at best, and this makes visual guidance of rapid actions difficult to achieve. Though it works beautifully for a watchmaker as he carefully assembles springs and wheels into precise alignment, it does not help the batsman in a game of cricket. One might think that he could use a direct feedback system to bring his bat up to the ball under visual guidance, using error information about the distance between bat and ball. But it is easy to calculate that the existence of this large visual delay makes this physically impossible, because visual information will be hopelessly out of date. If the bowler is delivering at 90 mph (40 m/s) the ball will travel nearly half the length of the pitch in the 200 ms it takes for any visual information about its position to be of use: thus the last useful visual fix on the ball is when it is still 8 m away. Clearly the bat cannot in any sense be guided on to it.[5]

Internal feedback: efference copy and virtual models

The final type of control system to be considered is at its best in this sort of situation. Though ballistic in the sense that it cannot immediately respond to errors, it has closer affinities with a guided, direct feedback, system than with a simple ballistic one, and uses *internal feedback*. The notion here is that if it is difficult to obtain feedback about *actual* results fast enough for them to be of use during an action, nevertheless it may be possible as the result of experience to *predict* what the result of a particular motor command is going to be, before the actual result is known.[6] Thus from a general knowledge of the mechanics of one's hand and arm, and information about the kinds of loads that are present, one can form an estimate in advance of what position the limb is going to adopt in response to any particular pattern of motor commands that is sent to it. Since this estimate is formed entirely within the brain, it can be available long before any feedback from the actual movement has found its way back from the periphery.

When things are happening fast, such an estimate – we can call it the *predicted result* – will at least be better than no information at all. This prediction is derived by sending a copy of the motor commands (an *efference copy* signal) to a neural *model* of the mechanical properties of the body, which is used to predict what will happen. So in an internal feedback control system (**Fig. 9.7**) it is this prediction, rather than the actual result, that is compared with the desired result to produce an estimate of the error. Thus an internal feedback system is a kind of virtual world – in the cricketer's brain, no doubt a virtual cricket pitch. A final, necessary, refinement is that this model must be updated all the time to ensure that it keeps in step with any changes in the actual muscles and bones and the objects in the real world with which we interact. How this is done is that the actual results are continually compared with the predicted results; any errors then represent faults in the model, which are then corrected by parametric feedback. In this way it continually improves the accuracy of its predictions.[7]

One excellent example of a physiological system that seems to work in this way is the one that controls saccadic eye movements. *Saccades* are the eye movements we make when we shift the direction of our gaze from one target to another. Large saccades are made with virtually constant velocity – which may be as much as 900°/s – whatever the separation between the targets; thus the duration of a large saccadic movement is a nearly linear function of its size (**Fig. 9.8**). At first sight, one might think that the eye was simply moving off towards the new target at a constant rate, and that as soon as the visual system senses that the target has been reached, the brakes are applied and the eye comes to rest. But it is easy to calculate – as in the case of the batsman and cricket ball – that this cannot possibly be the true explanation. Because the movement is so extremely rapid – most saccades last between 20 and 40 ms – and since visual processes in general take considerably

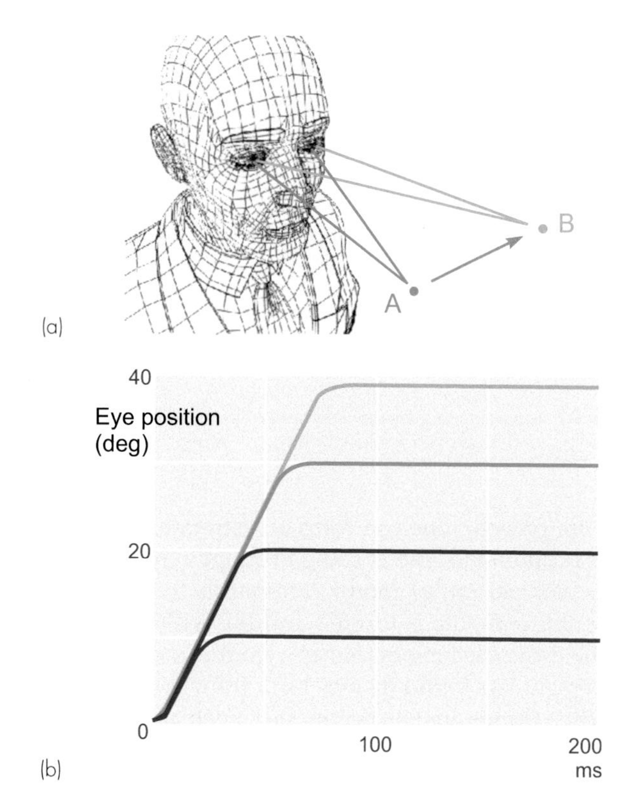

Figure 9.8 (a) A saccade is made in looking between two objects at the same distance. (b) The time-course of saccades of different amplitudes, showing the approximately constant velocity of the movements.

longer than this, by the time the brain had recognised that the target had been reached, the eye would have grossly overshot. Thus a *simple* feedback loop cannot possibly be used.[8]

However, there is a feature of the eyes that makes them ideally suited to control by internal feedback. Whereas movements produced by limbs in response to a given command vary with the load they are experiencing, this is not the case for the eyes, whose load is always constant. This means that the brain can form a very good idea of where the eye is pointing from knowledge of the commands it sends it. Thus an internal model using efference copy will work very well, and later we shall see good evidence that efference copy is indeed the means by which we normally sense the position of our eyes; more recent work suggests that the model that uses this information to work out where the eye is pointing is in the superior colliculus. In a saccade, when this calculated eye position is equal to the position requested by the visual system, the drive to the muscles is in effect switched off, and the eye comes to rest on the new target. In the long run, the visual system tells us whether the saccade was in fact successful in landing on the target, and this information is used to improve the internal model.[9]

There are several other types of eye movement apart from saccades and vergence. Oddly enough, between them they exemplify each of the different kinds of motor control that have been presented here, and they are summarised in **Table 9.1**.

The usefulness of internal models

On the face of it there is perhaps no obvious advantage in using internal feedback rather than ballistic control with parametric feedback. But because the model of the body that is embodied in the former system is a general one, not tied to any particular type of action, it means that experience in carrying out one kind of skilled motor function will benefit the performance of other ones in a rather more direct way than was the case for simple parametric feedback. This is particularly true when, as in the case of the eye, the expected result can be computed relatively easily from the motor commands. Learning motor skills then becomes a matter of learning to predict the behaviour of one's own body. Such a system can also cope much better

Table 9.1 The main classes of eye movements

Gaze-holding (maintaining the direction of gaze in space)		
Conjunct:	Optokinetic (OKR)	*Direct feedback*
	Vestibular (VOR)	*Ballistic, parametric feedback from vision*
Gaze-shifting (foveation of visual targets)		
Conjunct:	Saccades	*Probably internal feedback*
	Smooth pursuit[a]	*Ballistic, direct feedback*
Disjunct:	Vergence	*Disparity-driven: direct feedback*
		Blur-driven: ballistic
Spontaneous (Micro- or fixational)	Drift	*Central noise*
	Microsaccades	*Saccades correcting for drift*
	Tremor	*Peripheral noise*

[a] Small moving targets are tracked with a mixture of saccades, to move the eye to the right position, and smooth pursuit, to match the target velocity.

with changes of circumstances, for instance when forced to carry out with one's left hand an action for which you normally use your right.

A good analogy is using a map to get around in an unfamiliar town. A sequence of instructions such as 'first right, second left, then third right' is compact but not *robust*; were one of the streets blocked you would be helpless. A map can record information in such a way that experience gained while carrying out one action can be used to improve other actions, updating the internal model against which future efference copy will be evaluated: the blocked street can be avoided when making other journeys as well. It also allows rehearsal and planning. A complex manoeuvre – perhaps trying to unlock one's front door while encumbered with groceries, a bunch of flowers and an umbrella – can be tried out in advance within the virtual world before being put into operation.

The existence of parametric or internal feedback obviously makes the interpretation of the results of experimental stimulation or lesions no easy matter. For instance, it is not at all obvious what the effect would be of artificially stimulating a parametric feedback system at the 'modify' input. Equally, we would anticipate that lesions in such regions as the library of motor programs, or the neural model inside an internal feedback system, would result in complex and subtle effects: not just simple paralysis, but perhaps loss of *quality* of performance, of the ability to *modify* responses through experience, and perhaps the appearance of rigidly *stereotyped* patterns of behaviour not properly adjusted to their objects. As we shall see, defects of just these kinds are indeed characteristic of many types of clinical derangement of the higher levels of the motor system.

Box 9.1 Basic types of control system

Ballistic
The desired result is translated into a command, issued regardless of whether an error has occurred. However, subsequent commands may be modified in the light of errors that have been made (parametric feedback). In addition, feed-forward may be used: noise is monitored before it affects the output, and is used to modify the response.

Guided (direct feedback)
The desired result is compared with actual result at every moment, and any error results immediately in an alteration of the command so as to reduce the discrepancy.

Parametric feedback
Here the error signal is used not as direct input to the system, but to modify its parameters, so that the response is better next time.

Internal feedback
As direct feedback, except that a *prediction* of the result is used rather than its actual value. The commands that are sent out are monitored and used to calculate what result they ought to achieve, using a model of the way the system normally behaves. Any discrepancies between the predicted and actual result cause the model to be updated, using parametric feedback.

The hierarchy of control

Evolution by accretion

Another factor that makes both the anatomy and the physiology of motor systems alarmingly complicated is the way in which it has developed in the course of evolution. Whereas the behaviour of the very simplest organisms can largely be described in terms of simple local segmental mechanisms at a peripheral level – as for example the coordination of a centipede's legs when it walks – in ascending the evolutionary tree we find more and more domination of the special senses, and as a consequence of this, a corresponding degree of *encephalisation*: control by higher centres grouped near these sense organs, in the head.

It is important to appreciate that by and large this has been a process of *accretion*. Simpler mechanisms are not in general displaced by more recent ones: they are left essentially intact, but supplemented and controlled from above. They are, after all, carrying out useful functions. Man's walking movements are in essence not so very different from the centipede's and associated with rather similarly stereotyped sequences of muscle actions mediated by spinal mechanisms of the same general character: such sequences can often be evoked from spinal preparations, animals in which the higher levels of control have been surgically disconnected.[10] It would clearly be foolish for the brain to build its own neural circuits that merely duplicated what the spinal cord was already doing perfectly well, and it is important not to underestimate what the cord is capable of. Classical examples include the spinal dog wagging its tail after defecation, or the wiping reflex in the spinal frog: if a small piece of filter paper is moistened with acid and placed on its back, it will quite accurately use the nearest leg to wipe it off the skin; if that leg is held down, after a short delay another leg is used! It is clear that one should not think of the spinal cord merely as a sort of speaking-tube down which the brain shouts its orders to the muscles: rather, it provides a repertoire of fragments of action, 'party pieces' that can be called on when necessary by the higher levels. The main difference, in fact, between the spinal cord of a 'higher' and of a 'lower' animal is that the former in a sense expects to receive more in the way of commands from above; consequently, when isolated from the brain in a spinal preparation, it may appear less responsive. This phenomenon is known as *spinal shock*: immediately after making the cut that separates cord from brain, spinal reflexes are depressed or absent, because the usual 'permission' from above is not there. But after a period of time the cord becomes more lively, and may in the end actually show a greater degree of responsiveness than before the operation. This period of time depends markedly on the degree of encephalisation: in Man it may take many months; in a dog, days; and in a frog perhaps only a few minutes, reflecting the differing degrees of control normally descending from the brain.

In the end, one can never be sure that a spinal animal is really exhibiting all the things that the cord could do if the brain were intact: we always tend to underestimate what the spinal cord is capable of. A very clear example of the cord seeking permission from the brain to perform a routine task is urination. Clearly, our social conventions dictate that a full bladder does not automatically justify its emptying in any place at any time, and the brain performs a welcome task in suppressing such a reflex. However, the cord is so well trained in this respect that following a spinal injury above the area of cord controlling the bladder, voiding is suppressed to the point of causing enormous pain and sympathetic outflow – clinically, a sweating, tachycardic, hypertensive patient with a bladder that can be felt near the belly button! The back pressure can even be severe enough to cause kidney failure.

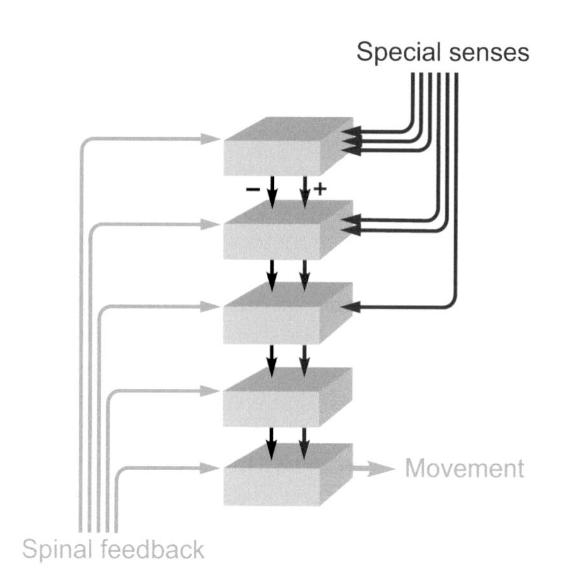

Figure 9.9 Schematic representation of hierarchy of levels in the nervous system. Higher levels have more access to information from diverse sources, lower ones have more immediate feedback. On the whole, levels act by controlling those immediately beneath them, rather than generating movements directly. Upper levels also tend to inhibit lower ones.

The functional characteristics of different hierarchical levels

Such observations lead naturally to the idea of a hierarchical organisation of the motor system into a series of functional levels (**Fig. 9.9**), the higher levels having more diverse kinds of sensory information at their disposal and therefore able to plan and anticipate more effectively than the lower.

Because of their ability to store experience through memory, they can also be more flexible in their responses and learn to conform to the outside world in a way the spinal cord cannot. It follows therefore that as well as being able to stimulate the spinal cord to generate particular patterns of output, these higher levels must also exert a tonic *inhibitory* influence on lower levels. Brain and cord may well often have conflicting ideas about what is the right thing to do in a particular situation – not flinching from a painful injection, for example – and this conflict needs to be won by the brain. Consequently, the effects of lesions in higher levels of the brain are usually two-fold: first, a *loss* of function, particularly of the more flexible and integrated kinds; and second, the *new appearance* of abnormal and more primitive modes of response. The latter phenomenon is often described by neurologists as *release*: the lower centres are released from the restraining influences of the higher, like schoolchildren when the teacher is called from their class.

But quite apart from what has happened in evolution, an engineer would recognise that from a purely functional point of view, hierarchies arise inevitably whenever something complicated has to be done, that naturally breaks itself down into relatively repetitive sub-tasks. Thus a well-written computer program will typically consist of simple subroutines that do very small tasks, called by other sub-routines which are in turn called by other more global sub-routines, and so on. A computer game might have one routine to draw a single dot on the screen, called by another routine that displays a set of dots to form a small pattern, called by another that draws a single object, called by another that displays an entire scene. Clearly at the level at which the programmer is thinking about the general organisation of the game, he does not want to be bothered with the repetitive detail of exactly how each dot in a picture is to be sent to the display.

Such hierarchies are particularly obvious in social organisations that are meant to get things done, most notably in how armies are organised.[11] When Napoleon decided to invade Russia he didn't himself give detailed instructions on how many pairs of boots to buy, or where to dig the latrines: he indicated the overall strategy to be followed – perhaps a little more detailed than *Envahissez la Russie!* – that then percolated down through all the hierarchical layers, getting elaborated as it went, until it eventually reached the soldiers in the firing line. A general has the advantage of integrated information available from a wide variety of sources, which he can use to develop wide-ranging strategies: the troops have the advantage of immediate and detailed experience of local conditions, with which they can modify their individual behaviour.

The existence of a hierarchical organisation carries very important implications for what happens if part of the system goes wrong. In military terms, the effects of blowing up a platoon are very different from those of shooting a general (**Fig. 9.10**). In the first case, the defect is obvious, immediate, and limited: very specific jobs no longer get done: there is a clear correlation between the 'lesion' and the 'symptoms'. In the second case, at first nothing may appear to be wrong at all: like a headless chicken, the army still functions. But gradually more subtle defects may begin

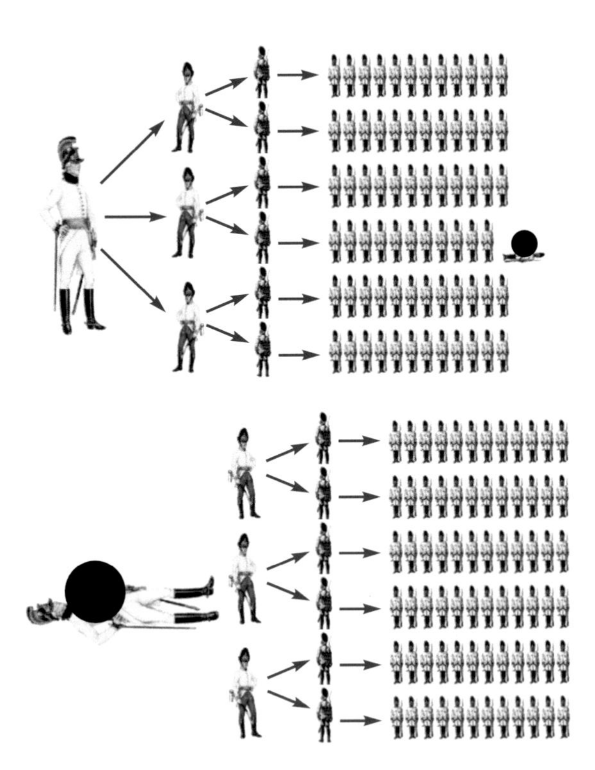

Figure 9.10 Different consequences of lesions at different levels in a hierarchy: localised at lower levels, diffuse at higher, and accompanied by release.

to show themselves, such as a lack of long-term planning or coordination. At the same time, new patterns of activity may start to become apparent as the general's subordinates begin to put their own ideas into practice without restraint: symptoms, in other words, of 'release'.[12]

These are precisely the kinds of disorders commonly described after damage to different levels of the central nervous system. In polio, for example, the loss of motor neurons causes total paralysis of a particular set of muscles, while others may be unaffected. With lesions at higher levels, we may see loss of some functions and release of others. A classical example is the *Babinski sign*, or 'up-going big toe'. If the plantar aspect of the foot of a normal adult is firmly stroked in the general direction of the big toe, the immediate response is an involuntary flexion of the foot and toes; but in damage to the corticospinal tract, as also in newborn children, the reaction is the exact opposite: the toes curl upwards (an 'extensor plantar' response).[13] It is clear in this case that cord and brain have different ideas about what to do when the foot is stroked; in the adult, the brain wins. Curiously, the Babinksi sign is only positive on the side of the body affected by an area of brain damage, for example the side affected by a stroke. A more teleologically tangible example is the *grasping reflex;* in children in their first few years of life, placing an object in the palm causes the fingers to close and grip it strongly – which makes sense considering how many mammalian offspring spend their first few months hitching a ride by gripping onto their mothers coats. This primitive grasp reflex is also found in adults with brain injuries, and is another example of 'release'.

Another, particularly alarming, reflex is the *Lazarus reflex*, sometimes seen in brain-dead patients in whom all neural structures above the spinal cord are destroyed. Such patients require mechanical ventilation in order to keep the remaining organs functioning (although, in the UK and Australia at least, they are legally dead) and usually stopping the ventilator is simply followed by absolute stillness and a swift loss of the heartbeat. However, it is a recognised, albeit rare, observation that when the ventilator is turned off patients may perform quite a complex set of movements, coordinated only by the spinal cord: they may roll their shoulders, extend their arms, turn their palms down and cross their arms across their chest. There may also be cough-like movements, grasping or extending of the fingers and even flexion and extension of the knees and ankles. The afferent stimulus for this reflex is unclear: whether it is the cessation of chest wall movement or a response to falling oxygen levels and climbing carbon dioxide is not certain. What is certain is the fear and distress caused to an unprepared family watching their 'dead' relative spontaneously adopt their funereal pose.

Another example is the coordination of walking movements. A newborn infant is actually able to walk after a fashion (**Fig. 9.11**), so long as its weight is supported. But one of the first things that the developing brain does is to suppress this primitive response, many months before it develops its own much more sophisticated patterns of walking, that make better use of integrated sensory information.[14]

With this notion of hierarchical control in mind, we begin the next chapter by considering what the lowest level of all, the spinal cord, can and cannot do, and the ways in which descending pathways from the brain may control and modify its activity. Later, in Chapter 12, we shall see how the control of movement by higher levels is dominated by its hierarchical organisation.

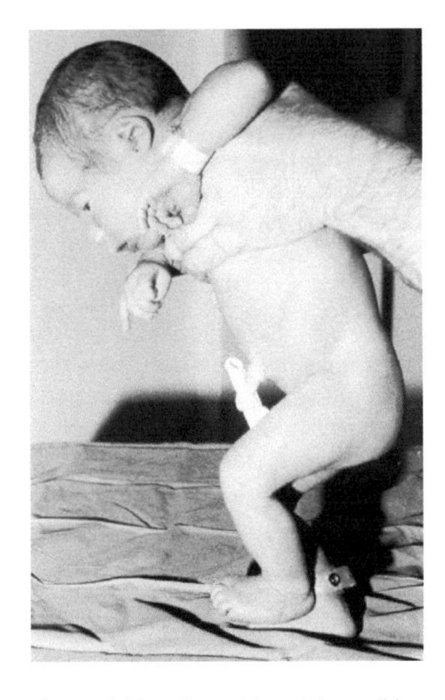

Figure 9.11 A newborn child walking. This ability will be suppressed by the developing brain, even though adult walking patterns will not appear until a year or so later.

Notes

1. Electrical stimulation While this may not be especially helpful for studying the motor system, it is invaluable in neurosurgical practice. For example, in the removal of low-grade gliomas patients are often kept awake during the operation so that the surgeon can detect areas involved in motor and language function by electrically stimulating the brain adjacent to the tumour. In this way, one can maximise its removal, while minimising permanent neurological deficits. A pioneer of this technique in the United Kingdom, Henry Marsh, has published two excellent accounts of his experience of a career in neurosurgery: *Do No Harm: Stories of Life, Death and Brain Surgery* (2014), and *Admissions: A Life in Brain Surgery* (2017), both published in London by Weidenfeld and Nicolson.

2. Konrad Lorenz For examples of similarly complex but apparently ballistic behaviour, see Lorenz's ***Studies in Animal and Human Behaviour.*** (1965; English translation, Robert Martin, 1970) (Cambridge, MA: Harvard). This area is intelligently and succinctly discussed in Marsden, C. D., Rothwell, J. C. & Day, B. L. (1984). The use of peripheral feedback in the control of movement. ***Trends in Neuroscience*** **7** 253–257. There is an interesting account of what it is like to lack motor feedback in Cole, J. (1991). ***Pride and a Daily Marathon***. (London: Duckworth).

3. Control systems Good books on specifically biological aspects of control systems that are not too technical are extremely hard to find. The appendix to Carpenter, R. H. S. (1989). *Movements of the Eyes.* (London: Pion) may be pitched at about the right level. Milsum, J. H. (1965). *Biological Control Systems Analysis.* (New York, NY: McGraw-Hill) and Stark, L. (1968). *Neurological Control Systems.* (New York, NY: Plenum) are both excellent but long out of print. A more recent account, but expensive, is Khoo, M. C. K. (2018). *Physiological Control Systems: Analysis, Simulation and Estimation.* (New York, NY: IEEE Press). A short but relatively accessible account is Carpenter, R. H. S. (2004). Homeostasis: a plea for a unified approach. *Advances in Physiology Education* **28** S180–187. Doucet, P. & Sloep, P. B. (1992). *Mathematical Modelling in the Life Sciences.* (Chichester: Ellis Horwood) is a more general account of modelling that may also be consulted, as is Brown, D. & Rothery, P. (1993). *Models in Biology.* (Chichester: Wiley). Some general aspects of motor control are discussed in Schmidt, R. A. & Lee, T. D. (2011). *Motor Control and Learning.* (Leeds: Human Kinetics Ltd).

4. Disparity vergence Two classic papers describing the relation between vergence and disparity: Rashbass, C. & Westheimer, G. (1961). Disjunctive eye movements. *Journal of Physiology* **159** 339–360; Westheimer, G. & Mitchell, A. M. (1956). Eye movement responses to convergence stimuli. *Archives of Ophthalmology* **55** 848–856.

5. Visual control of batting See for instance Bahill, A. T. & LaRitz, T. (1984) Why can't batters keep their eyes on the ball? *American Scientist* **May–June** 249–254, and Lacquaniti, F., Carrozzo, M. & Borghese, N. (1993) The role of vision in tuning anticipatory motor responses of the limbs. In A. Berthoz (ed.) (1993). *Multisensory Control of Movement,* (Oxford: Oxford University Press); Land, M. F & Mcleod, P. (2000). From eye movements to actions: how batsmen hit the ball. *Nature Neuroscience* **3** 1340–1345.

6. Predicting the results of actions A system of this type is the Smith Predictor, originally developed to control the thickness of the finished product in steel rolling mills: since there was inevitably a certain lag between the steel leaving the rollers and the point where it had cooled enough for the thickness to be meaningful, an ordinary direct feedback system would have led to unstable oscillations. See Smith, O. J. M. (1959). A controller to overcome dead time. *ISA Journal* **6** 28–33.

7. Two kinds of internal model In fact there are two different ways in which such a modal can be implemented. We can either use a *forward* model, which (as in the ballistic system) converts a desired result directly into an appropriate command, or an *inverse* model, which tells us what feedback to expect from a given action. On the whole the latter is more generally useful, since it contributes to perception as well as movement control.

8. Brainstem saccade circuits See for example Keller, E. (1992). The brainstem. In Carpenter, R. H. S. (ed.) (1992). *Eye Movements.* (London: MacMillan); Carpenter, R. H. S. (2004). The saccadic system: a neurological microcosm. *Advances in Clinical Neuroscience and Rehabilitation* **4** 6–8; Krauzlis, R. J. (2005) The control of voluntary eye movements: new perspectives. *The Neuroscientist* **11** 124–137.

9. Eye movements General accounts of eye movements include Carpenter, R. H. S. (1989). *Movements of the Eyes.* (London: Pion); Carpenter, R. H. S. (ed.) (1992). *Eye Movements.* (London: MacMillan); good sources of information on clinical aspects are Kennard, C. & Rose, F. C. *Neuro-ophthalmology: Volume 102 (Handbook of Clinical Neurology)* (2011), or Leigh, R. J., & Zee, D. S. (2015). *The Neurology of Eye Movements* (5th edn). (Philadelphia, PA: F A Davis). Very readable accounts of the relation between eye movements and vision are Findlay, J. M. & Gilchrist, I. D. (2003). *Active Vision: the Psychology of Looking and Seeing.* (Oxford: Oxford University Press) and Land, M. F. & Tatler, B. W. (2009). *Looking and Acting.* (Oxford: Oxford University Press, Oxford). An entertaining historical account is Wade, N. J. & Tatler, B. W. (2005). *The Moving Tablet of the Eye.* (Oxford: Oxford University Press).

10. Cleverness of the spinal cord The nineteenth-century physiologist Pierre Flourens has left a characteristic account of decerebrating a chicken: *'I removed the two cerebral lobes from a healthy chicken. The animal, thus deprived of its cerebrum, survived ten whole months in a state of perfect health and would in all probability have lived longer if I had not been obliged to leave Paris. I had scarcely removed the brain before the sight of both eyes was suddenly lost; the hearing was also gone and the animal did not give the slightest sign of volition, but kept itself perfectly upright upon its legs, and walked when it was stimulated – or when it was pushed. When thrown into the air, it flew; and swallowed water when it was put into its beak. It seemed entirely to have lost its memory, for when it struck itself against anything, it would not avoid it, but repeat the blow immediately'.*

11. Armies The analogy is a very old one: in one of his manuscript notebooks, Leonardo da Vinci (1452–1519) writes: *'The muscles and tendons obey the nerves as soldiers obey their officers; and the nerves obey the brain as the officers obey the general'.*

12. The benefits of hierarchies A thoughtful discussion, especially in relation to the oculomotor system (where hierarchical organisation is particularly obvious), is Berthoz, A. (2000). *The Brain's Sense of Movement.* (Cambridge, MA: Harvard).

13. The plantar response This important neurological sign can be elicited in several ways. The Babinski reflex is the most widely used, but the same response can be invoked by stroking the lateral aspect of the foot from the lateral malleolus ('Chaddock's reflex') or even along

the medial side of the tibia ('Opponheim's sign'). In each scenario the first movement of the big toe is used to define a normal (down-going) or abnormal (up-going) response.

14. Newborn walking The figure was very kindly supplied by Dr N. R. C. Roberton, Department of Paediatrics, Addenbrooke's Hospital, Cambridge. For a dissident view, that the loss of this stepping is due more to changes in body weight in relation to leg strength, see Thelen, E., Fisher, D. M. & Ridley-Johnson, R. (1984). The relationship between physical growth and a newborn reflex. *Infant Behavior and Development* **7** 479–493.

CHAPTER 10
LOCAL MOTOR CONTROL

The final output from the central nervous system to skeletal muscle is from the *motor neurons* in the ventral horn of the spinal cord or in the brainstem. Since every action we carry out is the result of the activity of these cells, we need to look in some detail at what they do and the way in which they relate to each other at the spinal level.[1]

Motor neurons

Motor neurons are large, reflecting the length of their axons. Each controls a scattered group of individual muscle fibres called a *motor unit*: a unit may comprise as few as a dozen fibres in some of the muscles of the middle ear and in eye muscles, or as many as 1500 in large and crude muscles such as gluteus maximus.[2] In the cat's leg, a single unit is capable of lifting some 10 g. Force is controlled partly by changes in the firing frequency of individual motor neurons, and partly by the process of *recruitment*, discussed in Chapter 2 (p. 39), in which more and more units are brought into play as the required muscle tension increases. Normally it is the smallest neurons, innervating the most 'tonic' and least fatigable muscle fibres, that are the first to be recruited.[3] Larger α motor neurons, innervating faster but fatigable muscle fibres, have higher thresholds and so operate for brief exertions requiring larger forces, such as in sprinting or jumping.

Although individual motor neurons may sometimes fire at very low frequencies, this does not normally cause discrete twitching of the muscles because different units fire out of synchrony with one another. One of the functions of the Renshaw cells, mentioned in Chapter 3, is probably to provide a kind of lateral inhibition between motor neurons that discourages synchronisation of this kind.[4]

Motor neurons have an orderly and systematic arrangement within the cord, in clumps not quite well enough defined to be called nuclei, which reflects the topology of the muscles they serve (**Fig. 10.1**). Medial neurons innervate the muscles of the trunk, the most distal parts of limbs are governed by the most lateral neurons, and flexors and extensors tend to be under the control of the more dorsal and ventral groups, respectively. Because these cells represent the ultimate funnel through which all nervous excitation must pass whenever a motor act is made, whatever its source, whether reflexive or voluntary, they form what is sometimes called the *final common path*. On these cells terminate all the afferents from interneurons, and in some cases receptors, that serve the various spinal reflexes and responses, as well as some of the descending paths from higher levels. The quantity of information that thus converges on a single ventral horn cell is enormous and is reflected in the huge size of its dendritic tree, which may often extend over a large part of the grey matter of the cord.

Sensory afferents entering the dorsal roots can give rise to *spinal reflexes*, by projecting – either directly or via interneurons – on to motor neurons (p. 55). It is never easy to

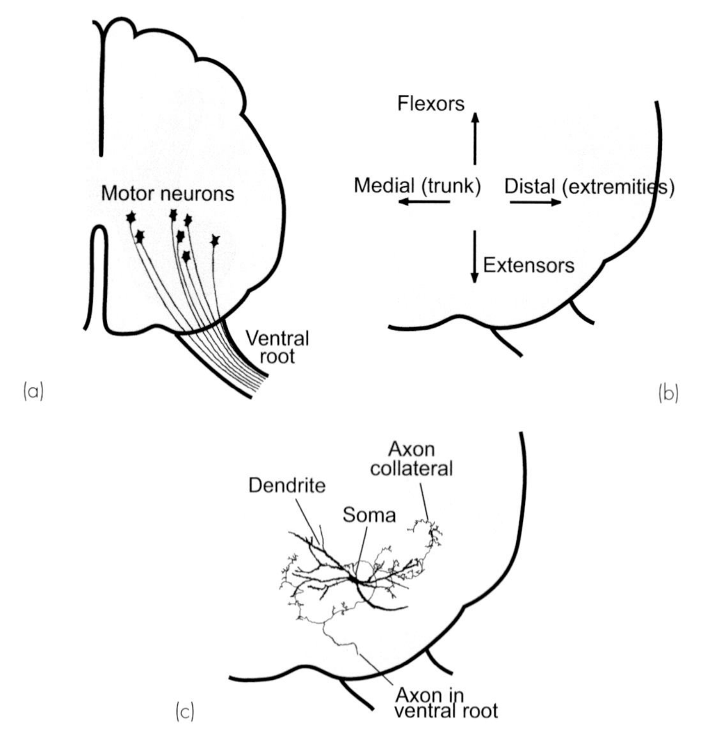

Figure 10.1 (a) Relation of motor neurons, spinal cord, and ventral root. (b) Motor neurons corresponding to various muscle groups are localised in different areas of the ventral horn. (c) Golgi preparation of a single ventral horn cell, showing the enormous extent of its dendritic field.

define exactly what is meant by a reflex. Any response that can be elicited from a spinal animal is certainly a spinal reflex.[5] But we have already seen that although a response may be essentially a spinal one, in the sense that its neural circuitry lies entirely within the spinal cord, one may not be able to elicit it in the spinal preparation because the usual facilitating, permissive, influences descending from the brain are absent. It is not quite enough to describe a reflex as an automatic, reproducible, response that is independent of the will, as one may readily influence one's own spinal reflexes by willed inhibition or facilitation from above. For example, there is a mechanism in the cord called the withdrawal or flexion reflex that causes a rapid flexion when the skin is touched by a hot object or other noxious stimulus: yet we all know from personal experience that in cases of necessity – when carrying a plate that proves to be hotter than we thought when we picked it up – we can (up to a point) inhibit the reflex completely. We shall see later on that what at first sight appears the simplest and most automatic spinal reflex of all, the monosynaptic stretch reflex, is actually under almost total control from other, mainly descending, influences. In fact, when we look at the mode of termination of the tracts descending from the brain, we find that the great majority of them end not on the motor neurons themselves, but rather on the interneurons that form part of these reflex arcs. Descending control is not so much of muscles as of *actions*, amounting to a selection from the cord's repertoire: the brain plays on the spinal cord not as one plays a piano, but rather as one selects a track from their music player.

Figure 10.2 Reticular formation: apparent chaos. (P. Brodal, 1998; permission pending.)

Descending pathways

Five important tracts descend from brain to spinal cord; four of these come from closely neighbouring parts of the brain, the brainstem and medulla. These are the *reticular formation*, the *vestibular nuclei*, the *red nucleus*, and the *tectum*; the fifth source of descending fibres is the *cerebral cortex*.

Reticular formation

If you take the brainstem, and remove all its sensory and motor nerve nuclei, all the fibre tracts that have to go through it, and all other well-demarcated structures, a considerable area of *terra incognita* is left unaccounted for. This is the *reticular formation* ('reticulum' = net; RF), described by Brodal as formed of *'diffuse aggregations of cells interspersed with fibres going in all directions'*,[6] full of axons and dendrites weaving inextricably between one another (**Fig. 10.2**), and with long branching axons going both upwards to the midbrain and forebrain, and also down to

Clinical box 10.1 Motor neuron diseases

There exists a large number of neurological diseases in which the principal site of pathology is the motor neuron and associated neuronal networks. The most common is a sporadic, late-onset, degenerative condition, termed motor neuron disease (MND), which is often synonymous with amyotrophic lateral sclerosis (ALS), or Lou Gehrig's disease (after the famous baseball player who succumbed to the disease), as this represents its most frequent variant.

In essence, MND is characterised by a progressive weakness (first affecting the lower limb in 35 per cent, upper limb in 30 per cent, and speech and swallowing in 30 per cent), in the absence of sensory symptoms. Examination, meanwhile, highlights the areas of pathology: there is evidence of dysfunction of the alpha (or 'lower') motor neuron (with weakness, wasting and spontaneous movements of denervated muscle units called fasciculations), but also of the descending (or 'upper') motor neuron pathways (manifest as increased tone and hyperreflexia, due to release of spinal reflex arcs). After starting with subtle symptoms such as foot drop or mildly slurred speech, this illness progresses relentlessly to affect all muscle groups, ultimately resulting in paralysis, inability to swallow, difficulty with communication, and, usually, as the respiratory muscles grow weak, the cough becomes inadequate to clear the chest, and most die of pneumonia and respiratory failure. Some patients choose to have a tracheostomy and percutaneous gastrostomy sited so they can be mechanically ventilated and fed directly into the stomach without the risk of choking. However, with no disease-modifying treatment available, much of the focus is on good quality palliative care alongside carer support.

The cause of MND remains elusive, but there is increasing evidence of a genetic predisposition; indeed 5–10 per cent have an affected first-degree relative. A plethora of potential genetic culprits have been identified, including superoxide dismutase (SOD), an enzyme that catalyses the transformation of superoxide to oxygen or peroxide. As superoxide is highly toxic, wantonly oxidizing and destroying sensitive and vital cellular structures, when it is dysfunctional, it is easy to see how a neurodegenerative disease can ensue.

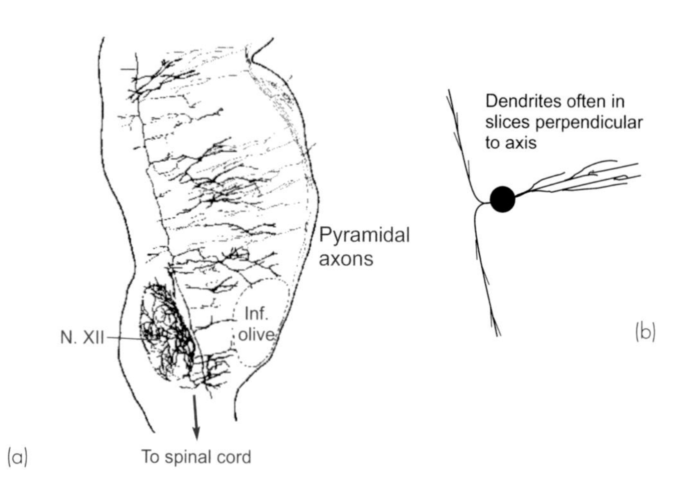

Figure 10.3 (a) Transverse organisation of brainstem reticular formation in the rat. (After Scheibel & Scheibel, 1960.) (b) Schematic representation of typical neuron with dendrites perpendicular to its axonal axis.

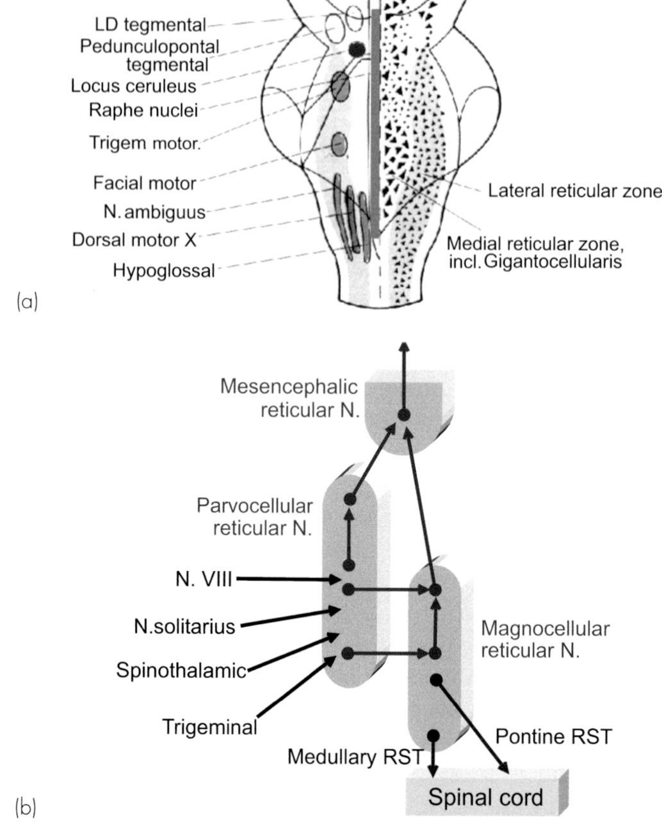

Figure 10.4 Reticular formation anatomy. (a) The position of some of the principal nuclei shown schematically. (After Fitzgerald, 1985.) (b) Highly schematic diagram of the main connecting pathways. RST, reticulospinal tract.

the spinal cord. However, one should not exaggerate its homogeneity and randomness. There *are* nuclei within it: although they can't always be made out by classical light microscopy, they can be defined by their transmitters or their connections or functions, rather than being apparent using conventional histological techniques.

The reticular formation is one of the most important and oldest structures in the brain, a direct descendant of the nerve nets controlling creatures like the sea anemone.[7] It stretches from the superior cervical spinal cord up to the intralaminar thalamic nuclei with which it merges at the top. One feature shared by cells in these primitive networks and in the reticular formation is the existence of long ascending and descending axons. Another defining feature of the reticular formation is that dendritic fields tend to lie in slices perpendicular to the axis of the brainstem: hence the term *isodendritic core*. These dendrites receive profuse axon collaterals from sensory fibres ascending to higher levels, as well as from descending motor pathways (**Fig. 10.3**).

Nevertheless, the reticular formation is not entirely shapeless and unstructured. Overall, it consists of essentially three longitudinal zones:

- *The raphe*, very close to the midline ('raphe' = seam);
- *The medial zone*, with cells that are particularly large because they are the origin of long ascending and particularly descending projections. Some of its subdivisions are shown in **Figure 10.4**: the most important are the *magnocellular (gigantocellular) nucleus* with unusually large cell bodies, and the *locus ceruleus;*
- *The lateral zone*, with smaller cells, and somewhat shorter axons; in particular, the *parvocellular nucleus*, which is more obvious at the bottom of the pons and top of the medulla, and has close association with cranial nerve nuclei: in a sense it is more sensory than motor, projecting predominantly upwards, partly via the *mesencephalic* reticular formation at the top (**Fig. 10.4**).

So what does all this *do?* Despite being so diffuse and apparently homogeneous, its functions fall into two fairly clear-cut categories, both a kind of integration. In its upwards projections, it is concerned with the regulation of the *level of activity* of the brain, including such functions as attention, sleep, arousal; these are dealt with in Chapter 14. Its downward-projecting systems are concerned with the generation of *patterns of response*, which can often be quite stereotyped. These in turn can be roughly divided into two kinds. First, the reticular formation provides in a sense the executive brains behind much visceral and vegetative activity: the control of heart rate and the generation of the rhythms of breathing, for instance. Second, similar circuits seem to underlie other timing functions that are more obviously motor in the conventional sense. One of the best understood is the generation of saccadic eye movements, and it's quite helpful to look at this in a little detail as an example of the rather sophisticated things the reticular formation is capable of.

Saccades were introduced in Chapter 9, and their visual control was discussed in Chapter 7. We saw that they are incredibly fast, as little as 20–30 ms in duration and with velocities around 900°/s. Yet the muscles are not intrinsically particularly fast, and in response to a step increase in motor neuron firing may take some half a second

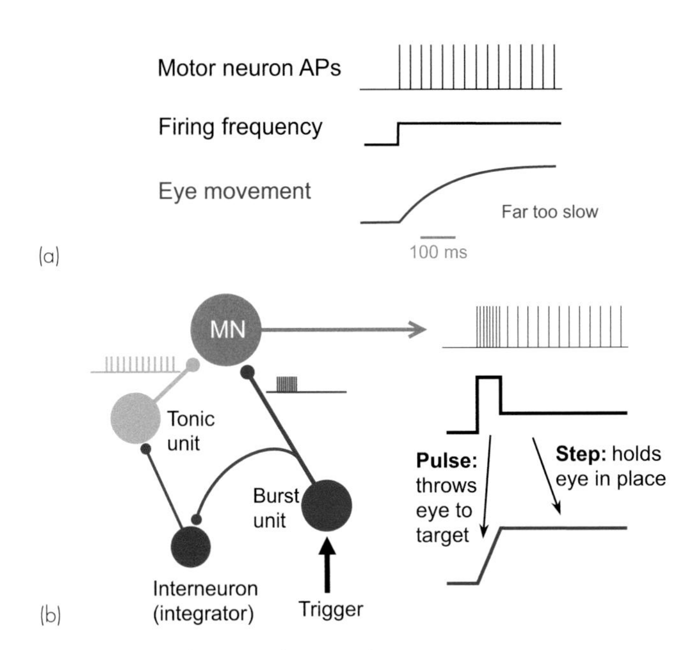

Figure 10.5 Reticular circuits for saccades. (a) A step increase in activity produces only a slow deviation of the eye. (b) The actual pattern observed is a combination of step and burst, throwing the eye rapidly to its new place; this is produced by summation of the activity of separate tonic and burst neurons. MN, motor neuron.

to settle down at a new length, which seems paradoxical. How can such sluggish muscles produce such extraordinarily fast responses? The answer comes from looking at the time-course of the signal that is sent to them by their motor neurons to create a saccade (**Fig. 10.5**). It consists of two components: a *step* of activity, designed to be just enough to hold the eye at its new position, and a very brief *pulse* of maximal firing, that kicks the eye as rapidly as possibly into place. It turns out that this very sophisticated pattern is generated by neural circuits in the prepontine and mesencephalic reticular formation, close to the motor neurons in nuclei III, IV and VI. Two kinds of neuron are found there, called tonic units and burst units, that provide the step and pulse components by summing together at the motor neuron. The burst neurons drive the tonic neurons through an indirect pathway and, descending from the superior colliculus and some other areas, trigger the whole thing off: a very neat neural mechanism. It is easy to imagine similar special-purpose neural circuits for all kinds of movements, and we know that the machinery for much of the very basic but beautifully coordinated patterns of behaviour like walking are in the reticular formation, as well as some even more biologically basic behaviours such as chewing, grooming, suckling (lateral and pontine regions), coughing, swallowing, sneezing (ventrolateral) and facial expression.

What are the routes by which all these movements are generated? In the case of eye movements, there are relatively direct projections on to the cranial nerve nuclei. But movements mediated by the spinal cord are brought about essentially by two large descending tracts that go down to the spinal cord and terminate on both alpha and gamma motor neurons, in some cases causing excitation

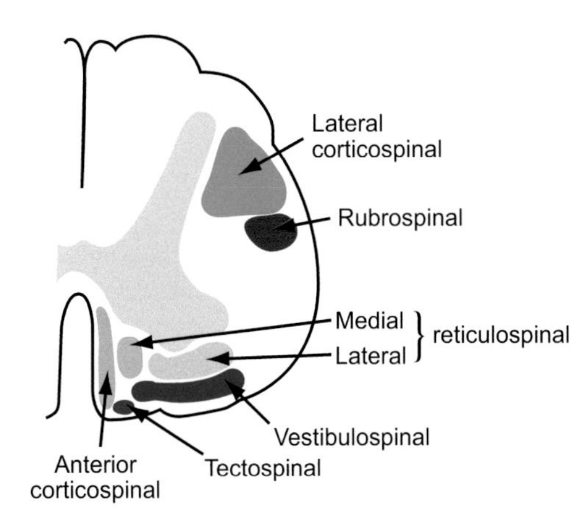

Figure 10.6 Section of cord showing approximate positions of the major descending tracts.

and sometimes inhibition: these are the two *reticulospinal* tracts (RSTs; **Fig. 10.6**):

- *Medial* RST: ipsilateral from pons (nucleus pontis caudalis);
- *Lateral* RST: mostly ipsilateral from medulla (nucleus gigantocellularis).

Most of the fibres terminate diffusely on interneurons rather than on the motor neurons themselves, except for certain fibres of the lateral tract; they influence muscles of the trunk and proximal parts of limbs rather than the extremities. Other fibres in the lateral tract contain enkephalin and are likely to modulate the spinal transmission of afferent responses to noxious stimuli (p. 96).

In addition to acting more or less directly on the spinal cord, the reticular formation also has extraordinarily widespread *indirect* influence on the central motor system. It projects to the cerebellum via the lateral and paramedian nuclei and to the basal ganglia via the mesencephalic reticular formation; and this is quite apart from the more diffuse ascending projections to basal ganglia and cortex that we'll be looking at later. Conversely, nearly every level of the motor system sends projections down *to* the reticular formation: basal ganglia, cerebellum, superior colliculus, vestibular nucleus, substantia nigra, and much of the cortex, including motor and somatosensory areas. It is clear that the reticular formation provides a *major* alternative output mechanism for the higher control of movement, as well as doing things on its own initiative.

The vestibular nuclei

In the course of evolution, areas of the reticular formation concerned with particular sources of sensory stimulation have tended to condense together to form separate nuclei and migrate towards their common source of excitation. The vestibular nuclei seem to have emerged in this way under the influence of afferent fibres from the vestibular apparatus, which, as we saw in Chapter 5, are concerned with sensing movements of the head and the direction of gravity. At least four nuclei make up the vestibular

Clinical box 10.2 Clinical aspects of the reticular formation

Because this area is anatomically and functionally diffuse, problems here rarely present as discrete, remediable lesions: any injury is likely to be accompanied by damage to structures with less subtle and more easily identified consequences, such as frank paralysis in the case of cord trauma affecting the corticospinal tract, or cranial nerve palsy if a nearby cranial nerve nucleus is affected by, for example, an inflammatory plaque of multiple sclerosis. Furthermore, injuries affecting the brainstem, as we shall see later, also tend to affect the conscious level, and identifying neurological deficits in comatose patients is very tricky. Because lesions to the reticular formation are difficult to identify, and even more importantly, there is little we can do about them, the reticular formation receives little formal attention from clinicians. However, injury to the brainstem can lead to several manifestations likely to be attributable to a damaged reticular formation.

Reduced conscious state

The reticular formation seems to be essential to maintain arousal and consciousness. With all higher centres intact, humans remain comatose if the reticular formation is injured. On the other hand, quite profound higher neurological injury is compatible with a degree of alertness if the reticular formation is intact, a sometimes ethically and philosophically challenging scenario.

The Cushing reflex

Raised intracranial pressure from a bleed, oedema or a tumour is associated with impaired blood flow into the cranium (which is, essentially, a non-expandable box). The resulting paucity of blood flow (ischaemia) stimulates the regions of the reticular formation responsible for augmenting blood pressure in a last-ditch attempt to raise the blood pressure enough to force some blood into the cranium and keep the neurons alive. This premonitory clinical picture mandates emergency decompression, usually involving surgical removal of part of the skull.

Abnormal respiratory patterns

The central genesis of normal respiratory pattern is a complex and poorly understood area but seems to reside largely in the reticular formation. Lesions of the reticular formation can yield a range of different breathing patterns depending on which precise region is injured – too fast, too slow, too shallow, too deep, or irregular and jerky. *Ondine's curse* is a rare condition which can be congenital or associated with brainstem injury, whereby the generation of breaths is no longer automatic. Patients suffer prolonged, potentially fatal, periods of apnoea when they fall asleep, and in some cases the only remedy is mechanical ventilation throughout the night. (Ondine was a jealous nymph whose rather foolish lover swore that his every waking breath would be a testimony to his love for her. He was then caught out in his infidelity and cursed by Ondine's king to forget to breathe should he fall asleep, which, the tale pre-dating Red Bull, he eventually did.)

complex, of which the lateral (Deiter's) nucleus is the origin of an important descending motor tract, the *lateral vestibulospinal tract* (**Fig. 10.7**). As might be expected from the reticular origin of the vestibular nuclei, the course of this tract is similar to those of the reticulospinal tracts: virtually all the fibres are uncrossed, most end on interneurons, but some excite motor neurons monosynaptically. A medial vestibulospinal tract also exists, projecting bilaterally from the medial vestibular nucleus mainly to cervical and upper thoracic regions; some of its fibres are inhibitory. The functions of these tracts are essentially to maintain posture and to support the body against the force of gravity; consequently they are mostly concerned with the control of extensors rather than flexors. The vestibular nuclei are also closely associated with the cerebellum, one of the largest and most important higher motor areas, and the descending tracts may provide one way in which the cerebellum may control the spinal cord: it has no direct projections of its own.

The red nucleus

The third descending pathway is the *rubrospinal tract*, which is derived from the red nucleus (Latin *ruber* = red), a well-defined region lying above the pons in the midbrain. This area, being highly vascular, is pinkish in fresh specimens, giving it its name. Like the vestibular nuclei, it forms all-important output relays from the cerebellum (interpositus and dentate), but is not associated with any particular sensory modality; it also receives descending fibres from the cerebral cortex. It is more prominent in animals other than humans, and projects somatotopically mainly to caudal rather than cervical parts of the cord, terminating on interneurons; it tends to produce flexion rather

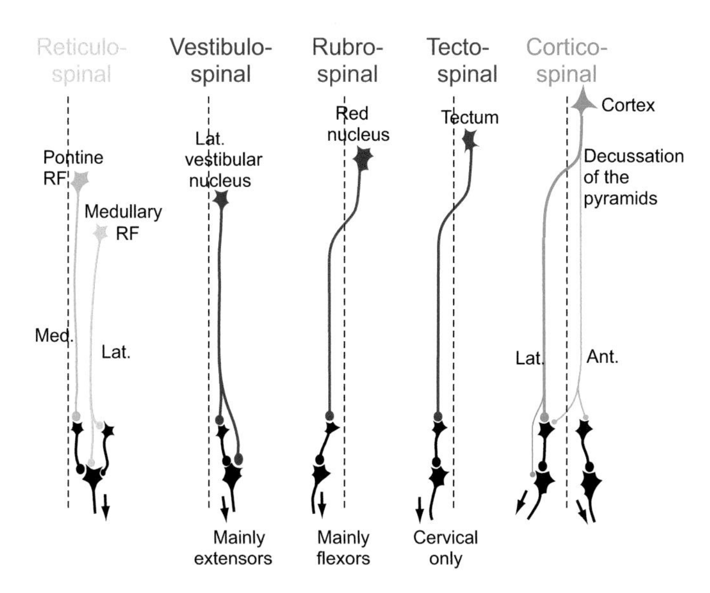

Figure 10.7 Diagrammatic representation of the arrangement of the descending fibres in the major motor tracts. RF, reticular formation.

than extension when electrically stimulated. Its fibres cross at a high level and then descend laterally; its functions are something of a mystery.

The superior colliculus

The fourth motor tract is the *tectospinal tract.* 'Tectum' means 'roof', and anatomically the tectum is simply the roof of the midbrain, comprising the superior and inferior colliculi in mammals. These are integrating centres for vision and hearing respectively, although as one ascends the evolutionary tree one finds their functions increasingly taken over, or at least supplemented, by the cerebral cortex. They seem to be concerned in particular with orientating responses, as for example in turning to look at the source of a sudden sound. The superior colliculus was described in Chapter 7 (p. 168), and you may recall that here one finds neurons that are responsive to visual stimuli in precise locations in the visual field, and when stimulated they cause the eyes to execute a saccade to the very same part of the field: large movements evoked in this way may involve the head as well. As one might expect from such orientating responses, the tectospinal tract projects no further than cervical segments, sending efferent branches to both areas of the reticular formation that trigger saccades and spinal regions controlling the neck. The fibres are crossed and appear to end on interneurons

The corticospinal tract

Finally we come to a tract that, because of its clinical significance, has been investigated to a degree perhaps somewhat out of proportion to its real importance in a broader view of the control of movement: the *corticospinal* or *pyramidal tract.* Its neurons are some of the longest in the body, since they run from the cerebral cortex in the top of the skull all the way down as far as the bottom of the spinal cord. Not much more than half come from frontal cortex, including primary motor cortex; many of the others come from somatosensory and parietal cortex. In Man, some 80 per cent cross in the medulla (in other species the proportion is greater: 100 per cent in the dog), and as they lie on the extreme ventral surface (the pyramids of the medulla, hence 'pyramidal'), the decussation can generally be seen with the naked eye. The crossed fibres descend as the lateral corticospinal tract, and the uncrossed ones as the anterior corticospinal tract. Both are relatively recent pathways, their development following that of the cerebral cortex itself, and consequently there is a good deal of species variation as to their size and disposition. In Man, there are about a million fibres in the pyramidal tract, forming some 30 per cent of the white matter of the cord: in the dog, 10 per cent. Only in Man and some other primates do any fibres terminate on motor neurons, presumably exciting them (they release glutamate); in many species they project no further than cervical segments, and in any case form a relatively small component of the total white matter of the cord; even in Man slightly more than half of them terminate cervically.

Their clinical importance is, however, very great indeed. At the upper end, the fibres fan out into a sheet called the *internal capsule,* in order to squeeze past the thalamus and basal ganglia; at this bottleneck they are almost entirely supplied with blood by small 'lenticulostriate' arteries, which are particularly susceptible to narrowing in the presence of hypertension, smoking or diabetes. Given the dense packing of fibres and their vulnerable blood supply, this site is a common site of ischaemic stroke, potentially paralysing one side of the body. The functions of the corticospinal tract will be considered in more detail both later in this chapter and also in Chapter 12; it predominantly controls the distal musculature and is by and large concerned with fine, skilled, voluntary movements of the extremities, and particularly with manipulation.

Effects of transverse sections

Some of the tonic actions of these various centres, and their inter-relations, can be deduced from classical experiments involving lesions in the brainstem that effectively disconnect them from the spinal cord, or from each other. Despite what might at first seem to be the crudity of the techniques, it is nevertheless possible to come to quite firm, if general, conclusions from them. The *spinal* preparation has already been mentioned; this is one in which all the tracts are cut, so that the cord is completely isolated from the brain.[8] Initially, there is a period of 'spinal shock' characterised by floppy or *flaccid* paralysis, in which there is loss both of voluntary movement and muscle tone and segmental spinal reflexes are depressed as the influence of the corticospinal, rubrospinal, vestibulospinal, tectospinal and reticulospinal tracts is lost. If the lesion is sufficiently high, this spinal shock is accompanied by a fall in blood pressure (confusingly termed 'neurogenic shock') as sympathetic outflow from the lateral columns is lost. Whereas a normal person's muscles fire tonically, excited by a steady

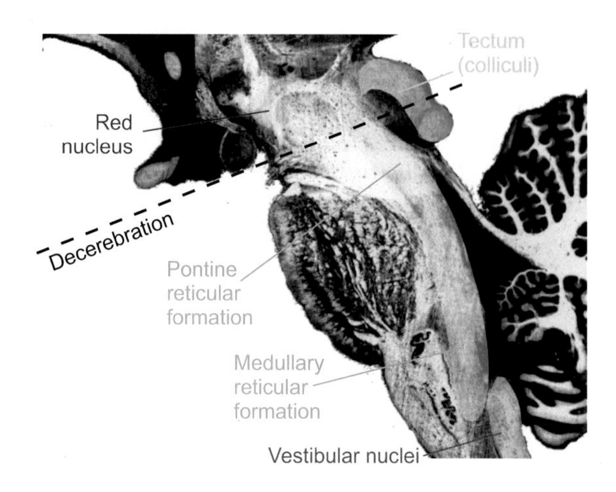

Figure 10.8 Sagittal section of human brainstem, showing reticular nuclei, and other areas that are sources of descending tracts. The dashed line shows the approximate position at which a cut is classically made in the decerebrate preparation.

level of motor neuron activity, and offer resistance to any movement imposed on the limbs from outside, in flaccid paralysis they are relaxed and offer no resistance at all: thus one may be able to pick up such a patient's arm and fling it in his face, something that cannot be done to a normal conscious subject. Spinal shock typically lasts from a day to a month before, in the absence of commands from above, spinal neurons regain intrinsic excitability and the classic findings of overexcitability and exaggerated reflexes appear.

Another preparation studied frequently in the laboratory is the *decerebrate* preparation, classically produced by a cut at the level of the colliculi (**Fig. 10.8**). The effect of such a transection, once the animal is allowed to recover, is utterly different from the floppiness of the spinal preparation: the animal now has muscles that, far from being flaccid and relaxed, are – especially extensors – tonically hyperactive, and the general picture is one of stiffness – decerebrate *rigidity*. The increased tonic activity of the decerebrate animal as compared with the spinal animal must presumably be interpreted as a release phenomenon of the kind discussed in the previous chapter, and due to unopposed activity originating in some structure that lies between the levels of the two cuts. A feature of decerebrate rigidity is that the pattern of stiffness in the legs depends markedly on which way up the animal is (**Fig. 10.9**), so it seems likely that the tonic over-activity is essentially due to the influence of sensory stimuli from the vestibular apparatus, normally held in check by centres lying above the brainstem: the rigidity is abolished by lesions in the lateral vestibular nuclei.

A different kind of rigidity develops, following initial flaccidity, after destruction of the cerebral cortex (giving a *decorticate* preparation) rather than decerebration: but in this case lesions of the vestibular nuclei have relatively little effect. This second type of rigidity is clinically called *spasticity*. Like rigidity, it increases resistance to passive stretch, but differs from it in that the resistance depends on the rate of stretch. In other words, the resistance to

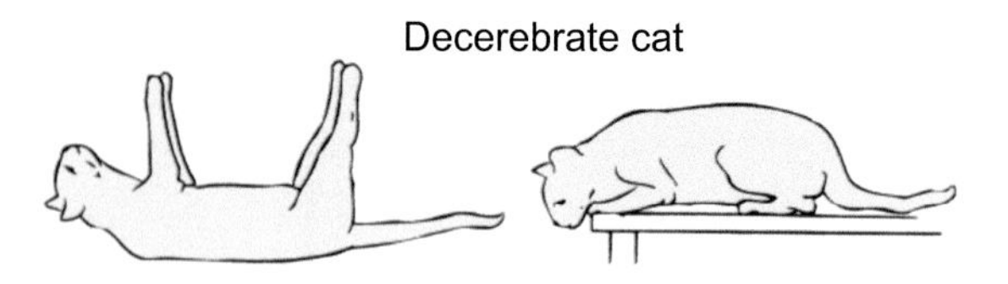

Figure 10.9 Tone in the limbs as a function of head position in the decerebrate cat.

movement in the spastic limb is greater if the limb is flexed or extended more quickly, and may even be normal if the limb is moved slowly enough. In contrast, the rigid limb generates the same, constant resistance to movement however it is applied. Spasticity is thought to be due to a tonic excitatory influence, predominantly on rate-dependent rather than tonic stretch reflexes, from upper areas of the reticular formation, which are disconnected from the cord in decerebration, and presumably normally inhibited by the cortex. Direct confirmation of this has come from electrical stimulation of the upper reticular formation, which results in a general facilitation both of spinal reflexes and of the effects of electrical stimulation elsewhere in the brain. In this respect, the lower or bulbar reticular formation is exactly the opposite: electrical stimulation causes not facilitation but depression of reflexes and evoked movements. The tonic relationships between these structures that may be deduced from such experiments are summarised in **Figure 10.10**; lesions and stimulation of the cerebellum

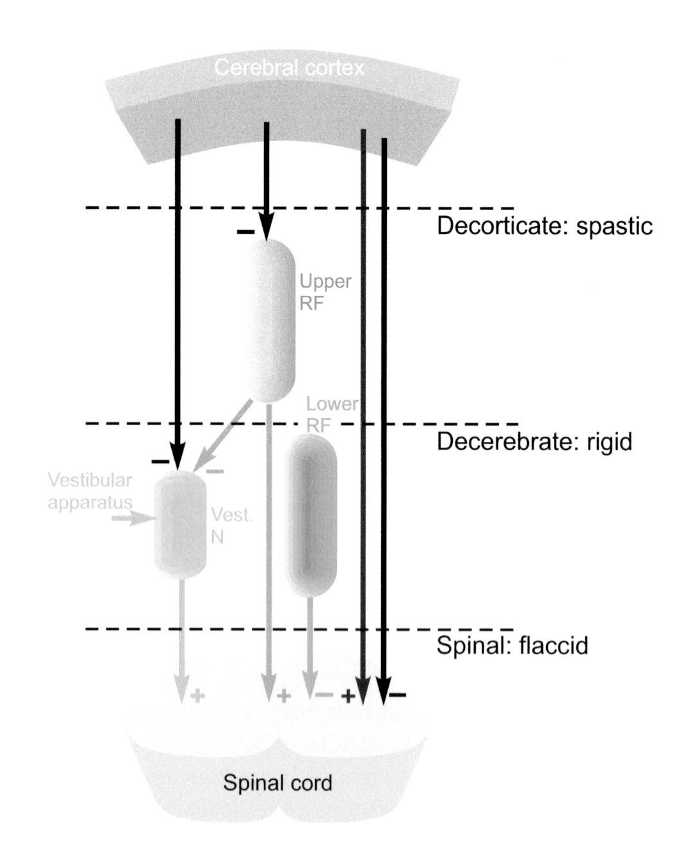

Figure 10.10 Highly schematic representation of apparent tonic facilitatory and inhibitory influences between various central regions, and their relationships to the levels of section in different experimental preparations. RF, reticular formation.

and basal ganglia also influence rigidity, but these effects are more complex and will be described later on.

A striking fact about the relation between the higher motor levels and the cord is that the two largest areas of all, the basal ganglia and cerebellum, have absolutely no direct projections to spinal levels: all that they do must be achieved at second hand, relaying through one of the five tracts described earlier. Furthermore, these tracts themselves act for the most part only indirectly, influencing spinal reflexes rather than motor neurons, underlining once again the essentially hierarchical nature of the motor system. The rest of this chapter will be concerned with one particular spinal reflex, the stretch reflex, which probably plays a more important part than any other in the control of movements, and illustrates something of the way in which the brain can make use of indirect control of this kind.

Sensory feedback from muscles

In the previous chapter, we saw that there is an intimate involvement of sensory feedback at every level of the nervous system and examined some of the ways in which this feedback might be used to improve motor control. Here we shall be concerned with the very lowest level of this feedback, that from the muscles themselves.

Figure 10.11 shows what an enormous quantity of this information there is. In a typical cat soleus muscle, compared with the 150 or so fibres that are truly motor, innervating extrafusal muscle fibres, there are some 150 sensory fibres, and another hundred or so γ fibres, which, as we saw in Chapter 5, do not directly cause contraction of the muscle itself, but rather modify sensory signals from the muscle spindles. In other words, some 250 fibres are concerned with afferent information, whilst only 150 are directly responsible for muscle tension.

Golgi tendon organs

Of the two types of sensory receptor within muscles, the spindle and Golgi tendon organ, the functions of the latter are less well understood. We saw in Chapter 5 that, being in series with the main contractile elements, it acts as a *force* transducer, signalling increases in muscle tension during active contraction of the parent muscle; what is not altogether clear is how this information about muscle tension is actually put to use. One reflex for which it is (partly) responsible is the clasp-knife reflex; if you take hold of someone's hand and then push on it in such a way as to bend his elbow – having told him to push back as hard as he can to resist you – there will come a point (assuming that you are the stronger!) at which the force he exerts suddenly seems to give way, and the arm folds up like a clasp-knife. This reflex is thought to be brought about by a neuronal circuit such as that in **Figure 10.12**, in which the incoming tendon organ fibres inhibit their parent motor neuron through an interneuron, with some kind of threshold; but Golgi tendon organs are not the only sensory input to what is now realised to be a complex response. It is usually claimed that this reflex is protective in function, preventing damage to tendons by pulling on them too hard; if so, it is not very good at its job, since athletes do of course frequently 'pull' their tendons despite the existence of the reflex. It is difficult to believe in fact that this is all the tendon organs do, and a role for them in the control of muscle movement is suggested later in this chapter: it is the spindles with which we are mainly now concerned.

Muscle spindles

We saw in Chapter 5 that spindles essentially signal both muscle length and also – especially in the case of the Ia fibres – rate of change of length, and that the messages they convey are also modified by the activity of the γ-efferent fibres to the intrafusal fibres: to a first approximation, their response is a function of the amount of external stretch plus the amount of internal stretch caused by γ activity. Or to put it another way, they signal γ activity minus the degree

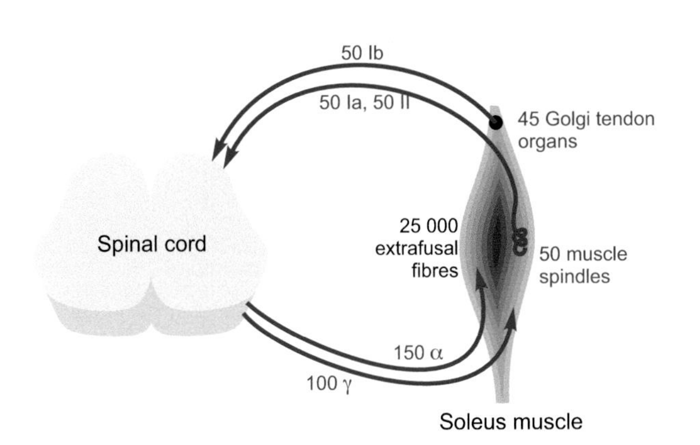

Figure 10.11 Flow of information to and from a typical cat soleus muscle, showing some 250 efferent fibres and 150 afferents. Of the total number of fibres, 250 are concerned with sensory information from the muscle, and only 150 are directly responsible for muscle tension. (Data from Matthews, 1972.)

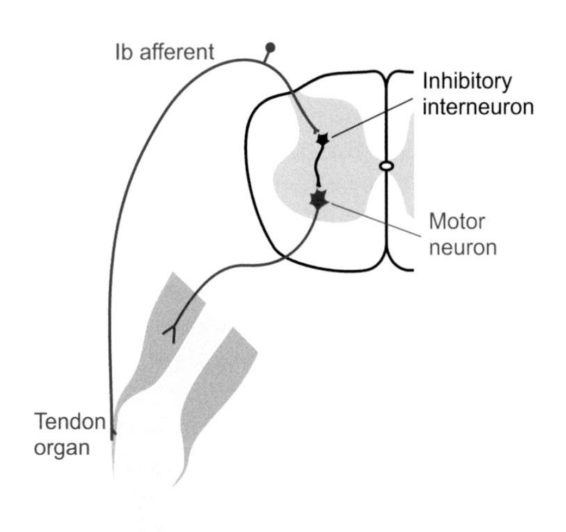

Figure 10.12 Schematic representation of the neural circuits thought to underlie the clasp-knife reflex: excessive muscle tension, sensed by the Golgi tendon organs, results in reflex inhibition of the muscle's motor neurons.

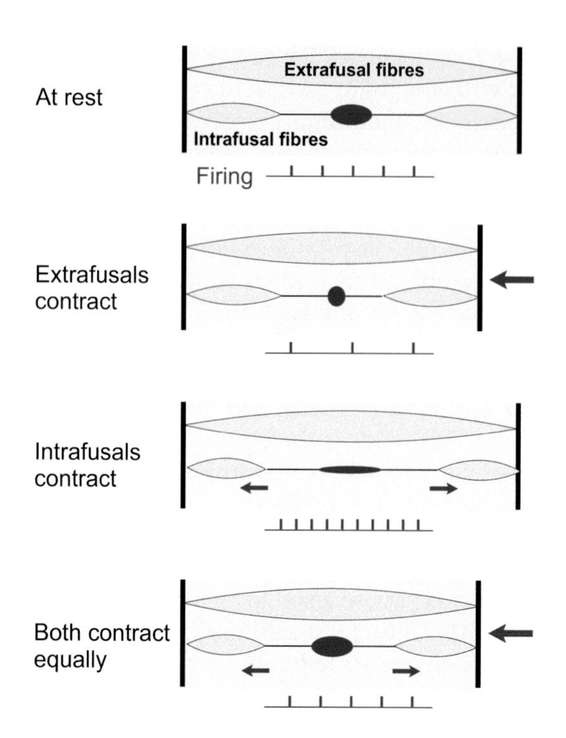

Figure 10.13 Muscle spindles signal the difference between intrafusal and extrafusal length. The frequency of firing of the afferent fibres is displayed below each scenario.

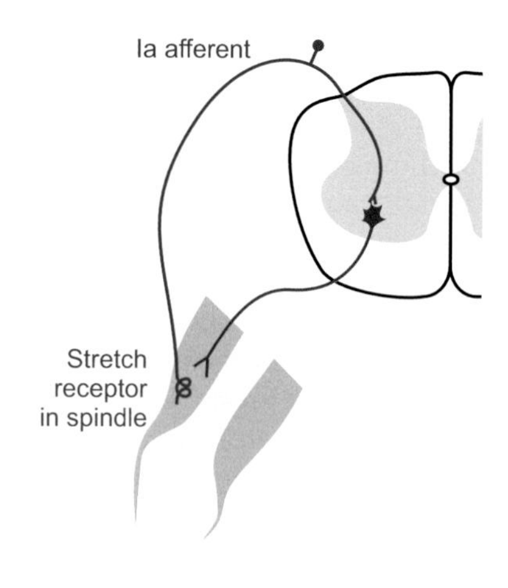

Figure 10.14 The monosynaptic phasic stretch reflex: rapid stretch activates the Ia fibres, which monosynaptically excite the motor neurons.

of muscle contraction (**Fig. 10.13**). What do these signals actually *do?*[9]

Spindle reflexes

One of their best known actions comes about because the Ia afferents monosynaptically excite motor neurons of the same muscle (**Fig. 10.14**), forming the classical *monosynaptic reflex arc,* the simplest imaginable kind of neuronal circuit that could link a stimulus to a response. The result is that any stretch of the muscle, but particularly a brief, phasic one that will preferentially excite the rate-sensitive Ia fibres, will stimulate the motor neuron and cause a rapid contraction of the muscle. An easy way to elicit such a response is in the familiar *tendon jerk:* tapping a muscle's tendon produces just the right sort of fast-rising stretch to elicit a brisk reflex contraction. The patellar tendon is convenient and gives an easily noticeable response, but other tendons such as the Achilles tendon will do just as well. Another effective way of stimulating the Ia fibres is by the use of massage vibrators; much of their 'exercising' effect is due to the fact that they induce tonic reflex contractions. In the normal person, these reflexes are rather feeble, but in certain experimental preparations – and pathological states – they are much increased, which is why they may often be valuable in clinical neurological diagnosis. In the decerebrate animal one may demonstrate not just a phasic reflex of this kind but also a tonic component, in which the muscle responds to steady stretch with a steady contraction – the *myotatic* or *tonic stretch reflex.* A record of this kind from a decerebrate cat is shown in **Figure 10.15**. In response to the 6 mm stretch shown by the lower line, the muscle responded with a steady tension of nearly 4 kg: some of this, but only a small part, was due simply to the muscle's intrinsic elasticity, which may be revealed if it is paralysed so as to suppress the reflex component. Since in this case a stretch of 6 mm generated about 3 kg of reflex tension, we can say that the *gain* of the reflex – the extra tension evoked per unit of stretch – was about 500 g/mm. The stretch reflex thus makes muscles appear stiffer – less elastic – than they naturally are. In fact, it is easy to show that the stiffness of decerebrate rigidity is entirely due to over-activity of the stretch reflexes, probably through excitation of the efferents, for if the dorsal roots are cut in such a preparation, preventing the Ia discharges from reaching the motor neurons, rigidity vanishes. Decerebrate rigidity might therefore be described as a sort of hyper-stretch-reflexia.

Muscle tone

Thus, the first notion about the function of muscle spindles that came to be accepted was that by acting through

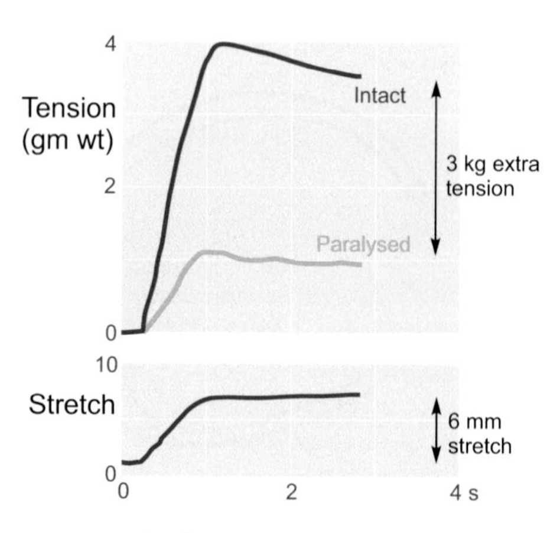

Figure 10.15 Tonic stretch reflex in decerebrate cat. The muscle was stretched with the time course shown in the bottom curve, and the resultant tension was measured (blue line). The green line shows the result of the same experiment when the muscle was paralysed. (After Liddell & Sherrington, 1924.)

the stretch reflex they were responsible for the generation of *muscle tone*, the constant muscular activity that is necessary as a background to actual movement in order to maintain the basic attitude of the body, particularly against the force of gravity. But tone is something that essentially *opposes* movement, that tends to keep muscles at pre-set lengths by making them resist any changes. Hence the idea arose that during movements one would have to alter the degree of tone in step with the movement if there was not to be a degree of conflict between the two, and that the γ fibres were ideally suited to doing just this. If every time a command was sent via the α-motor fibres (the ones innervating the extrafusal fibres) to make the muscle contract, the γ fibres were simultaneously activated, then all would be well. The internal stretch generated by the intrafusal fibres would make up for the reduced external stretch, so that the stretch reflex would still function and there would be no loss of tone.

The simple servo hypothesis

It then became apparent that one could carry this line of thought a stage further and envisage an even more active role for the γ fibres than this. Imagine for a moment that the γ fibres were stimulated without simultaneous direct activation of the α fibres. What would happen? By shortening the intrafusal fibres of the spindle, such a stimulus would result in excitation of the sensory afferents just as if an actual stretch had occurred (**Fig. 10.16**); consequently there would be a reflex activation of the α-motor neurons, and the muscle would contract automatically until the stretch receptors found themselves back at their original resting degree of stretch. In other words, γ stimulation could in principle initiate contraction, in exactly the same way as direct α stimulation. But would there be any point in such a roundabout way of making muscles contract?

There certainly would. We saw earlier that the sensory endings in the spindle are signalling something like the difference between the amount of γ activity and the shortness of the muscle. If we think of γ activity as telling the muscle how short it ought to be, then the spindle becomes a comparator generating an error signal, the 'error' here being the difference between the desired length of the muscle and its actual length. So if now we redraw the stretch reflex in a more formal way (**Fig. 10.17**), it is immediately apparent that what we have is a classic direct feedback or servo system. The γ fibres set the desired length; if this is shorter than the actual length, the spindle afferents stimulate the motor neurons to generate a force that makes the muscle contract. The advantage of such an arrangement, like all feedback systems, is that it automatically allows for noise, in this case the existence of unpredictable *loads* that have to be moved.

Load

The motor system faces a fundamental and frustrating dilemma. What we want to do is reach out for objects, pick them up, move them around: in other words, we want to achieve particular limb positions, joint angles, velocities. But unfortunately this is not the language that muscles understand: they think in terms of *force*, something that we do not usually want to know about.[10] In general terms, 'load' means the force against which a muscle has to operate. How much a muscle moves when a command is sent to it from α-motor neurons depends essentially on the difference between the resultant force of contraction and the load on the muscle. So we cannot tell how much movement will occur in response to a given command unless we know the load as well. Ultimately, what the motor system has to do is to create the best possible estimate of the force it needs to generate at every moment in order to achieve what it wants to do - lifting, moving, manipulating. To do this, it needs as much information as possible – and as quickly as possible – about its own muscles and joints, and – from the skin – about the object with which it is interacting: load and slippage (**Fig. 10.18**).

Now if all we used our muscles for was supporting our own weight and moving around, then these loads, if not constant, would at least be relatively predictable, and one could imagine the brain using feed-forward information to predict what α commands would be needed in any particular situation to achieve a particular movement. But – especially for humans, for whom these skills underpin their intellectual supremacy – if we are to manipulate objects in the outside world, lifting them, moving them, throwing them and so on, the loads we encounter will not

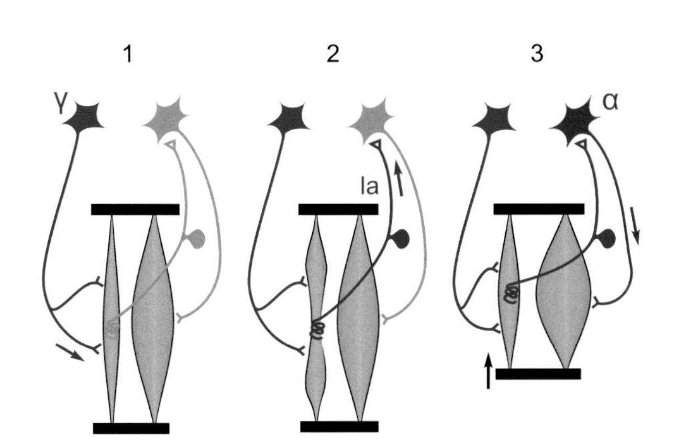

Figure 10.16 Reflex contraction via γ-fibre stimulation. (1) Activation of γ fibres stretches sensory endings in spindle, leading to (2) excitation of spindle afferents, which in turn cause (3) excitation of α-motor neurons and contraction of extrafusal muscle fibres.

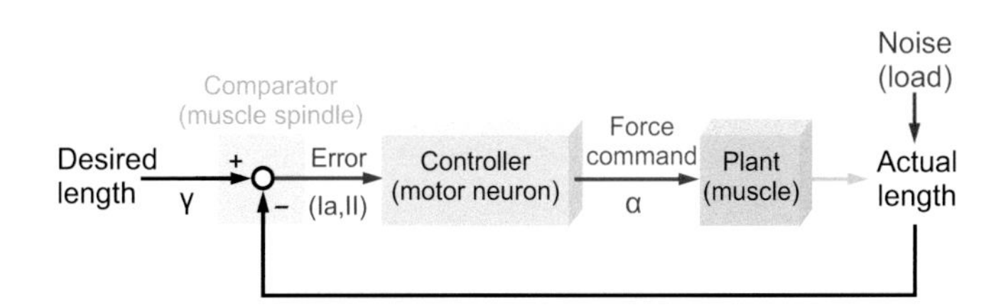

Figure 10.17 How the spindle might function as a comparator in a simple servo system in which a muscle's length is automatically made to conform to the desired length signalled by the γ-efferents.

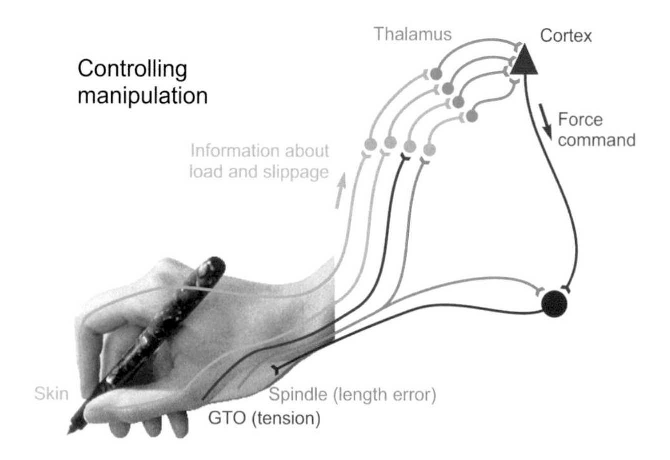

Figure 10.18 Manipulation relies on sensory information from muscle, joints and skin.

be under our control. While some feed-forward is still possible – we can estimate how heavy a case of wine is before we lift it – in general the need to cope with variable load creates a huge control problem.

Consider for example the challenge posed by holding out a cup while someone is filling it with tea: clearly, one's task here is to keep the various muscles concerned at a constant length, despite the fact that the force required to do this is continually increasing as the load – the amount of tea in the cup – gets bigger. Now one could of course imagine the brain continually monitoring the situation, and deciding at every instant exactly how much direct α-excitation to send down into the cord to keep the hand steady. But how much simpler it would be just to send, once and for all, a message indicating not the force needed but the *desired position of the hand*, and leave the spinal cord to get on with the job of adjusting the force to the load automatically, by sensing the extent to which the actual position of the cup matches the brain's command. In other words, such a feedback system would provide load compensation, by acting as what is sometimes called a follow-up servo: the main muscles simply act as slaves that follow any length changes signalled to the intrafusal fibres. The problem the motor system is faced with is how to work out what force is needed to produce a certain position with a given load: the beauty of the simple servo model is that it saves the brain having to worry about such matters at all. It can think simply in terms of the desired effect, and leave the lower levels to get on with the humdrum task of working out how to achieve it.[11]

Granted that the γ fibres could in principle initiate movements of limbs on their own, is there any evidence that they in fact do so? For certain types of movement, the answer is a clear yes. **Figure 10.19** shows an extremely elegant experiment that establishes this beyond dispute, at least in particular circumstances. Here recordings are made of afferent activity from muscle spindles in the leg of a decerebrate cat (black traces), while the head is moved rhythmically up and down. Through the vestibulo-ocular reflex, this causes reflex changes in the tension in the leg muscles (coloured traces). At the same time, the frequency

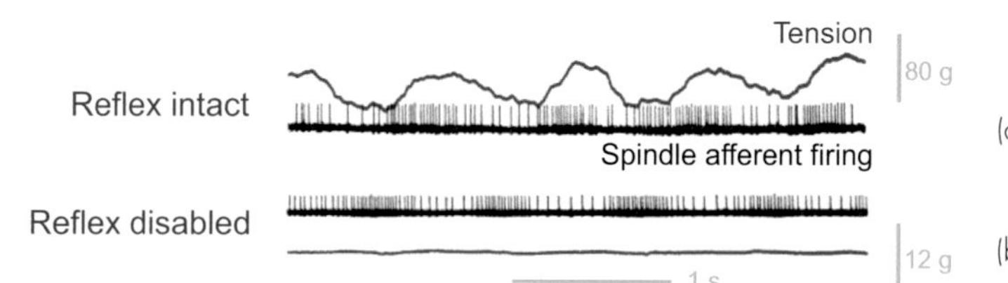

Figure 10.19 Demonstration that γ-fibres may be used to generate reflex movements. (a) Tension in decerebrate cat soleus muscle and associated firing of a spindle afferent during reflex contraction evoked by head movement. (b) After cutting the dorsal root, no contraction occurs, yet modulation of spindle discharge is much as before: this can only be through γ-activation. (After Eldred *et al.*, 1953; with permission from John Wiley and Sons.)

of firing is modulated in time with the head movement: this in itself proves nothing, since it could merely be the result of changes in length of the muscle, rather than activation of γ fibres. But if now the dorsal roots are cut, preventing the stretch reflex from operating, two facts are immediately obvious (bottom pair of traces). First of all, the limb activity is abolished. This shows conclusively that the original movement could *not* have been caused by descending pathways acting on a motor neurons, for these would not be affected by dorsal root section: the response must have been due to the stretch reflex. Second, the spindle discharge is still modulated by the stimulus. Since the muscle length is no longer changing, the only possible way in which this modulation can be taking place is by varying activation of the γ fibres. In other words, it is quite certain in this case that the movement is indeed initiated by γ-activation of the stretch reflex servo and not by descending a commands. Of course, a decerebrate cat with its dorsal roots cut is hardly in a physiological condition, and these observations concern only one kind of reflex activity, driven by the vestibular system. Nevertheless, it proves beyond doubt that some movements are indeed driven by a simple servo mechanism.

However, **Figure 10.20** shows a pair of representative records from an equally elegant and conclusive experiment, looking not at reflex vestibular activity but at the effects of a human voluntary contraction. Here, the activity in Ia fibres during voluntary movement of the wrist was recorded (black traces), together with the electromyogram from the muscle itself (colour). Again, there are two clear conclusions. First, the electromyogram starts at the same time as, or slightly before, the activity in the Ia fibres. This completely rules out the idea that – as in the simple servo – the contraction could be caused by the afferent Ia activity. Therefore this movement must have been initiated

Figure 10.20 Records of afferent Ia discharge and electromyogram (EMG) during voluntary human wrist movements. The Ia discharge clearly does not precede the muscle activity, as would be expected if the muscle were only driven by a servo system like that in **Figure 10.16**. (After Vallbo, 1971; with permission from John Wiley and Sons.)

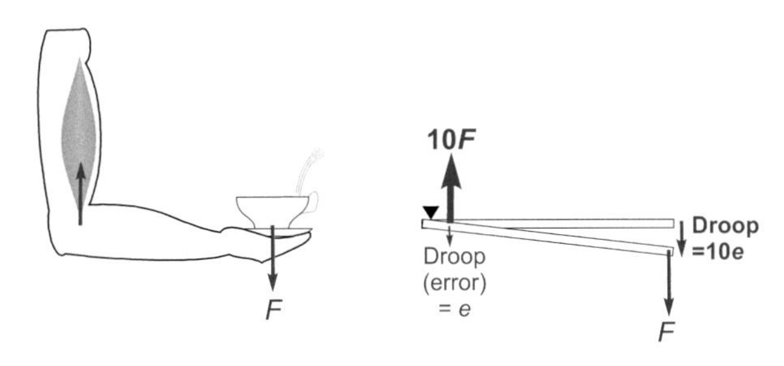

Figure 10.21 The problem of holding a cup. Taking the mechanical disadvantage of the lever system in the forearm as 10, a weight F in the hand must be balanced by a tension $10.F$ in the muscle. At the same time, an error e in muscle length will result in a descent $10.e$ by the cup. Thus, if the static gain of the stretch reflex is G, the cup will fall by $100/G$ for every unit of weight.

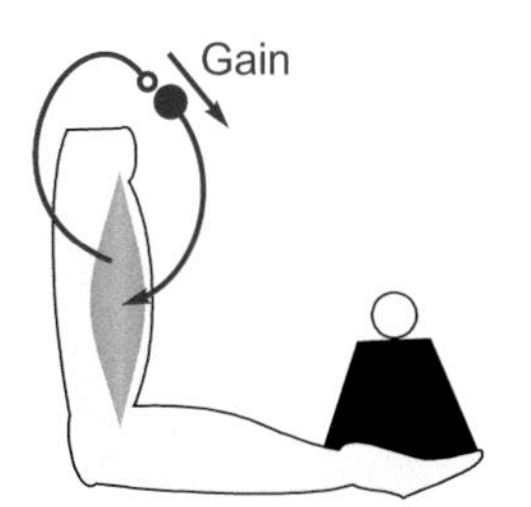

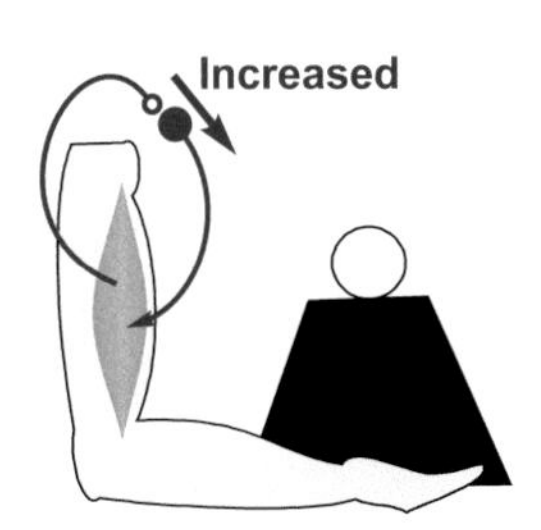

Figure 10.22

by direct activation of α-motor neurons. Second, although it is shortening that is taking place – which by itself would of course reduce Ia activity – nevertheless there is an increase in spindle discharge. The only possible explanation is that stimulation of γ fibres must be taking place as well as α. So while γ-initiation seems to occur for certain postural responses, this is almost certainly not the case for ordinary, willed movements.

Further, there is a theoretical consideration that casts doubt on the possibility of driving real movements solely by means of γ-activation, and that is the size of the *gain* of stretch reflexes; 'gain' here means how much extra tension is generated by a given error in length. Returning to the problem of holding out the teacup, it is a relatively simple matter to work out what gain this reflex would need to have in order to perform adequately (**Fig. 10.21**). From our definition of the gain, G, it follows that for every degree of error e (that is, for every value of the difference between actual and desired muscle length) there is a corresponding reflex force F that is developed by the muscle, where $F = e.G$. So when a muscle is supporting a load F, its actual length will be slightly greater than what is desired, by an amount $e = F/G$. In the case of the muscle in **Figure 10.15**, this means that a load of 500 g would cause an error of 1 mm, the stretch required to generate the 500 g needed to sustain the load. Now most muscles, because of the way they are attached to their bones, work under a considerable mechanical disadvantage: in the case of the human biceps, every kilogram of load on the hand results in some 10 kg of tension in the muscle; and conversely, a muscle movement of 1 mm moves the hand by some 10 mm. If for the sake of argument the gain of the stretch reflex being used to hold the teacup was also 500 g/mm, then every 50 g of load in the hand will cause a muscle extension of 1 mm, and the hand will move some 10 mm. In other words, the extra 300 g or so produced by filling the cup would be expected to make the hand droop by no less than 6 cm!

Somehow, the force generated by the system must be automatically increased in step with the changed requirements as the weight to be supported increases (**Fig. 10.22**). If, for instance, the force was always increased in such a way as to be *proportional* to the load, then the system would perform equally well, regardless of how big the load was. How could this information about load be obtained? It could, for instance, come from pressure receptors in the hand. Experiments carried out in human subjects suggest that something of the sort does indeed occur, and that pressure information does indeed modify the force generated in the stretch reflex. A subject was required to use his thumb to push a lever at a constant velocity against a fixed resistance: he was provided with visual feedback from the lever to tell him how well he was doing. Recordings of his electromyogram (**Fig. 10.23**) – averaged over a number of trials – show a steadily rising averaged activity during the course of the movement, as the muscle shortens. If now, without the subject's knowledge, a stop is introduced into the apparatus that prevents the lever moving past a certain point, one finds that after a short latency the EMG quickly rises, reflecting the subject's effort to overcome the unexpected obstacle. In fact, the latency is too short to be due to a conscious decision of this kind. A simpler explanation is that because of the stop there develops an increasing error between desired and actual thumb position – the former increasing steadily, and the latter having stopped – and this causes a stretch reflex. What is found is that if the experiment is repeated with different degrees

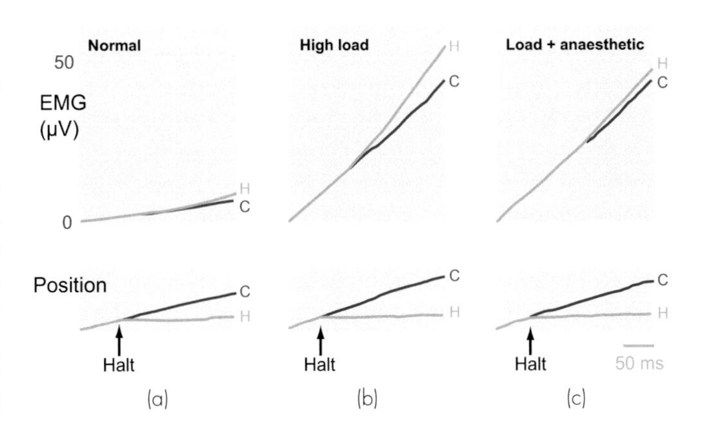

Figure 10.23 Evidence for variable forces produced by stretch reflex. (a) A subject moves his thumb (lower trace shows its position) against a steady load so as to track a uniformly moving target; the resultant steady increase in electromyogram (EMG) is shown above (C = control). If a stop is now introduced at the point marked Halt, after a latent period the EMG starts to rise more rapidly (H = halt trial). (b) If the load against which the thumb is pushing is increased by a factor of 10, the EMG in response to the error also increases by nearly the same factor. (c) If the hand is then anaesthetised, almost complete abolition of the stretch reflex results. (After Marsden *et al.*, 1972; with permission from Springer Nature.)

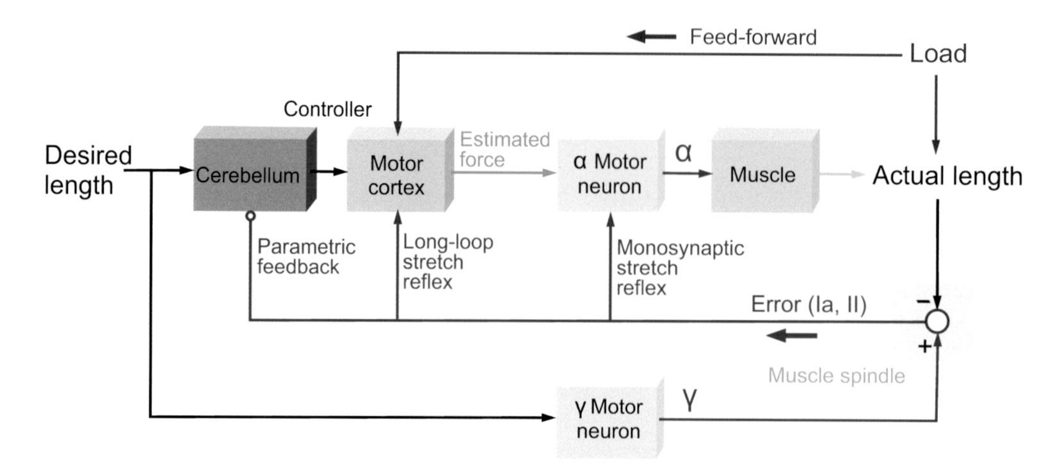

Figure 10.24 Hypothetical scheme of a servo-assisted control system for muscle length. Spindles provide an error signal (discrepancy between actual and desired length) that is used directly in the stretch reflex to correct the response, and also indirectly to modify the ballistic programs for future action. The brain signals to the cord not only desired length but also an estimate of the force needed to achieve that length, derived partly from stored programs embodying previous experience, and also partly from immediate information from sensory receptors about the load that is present.

of resistance to the thumb movement – with different loads, in other words – the force generated by a given error increases with increasing load in any particular trial (**Fig. 10.23b**). That this change is caused by pressure receptors in the skin is suggested by the fact that if the thumb is anaesthetised the extra force drops nearly to zero (**Fig. 10.23c**) – the stretch reflex has been almost completely abolished. The latency suggests that what is being modified is not the monosynaptic reflex but the slower long-loop reflex that probably passes through the motor cortex; it seems likely that the cutaneous signals that are also conveyed to the motor cortex are here used to modify the size of the response, perhaps by recruitment.

Servo-assistance

One is forced to conclude that in the control of movements there are two separate signals or commands that are sent to the spinal cord by the brain. One is a *position* command, that tells a muscle, via the γ fibres and muscle spindles, what length it is meant to be; the other is a *force* command, an estimate of the load that is going to be encountered. In the case of the teacup, load information could be obtained from receptors in the skin, or for that matter from Golgi tendon organs, but many tasks are more ballistic in nature and require anticipation in advance of what the load is likely to be. In such cases past experience and the more sophisticated use of special senses like vision may be brought into play as well. When we go to pick up a sack of potatoes as opposed to a sack of waste paper, or when we fling open a swing door with which we are familiar – too little force and we walk into it, too much and we smash it – we have clearly estimated beforehand what the likely load will be and thus what force is required. This notion of simultaneous force and length command is sometimes called *alpha/gamma co-activation;* if the estimate of force is an accurate one, then the system behaves, in effect, ballistically, and there is no error for the spindles to have to correct. So the job of the stretch reflex is now simply to deal with any residual errors left over after the estimated force is put into operation; it is not expected to provide the whole force necessary for the job, which we have seen it is too feeble to provide.[12]

A system of this kind is known as a *servo-assisted* system and may be represented by an arrangement like that of **Figure 10.24**. Spindles provide an error signal (discrepancy between actual and desired length) that is used directly in the stretch reflex to correct the response, and also indirectly to modify the ballistic programs for future action. The brain signals to the cord not only desired length but also its estimate of the force needed to achieve that length, derived partly from stored programs embodying previous experience, and also partly from immediate information from sensory receptors about the load that is present. It is really a ballistic system with a safety net provided by a rapid backup guided or direct feedback system. Thus a patient whose spindle afferents from the hand have been destroyed by disease may perform quite well in skilled movements where the loads are known in advance, for example in handling a coffee mug that he is familiar with, but will come to grief as soon as any kind of unexpected resistance or variation in load is encountered.[13]

In the servo-assistance model the spindles play two distinct roles. In the first place, through the stretch reflex, they provide *immediate correction* of any errors in the estimate of force. The Ia fibres are ideally suited to doing this, both because they are large and therefore fast, and also because they respond partly to length and partly to rate of change of length. As we saw in Chapter 9 (p. 193), this means that their firing effectively represents an estimate of what the length of the muscle *will* be in about 20 ms time, helping to prevent the hunting that would otherwise occur because of delay round the reflex loop. In the second place, spindles supply *parametric feedback* that in the long term can make future corrections of the estimates themselves. We shall see later that there is reason to think that the origin of the force command may be the cerebral cortex, via the corticospinal tract, and that it may be the cerebellum, which is richly supplied with afferents from muscle spindles (unlike the cortex) that is the site of the motor program store. These ideas will be discussed later, in Chapter 12.

Finally, it is perhaps worth mentioning that the muscles of the eye, though richly endowed with stretch receptors and γ-motor fibres, show no stretch reflexes whatever. It seems in this case that because of the predictable relation between force commands and resultant eye position emphasised earlier, errors arise so seldom that no short-term correction mechanism is needed. It seems very likely that the sole function of these spindle afferents is to provide parametric feedback in order that the essentially ballistic control of such movements as saccades may in the long run be performed accurately. Though ocular spindles fail to generate stretch reflexes, where they do project in great numbers is the cerebellum, a region – as we shall see in Chapter 12 – known to be associated with parametric feedback and other kinds of motor learning.

Notes

1. The final output *'To move things is all mankind can do, and for such the sole executant is muscle, whether whispering or felling a tree'.* (Charles Scott Sherrington, Man on his Nature.) But Bernard Shaw saw it differently: *'What made this brain of mine, do you think? Not the need to move my limbs; for a rat with half my brain moves as well as I'.* (George Bernard Shaw, Man and Superman.)

2. Muscle Discussion of the functional properties of muscle itself is beyond the scope of this book. A clear and stimulating account is McMahon, T. A. (1984). *Muscles, Reflexes and Locomotion.* (Princeton, NJ: Princeton University Press). There are also excellent accounts in Rothwell, J. C. (1994). *Control of Human Voluntary Movement.* (London: Chapman & Hall) and Cody, F. W. J. (ed.) *Neural Control of Skilled Human Movement.* (London: Portland Press, on behalf of the Physiological Society). Some clinical aspects are addressed in Shumway-Cook, A. & Woollacott, M. H. (2016). *Motor Control: Translating Research into Clinical Practice.* (London: Lippincott Williams & Wilkins).

3. Recruitment The result is that muscle tension is a sort of accelerating function of degree of afferent excitation, often referred to as the *size principle*; slow (S) motor units are the first to become active, followed by fast fatigue-resistant (FR) and finally the fast fatigable (FF) motor units. The ensuing increment of tension per unit extra-stimulation increases with the tension. This is obviously reminiscent of the Weber–Fechner law (see p. 150), that $\Delta I/I$ is constant.

4. Tetanus The importance of Renshaw cells in preventing synchrony in the alpha motor neuron pool is exemplified by this condition; they are preferentially targeted by the tetanus toxin. Spores from the soil-dwelling bacterium *Clostridium tetani* gain access via skin wounds, germinate, and then produce the toxin. The result is widespread activation of muscles, starting in the jaw ('lockjaw') before progressing to abdominal, limb and respiratory muscles. It can be treated with antibiotics and tetanus immune globulin (which mops up the circulating toxin), but prevention is better than cure, highlighting the importance of tetanus vaccination following potentially contaminated wounds.

5. Reflexes Two classic accounts of reflexes in general: Creed, R. S., Denny-Brown, D., Liddell, E. G. T. & Sherrington, C. S. (1932). *Reflex Activity of the Spinal Cord.* (Oxford: Oxford University Press), and Sherrington, C. S. (1906). *The Integrative Action of the Nervous System.* (New Haven, CT: Yale University Press).

6. Going in all directions The quotation is from Brodal, P (1998). *Neurological Anatomy.* (Oxford: Oxford University Press).

7. Antiquity of the reticular formation In the 5 cm long lancelet *Amphioxus* (a precursor of true vertebrates), all the descending control of swimming movements is via the reticular formation; however, it cannot really be said to have a brain, the central nervous system being only slightly enlarged at the front end (and it has no eyes).

8. Spinal preparation Sherrington's description of the state of spinal shock: *'There can hardly be witnessed a more striking phenomenon in the physiology of the nervous system. From the limp limbs, even if the knee jerks be elicitable, no responsive movement, beyond perhaps a feeble tremulous adduction or bending of the thumb or hallux, can be evoked even by insults severe in the extreme . . . a hot iron laid across thumb, index and palm remains an absolutely impotent excitant . . . A more impassable condition of block, or torpor, can hardly be imagined'.* In Denny-Brown. D. (ed.) *Selected Writings.* (1939). (London: Hamish Hamilton, p. 121).

9. Muscle spindles The classic account, thoughtful and comprehensive, is Matthews, P B. C. (1972). *Mammalian Muscle Receptors and their Central Actions.* (London: Arnold).

10. Not commanding position An analogy to this frustrating state of affairs is the driver of a horse and cart, who can tell the horse to start or to stop, but not *where*.

11. Thinking in terms of results A rather nice demonstration that broadly supports such a view is to have a subject first write their name and address in the usual way, with movement at the wrist, then (on a larger scale) with the wrist immobilised and using the elbow instead, and then finally with the whole hand and arm rigid, and the movement occurring at the shoulder. Despite the novelty of the second two tasks, and the fact that entirely different muscles are being used, the characteristics of the 'hand' writing are essentially retained.

12. Co-activation If commands are often being sent simultaneously to alphas and gammas, one might perhaps expect some efferent fibres to innervate both intrafusal and extrafusal fibres: and this is in fact the case. These fibres are called β-axons (though they are the same size as alphas), and in amphibia they appear to be the only motor innervation of spindles. This is not a useful arrangement when much variation of load can be expected, since the whole point of separate α and γ innervation is to be able to have separate control of force and length. But if the only load the muscle ever experiences is the frog's own weight, it could work.

13. Deafferentation See for instance Marsden, C. D., Rothwell, J. C. & Day, B. L. (1984). The use of peripheral feedback in the control of movement. *Trends in Neuroscience* **7** 253–257. Taylor, A. & Prochazka, A. (1981). *Muscle Receptors and Movement.* (London:

MacMillan) is a useful account of the use of muscle feedback in general. Cole, J. (1991). *Pride and a Daily Marathon*. (London: Duckworth) is a popular account of a patient with an unusually complete loss of afferent innervation, who has to carry out consciously, under visual guidance what most of us take utterly for granted: *'I'm going to buy a tin of beans in Sainsbury's. First, I have to pick a safe route down the aisle, because if someone brushes against me, and I haven't allowed for it, then I am thrown off balance. I reach the shelf and check my body position in space. I then lock my legs and focus on the tin. Then monitoring my arm movements all the time, I move it towards the tin. I grasp the tin – I don't know how much pressure to exert and I can't assess weight. The only way I know if something is too heavy is if I topple forwards'*. In addition, lack of sensory feedback can lead the owner to feel alienated from the part affected, as brilliantly described in Sacks, O. (1993). A *Leg to Stand on*. (New York, NY: Harper).

CHAPTER 11

THE CONTROL OF POSTURE

Movement begins and ends in posture: for most of the time, the motor system is not concerned with moving the body at all, but rather with keeping it still. This is especially true in humans, with our precarious twin supports needing constant motor commands to keep us upright against the force of gravity. One might perhaps think that this was merely a matter of keeping sufficiently rigid, once a stable balancing position had been found. But our centre of gravity is so high off the ground that this kind of passive stability is not enough: one need only compare the ease with which one can push over a tailor's dummy with the near impossibility of doing the same thing to a living person to realise that stability must involve active processes as well, which use proprioceptive feedback information.

The importance of support

In physical terms, whether someone – or something – falls over or not is entirely a matter of the vertical projection of his centre of gravity relative to their supports (**Fig. 11.1**). If this line of projection lies within the area defined by the points of contact with the ground (the support area), then all is well – small disturbances will result in a turning couple tending to restore the status quo. If it lies outside this critical area, then the system is unstable, and any further tilting will cause an ever-increasing rotational force that will make them fall over. Because the support area is much smaller in humans than in four-footed animals, maintaining an upright posture is correspondingly more difficult: a tilt of only a few degrees is sufficient to cause instability. Thus proper standing is not, as is often implied, just a matter of keeping upright: the man in **Figure 11.2** clearly has an excellent upright posture, but is equally clearly about to experience a postural disaster, because the vertical projection of his centre of gravity lies outside his region of contact with his support. Support is in every sense fundamental to posture, which must be controlled either by moving the centre of gravity relative to the feet or moving the feet relative to the centre of gravity.

Two sources of information enable us to do this. The first is the existence of pressure receptors in the feet themselves that provide information about the distribution of support. Knowledge of differences of pressure at different points of support (orange arrows in **Fig. 11.1**) tells us precisely what we need to know to determine our postural state, namely the position of the vertical projection of the centre of gravity relative to the body's supports. The second

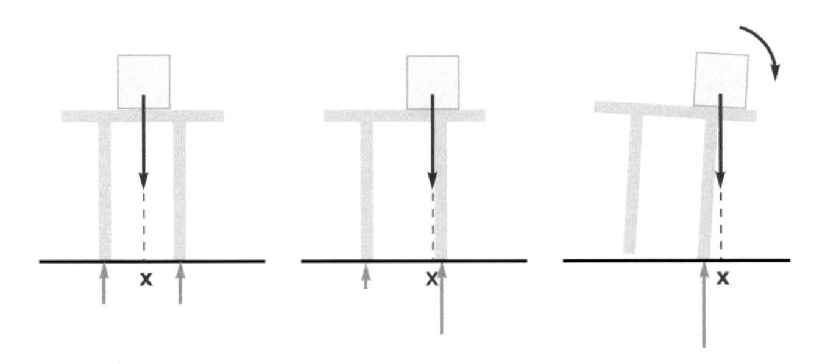

Figure 11.1 A sufficient condition for postural stability is that the vertical projection of the centre of gravity on the floor (X) should lie within the area defined by the supports. The distribution of pressure between the supports (small arrows) provides sufficient information to determine the position of X.

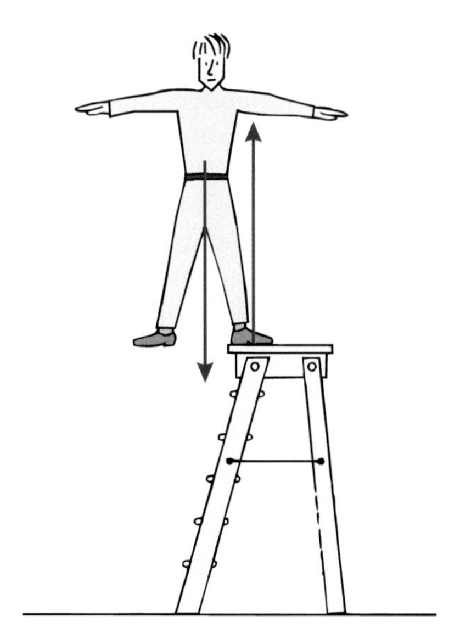

Figure 11.2 A splendidly upright but nevertheless unsatisfactory posture.

source of information is in the head: those special senses that can tell us about the position and motion of the head relative to the outside world, the *vestibular* and *visual* systems. It is convenient to consider the use of information from the feet first, since this is a more direct process than the contribution of the special senses.

Responses to pressure distribution

Imagine that we had to design an automatic system for keeping a lunar module standing upright on uneven ground. If the craft leans in the direction of one particular leg, then that leg will experience more force. So a good strategy would be to have a rule that any leg that experienced more pressure than the others was automatically extended, while any that experienced less was shortened. The result would be a system that would always bring the vertical projection of the centre of gravity to the middle of the critical area defined by the points of support. It is not difficult to demonstrate a precisely analogous mechanism in four-footed animals: if we suspend a decerebrate animal in the air with its legs hanging down, and push up on the sole of one of its feet, the animal responds by extending the corresponding limb. This tonic response is called the *positive supporting reaction;* there is also a transient component of the response called *extensor thrust,* and it is often accompanied by stiffening of the limb and arching of the back (**Fig. 11.3**). A variation is the *buttress reaction,* often observed on trying to pull a reluctant dog forward, increasing the differential pressure at the front: the extension of the front feet is obvious. The same mechanism may be seen at work in the *postural sway reaction:* if an animal's body is pushed from the side, the shift in pressure on the feet results in marked extension of the limbs on the opposite side and retraction of the others, so that the animal in effect leans against the experimenter.

It is easy to see how this mechanism would act to increase postural stability, as for example when the animal is standing on sloping ground (**Fig. 11.4**). On the level, there is a roughly even distribution of pressure amongst the four feet, and hence no tendency for one leg to lengthen more than another. But if the animal is facing up a slope, the pressure on the back feet is greater than on the front, and consequently the positive supporting reaction will result in rear limb extension and front limb flexion: the final result will be that the body adopts a more horizontal posture, and the projection of the centre of gravity is brought more nearly to the middle of the points of support. It seems likely that analogous mechanisms may be used when sitting, kneeling, lying down and so forth, involving information about differences of pressure on different parts of the body other than the feet. A blindfolded animal, whose vestibular system has been destroyed (leaving cutaneous receptors as the only remaining source of postural information) will nevertheless right itself when laid on its side on the ground. If a plank is laid on top of it that reduces the ratio of the pressures experienced by the two sides of the body, this body-righting reaction is inhibited,

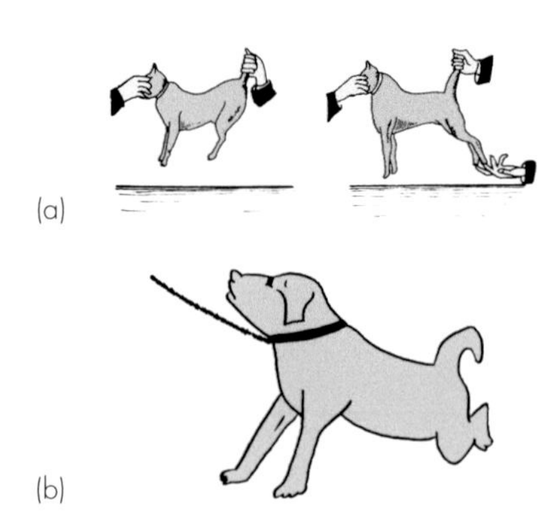

Figure 11.3 (a) Positive supporting demonstrated in a decerebrate dog suspended in the air, on making contact with the back paws. (After Walsh, 1964.) (b) Typical 'buttress reaction' on trying to pull a dog forward and thus increasing the differential pressure on the front foot; note obvious front-leg extension (From Rademaker, 1935.)

a phenomenon said to be used sometimes by veterinary surgeons to help restrain an animal for operation.

If for some reason the projection of the centre of gravity has moved outside the critical area, then automatic *stepping reactions* are elicited that in effect move the feet in such a way as to track the centre of gravity: if one happens to be standing on one leg, the result is a *hopping reaction.* One can demonstrate these responses quite easily to oneself by first standing upright and then trying to fall over deliberately by leaning over: at some point, however hard one tries not to, reflex stepping or hopping comes into play and actual falling is prevented. Incidentally, if you try to fall over backwards in this way, you will also observe an involuntary upward flexion of the feet, as expected from the positive supporting reaction – though in these circumstances it is of no use whatever! It is also interesting to note that whereas in walking one is for most of the time in postural equilibrium – in the sense that one may 'freeze' at nearly any point of the walking cycle without falling over – this is not the case in running, when the centre of gravity is normally ahead of the support area. One can think of running as being a series of regular and almost unconscious stepping reactions in response to the bent-forward posture that the runner maintains.

The mechanisms described so far assume that some sort of support is already present: there are other types of

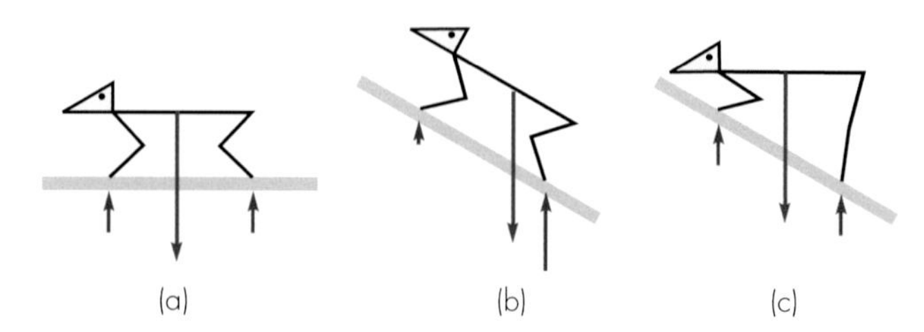

Figure 11.4 How the positive supporting reaction results in good posture. (a) Animal standing on level ground, with projection of centre of gravity centrally placed between the supports. (b) Facing up a slope: the weight is now unevenly distributed between the front and back legs, resulting in (c) front-leg flexion and back-leg extension, and a better position of the projection of the centre of gravity.

response that may be used to *find* postural support. Responses called *placing reactions* are used for acquiring postural support when none is present. If a blindfolded animal is suspended in the air and brought up to a table until the edge touches the backs of its paws, it will bring them smartly up to rest on the top of the table, in turn evoking a positive supporting reaction: a similar response may be triggered if the animal's whiskers are brought to touch the table. These responses are, however, rather more complex then those described above, and probably involve the cerebral cortex: unlike the stepping and supporting reactions, they cannot be shown in decerebrate animals.

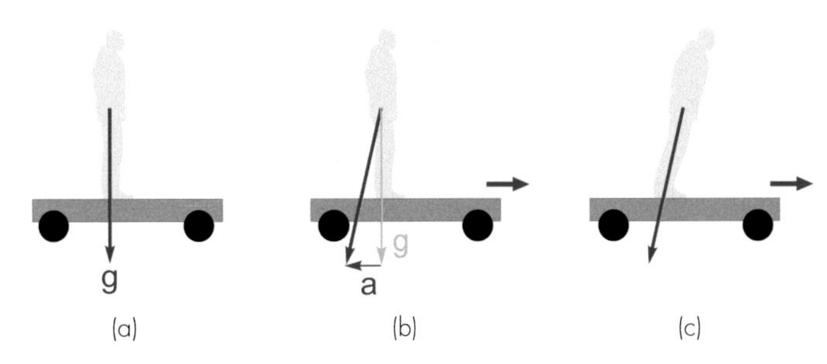

Figure 11.5 Effective direction of gravity, and true vertical, defines a good upright posture. (a) Man standing on a stationary platform, with vertical acceleration *g* due to gravity. (b) If the platform accelerates horizontally, the effective direction of gravity, as sensed by his otolith organs, is the vector sum of *g* and *a*, the acceleration of the platform. (c) It is the projection of the centre of gravity along this direction that must now be brought between his feet if he is not to fall over.

Vestibular contribution to posture

The vestibular apparatus was described in Chapter 5. We saw that it provided two separate types of information about the head: angular velocity from the semicircular canals, and attitude relative to the effective direction of gravity, from the otolith organs. From the point of view of the control of posture, it is, of course, the *effective* direction of gravity rather than its 'real' direction that matters. What determines, for instance, whether one falls over when standing in a bus that starts to accelerate is not the vertical projection of the centre of gravity vertically relative to the critical area, but rather its projection in the direction of the vector formed by gravity (g) and the horizontal linear acceleration (a) acting together (**Fig. 11.5**), and this is what the utricle and saccule tell us.

Each of these two divisions of the vestibular system gives rise to its own kinds of postural reactions, and so it is useful to distinguish between the *static* or *tonic* vestibular responses due to the otolith organs and the *dynamic* or *phasic* ones driven by the canals.

Static vestibular reactions

On the whole the otolith organs produce rather less powerful postural responses than do the canals, particularly in higher animals. But in the long run they are the *only* source of information about the absolute position of the head in space, since the canals essentially signal only changes of position. One of their main functions is in fact to keep the head upright despite changes in the position of the body, through appropriate changes in the tone of the neck muscles; these are the *head-righting reflexes*. If the head is forcibly tilted in different directions, so that the head-righting reflexes cannot operate, one can also observe compensatory *static vestibulo-ocular reflexes* that similarly help to maintain the normal attitude of the eyes with respect to the outside world. In humans these eye reflexes cannot easily be demonstrated, and if the head is tilted to one side the resultant counter-rolling of the eyes is seldom of more than a few degrees and so cannot maintain the correct orientation of the retinal image. But in animals such as the rabbit whose eyes essentially point sideways, tilting the head results in almost exact compensation over a wide range of angles of tilt (**Fig. 11.6**): they can continue to check the horizon while nibbling the grass at their feet. While both these types of response obviously aid the sensory organs of the head by providing a stable 'platform' from which to operate, they are not of course strictly postural in the sense of contributing to the maintenance of equilibrium of the body as a whole.

Reactions which are truly postural in this sense may be quite easily demonstrated in conscious animals, and are called *tonic postural vestibular reflexes*. If an animal is suspended in the air, and its head and body tilted nose downwards, one observes extension of the front legs and retraction of the rear (**Fig. 11.7**). Corresponding limb movements are found if the animal is tilted in other directions, for example to the side: in each case, there is extension of the limbs in the direction of downward tilt and retraction of the others. The function of this response is clear: if the animal is facing down a slope, this tonic vestibular response will assist the positive supporting reaction in shifting the centre of gravity backwards in relation to the feet, as may be seen in the human subject in **Figure 11.7**.

Dynamic vestibular reactions

Because the canals are velocity sensitive, they effectively give advance warning that one is *about* to fall over, possibly before the otolith organs have sensed that there is yet an

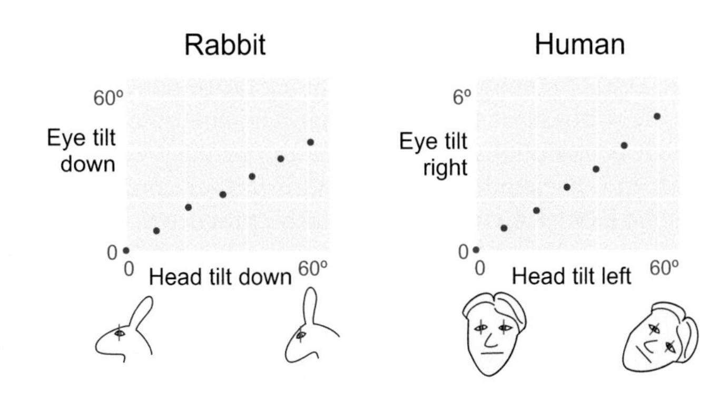

Figure 11.6 Static vestibulo-ocular reflexes. In the rabbit, with its sideways-pointing eyes, vestibulo-ocular compensation is substantial over a wide range of angles; but in humans, head tilt produces only a few degrees of ocular counter-rolling. (Data from de Kleijn, 1921; with permission from Springer Nature.)

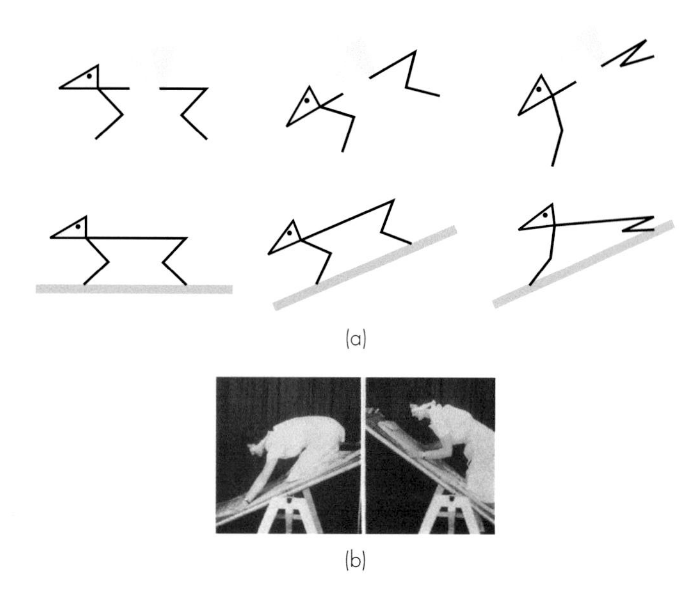

Figure 11.7 Static vestibular righting reflexes. (a) Tilting an animal's head and body nose-down results in front-leg extension and rear-leg retraction. This response helps to produce a good posture when standing on a slope. (b) A human subject under similar conditions. (Martin, 1967.)

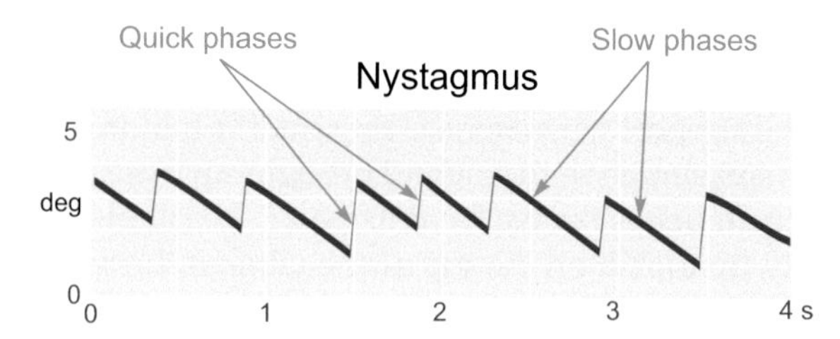

Figure 11.8 A record of human vestibular nystagmus, showing slow and quick phases.

actual error in head position. Perhaps for this reason, their responses are particularly fast, and generally bigger and more dramatic than the static vestibular reactions. The types of response they generate fall essentially into the same categories as the tonic ones: thus there is a dynamic component to the head-righting reflex that may be elicited by selective stimulation of the canals alone, and one may also demonstrate clear effects of canal stimulation on the eyes and limbs. If we seat someone on a rotating chair and record their eye movements (in the dark, so that there is no visual input to the oculomotor system), we find that the eyes move in the opposite direction to that of the head with a velocity that compensates almost exactly for the rotation, keeping the eye stationary with respect to the outside world. Clearly this dynamic *vestibulo-ocular reflex* (VOR) cannot go on forever, since sooner or later the eyes are going to reach the limit of their rotation in the orbit: what in fact is observed is that the smooth counter-rotation in one direction is interrupted at more or less regular intervals by a quick flick in the other direction, giving rise to a sawtooth-like eye movement called *vestibular nystagmus* (**Fig. 11.8**). The smooth, compensatory, movement is called the slow phase of the nystagmus, and the quick flick – which is essentially the same as an ordinary voluntary saccade – is called the quick phase (rather confusingly, it is the latter which is used in clinical practice to describe the direction of nystagmus, so that a subject turning to the right produces what would be called a nystagmus to the right, even though the more important, compensatory, component of the response is to the left). This allows the observer to maintain a stable visual field during the slow phase, with vision being only briefly compromised during the quick phase.

During rotation of this kind at constant angular velocity the response declines over a period of some 20 seconds or more, on account of the natural adaptation of the canals (see Chapter 5), and so does the velocity of the slow phase. If the chair is suddenly stopped, and with it the semicircular canals, the fluid tends to continue. As a result the cupula is deflected in the opposite direction, and a corresponding reversed nystagmus is seen, which in turn declines with a time-course of some 20 seconds as the fluid ceases its motion; these two types of nystagmus are respectively known as *per-rotatory* and *post-rotatory* nystagmus. Vestibular nystagmus provides a convenient means for clinical investigation of the functioning of the semicircular canals, since the eye movements produced by caloric stimulation (Chapter 5) of each of the two labyrinths may be examined separately to reveal imbalance of function on the two sides.

Finally, dynamic *postural vestibular reflexes* exist, functionally equivalent to the static ones (head moving down gives front-leg extension, and so on), but much more powerful. In humans, unnatural stimulation of the canals may produce inappropriate postural responses that are vigorous enough to throw the subject to the floor – despite the action of all the other postural mechanisms in trying to keep him upright – as for example if one attempts to stand up after having been rotated in a revolving chair with one's head on one side. Because they had adapted, the canals then falsely signal that one is falling over in the opposite direction: the consequent an extremely violent reflexes actually *make* one fall over.

Visual contributions to posture

The other receptors in the head that help maintain posture by providing information about head position are those of the retina. It turns out that there is a close parallel between the ways in which visual and vestibular information about head position are used in postural control, very probably because to a large extent they share common pathways. Once again, one can distinguish tonic effects from dynamic effects, and again one can distinguish effects on head and eye movements from truly postural responses involving the limbs.

Static visual responses

In so-called civilised surroundings, such as an urban street, our visual world is largely made up of horizontal and vertical elements, and our expectations about their orientation

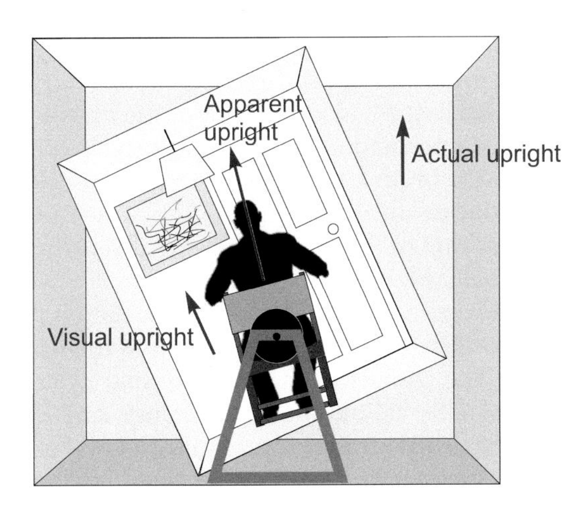

Figure 11.9 Experiment with tilting room and tilting chair, to investigate relative visual and vestibular contribution to the sense of being upright, (After Witkin, 1949.)

mean that we can in principle use our eyes to estimate head position. Many experiments have demonstrated that this tonic visual information is indeed used in making postural judgements and responses, and if subjects are seated on a tilting chair inside a dummy room, which can itself be tilted to various angles (**Fig. 11.9**), it is found that their sense of the upright direction generally lies somewhere between the true upright and the apparent upright of the room. However, it is not entirely obvious that information of this sort was very readily available in the more natural surroundings in which this ability presumably evolved: perhaps it is entirely learnt.

Dynamic visual responses

We have already seen in Chapter 7 that one division of the visual system, the *visual proprioceptive system,* has neurons that are specialised in responding to movement of the retinal image across the retina. If a sufficient number of these cells fire off simultaneously – in other words, if most of the visual field is in motion – the brain assumes, logically enough, that the world is in fact stationary and that it is the eye that is moving. This sense of self-movement is a very powerful one, as anyone who has had the experience of sitting in a train at a station while a neighbouring train moves off will agree. However, misleading circumstances like these are most infrequent in nature, and by and large when most of the visual field moves it is indeed because the head has moved relative to the visual surroundings, providing information that supplements what is provided by the semicircular canals, and is used in the same way to generate postural and oculomotor responses. Experiments in which victims are required to stand in seemingly normal rooms whose walls are then suddenly moved can be made to stumble and fall, even though the ground had remained still. A visit to a 'surround' cinema can be instructive: look at your neighbours during the sickening turns of the switchback ride and you will see them swaying about in helpless unison.

Movement of the visual field also generates eye movements that are extremely similar to the dynamic vestibulo-ocular reflexes. If we seat someone in front of a rotating striped drum, close enough for it to fill a substantial part of the visual field, the result is a sawtooth-like movement of the eyes called *optokinetic nystagmus* (or optokinetic response, OKR), in which the eyes follow the moving stripes during the slow phase, and flick back again during the quick phase. One might think that this was merely the result of a conscious decision by the subject to track the stripes with his eyes, but this is not so, as in fact if the subject tries very hard to keep his eyes still and ignore the movement, the nystagmus is nevertheless still observed, and indeed is in some ways more pronounced and more regular.[1]

How does the eye tell us about head movement?

Now movement of the retinal image does not really tell us about head movement. All it can tell us is that an image of an object in the outside world has moved relative to the retina: we cannot say whether this was because the object itself moved, because the eye moved relative to the head, or because head and eye moved together. We have already seen that movements of large areas of the visual field are generally interpreted by the brain as being due to movement of oneself rather than of the world around us. But from the point of view of controlling posture, it is obviously very important to be able to disentangle the effects of movement of the head from those of movements of the eye, since it is the former that we really want to know about. How is this done?

At one time there were two rival theories. One, due to Sherrington, was that spindles in the eye muscles sent signals to the visual system telling it where the eye was in the orbit; this estimate of eye position relative to the head – call it E_H – could then in effect be subtracted from the estimate of eye position relative to visual space (E_S) provided by the visual system, to generate an estimate of head position in space, H_S (**Fig. 11.10**). Helmholtz, however, argued that it was unnecessary to have receptors in the eye muscles to tell one the position of one's eyes. Since the eyes are never subject to external forces in the way that our limbs are, we can always deduce their position from the motor commands that we send to them, so Helmholtz suggested that a copy of the oculomotor commands – *efference copy* – was sent to the visual system to provide the required estimate of E_H. This theory, called the *outflow* theory to distinguish it from Sherrington's *inflow* theory, is supported by a number of pieces of experimental evidence. One of them is something you can do yourself: if you press on the side of your right eye with the left one shut, you will perceive an illusory movement of the outside world. This is exactly what would be expected from Helmholtz's model, but is not easy to explain with the inflow theory, since the muscle spindles ought still to be providing the necessary information to cancel the signal produced by the movement of the retinal image. More convincingly, if the oculomotor

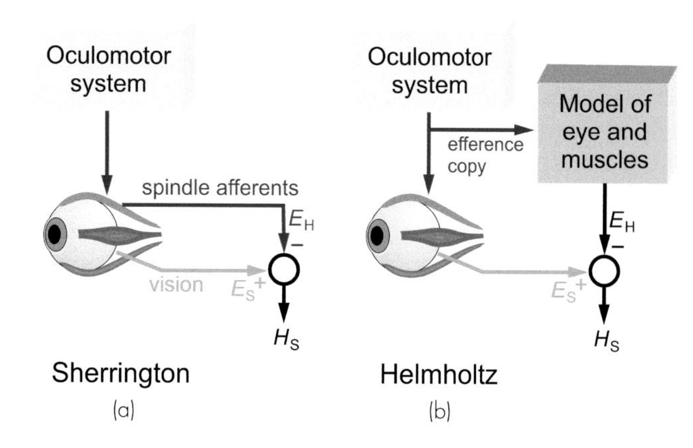

Figure 11.10 The sense of eye position. (a) Sherrington's scheme: proprioceptors in eye muscles provide information about the position of the eye in the head (E_H), and this is then compared with visual information concerning the position of the eye relative to the visual world (E_S) to give H_S, the position of the head in space. (b) In Helmholtz's scheme, a copy of the oculomotor commands is used to calculate, from knowledge of the behaviour of the eye on past occasions, the expected value of E_H; this is then combined with E_S as before, to give H_S.

commands are prevented from carrying out their usual effects on the eye muscles – for example, by some pathological condition affecting the oculomotor nerves or the eye muscles, or by artificial paralysis induced by local application of drugs like curare – then we would expect that every time the subject *tried* to make an eye movement, the E_H signal resulting from the efference copy would no longer be matched by the usual E_S signals from the retina, and the subject should therefore perceive an illusory movement of the visual world in the opposite direction; and this is precisely what is found. It seems therefore that efference copy is indeed the means by which the postural system can work out H_S, the position of the head in visual space, from the purely visual signal *ES*.

Vestibular and visual interactions

We have seen that we have two quite different estimates of the position of the head in space, one provided by the vestibular apparatus, and one by the visual system. These two signals complement one another almost uncannily. The vestibular signal is extremely fast, but not very accurate because the hair cells tend to suffer from low-frequency noise; the visual signals, on the other hand, are much slower (reaction times of the order of 150–200 ms) but potentially hugely accurate and unaffected by noise. How are these two signals combined? And what happens if they conflict with one another?

Recordings from the vestibular nuclei have demonstrated neurons that respond appropriately to optokinetic as well as vestibular stimulation, their response being roughly the sum of the two. In addition, under steady stimulation the optokinetic stimulus increases with a time-course that exactly complements the decline in the vestibular signal, due to adaptation of the canals. We saw earlier, in the tilting room experiment (**Fig. 11.9**), that in fact the brain seems to take something like the weighted mean of these two estimates of head position in arriving at a final answer. This in turn raises the interesting question of how one type of information is *calibrated* in terms of the other. How does the brain know that one particular rate of firing of certain fibres in the vestibular nerve is equivalent to some other rate of firing of a movement-sensitive neuron in the visual system?

The answer seems to be that this correlation of the two inputs is one that is *learnt* by the brain as the result of experience. The signals from the vestibular system are in fact continually being checked against, and calibrated by, the signals from the eye, providing a good example of *parametric feedback* of the kind described in Chapter 9. If the vestibulo-ocular reflex is not working properly, this will cause slippage of the image across the retina when the head moves; this error signal is then used to make corrective adjustments in the vestibular response. Wearing spectacles, for instance, causes a change in the magnification of the retinal image which means that vestibulo-ocular reflexes that were previously just the right size to keep the image stationary despite head movement are now incorrectly matched. But within a very short period of time the reflexes are modified by parametric feedback so that they operate correctly. Even more dramatically, if you wear prisms in front of your eyes that effectively reverse your visual field, so that an object moving from left to right appears to be moving from right to left, after a surprisingly short space of time – a matter of days – your vestibulo-ocular reflexes follow suit by reversing as well, even when measured in the dark.

A similar kind of long-term adaptation can be seen when there is unilateral impairment of the vestibular apparatus. You may recall that vestibular fibres are tonically active, but that central neurons in the vestibular nuclei are excited by afferents from one side and inhibited by those from the other, so that at rest in the normal subject the tonic activity cancels itself out. Unilateral damage upsets this balance between the vestibular signals coming from each side, for the tonic discharge from one side is now unopposed by that from the other. The result – as with unilateral infection (labyrinthitis) – is a false sense of rotation (vertigo) combined with postural reactions and spontaneous vestibular nystagmus (**Fig. 11.11**). However, these effects gradually decline, and in a matter of weeks may have disappeared altogether. This seems to be because the visual system has informed the central vestibular pathways that the signals are false, resulting in some kind of tonic shift of activity that once again cancels out the continual signals from the unaffected side. That this is so may be demonstrated by subsequently cutting the vestibular nerve on the good side. Despite the fact that there is now no vestibular system at all, the result is a nystagmus (*Bechterew* nystagmus) in the opposite direction to the first, presumably resulting from the central tonic activity that had been set up to correct the original imbalance, and this in turn gradually declines. This plasticity appears to be controlled by the cerebellum, the vestibular division of

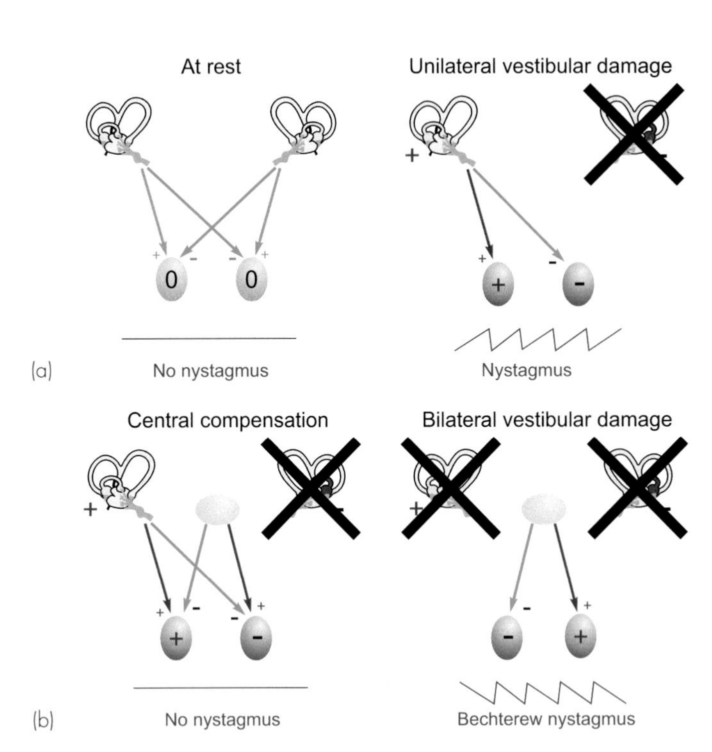

Figure 11.11 (a) Unilateral vestibular pathology results in a mismatch between afferent signals to the vestibular nuclei, simulating rotation in the direction of the intact side. The resultant vestibulo-ocular reflex drives a slow phase towards the side of the lesion and compensatory nystagmoid jerks away from the damaged side. (b) Central compensation gradually follows and eye movements and the sensation of rotation resolves. However, the effect of this compensation is obvious following damage to the remaining vestibular apparatus, manifesting as Bechterew nystagmus.

which (the flocculo-nodular lobe) is supplied with fibres both from the vestibular apparatus and from the visual system. The cerebellar contribution to vestibular adaptation is discussed more fully in Chapter 12, where it will be seen that it is a specific example of a general function of the cerebellum in learning to improve motor responses to stimuli.

Motion sickness

Conflicts between visual and vestibular information about head movement also tend to give rise to *motion sickness*. It is a matter of common experience that motion sickness is most likely to occur when our vestibular system tells us we're moving, but the visual field appears stationary. In a car one may feel perfectly well as long as one is looking out of the window, when visual and vestibular estimates of head movement match, but feel sick when trying to read, when the visual field moves with the head and the two sources of information conflict with one another. One may also feel motion sickness under precisely the opposite conditions. Sitting near the cinema screen, watching *Moby Dick* with its storm-tossed seas and heaving decks, our eyes tell us we are moving up and down, but our canals insist that we are stationary; the result is nausea. Removal of the vestibulo-cerebellum in dogs is said to eliminate motion sickness entirely, presumably because it is in this region that the correlation of the two kinds of input is made.[2]

Incidentally, it seems also that the sickness sometimes associated with alcohol intoxication is also, in effect, a kind of motion sickness. As we saw in Chapter 5 (p. 108), one of the effects of alcohol is to alter the relative density of the cupula and endolymph in the canals, so that they are no longer exactly equal. This means that the cupula is influenced by gravity – it becomes, in effect, an otolith – and is deflected by static head position as well as by angular velocity. Under these conditions one may see what is called a *positional* nystagmus, a nystagmus that occurs spontaneously when the head is held in different positions. It is a matter of common experience that one of the effects of over-indulgence can be a sense of giddiness, particularly on changing head position, and if one goes to lie down there may often be a sensation of the bed turning head over heels.[3]

Neck reflexes

All the kinds of information we have been looking at so far tell us about the position and movement of the head. But is that what we really want to know? It is not the *head* whose orientation we are trying to control, but that of the *body*. Postural control is all about keeping the centre of gravity of the body in the right relationship to its supports, and whether the head itself is upright matters very little. So just as we need to know how the eye is placed in the head before we can use vision to tell us about head position, in exactly the same way we need to know the position of the head relative to the body (H_B) before we can use information about the position of the head in space (H_S) to tell us how our body is situated (B_S).

This signal, H_B, which tells us the relationship between the head and the body, comes from proprioceptors in the neck, mostly joint receptors around the vertebrae. Their effects can be isolated either by destroying the vestibular apparatus or by moving an animal's body while keeping the head still. Since H_B must be subtracted from H_S in order to produce B_S, we would then expect the effects of these neck signals on their own to be as powerful as, and in the opposite sense to, the postural effects of changing the position of the head in space described earlier. This is found to be approximately true: if an animal's head is kept horizontal, dorsiflexion of the neck (by tilting the back up) results in front-leg extension and rear-leg retraction (**Fig. 11.12**). This, you will recall, was the vestibular effect of tilting the head down. These responses are the *tonic neck reflexes*. Thus corresponding to the vestibular mnemonic 'head down, front legs extend', we have the neck reflex mnemonic 'dorsiflexion, front legs extend'. As in the case of the vestibular reflexes, moving the head in other directions produces exactly analogous limb responses.

The use of these reflexes in maintaining posture is clear. An animal standing facing down a slope might have its head in the normal horizontal position so that there is no vestibular stimulation but the neck is dorsiflexed, or it might hold its head parallel to the slope, so that there is

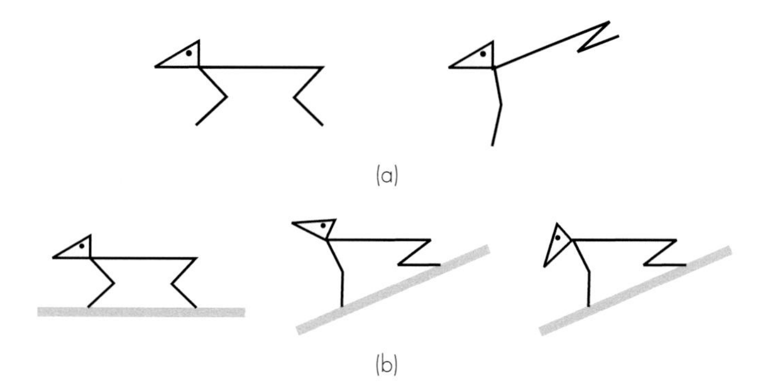

Figure 11.12 (a) Tonic neck reflex: if an animal is suspended with its head horizontal, dorsiflexion of the head (by tilting the back up) results in front-leg extension and rear-leg retraction (compare with **Fig. 11.7**). (b) This means that if an animal has its body tilted downwards, then whatever the position of the head, there will still be a postural drive to right itself.

vestibular stimulation but no activation of neck reflexes. Either way, the overall result will be the same. In fact, the animal will produce the same correct postural response *whatever* the position of the head. This consequence of the opposition of vestibular and neck reflexes is a very important one: if it were not so, then every time a cow tried to bend down to eat some grass, its front legs would extend, which would be frustrating. This is why, although the head is the single most important source of information about body position, nevertheless it can still move around freely without producing inappropriate postural responses. It is exactly analogous to the mechanism described earlier that permits us to move our eyes around without at the same time perceiving apparent shifts of the visual world.

Posture as a whole

Figure 11.13 is intended to summarise this chapter by bringing together the various sources of postural information and the way they interact in a hierarchical scheme that helps illustrate the parallels between the modes of operation of many of its parts. It is obvious that the scheme is highly redundant, in the sense that there are three distinct streams of information that maintain posture: from the eyes, from the vestibular system and from the support area. In fact it is clear that one can manage without any one of them, or even two, without completely losing one's ability to make postural adjustments. When swimming underwater, for example, only our vestibular system can provide us with information about body position; and in fact although patients with bilateral destruction of their vestibular apparatus can get about perfectly well in normal life by using the senses in their eyes and feet, it is precisely in circumstances like swimming underwater or running over a ploughed field that they may no longer be able to respond satisfactorily. But performance is always impaired by loss of one of the inputs. A simple sign of vestibular malfunction is simply to ask the subject to stand, and then to close their eyes. Most people sway around a

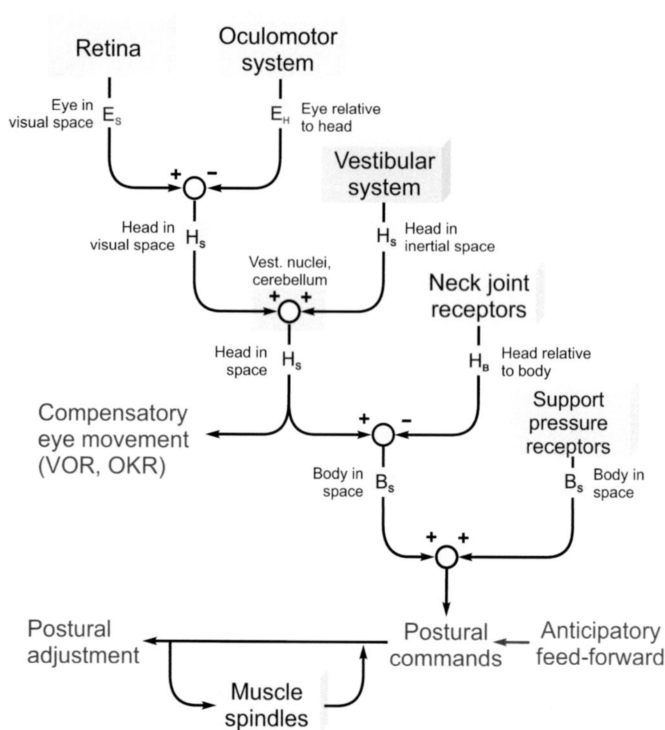

Figure 11.13 Summary scheme of the sources of information used in the control of posture, and their interactions.

little more as a result, especially if standing on something soft; with vestibular impairment, the additional instability is very marked.

One can get along without any one of these sources of information, but a sufficiently strong stimulation of one of the inputs can cause disequilibrium despite the correct functioning of the other two. As already mentioned, we may be thrown to the ground by our own vestibular reflexes after a spin in a revolving chair, and misleading visual information may similarly make us fall over if we stand looking up at the clouds.

At the output of this hierarchy is another kind of sensory receptor that has not yet been mentioned in the context of postural control, the *stretch receptor.* Though they provide feedback that helps to ensure that the commands issued by the system are correctly carried out by the limbs, as we saw in Chapter 10, it is important to realise that they do not generate useful information about the uprightness of the body in the sense that all the other inputs do: they are simply subservient to the postural system.

Finally, there is another source of information that is not sensory at all but the result of another kind of efference copy: feed-forward rather than feedback. Many of our voluntary actions are themselves threats to equilibrium, as for example if we pick up a heavy object and then throw it, or even for that matter as we simply walk along. In these cases the system can of course anticipate the threat, and can initiate postural compensations even before the disequilibrium has been sensed by the afferent pathways shown in the figure. A striking example of this kind

Table 11.1 Systems involved in maintaining an upright posture

System	Locates	With respect to	Confused by
Visual	Eyes	Visual environment	Movement of visual scene despite subject remaining stationary (e.g. 3D cinema) Static car interior despite movement
Vestibular	Head	Gravity	Acceleration of subject (e.g. on a bus)
Proprioceptors in neck	Head	Body	
Pressure receptors on body surface (feet, buttocks)	Body	Supporting surfaces	Externally applied pressure (e.g. plank placed on horse's flank during restraint for surgery)
Muscle stretch receptors	Actual muscle length	Desired muscle length	Vibration, muscle fatigue
Efference copy	Direction of current external force application (e.g. gravity)	Anticipated direction of external force application (e.g. gravity plus acceleration of train)	Unpredictable external events (e.g. sudden gust of wind, overzealous train driver)

of learnt postural response is the transient jolt one feels on stepping on to an escalator that has been turned off and is therefore stationary, which represents a temporary failure to suppress the normal, learnt, postural reaction to the sudden movement of the feet that would occur if the escalator were actually working.

Notes

A very intelligent account of posture, with many interesting examples (long out of print but not essentially out of date) is Roberts, T. D. M. (1967). *The Neurophysiology of Postural Mechanisms.* (London: Butterworths); Rothwell, J. C. (1993). *Control of Human Voluntary Movement.* (London: Chapman & Hall) has a more up-to-date, but less full account. Sherrington, C. S. (1906). *The Integrative Action of the Nervous System.* (New Haven, CT: Yale University Press) is a classic description of some basic responses, of some historical interest. Visual contributions to posture are thoroughly reviewed in Howard, I. P. (1982). *Human Visual Orientation.* (Chichester: Wiley). Developmental aspects of postural control – clinically a critical area – are discussed in Hadders-Algra, M. & Carlberg, E. B. (2008). *Postural Control: A Key Issue in Developmental Disorders.* (London: Mackeith). An excellent source of clinical reference is Bronstein, A. M., Brandt, T. Wollacott, M. & Nutt, J. (eds) (2004). *Clinical Disorders of Balance, Posture and Gait.* (London: Hodder).

1. Dynamic visual posture It is probably the main explanation for the postural instability associated with *heights*: if one is standing at ground level in a typical urban environment, most of one's visual field is filled with highly detailed visual texture, and the slightest head movement is likely to cause a brisk response from movement-detecting visual neurons. But on top of a mountain, things are very different. Most of the field is now occupied by clouds and sky, of low contrast and low spatial frequency, so that movements of the head are no longer so likely to be noticed by the visual system; the consequence is an increased instability, a tendency to sway about. Worse still, if one looks up, one's visual field is likely to contain nothing but the clouds moving past, which therefore give the visual system the illusion that one is falling over: the resulting postural compensation may well result in one actually falling over in the opposite direction.

2. Motion sickness The most comprehensive account is Reason, J. T. & Brand, J. J. (1975). Motion Sickness. (London: Academic Press). Nausea can occur even with quite mild visual/vestibular mismatch, for instance on putting on an unfamiliar pair of spectacles, when the slight degree of magnification or minification they produce temporarily upsets the natural relation between head movement and retinal slip.

3. Alcohol and the cupula It has been reported that the ingestion of heavy water may provide an instant antidote, by restoring cupular density – but it is obviously important to titrate accurately. See Money, K. E. & Myles, W. S. (1974). Heavy water nystagmus and effects of alcohol. *Nature* **247** 404–405.

PART 4

HIGHER FUNCTIONS

CHAPTER 12

HIGHER MOTOR CONTROL

Now we need to have a look at the levels of the motor system where we start to interpret and apply the detailed information delivered through the special senses. In ascending hierarchical order, these are the primary motor cortex, the cerebellum, the basal ganglia and prefrontal cortex.

The task of the motor system

Students often find the higher levels of the motor system difficult to understand: so do those who do research on them. As we have already seen (p. 189), there are peculiar experimental difficulties with investigating motor systems in conventional ways. At the same time the anatomy of the pathways is complex and open to many interpretations (compare it with that of the visual system, for instance). And finally, and perhaps most important of all, the complexity of the movements themselves is extraordinarily hard to grasp. Think what is involved in one of the most familiar sequences of actions of all, getting dressed in the morning. Faced in semi-darkness with a crumpled, partially inside-out, shirt, what astonishing feats of computation are needed to work out how to pick it up, manipulate into the right configuration, guide it over the correct parts of one's anatomy, deal with buttons and buttonholes. This is neither a uniquely human ability, nor particularly dependent on primate-sized cortex: you only have to see a bird elegantly swoop between the branches of a tree, killing its speed with exactly the right timing to come to rest on one particular twig, to realise that equally daunting motor tasks can be accomplished with remarkably little neural hardware.

At the end of Chapter 9, we introduced the concept of *hierarchical organisation*, and pointed out its utter inevitability in any system – a business enterprise, a military unit, a computer program – that is intended to carry out effective action. The areas of the motor system – cerebral cortex, cerebellum and basal ganglia – that we are about to look at are complex, ambiguous and poorly understood. The only way to make sense of them is – as always – to think *functionally*, and to be clear in one's mind about the problems that the brain faces in trying to generate meaningful movements. Thinking hierarchically turns out to be particularly enlightening, since the structures concerned happen to fall into a hierarchical order that has a natural, almost inevitable, relationship to different aspects of the movements themselves.

The three components of voluntary action

One way of thinking about movement that many students find helpful is first to consider a familiar, tangible, representative, action – and then to break it down into its components. A good example is reaching out to take a cake from a plate on a table. Starting at the end – always the best plan in neurophysiology – we have *grasping* (**Fig. 12.1**). The force in the fingertips must be carefully regulated to grip the cake safely without crushing it, despite its soft plasticity. There must be immediate sensing of any slip, and immediate adjustments must slightly increase the pressure of grip. Its weight must be estimated so that appropriate lifting forces are applied: it may perhaps stick to the plate. But before all these mechanisms of manipulation come into play, we must have succeeded in *reaching* it, an operation demanding an entirely different set of computations and control mechanisms. During the reaching the position of the cake must be monitored, using visual information supplemented by knowledge of eye position relative to the head, and head position relative to the trunk. Meanwhile, we need constant proprioceptive feedback from our limbs so that we can match the position of our hand to that of

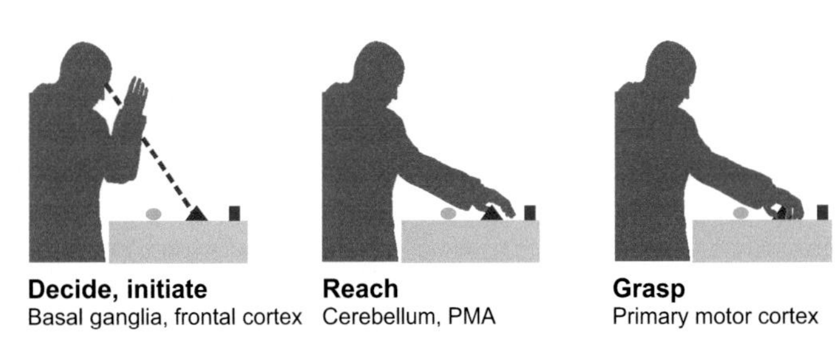

Figure 12.1 A sequence of events needed to achieve grasping an object, controlled by separate regions of the brain.

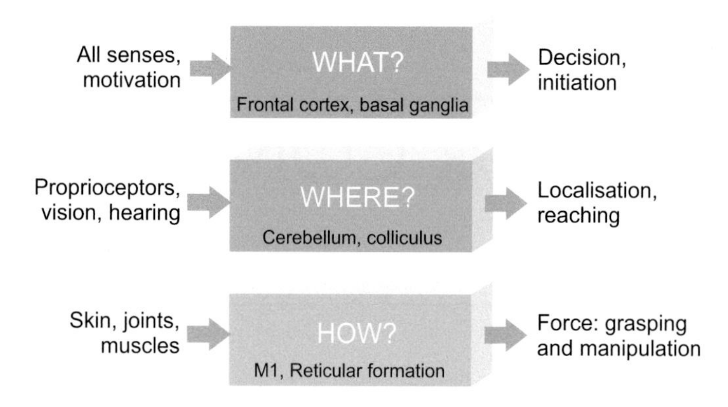

Figure 12.2 A hierarchy of motor control, with 'What' to do representing the highest level of control, followed downstream by 'Where' to move and 'How' to achieve this. Each level of the hierarchy requires separate inputs and has unique outputs to achieve its goal.

the cake. At the same time, we need to make slight postural adjustments to counteract the destabilising effect of extending the arm. And preceding both reaching and grasping come the processes of *decision and planning*. Are we actually hungry? Do we want the pink cake or the yellow one? Do social conventions either permit or forbid taking the cake at all? What route will avoid knocking over that milk-jug? These represent critical choices and plans needed to implement the subsequent movements.

By a happy coincidence, these logically distinct stages of planning, reaching and grasping – which amount to *what*, *where* and *how* – happen to correspond with precisely those three main areas that we shall be looking at in this chapter (**Fig. 12.2**). Grasping is the peculiar province of the *primary motor cortex*, which receives precisely the kind of information needed to manipulate objects, and has the correct descending connections to be able to control force and velocity with precision and speed. It is here that the desired positions of the limbs and so forth may be translated into detailed instructions about the forces required from moment to moment in order to achieve these actions. Patients with damage to this region, or indeed impairment of the sensory systems relaying the requisite information from the skin, can plan and reach perfectly well but are clumsy when it comes to the last stage of actually grasping and lifting, particularly when the thing being picked up is unfamiliar. For patients with damage to the *cerebellum*, on the other hand, the problem is mostly in the reaching: distances are misjudged, with overshoot and oscillation, and the trajectory may be broken down into a sequence of more tractable sub-components. The hand has to be guided on to the target with conscious effort. Finally, with hypokinetic disorders of the *basal ganglia* such as Parkinson's disease, movements may be well executed if they are executed at all, but the patient is conscious of the difficulty of getting them started, or sometimes of stopping them once they are under way. And prefrontal cortex damage may encourage cake-snatching, socially inappropriate but deftly executed – an error of decision rather than of planning, reaching *or* grasping. These behaviours highlight the distinct, yet equally important, roles of these various parts of the motor system which therefore deserve further discussion.

Primary motor cortex

By the middle of the nineteenth century, there was an increasing interest in the interactions between electricity and living tissues, and particularly in the kinds of responses that could be obtained by electrical stimulation of the central nervous system.[1] It turned out that there was a conveniently accessible region in the middle of the cerebral cortex where stimulation evoked reproducible movements of parts of the opposite side of the body; since no other area of the cortex generated movements on stimulation, this region was called the *primary motor cortex*. At about the same time, the English neurologist Hughlings Jackson had been studying the relation between paralysis of different parts of the body due to stroke and the postmortem location of the associated cerebral lesions. He was also interested in a special variety of epilepsy, subsequently named eponymously as 'Jacksonian epilepsy', in which there are characteristic motor signs such as spontaneous twitching, starting at one particular location – often an extremity such as a fingertip – and then moving progressively and systematically, for example up the hand and arm, until they culminate in a more generalised seizure. Putting these two types of observation together, he suggested that the progress of the epileptic signs over the body was the result of a 'march' of the region of pathological abnormality over the cortex itself. This is now known to be perfectly correct, as confirmed by later experimental recordings. An epileptic focus of overactivity in one cortical region tends to spread to the regions with which it communicates, a fact of importance in trying to treat epilepsy by surgical means. The orderly representation of parts of the body in the cortex was wholly confirmed when accurate motor maps were made by investigators such as Sherrington in the monkey, and later by Penfield and others in human patients, in the course of surgical operations intended to treat Jacksonian epilepsy. This area, the *primary motor area* (M1), lies just in front of the central sulcus and corresponds to Brodmann's area 4. It thus lies alongside the primary somatosensory projection area (areas 3, 2 and 1) and indeed shows a similar distribution of the representation of different parts of the body. Some regions of the body, such as the hands and mouth, have a disproportionately larger representation than others in this so-called 'motor homunculus': a possible reason for this will be considered later. Next to area 4 lies motor area 6, sometimes called the *premotor* area (PMA), and in humans this is considerably larger than area 4 itself. It is now recognised to have two or three subdivisions, with distinct functions, but the terminology is still confused. There is also a quite separate *supplementary* motor area (called M2, the first being M1; it is also known as the secondary motor area, SMA) which, like the corresponding

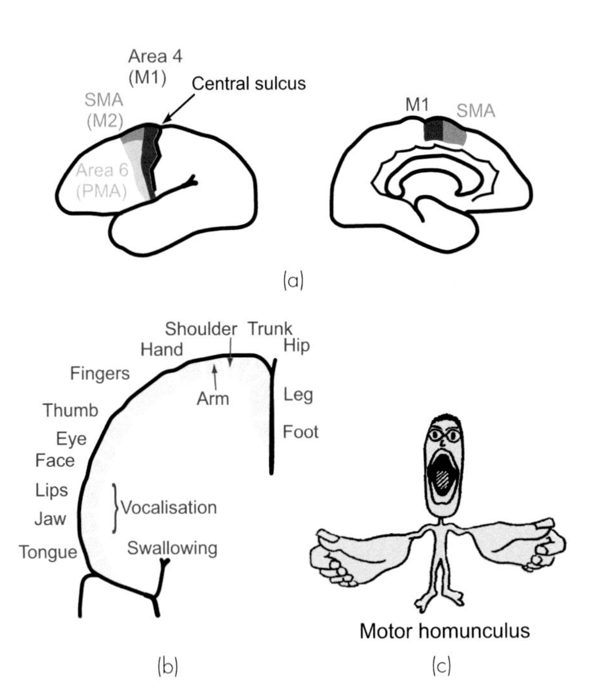

Figure 12.3 (a) Lateral and medial views of human brain, showing the approximate positions of primary motor cortex (M1), and the secondary areas, premotor area (PMA) and supplementary motor area (SMA). (b) Transverse section through M1, showing the distributions of areas devoted to different parts of the body. (c) Motor homunculus (compare with the corresponding sensory homunculus; see **Fig. 4.20**). (Partly after Penfield & Rasmussen, 1950; permission pending.)

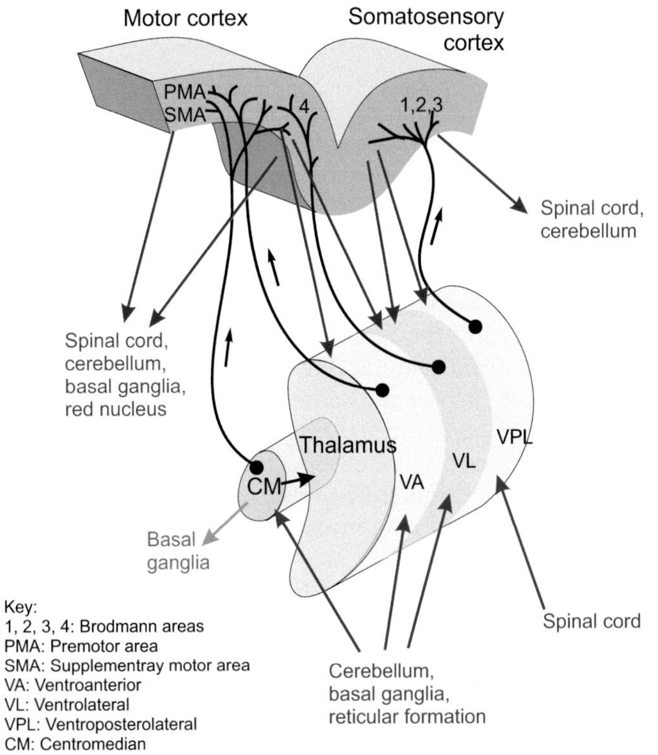

Motor cortex
Somatosensory cortex
PMA
SMA
4
1,2,3
Spinal cord, cerebellum
Spinal cord, cerebellum, basal ganglia, red nucleus
Thalamus
CM
VA
VL
VPL
Basal ganglia
Spinal cord
Cerebellum, basal ganglia, reticular formation

Key:
1, 2, 3, 4: Brodmann areas
PMA: Premotor area
SMA: Supplementray motor area
VA: Ventroanterior
VL: Ventrolateral
VPL: Ventroposterolateral
CM: Centromedian

Figure 12.4 Stylised representation of the main interconnections of motor cortex and thalamus.

sensory area S2, is more bilaterally organised than the strictly contralateral MI (**Fig. 12.3**).

Two thalamic nuclei are paired in this way with the cortical motor areas (**Fig. 12.4**): they are the *ventrolateral nucleus*, which projects mostly to area 4 and receives fibres back from both area 4 and the somatosensory cortex, and the *ventroanterior nucleus*, which projects to area 6 and receives returning connections from both area 4 and area 6. The ascending input to the motor nuclei of the thalamus is not sensory but comes partly from the basal ganglia and cerebellum and partly from the reticular formation; these regions also project to the more diffuse centromedial thalamic nuclei, which communicate more widely with both somatosensory and motor cortex, and other areas as well. The output from the sensorimotor regions is diverse: some of course makes up the corticospinal tract (CST, also called the *pyramidal tract*), some can reach the cord indirectly via the red nucleus, colliculus, and nucleus gigantocellularis of the reticular formation, but most project to the basal ganglia and – via relays in the pons and inferior olive – to the cerebellum. These indirect routes are sometimes lumped together under the heading of *extrapyramidal* pathways, those outputs that do not simply go straight down into the spinal cord.

Cortical influence on the spinal cord

Area 4 is differentiated from all other areas of the cortex by having a number of particularly big cells in layer V, the giant *Betz cells*. Their size suggests that they might be the origin of the corticospinal tract, and for a long time the notion was current that the essence of the voluntary motor system was something like what is shown in **Figure 12.5a**, with 'volition' somehow triggering off the Betz cells of Area 4, which in turn synapsed directly with spinal motor neurons: the former were called the 'upper motor neurons' and the latter the 'lower motor neurons', implying that the effects of stimulating the motor cortex were entirely due to stimulation of the cortical tract. These terms,

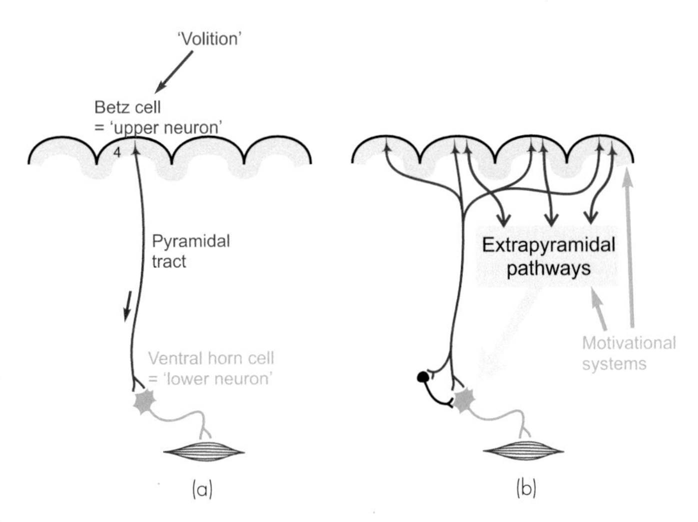

Figure 12.5 (a) Classical but misleading concept of 'upper' and 'lower' motor neuron. (b) A closer approximation to reality, highlighting the presence of extrapyramidal pathways and the influence of motivational systems.

upper and lower motor neurons, are still used in clinical practice to describe the location of neurological damage in patients.

This is an unhelpful and misleading picture in a number of ways. In the first place, it is clear that the Betz cells are not the sole origin of the corticospinal tract: for one thing, the latter contains about a million fibres, whereas there are only some 30 000 Betz cells. It is not even true that the tract comes only from the motor cortex: in primates about 30 per cent or less comes from area 4, another 30 per cent from area 6, and the rest from more posterior areas, somatosensory and parietal (**Fig. 12.5b**). Nor is it true that the effects of cortical stimulation are even mainly due to activation of corticospinal fibres. If in the monkey one traces the motor map by electrical stimulation both before and then after completely severing the pyramidal tract, the distribution of responses is found to be essentially unchanged (though they may tend to be a little slower and require a larger current to be evoked). And finally, as we have already seen, even in the primates only some of the corticospinal fibres end directly on peripheral motor neurons, while in other species none do: cortical control is probably rather of spinal circuits than of individual muscle units. The CST is not a particularly rapid conduction route: some 5 per cent are unmyelinated in Man, and of the myelinated fibres, 90 per cent of the fibres are quite small, with diameters around 1–4 μm.

Lesions of the corticospinal tract give similar, but not identical, effects to lesions of the cerebral cortex itself, no gross defect, but generally some hypotonia (reduced resting muscle tone) and weakness (paresis), and more specifically loss of skilled movements, particularly of extremities such as the fingers, such as in picking up a small object; lesions of area 4 are in general rather more severe, resulting at first in a wholly flaccid paralysis, and typically in greater functional loss than pyramidal section by itself, as the figure implies. When recovery has reached completion there is typically a degree of residual clumsiness, reminiscent of what is seen when feedback from the periphery is impaired (see Chapter 10, p. 212). There are certain difficulties of interpretation, however, both by reason of the marked species variation that is seen – a cat can still apparently run about after complete resection of its pyramids – and also because gradual (though variable) recovery is a feature of lesions of both cortex and pyramidal tract. In time, the immediate flaccidity resulting from a lesion in area 4 usually develops into a spasticity (with muscles continuously contracted), presumably through some kind of release phenomenon. Lesions that are large enough to encroach on area 6 as well as 4 also tend to produce spasticity, which might be interpreted as suggesting that area 6 has some kind of tonic inhibitory action, a notion reinforced by observations of the effects of stimulating area 6 during voluntary or evoked movements. In any event, the classical clinical description is that an 'upper motor lesion' gives a spastic paralysis, without muscle wasting, while a 'lower motor lesion' produces a flaccid paralysis. In summary, the CST is only partly from area 4; area 4 projects only partly to the CST.

Studying the relationship between motor neurons and pyramidal cells serves to exemplify how three techniques pivotal to studying neurophysiology complement each other:

- *Stimulation*: extracellular stimulation of cortical neurons reveals systematic representation in the cortex, analogous to that found in somatosensory or visual cortex, with groups of neurons corresponding to synergistic muscles found co-localised in columns. Stimulation of such a column generates coordinated activation of synergists (muscles that work together to generate a movement). Furthermore, stimulation with intracellular electrodes indicates clearly that each one of the pyramidal tract neurons excites only one muscle at a time, either directly (most obviously in Man) or via the specific facilitation (or sometimes inhibition) of one particular stretch reflex.
- *Reverse mapping*: peering back up at the cortex from an individual motor neuron in the spinal cord, we can identify the population of pyramidal cells making a monosynaptic connection with an individual peripheral motor neuron. In the monkey such a population can be mapped back to an area in the primary motor cortex of some 10 mm in diameter.
- *Recording*: conversely, recordings from individual pyramidal cells during spontaneous directed movements in monkeys show that movements are encoded by *populations* of neurons. The final direction of a movement is the result of a kind of weighted average of the activity in an ensemble of cortical neurons, an arrangement which – for reasons which are, as it were, the mirror image of the reasons given in Chapter 4, for the desirability of overlap of fields in sensory systems – gives robustness and sensitivity to small changes; conceptually, the ratio of activity of a 'move right' and a 'move left' neuron could elicit movement in any vector between 'right' and 'left', rather than having a neuron for every conceivable direction.

Somatosensory input and manipulation

One of the most interesting findings comes from using the same microelectrode for recording as well as stimulation. What is found is that monkey pyramidal cells have very wide, multimodal receptive fields, from joint receptors and tendon organs as well as skin receptors, and with a small representation of spindle afferents as well. If one looks at the relationship between what any particular cell does when it is stimulated and what it actually responds to, it is striking that frequently the cutaneous receptive fields represent those areas of skin that are normally brought into contact with one another when that particular muscle contracts (**Fig. 12.6**), and are found to be most active during precision grip rather than power grip. This strongly suggests that the cortex is the site of the kinds of sensorimotor

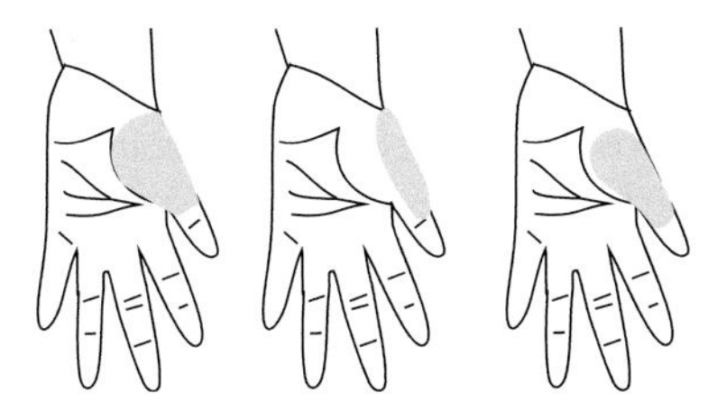

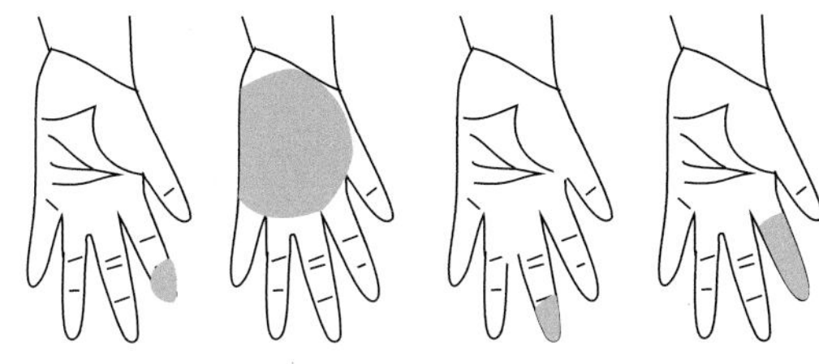

Figure 12.6 Cutaneous sensory fields associated with pyramidal cells of monkey causing (a) thumb flexion and (b) digital flexion in stimulation. The cutaneous fields correspond with areas that would be brought into contact by resultant movement. (After Rosen & Asanuma, 1972; with permission from Springer Nature.)

correlation required whenever we are grasping, touching or *manipulating* our environment.

That this is perhaps the most important single function of the primary motor cortex is suggested also by the relative sizes of different parts of the body in the motor map. In Man, only two areas of the body are used to any large extent for tasks of this kind that require accurate feedback from the skin, namely the hands and the mouth; and both of these together form by far the greatest proportion of the motor map. As shown in the highly schematic **Figure 12.7**, in the monkey, which makes more use of its feet for handling objects, they have almost as large a representation as the hands, and relatively much larger than in Man. In most four-footed animals it is only the mouth and lips that are used for grasping and exploration, and their representation is correspondingly enlarged; virtually the whole of the pig's motor cortex is said to be devoted to its snout! It is also perhaps significant that lesions in area 6 – which we saw to be predominantly inhibitory in effect – often produce uninhibited grasping of the kind frequently observed in very young children: any object touched against the palm results in immediate forceful grasp, with an unwillingness or inability to let go. Destruction of the motor cortex also abolishes the tactile placing reaction described in Chapter 11.

Figure 12.7 Motor 'homunculi' of rabbit, monkey and human, showing differences in the relative degree of representation of different regions. (Highly schematic: partly after Woolsey, 1958.)

One further function of primary motor cortex, of a related kind, is probably the generation of the 'expected force' commands that were described in Chapter 10. We saw there that these commands need to be adjusted to match the load that is to be encountered, and that one source of information about load that seems to be used to do this comes from pressure receptors in the skin; tendon organs, which as we have seen also project to pyramidal cells, are another likely source of information about load (see p. 207). Could it be that the pyramidal tract fibres are the route by which force commands are sent to the spinal cord? They certainly have the right kinds of information at their disposal to carry out such a function, and also appear to give the expected effects of facilitating stretch reflexes when stimulated. And if one records from the pyramidal tract of a conscious freely moving animal such as a monkey, and observes the circumstances under which pyramidal fibres actually fire during voluntary movements, one finds that whereas the correlation between pyramidal activity and muscle length or velocity is poor, there is a clear correlation between frequency of firing and the force that is being produced at any moment to generate the movement, suggesting very strongly that this tract does indeed provide force commands (as opposed to muscle length or velocity commands).

Finally, we have seen that the effect of lesions to motor cortex or pyramidal tract is not just a loss of the kinds of skilled movements that require close coordination between skin sensation and movement, but also a general weakness of the muscles, with a correspondingly increased sense of effort on the part of the patient. Sense of effort seems to be found in situations where the gain of stretch reflexes is insufficient, for example in the experiments described on p. 211 in which the pressure receptors in the subject's thumb were blocked by local anaesthesia, and may well be associated with voluntary excitation of motor cortex neurons. Partial paralysis tends to result in an apparent increase in the heaviness of a weight being lifted.

If it is true that the main function of the motor cortex is a relatively simple coordination between skin and other spinal afferents and the gain of stretch reflexes, one might well wonder why this function has to be carried out in the cortex and not in the spinal cord itself, where the inputs and outputs actually are. Why bother to send information from the periphery all the way up to the top of the head only to be sent back out again? There are probably two reasons why the spinal cord is essentially unsuited to this sort of coordination. The first is that most spinal reflexes seem to be organised in a segmental manner. But the feedback from skin receptors resulting from contraction of a particular muscle will not always return to the spinal cord via the dorsal roots of the same segment; in the cerebral cortex, however, information from different segments is brought into much closer approximation, along with information from other sources such as the visual

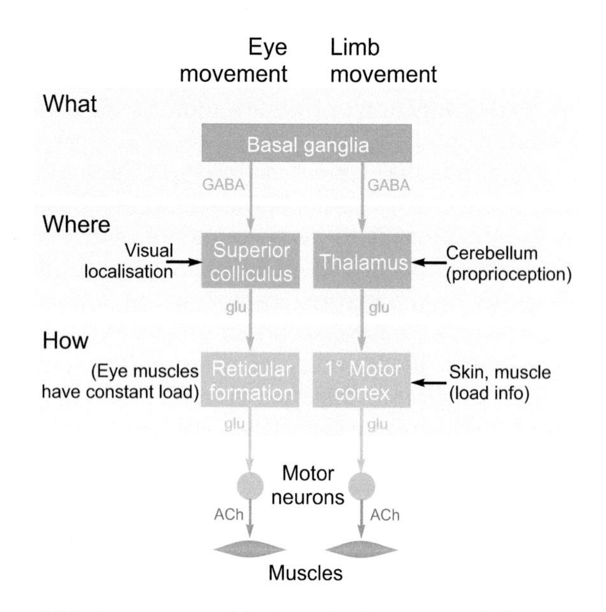

Figure 12.8 A comparison of the hierarchical motor control of eye movement with limb movements, noting different regions control the 'where' and 'how' components for the two types of movement.

and vestibular system. The large amount of neural convergence and divergence – in the monkey, each pyramidal cell receives some 60 000 synapses – means that connections from different inputs can presumably be allocated more easily to their appropriate muscles, rather as subscribers' lines converge on a telephone exchange. Thus, although the primary motor cortex is near the bottom of the motor hierarchy, physically it has to be near the top, a cortical area. It is instructive to compare its role in controlling the hand with what the reticular formation does in controlling the eye: because eye muscles have a constant load, they do not need the same range of feedback from mechanoreceptors, and the 'How' part of their control does not have to be cortical (**Fig. 12.8**).

The second point is that it is difficult to see how appropriate connections could ever be set up in the cord in the first place: how can an incoming sensory fibre from the skin *know* which of the hundreds of thousands of ventral horn cells are the ones to which they are supposed to make contact? It seems much more likely that such connections are established through experience, by frequent association of stimulation of a particular afferent with contraction of a particular muscle. This implies a richness and flexibility of connections, an ability to form functional contacts as the result of experience, which the cortex seems specifically adapted to perform, through the same types of memory-like mechanisms suggested in Chapter 7, in the case of the visual cortex. And just as depriving sensory input can cause large and rapid alterations in the relative areas of somatosensory cortex devoted to different regions of the body, similar changes can be demonstrated in primary cortex by stimulation or by reduction of use.

Finally, it should not be forgotten that if it is convenient for the control of movement to have cutaneous information readily available in areas 1, 2 and 3 right next door to the motor cortex, it is equally important in analysing cutaneous sensations to have information immediately on hand about what movements are being executed. It was emphasised in Chapter 4 that such common sensations as softness, resilience, roughness, stickiness, and the like rely on knowledge of what forces are being applied to the surface in question by the motor system, knowledge that is easily supplied by pyramidal tract neurons, through the rich interconnections between the two cortical areas. In human patients, electrical stimulation of the motor cortex produces sensory effects not very different from those resulting from stimulation of the somatosensory cortex: numbness, tingling, and a sense of movement which may or may not be accompanied by actual movement; if it is, the subject feels that he has been 'made' to carry out the action by the experimenter rather than 'wanting' to make it.

Secondary motor cortex

The functions of the two secondary motor areas, SMA and PMA, are much less well understood. Lesions in the supplementary motor area seem to cause a specific detriment in internally generated movements made from purely motor memory, though the animal can still learn the same movements in response to external signals. This is analogous, as we shall see, to what is observed in certain kinds of disorder of the basal ganglia (from which SMA receives input), and suggests a possible functional linkage between the two. Electrical stimulation of SMA can produce movement, and recording in conscious monkeys has shown a specific activation of SMA in more complex tasks, especially those involving carrying out a sequence of actions from memory. Stimulation in parts of PMA also produces movement, and lesions in certain parts of it suggest a clear distinction between SMA and PMA, that the SMA is on the whole more concerned with internally triggered movements and is related to the basal ganglia, whereas the PMA is more involved with movements guided by external stimuli, and related to the cerebellum (from which it receives much input). However, this is a very active area of research, complicated by difficulties of identifying and classifying the functional boundaries of the areas themselves, and by uncertainties in relating observations in monkeys to what happens in the human.[2]

Cerebellum

The cerebellum and basal ganglia are, very broadly speaking, at a higher hierarchical level in the motor system than motor cortex, in the sense that when they go wrong they tend to produce disorders of function, sometimes of a subtle kind, rather than discrete paralysis or weakness.[3] We have already seen that anatomically they are considerably further from the final output, in that they send no fibres directly into the spinal cord. They are also older structures than neocortex, prominent in all vertebrates; in

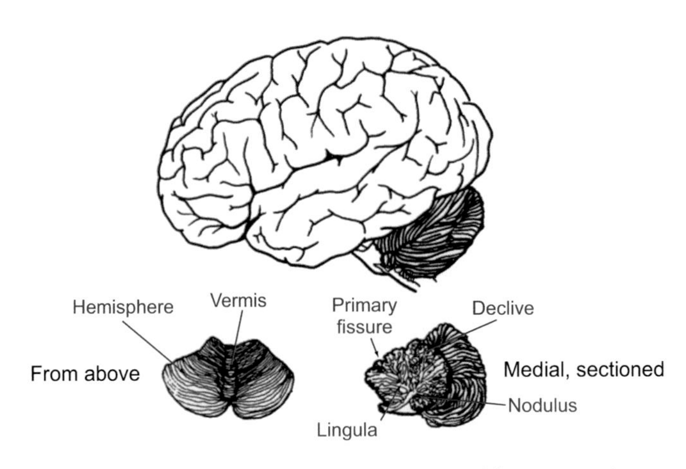

Figure 12.9 Gross structure of the cerebellum, viewed from various directions and in sagittal section.

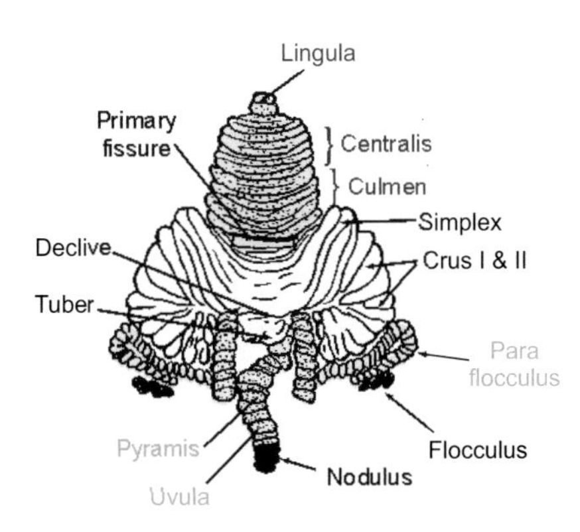

Figure 12.10 'Unrolled' cerebellum showing the main anatomical divisions. Functionally, it may be divided into vestibulocerebellum (black), spinocerebellum (stippled), and neocerebellum (unshaded), though the divisions are not as clear-cut as this diagram implies.

birds and reptiles the motor cortex or its equivalent does not appear to exist, so that one is forced to conclude that the cerebral cortex has developed more for *refining* actions than for generating them in the first place.

The cerebellum seems to have grown out of the brainstem as an adjunct to the vestibular system, in effect to provide it with some intelligence; but this oldest part of it, the *archicerebellum* or vestibular cerebellum, is now dwarfed by two newer areas: the *palaeocerebellum* associated with the spinal cord, and hence also sometimes called the spinocerebellum, and the large and central *neocerebellum* or corticocerebellum, whose development has been in step with that of the cerebral cortex from which most of its input is derived (**Fig. 12.9**). These areas are best appreciated when the cerebellum is anatomically 'unrolled', as in **Figure 12.10**.

The cortex of the cerebellum has a very regular and beautiful neuronal structure quite unlike the chaos seen practically everywhere else in the central nervous system, which has led to it being called 'the neuronal crystal' – so machine-like that it is impossible not to feel one ought to be able to work out what it is doing with just some concerted thought. Consequently it has provoked more speculation than anywhere else in the brain – particularly from mathematicians and computer scientists – about how its circuits might work. To put it another way, if we can't even work out what the cerebellum is doing, what hope is there for doing so anywhere else in the brain?[4]

Cerebellar neurons and their connections

The most conspicuous type of cell in the cerebellar cortex is the *Purkinje cell*, which with silver staining displays a huge dendritic tree that is confined to a plane perpendicular to the surface and roughly anteroposterior (**Fig. 12.11**). The cells are lined up in soldierly rows over the whole of the cortex, with their dendritic trees thus stacked like a pack of playing cards. There are about 15 million of them, and each has a dendritic surface area equivalent to two average-sized front doors.

Deep nuclei

The Purkinje cells form the sole output of the cerebellar cortex, for their fibres run downwards to the deep nuclei that lie in the core of the whole structure. In primates there are four of these nuclei on each side (fastigial, emboliform, globose and dentate) (**Fig. 12.12**); in other species the second and third of these are fused into one, called the interpositus. In one region, the flocculo-nodular lobe, the Purkinje output is not to a 'deep nucleus' but to part of the ipsilateral vestibular nuclei.

The projection from the cerebellar cortex on to these nuclei is a systematic one: medial areas (blue) project to the fastigial region, lateral ones (green) to the dentate, and the intervening region to the interpositus. A curious feature of the Purkinje cell is that although it is the only output from the cortex, it is entirely inhibitory (the transmitter is γ-aminobutyric acid (GABA)) on the deep nuclei. Finally, the deep nuclei themselves project to various other motor structures: the fastigial primarily to the vestibular nuclei, interpositus to the red nucleus, and dentate to the thalamus (mainly ventrolateral) and hence to the cerebral cortex; all probably also send fibres to the reticular formation of the brainstem. Thus there are several routes by which the cerebellum can influence the spinal cord.

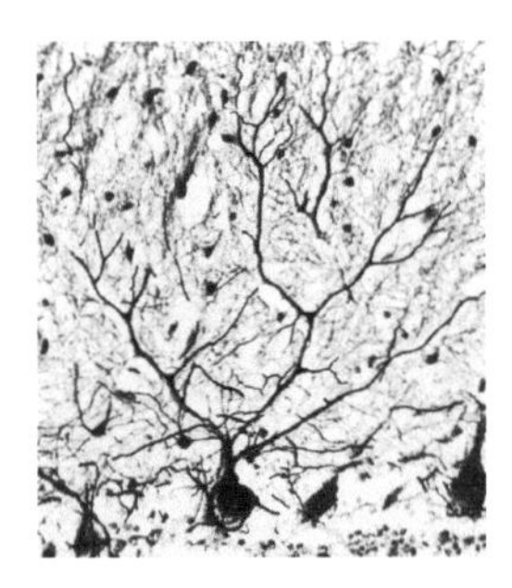

Figure 12.11 Cerebellar cortex, stained by the Golgi silver method, to show a large part of the dendritic tree of a single Purkinje cell.

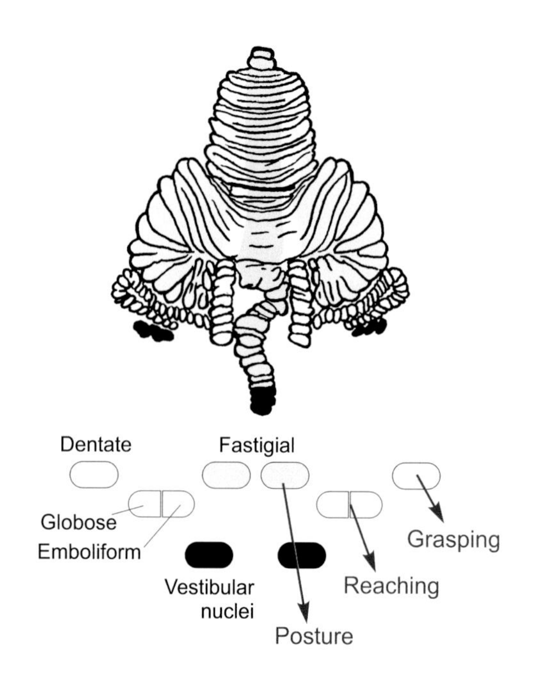

Figure 12.12 'Unrolled' diagram of cerebellum, as in **Fig 12.10**, showing approximate regions that project to particular deep nuclei.

Climbing fibres

Afferent information reaches the Purkinje cells by two quite distinct pathways (**Fig. 12.13**). In the first place, each Purkinje cell receives a unique single fibre that climbs up its efferent axon and then branches to clamber all over its dendrites like ivy on a tree, forming synaptic contacts that are believed to release aspartate and are very strongly excitatory. These *climbing fibres* come from the *inferior olive* in the brainstem, which receives its input in turn mostly from the cerebral cortex, but also from the spinal cord and to some extent from the special senses (as, for example, from the visual system, in the case of the vestibulocerebellum); as well as these contacts with Purkinje cells, they also excite the deep nuclei to which the Purkinje cells project. Such specificity of synaptic contact with a single cell is rather a rarity in the central nervous system, where wide neural divergence and convergence seems to be the general rule, and experimentally it is found that a single shock to a climbing fibre never fails to fire the associated Purkinje cell, sometimes repetitively. However, it is important to realise that Purkinje cells are normally extremely active at all times, including at rest, whereas climbing fibres seem to fire only infrequently (perhaps once a second on average). On the other hand, the extensive intimacy of contact between climbing fibre and Purkinje cell, and the latter's unusual electrophysiology (with voltage-gated calcium channels in its dendrites), means that a single climbing fibre impulse can have a highly significant effect on the Purkinje cell, triggering a 'complex spike' (actually a burst of spikes followed by a period of quiescence) that is capable of signalling to the whole dendritic tree, as is required if remote synapses from parallel fibres are to be capable of Hebbian plasticity (see p. 259). Such observations prompted neuroscientists to propose that these climbing fibre-Purkinje cell synapses have a role in motor learning as will be described later in this text.

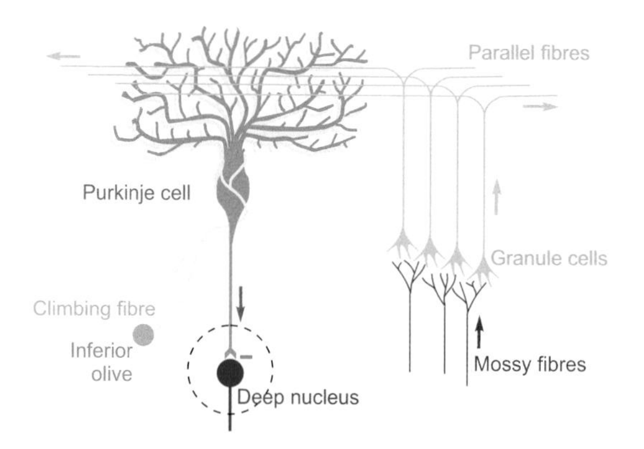

Figure 12.13 Diagrammatic representation of the connections between mossy afferents, granule cells, parallel fibres, Purkinje cells, and climbing fibres.

Mossy fibres

The other type of afferent system is utterly different. Here, the incoming fibres enter the lower layers of the cerebellar cortex and branch to form large terminal structures (giving them their name, *mossy fibres*) that synapse in *glomeruli* with a number of dendrites from the extraordinarily numerous *granule cells* in the cortex (there are some 10^{10} of them). These in turn send ascending axons to the surface, where they bifurcate and send two thin axons (called *parallel fibres*) in opposite directions, perpendicularly piercing the planes of the stacked Purkinje cells, to whose dendritic trees they form side connections, like telephone wires on a telephone pole. In this way, and in complete contrast to the highly specific climbing fibres, each parallel fibre can contact a large number of Purkinje cells; conversely, each Purkinje cell receives nearly *half a million* contacts from parallel fibres. Mossy fibres carry information from a very wide range of sources: directly from vestibular and spinal afferents, notably from muscle spindles, and visual, auditory and spinal afferents; indirectly through the *precerebellar* nuclei in the reticular formation (*lateral reticular nucleus*, the *nucleus reticularis tegmentum pontis* and the *paramedian reticular nucleus*); and finally from the cortex via relays in the pons. The receptive fields of Purkinje cells are very large indeed – sometimes extending over a whole limb – and are remarkably multimodal. This arrangement of parallel fibres piercing perpendicular sheets of flat Purkinje cells is an efficient way of providing the largest possible number of output channels with access to the largest number of input sources, within the smallest possible space.

It is quite instructive to compare this general arrangement with the other kind of cortex that is already familiar from Chapter 4, cerebral cortex (**Fig. 12.14**). In both cases there are large output cells (here in black), with a great deal of associative input (red) forming modifiable connections with them. But in cerebral cortex, the input projects to a relatively circumscribed region, the column, whereas

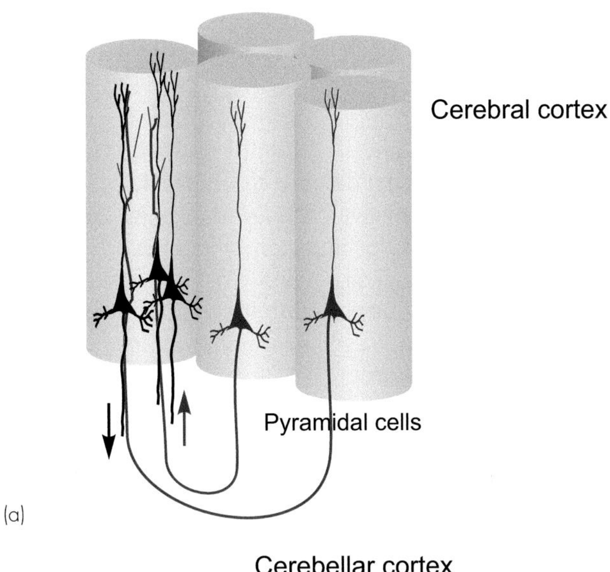

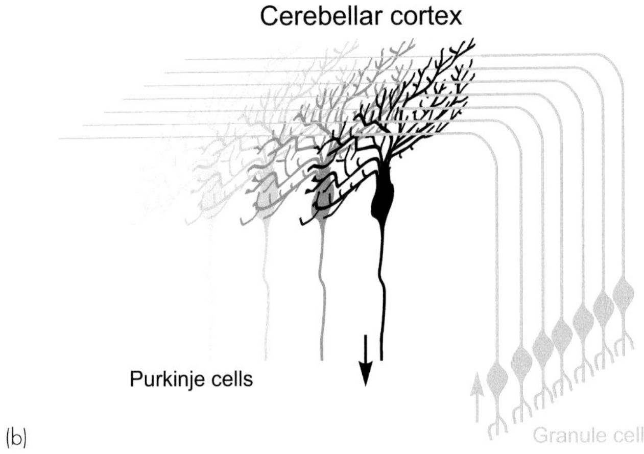

Figure 12.14 Comparison of cerebral cortex (a) and cerebellar cortex (b). Output cells are shown in black; associative afferents in red for cerebral cortex and green for cerebellar cortex.

the cerebellum seems designed for a given mossy input to reach the maximum possible number of Purkinje cells. Another very important difference is that in the cerebellum there is a second and very specific kind of input, the climbing fibre, which is not associative but forces the Purkinje cell to fire; as far as we know there is no exact equivalent to this in cerebral cortex, though one could draw a broad analogy with the specific thalamic afferents. Just as the job of the cerebellum could be said to associate general patterns of activity in mossy fibres with specific and real errors signalled by the climbing fibres, so as to be able to predict them, so in cerebral cortex one could regard one of the functions of a column as allowing general associative input from other columns, in order to learn to predict the activity in thalamic afferents that represents real events in the outside world.

Thus the basic structure is essentially simple: one system of inputs – the climbing fibres – which is extraordinarily precise, and a second system – the mossy fibres – which is equally extraordinarily copious, diffuse and non-specific (**Fig. 12.15**). It is only slightly complicated by the existence – as everywhere else in the central nervous system –

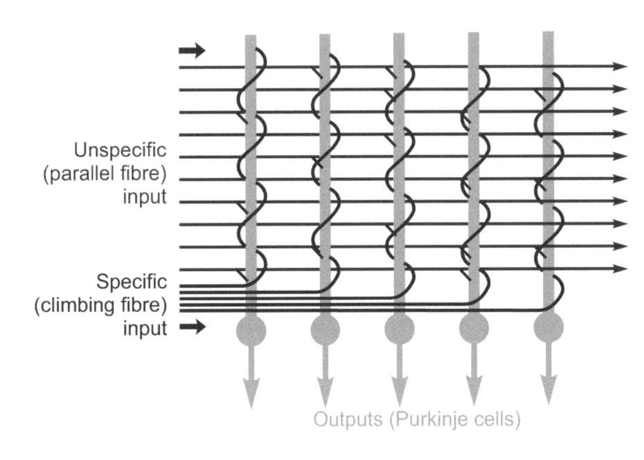

Figure 12.15 Highly schematic diagram of cerebellar inputs and output.

of various types of interneuron that mostly appear to provide lateral inhibition, sharpening up any spatial patterns of excitation that may be present. These include *basket cells*, excited by parallel fibres and inhibiting a parasagittal row of Purkinje cells, and *Golgi cells*, also excited by parallel fibres, but inhibiting granule cells instead and thus acting at the input rather than at the output (**Fig. 12.16**).

Though tucked unassumingly away at the back of the brain, overshadowed by the cerebral hemispheres, the cerebellum conceals a number of surprises. It is said to contain half the neurons in the brain, and the area of its cortex is almost 50 per cent that of the cerebral cortex; another significant fact is that roughly 40 times more fibres enter the cerebellum as leave it, suggesting it has an important role in deciphering inputting information prior to sending its output signals.

Disorders of the cerebellum

Although our knowledge of the neuronal structure of the cerebellum is quite precise, our knowledge of its function is less secure, and until recently was almost entirely limited to the effects of cerebellar lesions and other kinds of damage, observed most commonly in strokes or head

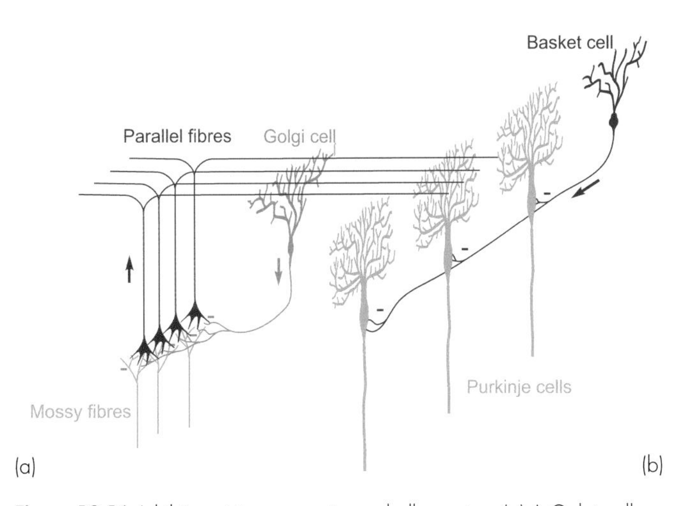

Figure 12.16 Inhibitory interneurons in cerebellar cortex. (a) A Golgi cell inhibiting a cluster of granule cells. (b) A basket cell inhibiting a row of Purkinje cells.

trauma. The deficits observed in these conditions are often quite specific and revealing (**Clinical box 12.1**). In general terms, the lack either of paralysis or of sensory loss led the earliest investigators to conclude that the cerebellum must be in the middle of the motor hierarchy, and concerned with *execution* of movement.

Damage to the vestibulocerebellum leads to difficulties of postural coordination that are similar to what is found with damage to the vestibular apparatus with which it is associated. There may be difficulty in standing upright, a tendency to dizziness and sometimes a staggering, unsteady gait when walking. There may be nystagmus (a jerky eye movement), partly when looking to one side, through faulty processing of vestibulo-ocular signals. A useful clinical observation is 'truncal ataxia': when asked to sit upright, unsupported in bed with arms folded, patients cannot maintain a still posture but jerk and squirm from side to side in attempt not to topple over. This test has the advantage that it eliminates contribution from limb weakness or clumsiness, and also avoids the risk of a patient falling and injuring themselves. It is very likely that the vestibulocerebellum acts as the centre of coordination for the various postural mechanisms described in the previous chapter. In some species, as was noted earlier, visual information enters this area through the climbing fibres, and vestibular fibres via the mossy fibres; it seems probable that it is here that the comparison and integration of postural information from these two sources takes place. As mentioned earlier, experimental cerebellar ablation in dogs leads not only to abolition of the effects of prism reversal on vestibular reflexes but also to freedom from motion sickness: these will be considered in more detail below.[5]

Difficulties of gait are also found after damage to other cerebellar areas, but the effects are then found to be more generalised and not just postural: a lack of coordination of all kinds of movement (asynergia), generally in association with a loss of muscle tone (hypotonia). It is worth considering some specific examples of these defects in more detail, since they reveal a good deal about the nature of cerebellar disability. Many can be explained in terms of the patient's motor system taking too long to respond to sensory information, of added delay round a feedback loop. Thus *dysmetria* or overshoot may be seen when the patient reaches out to touch something, their hand goes too far, presumably because the command to stop the movement is sent out too late. A consequence of this is *intention tremor*, in which the overshoot is subsequently corrected by a movement in the opposite direction which then itself overshoots, resulting in a new correction, and so on – the result being an oscillation or tremor around the desired position. The tremor is not seen at rest, but only when the patient is aiming to achieve a particular limb position. This slowness to react to changed circumstances is seen also in *rebound*: if, for example, the patient is asked to flex his arm against a force, which is then suddenly removed, whereas in the normal patient the resultant inward movement of the hand is quickly checked, in the cerebellar patient it is not, and may strike his body with considerable violence. A related defect is *dysdiadochokinesis*: the patient is unable to make rapid alternating movements, as for example rapid oscillatory rotation of his wrist between pronation and supination; he cannot apparently issue the command to reverse a movement sufficiently soon after having sent the command to start it (**Fig. 12.17**). In the same way, he may show *scanning speech*: whereas a normal person does not have to think about the sequence of mouth and tongue actions that he makes while speaking, the cerebellar patient seems unable to generate the series of commands sufficiently rapidly, and appears to have to think about the formation of each separate phoneme, like someone trying to speak an obscure foreign

Clinical box 12.1 Cerebellar impairment

The most important physical sign of cerebellar disease is ataxia, perhaps best defined as uncoordinated or inaccurate movement, which is not due to weakness, loss of postural sense or intrusion of involuntary movements. Neurologists elicit several classical signs to diagnose cerebellar impairment, generally revealing a deficit in the execution or automation of movement, rather than of initiation:

- Gait ataxia: patients typically do not fall, but walk with their feet widely spaced, reeling from side to side, even when supported.
- Truncal ataxia: a particular feature of midline cerebellar lesions, the patient is unable to sit or stand without support.
- Limb ataxia: when instructed to reach out with the index finger the subject may over- or under-reach (dysmetria) or demonstrate a non-rhythmic tremor that appears on action (intention tremor).
- Dysdiadochokinesis: rapid alternating movements, such as pronation and supination of the forearm, are fragmented, slow, and irregular.
- Scanning speech: the integration of bulbar muscle commands and breathing is lost resulting in a spluttering scanning dysarthria.
- Eye movement abnormalities: these can range from jerky nystagmus to more subtle impairments of smooth pursuit and saccades.

Causes of cerebellar ataxia include stroke, cerebellar tumours, and various inherited disorders. The cerebellum also seems particularly vulnerable to toxins, with alcohol and anti-epileptics being the main offenders. It is noteworthy that damage to a single cerebellar hemisphere, as in stroke, yields an ipsilateral ataxia; afferent connections to the cerebellum are ultimately derived from the contralateral cerebral hemisphere and ipsilateral proprioceptive afferent pathways.

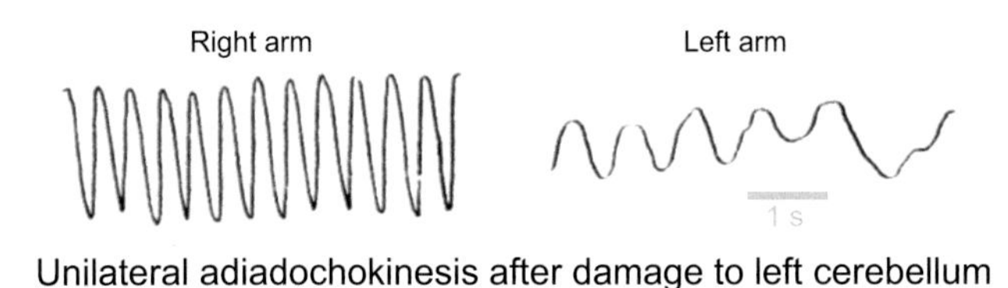

Figure 12.17 Adiadochokinesis: records from a patient with damage to his left cerebellum, who can make rapid alterations of pronation and supination with his right arm but not his left.

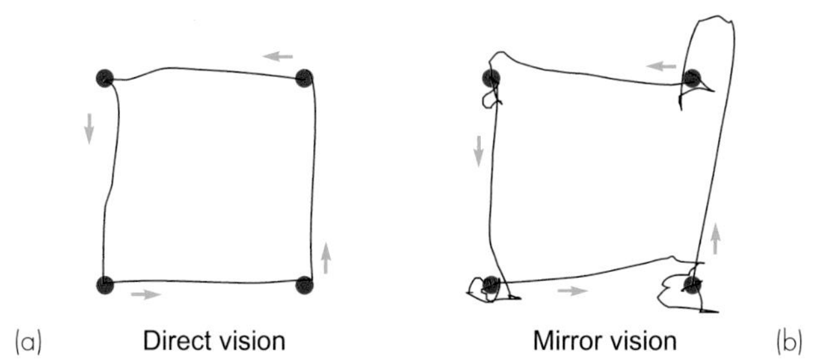

Figure 12.18 Illustration of signs of cerebellar damage. The subject was instructed to draw a pencil line linking the corner dots in the order shown, under direct vision (a) and looking in a mirror (b). Dysmetria, intention tremor and decomposition of movement, of a kind similar to the signs of cerebellar damage, are apparent.

language for the first time. The slurred end result is termed *dysarthria,* which should not be confused with *dysphasia* discussed in Chapter 13.[6]

Altogether, in fact, the patient has to bring enormously more conscious control into his movements, and it is the time required to think that slows things up. A normal person can walk along, pick things up and so on without thinking much beyond merely willing the final outcome; but a cerebellar patient has to plan and think about the details not just of what to do, but how to do it. This is perhaps most clearly demonstrated in another dysfunction called decomposition of movement: complex movements that require the temporal coordination of several different muscles are simplified by being broken down into their components, being executed in effect by one muscle group at a time. Normal people can see for themselves what it is like to be in this condition by the simple expedient of getting drunk: alcohol seems to have a particularly noticeable effect on the cerebellum, and in such circumstances one does indeed sense the need to think consciously about putting one foot in front of the other in order to walk, and one's conversation may begin to approximate to scanning speech.[7]

The difficulty that cerebellar patients have is essentially in using stored programs to carry out motor sequences that are usually automatic: their actions are performed as if they were learning them for the first time. In fact the execution of any new task by a normal subject is strikingly similar to what is seen in cerebellar patients all the time. A simple experiment you can do yourself is to try drawing while looking at what you are doing not directly, but in a mirror. If you attempt to move your pencil smartly towards a particular point on the page, you will see both dysmetria and intention tremor; more complex manoeuvres are only achieved by decomposition of movement, and all the time one is painfully aware of the need for continual thought about the details of the movements one is making (**Fig. 12.18**).With enough practice, of course, one would in time learn to execute mirror-drawing without these defects, and without the need for continual conscious intervention; presumably some part of the brain is then carrying out automatically motor sequences that previously required active thought, perhaps by means of some kind of internal model, as discussed on p. 193. It seems increasingly probable that it is the cerebellum which is the part of the brain that carries out this kind of motor learning – that it acts, in a sense, as the body's autopilot.

Theories of cerebellar action

What has always captured the imagination of neurophysiologists is its beautifully regular, near-crystalline, structure. In particular, the contrast between the tight, one-to-one coupling of climbing fibres and Purkinje cells and the grid-like arrangement of the connections with the myriad parallel fibres has suggested that, whereas the former are in a sense hardwired, the parallel fibre synapses might be of variable strength – programmable in a way that would account for the cerebellum's ability to learn. A plausible rule for such programming is the Hebbian rule (see Chapter 13, p. 259), that synaptic strengthening can occur if there is a coincidence of presynaptic and postsynaptic activity: if, in other words, a particular parallel fibre fires at the same time as the Purkinje cell. On this hypothesis, originally put forward by the late David Marr, it is possible to provide a ready explanation for the kinds of defects associated with cerebellar damage, as well as for more recent and precise observations.[8] He suggested that the cerebellum stores and executes specific sequences of actions by means of a gradual process in which the sequences are first generated consciously while the subject is attempting to master the task, but as the result of repetition are gradually taken over by the cerebellum itself. A crucial idea here is that every movement takes place in a sensory *context*, some of which will be due to feedback from the last movement that has been made. If we have a device that can learn to form an association between the feedback generated by one movement and the initiation of the next, then we can use it to generate each fragment of a movement automatically from its predecessor. In its final form the theory is a complex one, requiring complex mathematics to be fully appreciated: but a simplified example may help to convey Marr's basic argument.

Consider how we might learn to play a scale on the piano. To simplify things, let us begin by supposing that there is one Purkinje cell that corresponds to each of the fingers playing C, D, E, F (**Fig. 12.19**), and that, when the cell fires, it causes that finger to play the corresponding note. (The fact that the Purkinje output is inhibitory need not be an embarrassment: if we inhibit an inhibitory cell, the result is excitation, and neuronal circuits within the

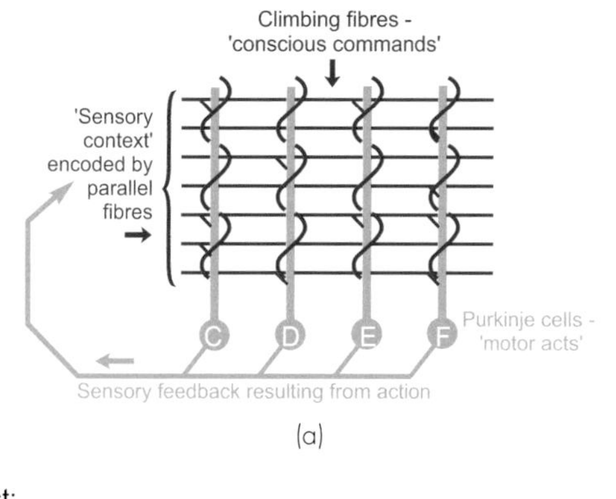

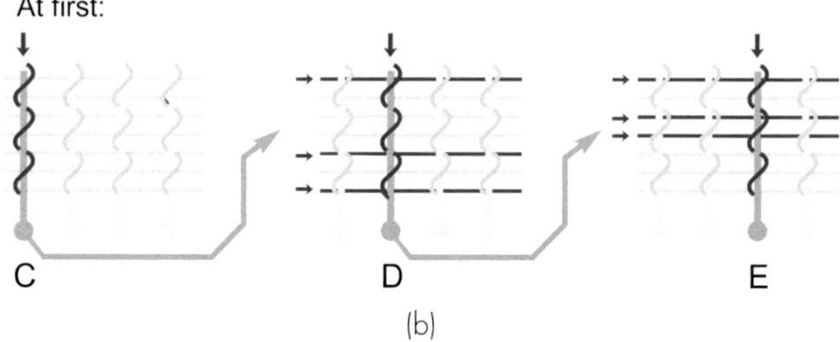

Figure 12.19 A model of how the cerebellum may learn sequences of motor actions. (a) A highly simplified diagram of cerebellar inputs and outputs, showing feedback from motor actions to the parallel fibres. (b) The hypothetical sequence before learning to play a piano scale C, D, E.

CNS can work equally well whether we consider either an increase or a decrease in firing rate to represent a 'positive' signal: one need only think of the hyperpolarisation of certain retinal receptors in response to light.) Now we have seen that the Purkinje cells are powerfully and specifically excited by their climbing fibres, and that one important source of these fibres, via the inferior olive, is the cerebral cortex. What is suggested is that during the initial learning phase, the Purkinje cells are driven by these climbing fibres – activated by some kind of volitional process – in the correct sequence: C, D, E, F; under these circumstances the cerebellum is doing nothing more elaborate than simply relaying this sequence of commands to some lower level.

But consider what meanwhile is happening to the parallel fibres: with their diffuse activation from every kind of sensory input, the pattern of their firing embodies the sensory context in which the action is taking place. Every time we carry out a motor act, it necessarily results in a kind of echo that comes back to us through our senses. When we play a note on the piano, we get feedback not only from proprioceptors, the muscle spindles, tendon organs and joint receptors, but also from endings in the skin, not to mention the visual stimulus of seeing the finger move and the note go down, and the auditory stimulus of hearing the result. Each note that is played consequently generates a particular pattern of sensory feedback that will be quite specific to that particular action and to no other. This pattern will be reflected in the pattern of activity of the parallel fibres, which we saw to convey information of the most diverse kinds to the dendrites of the Purkinje cells. Thus when we play the note D, having just played C, the Purkinje cell corresponding to D is activated by its climbing fibre during a sensory context that is quite specific to the state of having just played C, and is quite literally present at its dendritic branches in the form of a particular and specific pattern of parallel fibre activity. And if we now recollect our original, Hebbian, supposition that the condition for their synapses getting stronger is that the parallel fibre should often fire at the same time as the Purkinje cell, then we have a system that will learn to recognise the context associated with a particular action, and eventually respond to it automatically by generating the action itself. For if we play the sequence C–D over and over again, each time we do it the parallel fibres that are activated by the sensory feedback from C will fire at the same time as Purkinje cell D, so that their synaptic contacts with the latter will gradually get stronger and stronger. Eventually they will get so strong that they can fire D off even in the absence of a volitional command from the climbing fibre: D will then be produced spontaneously simply as the natural result of having played C, with no conscious intervention. In the same way, E will come to follow automatically from D, F from E, and G from F; and in the end all the subject needs to do is to initiate the sequence, and it will follow automatically, driven partly by sensory echo and probably increasingly by efference copy (**Fig. 12.20**). These theories have been confirmed experimentally in animals – blocking the climbing fibre-Purkinje cell synapses impairs animals ability to learn corrective eye movements in the face of forced head movements for example, implying these synapses are critical for teaching the cerebellum the eye movements needed in this context.

Such a model explains the phenomenon of dysdiadochokinesis particularly simply. If we imagine just two Purkinje cells, one for supination and one for pronation, then what we are doing when we learn to make the rapid alternation of hand position is to connect the two cells up reciprocally so that the context produced by one eventually comes to fire the other, resulting in almost automatic oscillation; those familiar with electronics will recognise that we have in effect built a multivibrator out of our cerebellar components. That these alternating movements have to be learnt in the first place is clear if you try to execute them with some less familiar part of the body: most people – unless they have been practicing – suffer from dysdiadochokinesis of the toes, as you can easily verify for yourself right now. Finally, it is perhaps worth mentioning

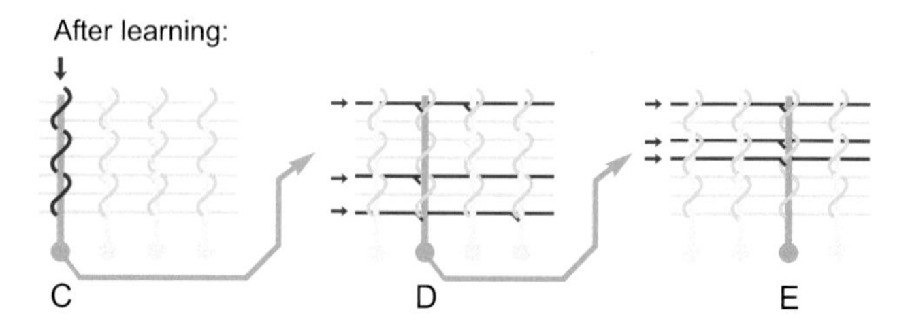

Figure 12.20 A model of how the cerebellum may learn sequences of motor actions. The hypothetical sequence after learning to play a piano scale C, D, E; compare with **Fig. 12.19**.

that Marr's model has already found a potentially useful application in the field of industrial robots: machines have been built that incorporate similar circuits and are equipped with sensors from the work area, and will learn to perform complex sequences of operations by first being driven 'consciously' (by a human operator, in fact) and then gradually recognising the patterns of sensory input that are to act as triggers for particular items of motor output.

With slight modifications, the same model can also form the basis of the other two types of motor learning discussed in Chapter 9, namely the storage of ballistic programs, and the prediction of expected results from copies of motor commands, by means of a stored model of the behaviour of the body (see p. 190). In the first case, we need only assume that a part of the mossy fibre input comes from other motor areas at a lower hierarchical level, rather than from sensory receptors (as indeed is the case, particularly in the neocerebellum). Then, instead of relying on actual feedback from the results of any particular item of a motor sequence, the command for one such item can trigger the next, producing ballistic sequences of motor acts that do not have to wait for actual feedback from results. Under these circumstances, the climbing fibre input (whose function in Marr's model is in effect to say to the Purkinje cells 'now learn this!') could be used to provide parametric feedback in order to improve ballistic performance through experience.[9]

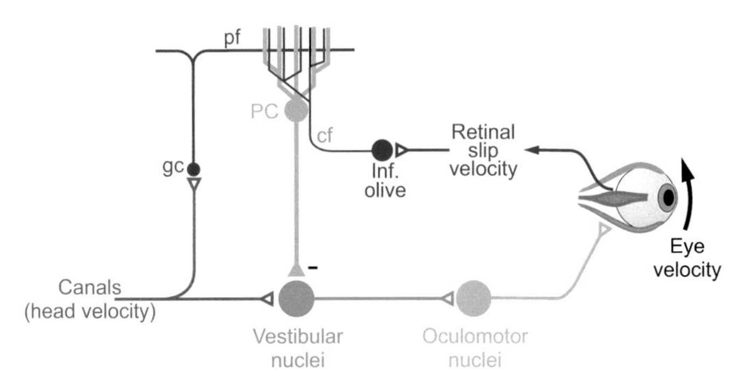

Figure 12.21 Postulated role of the flocculo-nodular lobe (FNL) of the cerebellum in adaptation of the vestibulo-ocular reflex (VOR). Head velocity signals from the semi-circular canals drive the oculomotor nuclei via the vestibular nuclei. In addition to this direct route, canal fibres also project to cerebellar cortex, where they impinge on Purkinje cells (PC) as parallel fibres (pf). Because the FNL Purkinje cells inhibit the VN pathway, this provides a route by which the gain of the VOR can be altered. Climbing fibres (cf) to the FNL come from the dorsal cap of the inferior olive (IO) and are driven by visual slip; because this is an error signal (it represents the difference between eye and head velocity), it is ideally suited to alter the strength of the connections between parallel fibres and Purkinje cells, and thus alter VOR gain to improve performance. However, experiments show that the learning in VOR adaptation is not confined to the cerebellum. Gc, granule cell.

Vestibular learning

For example, we have already seen that in the vestibulocerebellum there is evidence that visual information enters through climbing fibres and vestibular through the mossy fibres (**Fig. 12.21**). In Marr's model, this would imply that visual information would drive postural responses such as eye movements not only directly, through a direct 'reflex' route via the vestibular nuclei and oculomotor nuclei in the brainstem, but also via granule cells and thus through parallel fibres impinging on Purkinje cells that then modify the direct route, strengthen those vestibular connections to Purkinje cells that were appropriate, in the sense that they were in close correspondence with the visual signal. Such a mechanism could explain very nicely the way in which visual information appears to be capable of continually calibrating the vestibular input in such situations as the prism-induced reversal of vestibulo-ocular reflexes mentioned earlier. The synapses of those vestibular parallel fibres whose activity was in agreement with visual climbing fibre input would strengthen, and those in disagreement would weaken. When tested in the dark, this would lead to the kinds of alterations in response that are actually observed. (However, experiments show that the learning in vestibulo-ocular reflex adaptation is not confined to the cerebellum.)

Lastly, if we imagine the parallel fibres to convey copies of motor commands, and the climbing fibres to be activated by actual sensory feedback from the results of those commands, then we have a system in which Purkinje cell output will as a result of experience come to provide an estimate of the result of motor commands, of the kind required in internal feedback systems (see p. 193).[10]

Marr's model is thus a very versatile one, capable of embodying almost any kind of motor learning by defining its inputs and outputs in different ways. Many would say in fact that it explains almost too much, and is consequently difficult to test; and it has to be admitted that elegant and powerful though it is, there is as yet little direct neurophysiological evidence to support it. Its value at present is perhaps essentially explanatory, in that it helps to tie together in a coherent way both the observed effects of cerebellar dysfunction and what is known of cerebellar microanatomy and function. There are several details of it which certainly require correction in the light of subsequent work. For instance, while it is true that N-methyl-D-aspartate (NMDA) synapses showing the expected strengthening in response to coincidence of afferent and efferent activity are indeed found in the cerebellum, they are on granule cells rather than on the Purkinje cells where they ought to be. Synaptic plasticity of the connections from parallel fibres to Purkinje cells can be demonstrated, but it is weakening rather than strengthening (long-term *depression* rather than long-term potentiation). This in itself is not fatal, for shortly after Marr's work was published it was pointed out that the model would in some respects work better with just such a modification.[11] Much more seriously, in the case of modification of the vestibulo-ocular reflex, it is now clear that while some learning changes occur in the cerebellum, the most important ones occur in the *brainstem*, though they are certainly cerebellum-dependent. A picture is thus beginning to emerge of a cerebellum that learns, but also *teaches*: it learns to predict

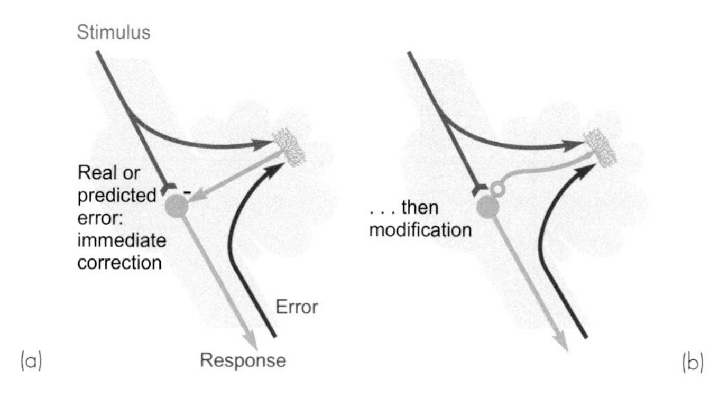

Figure 12.22 The cerebellum is posited to teach the brainstem circuits: (a) an error in movement (such as vestibulo-ocular reflex) requires an immediate correction; (b) the cerebellum learns the previous error and teaches the brainstem circuits to perform the correct movement.

errors before they actually occur, and these errors – whether real or virtual – are then used to modify the behaviour of the more primitive circuits in the brainstem (**Fig. 12.22**).

Error decorrelation

One way of trying to generalise what the cerebellum seems to be doing is something technically known as *error decorrelation.* We saw earlier that the cerebellum's origin seems to lie in supplementing the vestibular system by endowing it with intelligence. One particular example of this is that it learns to predict retinal slip before it happens, by monitoring head movement signals from the vestibular apparatus. But another way of looking at this arrangement is to say that it analyses the error signal from the eye – the retinal slip – into a component that can be explained by what we have done ourselves (derived from sensory feedback, or from efference copy) and a component that cannot be explained in this way, and must therefore genuinely be due to the object we are looking at moving in space. The explicable part of the error is what correlates with the vestibular signal: by decorrelating that component we are left with what is inexplicable or surprising. Quite apart from enabling us to distinguish different aspects of an otherwise ambiguous signal – as we saw earlier needs to be done with signals from the lateral line organ (p. 103), to sort out how much retinal movement is due to our own saccades (p. 219), or to compute albedo from luminance information (p. 128) – there are two profound reasons why we need to pay special attention to what is inexplicable. First, that novelty may signal a new opportunity – or, more likely, a threat – that requires deployment of the brain's full computational resources. Second, it may well mean that the stored model that we used to try to predict the error is out of date and needs modification. We shall see in Chapter 14 (p. 288) that a very similar mechanism associated with midbrain dopaminergic neurons appears to decorrelate *actual* reward from *expected* reward. And of course the adaptation that pervades every neural system in the brain can be viewed – in part – as a mechanism for discounting what is already familiar to reveal the unexpected.

Evolution, though innovative, can also be extremely conservative: when something really works, it tenaciously persists and is often multiplied and deployed to do things for which it was not originally developed. The cerebellum is a particularly good example: quite a complex modular neural circuit, developed originally to solve a very specific problem related to correlating vestibular and visual signals, was taken over virtually unmodified as the original archicerebellum burgeoned successively into the palaeo- and neocerebellum, that finally come to dwarf the vestibular archicerebellum where it all began.

Basal ganglia

Structure

Unfortunately, the functions of the basal ganglia are as uncertain in detail as their structure is complex.[12] Nevertheless, several observations have given us some insight into their function. A prominent component of the basal ganglia is the *corpus striatum* lying in the mesencephalon at the level of the thalamus; in higher animals it has suffered the fate that frequently falls to older structures in the brain – as we saw in the case of the archicerebellum – in that it has been elbowed out of the way by newer structures that have consequently distorted it into an even more tortuous shape than is altogether necessary (**Fig. 12.23**). Thus what was originally a relatively compact mass of cells has been disrupted by the arrival of the internal capsule – like a motorway through a village – with the result that it is now an elongated structure that twists its way round the newer ascending fibres. The stripes seen in cross section, which give it its name, divide it into a number of different regions: of these, the most important distinction is between the older part, the *globus pallidus*, which lies on the inside, and the outer *putamen*, which is continuous with the long arc of the *caudate nucleus*, the two together forming the *striatum*; combined with the globus pallidus it is the *corpus striatum.* Other nuclei are also conventionally considered part of the basal ganglia, though not everyone agrees as to what should or should not be included. They include the *subthalamus*, lying below the thalamus, and the *substantia nigra* (so called because certain of its cells are darkly pigmented with melanin). Both of these structures are highly developed in Man, but less so in other animals; and some authors also include the red nucleus.

Connections

The connections between these structures are very complex and their functional significance largely a matter for speculation. The important flow of information seems to be a projection from associational cortex to the putamen and caudate nucleus, thence to the internal and external layers of the globus pallidus (GPi and GPe) and substantia nigra pars reticulata (SNpr), both of which then project to ventroanterior and ventrolateral (VA and VL) areas of the thalamus, and thus back to motor cortex (mostly to the

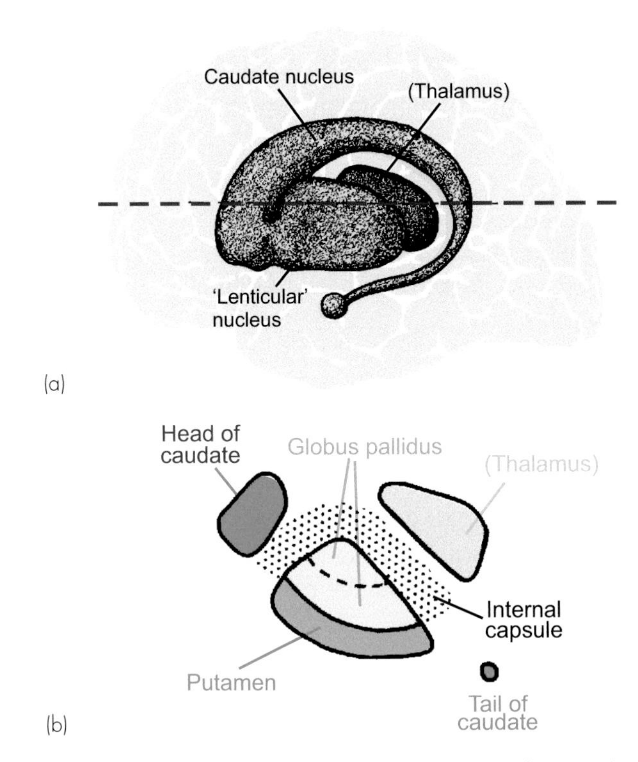

Figure 12.23 (a) Lateral view of the principal components of the basal ganglia, together with the thalamus. (b) A horizontal section at the level of the dotted line, showing in addition the fibres of the internal capsule pushing their way through. (Partly after Netter, 1962; copyright CIBA Pharmaceutical Company.)

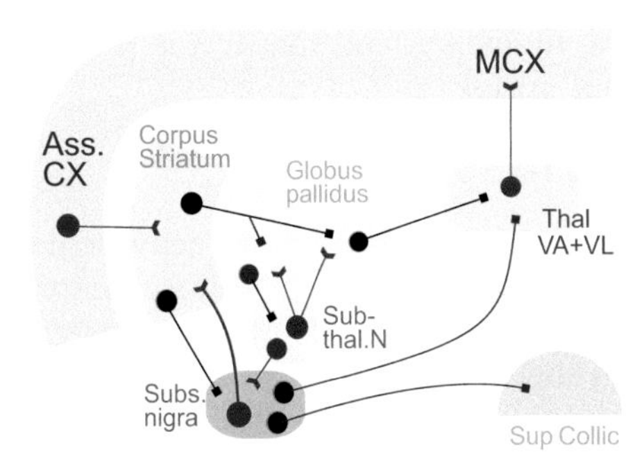

Figure 12.24 Schematic representation of the principal components of the basal ganglia. VA, VL, ventroanterior and ventrolateral thalamic nuclei. Inhibitory cells and endings in black; stimulatory ones are in red.

SMA, but also to motor and premotor areas; **Figure 12.24**). In addition, there is a second output route from SNpr to the superior colliculus, through which head and eye movements and possibly reaching can be controlled; in more primitive species this is the principal route. More is known about the transmitters here than in most parts of the brain: keen transmitter-spotters will be glad to know that the striato-pallidal and striato-nigral projections are GABAergic, and that the dendritic spines of efferent cells of the striatum receive glutamate-secreting fibres from the cerebral cortex (their action modulated by dopamine), serotonergic fibres from the raphe nucleus, dopaminergic fibres from the substantia nigra, endings releasing substance P and inhibitory cholinergic fibres from neighbouring interneurons. These connections are perhaps easier to make sense of if they are arranged functionally rather than anatomically, as in **Figure 12.25** (red arrows show excitatory projections, black are inhibitory). The subthalamus has connections to and from the globus pallidus, and the substantia nigra receives an input from the putamen, while sending fibres back to both parts of the corpus striatum (**Fig. 12.25**). Other connections are difficult to establish, though it is clear that there are indirect connections from the older areas of the brain such as the limbic system (which is concerned with motivation and emotion), possibly via the *nucleus accumbens*, and at least in fish one may demonstrate a functional projection from the olfactory system, which as we have seen is one of the oldest senses and one that is particularly associated with limbic functions. A suggestive finding is that within the striatum there appear to be two distinct populations of cells. One is in relatively dense islands or patches (*striosomes*) of spiny neurons containing opiates, projecting predominantly to substantia nigra pars compacta; the other forms a looser background (*'matrix'*) and projects more to the pars reticulata. The input to the striosomes appears to be mostly from motivational

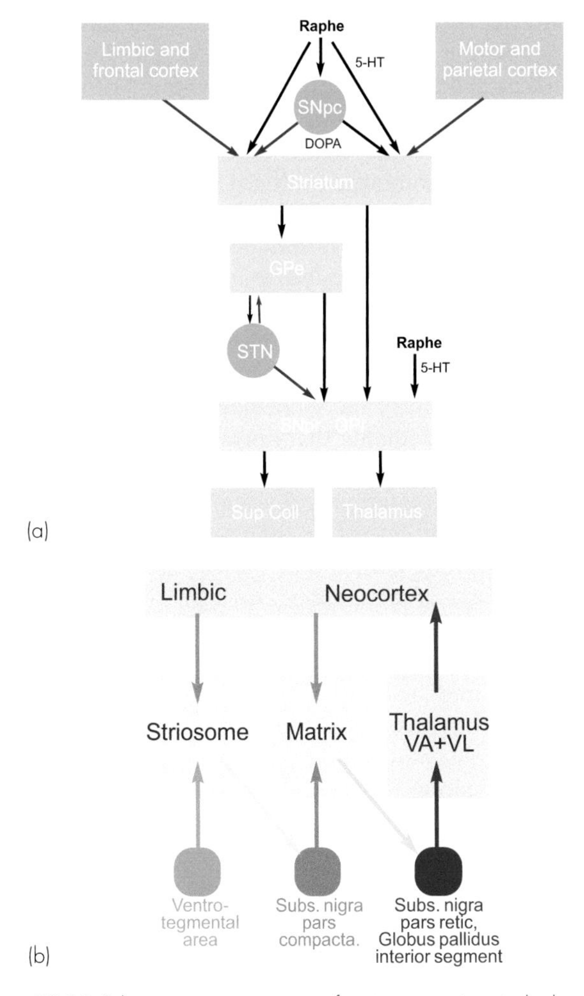

Figure 12.25 Schematic representations of main connections in the basal ganglia. The upper panel (a) demonstrates the circuitry and their outputs, while the lower panel (b) is an alternative schematic of the basal ganglia that distinguishes between the striosomal and matrix areas of the striatum.

areas of the limbic system, that to the matrix is mostly from cerebral cortex.[13]

Neural activity

Not a great deal is known about the activity of neurons in the basal ganglia. The final, inhibitory, output neurons in SNpr and GPi fire tonically at high frequency, so that they act on their targets through a process of disinhibition. Some striatal and pallidal neurons respond to joint movement and other proprioceptive feedback from motor activity; rather little activity is driven by cutaneous stimulation. Although many neurons fire in relation to limb movements, in general the relation of this activity to the movement itself is not understood. As yet, recording from basal ganglia has – apart from a specific role in the control of eye movements – produced only equivocal findings. In the case of the putamen, it is not even at all clear that the neurons are motor and not sensory (if that distinction has meaning at this level): most are best described as responding to a stimulus if it has some behavioural significance for the animal.

Clinical box 12.2 Classical signs of basal ganglia impairment

Hyperkinetic:
- Ballismus (subthalamus)
- Chorea (corpus striatum)
- Athetosis (corpus striatum)

Hypokinetic: Parkinsonism (nigro-striatal)
- Rigidity
- Bradykinesia
- Tremor at rest: typically unilateral, often described as a 'pill-rolling tremor'
- Akinesia: loss of associated movements; loss of expressive movements; difficulty of initiation

In general, difficulties are at a higher level: of initiation rather than of execution. In the right circumstances, movements may be performed relatively normally.

Effects of lesions

As with the cerebellum, such meagre information as we have about the functions of the basal ganglia is almost entirely derived from clinical observations of the effects of damage in Man. The best known of these disorders is *Parkinson's disease* (PD), associated especially with damage to the dopaminergic pathways linking the substantia nigra and the putamen.[14] The main feature of classical PD is a general poverty of movement (*akinesia*, meaning lack of movement, or *bradykinesia*, meaning slowness of movement).

Expressive movements, such as the normal mobility of the face, may be absent, giving the patient a lifeless and apathetic appearance, and there may be a loss of associated movements, movements that normally occur in conjunction with a particular primary activity, but are not strictly necessary (such as swinging the arms when walking). The patient may blink less often than a normal subject; there is often a shuffling gait, and he may be very slow in walking about. None of these things are defects of the peripheral motor apparatus, for under the right circumstances, especially under strong emotional stimulation, quite normal movements may be made. Thus a Parkinson patient may be shuffling his way across the road when a car comes: he then runs briskly to the other side, only to continue his slow shuffle along the pavement. The difficulty, in other words, is not in the *execution* of the movement, but in its *initiation*. A patient may find it very difficult to start to walk, but some martial music, or even a few lines drawn on the ground to act as a visual stimulus, may be sufficient to get the movement going.[15] Equally, once started it may be difficult to stop: *perseveration* of movement. It is clear that these are difficulties at a very high level of the motor system, and it is significant that Parkinsonian patients often show emotional disorders of a related kind: a more general apathy and immobility of mind as well as body. Other common features of Parkinsonism are a general rigidity and slowness (*bradykinesia*), vague postural difficulties, and a tremor (typically on one side) that is the exact opposite of the intention tremor of cerebellar damage, being present only at rest, and disappearing as soon as some voluntary action is attempted.

Parkinsonism, with its poverty and slowness of movement, is called a *hypokinetic* disorder. Other disorders of the basal ganglia, by contrast, result in the spontaneous production of *unwanted* movements: *hyperkinesia*. Lesions of the subthalamus in particular give rise to *ballismus*, in which the patient may throw his limbs about in a violent manner. Huntington's disease (HD), a progressive and fatal inherited condition affecting the basal ganglia, is characterised by spontaneous movements called *chorea* ('dancing'), for example continual shaking or twitching; other kinds of basal ganglia impairment my give rise to slow writhing movements known as *athetosis*. Some of these symptoms get worse as the patient tries to reach a particular goal. There may also be an exaggeration of associated movements. With the exception of the ballismus that can be produced in monkeys as well as humans by damage to the subthalamus, these effects are not clearly associated with lesions of specific areas of the basal ganglia, and explanations of their origin are correspondingly vague. For instance, in HD the inhibitory projection from the external globus pallidus to the subthalamus is increased, and if you mentally follow round what the effects of this ought to be, in **Fig. 12.25**, you can see that this ought to result in reduced activity in SNpr and GPi, and thus hyperkinesia.

Consequently, treatment of basal ganglia disorders is generally of a rather rough-and-ready kind: it is sometimes found, for example, that a Parkinson patient's rigidity and tremor may be alleviated by actually making further lesions in other parts of the basal ganglia. Another treatment for parkinsonism that is often helpful and has the appearance of being more scientific is to treat the patient with levodopa, a precursor of dopamine, the transmitter in the projection from substantia nigra to putamen; it is possible that it is some defect in the production of transmitter here that gives rise to the condition in the first place. More recently, many PD patients have been shown

Clinical box 12.3 Clinical aspects of Parkinson's disease and Huntington's disease

The mainstay of therapy for PD is to use drugs designed to increase the presence or reduce the removal of dopamine. The most prominent example is levodopa, which is a dopamine precursor. It is incredible that so blunt an approach as taking a dose of neurotransmitter precursor is remotely beneficial, evoking images of analogies of throwing petrol at a car or throwing sand on a circuit board.

Nevertheless, such an approach *does* work, for a while anyway. The efficacy does tend to drop off with time, requiring increasing doses for effect and sometimes necessitating the use of a 'dopamine holiday' in which medication is briefly stopped. However, escalating doses are associated with the generation of unwanted movements as well as hallucinations and the loss of impulse control as dopaminergic pathways which had remained intact are flooded with excessive levels of poorly targeted dopamine activity. The evocation of thought disorder (including hallucinations reminiscent of schizophrenia) by agents effective for movement disorder may be another indicator of the common mechanisms underlying the regulation of thought and action.

An alternative approach is to try to manipulate the dysfunctional neural pathway surgically: the atrophy of the substantia nigra and dopamine deficiency characteristic of PD lead to excitation of the internal globus pallidus by the subthalamic nucleus, and in turn the globus pallidus inhibits the thalamus. The subsequent suppression of thalamocortical activity is thought to underlie the paucity of movement and psychomotor retardation of the disease. Two surgical approaches have been tried: pallidotomy lifts the inhibitory influence on the thalamus, while deep brain stimulation of the subthalamic nucleus (which despite being 'stimulatory' appears to interrupt the outflow of the nucleus) prevents excessive activation of the globus pallidus. Both techniques are increasingly recognised as effective means of symptomatic improvement, with deep brain stimulators being increasingly favoured as they are reversible and modifiable as the disease manifestations change with time. If the bradykinesia worsens, the stimulator settings can be altered to improve the symptoms; if unwanted side effects are too problematic, the stimulator can be switched off. Alternative targets for deep brain stimulation other than the subthalamic nucleus in PD are being explored. However, deep brain stimulation is not effective for all patients with PD; it does not alter the inevitable degenerative nature of the disease, and the costs associated with the operation can be prohibitive in some cases. PD is also a potential target for stem cell therapy, with efforts at implanting stem cells to regenerate atrophic nigrostriatal tissue. This approach however carries with it ethical dilemmas of foetal tissue use for medical purposes, and so clinical trials seem more likely to focus on implanting adult stem cells that have been reprogrammed to differentiate into dopaminergic neurons.

Huntington's disease is an autosomal dominant, inherited condition involving a mutation of the gene coding for the 'huntingtin' protein; the primary function of huntingtin is unclear, although it is known to be essential for many cellular 'housekeeping' processes. The mutation is expansion of the CAG triple repeat, resulting in abnormal protein that is expressed throughout the brain, but the neuropathological hallmark is neuronal loss in the basal ganglia, particularly the caudate and putamen. Disease progression is accompanied by progressive generalised brain atrophy and loss of cerebellar Purkinje cells.

The clinical manifestations are dominated by motor abnormalities, particularly the development of rapid, involuntary non-repetitive movements of the limbs, face and trunk termed *chorea* (Greek *khoreia* – dance). What begins as an apparent mild restlessness becomes more florid, descending into a distressing, disabling movement disorder. Towards the end stages of the disease the generalised hyperactivity can be replaced by a state more akin to a parkinsonian picture, marked by akinesia and rigidity. Eye movements are involved; in particular there can be a delay in the initiation of saccades.

Consistent with the lofty place of the basal ganglia in the neurological hierarchy, the motor abnormalities are accompanied by deterioration of 'higher functions', and it is striking that the cognitive decline characteristic of these unfortunate patients predominantly affects their executive function, their ability to make decisions or initiate and switch tasks. This correlates with our neurophysiological model of the basal ganglia as the engine for the initiation and goal orientation of thought as well as movement. Similarly, the psychiatric manifestations ubiquitous among Huntington patients are marked by agitation and apathy, psychosis and depression, perhaps mirroring the motor hyperkinesis and the progressively subdued cognition.

to benefit considerably from continual electrical stimulation of the subthalamic area, through chronically implanted electrodes – *deep brain stimulation*, or DBS; yet we have practically no idea at all how this procedure leads to the observed improvement. Experimentally lesions give similar results to electrical stimulation (as for instance in GPi, where both give rise to bradykinesia), which makes interpretation difficult.

At all events, it is certainly not yet possible to use the existence of these clinical disorders to deduce the detailed functioning of the basal ganglia; and single-unit recording is only just beginning to make a contribution to our understanding of what they do. To suggest, as some of the older accounts do, that because damage to a certain area gives rise to tremor or to sudden violent movements, the function of that area is to reduce tremor or smooth movements out in some way, is only slightly less absurd than the analogy presented in Chapter 1, of removing a circuit board from a radio and deducing that its function was to inhibit humming. Clearly, deciphering the role of the basal ganglia will require more in-depth neurophysiological experimentation.

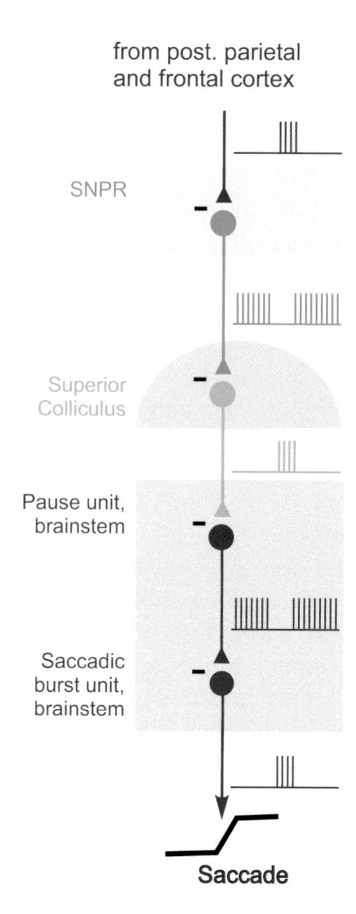

Figure 12.26 Cascade through basal ganglia causing the initiation of saccades. Saccadic burst units in the brainstem drive oculomotor neurons; they are tonically inhibited by pause units, which stop firing to permit a saccade. The pause units are in turn inhibited by burst units in the superior colliculus, under visual control, but these too are held tonically in check by inhibitory pause units in substantia nigra pars reticulata (SNPR), which in turn appear to be gated by higher areas, including parietal and frontal cortex. The '–' sign implies an inhibitory neuron.

What do the basal ganglia do?

So do we know *nothing* about the function of the basal ganglia? Not quite, for one can deduce a great deal about the hierarchical level at which the basal ganglia operate. In all these cases, whether there is loss of voluntary initiation of movements that can be evoked involuntarily, or the intrusion of unwanted movements that are, in their way, quite well executed (or even elegant, as in athetosis), it is clear that we are at a very high level in the motor system. Lesions of the cerebellum give rise to defects of execution but not of initiation, and lesions of the cortex lead to even 'lower' defects like weakness or frank paralysis. But damage to the basal ganglia clearly interferes with the level at which movements are *strategically planned and initiated*, and in some cases at least appears to work by disinhibition: movements occur when their tonic inhibition is removed. This mechanism has been particularly clearly demonstrated in the saccadic system. We saw in Chapter 7 that oculomotor neurons are driven by burst units in the brainstem, which are tonically inhibited by pause units, which stop firing to initiate a saccade. This inhibition comes from saccadic burst units in the colliculus, under visual control, but these too are held tonically in check by inhibitory pause units in the SNpr; these in turn appear to be gated by higher areas, including parietal and frontal cortex. Thus the whole system constitutes a remarkable cascade of serial inhibition (**Fig. 12.26**).

Another general kind of feeling, engendered by the predominance of inhibition in the circuits of the basal ganglia, is that at some deep level they are as much about stopping movements as starting them. A feature of movement in general that has tended to be overlooked is that our activity tends to occur in bursts, separated by periods of complete stillness: in a phrase often (wrongly) attributed to Sherrington, *Movement begins and ends in posture*. This alternating pattern – a kind of systole and diastole to make an analogy with the heart beating – is something we have already seen in saccades, but it is equally obvious in the scampering interspersed with freezing so characteristic of squirrels and other timid animals. It should be obvious by now that keeping still is not simply a matter of turning on global inhibition: since continual neural commands must be sent to striated muscle just to keep it at the same length (because of the shamefully wasteful spontaneous breaking of actin–myosin bonds), 'freezing' is just as computationally demanding as moving.[16]

One might well ask what, if anything, lies *above* the basal ganglia in such a scheme. The problem here is partly one of terminology. In the general representation of the brain presented at the beginning of Chapter 9, it was emphasised that there is in effect a gradual series of neuronal levels that convert sensory information into motor movements. When we are considering areas near the centre of such a scheme, at the highest hierarchical levels, the distinction between 'sensory' and 'motor' becomes somewhat arbitrary. (Of course, one might cut through this problem at one stroke by introducing a 'ghost in the machine' which is both conscious of incoming sensory information and also capable of willing volitional movements: by definition, any structure upstream of such an entity is sensory, and downstream, motor. The question of whether such a notion is necessary in understanding human behaviour is postponed until Chapter 14.)

Meanwhile, all one can usefully say is that it is simply a matter of convention that the inputs that provide the drive for motor acts are normally reckoned not to be part of the motor system. There are in fact two distinct types of input that must be considered: first, the sensory information about the environment, without which one clearly cannot make decisions about how to plan one's motor acts, even on a large scale; and second, the neural mechanisms that decide what is to be done, that choose between all the various possible courses of action which are open to the brain at any particular moment. The first type of input, that of high-level integrated sensory information about the environment, forms the subject of Chapter 13; the second type, which is what motivates the motor system, and requires information not just about the outside world but also about one's internal environment as well, one's state of need, forms the subject of Chapter 14.

Prefrontal cortex

It is convenient at this point also to consider a region that is not often considered 'motor' in the sense that the cerebellum and basal ganglia are normally regarded, but is nevertheless clearly concerned with the decision to initiate movement.[17] Prefrontal cortex forms the largest single division of the cortex in humans, with a diverse output that extends to the hypothalamus as well as to the striatum, subthalamus and midbrain. It receives afferents from the correspondingly large *dorsomedial nucleus* of the thalamus, which in turn receives fibres not only back from the frontal lobe but also from the hypothalamus and other parts of the limbic system, an area predominantly associated with such functions as emotion and motivation, to be discussed in Chapter 14. These are very old parts of the brain indeed, found practically unchanged throughout the animal kingdom, and it is striking that they have developed such an integral relationship with the prefrontal cortex, the most recently developed region of neocortex – itself the newest part of the brain. Humans are in fact distinguished most from primates by the absolute and relative size of our frontal lobes; until a century or so ago it was assumed that these must therefore be the seat of the very highest functions – intelligence, morality, religion, etc. – those that were thought to differentiate 'Man' most clearly from the apes.

The case of Phineas Gage in 1848 therefore came as something of a shock. He was an American mining engineer, who was one day tamping down explosive with an iron bar. The iron struck a spark, setting off the explosive and converting the bar into a ballistic missile. The consequence can be imagined (**Fig. 12.27**). Astonishingly, when one considers contemporary standards of medical care, he survived, and made a precarious living for some years by exhibiting himself in public, together with the bar. Post-mortem examination showed that a very large portion of the frontal cortex had been destroyed; yet what was extraordinary was how little effect this seemed to have on him. Far from turning into an ape, or losing his powers of reason, he seemed to suffer little more than a slight change in personality. His doctor described him as *'fitful, irreverent, indulging at times in the grossest profanity (which was not previously his custom), manifesting but little deference for his fellows, impatient of restraint or advice when it conflicts with his desires, at times pertinaciously obstinate, yet capricious and vacillating'* – what, in fact, some would now call perfectly normal in modern society. Indeed his friends said that in some ways he was actually happier – more carefree and less inhibited – after the accident than before.

As a result of this dramatic demonstration, controlled experiments were performed on animals, with the same conclusion: that lesions in the frontal lobes seem generally to reduce anxiety – monkeys worry less when they make mistakes in learning tasks – and inevitably the idea developed that such a procedure might even be of benefit to depressive patients or anxious schizophrenics. This operation, *frontal leucotomy* or *lobotomy*, began to be practised around 1935 and remained popular until the introduction of pharmacological agents doing much the same thing in a more reversible manner, in the early 1960s. **Figure 12.28** shows the extensive area of sectioning of the fibres communicating with frontal cortex, severed by means of a spatula inserted through holes on each side. There is no doubt that operations of this kind could give a great deal of relief – not only alleviation of tension and anxiety but better adjustment to work and increased weight and energy. A difficulty was that the changes in personality might go too far, developing into euphoria, tactlessness, a lackadaisical approach to life, and a lack of such social inhibitions as those that discourage urination in the fireplace. One circumstance where these side effects seemed worth putting up with was in the treatment of intractable pain, not easily dealt with by other measures. The result in this case was not so much loss of the objective knowledge of the pain – not, in other words, analgesia – but rather a loss of the *'affekt'* of the pain, its unpleasant or emotional quality.

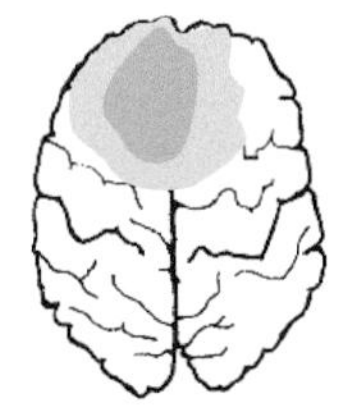

Figure 12.27 Prefrontal damage: Phineas Gage's skull, with tamping iron. The horizontal section on the right indicates the subsequent area of destruction (red) and probable neural damage (shaded). (After Cobb, 1946.)

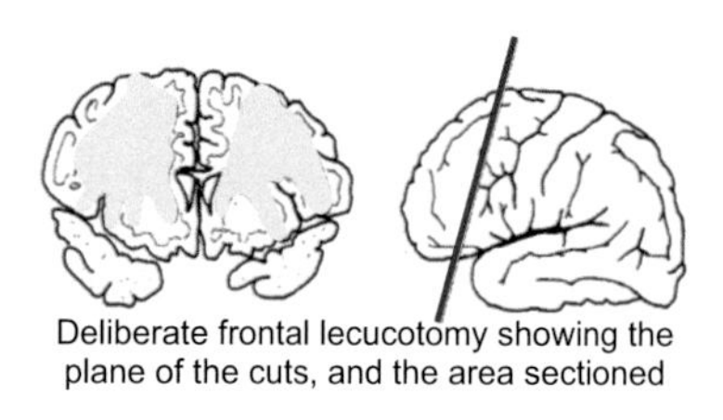

Figure 12.28 Deliberate frontal leucotomy, showing the approximate plane of the cut (right), and cross section of an actual cut of this kind. The operation is performed by means of a spatula inserted through skull openings. (After Freeman & Watts, 1948.)

Thus when asked what the pain was like, a patient might reply 'Oh doctor, it's absolutely appalling, unbearable', yet would be smiling as he said it and not – apparently – really feeling it despite being able to sense it.

What is perhaps most notable about the effects of frontal lesions is how little defect of ordinary intelligence occurs, with one exception: there are almost always difficulties in carrying out two programs of activity simultaneously. Thus a patient may be asked to recite the letters of the alphabet 'A, B, C, D', and then be interrupted and asked to add together 13 and 15: '28'; if then told to carry on the response may be '29, 30, 31, . . .' – the original task is forgotten. Similarly, there may be an inability to organise actions in proper temporal sequence; this may become apparent when trying to prepare a meal, where one has of course to plan well ahead, bearing in mind the different times that different things take to cook, so that everything is ready simultaneously. This impairment of the sense that something has to be done can be demonstrated in monkeys by means of the *delayed reaction test* (**Fig. 12.29**). Here, as shown on the left, a monkey sits behind a glass partition in a cage, in front of which are two boxes, one of them containing a reward such as a banana. The doors of the boxes are first opened to show what is in them, then closed again; after an interval of perhaps 10 minutes, the partition is raised, allowing the monkey to go and open the correct box and receive his reward. Normal animals can do this very well: animals with prefrontal lesions cannot, unless they spend the waiting period doing nothing except sitting and concentrating single-mindedly on the correct door – as soon as they are distracted, they are no longer able to recall which box holds the reward. On the right is a similar delayed saccadic task. Recordings from neurons in prefrontal areas during delayed response trials confirm the idea that these areas are in some sense to do with *waiting to do something*, with activity in many units starting up on receipt of the command, then firing in a sustained way until the response is finally made.

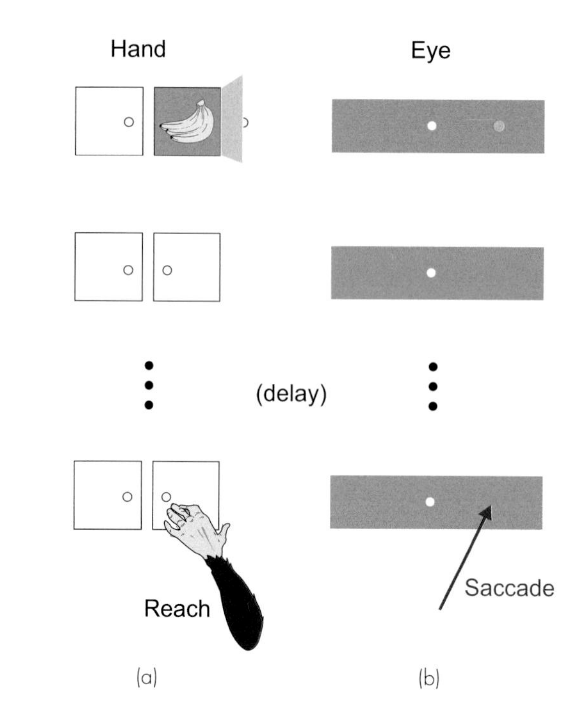

Figure 12.29 Two kinds of delayed reaction tasks. (a) Reaching for a reward after an enforced wait. (b) Making a saccade to a briefly glimpsed target, again after a wait.

Deferred action

These miscellaneous observations concerning the prefrontal areas can be unified quite satisfactorily once it is appreciated that each involves a defect in remembering to do a *deferred action* – as if there were some sort of prefrontal supply of Post-it notes. Anxiety is of course a side effect of the sense that something has to be done in the future, and lack of anxiety may sometimes merely indicate a lack of forethought: worry, if rational, is a thoroughly good thing. Thus it was anxiety that presumably made our ancestors save some of their seed harvest to plant for next year, despite their immediate needs, and it is perhaps not going too far to suggest that the enjoyment of *useful anxiety* of this kind is indeed what separates us most from the other primates. Finally, it may also be that the unpleasantness of pain, particularly when it results from terminal illnesses (it is this type of pain that frontal leucotomy seems best at alleviating) is at least in part due to the anxiety it causes us by reminding us of our impending death: the painfulness of an injury depends very much on the significance that we attach to it (see Chapter 4). By stripping pain of its meaning for the future, we also relieve its emotional threat.

Notes

Higher motor areas Some excellent general accounts: Evarts, E. V., Wise, S. P. & Bousfield, D. (1985). *The Motor System in Neurobiology.* (Amsterdam: Elsevier); Rothwell, J. C. (1994). *Control of Human Voluntary Movement.* (London: Chapman & Hall); Brooks, V. B. (1988). *The Neural Basis of Motor Control.* (Oxford: Oxford University Press); Jeannerod, M. (1997). *The Cognitive Neuroscience of Action.* (Oxford: Blackwell); Berthoz, A. (1997). *The Brain's Sense of Movement.* (Boston, MA: Harvard University Press). A recent and quite comprehensive review of clinical aspects is Watts, R. L., Standaert, D. G. & Obeso, J. A. (2011). *Movement Disorders.* (New York, NY: McGraw-Hill Medical);

an outstandingly comprehensive but correspondingly expensive clinical book is Donaldson, I., Marsden, C. D., Shneider, S. & Bhatia, K. (2011). *Marsden's Book of Movement Disorders.* (Oxford: Oxford University Press).

1. Motor cortex Useful accounts include: Passingham, R. (1993). *The Frontal Lobes and Voluntary Reaction.* (Oxford: Oxford University Press); Asanuma, H. (1989). *The Motor Cortex.* (Philadelphia, PA: Lippincott Williams & Wilkins); Porter, R. & Lemon, R. (1993). *Cortico-Spinal Function and Voluntary Movement.* (Oxford: Oxford University Press); Schmitt, F. O., Worden, F. G., Adelman, G. & Dennis, S. G. (1981). *The Organization of the Cerebral Cortex.* (Boston, MA: MIT Press); Passingham, R. E. (1996). Functional specialization of the supplementary motor area in monkeys and humans. *Advances in Neurology* **70** 105–116; Riehle, A. & Vaadia, E. (2004). *Motor Cortex in Voluntary Movements.* (Boca Raton, FL: CRC Press).
2. Secondary motor cortex An excellent review is Nachev, P., Kennard, C. & Husain, M. (2008). Functional role of the supplementary and pre-supplementary motor areas. *Nature Reviews Neuroscience* **9** 856–869.
3. Cerebellum Two comprehensive accounts: Ito, M. (1984). *The Cerebellum and Neural Control.* (New York: Raven Press); Palay, S. L. & Chan-Palay, V. (1974). *Cerebellar Cortex.* (Berlin: Springer). A recent and imaginative book is Ito, M. (2011). *Cerebellum: The Brain for an Implicit Self.* (Upper Saddle River, NJ: Prentice Hall).
4. Cerebellar speculation It is odd how *certain* people always seem to have been about what the cerebellum does. A passage from an 1853 textbook of phrenology: '*When the cerebellum is really large, and the temperament active, the individual becomes distinguished from his fellows by the predominance of his amorous propensities. In all his vacant moments, his mind dwells on objects related to this faculty, and the gratification of it is the most important object of his thoughts. If his moral and intellectual organs be weak, he will, without scruple, invade the sanctity of unsuspecting innocence and connubial bliss, and become a deceiver, destroyer, and sensual fiend of the most hideous description*'.
5. A cerebellar urban myth Someone is sure to tell you that the cerebellum can't be *that* important since just the other day a body turned up in the dissecting room where the cerebellum had been totally absent from birth, yet the man had made his living as a steeplejack (or ballet dancer or pole-vaulter or something similar). Mitchell Glickstein has traced the origin of this extraordinarily resilient urban myth in Glickstein, M. (1994). Cerebellar agenesis. *Brain* **117** 1209–1212.
6. Effects of cerebellar damage Much of this detailed characterisation of the effects of cerebellar damage is thanks to the work of the great neurologist Gordon Holmes during the First World War. The newly developed high-velocity rifles produced clean, precise wounds that were less likely to lead to death through sepsis; head injuries were frequent when trying to peer over the parapet, and the British helmet gave inadequate protection to the cerebellum. See Holmes, G. (1917). The symptoms of acute cerebellar injuries due to gunshot injuries. *Brain* **40** 461–535.
7. A cerebellar mnemonic A popular mnemonic used by medical students to recall the signs of a cerebellar injury (from, for example, trauma, tumour or stroke) is Danish P:

   Dysdiadochokinesis
   Ataxia
   Nystagmus
   Intention tremor
   Scanning speech
   Hypotonia
   Past pointing (dysmetria)
8. Marr's theory Marr, D. (1969). A theory of cerebellar cortex. *Journal of Physiology* **202** 437–507.
9. Cerebellum: a motor learner And perhaps not just motor: some believe that it may play a part in purely cognitive functions, but this is somewhat controversial. See Leiner, H. C., Leiner, A. L. & Dow, R. S. (1993). Cognitive and language functions of the cerebellum. *Trends in Neuroscience* **16** 444–454, with its lively commentaries and discussion afterwards.
10. Cerebellar predictive models? There is an excellent discussion of this possibility in Miall, R. C., Weir, D. J., Wolpert, D. M. & Stein, J. F (1993). Is the cerebellum a Smith predictor? *Journal of Motor Behaviour* **25** 203–216.
11. Depression and not potentiation See Albus, J. S. (1971). A theory of cerebellar function. *Mathematical Biosciences* **10** 25–61.
12. Basal ganglia Some specialised accounts in this difficult area: Steiner, H. & Tseng, K. T. (2010). *Handbook of Basal Ganglia Structure and Function.* (London: Academic Press); Schultz, W. (1998). Predictive reward signal of dopamine neurons. *Journal of Neurophysiology* **80** 1–27; Edwards, M., Quinn, N. & Bhatia, K. (2008). *Parkinsons Disease and other Movement Disorders.* (Oxford: Oxford University Press). A brief overview of basal ganglia circuitry for decision-making in eye movements is given in Noorani, I. & Carpenter, R. H. (2014). Basal ganglia: racing to say no. *Trends in Neuroscience* **37** 467–469.
13. Striosomes See Graybiel, A. M. (1990). Neurotransmitters and neuromodulators in the basal ganglia. *Trends in Neuroscience* **13** 244–253.
14. James Parkinson It is worth reading Parkinson, J. (1817). *An Essay on the Shaking Palsy.* (London: Whittingham and Rowland) partly as an admirable example of clear clinical description, and partly to see how excellent observation and plausible deductions can lead to an utterly false conclusion, in this case that the cause was '*some slow morbid change in the structure of the medulla, or its investing membranes, or theca, occasioned by simple inflammation, or rheumatic or scrophulous affection*'.
15. The need for external cues For a description of such a patient from that unusually thoughtful and wise analysis of what is really happening in parkinsonism, see Sacks, O. (1982). *Awakenings.* (London: Pan): '*Once a first step was taken – and walking could be inaugurated by a little push from behind, a verbal command from the examiner, or a visual command in the form of a stick, a piece of paper, or something definite to step over on the floor – Miss D. would teeter forward in tiny rapid steps. . . . In remarkable contrast was her*

excellent ability to climb stairs stably and steadily, each stair providing a stimulus to a step; having reached the top of the stairs, however, Miss D. would again find herself 'frozen' and unable to proceed. She often remarked that 'if the world consisted entirely of stairs' she would have no difficulty in getting around whatever'.

16. Holding This much-neglected topic is touched on in: Carpenter, R. H. S. (2011). What Sherrington missed: the ubiquity of the neural integrator. *Annals of the New York Academy of Sciences* **1233** 208–213. Carpenter, R. & Nooran, I. (2017). Movement suppression: brain mechanisms for stopping and stillness. *Philosophical Transactions of the Royal Society of London B Biological Sciences* **372** 20160542. Noorani I. (2017). Towards a unifying mechanism for cancelling movements. *Philosophical Transactions of the Royal Society of London B Biological Sciences* **372** 20160191. Noorani, I. & Carpenter R. H. (2017). Not moving: the fundamental but neglected motor function. *Philosophical Transactions of the Royal Society of London B Biological Sciences* **372** 20160190

17. Prefrontal cortex Three excellent and comprehensive accounts are Fuster, J. M. (2008). *The Prefrontal Cortex*. (New York, NY: Raven Press); Levin, H. S., Eisenberg, H. W. & Benton, A. L. (1991). *Frontal Lobe Function and Dysfunction*. (New York, NY: Oxford University Press); and Roberts, A. C., Robbins, T. W. & Weiskrantz, L. (1998). *The Prefrontal Cortex: Executive and Cognitive Functions*. (Oxford: Oxford University Press).

CHAPTER 13

ASSOCIATIONAL CORTEX AND MEMORY

The processes we look at in this chapter are those that we humans tend to be inordinately proud of, and to which we owe any temporary biological success that we have managed to achieve. Many animals are more agile and better coordinated than we are, can hear better or see better, and nearly all of them have more sensitive noses. Our special virtue is that we are quite good at *storing* and *processing* such sensory information as gets through to us, so that we use it to respond more effectively to our environment. But this difference is quantitative, not qualitative: there is no very sharp distinction between us and other creatures in this respect. It is simply that the parts of their brains that carry out these functions in a rudimentary way have in us been very greatly expanded and developed, above all the cerebral cortex

The organisation of associational cortex

If we compare the cerebral hemispheres of a series of animals from different evolutionary stages, what is striking when we get to humans is not just the expansion in the absolute mass of neural tissue, but the dramatic changes in the relative proportions of the cortex devoted to different functions (**Fig. 13.1**). Very little of a rat's cortex is not either primary motor or a projection area for one of the senses; in Man, by contrast, most areas of the cortex neither respond in an obvious way to simple sensory stimulation nor produce movements when electrically activated: they are what have sometimes been called *silent* areas.

Now these are precisely the properties we would expect from neural levels in the middle of the model of the brain shown in **Figure 13.1**, first presented in Chapter 1. Because a neuron in any level is activated only by a particular pattern of activity in the preceding layer, as we penetrate deeper into the sensory side we find that individual neurons become fussier and fussier about what they respond to, and eventually the chance of our finding out, in an experiment of finite duration, what they *do* actually do becomes vanishingly small. Stimulation is equally frustrating: unless we happen to stimulate them in a pattern that makes some kind of neural sense, corresponding to what is needed to activate the next layer along, nothing will happen at all. Both these problems are accentuated in

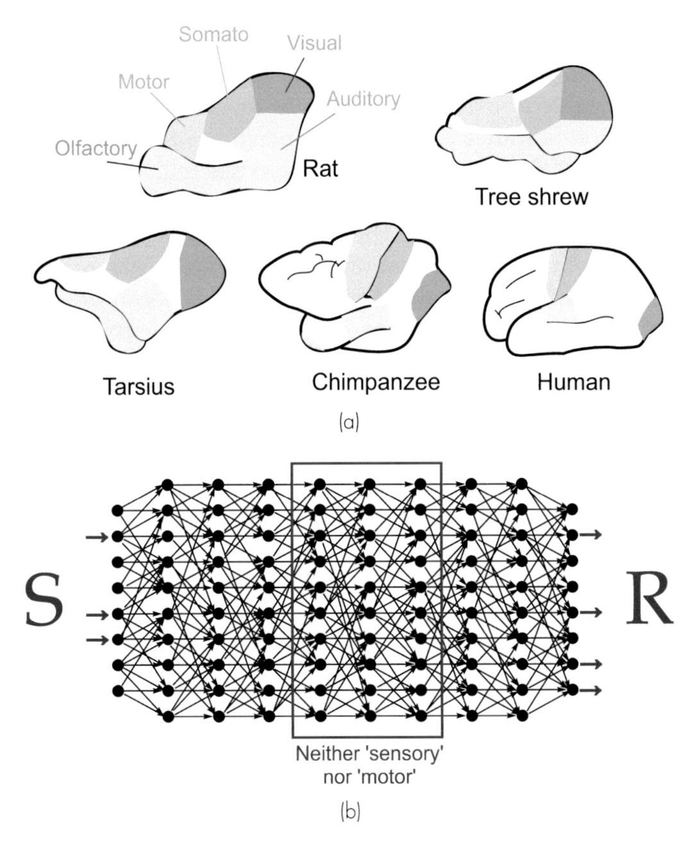

Figure 13.1 (a) Series of brains of different species, showing increase in extent of 'silent', associational, cortex (unshaded) in the course of evolution. (After Stanley Cobb, in Penfield, 1967.) (b) Representation of the brain as a series of neuronal levels.

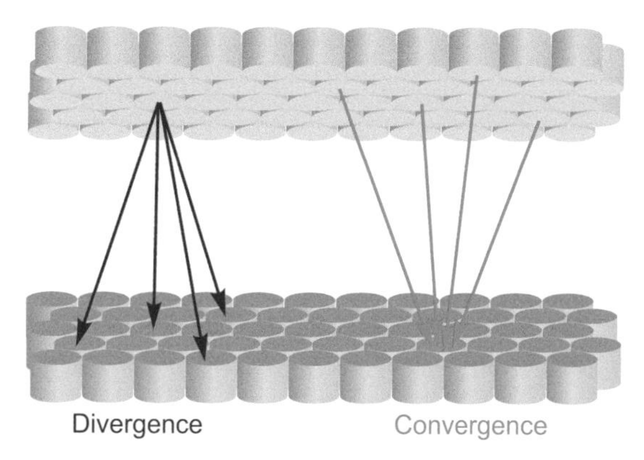

Figure 13.2 Convergence and divergence. (Reproduced with permission of Cambridge University Press from Cotterill, R. M. (1998). *Enchanted Looms: Conscious Networks in Brains and Computers.*)

cortical areas, where there is an immense degree of convergence and divergence from neurons in one column to those in others (**Fig. 13.2**).

The third weapon in the neurophysiologist's armoury, lesioning, is similarly blunted. Because these areas integrate or associate information from diverse sources which cut across the conventional divisions of sensory modality (which is why the corresponding cortical areas are called *association areas*), the effects of lesions often lack the functional specificity found, for example, in damage to the primary visual or motor cortex: we are at a high hierarchical level, in the sense discussed in Chapter 9. Lesions can often lack spatial specificity as well, and there is little of the topological orderliness found at more peripheral levels. This situation is rather like what happens in a telephone exchange: at the periphery – the region where the incoming cables arrive – there is a systematic relationship between a subscriber's number and the position of his particular connection, but the circuits in the heart of the exchange that set up the connections and form, in effect, associations between different subscribers are shared by all of them and used to set up different circuits on different occasions, and therefore have no obvious spatial organisation.

The analogy of the telephone exchange, introduced in Chapter 1, is a suggestive one. Just as an exchange is capable of connecting any subscriber to any other, so cortical convergence and divergence guarantees to provide a neural pathway from any sensory stimulus to any motor response, clearly a necessity for complete flexibility of behaviour. Similarly, the capacity of an exchange – the number of associations it can make at any one time – is simply proportional to the quantity of the common switching equipment it contains. Might the neural elements of associational cortex also be in some sense shared in this way? Such a notion, of associational cortex being uncommitted to any particular task, but providing a reserve of computing power that can be applied to whatever job is on hand, was originally suggested by the experiments of Karl Lashley described on p. 15. Lashley's extreme statement, his 'Law of Mass Action' – that the effect of lesions in associational cortex depends more on how large they are than on their exact location – is now less in favour. Recording from units in associational areas shows in many cases that they are 'silent' because we are using unnatural or boring stimuli; with adequate sensory patterning they can often be made to respond, in a way that may be complex and highly time-dependent, but does not alter radically from one experiment to the next. And it is clear from clinical observations in particular that discrete lesions in associational cortex can often lead to relatively specific functional defects, rather than something like a generalised loss of 'intelligence'. What *is* true, as we shall see, is that these functional defects may be of the wide-ranging and subtle kind that is characteristic of damage at a high hierarchical level – for instance the loss of the ability to speak French, while spoken English is unimpaired – and also that there is little reproducibility from subject to subject, in the sense that a lesion in a particular place in one person may have a completely different effect in another person. Such nuanced deficits are not readily appreciated by the busy non-specialist clinician, for several reasons. First, the commonest cause of discrete cortical lesion is a cerebral infarct, usually as a result of occlusion of a large cerebral artery or even a carotid. Among the devastation wrought by such a large lesion – paralysis, hemianopia, aphasia, hemineglect – impaired performance of higher, more specialised areas is often obscured; recognition is made particularly difficult as patients often appear curiously unaware of their loss. Second, despite advances in our diagnostic techniques and our understanding of the pathogenesis of stroke, our ability to do much about it is frustratingly rudimentary, and, certainly in the acute and early management of a stroke, any variation in the treatment administered is guided by the grossest of clinical signs, such as the presence of significant limb weakness or a marked visual deficit. Subtle difficulties with mental arithmetic or recognising street names are not easily identified because they are not intentionally sought, and not sought because, at this stage, they have only a limited bearing on clinical management.

In addition, we shall see clear evidence that cortical neurons can change their function to help cope with a change in functional demand. Thus the idea of a completely uncommitted pool of 'brain power' – rather like the 'cloud computing' that enhances the capabilities of smartphones – is an over-simplification; at a given moment the neurons are specialised in their function, but at a high level in the hierarchy: it is this that gives lesions in associational cortex their subtle and unpredictable quality. Cortex might be described as a community of specialists, acting in concert through an astonishing communication network by which they can share their ideas.

Structure

In Chapter 4 (p. 84) we saw how the cerebral cortex can be parcelled up into the Brodmann areas on the basis of

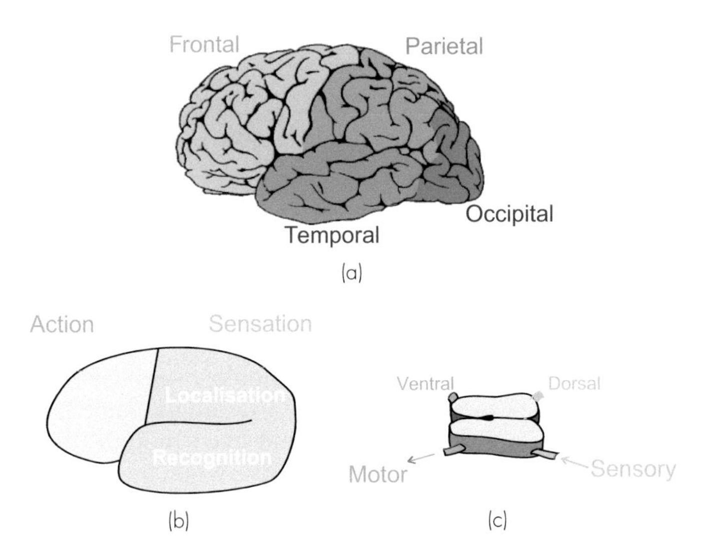

Figure 13.3 Gross divisions of the cerebral cortex. (a) The four major conventional regions. (b) The front half is essentially motor, the back half sensory (and further divided into localisation and recognition); this reflects the similar division of the spinal cord (c).

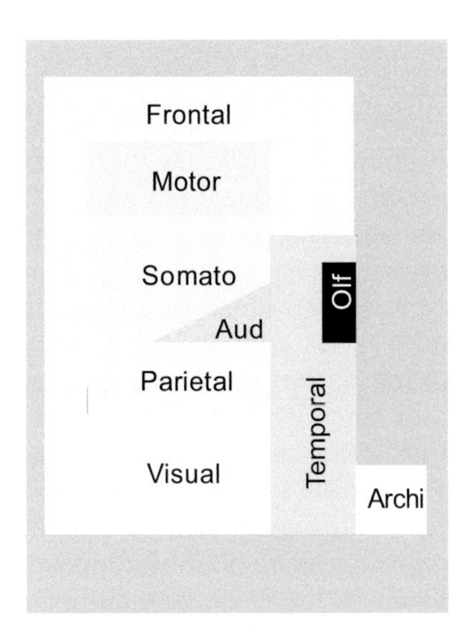

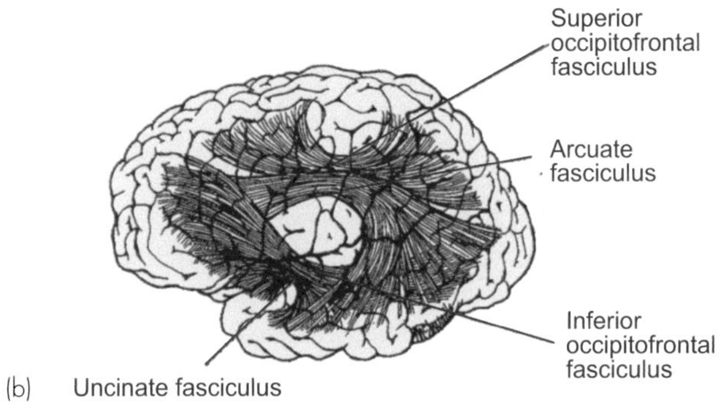

Figure 13.4 (a) A highly stylised representation of cortical topology, based on contiguity. (b) Major fibre bundles (fasciculi) of the human cerebral cortex. (Nolte, 1999.)

variations in the size and composition of its six layers. Conversely, it can be useful to group them together into larger functional units. In terms of gross anatomy, primate cortex is classically divided into four areas: frontal cortex (FCX) anterior to the central sulcus, temporal cortex (TCX) along the thumb of the cerebral boxing glove, occipital (OCX) at the back, and parietal (PCX) in between. In fact, the functional boundaries between PCX and OCX and TCX are not very distinct, and many neurologists prefer to lump them all together as parieto-temporo-occipital cortex (PTO CX, or POT). What we are then left with is a binary division of cerebral cortex by the watershed of the central sulcus into just two areas, front and back. Just as in the spinal cord, where the ventral half is essentially motor and the dorsal half sensory, so, broadly speaking, everything anterior to the central sulcus is, in a deep sense, motor, everything behind it sensory. Furthermore, the posterior half is itself divided into an upper part concerned largely with localisation and movement (the 'where' stream; see p. 160) and a lower part concerned with recognition ('what') (**Fig. 13.3**).[1]

Another classic way of dividing it up is into *primary sensory* and *primary motor* areas – and the rest. As we have seen, it is this remnant, *associational* cortex that has grown most in evolution, particularly the frontal associational area. Nowadays, more of the associational cortex tends to get called *secondary* or *tertiary* cortex, for example the large number of secondary visual areas. **Figure 13.4** shows human cerebral cortex in a highly stylised, topological form, based on contiguity. It demonstrates how these associational areas seem to form bridges between primary cortex devoted to different sensory modalities, and between primary sensory and primary motor cortex.[2] Right in the middle is the *posterior parietal cortex* (Brodmann areas 5, 7, 39 and 40), and one might therefore expect it to be concerned with the coordination of information from the visual, auditory, somatosensory and motor areas which surround them; and on the whole this seems to be true. There are massive fibre bundles connecting these neighbouring cortical regions with the parietal region, and it also receives a projection from the pulvinar and lateral posterior nuclei of the thalamus. The pulvinar in turn receives sensory information from visual areas 18 and 19, and from the colliculi and lateral and medial geniculate bodies; in addition it receives the usual reciprocal fibres from the parietal cortex itself. The lateral posterior nucleus obtains its input partly from the pulvinar and partly from the (somatosensory) ventroposterolateral thalamic nucleus. Efferents from parietal cortex go to the premotor and supplementary motor areas, to the frontal eye fields, to basal ganglia (and hence to colliculus) and indirectly to the cerebellum. In addition to connections between neighbouring areas, there are also bands of fibres (*fasciculi*) that link distant areas, and can be big enough to see easily with the naked eye in gross dissection (**Fig. 13.4**). Meanwhile, the corpus callosum with its staggering 100 million fibres shuttles information backwards and forwards between the two hemispheres.

Hence, by receiving a range of sources of sensory information and enjoying reciprocal communication with motor planning and execution, posterior parietal cortex mediates bidirectional sensorimotor communication: the posterior parietal cortex is ideally placed to provide a working model of our environment, which may then guide an appropriate motor response. Indeed, patients suffering vascular lesions to the posterior parietal cortex exhibit difficulty in tasks requiring sensory guided activity such as reaching for visually or somatosensorily presented objects, clinically manifesting as apraxia (see p. 264). On the other hand, receiving output reports from motor areas allows one to account for any motor activity that might influence interpretation of the sensory data reaching the posterior parietal cortex – did I walk into the chair, or did the chair walk into me? Consistent with this idea, patients with posterior parietal lesions often suffer from tactile agnosia (discussed later (see p. 262), the inability to recognise objects based on purely tactile information, a feat usually accomplished by interpreting the somatosensory consequences of manipulating an object. 'Feeling' is, after all, a very active process.

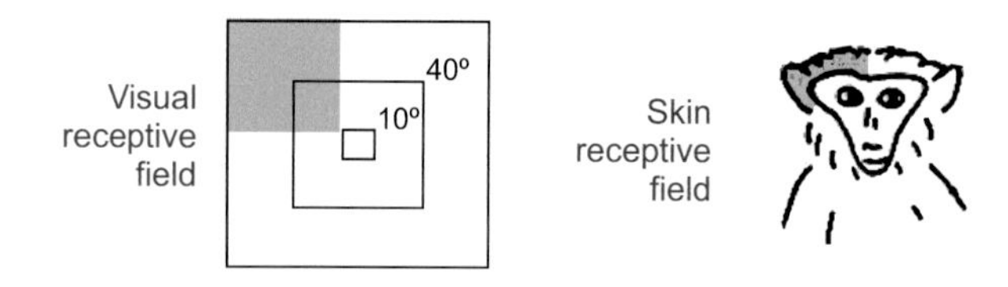

Figure 13.5 A neuron from monkey parietal cortex responding to vision as well as touch. Receptive field in one quadrant, with typical responses in a number of trials.

Neuronal responses

The responses of units in associational cortex show progressively more specific recognition of specialised features of the outside world, especially those that are *behaviourally* important such as hands and faces and eyes, compared with those of primary projection areas: some visual examples were noted in Chapter 7. In addition, as might be expected from such a diversity of input, neurons in PTO cortex often show complex responses to stimulation of more than one modality. **Figure 13.5** shows an example from such a neuron in the monkey, responding to both visual and somatosensory stimulation. Visually, it responds best to objects close to the eye, in one quadrant; it also responds to touching the skin in a corresponding quadrant near the orbit. Sometimes one observes responses when an animal *expects* a stimulus even though it is not present: in a sense the 'near the eye' unit could be thought of one that fires in response to a very near visual target because through experience it has come to expect contact to be made with the skin after visual stimulation.

Responses are also often greatly influenced by context and attention. For instance, many parietal units are visually driven, with receptive fields that can be mapped out; but unlike visual cells in visual cortex itself, they may or may not fire when a stimulus appears within the field, depending on whether or not the stimulus is sufficiently interesting to evoke a subsequent motor response such as an eye movement. For this reason, this area is better regarded as sensorimotor than purely sensory (if indeed such a distinction has much meaning).

A compelling example of the intimate association of 'sensory' and 'motor' areas is the coupling that has been shown between a region of the premotor area that is active in monkeys when doing; precision grip with the fingers, and another area in parietal cortex that is visually driven, and only active when the monkey sees itself – or indeed another monkey or human – doing the same task. Recording from PMA, we find that even though it is a motor area, its neurons are more active when the monkey is looking at the same task being performed even in a video – and conversely the sensory area is activated when the monkey carries out the task, even if it can't in fact see it.[3] The functional importance of such 'mirror neurons' and their system of mutual association is twofold: partly to enable one to learn tasks by looking at them, but also to understand what someone else is doing when you see them doing something – *prediction.* All of this appears to be due to a rather precise set of fibre connections that links the two areas together – both ways – in a remarkably specific way. Presumably there are many such mutual systems of association forming hierarchical ladders of predictive links between stimulus and response (**Fig. 13.6**), extending from the most basic, semireflex level (for instance, the links between primary somatosensory and primary motor cortex) up to the most subtle conceptual relationships, many stages removed.

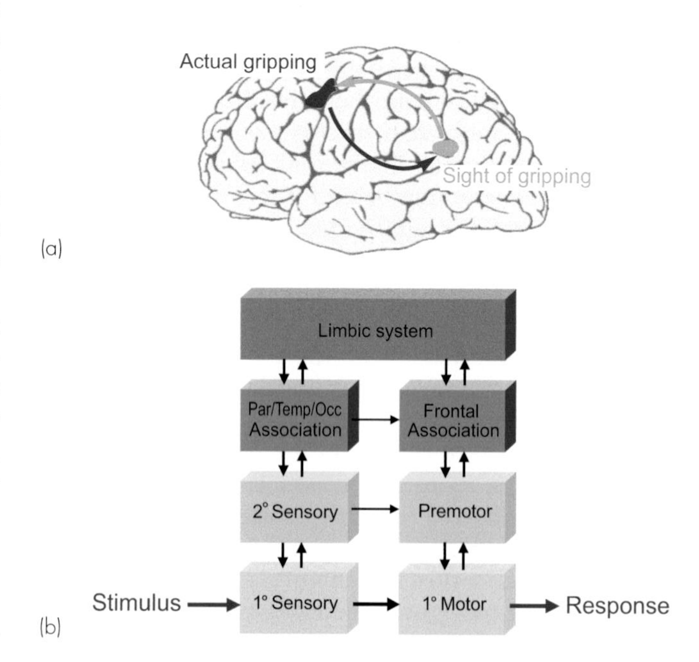

Figure 13.6 (a) In order grip an object, there is a communication between the parietal cortex that is visually driven and the premotor area that actually plans the movement. (b) Association of chains, ascending ladder of more and more indirect routes by which stimuli may cause responses. Par/Temp/Occ, parieto-temporal-occipital association cortex.

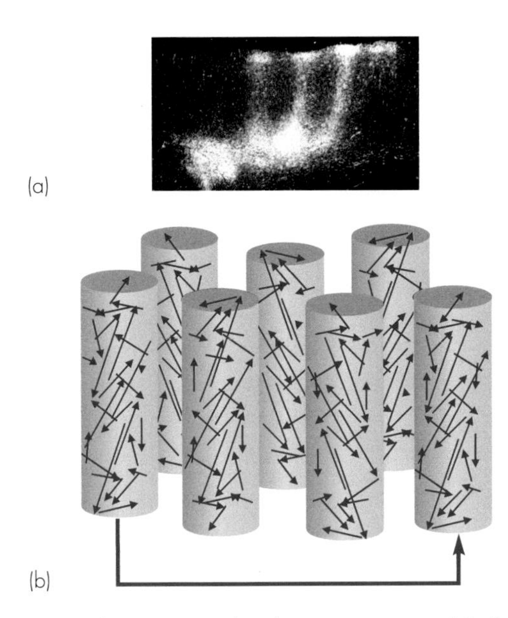

Figure 13.7 (a) Columns in retrosplenial association cortex, labelled by using local injection of tritiated leucine into frontal association cortex. (From *Fundamental Neuroanatomy* by Walle J. H. Nauta and Michael Feirtag, 1986.) (b) Columns debate within themselves, then send their conclusions to other columns.

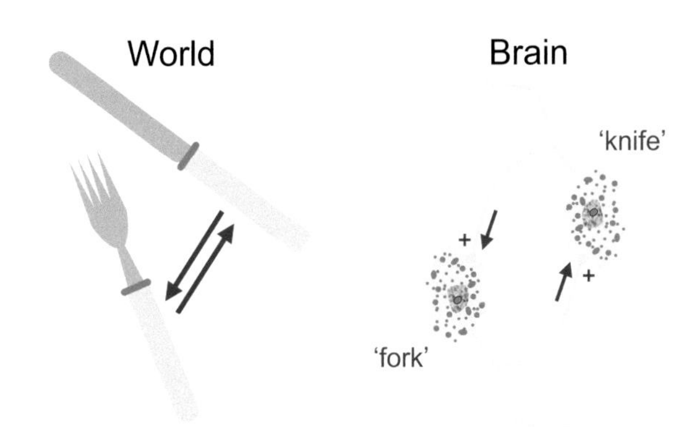

Figure 13.8 An illustration highlighting that the relationship between neurons in the brain mimics real-world relationships such as that between a knife and fork being used.

Association

These associational connections, so obvious even with the naked eye, are the sole reason that we have a cortex at all. We saw in Chapter 4 that the basic microstructure of the cortex seems to be of columns in which direct input from subcortical structures, especially the thalamus, mingles with associational projections from the pyramidal cells of other columns, the result being sent back down again through the projection efferents. This columnar arrangement appears to apply to all areas of cortex, association and motor as well as sensory: in the section of retrosplenial association cortex in **Figure 13.7**, fibres have been labelled by using local injection of tritiated leucine into a different area, frontal association cortex. It is important to emphasise that cells in any one column talk a great deal to *each other* but essentially turn their backs on their neighbours, like people sitting gossiping at different tables in a pub. The intracortical associational links are rather as if they keep ringing the other tables on their mobile phones to tell them the conclusions of their discussions.[4]

Why is this useful? Because the resultant network of links between neurons mimics the relationships between things in the outside world (**Fig. 13.8**), what we end up with is a probabilistic model of the world in the brain, a *predictive* model that can be used to anticipate what is likely to happen next: when we see a knife we think of – and expect – a fork.

But to make such models work, it is not enough to provide the largest possible number of potential connections between all the neurons representing stimuli and actions, things in the outside world that we might want to link together. We must also have a mechanism for making the strength of these connections *change* to reflect the associations that are actually observed in the outside world. And this is the secret of cerebral cortex: it provides a mechanism for creating physical connections between neurons that are often active simultaneously (*fire together, wire together*). Artificial neural networks, programmed with this rule, can be implemented in computers and are increasingly being used to perform complex tasks of pattern recognition that must by any criterion be regarded as 'intelligent'.[5]

From the very earliest days of experimentation on cortex by investigators such as Sherrington, it became clear that even non-associational cortex like motor cortex had extraordinary properties of plasticity, with functions rather *diffusely* represented and maps that could expand or contract as the result of stimulation or disuse. Simply having an arm in plaster for a few days leads to the arm 'losing its place' in the sensory cortex. Conversely, relatively short periods of localised stimulation (such as monkeys receiving tactile stimulation from their fingertips by means of a rotating textured wheel) lead to very obvious expansion of the corresponding areas of the cortical map at the expense of neighbouring areas (**Fig. 13.9**). Cortex is a turbulent region: stimuli jostle each other in a desperately competitive way.[6]

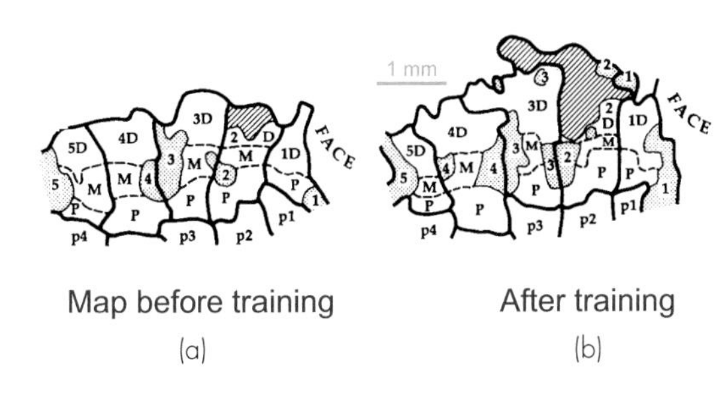

Figure 13.9 Effect of training on map in monkey somatosensory cortex. (a) Monkeys received continual tactile stimulation of their fingertips by a rotating textured wheel. (b) The result was expansion of the corresponding parts of the digital map, partly at the expense of neighbouring areas. (From Merzenich et al, 1990; permission pending.)

Figure 13.10 How specific can feature detectors be?

Figure 13.11 What is 'A-ness'?

Functions of association

When Hubel and Wiesel published their amazing findings – that cells in the visual cortex code for a wealth of information about the visual world, looking for spots and edges and lines of a certain orientation, of a particular length and moving in a particular direction and so on – people thought *At last we understand how the brain works*! Clearly, having once seen the principle of what is sometimes called feature extraction, we could go on like that forever. We can join line-detectors together to make detectors for squares, for numbers, for letters of the Russian alphabet, for *teacups* and so on ad infinitum (**Fig. 13.10**). Is there any limit to this? Is there a special cell for recognising *every single thing* we see? It is not at all clear that even with the countless numbers of neurons at our disposal we could *actually* do something like that. For teacups, just possibly: but for *porcelain* teacups, for *eighteenth-century French* porcelain teacups? It doesn't sound very likely, if only because the number of cells required would soon exceed even the million million that we are provided with. So how is it done?

Recognition implies classification. A classification based on only one criterion is trivial to implement: it is easy for example to build a machine that will sort peas according to size before stuffing them into appropriate tins. The problems start when there is a large number of attributes to be taken into account before a stimulus can be assigned to one or another category when they are subject to random variation and when it is the *relation* between them that is the crucial factor, rather than exact correspondence with some kind of template. If the objects to be recognised are highly stereotyped, like £10 notes, it is not difficult to make a device that looks for a match between a stored 'ideal' bank note in the machine's memory, and the actual specimen that is presented. But how, for instance, do we recognise that a set of objects as different as this set of As actually belong in the same category? It is hard to define an 'ideal' letter A, or say what essentially is the A-ness that all these particular examples have in common (**Fig. 13.11**). And we recognise a rose not because it is identical to some archetypal rose but because some aspects of it are similar to other specimens that we have seen. Although in other respects – perhaps its size or its colour – it may be different, we know these aspects are irrelevant and can be ignored. Thus there are two components to recognition: one is to do with associating together those attributes of a stimulus which define what it is; the other is the *filtering out* of aspects of the stimulus that are irrelevant.[7]

Filtering out irrelevance

A stimulus is a function not only of the object but also of quite accidental things like how it happens to be illuminated, what angle you are looking at it, whether it is heard through a blocked ear or felt with numbed fingers, and so on. So while it is easy to make a machine recognise banknotes fed into a slot in a fixed position and with constant illumination, *real* recognition means being able to do this when one is just waved briefly in front of you, or felt crumpled in one's back pocket. So the job of any sensory system is to take the stimulus and filter out those aspects of it that are *accidental* in this sense, and leave behind those that are intrinsic to the object, or *essential*. This is what we mean by recognition.

People often get muddled about the difference between the *stimulus* – the pattern of energy falling on receptors – and the *object* that gave rise to that pattern in the first place. Of course it is the object that has to be recognised, not the stimulus: stimulus is, in a sense, a coded version of the object that has to be decoded again. And this is the essential problem in recognition, because the same object can give rise to very different stimuli on different occasions. Objects in the real world are perceived at different times under lighting of different intensities and colours, and from different distances and directions. The stimulus is a coded version of the object that gave rise to it, some aspects being *essential*, and due to the object itself, and some being merely *accidental*, and nothing to do with the object at all. A particular retinal image of a cube under particular conditions is as much a coded version of the cube, that has to be deciphered, as are the four letters CUBE: in many ways the latter presents an easier task. So the job of the visual system (and indeed any sensory system) is to separate off those aspects of an object's image which are its essential attributes in the sense of defining its essence, and those which are merely accidental and the result of temporary circumstances (**Fig. 13.12**).

This mechanism of filtering out is something we have already met when considering the functions of adaptation,

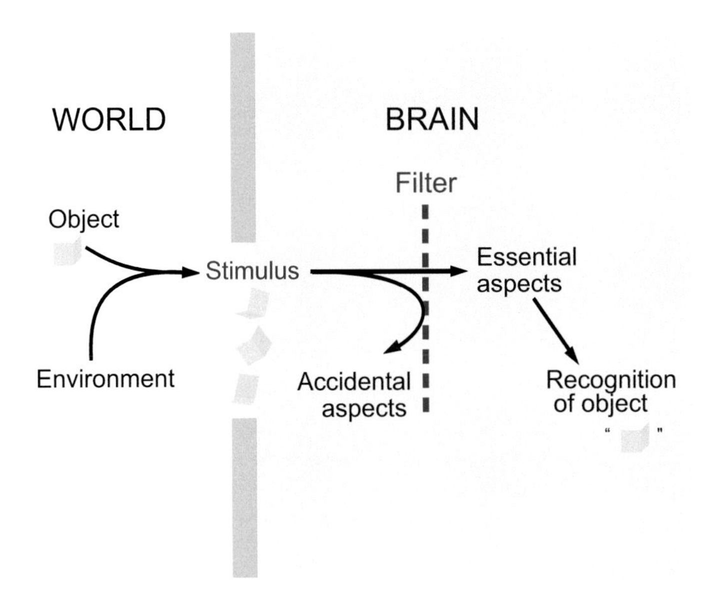

Figure 13.12 A stimulus is a coded version of the object in the environment that causes it, some aspects of it being due to the object itself, and some to accidental factors. It has to be decoded by the brain filtering out the accidental properties to leave behind those that are essential to the object itself.

more particularly in the eye. We saw that one of the functions of dark adaptation is to enable us to perceive the intrinsic albedo of an object despite the fact that on different occasions it may be brightly or dimly illuminated. It may seem strange that whether something is perceived as white or black should apparently bear so little relation to how much light the eye gets from it, but if you think about it for a moment you can see that it is exactly what is needed in order to *recognise* objects. What the brain has to do is to make allowance for different levels of illumination, so you actually perceive albedo rather than luminance. Albedo is what tells you *what sort of object it is* – a snowball or a lump of coal – whereas luminance is messed up with how brightly the sun happens to be shining, quite irrelevant to deciding what you are looking at. By filtering out the accidental aspect of the stimulus – illumination – we are left with the essential property, albedo, which is characteristic of what we are trying to recognise.

It is not hard to extend this idea – of discarding accidental properties to leave behind what is most characteristic of an object – to much higher levels of perception. When we look at a coin on a table, it looks circular even though its retinal image is in fact an ellipse. It is easy to see how a mechanism of associative memory will do it: as we move our head, different ellipses tend to occur together and the corresponding neurons get wired together; as a result, we are filtering out the accidental aspects of the stimulus that depend on us and not on the coin (**Fig. 13.13**). Or again, when we look at that set of As, we are filtering out the accident of the way in which they happen to have been designed. At a higher level, we recognise that *arbre, baum* and *tree* are all essentially the same by disregarding the accident of what language they happen to have been expressed in. The same principle operates at the very highest levels of thought: in chess, rather than laboriously calculating all the

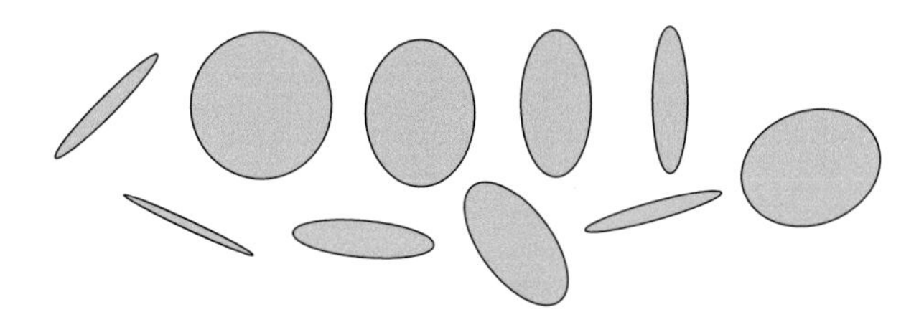

Figure 13.13 The brain filters out the accidental aspects of the shape of a coin that depend on our perspective in order to correctly interpret the circular shape of the coin.

possible consequences of a given move, a good player will perceive that the position he is faced with, though never experienced before, is in some *essential* way the same as one he knows how to win. Intelligence in fact is the ability to discern *an underlying similarity between things when they are obscured by irrelevant detail.*

Thus filtering, classification, recognition and association are simply different aspects of the same underlying phenomenon. Our brains are not *genetically* programmed to recognise lines and faces and letters of the alphabet and so on: they get wired up through experience. We recognise a figure '3' because it has certain topological features that are found in association together: a single continuous line with a cusp in the middle to the left and a couple of bulges to the right. If we imagine individual neurons that respond to each of these features, we can see in general terms how, with sufficient repetition, they would tend to strengthen their mutual connections and form a functional cluster corresponding to the existence of 3s in the outside world.

Learning to classify

Consider, for example, line-detectors in the visual cortex discovered by Hubel and Wiesel, which are essentially created by wiring together retinal ganglion cells that lie in a row. Experiments have shown that even these relatively simple detectors require the animal to have actually *experienced* lines for them to wire themselves up properly. If, for example, you rear kittens from birth in such a way that they never experience horizontal lines, then one finds on testing their cortical units that the great majority of them respond only to vertical or near-vertical lines and not to horizontal lines at all. It seems therefore that the *stimuli themselves wire the units up.*

It is not difficult to think of a plausible mechanism by which this could happen (**Fig. 13.14**). Imagine a 'naive' or untrained cortical cell, that starts with quite random retinal connections as shown in the figure. If we then present a line at a particular orientation to it, it is clear that though the field is initially random, nevertheless the line may well stimulate it enough to make it fire. If now we make the usual Hebbian assumption, that those synapses where the pre- and postsynaptic cells are active simultaneously get strengthened, you can see that the *useful* inputs to that cell will get stronger and the others presumably relatively weaker, so that eventually it builds its own linear receptive field, and is also capable of responding to incomplete

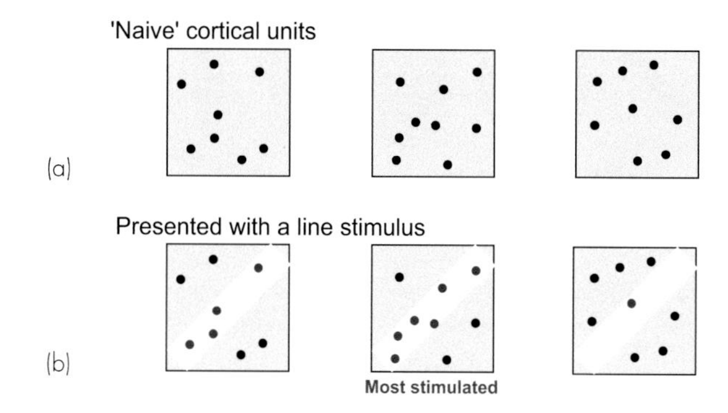

Figure 13.14 Hypothetical mechanism by which the specificity of central visual neurons might grow from experience. (a) Receptive fields of three 'naïve' neurons, indicated by the dots. (b) On stimulation with a slit of light at a particular orientation, only the middle neuron fires: its afferent fibres grow stronger, while the others decay.

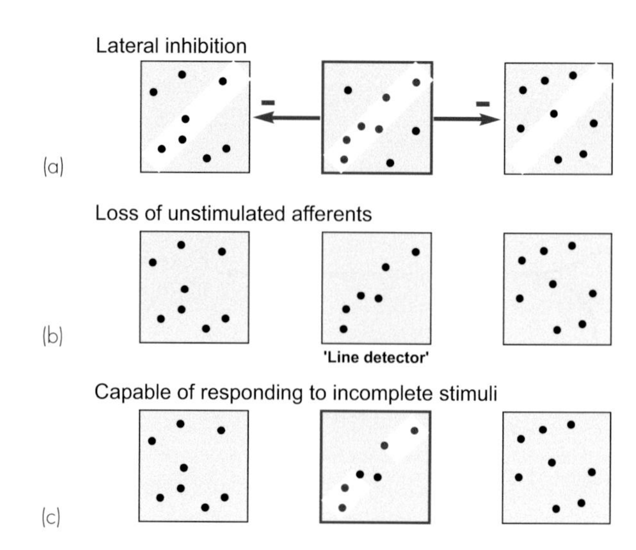

Figure 13.15 (a) Through the mechanism of lateral inhibition, a neuron selectively responding to a line of a certain orientation will suppress the response of its neighbouring neurons that are less responsive to that stimulus. (b) Unstimulated afferents will not respond to a stimulus that another neuron becomes selectively responsive to. (c) The middle neuron that selectively responds to the stimulus is capable of responding even if only part of the appropriate stimulus is present.

stimuli. (Such a model can be extended to cover the generation of inhibitory surrounds as well; but it must be said that recent work has indicated that the true mechanism, though not understood in full, is not quite as simple as the one presented here.)

Extending such a notion to yet higher stages of cortical processing, we can easily imagine units that could learn in exactly the same way to respond to the more complex sets of essential features that make up things like teacups and human faces. It is certain that some such mechanism of learned connections must exist, for we know that the cortical units of young kittens brought up in a visual environment consisting entirely of lines having a single orientation will on subsequent testing respond only to lines of that same orientation. Once such a set of features have been associated together in this way, the detector may not mind very much if some of its inputs are missing on a particular occasion: so long as it fires more actively than any of its neighbours in response to a particular object, then it will in effect form a hypothesis about what is present (**Fig. 13.15**). We not only recognise the A, but are convinced we can actually see phantom contours that we expect but are not actually there. In this way we can deal with situations where – as is almost always the case in real life – parts of objects are obscured because they are hidden behind other objects. So our perceptions are conditioned all the time by our expectations, and in fact the cortex's job is to be a sort of Sherlock Holmes, jumping to conclusions on minimal evidence: *having* recognised, we perceive associated elements we *expect* to see but not actually present in the stimulus at all. Decision-making columns, receiving fragmentary clues from the outside world, but also exchanging gossip with each other through their associational fibres, provide an ideal mechanism for doing this. **Figure 13.16** shows how this could operate in a particular case. An ambiguous letter (A? H?) is perceived as A because of the predictions from neighbouring columns that respond to part of the local *context*.

Recent work has revealed something of the dynamic properties of associative networks of this kind. Many individual units all over the cortex and in some subcortical regions show electrical activity that appears to rise steadily before an action is about to be performed, collapsing again once the movement actually starts. A well-studied example is the saccade (**Fig. 13.17**). In the superior colliculus, as well as in the frontal eye fields and posterior parietal cortex, there are units whose firing starts to accelerate as long as two or three hundred milliseconds in advance of a saccade made to fixate a visual target, and similar early responses can be seen in the electroencephalogram (EEG) long before other kinds of movements. When this build-up of activity reaches a particular level, it appears to trigger the movement itself, and variation in the rate of rise from trial to trial seems to be reflected in variation in reaction time. (These mechanisms of decision are discussed further in Chapter 14, p. 284.) It is not difficult to see how this kind of behaviour might arise through associative circuits, with information from the stimulus being combined with associative links representing the probability of the target existing (given the context), to form a crescendo of agreement between widely scattered units that nevertheless act in synchrony and finally decide to initiate action. The parallels with how decisions are reached in human organisations are obvious: a company board of directors would have to vote with a majority in order to take a major action affecting the entire company.[8]

Lateral inhibition

Lateral inhibition is an important component of models of this kind. It plays an important role in discarding accidental information when distinguishing the aspects of visual, auditory or tactile stimuli that allow recognition; but it also plays a part in association. As we saw when discussing the process by which line-detectors might be generated, during the learning process it will ensure that only the

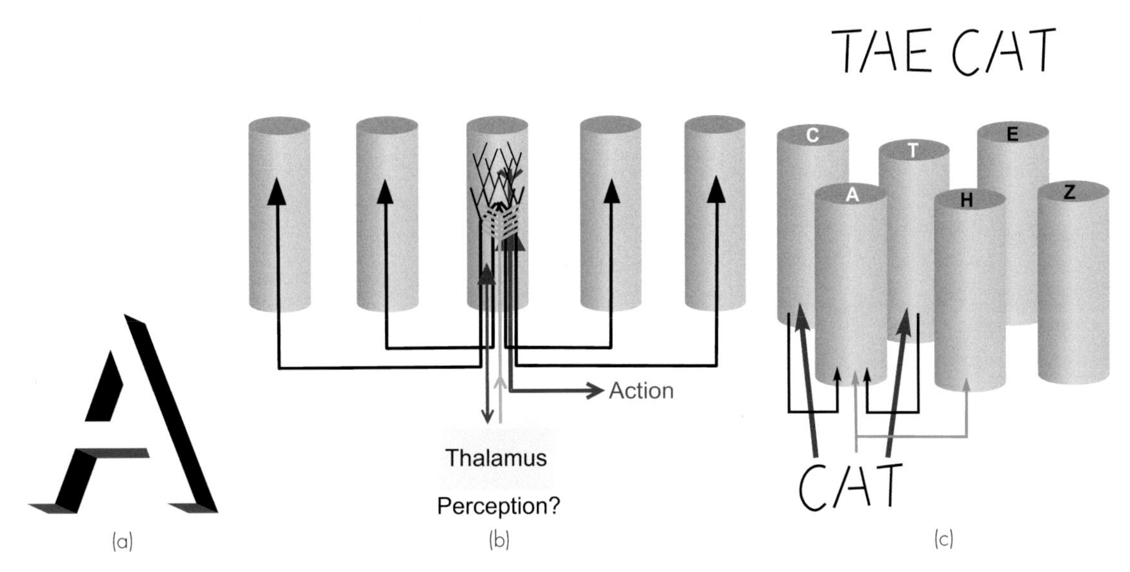

Figure 13.16 (a) Phantom contours. (b) Neural circuits for association: 'real' input is brought into association with predictions from other columns, resulting in activity of pyramidal cells generating action and perhaps perception. (c) The mechanism in action – an ambiguous letter (A/H) is perceived as 'A' because of the predictions of neighbouring columns responding to parts of the local context.

cells that are most stimulated by a particular pattern will be activated enough to increase the strength of their afferent synapses. And subsequently it will help to sharpen up the discrimination between stimuli that differ only slightly by enhancing any differences in the patterns of neural activity that they evoke. Lateral inhibition of this kind is not strictly spatial, of the kind introduced in Chapter 4, but operates rather along what might be called an abstract sensory variation dimension. For example, the mutual inhibition between red- and green-sensitive channels in the retina that generates colour-opponent responses can be thought of as lateral inhibition along a wavelength axis, that sharpens up colour discriminations. In the same way, lateral inhibition between line-detectors in the visual cortex acts along a dimension of orientation, improving angle discrimination. A consequence of this is that if two lines are presented at once, forming an angle, the effect of the lateral inhibition is to exaggerate the difference in their orientation, and thus make the angle seem larger. Many well-known optical illusions can be explained by angle expansion of this kind (**Fig. 13.18**).[9] In the Zöllner illusion the thick zigzags twist the apparent orientation of the thinner lines away from their true orientation. The same mechanism is at work in the 'café wall' illusion. In the same way, the square (below left) becomes distorted in shape because of the tilted lines behind it, and on the right (the Poggendorf illusion) angle expansion produces the strong impression that the two parts of the oblique line will not in fact line up. While these illusions appear merely to distort our interpretation of the visual environment, in less artificial circumstances the same mechanisms serve to exaggerate differences and facilitate distinction between objects.

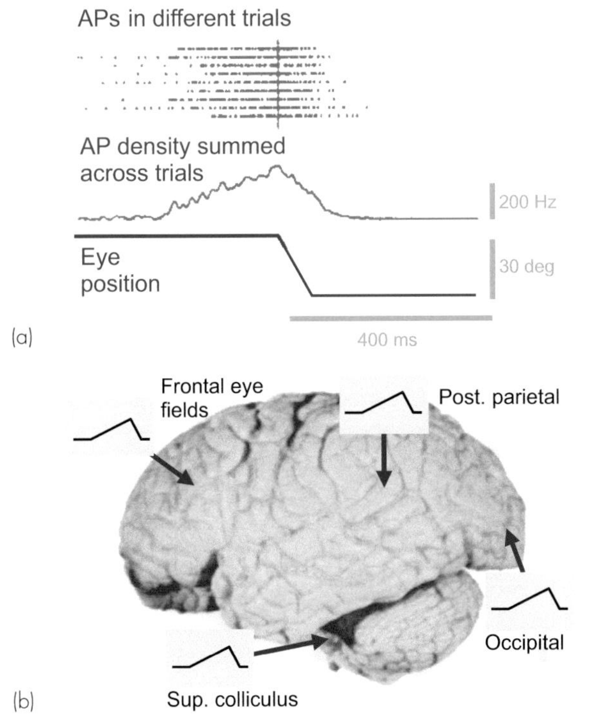

The most abstract example of all is perhaps in the olfactory bulb. We saw in Chapter 8 that olfactory receptors are very unspecific as to the chemical stimuli they respond to: to take a simplified example, while one receptor might respond to substances A, B, C and D, its neighbour might respond to A, B, C and E. But the effect of lateral inhibition between second-order neurons within the olfactory bulb will be to eliminate the overlap between the two

Figure 13.17 (a) Before a saccadic eye movement to a particular target, neural activity as shown by action potentials (APs) rises steadily in related neurons in the colliculus. (From Munoz & Wurtz, 1995; with permission.) (b) Similar rises occur simultaneously over several areas of cortex, presumably linked by associational connections.

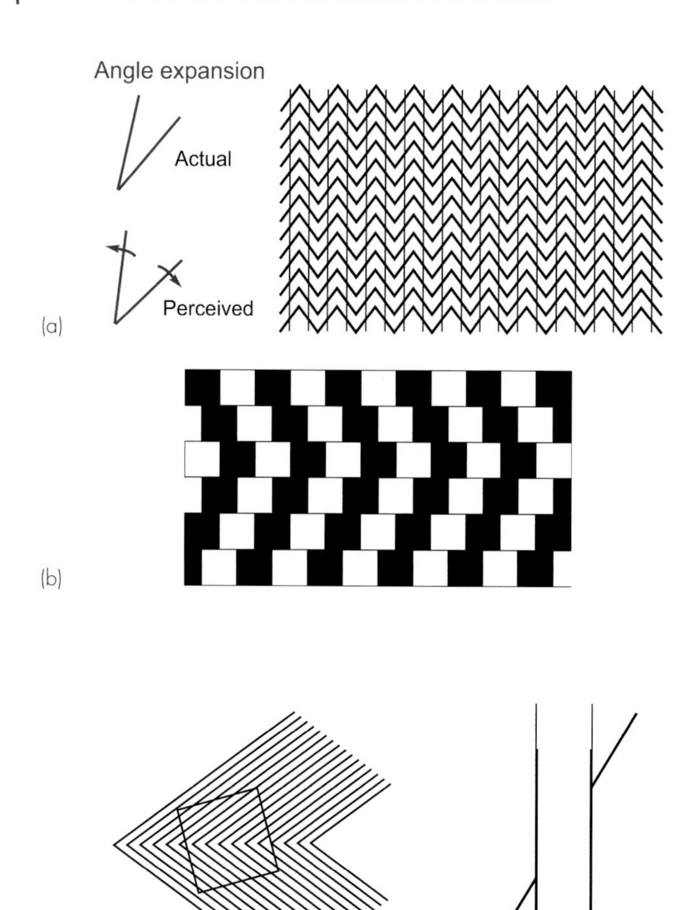

Figure 13.18 Some illusions caused by lateral inhibition amongst orientation detectors; in each case, the apparent orientation of a line twisted away from a neighbouring line of different orientation. (a) The thin lines are in fact parallel (the Zöllner illusion). (b) The 'café wall' illusion: the zigzag lines formed by the darker bricks cause the apparent orientations of the thinner lines to alter. (c) The figure on the left is actually a square; on the right the angled lines are in fact aligned, though the expansion of the angle they make with the vertical lines makes it look as though they would not meet if extended (the Poggendorf illusion).

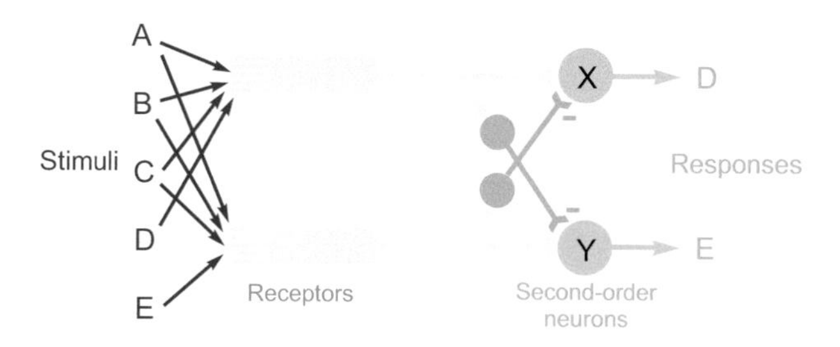

Figure 13.19 Lateral inhibition as a means of increasing modality specificity. Left: two idealised olfactory receptors, each responding to its own list of substances (A, B . . . E), but with considerable similarity of response. The effect of mutual inhibition at the level of the second-order neurons (right) is to increase the specificity of the response by eliminating responses that are common to the two receptors: neuron X will only respond to D and neuron Y only to E.

'receptive fields' and thus considerably sharpen up their modality specificity to particular stimuli – X responds only to D and Y only to E in **Figure 13.19**. Here we have lateral inhibition that is in effect operating in a multi-dimensional stimulus space.

The final goal of recognition is not of course simply the identification of individual objects, but of attaching *meaning* to them. This implies, in effect, associating them not only with each other, but with words, actions, and above all with emotional states and with the satisfaction of physiological needs. The only way to make sense of what the brain does is to take a firmly pragmatic line, to insist all the time on asking what *use* things are. Our senses are not after all merely there to provide some sort of in-flight entertainment for the soul: they have evolved because they help us to survive. They are required by the motor system both in the planning and execution of actions, and also by the *motivational* systems that decide what action to take: whether an object is nice or nasty, whether it is something to eat, or something that will eat us. It is not difficult to see how this can be done, by an extension of the mechanism for forming associations by means of synaptic strengthening, and this is what the next section will discuss.

Neural mechanisms of association

Psychologists prefer classifying learning and memory in elaborate ways. One which has been popular is to distinguish between procedural memory – learning *how* to do things – and declarative memory – learning *what*, the latter essentially conscious recognition, often verbal: 'facts'.[10] The latter may be further divided into *semantic* memory – such as knowing that Copenhagen is the capital of Denmark – and *episodic* memory – that last night I went to a concert. A difficulty here is that a lot of recognition is not conscious at all, and is part and parcel of how we control our actions, and is therefore sort of procedural (for example lane markings on the road guide our steering – we remember what they mean without consciously attending to them). Another classification in some ways more satisfactory from a physiological point of view is a functional one based partly on the structures in question: in these terms there are essentially three kinds of memory. The first and perhaps most fundamental kind is the learning of secondary motivation, introduced in Chapter 8 (p. 177), which can be regarded as associated with the limbic system, and especially the *hippocampus*. The second kind of memory is obvious in the motor system, in parametric feedback, as for example when one learns to recalibrate the vestibular responses to head movement after wearing reversing prisms (see Chapter 11) or to learn to produce ballistic sequences of actions, using sensory feedback to modify the responses if they do not lead to success, and is associated with the *cerebellum*. The third kind is for us the most important of all, the associational function of our grossly inflated *cerebral cortex*: sensory learning, which is the basis of recognition. We have, for instance, the kind of *sensory* memory which is implied by the ability of higher-order sensory cells – beginning in primary visual cortex – to develop for themselves a selectivity to those particular patterns of input that actually occur in the environment (p. 254). Can we hope to find any common mechanism for each of these types of memory, performed by different areas of the brain?

Certainly the first and third of these varieties are not conceptually very different: one can imagine a continuum of types of learning between, say, learning to associate patches of light on the retina into lines and edges, learning letters of the alphabet, learning a song composed of the same letters, and learning that singing it on television culminates with a large cheque. In each case, the key operation is one of forming *associations* between those elements of the stimulus that tend to recur together: if retinal units tend to fire in rows, we learn to recognise lines; if lines quite often lie in a certain relation to one another, we learn to recognise an 'E'; after we have seen a particular configuration of such letters a few times, we have learnt 'Mr Tambourine Man'; and finally we learn that singing it is associated with huge monetary reward. Similarly, in the case of Marr's plausible model of motor learning by the cerebellum, examined in Chapter 12, it is the associations formed between sensory feedback patterns and ensuing fragments of action that ultimately result in the learning of motor sequences. Of course, repeated association of stimuli is not always sufficient or necessary to generate memories, you may well fail to recall billboard advertisements or street signs that you pass by every day, while it does not take repeated trips to hospital before you learn which of the neighbours' dogs will bite the hand that strokes it. Clearly, higher centres contextualise the associations, a descending influence that coincides with and more strongly 'associates' those connections which are worth remembering while neglecting those which are less important. It is, of course, the student's dream to be able to control this process: at some time we have all lamented our ability to recall the words of an irritating pop song at the seeming expense of being unable to retain the information demanded in an exam.

Nevertheless, all learning by the brain must amount, in the end, to the formation of physical connections between neurons in such a way as to mirror the associations that exist in the real world between the stimuli that those same neurons code for. Memory, the process that models the world within our heads, must operate through *synaptic plasticity*.[11]

Conditioning

Consider a classic example: Pavlov's famous experiments on dogs, which for the first time showed that learning could be quantified and treated as a thoroughly scientific phenomenon. A dog is trained by frequent association of sound and food to salivate when a bell is rung (**Fig. 13.20**). Since he didn't do it before, there must have been a change in his neural connections. What can we deduce about what must have been going on in his brain? Here there are, in simplest terms, two stimuli or inputs (the *unconditional stimulus* (UCS), the sight of food; the *conditional stimulus* (CS), the sound of the bell) and one output or response

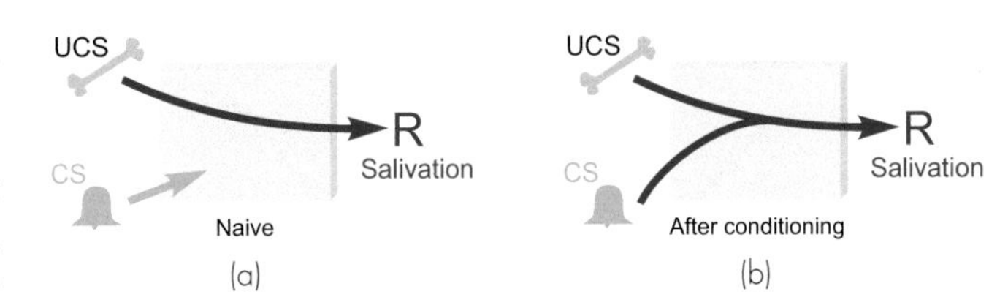

Figure 13.20 Memory and neuronal connections. Schematic representation of functional pathways before (a) and after (b) Pavlovian conditioning of salivation to the sound of a bell (CS, conditioned stimulus) by frequent pairing of bell and food (UCS, unconditioned stimulus).

(R, salivation). In the end, since either input will produce the output, there must be at least one chain of neurons forming a functional pathway from UCS to R, and another from CS to R. Before the period of training, the second pathway either does not exist, or perhaps exists in the structural sense but is functionally incapable of initiating salivation. It follows that learning the association between CS and R is brought about either by growth of new neuronal connections, or by the activation of pre-existing ones. The only questions that remain are, first '*What are the conditions under which such growth or activation occurs?*' and second, '*What is the neuronal mechanism of these processes?*'

Now there is one further point that may be deduced about the Pavlov dog's brain when it has finally learnt to make its conditioned response. There must be *at least one* neuron – the one that actually innervates the salivary gland, if none other – that is common to both pathways and where they first come together; this is the cell X shown schematically in the diagram, and in the simplest case of all might have exactly one synapse (A) driven ultimately by UCS, and one (B) driven by CS. Let us for the moment consider only the second and more likely of the two possibilities mentioned earlier, namely that both synapses are structurally in existence before the training period, but that the synapse B is in some kind of inactive, dormant state; we assume that synapse A on the other hand is always capable of firing X and hence producing salivation. What we observe is that after sufficient pairings of food with bell, the bell alone eventually produces salivation. Translating this into what is happening in the region of X, this means that the more often A (and hence X) fires at the same time as B, the stronger becomes the connection from B to X, until in the end B is able to fire X all by itself: the bell produces salivation (**Fig. 13.21**).

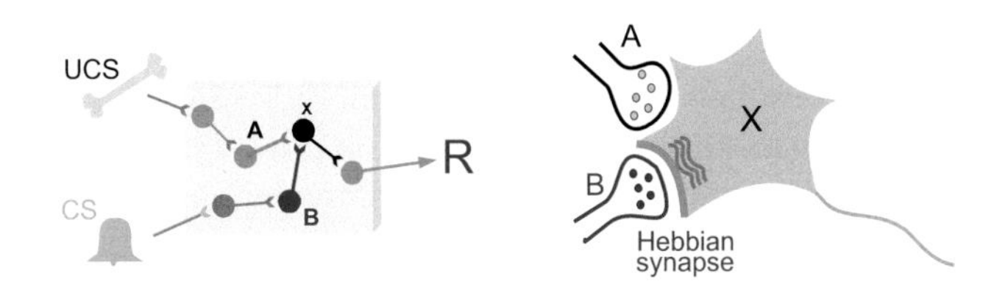

Figure 13.21 Simplified representation of the functional chains of neurons that must exist after conditioning: X (shown in more detail on right) is the first neuron common to both paths. It has an afferent, A, that is driven by UCS and an afferent B, driven by CS, whose effectiveness has increased because of frequent joint activity of X and B.

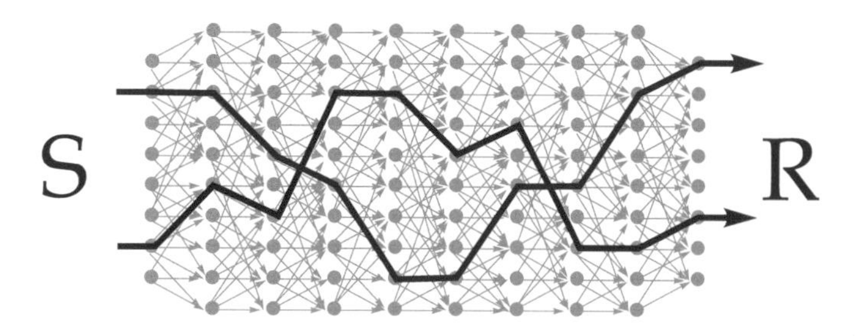

Figure 13.22 A schematic diagram illustrating how multiple routes can be taken from a sensory neuron to an output motor neuron, given the large amount of divergence and convergence present in the brain.

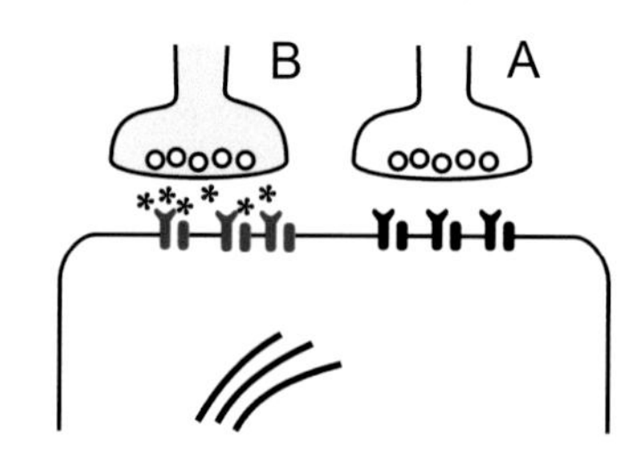

Figure 13.23 NMDA synapses and learning. Synapses A and B both release glutamate, but B has NMDA receptors (purple), whereas A has only AMPA receptors (black).

Note that it is the *associated* firing of B and X that is necessary to strengthen the synaptic connection: mere overactivity of B alone (if, for example, the unfortunate dog was to be subjected to continual bell-ringing) is not a sufficient condition. In other words, it is the conjunction of presynaptic and postsynaptic activity that is postulated to cause synaptic strengthening. What it amounts to is *fire together, wire together:* neurons representing things that tend to happen together get physically linked together, so that brain eventually embodies a model of the outside world (cf. the knife and fork on p. 253). That this must be so was deduced – as we shall see, with extraordinary prescience – more than 50 years ago by Donald O. Hebb, and synapses with these properties are called *Hebbian synapses.*[12] You may recall that exactly the same hypothesis was used in Marr's model of cerebellar learning (Chapter 12): once again, it is the paired association of Purkinje cell firing with parallel fibre activity that results in strengthening of the connection from one to the other. A is the climbing fibre, B is the parallel fibre, and X the Purkinje cell itself.

It may of course be objected that the notion that the connection from B to X already exists structurally before the period of training is an implausible one. But given the amount of convergence and divergence of pathways that occurs in the brain, and the bringing together of diverse sources of information in such regions as the hippocampus, one can appreciate that there *must* be multiple pathways from any given sensory receptor cell to any given motor neuron (**Fig. 13.22**). In any case, the model will still work without that assumption, if *we* imagine that paired firing of B and X results in some way in growth of B towards X and eventual functional contact (or alternatively, in growth of dendrites of X towards B).

Synaptic learning

Hebb's formula is more than 60 years old, and is an extraordinary example of a prediction on purely theoretical grounds that suddenly turned out, many decades later, to be absolutely correct: synapses having exactly the properties predicted by Hebb were actually discovered. These are the *NMDA* (N-methyl-D-aspartate) *synapses*, with their long-term potentiation or LTP. The principle of their operation is simple, and it is perhaps surprising that it had not been proposed earlier. It is simply that whereas conventional ionic channels are either voltage- or ligand-gated, the NMDA receptor is *both.*[13] The condition for it to open is both that the postsynaptic cell is depolarised, and also that the transmitter, glutamate, is present. If both conditions are met, calcium enters the postsynaptic cell, where it appears to turn on cellular machinery for the manufacture of more glutamate receptors: not NMDA ones, but conventional α-amino-5-hydroxy-3-methyl-4-isoxazole propionic acid (AMPA) ones that require only the presence of glutamate to produce depolarisation; existing AMPA receptors are also potentiated. Eventually, if this sequence of events is repeated, the synapse will be strong enough to fire the postsynaptic cell on its own. How would this produce useful learning?

Figure 13.23 shows – very schematically – examples of the A and B synapses for Pavlovian conditioning. Both release glutamate, but the receptors under A are of the AMPA type, and thus always cause excitation, whereas those under B are – initially – only of the NMDA type. Thus, in the naïve state, only the UCS, activating A, will cause salivation. Now imagine what will happen during conditioning: A and B will frequently fire together, so that the conditions are met (activity of B combined with postsynaptic depolarisation) for calcium to enter, triggering the production of AMPA receptors under B. After sufficient training, there will be enough of them for the CS to be able to generate the response all by itself (**Fig. 13.24**).

There is a peculiarity of the dendrites of neurons in those regions of the brain that are particularly associated

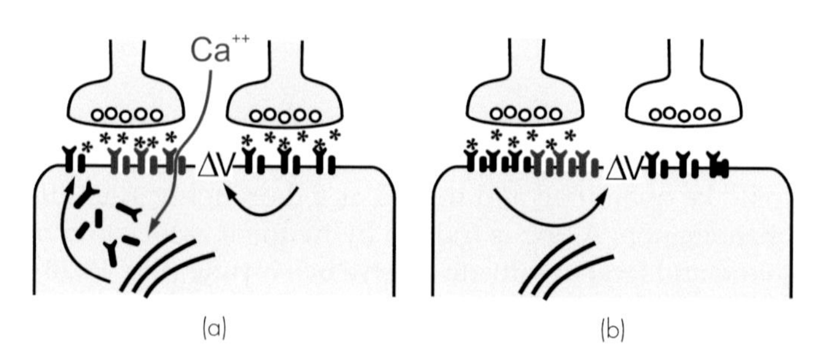

Figure 13.24 (a) If A is active as well, causing post-synaptic depolarisation, the NMDA receptors now open in response to the transmitted from B, allowing calcium to enter. This calcium then improves the effectiveness of B, either (as here) by increasing the number of AMPA receptors, or possibly through pre-synaptic mechanisms. (b) As a result, B is now effective on its own.

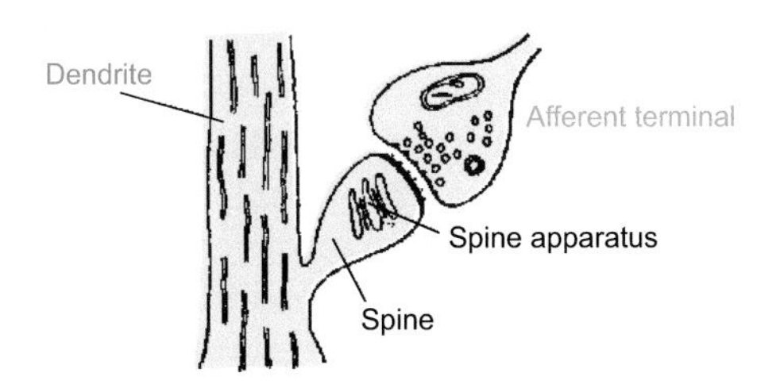

Figure 13.25 An afferent neuron synapses with the dendritic spine of another neuron.

with learning of one sort or another – the pyramidal and stellate cells of neocortex and of hippocampus, and the Purkinje cells of the cerebellum – which supports the idea of local postsynaptic change. Certain classes of afferent in each case terminate not directly on the soma or dendrite surface but rather on a sort of bud sticking out from it (the *dendritic spine*), which contains a prominent Golgi apparatus, implying a specifically localised production of protein, presumably of new AMPA receptors; the number of spines in visual cortex is greatly reduced by visual deprivation (**Fig. 13.25**). It is of course essential that synaptic strengthening should be strictly limited to only the one particular synapse and not over the whole cell. At many sites there is also evidence for *presynaptic* changes, an increase in the amount of transmitter released being triggered by nitric oxide diffusing from the postsynaptic cell, generated in response to the entry of calcium through NMDA receptors.

To summarise, the NMDA receptor (and others similar to it, more recently discovered) is the cellular mechanism for 'fire together, wire together'. The structure of neocortex provides an ideal way of enabling the most diverse sources of input and output to be brought together for association of this kind, and it is not surprising that both spiny stellates and pyramidal cells themselves are covered with an extraordinary density of dendritic spines; and similar spines are found in profusion in the two other regions associated with learning, cerebellum and hippocampus.

Learning to inhibit

Another potential function of Hebbian synapses is at first sight the exact opposite of what happens in Pavlovian conditioning, though it is still 'fire together, wire together'. A prominent feature of the way the brain is wired up that we first encountered in earlier chapters in a 'sensory' context (p. 88, p. 149), but is universally found in motor systems as well, is *lateral inhibition.* Neighbouring neurons tend to inhibit one another, with the desirable effect of sharpening up patterns of activity, and minimising redundant neural firing. But who is my neighbour? Functionally, a neuron is a neighbour if it often fires at the same time as you do yourself, though it may be physically quite distant. Therefore a good rule for learning to laterally inhibit is for a neuron that often fires at the same time as another to strengthen its inhibition of it. To put it another way, a neuron that predicts the activity of another neuron earns the right to suppress it, another example of the vicious, jostling neural competitiveness that we noted earlier (p. 253). Such a mechanism will not only guarantee to reduce redundancy, but will also automatically result in the process of error decorrelation that we saw (p. 240) to be an essential aspect of how sensory signals are processed. We saw earlier in this chapter how it sharpens up the discrimination of colours, of orientations, and of odours, but it can operate at much higher levels as well. In language, for example, it helps us to learn finer and finer distinctions between words of overlapping sound or meaning: hit/hat/hot, or irritated/vexed/furious.

Generalising yet further, one could argue that a grand organising principle of the brain is a principle of *least neural effort,* minimising the total amount of neural activity by laying down pathways of inhibition that stop it being disturbed by what is perfectly predictable. For instance, it prevents a startle reaction to the switching on of a light when one has already heard the key in the front door. A telling analogy is perhaps the way that bone is shaped by external forces: as d'Arcy Thompson first pointed out, the trabeculae in bone represent lines of stress – they form where stress is greatest, and by doing so relieve that stress.[14]

Short- and long-term storage

Finally, it is clear that whereas a stimulus that one remembers for a lifetime may only be present perhaps for less than a second, growth or strengthening of synapses must take some time to implement. There must therefore be a period of consolidation during which the event to be remembered is actually converted into some kind of semi-permanent structural change. A number of observations suggest that there are really two distinct memory stores in the brain: a *long-term memory* (LTM), which takes the form of the kind of synaptic changes that we have been considering, and a *short-term memory* (STM: there is evidence that STM has a number of separate components, including what psychologists call 'working memory'), which retains information temporarily to cover the period – probably of the order of minutes – during which consolidation takes place. It appears that STM is much more vulnerable than LTM, suggesting that the short-term store is a dynamic one, perhaps consisting of impulses continually circulating round looped chains of neurons. We all know how easily when engaged in difficult mental arithmetic, the numbers stored in working memory can vanish when someone asks if we want a cup of coffee. Clinically, sudden shocks of any kind – a blow on the head, or the passage of a large electrical current across the skull, as in electroconvulsive therapy – are a potent way of disrupting STM. They cause a characteristic type of amnesia called *retrograde amnesia,* in which the ability to recall events that occurred either after the shock, or long before it, is unimpaired (shaded blue), but a period of some 20 minutes or so before the shock remains more or less blank (see **Figure 13.26**). It seems as though memories need to be stored for

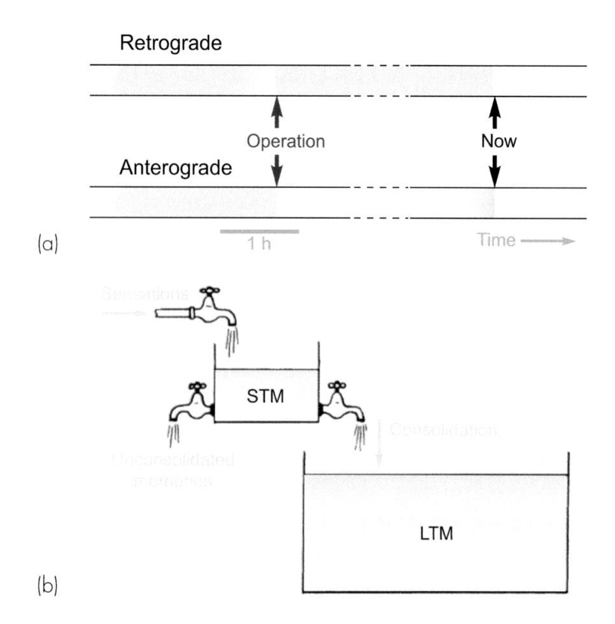

Figure 13.26 (a): two kinds of amnesia. Retrograde amnesia: there is a loss of recall of occurrences just before the operation of precipitating event (shaded areas represent the stretches of past experience that can be recalled). Anterograde amnesia: only very recent events or those before the operation can be recalled. (b): two-tank analogy of short-term and long-term memory (STM and LTM).

a certain time in STM in order to make as it were a sufficient impression on the permanent memory trace, as suggested by the two-tank analogy, and that violent disruption of the brain's activity through electroconvulsive therapy or some other shock simply empties the STM of its contents (**Fig. 13.26**).

This interpretation is strengthened by the existence of a related condition, *anterograde amnesia*,[15] in which the patient can recollect very recent events, within a timescale of some 30 minutes, as well as events that occurred before the condition began, but cannot transfer memories from STM into LTM. It is as if the flow from the upper to the lower tank, from STM to LTM, had been permanently disconnected, leaving the patient with a functional STM but fossilised LTM. Finally, the leak in the STM tank is a reminder that not everything in STM – perhaps fortunately – finds its way into permanent memory, and there is little conscious control, if any, over what is or is not permanently stored. Some unconscious control certainly does occur, since experiences with a strong emotional significance are almost always transferred to LTM. (A striking instance of this is that so many remember with unusual vividness exactly what they were doing when they heard the news about the death of Princess Diana, or the terrorist attack in New York on 11 September 2001.)

One complicating factor is that some patients with short-term memory defects may nevertheless be able to lay down long-term memories; another is that things may have been stored perfectly well in LTM that may not be recalled because the mechanism for *retrieval* is not working properly. This is particularly obvious in the case of experiences that are unpleasant, and psychiatric help may be required in order to bring such repressed memories to consciousness. In other cases, forgetting may be the result of learning new material. Since retrieval is essentially by association, memories that are linked together by too many associations may become irretrievably entangled. Unique and strange events are easy to recollect; boring things like telephone numbers are much more difficult, because of the vast number of pre-existing associations in our minds between each of the digits, the result of having remembered many other numbers in the past. Methods for improving one's memory that are commonly advertised generally work by translating each digit into a unique and vivid mental image: thus if 7 is 'elephant', 3 'cigar' and 4 'bicycle', the number 734 could be recalled by picturing an elephant smoking a cigar and riding a bicycle. The snag is of course obvious: after a while, there will be such a tangled knot of connections between elephants, bicycles and cigars and so forth as a result of learning one's friends' telephone numbers that new numbers will be just as difficult to remember as ever.[16]

Disorders of association

This idea of associational cortex putting together fragments of information in order to recognise a complex object is reinforced by studies of the effect of clinical lesions in these areas. They can often give rise to conditions in which the patient's vision, hearing or tactile sensitivity is in one sense perfectly *normal* – as measured by Snellen charts, audiometry, two-point discrimination and so on – but what he cannot do is somehow put this information together in a coherent way.[17] These clinical disorders fall broadly into three groups:

- *Agnosia* – disorders of high-level sensory analysis;
- *Apraxia* – disorders of high-level motor coordination and appropriateness;
- *Aphasia* – disorders in communicating and using symbols.

Agnosia

Agnosia is the technical name for a condition where a patient's peripheral nervous system is working fine but they have difficulty in recognising things because they can't put the component bits together in an associative way. One kind of agnosia has already been mentioned in Chapter 4: lesions of parietal cortex near the somatosensory region may give *tactile agnosia*. Here there is no appreciable peripheral disorder – the subject has normal sensitivity to touch or temperature, and his acuity as measured by the two-point discrimination test may be unimpaired – but what is lacking is the ability to *use* this sensory data properly in order to recognise and respond to objects that are sensed by the skin. Such a patient may not recognise a matchbox when he is given one to hold, but can do so if he is allowed to see it; such difficulties in feeling the shape of an object in the hand are sometimes called *astereognosis*. (The -gnosia root, incidentally, means 'knowledge': 'astereognosia' means 'no-shape-knowledge'. A little Greek helps make some sense of the forbidding clinical jargon for parietal lobe defects.) A similar state of affairs, involving the visual system, can be seen with lesions nearer the visual cortex and the recognition stream, called visual agnosia. Again, simple tests of visual performance reveal no abnormality – acuity, colour vision and sensitivity may all be normal – but the subject cannot always *appreciate* what he sees, and recognition of objects and places may be difficult. One highly intelligent patient described by Oliver Sacks, when asked to identify a flower, described it as 'a convoluted red form, with a linear green attachment' but only recognised it as a rose when allowed to smell it. As in all agnosias, it is generally the most difficult tasks that are most affected, for example a difficulty in recognising people's faces (*prosopagnosia*) may be the first sign that something is wrong. There is no lack of intelligence, not of the concept of the objects, nor mechanisms for perceiving basic forms and colour – it is putting it all together that is difficult. A characteristic in such cases is that though the victims cannot immediately recognise things, they can do a sort of work-around by checking things off on a list. Normal people find themselves doing the same, when faced with recognition tasks (for example, identifying obscure birds) that is beyond their learnt experience.

A related but distinct defect is *spatial agnosia:* the subject has difficulty in appreciating the spatial relationships between objects, tends to get disorientated more easily, or may have difficulty in drawing or using a map or in sketching a complicated object like a bicycle; even though he knows *intellectually* what all the bits are, he can't assemble it as a whole. Very commonly the defect is unilateral, as a result of one-sided brain damage, normally to inferior parietal cortex, and the disability is then confined to half of the visual field, which may often show lack of use or *neglect.* Asked to draw a map of Piccadilly Circus, the patient may ignore the whole of one side, cramming the streets into the other: yet the fact that all the streets are included shows that there is no lack of knowledge of *what* ought to be there, only of *where* they should be (**Fig. 13.27**). This unilateral neglect may be reflected in their everyday activities, for instance when only food on one side of the plate is eaten. There is in fact increasing neurophysiological evidence that cells in parietal cortex may be specifically involved in the representation of visual and somatosensory space and associating them with knowledge of ongoing movements. Every time we make a movement, sensory transformations and re-orientations are necessary that are mathematically complex, but capable of being carried out by neural networks working through association.[18]

A curious feature of many of these high-level defects is that the subject may often be strikingly unaware that anything much is wrong, and recent suggestions to the contrary; the defective field is simply ignored – much as we ignore our own blind spot – and it may require specially designed neurological tests to reveal the disorder. It is important to remember that this neglect of one side of visual space or the body is not the same as a hemianopia or hemisensory loss: visual acuity and fields are intact when an isolated object is presented to the neglected part

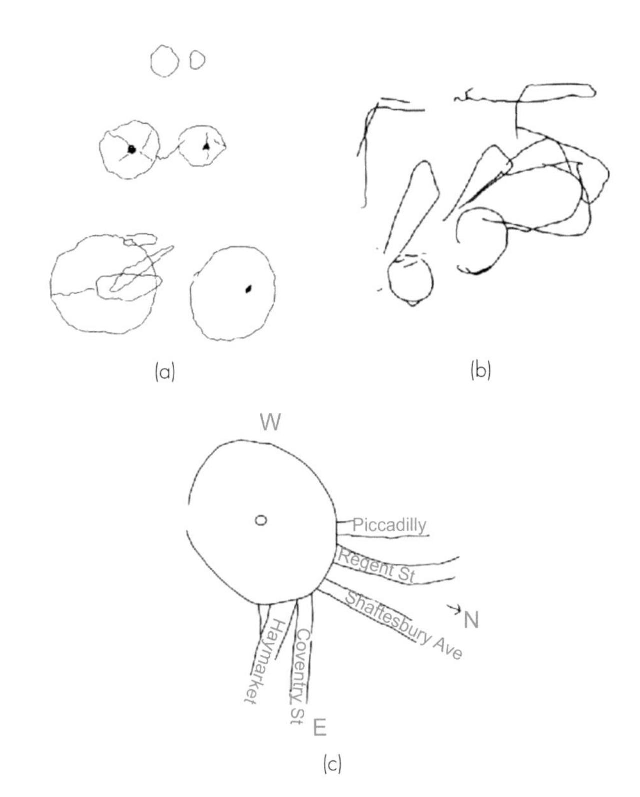

Figure 13.27 (a) Attempts at drawing a bicycle: left parietal lesion. (b) Another attempt: biparietal vascular lesion. (c) An attempt at a map of Picadilly circus, showing neglect of top and left: right parietal lesions. (From Critchley, 1971.)

of a visual scene, and tactile stimulation is appreciated normally when applied to the neglected side of the body in isolation. The deficit is only apparent when multiple stimuli are presented together as in the normal visual and tactile environments. Clinically, we identify neglect by eliciting a phenomenon called extinction: an examiner will face the patient and raise their hand on each side. When presented on either side, the patient can identify which hand has been raised, and using a more complex stimulus, acuity, visual fields and even recognition can be demonstrated. However, when both hands are raised simultaneously, the hand on the neglected side will be ignored and the patient only acknowledge the hand raised on the non-neglected side, i.e. the latter has 'extinguished' the former. Exactly the same can be demonstrated using tactile stimuli applied to one or both sides of the body. Deficits usually follow lesions to the non-dominant parietal lobe and neglect of the contralateral side is typical. Patients may look to the ipsilateral side when approached from the contralateral side or consistently steer their wheelchair towards the ipsilateral side.

One particularly bizarre example of this is when the agnosia takes the specific form of defects in the perception of one's own body image (*anosognosia*). Such a subject may emphatically deny that a particular part of his body such as a leg actually exists, and disown it or even express distaste for it when it is forcibly brought to his attention.[19] This is not merely a conscious fabrication on the part of the patient: he may be completely consistent in his attitude

Clinical box 13.1 Examples of specific agnosias and apraxias

The term agnosia describes a modality-specific inability to access semantic knowledge of a stimulus or object, which cannot be attributed to impairment of basic perceptual processes – that is, a percept is stripped of its *meaning*. For example, one might elicit visual agnosia by testing the ability to name objects that would have previously been known (perhaps a selection of toy animals) and to provide semantic information about them (such as the ability to identify which of the given selection are ferocious); in a pure visual agnosia semantic knowledge should be demonstrable through other modalities such as somatic sensation. It is not uncommon for a patient to suffer a more general loss of semantic knowledge, not restricted to any one sensory modality, such as in the semantic dementia variant of frontotemporal dementia.

Apraxia refers to the difficulty in performing skilled movements, which cannot be accountable by lack of understanding of the task, weakness, or proprioceptive loss. The most common causes are stroke and neurodegenerative diseases – such as Alzheimer's or cortico-basal degeneration.

Name	Area of difficulty
Astereognosia	Tactile recognition
Visual agnosia	Visual recognition
Auditory agnosia	Auditory recognition
Spatial agnosia	Orientation, drawing, maps, etc.
Anosognosia	Appreciation of body topography
Prosopagnosia	Recognition of faces
Ideomotor apraxia	Execution of skilled sequences
Constructional apraxia	Assembling components into a whole
Ideational apraxia	Formulation of plans of action

to the limb, not drying it after a bath, not bothering to dress it, tending to bump it against door frames, and the like. Nor can one talk about any lack of intelligence in the normal sense: rather the lack of a certain kind of synthesis between somatosensory and visual inputs.

Apraxia

Apraxia implies clumsiness, but of a kind that is much more specific for *particular* tasks than the more general impairment associated with lesions of the cerebellum or motor cortex. It may be especially noticeable when the subject had previously been extremely skilled at using a particular tool or carrying out a highly trained sequence of actions, as in the case of a fish-filleter whose biparietal lesion led her to forget how to do it: although she 'knew in her mind' how to set about filleting a fish, she was unable actually to execute the manoeuvres that she wanted, and was sent home by the foreman for 'mutilating fish'. Sometimes a patient cannot produce specific actions on command – for example, gestures such as beckoning or saluting – but can do so spontaneously in appropriate circumstances, illustrating the high hierarchical level of the deficit. A more specific variety of apraxia is *constructional apraxia*, a sort of motor version of spatial agnosia: though the patient may seem to perceive spatial relationships quite readily, they may for example find it difficult to put building blocks together to make a particular shape, or construct a simple jigsaw puzzle. Curiously, constructional apraxia is particularly associated with the encephalopathy that accompanies liver failure and its progression can be used to gauge the clinical progression of the underlying hepatic dysfunction and recovery. A dressing dyspraxia can complicate injury to the non-dominant parietal lobe and manifests as difficulty performing the complex sensorimotor integration required to dress oneself. Resourceful clinicians make use of an inside-out pyjama top as a diagnostic challenge as one watches a patient try to turn it the right side out using visual cues to guide precise bimanual responses before guiding the hidden limbs through convoluted flannel tunnels using tactile and proprioceptive information.

Aphasia

There are many different ways in which aphasia may be manifested: a useful classification is into *sensory aphasia*, *motor aphasia* (these being in effect agnosia and apraxia in the particular field of language and communication; an older nomenclature still clinically current is to call them *receptive* and *expressive* aphasia), and *central aphasia*. (Strictly speaking, disorders of these kinds should be called *dysphasias* since there is not usually complete loss of function.) Some of these classifications are ambiguous and subject to change with fashion. For instance, the more peripheral problems that can impair communication are in clinical practice not usually called aphasias. Like any such clinical classification, it is much more clear-cut than the symptoms that actual patients typically present with. Nevertheless, the reception and generation of speech and other forms of communication are clearly a microcosm of what the brain does as a whole, and a systematic approach can at least help in trying to understand a complex area.[20]

It is helpful to think of the processing of language by the brain in the same hierarchical terms as the generation of other kinds of movement (**Fig. 13.28**). Raw sensory information enters through the eyes or ears, and is analysed by successive levels to the point at which letters, words, and larger syntactic units are recognised. At the highest level, *meaning* comes about by association of these symbols with other kinds of sensory information to form concepts;

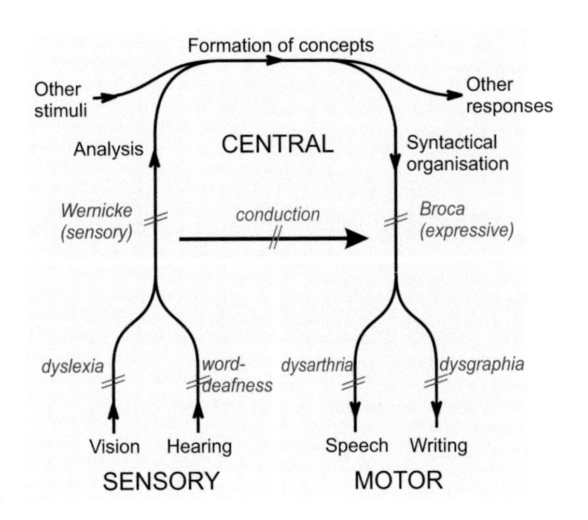

Figure 13.28 The hierarchical organisation of the neural processing of language.

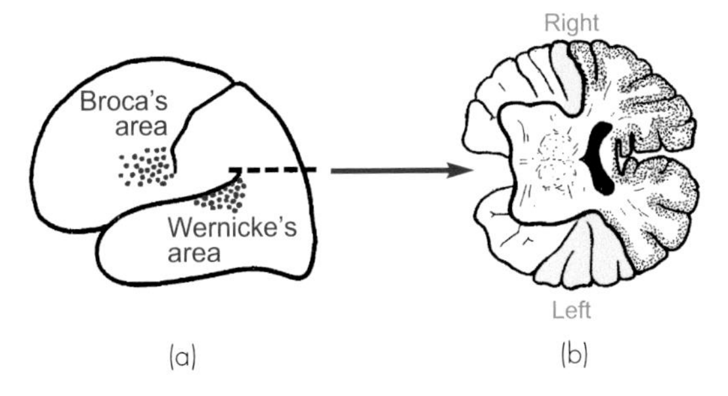

Figure 13.29 (a) Lateral view of left hemisphere, showing approximate location of Broca's and Wernicke's area. (b) Section along the dotted line showing relative enlargement of the left planum temporale (approximating to Wernicke's area, shaded) in comparison with the right. (After Geschwind & Levitsky, 1968; copyright AAAS, with permission.)

these in turn may result in speech or writing by an exactly converse process of elaboration down the motor side, ending up with the firing of motor neurons in appropriate patterns to form phonemes or fragments of writing or typing. Defects at the most peripheral levels – simple blindness, or paralysis of the writing arm – will prevent certain kinds of communication, but do not count as aphasia because the effects are unspecific.

In *sensory aphasia*, the patient's sense of hearing may for example be perfectly normal, and the sounds of speech are heard, but they make no sense: he may complain that everything sounds like a foreign language. This '*word-deafness*' is in effect an agnosia specific to speech, and may indirectly lead to defects in *producing* meaningful speech as well, since the patient can no longer monitor effectively the words he is producing. A similar disability specifically affecting reading is word-blindness or *alexia* (*dyslexia* in milder forms). These sensory aphasias are generally associated with a relatively localised region that borders on both visual and auditory cortex, called *Wernicke's area* (**Fig. 13.29**). In a horizontal section at an appropriate level, this area is normally enlarged on the left relative to the right.

In *motor aphasia* the subject can show by his actions that he understands what is said to him or what he reads, but has difficulty in initiating such communications himself. Such an expressive or *Broca's* aphasia can range in severity from a total absence of speech, through non-fluent speech with word finding difficulty, increased effort and grammatical mistakes to a mild stutter. Interestingly, in other cases emotional expression may be unaffected – swearing may continue unabated – and stutterers often find that under sufficient duress, when they are not thinking consciously about what they are saying, the disability may suddenly vanish.[21] Lesions specifically affecting speech and articulation are associated with an area of cortex called *Broca's area*, which is close to the tongue and mouth regions of the motor cortex, and is not actually in the parietal lobe at all, but in the posterior frontal lobe. Damage here can frequently result from stroke due to a vascular occlusion, usually by a blood clot (embolus /thrombus). Physicians distinguish dysphasia from *dysarthria*, the latter being a defect in articulation which typically arises following a lesion to the cerebellum, cranial nerves or their nuclei, neuromuscular junction, muscles or structures required to generate speech, i.e. a defect in areas generally lower in the neurological hierarchy. It can, however, be difficult to distinguish dysarthria of this nature from dyspraxic speech, which arises from lesions of those 'higher' regions of the cortex usually associated with dysphasic syndromes. *Agraphia* describes an inability to write despite otherwise normal praxis and sensorimotor function: in the late nineteenth century it was elegantly described as aphasia of the hand. The ability to spell and type is usually preserved; in different forms it can result from injury to the non-dominant parietal cortex or frontal cortex.

Interestingly, and perhaps entirely logically, there is some evidence that the cerebellum has a linguistic role and contributes to grammatical correctness, beyond that of simply coordinating the motor patterns of oral muscles. One can certainly imagine how formulaic rules of grammar might be engrafted onto the cerebellar circuitry in the same way as any other frequently employed motor sequences.

Central aphasia

This term covers a number of miscellaneous conditions in which the defect is not primarily either sensory or motor, but involves the mental mechanisms for forming concepts, for understanding symbols, and making sentences. A patient may be shown a common object such as a knife and be unable to name it (*anomia*). Yet he can use it, and by employing a paraphrase – 'what you use to cut with' – he may show that he can designate it in speech. Nor is the defect simply motor, since he can repeat the word 'knife' when told to do so; what seem to be at fault are the normal central *connections* that ought to link the sight of the object to the utterance of its name. Sometimes such a patient may use the wrong word for something without realising it (*paraphasia*): given a pair of scissors he promptly describes them as a nail-file, and on being corrected may say 'no, of course it's not a nail-file, it's a nail-file'. Such errors

Clinical box 13.2 Clinical localisation of aphasias

	Localisation	Clinical picture
Receptive aphasia	Wernicke's area at the temporoparietal border, associated with auditory and visual areas	Poor comprehension Unable to follow three-stage commands, e.g. 'Put your left index finger on your right ear'
Central aphasia	Regions of temporal, occipital, and parietal cortex that support knowledge of words and their meanings	Meaningless, often voluminous speech, 'word salad' Neologisms and paraphasic errors Inability to name objects presented to patient
	Areas of frontal and temporal lobe that support sentence-level processing	Errors of syntax and sentence construction
Conduction aphasia	Arcuate fasciculus connecting Wernicke's and Broca's areas	Unable to repeat a spoken phrase, e.g. 'No ifs, ands or buts'
Expressive aphasia	Broca's area of the posterior frontal lobe, near mouth and tongue motor representation	Sparse, non-fluent, effortful speech. Difficulty with word-finding when asked to describe a scene or answer questions. Agrammatical Preserved comprehension

can be semantic, substituting cat for dog, or phonemic, substituting cat for hat. Other times he may simply make up words: *neologisms*. He may produce speech sounds that are correctly executed and sound grammatical but actually make no sense; one such patient, for instance, shown a bunch of keys, came out with: 'Indication of measurement or intimating the cost of apparatus in various forms'. Such a response, often with much repetition of meaningless phrases, is described technically as *jargon*: in some cases it appears to be related to a sensory aphasia (Wernicke's aphasia) that interferes with normal feedback of what is spoken.

One remarkably specific problem arises when there is damage to the prominent associational projection from Wernicke's area to Broca's area, the arcuate fasciculus. The victim can understand and generate speech fairly well, but there is an almost specific inability to do what might seem a much easier task, namely to *repeat* what has just been said; this condition is called *conduction* aphasia.

All these defects may be quite specific for only one category of symbolisation: thus in bilingual patients, only one of the languages may be affected, and morse aphasia, or aphasia of sign language in deaf and dumb patients, has also been described. An interesting case of specificity of this kind occurred in the composer Maurice Ravel, who was affected by aphasia in later life, yet though unable to speak or write, could still sing and play and compose music. Other specialised aphasias of the central kind that have been described include *acalculia*, an inability to perform arithmetical operations, and *amusia*, an inability to appreciate music. Conversely, individuals are not infrequently found with extraordinary development of these same faculties, the *idiots savants* or calculating prodigies, infant musicians and those remarkable people who seem to find it no trouble at all to learn 20 or 30 different languages, but these are not normally reckoned to be disorders.[22]

It is important to appreciate that normal people suffer from all types of dysphasia on occasion. Not everyone can guarantee to complete *The Times* crossword puzzle; we all sometimes stutter or stumble over words; we are often at a loss for the name that goes with a face we know well, or for something that a moment ago was 'on the tip of our tongue'; and all of us are guilty from time to time of generating jargon, especially in social situations where we are compelled to speak, but have nothing to say. In people of limited education, one may observe a tendency for remarks to be repeated endlessly with only slight variations, with disproportionate use of a small number of concise monosyllables; at a more exalted level we find 'win-win situation', 'singing from the same hymn sheet' and so forth. Indeed, short stretches of dialogue with patients who really do have a high-level central aphasia can sound remarkably normal. Equally, we all suffer at times from more or less severe attacks of agnosia: few normal people are really very proficient at, say, drawing a map of the town where they live, and most people have difficulty with tests like the one above. Similarly, untrained drawings of the human face reveal obvious distortions of the proportions between the various parts that amount to a kind of disorder

Clinical box 13.3 Examples of specific speech disorders and dysphasias

Name	Area of difficulty
Dysarthria	Articulation – a motor disorder affecting motor aspects of speech
Aphonia	Speaking – due to damage to the mouth or larynx.
Dyslexia	Reading – possibly associated with abnormal processing in inferior frontal gyrus
Dysgraphia	Writing – linked to a deficit in working memory
Broca's (expressive) aphasia	Expression of communication – due to damage to inferior frontal gyrus (usually left)
Wernicke's (sensory) aphasia	Understanding – due to damage to posterior superior temporal gyrus (usually left) communication
Conduction aphasia	Repeating – caused by lesions to the arcuate fasciculus
Nominal aphasia	Recalling names – associated with damage to left parietal lobe
Global aphasia	All aspects of communication – caused by damage to Broca's and Wernicke's areas
Amusia	Music – associated with damage to various brain regions including Broca's area
Acalculia	Arithmetic – a symptom of Gerstmann's syndrome (posterior parietal lobe disease)

of body-image perception: the mouth and eyes tend to be too big, the cranium too small (to an extent reflecting the sensory homunculus, as it happens).[23]

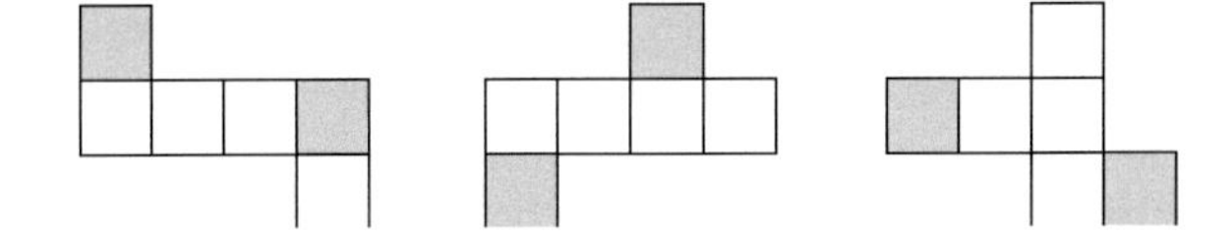

If folded to make a cube, for which shape would the areas not touch?

All this suggests that the parietal lobes are a fruitful area for human improvement, that might well become better developed in the course of future evolution, and sheds some light on what we mean by 'intelligence'. We would all like to be able to speak five foreign languages, to play the violin to diploma standard, to remember the name of everyone we meet, and be quick at doing mental arithmetic, but it seems that our cerebral cortex is just not up to doing all these things well at once: all too often truly extraordinary ability in one field seems to be associated with regrettable defects in others. Intelligence is perhaps no more than a relative freedom from the more obvious kinds of dysphasia.

Left–right asymmetry in the brain

One important point of interest in connection with the dysphasias is that they show a functional asymmetry between the left and right halves of the brain. Although the agnosias can on the whole be found with lesions of either hemisphere, neglect is more common after damage to right hemisphere, and aphasia is nearly always associated with lesions of the left hemisphere (that governs the right side of the body), at least in right-handed people. This asymmetry is reflected in the relative anatomical size of certain parts of the cerebral cortex on the two sides, notably in Wernicke's area, as can be seen on p. 265. In living subjects this cerebral *dominance* (the dominant hemisphere being the one associated with aphasia) may be demonstrated by the Wada test: a substance such as sodium Amytal (amobarbital sodium) is injected into the carotid artery on one side or the other, while the subject is carrying out some such task as reciting the letters of the alphabet. If the injection is on the dominant side, the recitation is interrupted for a short time and then continues; on the non-dominant side, very little is observed or felt by the subject. The various types of brain scan described in the Appendix enable both dominance and other aspects of cerebral localisation to be shown in a dramatic manner, with graphic pictures of the changing patterns of activity associated with different types of mental process; in this instance, the differences in regional blood flow for different kinds of tasks are obvious (**Fig. 13.30**).

In most people with left-hemisphere dominance, one finds that not only is the right hand used preferentially for writing and other skilled tasks, but often the subject is right-legged and right-eyed as well: one may discover this by observing which foot is used to kick a ball, or which eye looks through a peep-hole. Other, more unconscious, actions may be revealing: thus the right leg may be crossed over the left when sitting, or if asked to fold his arms, the right arm may be placed on the left. But in the 7 per cent or so of the population who have right- hemisphere dominance, most (but not all) are found to be left-handed. Some statistics relating to this correlation between dominance and handedness are shown in **Figure 13.31** (derived from observations of the incidence of aphasia after unilateral brain lesions in right- and left-handers, and therefore necessarily somewhat approximate). It is clear that although there is a strong correlation between the two, it is

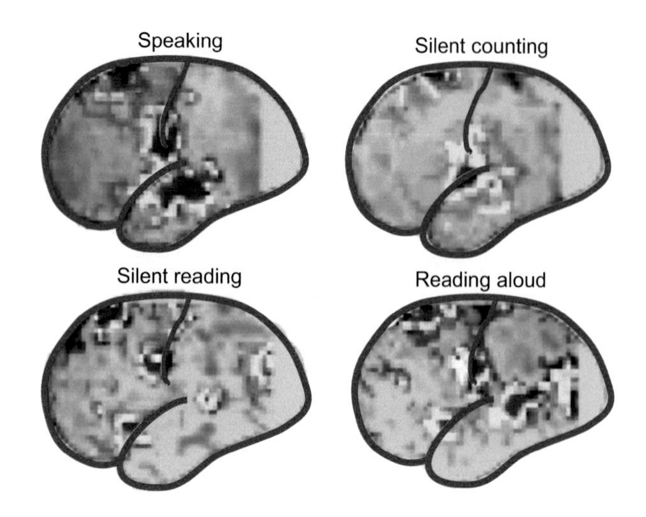

Figure 13.30 Regional blood flow in the cerebral cortex of a conscious human subject, revealed by a radioactive marker, under the various conditions shown. (After Lassen et al., 1978; Copyright © 1978, Scientific American, Inc.; permission pending.)

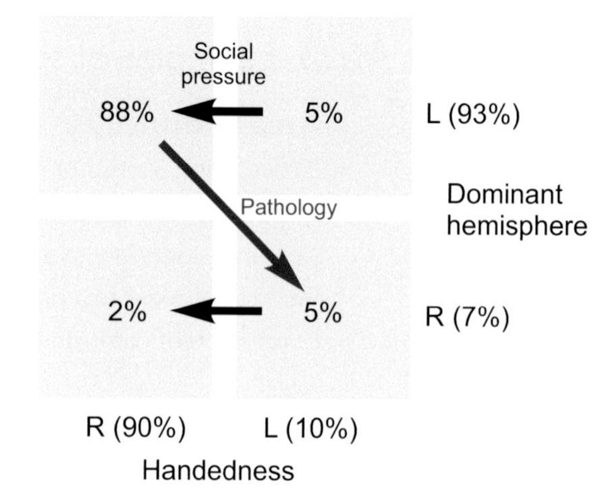

Figure 13.31 Incidence of right-handedness and left-handedness and brain dominance, expressed as percentages of the whole population. The horizontal arrows indicate the social pressures tending to turn natural left-handers into apparent right-handers; the red arrow indicates the likely effect of early damage to the left hemisphere. The data are derived from observations of the incidence of aphasia after unilateral brain lesions in right-handers and left-handers (Zangwill, 1967 © 1967 American Psychological Association) and are therefore necessarily somewhat approximate.

not an absolute one; one factor that tends to distort such figures is that there are considerable social pressures from school and family for 'natural' left-handers to learn to use their right hands in preference, producing an artificial shift of the distribution towards right-handedness, shown by the horizontal arrows. It is likely in fact that in the absence of such pressures the number left-handers in the population would be rather more than the 10 per cent or so usually reported, though this percentage has remained essentially unchanged throughout recorded history. It also appears from these statistics that there are essentially two distinct ways in which left-handedness can come about. The first is what might be called 'normal' left-handedness, and is probably genetically determined and essentially independent of dominance. The second type may be the result of slight brain damage to the left hemisphere early in development, which causes both speech and handedness to shift to the other hemisphere, as indicated by the downward diagonal blue arrow, or become to some extent bilateral; of these, some are again converted to apparent right-handedness by social pressures. Left-handers in this category may often show vague disabilities of speech such as stuttering, or mild forms of apraxia or agnosia, a fact that has given left-handers as a whole – including the 50 per cent of left-handers who are in every way perfectly normal – a bad name: literally so, when one considers the etymology of words such as 'dextrous' and 'sinister', not to mention 'right'! (**Fig. 13.31**).[24]

One interesting consequence of the lateralisation of speech occurs in patients who have undergone surgical section of the fibres of the corpus callosum (see p. 8), producing a so-called 'split-brain' patient. This controversial procedure has been performed as a last resort on people suffering from intractable forms of epilepsy in an attempt to prevent focal seizures spreading across the corpus callosum and involving both hemispheres – creating a sort of neurological firewall. What is most surprising in such cases is the apparent lack of ill-effects, despite the severance of a massive fibre bundle containing more than 100 million fibres (a fact that has prompted the facetious suggestion that the function of the corpus callosum is to allow the spread of epilepsy across the brain). In normal everyday life such patients seem perfectly normal; it is not until you set up experiments that restrict incoming visual information to one visual hemifield, L or R, that an extraordinary state of affairs is revealed: each half of the brain now appears to act independently, receiving information from the opposite half of the visual field, and controlling the opposite half of the body. Thus a patient can be shown a picture of something on one side, and use the *corresponding* hand to point to an object on the same side that matches it, but cannot do so with the *other* hand; and tests involving a comparison of stimuli on the left and right cannot be done (**Fig. 13.32**).[25]

Because only the dominant hemisphere can speak, if the subject is shown an object on the left he cannot say what it actually is, but he can when it is on the right. But the non-dominant side is not entirely aphasic, for it can understand speech and also read: the left hand will pick

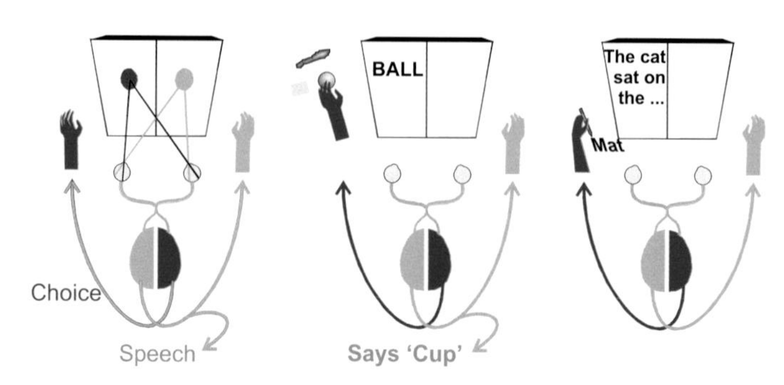

Figure 13.32 Only the left hemisphere can activate speech; but the right hemisphere can apparently read sufficiently well to be able to direct the left hand appropriately, or perhaps complete simple sentences using the left hand.

up a ball to correspond with the word 'BALL' presented in the left visual field, but the patient will be unable to name the object he has just selected, because the specific function of speech is wholly localised in the other hemisphere. In general, the non-dominant hemisphere has great difficulty with things like selecting from a list of alternatives the word needed to complete the sentence 'The cat sat on the . . .', indicating that it suffers from central as well as motor aphasia. It may, however, have a specific role in the *prosody* of speech – the rise and fall of pitch that can have emotional connotations. However, recently some doubt has been expressed about the validity of some of these claims, which were necessarily based on very few subjects.

Some communication appears to be possible between the hemispheres, but of a subconscious, emotional kind rather than of 'facts'. Thus a patient whose non-dominant hemisphere is allowed to see a pornographic picture may blush or giggle, and asked what was there may indicate the awareness of the emotion without being able to describe exactly what was seen. Some functions – for example spatiovisual tasks such as drawing, and probably musical appreciation as well – appear to be performed better by the non-dominant hemisphere. **Figure 13.33** shows right-handed and left-handed attempts by a split-brain patient to copy the series of drawings on the left. Though the drawings generated by the right hand (left hemisphere) are technically better in the motor sense, the others (right hemisphere) are *artistically* better (in the sense of conveying the three-dimensionality of the scene). Similarly, when asked to choose one of a set of drawings to go with another drawing, although the dominant side may select according to similarity of function, the non-dominant side uses similarity of appearance. Findings of this sort have led to a certain amount of semi-mystical speculation about

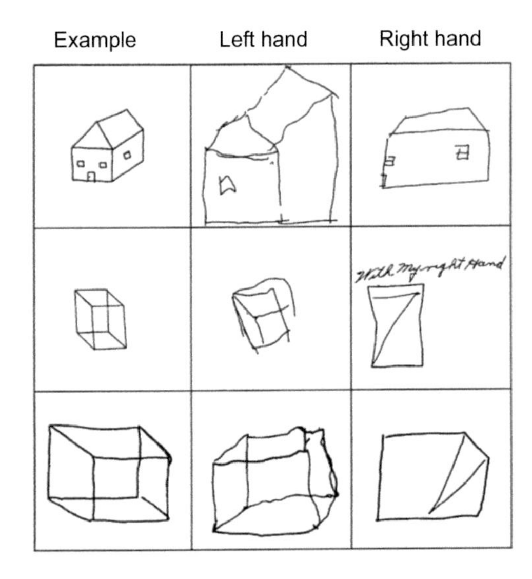

Figure 13.33 Right-handed and left-handed attempts by a split-brain patient to copy the series of drawings on the left; the superiority of the left hand in dealing with the implied three-dimensional relationships is evident, although the right hand is slightly better at carrying out the drawing movements. (From Gazzaniga, 1967; copyright Scientific American, Inc.; permission pending.)

the possibility of a fundamental split in Man's psyche between the rational, nerdish and factual left hemisphere and the intuitive, bohemian and artistic, right hemisphere, and the importance of not allowing one hemisphere to develop at the expense of the other.[26]

Finally, one may sometimes observe in such split-brain patients the effects of evident struggles between the two sides about what should be done, one hand perhaps trying to tie up the patient's shoelaces while the other unties them. Such observations raise rather difficult problems concerning the nature of consciousness and its relation to the brain: do we have here two minds in one body? The one thing we cannot do, of course, is to ask the patient what *he* thinks is going on: only the dominant hemisphere will reply.

Temporal archicortex and memory

There is no very clear distinction between temporal and parietal cortex, and we have already seen that some of the areas mentioned in the preceding section – Wernicke's area, for example – lie partly in the temporal lobe. There is, however, a specific type of disability associated with damage to the temporal cerebral hemispheres, *amnesia*, which is quite different in kind from anything seen with damage to frontal or parietal cortex. It is now recognised that these and other classical 'temporal lobe' disabilities are probably more related to structures forming part of the *limbic system*, lying within the temporal cortex itself. To make clear the distinction between these two quite different areas that are sometimes lumped together as 'temporal lobe' it is necessary to begin by outlining the structure and development of those deeper and older structures that form the limbic system, and which on account of their close association with the olfactory sense, discussed in Chapter 8, are sometimes classed together as the *rhinencephalon* or 'nose-brain'.

We have noted several times before that older structures of the brain tend to get elbowed out of the way by newer ones, and thus often become twisted into complex and at first sight incomprehensible shapes; this is markedly true of the limbic system. It consists of nuclei (notably the *amygdala, septal nuclei, mammillary body* and *hypothalamus*) and areas of cortex (in particular the *hippocampal gyrus, cingulate gyrus*, and *entorhinal*, *periamygdaloid* and *prepyriform cortex*, the latter two having especially important olfactory connections), all joined together by fibre tracts (for instance the *fornix*, *medial forebrain bundle*, and projections from the mammillary body to *anterior thalamus*, and from there in turn to the cingulate gyrus); these are all shown schematically in **Figure 13.34**.

Originally, it seems that the two main areas of limbic cortex, hippocampal and cingulate, formed almost the entire cortical surface of the brain, lying side by side immediately over the relatively compact group of their associated nuclei. But in the course of time, they were infiltrated by the newer six-layered *neocortex*, which by expanding almost

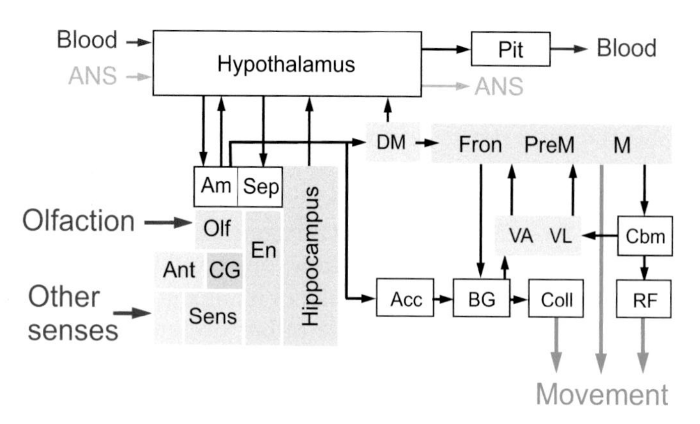

Figure 13.34 Highly schematic and simplified representation of the principal areas of the limbic system and their connections to other structures. Key: Neocortex (pink): sensory and associational (Sens), cingulate (CG), olfactory (Olf), frontal (Fron), premotor (PreM), motor (M); archicortex (grey): entorhinal (En), cingulate (CG); thalamus (blue): dorsomedial (DM), ventroanterior (VA), ventrolateral (VL). anterior (Ant); Other (white): N. accumbens (Acc), basal ganglia (BG), colliculus (Coll), cerebellum (Cbm), reticular formation (RF), pituitary (Pit), autonomic nervous system (ANS).

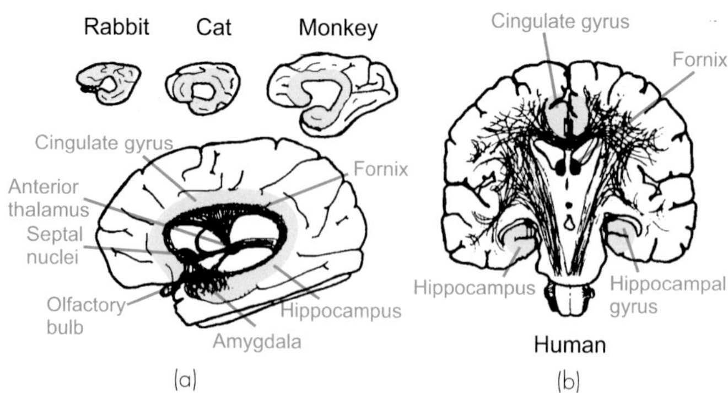

Figure 13.35 The eclipse of the limbic cortex. (a) Approximate area of limbic cortex (shaded) in rabbit, cat, monkey and human, showing the relative growth of neocortex and consequent relegation of limbic structures to medial and central regions. (Partly after Ochs, 1965.) (b) Transverse section of human brain, showing limbic cortex (shaded) and fornix, in relation to massive fibre bundles serving the neocortex.

explosively in the region separating the two areas, swelled into the modern balloon-like cerebral hemispheres, leaving the now dwarfed limbic cortex round the edge ('out on a limb'; limbus = edge or border; **Fig. 13.35**), and tucked away out of sight. Meanwhile, the massive fibre tracts required to link all this bulk of new cortex to the thalamus and other subcortical structures have forced their way between the older nuclei, cutting them off from one another and making their communicating nerve fibres wind their way right round the outside in circuitous fashion. At the same time, the amygdala, originally a structure on the wall of the hemisphere, became – like the corpus striatum – submerged beneath the incoming tide of neocortex, and ended up as an additional subcortical nucleus.

All this makes the neuroanatomy of the limbic system look a great deal more complex than it really is, and a schematic representation of the functional connections is in many ways more helpful than trying to reproduce in one's mind all the three-dimensional muddle of its actual form. Most of the limbic system appears to be concerned with such functions as emotion and motivation, with the neural control of the body's internal environment, and to some extent with olfaction; these aspects will be dealt with in Chapter 14. The cortical regions, and especially the hippocampus, seem to be more concerned with the contribution of learning and memory to these functions. In Man, electrical stimulation in this region has been undertaken occasionally as a preliminary to the surgical treatment of epileptic foci. What has sometimes been reported is that stimulation at particular sites gives rise not to the discrete flashes and spots of light characteristic of electrical stimulation of the human visual cortex, but rather to complex and repeatable *hallucinations* of an unusually realistic kind, sometimes apparently not static but moving in 'real time' (for example, of a tune played by an orchestra, to which the patient could beat time), and often producing an experience which is a synthesis of many sensory modalities at once. In one case a patient described a sense of it being Sunday morning, a bright summer day, the car being washed, children shouting, and so on. Such experiments have naturally been rare, and there must be some doubt as to how they should be interpreted: necessarily, they have not been carried out on normal people, and what makes these results less exciting than they might appear at first sight is that these hallucinations – only found in 8 per cent of patients – were in practically every case part of the aura that the patient felt anyway at the beginning of the attack; while the best responses were found on the whole when nearest the epileptic focus, excision of the region didn't always prevent the same hallucination being evoked later from a different spot. Nevertheless, it does seem probable from electrical recording in animals that some kind of progressively more detailed analysis of sensory information, and its integration across sensory modalities, does occur on its way down to the archicortex that runs round the bottom edge of the temporal cortex, and that this analysis – unlike what is seen in the parietal lobe – is of *recognition* rather than *localisation*.

Hippocampus

As we have seen, the hippocampus lies along the bottom edge of the temporal neocortex, and it is perhaps not too fanciful to think of it as a kind of cortical *gutter*. Sensory information is increasingly analysed and refined as it trickles from neuronal level to neuronal level down from sensory projection areas, through the complex associational networks of parietal and temporal cortex, and finally drains into the hippocampus itself. Certainly in the more posterior region of the temporal neocortex, which borders on visual areas, the logical progression already noted in the visual cortex by which cells are found first with simple concentric fields, and then with more and more specificity in terms of such parameters as line orientation, colour movement, or width, appears here to be continued; in Chapter 7 (p. 160) we saw that cells have been found in the temporal cortex which respond to quite specific objects

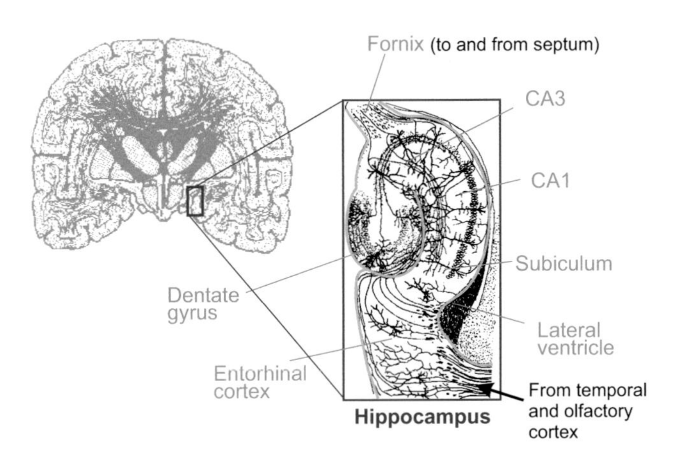

Figure 13.36 Simplified transverse section of the hippocampus, showing the main subdivisions, and its location in the brain.

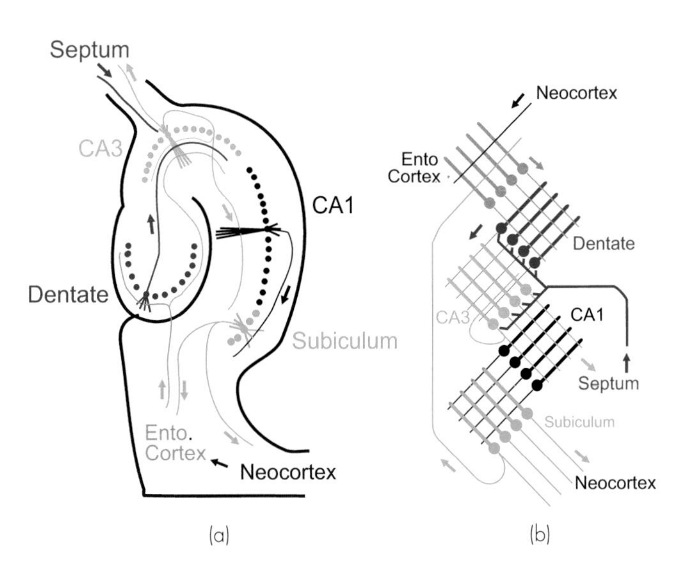

Figure 13.37 (a) Typical connections of neurons in different areas of the hippocampus. (b) Highly stylised representation of neural circuitry of the hippocampus, showing the sequence of cerebellar-like learning grids.

– faces, hands and so on – with significant behavioural implications.

A cross section through this gutter – its seahorse-like shape giving rise to the name 'hippocampus' – reveals a surprisingly regular neuronal structure, a neuronal crystal almost as machine-like as the cerebellum (**Fig. 13.36**). Archicortex differs from neocortex in having only three layers instead of six; there is only one layer of pyramidal cells, with fibres running predominantly transversely above and below them, and making convergent and divergent synaptic contact with the pyramidal cell dendrites. Pyramidal cells in the entorhinal cortex project to long rows of pyramidal cells in the dentate gyrus, these in turn project to rows of cells in the CA3 region of the hippocampal gyrus, and these cells in turn projecting to the pyramidal cells of the CA1 region. Branches of the CA3 cells form the large fibre bundle called the *fornix,* which projects to the mammillary bodies and septal nuclei, and thence indirectly to the hypothalamus and amygdala. In Man, the fornix contains more fibres than either the pyramidal tracts or optic nerves. The CA1 cells project to the neighbouring *subiculum* and thence, among other areas, to the anterior and central regions of the thalamus, by which route they may ultimately influence the basal ganglia and (prefrontal) neocortical areas.

Through all these routes the hippocampus can influence both overt behaviour and also internal responses via autonomic and endocrine mechanisms. Apart from receiving fibres from temporal neocortex, the entorhinal region has projections also from the neighbouring olfactory areas, prepyriform and periamygdaloid cortex, and septum. Thus the entire structure can be represented in the highly schematic form shown in **Figure 13.37**, a strongly hierarchical arrangement – a bit like four or five cerebellums in series – well adapted to integrating together information from neocortex and from the olfactory system, recognising specific patterns of activity, and producing both motivational responses through the motor system and emotional responses via the limbic nuclei. The question of what its output actually does must wait until the next chapter; what we are concerned with here is how the output is derived from its input.

That the hippocampus does indeed form, in a sense, the final output from the sensory analysers of the neocortex is clear from electrical recordings from its pyramidal cells. Ninety-five per cent of the pyramidal cells of area CA3 have been described as totally multimodal, responding to almost any combination of sensory modalities, and they are also described as being 'novelty-conscious'; that is, that they tend to show *habituation* to a stimulus if it is repeatedly presented, and respond more readily to things that are new; this in itself represents a kind of memory process. The results of lesions, particularly in the anterior and inferior parts of the temporal lobe, have classically also been conclusive in associating this area with the function of memory. The famous case is that of a patient called H. M., who was treated for severe bilateral temporal epilepsy by the drastic expedient of cutting away the major part of both temporal lobes, including most of the hippocampus, with the unfortunate result that although the patient's memory for things that had happened *before* the operation was good, he was effectively unable to lay down *new* memories for periods of longer than a matter of some minutes, an example of *anterograde amnesia* (see p. 260). H. M. woke up every day for the next 50 years of his life believing he was just over 30 years old. In some ways he was an ideal patient to study – despite being interviewed thousands of times and the subject of countless papers, he never grew bored with the attention. Subsequent work in rats has confirmed that it is damage to the hippocampal region that is responsible for this defect, and that it only happens when the lesion is bilateral, in which case the deficit is quite unspecific as to the nature of the material to be learnt: it can be recalled for some 5–10 minutes, but after that time, unless the subject can in some way rehearse it in his mind – as for example when trying not to forget a

Clinical box 13.4 Korsakov syndrome

A quite similar kind of anterograde amnesia is also seen as the result of chronic alcoholism; it is called the *Korsakov syndrome.* It is not so much the effect of alcohol itself as the result of thiamine deficiency brought on because a liquid diet is somewhat imbalanced. After death, one may see degenerative changes to various areas in the limbic system, notably the mammillary bodies and anterior thalamus, both of which lie on output routes from the hippocampus; however, one cannot of course be sure whether or not other regions, such as the hippocampus itself, may be functionally deranged even though their gross visual appearance may be normal. As with temporal lobe damage, motor learning is unimpaired, and stimulus recognition is still normal. But long-term storage of memories is impaired and things cannot be remembered for more than a few minutes without conscious rehearsal. The victim consequently often seems to be stuck in a past era: if asked who the prime minister is, he might reply 'Mrs Thatcher', and when told to describe the latest fashions, he might talk about bell-bottoms and micro-skirts. As with the agnosias associated with parietal cortex, the patient is often strikingly unaware that anything is wrong, and if confronted with facts that don't fit into his private time-warp, may start to *confabulate,* making up elaborate fantasies to explain the discrepancies; he may also become paranoid and aggrieved, believing that there is some kind of global conspiracy directed against him.[27] In order to avoid this, it is of clinical importance that patients suspected of alcohol dependency presenting to hospital are treated with thiamine supplementation as a preventative measure.

telephone number in the interval between looking it up in the book and dialling it – it is lost for good. Significantly, purely *motor* skills are unaffected: previously learnt ones are not lost, and new ones (like learning to type or to ride a bicycle) may be acquired, as demonstrated by patients such as H. M. Motor skills are learnt elsewhere, presumably either in the cerebellum or neocortex. Unilateral lesions do not have the same dramatic effect, although some difficulty has been reported in learning verbal material if the lesion is on the dominant side, and scans similarly demonstrate an asymmetry in learning verbal and non-verbal material.

Notes

1. Does size matter? The effects of evolution are not necessarily beneficial, and in some particularly dramatic cases – the peacock's tail, the stag's antlers – can be downright silly. The cortex that we are so proud of may come in the same category, and is perhaps better perceived as a grotesquely overblown excrescence, its size being selected for because it generates wit and art and music, and makes its owner attractive to the opposite sex.
2. Parietal cortex Good general accounts of parietal cortex in particular may be found in Beaumont, J. G. (2008). *Introduction to Neuropsychology.* (Oxford: Blackwell); Critchley, M. (1971). *The Parietal Lobes.* (London: Hafner); Stein, J. F. (1991). Space and the parietal association areas. In Paillard, J. (ed.) *Brain and Space.* (Oxford: Oxford University Press); and Walsh, K. W. (1978). *Neuropsychology, a Clinical Approach.* (Edinburgh: Churchill Livingstone). Discussions of association cortical function in general, and its theoretical and quantitative basis, include Wilson, H. R. (1999). *Spikes, Decisions and Actions.* (Oxford: Oxford University Press); Abeles, M. (1991). *Corticonics.* (Cambridge: Cambridge University Press); and Fuster, J. M. (2005). *Cortex and Mind: Unifying Cognition.* (Oxford: Oxford University Press).
3. Mirror neurons As a matter of fact there is an example of this on p. 252: when performing 'silent reading' there is nevertheless considerable activity in the motor cortex.
4. Café tables And to pursue the analogy a little further, the diffuse inputs ascending from the brainstem that regulate attention, sleep and other large-scale states function like the different kinds of drinks and other substances that are being consumed: see Chapter 14. An original and imaginative viewpoint on the functions of cortical and columnar organisation is presented in Calvin, W. H. (1998). *The Cerebral Code.* (Cambridge, MA: MIT Press).
5. Neural networks A readable account of neural networks is Levine, R. & Drang, D. (1988). *Neural Networks: The Second AI Generation.* (New York, NY: McGraw-Hill). A more technical but comprehensive source is Rojas, R. (1996). *Neural Networks: A Systematic Introduction.* (Berlin: Springer). A stimulating, biologically oriented but controversial book is Edelman, G. M. (1989). *Neural Darwinism: The Theory of Neuronal Group Selection.* (Oxford: Oxford University Press). Other accounts include Alexander, I. & Martin, H. (1995). *An Introduction to Neural Computing,* 2nd edn. (London: Thomson); Callan, R. (1999). *The Essence of Neural Networks.* (London: Prentice-Hall); and Gurney, K. (1997). *An Introduction to Neural Networks.* (London: UCL Press).
6. A turbulent area See for instance Carpenter, R. H. S. (2004). Supplementary eye field: keeping an eye on eye movement. *Current Biology* **14** R416–418.
7. Nature of perception See, for example, Kaufman, L. (1979). *Perception.* (Oxford: Oxford University Press). Some of the most thoughtful and penetrating insights in this area have been voiced by the distinguished art historian, E. H. Gombrich, who frequently showed so much better an understanding of perceptual mechanisms than many neurophysiologists: see for instance Gombrich, E. H. (1982). *The Image and the Eye.* (London: Phaidon).
8. Diffuse build-up to action No one has written more thoughtfully about how decisions are made in diffuse networks than Leo Tolstoy, in *War and Peace*: *'and in the same way, the innumerable people who took part in the war acted in accord with their personal*

characteristics, habits, circumstances and aims. They were moved by fear or vanity, rejoiced or were indignant, reasoned, imagining that they knew what they were doing and did it of their own free will, but they all were involuntary tools of history, carrying on a work concealed from them but comprehensible to us. Such is the inevitable fate of men of action, and the higher they stand in the social hierarchy the less they are free' (Book X).

9. Angle-expansion See Carpenter, R. H. S. & Blakemore, C. B. (1973). Interactions between orientations in human vision. *Experimental Brain Research* **18** 287–303. An excellent and comprehensive source of visual illusions in general is Robinson, J. O. (1999). *The Psychology of Visual Illusion.* (London: Hutchinson). Other, more popular, compilations include Seckel, A. (2011). *Incredible Visual Illusions.* (London: Arcturus); and Sarcone, G. A. & Waeber, M. J. (2011). *Amazing Visual Illusions.* (London: Arcturus).

10. Memory A well-written popular account is Baddeley, A. (1983). *Your Memory: A User's Guide.* (Harmondsworth: Penguin); another approachable account is Gluck, M. A., Mercado, E. & Myers, C. E. (2010). *Learning and Memory: Brain and Behavior.* (New York, NY: Worth); Dudai, Y. (1989). *The Neurobiology of Memory.* (Oxford: Oxford University Press) is more physiological.

11. Memory and development Some accounts of neuronal memory mechanisms, particularly in relation to development: Abeles, M. (1991). *Corticonics: Neural Circuits of the Cerebral Cortex.* (Cambridge: Cambridge University Press); Byrne, J. H. & Berry, W. O. (1989). *Neural Models of Plasticity.* (New York, NY: Academic); Gaze, R. M., Sharma, S. C. & Fawcett, J. W. (1992). *The Formation and Regeneration of Nerve Connections.* (London: Academic); Hopkins, W. G. & Brown, M. C. (1984). *Development of Nerve Cells and their Connections.* (Cambridge: Cambridge University Press); Lund, R. D. (1978). *Development and Plasticity of the Brain.* (Oxford: Oxford University Press); Hopkins, W. G., Keynes, R. J. & Brown, M. C. (1991). *Essentials of Neural Development.* (Cambridge: Cambridge University Press).

12. Hebb The postulate was most clearly stated in Hebb, D. O. (1949). *Organisation of Behaviour.* (London: Wiley). *'When an axon of cell A is near enough to excite a cell B and repeatedly or persistently takes part in firing it, some growth process or metabolic change takes place in one or both cells such that A's efficiency, as one of the cells firing B, is increased'.*

13. NMDA receptors More specifically, the channel is normally blocked by external Mg^{2+} ions: depolarisation removes the block. When open, the channel is permeable to sodium and potassium ions as well as calcium, with an equilibrium potential near zero. To open, glycine must also be present as a co-agonist; clearly this could have immense significance in terms of providing a possible mechanism by which one might control whether certain associations are learnt or not.

14. Trabeculae The reference is d'Arcy Wentworth Thompson (1917). *On Growth and Form.* Cambridge: Cambridge University Press); reprinted 1992 by Dover books.

15. Anterograde amnesia A moving account of such a case (*The Lost Mariner*) can be found in Sacks, O. (1985). *The Man who Mistook his Wife for a Hat.* (London: Duckworth).

16. Memory methods The technique of bizarre association is of very great antiquity, used for instance by the great Roman orators. In its original form it involved the mental placing of things to be remembered into a fixed sequence of locations in a real or imagined building: hence the expression 'in the first place . . . in the second place.'. . . See the extraordinarily stimulating Yates, F. A. (1969). *The Art of Memory.* (Harmondsworth: Penguin); and also Rossi, P. (1990). Creativity and the art of memory. In Shea, W. R. & Spadafora, A. (eds) *Creativity in the Arts and Science.* (Canton, MA: Science History Publications).

17. Symptomatology Neurological anecdote has now – deservedly – achieved the status of a recognised literary genre, and sometimes staged as well; some of the best are: Critchley, M. (1979). *The Divine Banquet of the Brain.* (New York, NY: Raven Press); Klawans, H. L. (1989). *Toscanini's Fumble.* (London: Bodley Head); Klawans, H. L. (1990). *Newton's Madness.* (London: Bodley Head); Sacks, O. (1985). *The Man who Mistook his Wife for a Hat.* (London: Duckworth).

18. The representation of space See de Renzi, E. (1 982). *Disorders of Space Exploration and Cognition.* (Chichester: Wiley); de Renzi, E. (1988). Visuo-spatial agnosias. In Kennard, C. & Rose, F. C. (eds) *Physiological Aspects of Neuro-ophthalmology.* (London: Chapman and Hall).

19. Denial of body parts As in the following Pinter-like dialogue:

    Doctor: Is this your hand?
    Patient: Not mine, doctor.
    Doctor: Yes it is. Look at that ring: whose is it?
    Patient: That's my ring. You've got my ring, doctor!

    (Sandifer, P. H. (1946). Anosognosia and disorders of the body scheme. *Brain* **69** 122–137.)

20. Aphasia Brain, L. (1975). *Speech Disorders.* (London: Butterworth) is a classical account; see also Rose, F. C., Whurr, R. & Wyke, M. A. (eds) (1988) *Aphasia.* (London: Whurr Publishers); Basso, A. (2003). *Aphasia and its Therapy.* (Oxford: Oxford University Press) combines science and clinical application.

21. Aphasia after a stroke A classic example is that of Dr Samuel Johnson, who has left a vivid account of what it is like to experience such a stroke. Johnson was a stutterer before this episode, and was notoriously clumsy: it is possible that he may in fact have suffered some slight brain damage in early life: *'I went to bed, and in a short time waked and sat up. I felt a confusion and indistinctness in my head that lasted, I suppose, about half a minute. I was alarmed, and prayed God, that however he might afflict my body, he would spare my understanding. This prayer, that I might try the integrity of my faculties, I made in Latin verse. The lines were not very good, but I knew them not to be very good: I made them easily, and concluded myself to be unimpaired in my faculties. . . . Soon after I perceived that I had suffered a paralytick stroke, and that my speech was taken from me. Though God stopped my speech, he left me my hand. My first note was necessarily to my servant, who came in talking, and could not immediately understand why he should read what I put into his hands. In penning this note I had some difficulty; my hand, I knew not how nor why, made wrong letters. My physicians are very friendly, and give me great hopes; I have so far recovered my vocal powers as to repeat the Lord's Prayer with no very imperfect*

articulation'. A number of features of this account are interesting: the aphasia was clearly predominantly expressive, with a little disturbance of writing, but no disorder either of the ability to comprehend speech, or to formulate it in the mind – even in Latin verse! – and certainly no evidence of any general impairment of intelligent thought.

22. Prodigies See for instance Treffert, D. A. (1989). *Extraordinary People: An Explanation of the Savant Syndrome.* (London: Bantam). One example from Treffert: *'Leslie has never had any formal musical training. Yet as a teenager, on hearing Tchaichovsky's 1st Piano Concerto for the first time, he played it back flawlessly and without hesitation, and can do the same with any piece of music however long or complex. Leslie is blind, severely mentally handicapped and has cerebral palsy. He cannot hold a spoon to eat and merely repeats in monotone fashion what is spoken to him'*. There are also some good examples in Sacks, O. (1985). *The Man who Mistook his Wife for a Hat.* (London: Duckworth), and in his (2011). *Musicophilia.* (London: Picador).

23. Face perception An excellent account of this fascinating area is Bruce, V. & Young, A. (1998). *In the Eye of the Beholder: the Science of Face Perception.* (Oxford: Oxford University Press).

24. Dominance and handedness There is a good discussion in Morgan, M. J. & McManus, I. C. (1988). The relationship between brainedness and handedness. In Rose, F. C., Whurr, R. & Wyke, M. A. (eds) *Aphasia.* (London: Whurr Publishers). There is also much useful material in Hellige, J. B. (1993). *Hemispheric Asymmetry.* (Boston, MA: Harvard University Press); Annett, M. (1998). Handedness and cerebral dominance: the right shift theory. *Journal of Neuropsychiatry* **10** 450–469 makes special reference to possible genetic influences.

25. Split brains A useful survey is Gazzaniga, M. S. (2005). Forty-five years of split-brain research and still going strong. *Nature Reviews Neuroscience* **6** 653–659.

26. Abilities of the non-dominant side A full account is Springer, S. P. & Deutsch, G. (1993). *Left Brain, Right Brain.* (San Francisco, CA: Freeman). With special exercises to encourage the right hemisphere, one can learn to draw better (it works – I've tried it): see Edwards, B. (2012). *Drawing on the Right Side of the Brain.* (New York, NY: Tarcher).

27. Confabulation When something in our memory or perceptual world is missing, we do not perceive a hole: the brain makes strenuous efforts to fill it in. Think of the blind spot (see p. 142) interpolating a surrounding pattern, or the fact that with partial deafness, one mishears words – inventing them – rather than sensing gaps. An ancient joke: two elderly men in a train: *'Is this Wembley?' 'No, Thursday.' 'Yes, so am I. Let's have a drink'*. It's featured by P. G. Wodehouse in *The Mating Season,* told by Gussie Fink-Nottle to a dinner table full of aunts; it is not a success, not least because one of them is deaf. *The stout aunt spoke into her ear, spacing her syllables carefully. 'Augustus was telling us a story about two men on a train. One of them said 'To-day is Wednesday', and the other said 'I thought it was Thursday', and the first man said 'Yes, so did I'.' 'Oh?' said Aunt Charlotte, and I suppose that about summed it up.*

CHAPTER 14

MOTIVATION AND THE CONTROL OF BEHAVIOUR

We first came across the concept of the hierarchical structure of the brain, its organisation in an ascending series of levels, in Chapter 9. Now, in this final chapter we look at the very highest levels of all, those that determine *what* we do rather than how we do it.[1]

Actually, there is no very clear or logical distinction between 'what' and 'how' in this sense: the task of deciding what to do amounts in the end to deciding *how* to stay alive or, at worst, how to immortalise our genetic instructions. It is for this reason that the sensory inputs of this highest level come – paradoxically – not just from the special senses that tell us about the outside world, but also from interoceptive, self-monitoring senses that are usually considered so 'low' as to be beneath our conscious notice. What they provide is information about the physiological well-being of the body, the state of the milieu intérieur, and our distance from that final condition that awaits all of us.

Motivation

Why, in fact, do we ever bother to do anything at all?

The answer is basically to do with income and expenditure – of energy. Even at rest, we are remorselessly expending energy: if we don't replace this energy, we die. If like corals or sea anemones we were lucky enough to live in an environment where we were bombarded by food, we could just glue ourselves to a rock and keep our mouths open. But for the big spenders, warm-blooded animals like us, the only way of keeping in surplus is to *gamble*. We expend a lot of energy as a *stake*, in order to perform actions from which we hope to get more in return, rather like a business investing some of its profit in the hope of even huger profits in the future. In a sense this decision-making – *to do or not to do* – is the most difficult task an organism has to undertake. As we shall see, the whole of the brain can usefully be thought of as a mechanism for reducing the risk, by making more and more accurate *predictions* about the likely result of any particular course of action, on the basis of past experience, stored not just in our brains, but in our books.

To put it another way, we need to apply the principles of *homeostasis*, which loom so large in general physiology, not just to the milieu intérieur but to the outside world as well. In addition to *internal* homeostasis, controlled by hormones and the autonomic nervous system (ANS), we have to add *external* homeostasis, controlled by the brain, achieved sometimes by literally altering our environment (wearing a pullover, for instance), but more often by moving to somewhere nicer, or by engulfing or penetrating things we like.

Motivational maps

But the decision process need not be as complex as this. In a simple creature – an amoeba is an extreme case – the nature of the fundamental mechanism is particularly obvious: its motivation is entirely a function of its immediate environment, sensed chemically. Consequently we see *tropisms* in response to gradients of things like food (positive) or poisons (negative). On the one hand, attractive stimuli set up a positive gradient down which the animal moves; on the other hand, threatening conditions create a negative gradient, and it moves away (**Fig. 14.1**). So the amoeba's environment is a sort of motivational potential field or contour map, and the amoeba is like a little charged particle that moves around in response to local gradients, the path it traces out being a direct function of its environment. Very simple tropistic mechanisms like these give rise to surprisingly life-like behaviour, as in Dr Grey Walter's pioneering electromechanical tortoise Elsie which ran amiably around the floor looking for light. Though higher animals produce more complex behaviour, the mechanism is essentially the same. The added complexity comes about for two reasons.

First, because there are many more types of desirable and undesirable stimuli to which they may react, and

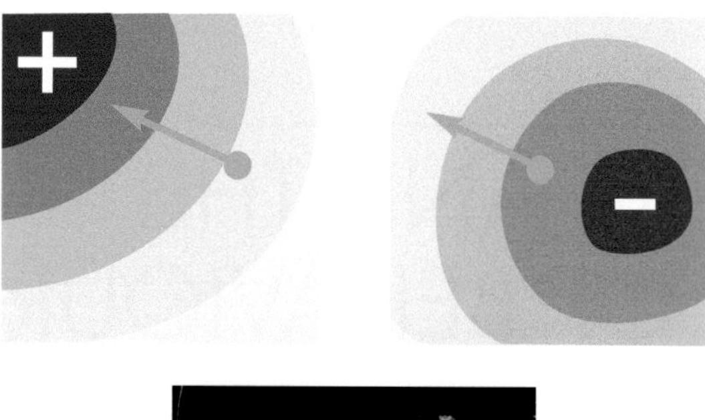

Figure 14.1 Positive and negative motivational gradients.

many of them – perhaps most – are *learnt*: these are the secondary motivators (like money) that through experience become associated with other more self-evidently desirable goals (Chapter 8, p. 177). Consequently each individual has its own classification of stimuli into desirable and undesirable categories, unique because it is the result of that individual's own personal experience.

Second, because whether a particular stimulus like food *is* a motivator or not depends also on one's own *need* – in this case, whether or not one is hungry. Motivation, in other words, is something like the *product of gradient and need*, so changing patterns of need give rise to changing patterns of activity even though the environment itself is the same, and to an outsider the resulting behaviour may appear to be complex or even unpredictable. Thus Cambridge for me consists of a large number of separate gradient or contour maps, each corresponding to a different need: one for food, with high points at all the food shops and restaurants, one for money, centred on my bank and supplemented by cash machines, one for newspapers, one for avoiding rain, and so on (**Fig. 14.2**). Which one is operative at any particular moment depends on my need at that moment, rather like those electrical maps sometimes seen at the more down-market tourist resorts, with

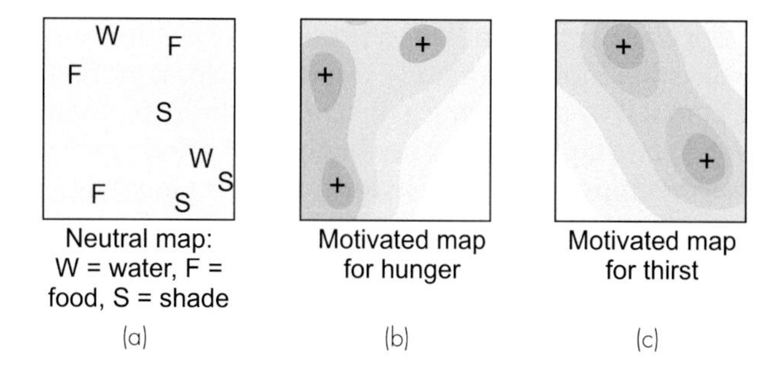

Figure 14.2 Motivational maps. (a) A neutral environmental map showing the location of food (F), water (W) and shade (S). The corresponding motivational maps when the animal is hungry (b), or thirsty (c).

bulbs that light up when you press one of a set of buttons marked 'parking', 'pubs', 'post offices' and so on.

So the fundamental limbic motivational computer has to be a sort of phone directory connecting particular needs with a kind of library of motivational maps of the outside world: like the tourist map, it translates information about need into the kind of tropistic data that can in turn be changed by the higher levels of the motor system into actual patterns of activity (**Fig. 14.3**). Here A, B, C, etc. are separate needs such as hunger, thirst, etc., and each has its own stored motivational map that is activated in appropriate physiological circumstances. Some evidence suggests that these 'yellow pages' or motivational maps are embodied in the *hippocampus*. Hippocampal neurons have been found in the rat that respond specifically when the animal is at a particular point in its environment, for example within a maze that the rat has learnt; its involvement in certain kinds of learning was discussed in Chapter 13.

In **Figure 14.3b**, the shading represents average firing frequency of a unit in rat hippocampus at different points within an enclosure, showing that it seems to be associated with a particular location. Similarly, brain scans of

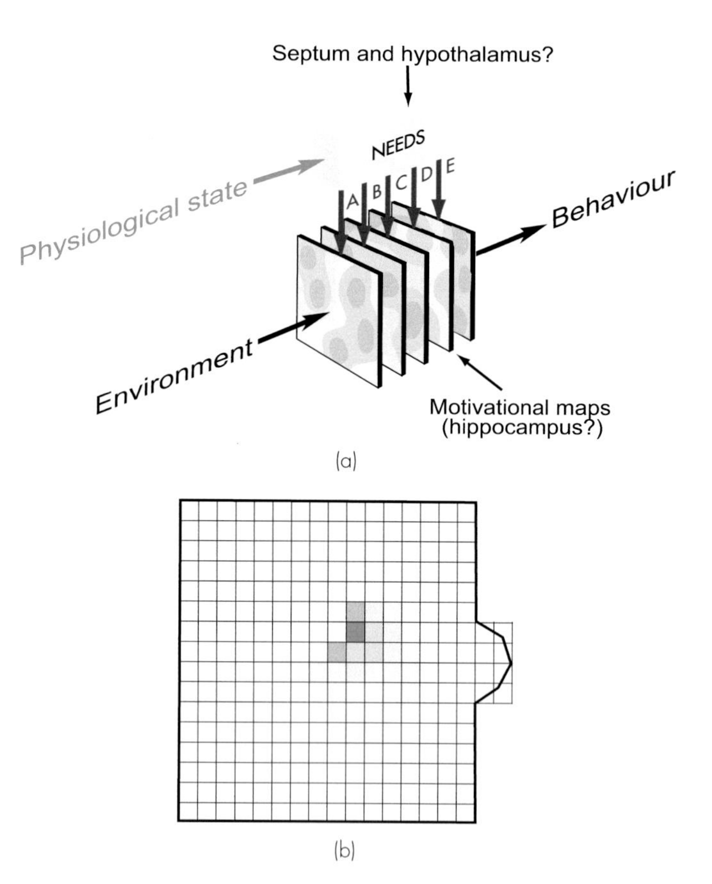

Figure 14.3 (a) Hypothetical model of a mechanism for computing directed behaviour from needs sensed by monitoring the body physiological state. A, B, C etc. are separate needs, such as, for example, hunger, thirst etc., and each has its own stored motivational map that is activated in appropriate circumstances. (b) Hippocampal mapping: plan view of an enclosure in which a rat with an electrode implanted in the hippocampus was free to move around. The shading represents the average firing frequency of a hippocampal unit in each of the squares, showing that the unit appears to code for a particular area within the enclosure. (Data from Wiener et al., 1989, ©1989 Society for Neuroscience, with permission.)

London taxi drivers, who have to undergo a period of rigorous spatial learning, demonstrate an enlarged posterior hippocampus relative to matched controls.[2] Equally, it is the *hypothalamus* that provides information about need, about the state of the body, and projects to the hippocampus via the septal nuclei. It is the centre to which autonomic afferents project, and its neurons monitor such physiological states of the blood as glucose concentration, temperature and osmolarity, as well as levels of circulating hormones, that decides one's state of need. It is also in the hypothalamus that primary consummatory responses such as eating and drinking may be triggered off by electrical stimulation. The hypothalamus is thus utterly at the heart of the neural mechanisms that generate motivation: it is discussed in detail later in this chapter (p. 279).

It is natural to feel a certain resistance to the notion that our own richly complex lives, the apparent wealth of choices open to us, and our sense of liberty to choose among them, could possibly be determined by so simple a mechanism. But as Herbert Simon has said, human behaviour is really rather simple, but because most people live in very complex physical, man-made and social environments, their actual behaviour *appears* extremely complicated; thus the path traced out by an ant moving over rough ground may be very complex in appearance, even though its behaviour is simply directed at getting back to its nest.[3] To some extent it is in fact possible to plot motivational maps in humans: by averaging over large numbers of individuals, it is not difficult to measure quite directly the same kinds of tropistic gradients for us humans, that work so well in describing what an amoeba does. If you take a group of people and ask them the very simple question *'Where in Britain would you like to be?'*, it is possible to obtain contour maps of average preferences, in this case of the relative desirability amongst a group of school-leavers of different parts of the country (**Fig. 14.4**). These are certainly motivational maps, in the sense that if the individuals had the means to do it, they would be translated into actual migratory behaviour not very different in essence from our amoeba moving blindly down its tropistic gradient.

Figure 14.4 Motivational maps for humans are not too dissimilar than those of other organisms.

Emotion

But motivational tropisms are not the only kind of behaviour. Some of our activity is not aimed directly at achieving particular goals in the way that the tropisms are; instead it is *preparatory* to the directed behaviour itself. The release of adrenaline associated with the need for sudden exertion is a classic example: it has the obvious effect of preparing the muscles and circulatory system for action. This sort of thing is what is meant in the widest sense by *emotional behaviour*; the *emotions* that we may feel at the same time are the sensory side effects of this undirected behaviour (**Fig. 14.5**). There are as many types of emotional behaviour as there are types of motivational goal, and they include some kinds of activity that are not regarded as 'emotional' in common parlance. Salivation, for example, is in this sense an emotional response accompanying the directed behaviour of getting food and eating it; and penile erection is an obvious preparatory response to another kind of goal. Much of the release of hormones falls in this general category, as for example the surge of luteinizing hormone (LH) that triggers ovulation in response to copulation in some species. Thinking of emotion and emotional behaviour in this more inclusive way helps to dispel some of the misunderstanding and muddle that tend to surround this topic.

As Man has a richer set of possible needs and goals, including abstract or even spiritual ones, so his types of emotional behaviour and emotional sensations are more varied and complex. But there are two absolutely basic emotional patterns found throughout the animal kingdom, and perfectly evident in Man as well, associated with tropisms of any kind: these are *arousal* and *conservation*.

Two basic emotional states

Arousal signifies the emotional state associated with a steep tropistic gradient, which may be either towards a desirable goal or away from a source of threat (p. 275): the state

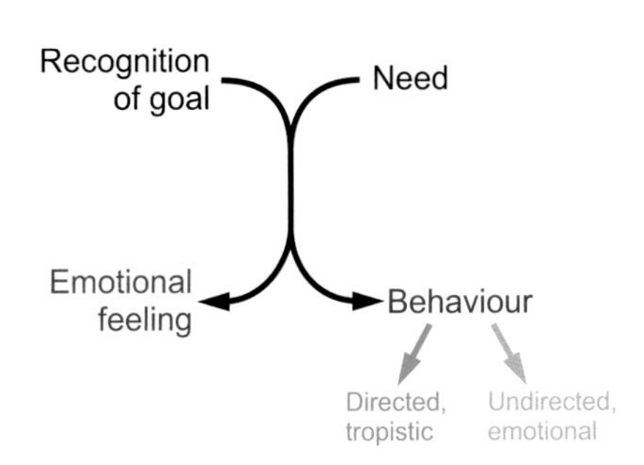

Figure 14.5 The relationship between emotion and behaviour.

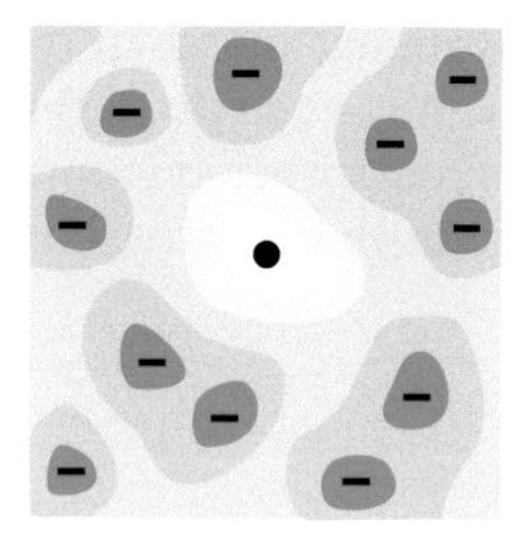

Figure 14.6 The kind of motivational map for which the only appropriate response is conservation.

Figure 14.7 Features of acute conservation. (From Engel and Schmale, 1972; © 1972 Ciba Foundation; with permission from John Wiley and Sons.)

often described by physiologists as 'fight, fright or flight', that results in an increase in the general activity of the sympathetic system, and the release of adrenaline. The consequent bodily responses are all of more or less obvious use in preparing the body for the expenditure of the energy that will be used to achieve the goal: blood flow through the muscles is increased, the heart rate is raised, glucose is released into the blood, the bronchioles and pupils dilate, the electrical activity of the brain increases, reaction times get quicker, and there is an associated feeling of general excitement. All of this of course involves a certain expenditure of energy, and would be a drain on the body's resources if kept up for a long time: but much is now at stake, and the gamble is one worth taking.

Conservation or withdrawal is in a sense the opposite of arousal. In a situation like the one in **Figure 14.6**, when every possible action is unpleasant – like standing in the middle of a minefield – the only sensible response is to conserve one's resources and do nothing at all, in the hope that the difficulties will go away of their own accord. The result is inactivity and stupor, a loss of muscle tone, sleep or even hibernation; if the situation is a sudden one, there may be abrupt immobility or freezing – the animal thus incidentally making itself inconspicuous and feigning death (a common response to oncoming motor-cars, but not a particularly helpful one). By all these means the rate of energy expenditure is greatly reduced, enabling the animal to ride out what may be only a temporary state of siege. The associated feelings are of apathy, tiredness and weakness: because of the reduction in muscle tone, one may actually feel heavier, pressed to the ground – the origin of the word 'depression'. Loss of muscle tone in the face produces a characteristic sagging of the lower jaw and of the corners of the mouth, and bowed head. In mild forms, the conservation state occurs only too commonly when a person feels that nothing is worth doing and circumstances are against him, giving rise to reactive depression. The more acute form of conservation is fortunately only rarely seen, except in response to cataclysmic disasters. The woman shown in **Figure 14.7** has just emerged from shelter after an earthquake that has destroyed most of the town in which she lived. The objective signs of conservation are obvious: the stooped posture, the hand lifted to the face to support the dropped jaw, the immobile staring eyes. In such circumstances one may find a general state of apathy and inactivity that continues for a long time and is not conducive to survival. A curious feature of such chronic depression, though one that is readily understandable in terms of motivational maps, is that in times of severe and particular stress, as in war, the incidence of this kind of emotional state actually decreases, perhaps because collective activity is required. *Sleep* can usefully be regarded as a variety of conservation, but is considered in a separate section below (p. 289).

We seldom see either of these two kinds of emotional state in their pure forms. Real objects tend to be both attractive and repellent: a hunting animal's prey may both be desirable as food and also dangerous, and in many species even the sexual act is a risky undertaking for the male. It can be illuminating to think of emotions in terms of a continuum of mixed types distributed around the two primary axes of arousal and conservation, emphasising the ambivalent nature of such states as rage and fear, the knife-edge between attack and retreat (**Fig. 14.8**). Food does not usually have quite this effect on humans – dining is rarely a frightening experience in modern society – but rage can easily be elicited in situations of frustration, when the positive and negative aspects of a possible goal are

Figure 14.8 Mixed emotional state as a result of stimuli that are partly attractive and partly repellent.

nicely balanced. The nature of the goal clearly affects the precise response that is made, yet there can often be a curious generality about arousal, most obvious perhaps in human sexual behaviour: sexual aggression shades off imperceptibly into sadism, and affection is expressed by licking and biting and other responses more appropriate to an edible goal.

Neural mechanisms

Of course such schemes are over-simplistic; for one thing, they ignore the important part that *memory*, especially the kind of anticipatory memory discussed in Chapter 13, in connection with the frontal lobes, may play in introducing an extra temporal dimension into our emotions: such emotional states as hope, worry, confidence and regret clearly involve an element of this kind. But it may help us to remember that there is nothing particularly recherché or high-falutin' about human emotional responses, and no reason to suppose that they are produced by fundamentally different mechanisms from those generating the remarkably similar patterns of behaviour seen in other animals.

In short, keeping alive is a matter of monitoring the milieu intérieur and making homeostatic adjustments to it, adjustments that are partly neural and autonomic and partly hormonal: internal responses to internal stimuli. But this process is made much more effective by reacting to external stimuli as well, and by generating external responses. The development of the brain has permitted more and more sophisticated analysis of external stimuli, and greater and greater elaboration of patterns of external response, so that most of its bulk is concerned either with sensory analysis or motor coordination. But in the end, the only part of it that really matters is the region where the four fundamental signals – internal and external inputs and outputs – actually come together. That region is the *hypothalamus*.

The hypothalamus

Control of the pituitary

The hypothalamus lies on either side of the third ventricle, immediately above the pituitary (hypophysis) and below the thalamus, and consists of several fairly distinct nuclei (**Fig. 14.9**).[4] At its lowest level, it controls the pituitary (hypophysis), through two distinct mechanisms: one is direct, the other indirect.

- *Direct:* the axons of magnocellular neurons (black, p. 279) in the supraoptic and paraventricular nuclei pass right down into the pituitary stalk to terminate in the posterior lobe (neurohypophysis) (**Fig. 14.10**). Here they release their transmitters, not at synaptic junctions but directly into the bloodstream; thus these neurons are acting directly as endocrine cells, and their transmitters are actually hormones. Neurons of the supraoptic region predominantly release *antidiuretic hormone* (ADH), those of the paraventricular region mostly *oxytocin*. Both these hormones are nonapeptides of very similar structure, but their effects are entirely different. ADH helps control the osmolarity of the blood by stimulating the retention of water in the kidney; in large doses it may also increase blood pressure through arteriolar constriction (hence its alternative name of vasopressin). Oxytocin stimulates the smooth muscle of the uterus in labour, and also causes milk ejection during lactation; in both cases the stimulus to its release is essentially neural, predominantly from mechanoreceptors in the regions concerned.
- *Indirect:* the other route by which the hypothalamus controls the secretion of hormones from the pituitary is quite different. Axons from other, parvocellular, hypothalamic neurons terminate in a region on the ventral surface (the median eminence) where a system of fenestrated capillaries carries arterial blood down to the anterior pituitary through a portal system.[5] The substances released from their terminals (*releasing* or *inhibiting hormones*) enter this portal system and are transported to

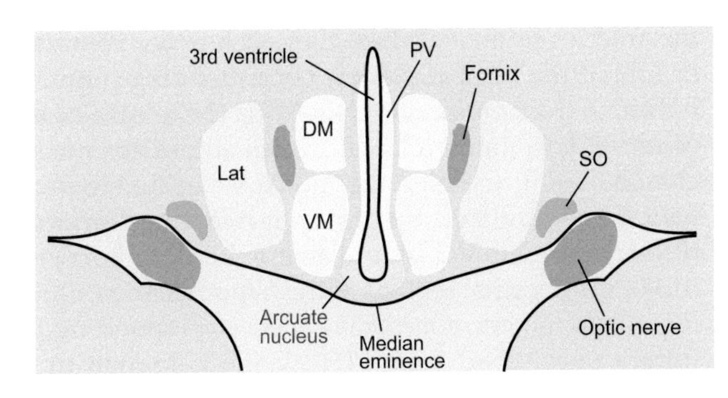

Figure 14.9 Highly schematic representation of the main hypothalamic nuclei (transverse section): Lat, lateral; PV, paraventricular; DM, dorsomedial; VM, ventromedial; SO, supraoptic.

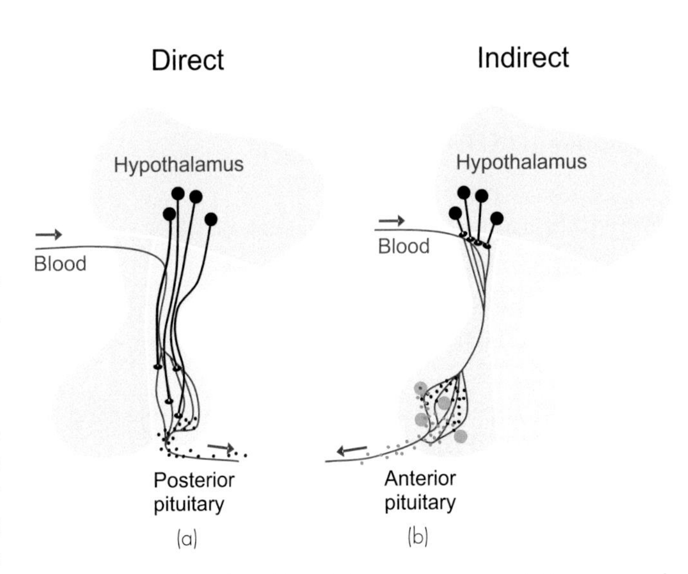

Figure 14.10 Relationship of hypothalamus with pituitary. (a) Innervation of the posterior lobe by hypothalamic fibres releasing pituitary hormones at their terminals. (b) Releasing hormones from hypothalamic neurons being carried by a portal system to the anterior lobe, where they control the release of pituitary hormones from pituitary endocrine cells.

the anterior pituitary where they each either stimulate or inhibit the release of some corresponding pituitary hormone from endocrine cells. Thus the release of the pituitary hormone prolactin, which stimulates the secretion of milk and has other functions related to pregnancy, is stimulated by prolactin-releasing hormone (PRH) and inhibited by prolactin-inhibiting hormone (PIH), from medial regions of the hypothalamus. Other hypothalamic hormones have their corresponding pituitary ones: apart from PIH (which is known to be dopamine) they are all small peptides.

Homeostatic functions

Though scarcely larger than a peanut, the hypothalamus has an absolutely fundamental role in the control of behaviour. The reason is that – uniquely – it straddles the blood and the brain. At its lowest level – as we have just seen – it directly oversees the endocrine mechanisms that dominate hormonal and metabolic functions. In doing this it relies on the fact that it is itself capable of monitoring critical aspects of the blood, notably blood glucose, temperature and osmolarity: the hypothalamus is a death predictor. At an intermediate level it acts as what Sherrington called the *head ganglion of the autonomic system*, effectively being in charge of the whole of the milieu intérieur, through its autonomic efferents that control the heart, digestive tract and other vital organs. Again, it is able to do this effectively because it has feedback from the body through autonomic afferents. In both cases, sensory information about the internal state of the body – S_{int} – is used as feedback to create homeostatic internal responses – R_{int}. The intermingled coordination of these two kinds of feedback control – partly hormonal, partly neural – is what keeps us alive.

In the course of evolution what was already a beautifully functioning system became dramatically better. Partly this was a matter of *prediction*: From the special senses, sensory cues about the outside world could be interpreted through experience to anticipate what was *going* to happen to the milieu intérieur – think of temperature receptors in the skin, for example, or information about general light level telling you about the time of day, or taste receptors telling you that you're about to experience an elevation of blood glucose. Thus the internal sensory signals, Sint, came to be supplemented by external ones, S_{ext}. At the same time, on the output side, the hypothalamus developed the ability to generate actual behaviour – R_{ext} as well as R_{int} – through its projections to limbic and motor pathways: a sensible response to cold is putting more clothes on; a sensible response to a river if you're thirsty is to bend down and drink (**Fig. 14.11**). The hypothalamus is both the ultimate detector of need and the ultimate generator of behaviour: the rest of the brain is really just a *way of making the hypothalamus work better.*

It may be helpful to reinforce this fundamental point by examining a number of specific in stances of homeostatic systems in which hormonal and neural signals are integrated by the hypothalamus to produce external, behavioural, responses as well as internal ones.

Hypothalamus and glucostasis

The control of blood glucose is a particularly clear example of the interplay between internal and external homeostasis (**Fig. 14.12**). On the one hand, there are the well-known

Table 14.1 Hypothalamic control of the pituitary

Pituitary hormone	Control of release	Actions
Oxytocin	Neural	Milk ejection Uterine contraction
Vasopressin (ADH)	Neural	Water retention Vasoconstriction
Growth hormone (GH)	GHRH, GHIH	Medium-term provision of metabolic energy Promotion of growth
Thyroid-stimulating hormone (TSH)	Hypothalamic thyrotropin-releasing hormone (TRH)	Stimulates thyroid; its hormones raise body temperature and have other miscellaneous effects
Adrenocorticotrophic hormone (ACTH)	Corticotropin-releasing hormone (CRH)	Regulates levels of cortisol and androgens from the adrenal cortex; some effects on aldosterone as well
Luteinizing hormone (LH)	LH-releasing hormone (LHRH)	Stimulates ovulation or testosterone secretion
Follicle-stimulating hormone (FSH)	GRH	Stimulates follicular growth or spermatogenesis
Prolactin	PRH, PIH	Stimulates milk secretion and maternal behaviour
Melanocyte-stimulating hormone (MSH)	MSH-releasing and -inhibiting hormones (MRH, MIH)	Controls skin colour in some species; in Man, function unclear

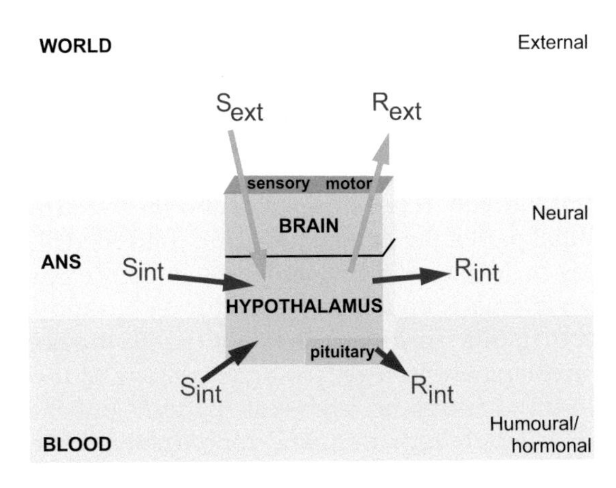

Figure 14.11 The hypothalamus as an interface between blood and brain linking internal and external stimuli (S_{ext}, S_{int}) to internal and external responses (R_{ext}, R_{int}).

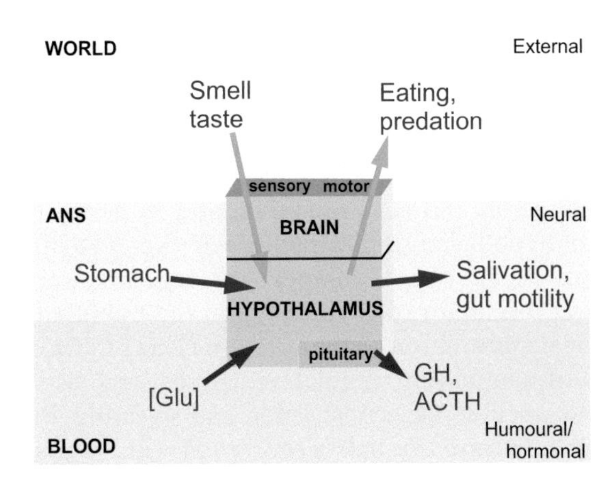

Figure 14.12 Hypothalamic glucostasis, showing a specific example of the interplay between internal and external stimuli and responses.

hormonal mechanisms that ensure that glucose flooding in from the gut during a meal is quickly stored away, and later distributed to longer-term depots; in addition, there are several hormonal mechanisms that liberate glucose from these stores, during acute or chronic periods of need. On the other hand, it is only through behaviour – predation and ingestion – that glucose or its precursors will ever arrive in the gut at all. And in the hypothalamus we find neurons that are concerned with all aspects of both internal and external glucostasis.

The *ventromedial* and *lateral* areas regulate feeding behaviour by monitoring the level of blood glucose, and this information is also used in a negative feedback loop that regulates pituitary growth hormone release by means of GHRH and GHIH (somatostatin) in response to fluctuations in blood glucose. The ventromedial hypothalamus has also long been known to be associated with the control of *eating*. An animal with a lesion in the ventromedial area develops a voracious appetite, as if unable to sense when it has had enough, and as a consequence it becomes obese. Lesions in the lateral hypothalamus have exactly the opposite effect: appetite is reduced, the animal displays little interest in food and loses weight. For this reason, the lateral area is often described as a 'feeding centre', and the ventromedial area as a 'satiety centre'. Cells of the ventromedial area take up glucose at a particularly high rate; as a consequence, injections of the poisonous glucose derivative gold thioglucose cause specific localised lesions that result in hyperphagia. In addition, the short-term control of eating is of course also dependent on sensory information coming both internally from the digestive tract and externally from smell and taste. Thus we have here a clear example of a system in which internal information from both the blood and viscera is used in conjunction with external stimuli to produce an integrated response that is partly internal – the regulation of growth hormone, and also the production of saliva and other digestive secretions, and other autonomic effects – and partly external – the eating itself.[6]

Hypothalamus and fluid balance

The control of the concentration and volume of the body fluids is equally clearly a matter of cooperation between internal and external mechanisms of homeostasis, between hormonal regulation and drinking behaviour, and is associated particularly with the *supraoptic* region. Certain cells in this region act as *osmoreceptors*, stimulating the release of ADH when the blood becomes too concentrated; autonomic afferents carrying information about blood volume from stretch receptors in the venous circulation also appear to contribute to the control of ADH by the hypothalamus. Other information that is relevant to the regulation of water balance comes from receptors in the subfornical region, just above the hypothalamus; they respond to the hormone angiotensin II that essentially signals a low average blood pressure, but are more concerned with the regulation of drinking than with the control of ADH. Again, autonomic afferents from the oesophagus and stomach are also believed to contribute to thirst and to the initiation and especially the termination of drinking: animals stop drinking long before their body fluids have yet become fully rehydrated – if they didn't, they would drink far too much. The effects of hypothalamic lesions suggest that like eating, drinking is controlled by two opposed systems located in different areas. Lesions in the supraoptic region produce excessive drinking (polydipsia), while those in the lateral hypothalamus reduce drinking as well as eating; electrical stimulation of the lateral nuclei, on the contrary, causes an animal to take in enormous amounts of water. The fact that lateral lesions affect both eating and drinking, and in rats have more generalised effects on other kinds of directed behaviour, suggests either that it has in some sense a more global role in mediating motivation, or perhaps that it is a mosaic of sub-regions devoted to specific varieties of directed behaviour.

Hypothalamus and temperature regulation

Temperature regulation is another example of homeostasis achieved through a mixture of internal and external responses: autonomically, hormonally, and also through overt behaviour such as curling up in the cold and seeking warmth (not to mention putting on or taking off one's clothes). Once again, the input to this system is partly neural and partly humoral; afferent signals from somatosensory warm and cold receptors, and from cells in the anterior hypothalamus that themselves respond to the temperature of the blood. Temperature regulation appears to be represented rather diffusely in the hypothalamus. Electrical stimulation at many points can produce fragments of temperature-regulating activity such as shivering, piloerection, vasoconstriction and sweating. Broadly speaking, the anterior half is concerned with mechanisms for losing heat in a hot environment and the posterior half with conserving heat when it is cold. More generally, there is a tendency for sympathetic responses to be found in the posterior half and parasympathetic in the anterior half. The fact that none of these effects is sharply localised simply reflects the high degree of inter-relationship that exists between different homeostatic functions: a given response such as vasoconstriction may be caused by many diverse kinds of stimulus (fear, low temperature, low blood pressure), and will in turn have a disturbing effect on several different homeostatic systems. The *thyroid-stimulating hormone* is controlled by TRH, associated with more ventral parts of the hypothalamus. The thyroid hormones have many interrelated effects on metabolism and growth, which are not well understood. One of its functions appears to be to cause a general increase in metabolic rate, helping to maintain body temperature under conditions of chronic cold.

Hypothalamus and reproduction

The endless multiplication of examples can soon become wearisome, and in any case leads far outside the scope of this book; the remaining pituitary outputs will be only mentioned very briefly. The *gonadotropic hormones* LH and FSH, jointly controlled by a single releasing factor (gonadotrophin-releasing hormone, GRH), together with the pituitary hormone *prolactin* are the means by which the brain influences the reproductive systems. What is particularly interesting about them is that gonadal steroids feed back on to receptors in the hypothalamus, not only influencing the production of the hormones themselves (and thus generating reproductive cycles) but also controlling sexual and maternal behaviour through its descending control of reticulospinal neurons. They also illustrate very clearly how an essentially hormonal control system can be influenced by a host of different types of stimuli from the special senses. One need only think of the effects of light on the timing of ovulation and breeding seasons, of the effect of skin and other kinds of stimulation on sexual arousal, of the influence of the sound or smell of offspring on maternal behaviour, of pheromones on ovulation and mating, and so on. Similarly, the effects of the various kinds of internal and external stimuli that constitute 'stress' on the secretion of ACTH, which regulates the secretion of corticosteroids from the adrenals and is controlled by CRH, are well known: ACTH is actually quite widely distributed in the brain, including the superior colliculus and substantia nigra and amygdala. In addition, there are the *melanocyte-stimulating hormones*, regulating the pigmentation of the skin in certain species and released from the interstitial part of the pituitary, that are controlled in a similar way by MRH and MIH, under the influence predominantly of visual stimulation (see **Table 14.1**). And finally, a hypothalamic function that is not exactly homeostatic is its involvement in registering the passage of *time*, and thus in turn controlling large-scale patterns of behaviour such as sleep, hibernation, and reproductive behaviour in animals where this has a seasonal basis. The *suprachiasmatic nucleus*, which receives input directly from the retina informing it about the time of day but also has intrinsic circadian clocks, appears to play an essential part in this: it is discussed further on p. 289.

Relation to limbic system

The ancient neural pathways to and from the hypothalamus are not well understood.[7] In particular, although it is evident that the hypothalamus is indeed 'the head ganglion of the autonomic system', it has not been possible to identify conclusively by what routes its autonomic functions are mediated. Its connections with the limbic system are clearer: it receives afferents from the hippocampal region through a massive fibre bundle (the *fornix*), and is interconnected through the *medial forebrain bundle* with many parts of the limbic system including the amygdala, with orbitofrontal cortex, and with reticular formation (RF), including the areas controlling respiration and the cardiovascular system (see p. 204). The medial forebrain bundle also carries afferent olfactory information. Some hypothalamic nuclei – notably the supraoptic and paraventricular – project to extraordinarily diverse areas, including substantia nigra and the substantia gelatinosa of the dorsal horn. Corresponding to this diverse output, stimulation of many regions of the hypothalamus can give rise to fragments of emotional behaviour such as sweating, piloerection, freezing, sexual responses, and so on.

The amygdala

The *amygdala* ('almond' – from its shape) appears to be an important structure on the output side.[8] It projects partly to the hypothalamus (generating autonomic and hormonal effects) and is also capable of generating actual *behaviour* by two routes to the motor system: to dorsomedial nucleus of the thalamus, which projects in turn to frontal cortex; and to the nucleus accumbens (ventral striatum), which projects to parts of basal ganglia, particularly the globus pallidus and substantia nigra. It can be broken down into three major divisions (basolateral, cortical and centromedial), further subdivided into 10 or so nuclei. The projection from the centromedial region to the hypothalamus through the medial forebrain bundle is particularly interesting, as

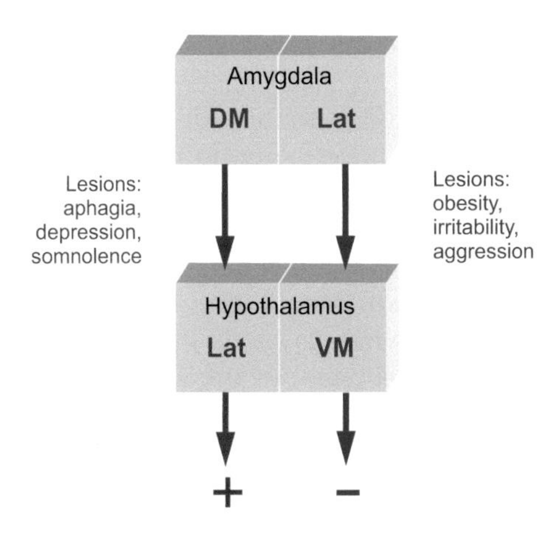

Figure 14.13 Corresponding regions of amygdala and hypothalamus, broadly associated with arousal (+) and conservation (–). DM, dorsomedial; VM, ventromedial; Lat, lateral. (From data of Fonberg, 1972, © 1972 Ciba Foundation; with permission from John Wiley and Sons.)

its organisation seems to correspond rather nicely with the two fundamental types of emotion described earlier in this chapter, conservation and arousal (**Fig. 14.13**).

- The *lateral area* projects directly to the ventromedial hypothalamus, and broadly speaking both seem to correspond with conservation, the horizontal axis of **Figure 14.8**. Stimulation in this region of the amygdala produces passivity, and even sleep or stupor. Lesions in the ventromedial hypothalamus, the 'satiety centre', produce over-eating and obesity; again, similar effects are produced in the lateral amygdala, but with a more general affective change as well – an increase in irritability and aggressiveness.
- The *dorsomedial area* projects directly on to the lateral hypothalamus, and both regions seem, broadly speaking, to be concerned with arousal, the activation of a positive motivational drive. We have already seen that the lateral hypothalamus is associated with the initiation of eating and drinking behaviour, in the sense that lesions give rise to aphagia and adipsia. But they also produce a more general depression: dogs with lateral hypothalamic lesions are described as having a sad appearance, and are listless and somnolent. Lesions of the dorsomedial amygdala have similar effects, with perhaps more of the general, affective, component; stimulation in this region may produce hissing and growling and other signs of positive arousal.
- Large bilateral lesions that include the amygdala create an animal in which tropistic behaviour is greatly exaggerated (the *Klüver–Bucy syndrome*): everything in the environment seems indiscriminately attractive, and such a monkey will compulsively examine and try to eat such things as the bars of its cage and its own faeces, and even things like snakes that would terrify a normal animal. The same kind of hypertropism is seen in its sexual activity: the animal is markedly hypersexual and may try to copulate with members of its own sex, as well as inanimate objects. Again, it is as if objects do not receive their normal emotional colouring, and inappropriately directed behaviour results, though a complicating factor is that such widespread lesions also damage regions of temporal cortex involved more generally in visual recognition. Studies in humans have confirmed the general sense that the amygdala is in part involved in *fear*, aversion from stimuli that have been experienced as harmful. Early studies showed that electrical stimulation of the amygdala in conscious patients (as part of preliminary investigations of possible sites of epileptic disturbances) elicited feelings of fear, and lesions in this area are reported to interfere with the formation of associations with unpleasant stimuli. Some other studies suggest that the amygdala is concerned not just with aversive responses but also with positive responses to stimuli associated with reward.

More specifically, sexual responses (as well as more general items of emotional expression such as pupil dilatation or changes in facial expression, and also respiratory responses) may be elicited from the *cingulate gyrus*, a primitive mesocortical region of the limbic system that receives a projection from the hypothalamus by way of the mammillary bodies and anterior thalamus, as well as from neocortex. Together with the nucleus accumbens, which appears to provide a link from the amygdala to the basal ganglia, it may provide another route by which the hypothalamus generates overt behaviour, though the functions of the cingulate remain unclear. Other regions of the limbic system are described as *pleasure centres*, in the sense that if an electrode is implanted in, for instance, the septal nuclei, and connected up so that when an animal presses a lever in its cage it receives a pulse of electrical stimulation through the electrode, then as soon as the animal discovers what the lever does, it will go on pressing it repeatedly, often in preference to 'really' pleasant stimuli such as food or sex. Of course one cannot tell whether it is *feeling* pleasure as a result, but it is clear that the electrode must in a sense be bypassing the normal motivational mechanisms of the hypothalamus and in some way activating the tropistic input to the motor system directly. Other sites that have been found to produce direct motivation of this kind include parts of the amygdala and the hypothalamus itself. In some locations (dorsomedial thalamus, amygdala, hypothalamus) electrical stimulation has exactly the opposite effect: once the lever is pressed, it is never pressed again, presumably because the stimulus is evoking avoidance rather than positive tropism; but one has to be sure in such cases that the animal is not merely feeling pain.

Decision

At the beginning of this chapter it was pointed out that the whole purpose of the evolutionary development of the brain is to help the hypothalamus in carrying out homeostasis, by making *predictions*: predictions about what we are about to experience, predictions about what will be the result of doing this rather than that.[9] If the world behaved with clockwork regularity, and if we had unlimited information about it, making decisions about how to respond to a particular stimulus would be entirely trivial, like playing noughts-and-crosses. But that is not the case; and just as a successful business makes the best estimates of the risks and potential gains from taking one course of action rather than another, so that its gambles tend to pay off, so the neural mechanisms that determine what we do must similarly compute the probabilities of different outcomes, and the rewards likely to accrue from different actions. The rational way to proceed is to calculate, over all the things we might do, the one for which the expected return (the size of the reward multiplied by the probability of getting it) is maximum. This would be relatively straightforward, if our sensory systems were fast, reliable and capacious, but they are not. As we have seen again and again, they are subject to noise (some originating in the outside world, some internally), and – by computer standards – operate with a very low bandwidth. The result, as has been repeatedly emphasised, is that we are continually making *guesses* about what's going on, based mostly on our experience of what has happened in the past (embodied in our internal models) and only partly on real information trickling in through our senses (p. 255). The situation is not entirely unlike making a clinical diagnosis: on the one hand, the doctor has a lifetime's experience of the probability of different kinds of disease, and this is supplemented by the often rather meagre information afforded by the patient in front. The question then arises as to how to combine these two kinds of information – actual evidence and prior expectation – in order to arrive at the most likely assessment of the underlying cause.

It turns out that there is a simple rule for doing this, called Bayes' law.[10] To see how it works, imagine playing a little game. Suppose you are presented with two identical bags: one – call it A – contains nine red balls and one black, the other – B – has nine black balls and one red. One is on the left, the other on the right, but you have no idea which of them is A. So initially you could say that the probability P_L of A being on the left is the same as the probability P_R of it being on the right, so the *prior odds* $P_L / P_R = 1$. Now you are invited to reach into the left-hand bag, choose a ball with your eyes shut, see what colour it is, and replace it. It turns out to be red. What can you now say about the odds of A being on the left? Intuitively, it now seems more likely, but *how much* more likely? Bayes' law tells us the answer. What we have to do is calculate what the chance of drawing a red ball would have been from each of the two bags: from A it is obviously 0.9, from B, 0.1: the ratio of these numbers, 0.9/0.1, is called the *likelihood ratio*. Then Bayes tells us that the new estimate of the odds – the posterior probability – is given by the prior odds multiplied by the likelihood ratio, in this case $(1/1) \times (0.9/0.1)$, or 9:1. In other words, it is now nine times more likely that A is on the left. If we wanted to be more certain, we could repeat the sampling. Suppose the ball is once again red: then the odds are once again multiplied by 9, so they are now 81:1. Clearly, the more samples we take, the more certain will be our conclusion about which bag is which.

This is a general model for how decisions must be made under conditions of uncertainty, and results in a situation in which some kind of decision signal, representing the extent to which we favour one hypothesis rather than other, rises steadily as a result of acquiring more and more information from our senses, until some particular outcome is so overwhelmingly likely the expenditure of energy on an actual response is justified. Many experiments – behavioural in humans, electrophysiological in animals – have demonstrated this kind of 'rise-to-threshold' activity in the period leading up to a behavioural response (see p. 256). In humans, measuring reaction times is a good way to gain insights into these mechanisms of decision. Although reaction time is in part a result of such 'physiological' factors as conduction velocity and synaptic delay, in practice most of it appears to be due to the time taken for these growing decision signals to rise from their starting levels (that reflect prior probability) to the threshold at which they initiate a response. Because reaction times can be measured cheaply and non-invasively, this technique is beginning to find a range of clinical applications, especially in neurodegenerative disorders.[11]

Cortical arousal

There is a further aspect of the general emotional states of arousal and conservation that has yet to be considered. In addition to altering the visceral functions of the body, and determining patterns of overt behaviour, they also regulate the activity of the thinking, cortical, areas of the brain. Any system that works through association must have some way of regulating its sensitivity. Too sensitive, and it will tend to jump to conclusions, recognising objects on insufficient evidence and wasting energy by making inappropriate responses; too cautious, and it may fail to respond to sensory signals hinting at the presence of predator or prey. In states of emotional arousal we generally find low cortical thresholds and over-excitability; in states of conservation (of which sleep is the most familiar example) thresholds are high and responses are hard to elicit. The control exerted by these systems may on some occasions be relatively local, acting rather like a spotlight to focus attention on a particular cortical region – as when a sudden sound at night alerts our auditory system – or more widespread, in the generalised states of arousal described earlier. They may well also have a role in preventing cortical activity from getting out of hand. Complex networks like those of the cortex, where the huge amount of convergence and divergence (the average number of synapses on a neuron in the monkey's motor cortex is around 60 000) tends to create positive feedback loops, are inherently unstable. We need the equivalent of the damping rods in a nuclear reactor, altering the thresholds of the neurons in step with the level of incoming sensory activity, in such a way as to maintain a sufficient degree of sensitivity without triggering off the kind of neural explosion that is seen in epilepsy (see p. 292).

One way to monitor the overall level of activity of the brain is to attach electrodes to a subject's scalp, which pick up the average electrical activity of very large numbers of cortical cells at once; a record of these potentials is called the *electroencephalogram* (EEG).[12] With lots of electrodes, one may get some idea of the spatial pattern of activity (see Appendix, p. 304). Paradoxically, the largest potentials are not those recorded when the brain is active, but when it is at rest. The reason seems to be that because cortical cells are so richly interconnected with one another, with multiple opportunities for feedback circuits that loop back on themselves, if left to their own devices they tend to lock into rhythmic oscillation, giving rise to waves of potential that run across the cortical surface like ripples on a pond. In a state of conscious quiet relaxation, these waves have a frequency of around 10 Hz, and are known as *alpha waves* (**Fig. 14.14**), most prominent near the occipital region. In the two phases of sleep, dreaming (REM) and dreamless (NREM), there are characteristic alterations in the pattern of activity. If the brain is aroused as a result of outside stimulation, this idling pattern is broken up into essentially random fluctuations of no particular frequency and small amplitude, just as the even flow of waves across a pond is disrupted by a shower of rain; the resultant EEG is then described as *desynchronised*, as can be seen in the **Figure 14.14b** record of EEG in an isolated cat's brain after electrical stimulation of the RF. Conversely, the more profoundly inactive the brain is – as for instance in sleep or in certain kinds of pathological condition – the larger and slower are the waves of electrical activity.

Ascending regulation and activation

Using the EEG, one can discover, by using lesions and stimulation, which parts of the brain seem to be concerned with the control of the overall activity of the cortex. Over 60 years ago, Moruzzi and Magoun made some observations on cats using a preparation called the *encéphale isolé*, in which the brain, isolated from the spinal cord, shows essentially normal cycles of waking and sleeping, easily detected in the EEG. What they showed was that electrical stimulation of part of the RF converted the resting state of alternating sleep cycles into one of almost instant EEG alertness. The same effect was produced by a mid-pontine section, creating a cortex that was permanently awake. Conversely, lesions in the top half of the RF, or disconnection of this region from the cortex by a transverse cut at a higher level (creating a preparation called the *cerveau isolé*), caused permanent reversion of the EEG to a somnolent condition (**Fig. 14.15**). Finally, a mid-pontine section had the opposite effect, creating a permanent state of alertness, suggesting that there were relatively specific regions within the RF that were capable of determining the state of arousal of the cortex.

This was a dramatic demonstration: until then, the RF, with its apparently chaotic structure, was a complete

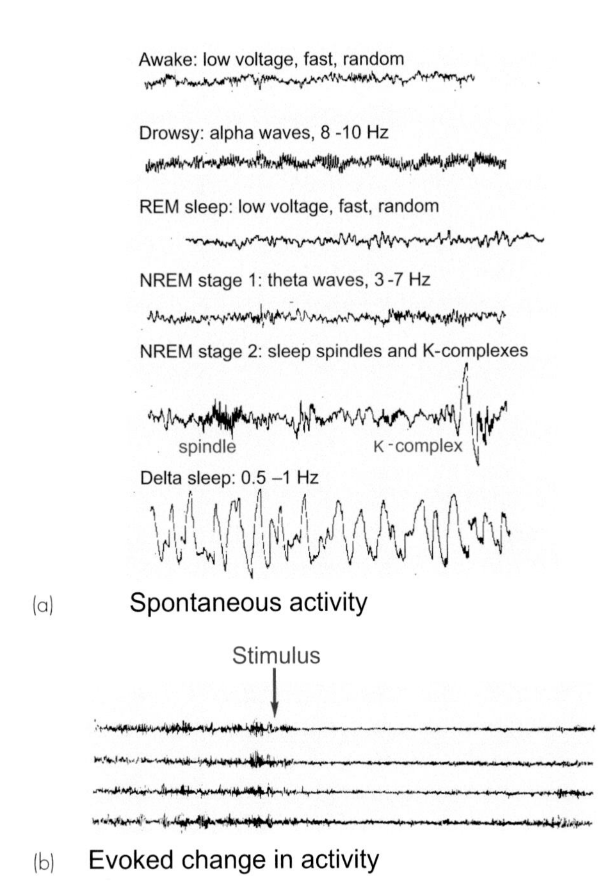

Figure 14.14 The electroencephalogram (EEG). (a) spontaneous human EEG in different arousal states. (From Penfield and Jasper, 1954.) (b) Desynchronisation resulting from electrical stimulation of reticular formation in cat. (From Moruzzi and Magoun, 1949; with permission from Elsevier.)

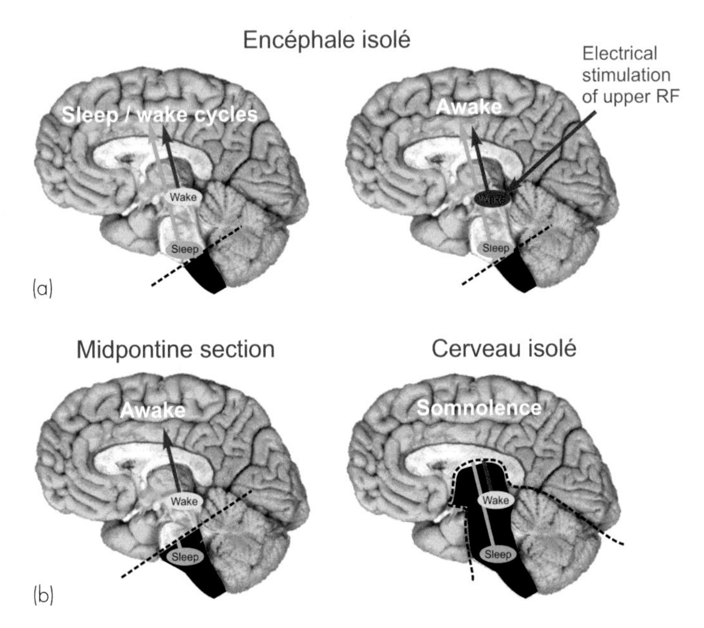

Figure 14.15 Experiments in cat demonstrating the role of parts of reticular formation (RF) in controlling arousal. (a) In encéphale isolé, stimulation of upper RF causes awakening. (b) A mid-pontine preparation is awake, but a higher section (cerveau isole) is somnolent.

Clinical box 14.1 The EEG

The EEG is acquired by applying a series of electrodes across a patient's head to register patterns of electrical activity.

	Frequency (Hz)		Normal *Abnormal*
Alpha	8–13		More prominent posteriorly and pronounced on the non-dominant side. Accentuated by relaxation with the eyes closed; alpha activity is suppressed when concentrating or performing tasks *Severe, destructive brain injury such as haemorrhage, trauma or hypoxia can be accompanied by alpha coma – the patient is unresponsive to stimuli and exhibits widespread, uniform alpha wave activity on EEG*
Beta	>13		Normally present when active and thinking, particularly frontal areas *Exaggerated by sedative medications such as benzodiazepines and barbiturates*
Theta	4–8	Together known as slow waves	Seen in sleep, may be associated with deliberate suppression of a response *Abnormal if seen in excess in awake adults, e.g. during metabolic encephalopathy (hypoxia, hypoglycaemia, drug overdose, electrolyte imbalance, nutritional deficiencies, hypothyroidism)*
Delta	<4		Seen in deep 'slow-wave' sleep *In awake adult associated with metabolic encephalopathy or damage to subcortical structures*

Hypoxic brain injuries are associated with excessive theta activity, diffuse slowing and, ultimately, alpha coma. *Herpetic encephalopathy* (herpes simplex encephalitis), usually leading to necrosis of the temporal lobes, is characterised by periodic sharp spikes. *Hepatic encephalopathy,* a special cause of metabolic encephalopathy, is associated with paroxysmal triphasic waves. It occurs during acute or severe chronic liver failure and is thought to arise as the liver fails to filter aminergic substances absorbed from the gut, allowing them to reach the brain where, among other actions, their similarity to endogenous neurotransmitters may allow them to behave as 'false' neurotransmitters causing delirium, coma and brain swelling. (It is probable that the pathogenesis is less direct than this, involving altered glutamine metabolism and activity.)

The EEG is also particularly important for diagnosing *non-convulsive status epilepticus* (NCSE) – persistent seizure activity on EEG not accompanied by any motor manifestation. Clinically, patients may have unexplained confusion, bizarre behaviours or reduced conscious state and coma. Because NCSE often occurs in patients who are already critically unwell, seizures may not be recognised as such, drowsiness being attributed to head injury, infection and so on. In less obviously unwell patients, the peculiar behaviour may be attributed to psychiatric disturbance. In the meantime, it is thought that uncontrolled seizure activity, which may endure for days, continually damages the relentlessly activated neurons. NCSE is hard to diagnose without EEG.

mystery – people didn't really want to know about it. Suddenly there was this revolutionary idea, that our wonderful cortex might actually be totally under the control of the despised and apparently primitive RF. This control system came to be called the *ascending reticular activating system* (ARAS), though this is something of a misnomer as important parts of it are not in the RF, and some of it projects not upwards but sideways or downwards. It is a system with extremely widespread inputs from all over the brain and from collaterals of ascending fibres (**Fig. 14.16**), integrated over neurons with astonishingly large fields of projection as can be seen in this single cell (soma arrowed) in a rat nucleus magnocellularis. Two specific areas in the RF form the origin of part of this system, the *raphe nucleus,* which is serotonergic, and the *locus ceruleus,* which is noradrenergic. They have endings distributed to nearly all parts of the brain, with extensive diffuse projection to the cerebral cortex. Another component is cholinergic, and has various functions. One group of very large cholinergic neurons, in the pontine RF, project on to a rather diffuse set of neurons, the *thalamic reticular nucleus* (TRN), that we encountered earlier, in Chapter 4. It encircles the thalamus and sends inhibitory fibres into it that in effect gate the flow of afferent information to the cortex, possibly mediating focused attention (**Fig. 14.17**). There are also diffuse cholinergic projections to the *intralaminar thalamic nuclei* (centromedian and parafascicular): they can be regarded as a diffuse diencephalic continuation of the RF, but they also receive information from a surprising variety of other areas as well (spinothalamic and spinoreticular branches –

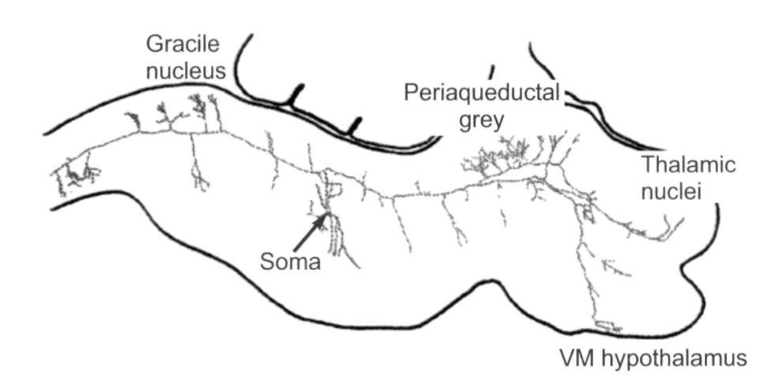

Figure 14.16 Single cell (arrowed) from the nucleus magnocellularis of rat reticular formation, showing the distribution of branches of its axon to widespread areas of brain. (From Scheibel & Scheibel, 1957.)

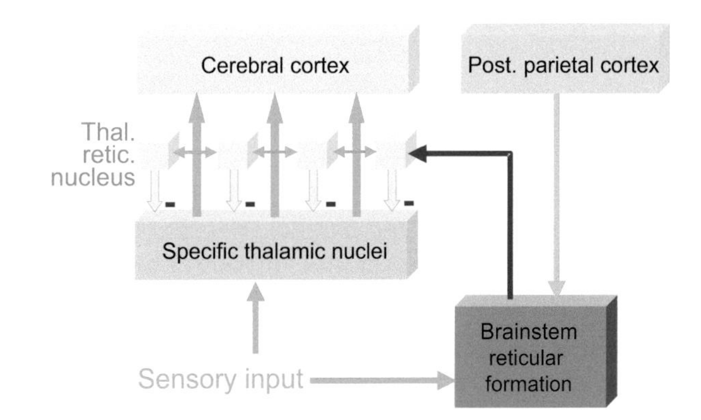

Figure 14.17 Gating of cortical input via the thalamus by the thalamic reticular nucleus, acted on by part of the brainstem ascending reticular formation; this is in turn ultimately driven by collateral sensory activity, and by focal activity from posterior parietal cortex, possibly mediating focused attention.

nociceptive, for example – basal ganglia, and cerebellum). Their output is to widespread areas of the cortex, but unusually also to the putamen and to ventral thalamus and thence to motor cortex.

The RF's overall regulation of the cerebral cortex takes place in two distinct ways: by controlling the *access* of sensory information to higher levels, and by *direct* regulation of the activity of the higher levels themselves.

Regulation of access

In the first of these functions, which can be called *gating*, the relationship between thalamus, RF and cortex seems to play a central role. There are three kinds of afferent to the thalamus: *specific*, including the primary sensory projections; *reciprocal* from cortex; and *regulatory* from the RF and TRN (**Fig. 4.18**). The role of the reciprocal connections is not entirely clear, but may well represent the coming together of 'real' stimuli and those that are predicted by the associative networks of the cortex. The reticular nucleus forms a sheet of neurons with dendrites in the plane of the sheet. Thus all connections between thalamus and cortex have to pass through it, and in fact they send collaterals to the TRN cells, which in turn project back to the thalamus itself and are inhibitory (γ-aminobutyric acid (GABA)). This probably provides a mechanism for regulating the overall rate of information exchange, and perhaps a kind of lateral inhibition (compare for instance basket and Golgi cells in the cerebellum). The TRN also receives input from parts of the RF, which are in turn driven by projections from parietal cortex – the cortical area concerned with the localisation of objects around us and may represent a pathway for directing attention to particular regions of the outside world.

The action of the regulatory input has become a little clearer thanks to studies of the electrical behaviour of thalamic neurons in different states of alertness.

Thalamic neurons seem to have two physiological states. In their *tonic mode*, they are slightly depolarised and respond to stimuli in a normal way with tonic changes in frequency depending on the degree of stimulation (**Fig. 14.19**). However, hyperpolarisation causes them to enter *burst mode*: they then generate short bursts of action potentials through the opening of calcium channels in response to small depolarisations. In tonic mode these channels are inactivated by the steady depolarisation. When the brain is asleep, most of the neurons are in burst mode; when awake, some are tonic but some are still in burst mode, perhaps depending on the state of attention. It is as if in burst mode there is a kind of hurdle to discourage information from getting to the cortex, but if it manages to overcome the hurdle, these large spikes have much more effect than the same stimulus would in tonic mode: it has been suggested that this is meant to wake the cortex up and make it attend to something important or novel.

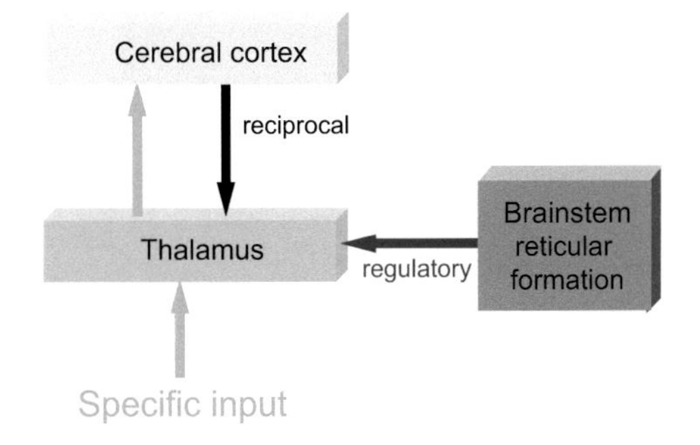

Figure 14.18 Thalamic gating. Reticular input to the thalamus regulates the specific ascending and reciprocal cortical input.

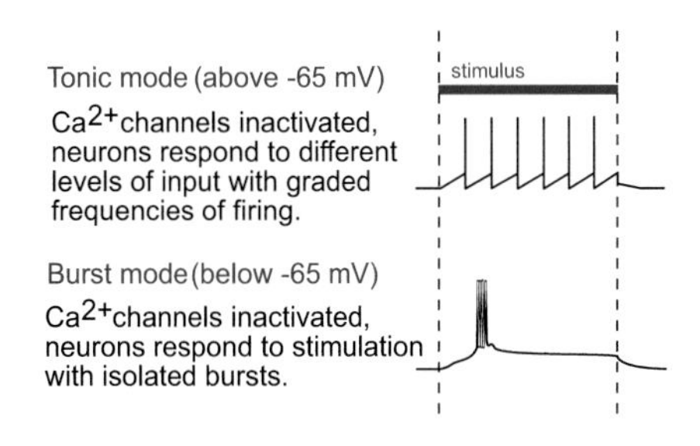

Figure 14.19 Two modes of action of thalamic neurons, mediated by voltage-gated calcium channels.

Regulation of activity

Apart from regulating the thalamic gating of input to the cortex via the thalamus, the RF also regulates the overall level of cortical activity by means of curiously pervasive ascending projections to the cortex itself. There appear to be at least four separate and parallel systems, each associated with a particular transmitter and one or more particular sites of origin (**Fig. 14.20**):

- *Noradrenergic*, from the locus ceruleus and other parts of the lateral RF;
- *Serotonergic*, from the raphe nuclei;
- *Dopaminergic*, from the ventral tegmental area (mesencephalic);
- *Cholinergic*, from various reticular and related sites, for example the substantia innominata.

Before considering each in turn, it is important to bear in mind that the monoamine systems in particular, though certainly mostly genuinely neural, have a sort of autonomic function as well, in that they innervate blood vessels and therefore can have an indirect effect on neural function.

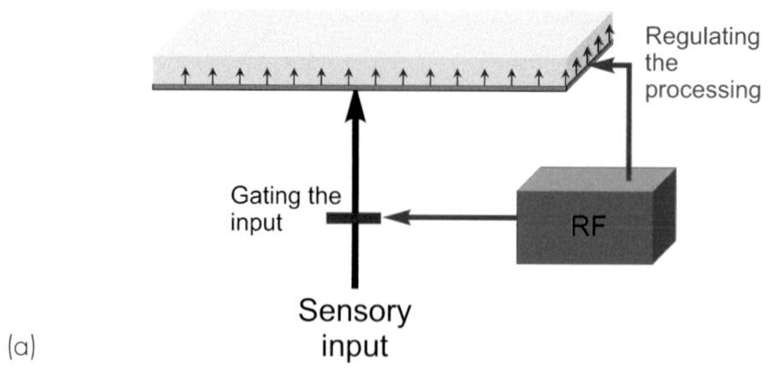

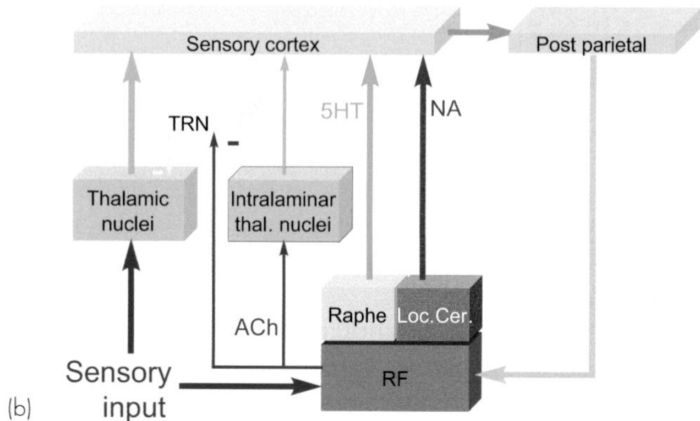

Figure 14.20 Reticular control of cortical activity. (a) Control can be by gating cortical input, or by regulation of the cortical processing itself. (b) Some pathways by which these effects may be mediated. Loc.Cer, locus ceruleus; RF, reticular formation; TRN, thalamic reticular nucleus.

Noradrenergic

There are two main sources of ascending noradrenergic control: the *lateral reticular nucleus* and, more importantly, the *locus ceruleus*.[13] This excitatory noradrenergic innervation covers practically the whole central nervous system: thalamus, basal ganglia, hypothalamus, cerebral and cerebellar cortex, spinal cord – no cell in the cortex is more than 30 μm from a terminal from the locus ceruleus. Most of its input seems to come from the medullary RF. This system may be something to do with sensory alerting or arousal: the general idea is that an incoming stimulus, destined for some localised cortical area, also activates the RF en route, which wakes up the cortex and prepares it to receive the stimulus, not only in the particular area in question, but in other areas that may have to deal with the computational consequences of the stimulus (note that it projects particularly to layer IV, where thalamic input arrives). Stimulation of the locus ceruleus in monkeys appears to cause noise levels in cortical neurons to drop and the response to incoming stimuli to increase, enhancing the signal-to-noise ratio. It is quite interesting that cortical noradrenaline levels are enhanced by some antidepressants.

Serotonergic

The top end of the *rostral* raphe nucleus projects up to cortex and basal ganglia, perhaps a little more diffusely even than the locus ceruleus. A rather interesting difference lies in the cortical layers it projects to: not so much to IV as to V and VI, where the pyramidal neurons are, suggesting a control of output rather than input (**Fig. 14.21**). It may be that the serotonin system acts as a censor, suppressing immediate or impulsive responses to stimuli. For instance, obsessional behaviour is said to be associated with reduced 5-hydroxytryptamine (5-HT); on the other hand, fluoxetine (Prozac), which increases 5-HT and is primarily used to treat depression, actually tends to increase spontaneous movements. One should perhaps be wary of psychopharmacological generalisation. The function of the raphe nucleus in the control of pain was discussed in Chapter 4 (p. 96); that too might be regarded as a process of suppressing instinctive reactions to a stimulus. The *caudal* raphe nucleus is the origin of another serotonergic projection, to the spinal cord and cerebellum and also to the hypoglossal nucleus: its functions are unknown.

Dopamine

We have already encountered (in Chapter 12) one dopaminergic projection, from substantia nigra pars compacta

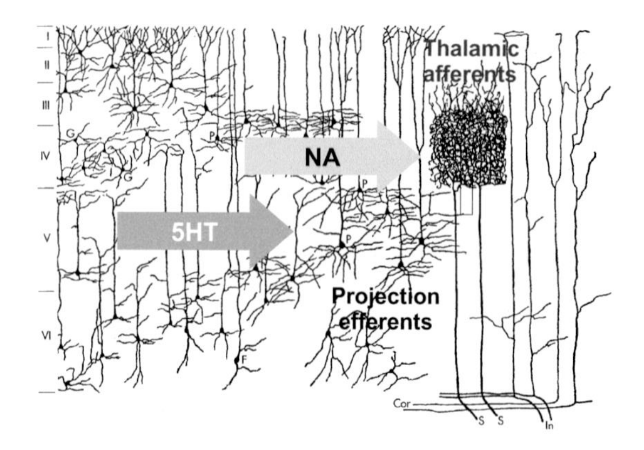

Figure 14.21 Diffuse monoamine projections are concentrated in particular layers: noradrenergic in layer IV, where inputs from the thalamus terminate; 5-HT in layers associated with the pyramidal cells that form the cortical output.

upwards to the caudate and putamen. The ventral tegmental area provides a similar rostral dopaminergic projection, but to parts of the limbic system, especially the amygdala, and to the cortex, the front (motor) half rather than the back. What does it all mean? Dopamine levels are often raised in schizophrenia, and dopamine blockers tend to be antipsychotic; but lesions of dopaminergic systems also lower spontaneous activity, and depress such behaviours as eating and drinking. Dopamine seems bound up in some way with the higher levels of motor control and response preparation and reward and decision and so on, nobody really has much idea what. One specific finding is of midbrain dopaminergic neurons that seem to be performing the equivalent of error decorrelation in respect of rewards: their activity reflects the difference between expected or predicted reward and what is actually received.[14] Clearly such a signal is important in updating internal models so that they accurately reflect the way the world actually works, and it has long been suspected that dopamine may have something to do with what psychologists call *reinforcement*: in this context, the demonstration that dopamine seems to have a specific influence on the modification of synaptic plasticity at *N*-methyl-D-aspartate (NMDA) synapses may or may not be significant. Again, like much psychopharmacology, it's just a little bit vague. Dopamine is of course a local neurotransmitter, and we have already seen examples of its specific and unemotional role in such functions as re-wiring the retina in dark adaptation (p. 152). There is perhaps a danger of a kind of conspiracy theory, of seeing a global meaning in all the things it does that does not really exist.[15]

Acetylcholine

There are two particularly prominent reticular output systems that are cholinergic. One has already been discussed: the projection to the thalamic reticular nucleus, inhibitory and therefore promoting access to the cortex. The other forms yet another diffuse cortical system, which originates from a group of neurons in the substantia innominata called the *basal nucleus* (of Meynert). We shall see shortly that both these cholinergic systems have an important part to play in sleep, particularly in what might be called the control of dreaming. Is there anything useful to be said about cholinergic systems in *general*? In Alzheimer's disease there is a specific loss of cortical cholinergic neurons and especially those in the basal nucleus. The basal nucleus is said to increase its activity when animals are rewarded, so in some very general sense forebrain cholinergic neurons are perhaps something to do with learning and memory; but once again it is very unclear at this stage.

These four diffuse systems are undoubtedly intriguing. Clinically they are of profound importance: practically all psychoactive drugs work by influencing one or other system, and in various diseases – Parkinson's, Alzheimer's, progressive supranuclear palsy (PSP) – some are characteristically affected more than others. It is all obviously trying to tell us something, but rather frustratingly, nobody is quite sure what.[16]

Sleep

Sleep is of course governed by a circadian cycle, which is normally synchronised to the alternation of day and night, the light acting as what is known as a *Zeitgeber* or time-giver; without it, the cycle in Man tends to be around 25 hours – in mice a little over 23 hours.[17] **Figure 14.22** shows the effect of isolation from daylight on the human sleeping cycle. The pale blue bars show successive periods of waking in an experiment in which the subject was isolated from natural light for the period shown. His cycles then fell behind the normal 24-hour period, but resynchronised once normal conditions were resumed. This is typical of diurnal animals; for nocturnal animals the natural period is usually less than 24 hours.[18] You don't need complete day and night to act as a Zeitgeber: just a 10-second pulse of light is enough; nor does it have to be light – in mice, the existence of regular 24-hour food patterns is enough even under continuous lighting. The neural clock that seems to do all this appears to be located in the suprachiasmatic nucleus of the hypothalamus (SCN: see p. 282), and isolated cultured SCN cells continue to show circadian rhythms: how they do it is still a mystery. Apart from the gross behavioural effect of sleep, other less noticeable physiological functions such as core body temperature also show circadian rhythmicity.

A further cyclical process occurs within sleep itself. One of the rather few things that the EEG is useful for is monitoring and in a sense quantifying the various gradations of sleep and sleep-related states such as coma (see p. 286). Two completely different patterns of activity alternate with one another during sleep. In one, which can be called *slow-wave* sleep, the oscillations are larger and slower and the muscular tone of the body is much reduced; the lack of cortical control evident from the fact that primitive spinal responses like the Babinski sign may sometimes be evoked. This kind of sleep can be further classified by means of the frequency of the waves – the slower they are

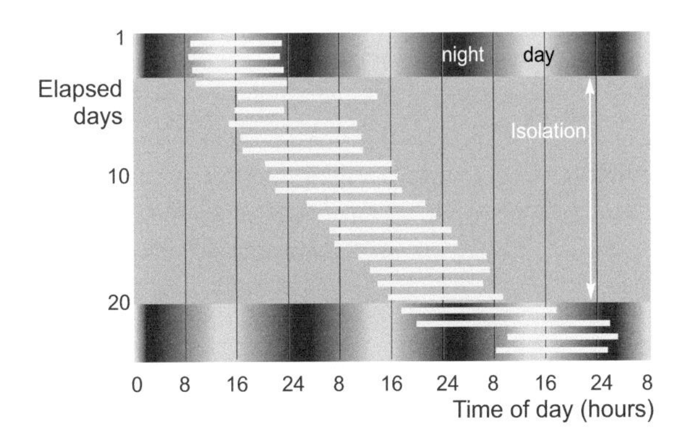

Figure 14.22 Effect of isolation from daylight on sleeping cycle in humans. The bars show successive periods of waking in an experiment in which the subject was isolated from natural light for the period shown. His cycles then gradually fell behind the normal 24-hour period, but resynchronised once normal conditions were resumed. (After Aschoff, 1965; with permission.)

Table 14.2 Activity during waking and different stages of sleep

	Awake	Asleep	
		NREM	REM
EEG	Desynchronised	Slow waves	Desynchronised
EMG	+++	++	+
Ease of waking	(Already awake)	Easy	Hard
Eye movement	++	–	++
Dreaming	–	–	+
Raphe (5-HT)	+	–	
Locus ceruleus (NA)	+	–	
Pontine RF (ACh)	+	–	+
Nucleus Magnocell. (ACh)	–	–	+

the deeper the sleep – and also by the occurrence of various minor features such as bursts of high-frequency activity called sleep spindles, and so on. The other kind of sleep is totally different: now what we have is desynchronisation, essentially very similar to the alert waking state; at the same time we get rapid eye movements, which gives rise to the usual name for this kind of sleep, *REM* sleep (slow-wave sleep is nowadays often called non-REM or NREM sleep). At the same time, although the EEG looks more alert, paradoxically the subject is actually more difficult to wake up, so that another name for this type of sleep – less used nowadays – is *paradoxical sleep.* All this is summarised in **Table 14.2**, which shows that muscle activity drops steadily as you go from waking to NREM to REM sleep. During a night's sleep, REM and NREM sleep alternate in cycles of about 90 minutes in humans, getting steadily shallower until one wakes up (**Fig. 14.23**).

A number of peculiarities of REM sleep are worth mentioning: reptiles don't have it, in birds it forms only half a per cent of total sleep, in rats 15 per cent, and for us about 20 per cent. As you get older, your total sleep time gets steadily smaller, and so does the proportion of time in REM (shaded in colour): in late middle age we rapidly approach the rat level (**Fig. 14.24**). The final and very interesting thing about REM sleep is that it is only then that we *dream*; or to put it another way, it is only in NREM sleep that we are unconscious.

Sleep can be thought of as an emotional behavioural state, a variety of conservation akin to hibernation or freezing. Clearly there is no point in an animal wasting energy by being awake and moving about if it can't see what to eat. If the environment is adverse, and you are more likely to *be* food than find food, then much better batten down the hatches, make yourself inconspicuous and save on energy expenditure. However, that can't be the whole story, as in some aquatic species – for example, the porpoise – the two halves of the brain take it in turns to sleep, and the animal itself is perpetually alert and swimming. So it seems as though sleep is needed in some sense to restore brain cells, perhaps to consolidate memory, perhaps to perform some sort of tidying up operation on our associations. On the other hand, extensive studies

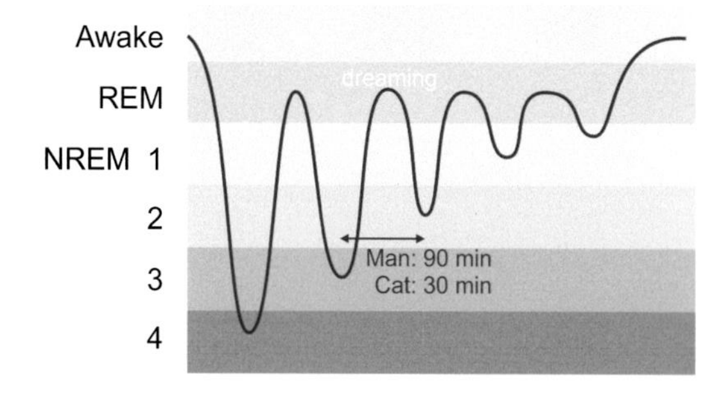

Figure 14.23 Cyclic changes in a human subject during three different nights' sleep, showing fluctuations in the depth of sleep, as determined by the EEG, of increasing shallowness of the cycles as the night progresses, and periods of rapid eye movement (REM) associated with dreaming. (After Dement & Kleitman, 1957; with permission from Elsevier Ireland Ltd) Note that, since these experiments were first performed, stages 3 and 4 have now been amalgamated into a single – slow-wave – stage 3.

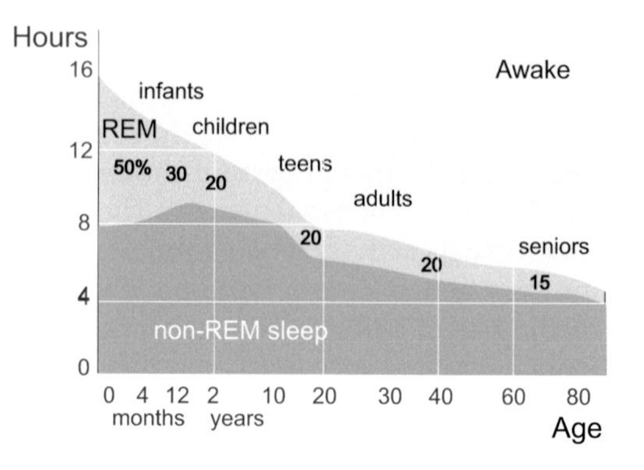

Figure 14.24 Sleep and age. Average proportions of time spent in REM and non-REM sleep as a function of age; note the non-linear time axis. (Reprinted with permission from Roffwarg HP, Muzio JN, and Dement WC, 1966, Ontogenetic development of the human sleep-dream cycle. *Science* 152, 604–19, Copyright 1966 AAA; with permission.)

have never found anything very specific that is impaired by *moderate* lack of sleep. Although one may feel dreadful, objective measures of performance are usually pretty normal, provided the tasks are not so boring or repetitive that you simply fall asleep. There are several well-attested instances of people who have succeeded in ridding themselves of the habit of regular periods of sleep, though it is generally believed that in such cases, instead of having all his sleep in one daily dose, the subject tends to drop off continually for periods of perhaps a few seconds without noticing it. But when sleep deprivation is properly enforced, marked irritability ensues and eventually a state resembling psychosis results.

What is particularly unclear is the particular function of REM sleep, and the dreaming that goes with it: it is evident that less conservation of energy is taking place during it. It may be that dreaming is in some way *needed* by the brain, for if a subject is specifically deprived of REM sleep – by waking him up as soon as his EEG shows desynchronisation – it is found that for several subsequent nights the proportion of time spent in REM sleep is increased to make up for it.

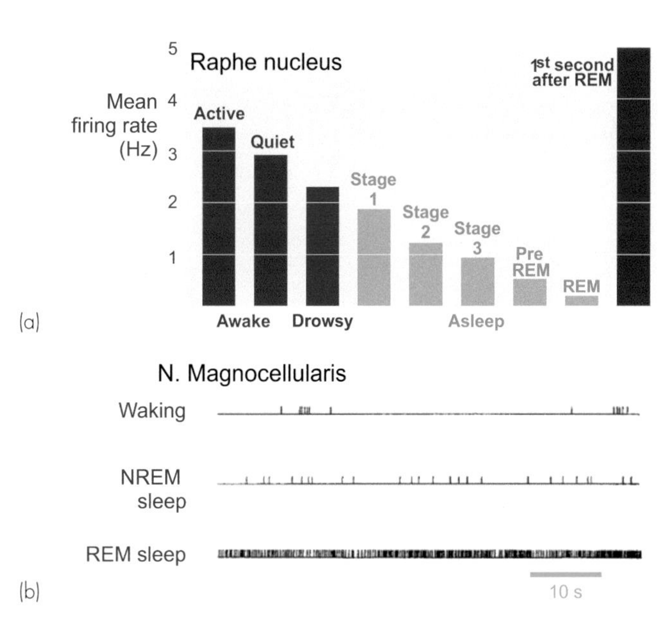

Figure 14.25 Neurons in sleep. (a) Mean firing rate of a neuron from the raphe nucleus at different stages of sleep. (From Trulson and Jacobs, 1979.) (b) Firing patterns of a neuron from nucleus magnocellularis in waking, NREM and REM sleep. (After El Mansari *et al.*, 1989; with permission from Springer Nature.)

Neural mechanisms of sleep

These various states are reflected in the activity of neurons in the ascending activating systems described earlier (p. 284). We saw that the classic *encéphale isolé* preparation shows alternating wake–sleep cycles, but when the isolating cut is higher up, in the *mid-pontine* preparation, instead of having such cycles, the animal is permanently awake: this suggests that there is some kind of 'sleep centre' in the region below the cut. In the *cerveau isolé* preparation, on the other hand, the animal is permanently asleep, suggesting a sort of 'waking area' higher up the brainstem. Combining this with what we know about the role of the RF in regulating cortical activity, it suggests that different parts of the RF influence the arousal of the cortex in different ways. Many general anaesthetics and sedatives act primarily on the RF, as do such stimulants as amphetamine.

More clear-cut information has come from recording from various reticular regions during the stages of sleep (**Fig. 14.25**). Recording from the *dorsal raphe*, which you remember is serotonergic, or from the *locus ceruleus*, which is noradrenergic, an obvious pattern emerges which seems to be essentially the same for both: the deeper the degree of sleep, the lower the mean firing rate. Given what was noted earlier about noradrenaline enhancing sensory input and 5-HT possibly inhibiting pyramidal cell output, this seems to suggest that in sleep the input is being shut down, but the associational connections are if anything being encouraged. At the same time the various *cholinergic* cells are behaving quite differently. Some of the large cells in nucleus magnocellularis are essentially inactive except in REM sleep, where – as can be seen above – they dramatically come to life; given that these have large descending connections, it seems likely that they are part of a mechanism for paralysing the body, and are themselves apparently under the control of noradrenergic cells just below the level of the locus ceruleus. In waking cats, stimulation of this area is said to inhibit spontaneous movement; and if the nucleus is lesioned in cats, they apparently act out their dreams; sleep-walking in humans may be associated with pathological changes in this area. Other cholinergic cells in the pons are even more interesting, as they are specifically active during waking and also during REM sleep – in other words the two periods of actual consciousness.

Although the details of cause and effect in all this are not at all certain, the general sense of what is going on is fairly clear. A continuing theme of this chapter has been the way in which the cortex constructs a model of the outside world through its associational predictions, and our waking perceptions are a mixture of what genuinely arrives through our senses and what our cortex *expects* to happen (**Fig. 14.26**). In NREM sleep the whole thing more or less shuts off, partly through thalamic cells going into burst mode, partly through lack of diffuse stimulation of the cortex, and partly through specific cholinergic mechanisms that paralyse movement – at least, all movement except the eyes. In REM, the shutting off of input and output remains and is even intensified (except that eye movements are spared the general paralysis), but the cortex is permitted to come to life under reticular control – safely, because of the paralysis of actual movements. The internal model then runs all by itself, generating the self-sustaining *dreams* that we are all familiar with: we imagine doing something, the model then predicts what the consequence will be, we respond to that and so on.[19] It seems very likely therefore that dreams are the result of the associational mechanisms being allowed to free-wheel, without the check to fantasy that is imposed in the waking state by incoming

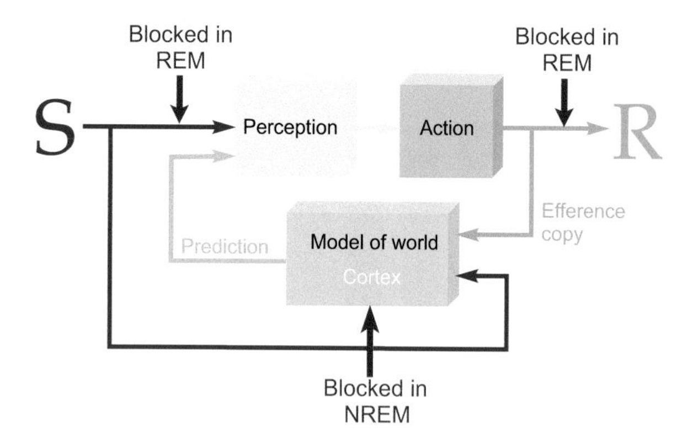

Figure 14.26 A model that explains some aspects of sleep and dreaming. Stimuli S are combined with predictions based on a cortical model of the outside world to provide a percept that is used as a basis for making response R. In non-REM (NREM) sleep, both the model and the inputs and outputs are disabled; in REM sleep, the input and output gating is further increased, but the cortex is permitted to be active. This results in the system free-wheeling, with perceptions being entirely the result of predictions from imaginary actions: this is dreaming.

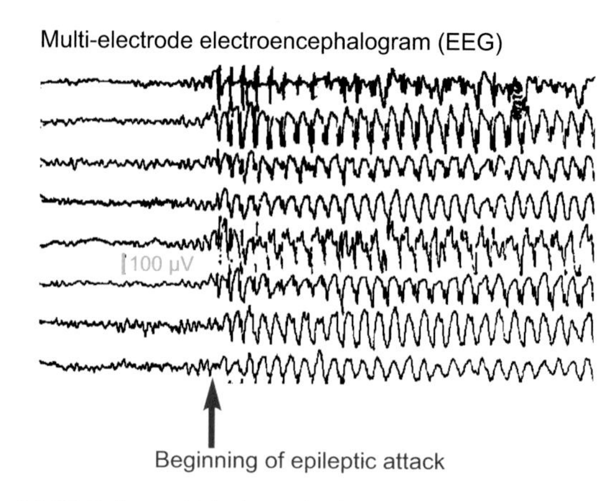

Figure 14.27 Epilepsy. Multi-electrode electroencephalogram (EEG) recording of a brief generalised epileptic attack. (From Lothman & Collins, 1990.)

messages about what the world is *really* doing. A characteristic of dreams is that everything seems very logical over short stretches of time – presumably because of the accuracy of the immediate predictions – but wanders off course on a longer time-scale. Certain kinds of psychiatric disorder can also usefully be thought of as free-wheeling of associative mechanisms, insufficiently checked by sensory input from the real world. It is also significant that waking subjects deprived of sensory input for sufficiently long periods often report dream-like hallucinations.[20]

Disorders of arousal

There are basically two ways in which these regulatory systems can go wrong, resulting in over-activity or under-activity. *Epilepsy* is an explosion of synchronous activity by lots of neurons at once, as can be seen in the multi electrode EEG recording of a generalised epileptic attack (**Fig. 14.27**). It has a tendency to spread throughout the cortex, and seems to represent a failure of inhibitory regulation, in that it can be simulated with strychnine or picrotoxin or other agents that block local inhibition. Epilepsy is in fact the second commonest neurological disorder after stroke, and comes in various forms:

- *Focal* epileptic seizures involving motor areas are heralded by signs such as twitching of thumbs and illusory sense of movement. They often appear to move across the body as the synchronised neural discharges spread over the motor and somatosensory cortex. In the same way, fits involving temporal cortex often start with an aura that combines intense hallucinations with strong emotional feelings and occasionally also olfactory illusions, and disturbances of memory such as *déjà vu*, the illusion that what you are experiencing has happened to you before (*psychomotor epilepsy*; see **Clinical box 14.2**). Focal seizures can occur both with and without loss of awareness, and can also progress to a generalised seizure.
- *Generalised* epilepsy on the other hand tends to start simultaneously over the whole cortex rather than spreading. In a milder form called *petit mal* (more properly termed absence) there is transient loss of consciousness – the subject may suddenly appear completely distracted – though muscle tone is often maintained. *Grand mal* is much more dramatic, and is what is popularly understood by an 'epileptic fit': there is sudden loss of consciousness, the subject falls to the floor rigid (*tonic*) followed by a period of symmetrical jerking movements (*-clonic*).

Disorders of sleep include *narcolepsy*, a fairly rare condition in which the subject suffers from 'sleep attacks',

Clinical Box 14.2 Two examples of focal epilepsy

1. A woman developed temporal lobe epilepsy as a result of head injuries caused by being knocked down by a car. The attacks were all heralded by the appearance of a human face and shoulders clothed in a red jersey. The hallucination would then topple sideways and disintegrate into discrete fragments like a jigsaw puzzle, the patient meanwhile experiencing extreme fear with an unnatural quality to it, followed by amnesia in which a generalised (tonic-clonic) convulsion occurred. This would be described as a focal seizure with secondary generalisation.
2. 'For me there is often a characteristic smell and a terrible sinking feeling in the pit of the stomach. I get a tremendous sense of *déjà vu*. I know that I have been in the same place, I have heard all the same words and seen the same people saying them many times before, like going to a film for the fiftieth time. I believe I know what will happen next. I have just enough time to take a capsule, and then I black out.' This would be termed a focal seizure with loss of awareness.

going to sleep with extraordinary rapidity, typically for just a few minutes at a time. It is so sudden that often there is no time to prepare, and subjects may slump to the floor (*cataplexy*). Paradoxically, it can be triggered by high levels of excitement, and often the subject goes straight into REM without first traversing NREM sleep.

A cluster of conditions can probably be explained by lack of synchrony of the different neural components of sleep control. In *sleep paralysis*, the subject may wake up – terrifyingly – unable to move, presumably because the cholinergic paralysis mechanisms have not yet turned off. Even more terrifyingly, REM dreaming may still be going on, giving rise to appalling half-waking nightmares of peculiar realism. *Sleep-walking*, enuresis and so on can easily be explained in terms of these cholinergic mechanisms being slow to turn *on* rather than slow to turn off. Finally, and much the most common, is *insomnia* – arguably not so much a sleep disorder as a symptom of psychological disturbance, stress or anxiety.

To end on a less cheerful note, we have coma and of course death, which despite what Hamlet says is not a kind of sleep; coma is a completely different biological state with a much lower metabolic rate and much greater difficulty of arousal. It can be caused by all sorts of things – anoxia, hypoglycaemia, or conditions affecting the brainstem. There is a kind of official progression here, which starts with *stupor*, in which the subject can be made to respond but only with extreme measures like shouting and hitting and causing pain generally; then in *coma*, by definition, the subject is completely unresponsive; finally *brain death*, diagnosis of which requires unresponsiveness plus a number of other criteria including electrical silence in the brain, all of which need to be spelled out in detail, for obvious legal reasons.

Clinical box 14.3 Death and brain death

One might think that diagnosing death ought to be a fairly straightforward matter, but in fact it has challenged ethicists and legislators, not to mention perplexed clinicians and distraught relatives for decades. The problem is that since the advent of modern intensive care units, bristling with ventilators, mechanical circulatory devices and dialysis machines, we have become really rather good at keeping oxygenated blood flowing around the body almost indefinitely, supporting or even replacing organ function. This is good news for the patient with lung, heart or kidney failure which, following a bit of support, may get better. Unfortunately, our ability to support viable tissues has greatly outstripped our ability to regenerate and repair dead brain tissue and it is not uncommon that, following a major brain injury (such as a bleed, stroke or trauma), the injured brain swells, raising the pressure in the skull such that blood can no longer get in to perfuse the brain. At this point the brain and brainstem die – *brain death*. However, patients can still have a heartbeat, a pulse, feel warm and even demonstrate quite complex spinal reflexes. They will absorb food and pass urine. Are they alive?

There are differences between the definitions of death between countries and, in the USA and Australia, even between states. In the UK and in Australia a diagnosis of brain death (sometimes considered as brainstem death in the UK) is equivalent to death, and this can be a difficult concept to impart to a family member holding their mother's warm hand and watching a steady heartbeat on a monitor.

In the UK brain death is defined as the 'irreversible loss of capacity for consciousness, combined with irreversible loss of the capacity to breathe'. The Australian and New Zealand Intensive Care Society (ANZICS) determined that 'brain death requires unresponsive coma, the absence of brain-stem reflexes and the absence of respiratory centre function, in the clinical setting in which these findings are irreversible. In particular, there must be definite clinical or neuro-imaging evidence of acute brain pathology (e.g. traumatic brain injury, intracranial haemorrhage, hypoxic encephalopathy) consistent with the irreversible loss of neurological function.' How do we establish this?

- Is there evidence of sufficient intracranial pathology to account for brain death? Usually confirmed on computed tomography (CT) or magnetic resonance imaging (MRI).
- Are there reversible causes? The clinician should ensure that the temperature, blood pressure, electrolytes and oxygen levels are not causing the coma, and exclude hypothyroidism or drug toxicity as suppressing the conscious state.
- Is the patient comatose? There should be no motor response to painful stimuli applied to any limb or any area of cranial nerve distribution. Spinal reflexes occasionally occur and are compatible with a diagnosis of brain death (see p. 206).
- Cranial nerve (CN) function:
  - No pupillary response to light (CN II/III);
  - No blink reflex to touching the cornea (CN V/VII);
  - No response to pain over the distribution of the trigeminal (CN V) nerve;
  - No eye movement in response to ice cold water against the tympanic membrane (causes convection currents in vestibular fluid, simulating head movement, i.e. vestibulo-ocular reflex) CN III/IV/VI/VIII);

  - No gag reflex when pharynx touched (CN IX/X);
  - No cough when trachea touched by catheter passed through endotracheal tube (CN X).
- Will the patient breathe when disconnected from the ventilator? The patient is left unventilated until there has been sufficient carbon dioxide build up to stimulate breathing if the respiratory centres of the brain are intact.

Only if all of the above criteria are met can a patient be termed brain dead, and thus dead. This holds enormous import in the field of organ donation, because once the patient is dead, they can legally donate their organs. The brain dead donor can maintain perfused and oxygenated organs right up until the moment of transplant and such donations can be more successful than emergency retrieval from donors who have died with circulatory arrest and consequent interruption of blood flow to the organ. More importantly perhaps, the whole scenario is more controlled and less demanding of the grieving family making desperately hard decisions in a short timeframe.

A difference between the UK and Australia is the requirement by the latter that blood flow and viability of the entire brain must be excluded, whereas in the UK brainstem death, even in the presence of cerebral blood flow, is sufficient to determine death. Brainstem death is incompatible with liberation from coma or self-ventilation, so clinically the distinction is perhaps moot: even if not legally 'dead' in Australia, most families and clinicians would be comfortable to cease aggressive 'life-prolonging' interventions in such a case.

Even more ethically thorny is the *persistent vegetative state,* where profound cerebral injury with relatively intact brainstem leaves a patient unable to communicate, unaware of other people or the environment but able to generate spontaneous limb and eye movements, grimacing and yawning. Patients in such a state are the greatest source of heartbreak and dilemma because of uncertainty about prospects for recovery and differing opinions as to evidence of awareness (see **Clinical box 14.4**). The antithesis to this scenario is the *locked-in syndrome,* where a bilateral pontine lesion separates an alert and functional brain from all forms of motor expression other than blinking and vertical eye movements. People in this apparently hopeless state have gone on to lead fulfilling and indeed inspiring lives.

Occasionally there is ethical, religious or philosophical discord among families and even clinicians, and this applies to severe brain injuries in all their forms. This can be a spiritually and intellectually challenging time, and vigorous debate is the healthy and vital way to explore the ethical minefield of brain, mind and machine.

A last look at the brain

Our journey of exploration is almost ended. When it began, the goal of our ascent seemed unattainable, its heights unscalable and swathed in clouds of mystery. Yet here we are at the summit: do things look different now?

If distance lends enchantment to the view, it is because it eliminates messy details – in this case the horribly numerous neurons that produce that rather queasy feeling we tend to get when we try to think about the brain. A glass of water contains far more molecules than the brain has neurons, but it seems quite simple to us because we ignore them. If we are prepared to take the neurons for granted, to lump them all together and look at the brain in terms of the flow of patterns of information from one part of it to another, then trying to grasp what exactly it does becomes a much less daunting prospect. But we do of course have to *earn* the right to look at it in this way, by first mastering the intricacies of synaptic integration and the principles on which patterns of neuronal activity are recoded as they filter through from sensory input to motor output.

Figure 14.28 shows an attempt of this kind: the brain as seen from a very distant viewpoint – fuzzy and oversimplified, to be sure, but it can at least help us get our bearings should we wish to examine it more closely. The left-hand side is sensory, the right-hand motor; its hairpin shape comes about because in going from sensory input to motor output we have first to climb to higher and higher hierarchical levels, and then descend again as motor patterns are progressively elaborated into their component parts.

At the top of the hierarchy comes the hypothalamus, which ultimately determines what we do; and it is here, as we have seen, that internal stimuli and responses are brought into contact with external ones. But side by side with this grand strategic planning that requires the identification of possible goals through mechanisms of sensory

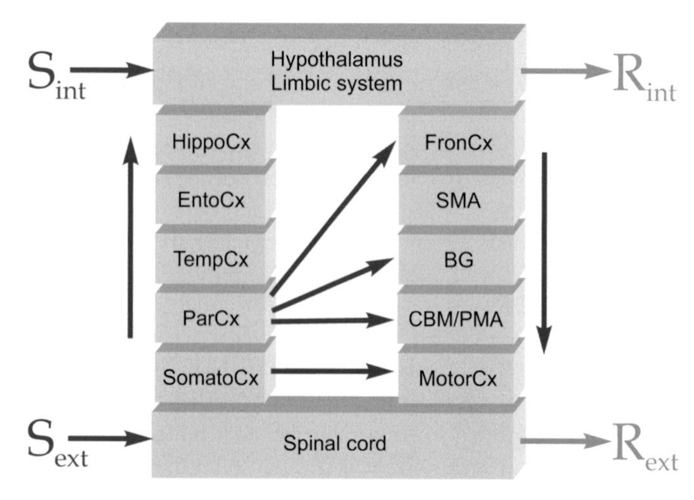

Figure 14.28 A very distant view of the brain.

recognition, we also require the humbler and simpler kinds of tactical coordination that are to do not with *what* objects are, but *where* they are, where *we* are, and the effective execution of the generalised commands sent down from above. This requires pathways for localisation and proprioception, often anatomically separate from the cortical mechanisms for recognition, that bypass the higher levels by cutting across from one side of the hairpin to the other. Some of these short-circuits – the stretch reflex, for example, which helps to ensure that limbs really are where we intend them to be – are at the lowest level of all, the spinal cord. Others, because they demand the integration of many different sources of information, or because they imply a certain degree of learning, have to take place at higher levels such as the cerebellum. For simplicity, only input from the spinal cord is shown, but vision and hearing are organised in essentially the same general way. Olfaction is different: because it does not demand the analysis of patterns in the way that visual and auditory recognition do, and also because it provides purely motivational information with no localisation component, it enters very near the top of the sensory hierarchy.

As a blueprint for a biological machine, such a figure may seem plausible enough. But one may feel a little uneasy at the idea that such a scheme is a picture of oneself. Hasn't something been left out?

'Mind' and consciousness

'Nothing puzzles me more than time and space; and yet nothing troubles me less, as I never think about them' (Charles Lamb) – a reaction not very different from that of most neurophysiologists to problems of mind, brain, and consciousness.[21] This is of course a field that has been thoroughly dug over since the days of Descartes and Hume and indeed long before, and philosophers have every right to question whether mere empirical physiologists can add much to such a hoary debate, in which the various arguments have been rehearsed so exhaustively. But recent developments both in neurophysiology and in computer science – for £20 I can purchase an electronic device hardly bigger than a pack of cards that is the intellectual superior of half the animal kingdom – have so enlarged our notions of what classes of operation a physical system may in principle be capable of, that a great deal of earlier thought on the subject is now merely irrelevant. In a nutshell, 'brain versus mind' is no longer a matter for much argument. Functions such as speech and memory, which not so long ago were generally held to be inexplicable in physical terms, have now been irrefutably demonstrated as being carried out by particular parts of the brain, and to a large extent imitable by suitably programmed computers. So far has brain encroached on mind that it is now simply superfluous to invoke anything other than neural circuits to explain every aspect of overt human behaviour. Descartes's dualism proposed some non-material entity – the 'ghost in the machine' – that was provided with sense data by the sensory nerves, analysed them within itself, and then responded

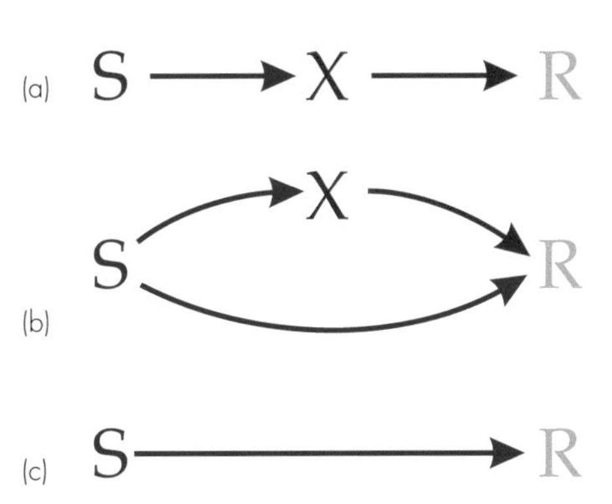

Figure 14.29 The relationship between stimulus and response. Please see text for explanation.

with appropriate actions by acting on motor nerves (the mind thus having the same relation to the body as a driver to his car; **Fig. 14.29a**). Clearly one must modify such a scheme to include the existence of certain automatic reflexes that clearly do not pass through the mind (**Fig. 14.29b**); and in fact modern neurophysiology goes further still, admitting of no other path between stimulus S and response R than unbroken chains of neural connections (**Fig. 14.29c**): X, the ghost in the machine, has finally been laid to rest.

So is there still a problem, or have the philosophers been wasting their time? Indeed there is: that problem is *consciousness*. However sure I may be that (c) is a fair representation of *your* brain, there remains the obstinate and unshakable conviction that *my* brain is like (a). Though – after reading Freud – I might reluctantly agree that a great deal of what I do is not consciously willed, and that (b) is perhaps nearer the truth, nevertheless that there is no X at all is simply inconceivable. Now philosophers can have a great deal of fun with beliefs such as these, since the existence of my own consciousness is not something I can prove to other people in the way I can, for example, prove that I have hands. Clearly its outward manifestations could easily be imitated by a machine (like Hebb's example of the calculator programmed to say 'I am multiplying' every time it multiplies). But this kind of scepticism is so self-consistent as to be utterly tautological: if, like Wittgenstein, we decide that the only criterion for consciousness must be overt, public, behaviour, then of course we have nothing to say about it, because we have defined it out of existence. And for lazy neurophysiologists it provides a veneer of philosophical respectability for their unwillingness to think about the subject at all: Pavlov used to fine his students when he caught them using the words 'voluntary' or 'conscious'.

Yet to evade the problem by such specious materialism is perhaps no worse than to take the opposite extreme and accept consciousness as something much too mysterious and wonderful for a scientist even to be able to begin to think about. Both attitudes contribute to the evident intellectual muddle that surrounds the whole subject. So how would a brash and simple-minded physiologist proceed? Once he had accepted the reality of the phenomenon, he might go on to relate it to the fabric of the brain in much the same way as he would in the case, say, of the sense of sight. It is clear, for example, that loss of a limb does not lead to blindness, whereas loss of the eyes does; and by the use of inductive reasoning hardly more sophisticated than that, one may proceed into the brain itself and map out, almost neuron by neuron, the mechanism of the visual pathways. This kind of work has not of course been carried out systematically in the case of consciousness, if only because experiments of this sort on animals are useless to us. All the same, it is clear that we do in fact already know quite a lot about the functional anatomy of consciousness, even if we have little idea what consciousness actually is. We know, for instance, that while massive lesions of the cerebral cortex and its underlying fibres may blunt our perceptions, paralyse our limbs, impair our intelligence or even – as in the case of Phineas Gage – our morality, they have little effect on consciousness itself. Conversely, relatively slight injuries – perhaps a blow on the chin – that affect an area in the core of our brainstem (the same region of the RF that is associated with arousal and sleep) can produce complete unconsciousness even though the whole of the rest of the brain is unimpaired.

At a different level of description, it is clear that we are conscious of some kinds of brain activity but not others, and that the boundaries of this zone of awareness are not fixed, but vary on occasion. By and large it is what goes on at moderately high hierarchical levels that we are conscious of, although by introspection we can often learn to increase our awareness of lower levels. Curiously enough, we also tend to be relatively unconscious of the highest levels of all, those that control our motivations; and further, even the most complex mental processes can sometimes be carried out without being conscious of the fact at all. While reading through a difficult score at the piano, I have suddenly had the realisation that for several bars I have been thinking about something entirely different, yet my brain had been getting on with the complex task of translating printed notes into finger movements perfectly well without me. Often, quite suddenly and unexpectedly when we were not thinking of it at all, we may find the solution to a problem that has baffled the most energetic conscious cerebration – perhaps, like Archimedes, in the bath; or, like Coleridge with Kubla Khan, asleep![22] L. S. Kubie has gone as far as to say that there is nothing we can do consciously that we cannot also do unconsciously. The natural conclusion must surely be that since consciousness is more associated with 'higher' functions than 'lower', yet not particularly affected by damage to precisely those regions of the brain that we know to carry out

Clinical box 14.4 The vegetative state

Following brain injury, some patients emerge from coma into a sort of twilight state of wakefulness without conscious awareness of self or environment. Their sleep–wake cycle is preserved, yawning and spontaneous eye and limb movements can be observed; however, there is no attempt to communicate, no response to commands and patients do not orientate to, fixate or track visual or auditory stimuli, nor localise to pain. This state is referred to as the 'vegetative state'. A concern with making conclusions about the conscious state in this way, however, is that it depends on observing behaviour, i.e. motor expression, in a severely injured brain. Might consciousness be preserved but obscured by the inability to express it?

This conundrum has been explored over the past decade using the techniques described above to circumvent the need for intact motor pathways. In a landmark study by Owens, a 23-year-old woman with severe traumatic brain injury 'awoke' from a coma into such a state. After 5 months, she would open her eyes spontaneously and exhibit inconsistent pain, startle or olfactory reflex responses but would not follow commands, try to communicate or look at or follow objects placed in her line of sight. Extensive clinical examination concluded that she was unaware of her surroundings, and showed no evidence of willed response, i.e. she was in a vegetative state.

Owens and his team proceeded to look at her brain more closely using functional MRI (fMRI). They instructed her to imagine two different scenarios, either playing tennis or navigating through her own house. The fMRI scan showed that both tasks elicited cerebral activity comparable with that in normal, healthy controls. The instruction to imagine playing tennis was associated with activity in the supplementary motor area, while exploring her house was associated with activity in the parahippocampal gyrus, posterior parietal lobe and lateral premotor cortex. The investigators concluded that a clinically diagnosed vegetative state could in fact be consistent with comprehension of commands, generation of neurologically appropriate responses and, perhaps most controversially, an active 'decision' to cooperate with the investigators. An extension of this study concluded that communication with a number of patients in a clinically vegetative state was possible by asking them to generate yes/no responses; imagining playing tennis for 'yes' and walking through their houses for 'no'.[23]

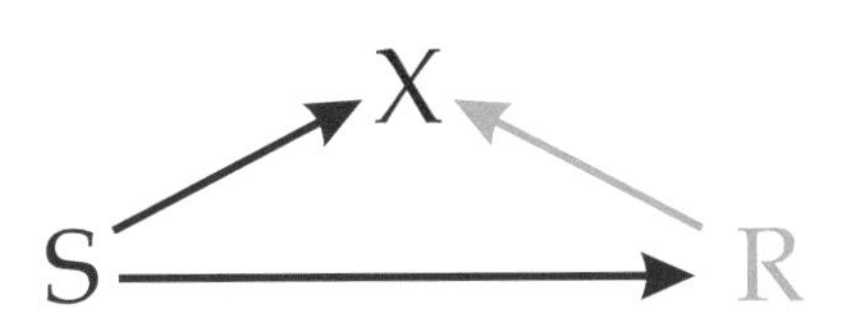

Figure 14.30 Consciousness may be like a spectator, possibly able to influence where attention is to be paid but unable to influence the response.

those higher functions, nor is it necessary to be conscious for those functions to be carried out perfectly adequately, that the ghost in the machine is not an executive ghost, as it is in (a) and (b), but rather a *spectator*, watching from its seat in the brainstem the play of activity on the cortex above it, perhaps able in some way to direct its attention from one area of interest to another, but not able to influence what is going on (**Figure 14.30**).

But what about *free will*? The ghost in such a scheme would observe the body's actions being planned, and see the commands being sent off to the muscles before the actions themselves began, and so one can well imagine how it might develop the illusion that because it knew what was going to happen, that it was itself the cause. For X, the distinction between 'I lift my arms' and 'My arms go up', in which Wittgenstein epitomised the notion of voluntary action, would amount simply to the distinction between those actions which it observed being planned, and those – such as reflex withdrawal from a hot object – which it did not. There is no implied necessity here for us to be deterministic in our actions – to an outsider we may appear to have free will – since the physical processes linking S and R can be as random and essentially unpredictable as we please.[24] Such a scheme seems more intellectually satisfying than (a) or (b) without conflicting with our own feelings about ourselves; and unlike (c), does not merely evade the issue. The most serious objection to it is perhaps that it is difficult to see what on earth X is for, since it can't actually do anything. Perhaps it does just occasionally intervene. But in any case, what is the audience at a concert for? Or the spectators at a football match? The idea that I am being carried round by my body as a kind of perpetual tourist, a spectator of the world's stage, is not – on reflection – so very unattractive. The moral is clear: *enjoy your trip!*

Notes

1. Neuropsychology Four excellent general books dealing with the topics covered in this chapter: Beaumont, J. G. (2008). *Introduction to Neuropsychology.* (New York, NY: Guildford); Walsh, K. W. (1994). *Neuropsychology, A Clinical Approach.* (Edinburgh: Churchill Livingstone); Robbins, T. W. & Cooper, P. J. (1988). *Psychology for Medicine.* (London: Edward Arnold); Carlson, N. R. (1999). *Physiology of Behaviour.* (Boston, MA: Allyn & Bacon). The neurophysiology of emotion thoroughly discussed in Rolls, E. T. (1999). *The Brain and Emotion.* (Oxford: Oxford University Press).

2. Hippocampal maps See for instance O'Keefe, J. & Nadel, L. (1978). *The Hippocampus as a Cognitive Map.* (Oxford: Clarendon); Lopes da Silva, F. M. & Arnolds, D. E. A. T. (1978). Physiology of the hippocampus and related structures. *Annual Review of Physiology* **40** 185–216; O'Keefe, J. (1990). The hippocampal cognitive map and navigational strategies. In Paillard. J. (ed.) *Brain and Space.* (Oxford: Oxford University Press). The study of London taxi drivers was by MaGuire, E. A., Gadian, D. G., Johnsrude, I. S., Good, C. D., Ashburner, J., Frackowiak, R. S. & Frith, C. D. (2000). Navigation-related structural change in the hippocampi of taxi drivers. *Proceedings of the National Academy of Sciences of United States of America* **97** 4398–4403.

3. Herbert Simon In *The Sciences of the Artificial* (1981). (Cambridge, MA: MIT Press).

4. Hypothalamus Some useful accounts in this general area: Morgane, P. J. & Panksepp, J. (1979). *Handbook of the Hypothalamus.* (New York, NY: Marcel Dekker); Bloom, F. E., Lazerson, A. & Hofstadter, L. (1985). *Brain, Mind and Behaviour.* (New York, NY: W.H. Freeman); Donovan, B. T. (1985). *Hormones and Human Behaviour.* (Cambridge: Cambridge University Press); Brown, R. E. (1994). *Introduction to Neuroendocrinology.* (Cambridge: Cambridge University Press); Lovejoy, D. A. (2005). *Neuroendocrinology: an Integrated Approach.* (Chichester: Wiley).

5. Portal system A portal system is a vascular pathway connecting two separate capillary beds in series: the best-known example is the portal vein carrying nutrients absorbed from the gut to the liver.

6. Control of eating A relatively recent discovery is *ovexin*, a neuropeptide hormone found in lateral and posterior regions of the hypothalamus that stimulates hunger (as well as wakefulness); its release is inhibited both by blood glucose level and also by *leptin*, a hormone released from adipose tissue when fat levels are high. This hormonal system has considerable promise in relation to attempts to create drugs to counter both obesity and sleep disorders.

7. Limbic connections See for instance Isaacson, R. K. (1982). *The Limbic System.* (New York, NY: Plenum); Bloom, F. E., Lazerson, A. & Hofstadter, L. (1985). *Brain, Mind and Behaviour.* (New York, NY: W.H. Freeman); Mogenson, G.J., Jones, D. L. & Yim, C. Y. (1980). From motivation to action: functional interface between the limbic system and the motor system. *Progress in Neurobiology* **14** 69. Aggleton, J. P. (2001). *The Amygdala.* (Oxford: Oxford University Press) is a useful recent source of information, and Rolls, E. T. (1999). *The Brain and Emotion.* (New York, NY: Oxford University Press) is a stimulating account of some recent ideas in this general area.

8. Amygdala A useful review is Sah, P., Faber, E. S. L., Lopez de Armentia, M. & Power, J. (2003). The amygdaloid complex: anatomy and physiology. *Physiological Reviews* **83** 803–834.

9. Decision Now a trendy area, especially now that it has been taken up by the financial community, who hope that 'Neuroeconomics' will result in ever bigger bonuses. There are many stimulating books in this area, including: Glimcher, P. W. (2003). *Decisions, Uncertainty and the Brain.* (Cambridge, MA: MIT Press); Wilson, H. R. (1999). *Spikes, Decisions and Actions.* (Oxford: Oxford University Press);

Lauwereyns, J. (2010). *The Anatomy of Bias: How Neural Circuits Weigh the Options*. (Cambridge, MA: MIT Press). More specific accounts include: Gold, J. I. & Shadlen, M. N. (2007). The neural basis of decision-making. *Annual Review of Neuroscience* **30** 525–574; Reddi, B. A. J. (2001). Decision making: the two stages of neuronal judgement. *Current Biology* **11** 603–606; Carpenter, R. H. S. (1999). A neural mechanism that randomises behaviour. *Journal of Consciousness Studies* **6** 13–22; Schall, J. D. (1999). Weighing the evidence: how the brain makes a decision. *Nature Neuroscience* **2** 108–109; Yang, T. & Shadlen, M. N. (2007). Probabilistic reasoning by neurons. *Nature* **447** 1075–1080.

10. Bayes The Reverend Thomas Bayes (1701–1761) was remarkable for being both a dissenting clergyman and also a Fellow of the Royal Society. His law was published posthumously in *An Essay Towards Solving a Problem in the Doctrine of Chances* by a friend, and it is only within the past few decades that its all-pervasive nature has come to be realised. Among other applications, it was instrumental to Turing's unravelling of the Enigma codes at Bletchley Park, which is often said to have been a decisive factor in the battle for naval supremacy during the Second World War. See, for instance, McGrayne, S. B. (2011). *The Theory that Would Not Die*. (New Haven, CT: Yale University Press). Swinburne, R. (ed.) (2002). *Bayes's Theorem*. (Oxford: Oxford University Press) has examples of interesting applications in different areas.

11. Reaction time Accounts of the relation between reaction times and decision processes include Shadlen, M. N. & Gold, J. I. (2004). The neurophysiology of decision-making as a window on cognition. In Gazzaniga, M. S. (ed.) *The Cognitive Neurosciences*. (Boston, MA: MIT Press, pp. 1229–1241); Carpenter, R. H. S. & Williams, M. L. L. (1995). Neural computation of log likelihood in the control of saccadic eye movements. *Nature* **377** 59–62; Reddi, B. & Carpenter, R. H. S. (2000). The influence of urgency on decision time. *Nature Neuroscience* **3** 827–831; Noorani, I. (2014), LATER models of neural decision behavior in choice tasks. *Frontiers in Integrative Neuroscience* **8** 67; Noorani, I. & Carpenter, R. H. (2016). The LATER model of reaction time and decision. *Neuroscience & Biobehavioral Reviews* **64** 229–251.

12. EEG First described over 60 years ago: Adrian, E. D. (1944). Brain rhythms. *Nature* **153** 360–367.

13. Locus ceruleus Ceruleus means 'blue', and its blue colour – like the black of substantia nigra – is due to small amounts of melanin that get deposited as some kind of by-product of the production of catecholamines.

14. Reward prediction See, for instance, Tobler, P. N., Florillo, C. D. & Schultz, W. (2005). Adaptive coding of reward value by dopamine neurons. *Science* **307** 1642–1645; or Schultz, W. (2007). Multiple dopamine functions at different time scales. *Annual Review of Neuroscience* **30** 259–288.

15. Dopamine and addiction The dopaminergic pathways may also play a part in some forms of drug addiction; the feeling of satisfaction associated with achievement of a goal is thought to be somehow signalled through dopaminergic activity in the limbic system. Drugs such as amphetamine (speed) and cocaine directly or indirectly stimulate this dopaminergic activity, recreating the euphoria of success without one actually having to go to the trouble of finding food, mating or winning a race. A downside of this neurophysiological 'cheat' (aside from the potentially catastrophic extra-limbic side effects such as heart attack or stroke) is that, like all neurophysiological systems, there appears to be an element of adaptation: with continual stimulation of the reward pathway, the response to the drug is diminished, requiring ever-greater doses of the substance to re-create the pleasant experience. Meanwhile, natural sources of pleasure are no longer sufficient to activate pleasure centres, leaving users *dysphoric*, no longer able to derive pleasure from previously enjoyable hobbies and activities.

16. Multiple ascending systems A thoughtful discussion is Robbins, T. W. & Everitt, B. J. (1995). *Arousal systems and attention*. In Gazzaniga, M. S. (ed.) *The Cognitive Neurosciences*. (Cambridge, MA: MIT Press). More recent accounts include Björklund, A. & Lindvall, O. (2011). *Catecholaminergic Brain Stem Regulatory Systems*. (Chichester: Wiley); Cools, R., Roberts, A. C. & Robbins, T. W. (2008). Serotoninergic regulation of emotional and behavioural control processes. *Trends in Cognitive Sciences* **12** 31–40; and Robbins, T. W. (2005). Chemistry of the mind: neurochemical modulation of prefrontal cortical function. *Journal of Comparative Neurology* **493** 140–146.

17. Slow body clock The fact that our body's natural day is longer than 24 hours might just explain what is well known to transatlantic travellers, that jet-lag is worse going east than west; presumably for jet-set mice it should be the other way round.

18. Sleep The classic account is Kleitman, N. (1939). *Sleep and Wakefulness*. (Chicago, IL: University of Chicago Press); a recent and comprehensive book is Cooper, R. (1994). *Sleep*. (London: Chapman & Hall); Orem, J. & Bernes, C. D. (1980). *Physiology in Sleep*. (New York, NY: Academic) is a useful source of reference; Hobson, J. A. (1990). *The Dreaming Brain*. (Harmondsworth: Penguin) is a popular account, with an extended historical review of the study of dreaming: a similar book is the same author's (2002). *Dreaming: An Introduction to the Science of Sleep*. (Oxford: Oxford University Press). Another, more recent, popular account is Horne, J. (2007). *Sleepfaring: A Journey through the Science of Sleep*. (Oxford: Oxford University Press).

19. Dreams In popular folklore, dreams are supposed only to occupy a fraction of the *time* the events they portray would really take, but experiments on what are called lucid dreamers suggest that this isn't so. A lucid dreamer is someone who has learnt to exert their will during dreams, consciously influencing what happens in them and also being able to move their eyes voluntarily. In one experiment, a lucid dreamer agreed that while dreaming he would move his eyes in a particular pattern, with what seemed to him like one-second intervals. This he did, and the timing of the eye movements were essentially the same as when he did the same manoeuvre awake. However, it is possibly debatable whether lucid dreamers are really dreaming in the normal sense: they may be half awake. See LaBerge, S. (2000). Lucid dreaming: evidence and methodology. *Behavioral and Brain Sciences* **23** 962.

20. Dreaming and insanity This idea is beautifully expressed in one of Dickens' less-known passages: *'I chose . . . to wander by Bethlehem hospital; partly, because it lay on my road round to Westminster; partly, because I had a night fancy in my head which could be best pursued within sight of its walls and dome. And the fancy was this: Are not the sane and the insane equal at night as the sane lie a dreaming? Are not all of us outside this hospital, who dream, more or less in the condition of those inside it, every night of our lives? Are we not nightly persuaded, as they daily are, that we associate preposterously with kings and queens, emperors and empresses, and notabilities of all sorts? Do we not nightly jumble events and personages and times and places, as these do daily? Are we not sometimes*

troubled by our own sleeping inconsistencies, and do we not vexedly try to account for them or excuse them, just as these do sometimes in respect of their waking delusions?' Dickens, C. (1860). *Night walks*. In *The Uncommercial Traveller.*

21. Consciousness By far the best of the more physiologically oriented books in this area is Honderich, T. (1988). *A Theory of Determinism: the Mind, Neuroscience and Life-hopes*. (Oxford: Clarendon); some of the issues have also been addressed in a stimulating article: Cotterill, R. J. (1994). Autism, intelligence and the brain. *Biologiske Skrifter det Kongelige Danske Videnskabernes Selskab* **45** 1–93. Walshe, F. M. R. (1972). The neurophysiological approach to the problem of consciousness. In Critchley, M. (ed.) *Scientific Foundations of Neurology*. (London: Heinemann) is also well worth reading. Cotterill, R. M. J. (1989). *No Ghost in the Machine*. (London: Heinemann) very clearly states the argument for pure mechanism, while a clear analysis of the basic problem can be found in Lewes, G. H. (1859). *The Physiology of Common Life*. (London: Blackwood); Lewes was George Eliot's husband. Discussions of what functions are possible *without* consciousness can be found in de Gelder, B., de Haan, E. & Heywood, C. (2001). *Out of Mind*. (Oxford: Oxford University Press). Possible relations between consciousness and deep problems in physics are discussed in Penrose, R. (1994). *Shadows of the Mind*. (Oxford: Oxford University Press). A recent, highly analytic and careful approach can be found in Sommerhof, G. (2000). *Understanding Consciousness*. (London: Sage). Finally, the musings of a scientific mystic (and poet): Sherrington, C. S. (1951). *Man on his Nature*. (Cambridge: Cambridge University Press).

22. Kubla Khan In the preface to the poem, written in 1797, Coleridge writes: *'The Author continued for about three hours in a profound sleep, at least of the external senses, during which time he had the most vivid confidence, that he could not have composed less than from two or three hundred lines. . . . On awakening he appeared to himself to have a distinct recollection of the whole, and taking his pen, ink, and paper, instantly and eagerly wrote down the lines that are here preserved. At this moment he was unfortunately called out by a person on business from Porlock, and detained by him above an hour, and on his return to his room, found, to his no small surprise and mortification, that though he still retained some vague and dim recollection of the general purport of the vision, yet, with the exception of some eight or ten scattered lines and images, all the rest had passed away like the images on the surface of a stream into which a stone has been cast, but, alas! without the after restoration of the latter!'* However, scholars are not agreed that this account is actually to be taken literally,

23. Vegetative state Owens, A. M., Coleman, M. R., Boly, M., Davis, M. H., Laureys, S. & Pickard, J. (2006). Detecting awareness in the vegetative state. *Science* **8** 1402. A recent and similar experiment used the cheaper and more portable technique of EEG. Patients meeting clinical definitions of vegetative state were commanded to imagine either clenching a fist or wriggling their toes. In three patients the instructions met with distinct EEG responses similar to those generated by healthy controls and this was again interpreted to mean that these three patients in a clinically vegetative state might well retain awareness, comprehending instructions and consciously generating the appropriate response (Cruse, D., Chennu, S., Chatelle, C., Bekinschtein, T. A., Fernândez-Espejo, D., Pickard, J. D., Laureys, S. & Owen, A. M. (2011). Bedside detection of awareness in the vegetative state: an observational cohort study. *Lancet* **378** 2088–2094). But do such studies really prove that these patients are conscious? With no means of external validation or measurement of the internal experience of receiving command and generating response, have brain scans and EEGs really enlightened us? Some would vehemently argue yes, others are more sceptical. An editorial to this paper points out that complex types of behaviour, such as task switching, response inhibition, error detection and command following, have all been described in the absence of conscious experience. Both split-brain and blindsight patients can respond to commands without reported awareness. In the case of blindsight, a distinction has been made between vision for action and vision for perception. Following this train of thought, it is interesting that three of Cruse's 12 healthy controls failed to generate the expected EEG response despite being 'aware' of the instruction and 'deliberately' trying to respond appropriately, immediately raising the question of whether activity at the purported site of response generation necessarily equates to consciousness of instruction and response. Is the conscious experience occurring elsewhere? (Overgaard, M. & Overgaard, R. (2011). Measurements of consciousness in the vegetative state. *Lancet* **378** 2052–2054.)

24. Free will For a particularly compelling argument that 'free will' is an illusion, see Wegner, D. W. (2002). *The Illusion of Conscious Will*. (Cambridge, MA: MIT Press); helpful discussions can also be found in Pocket, S., Banks, W. P. & Gallagher, S. (eds) (2009). *Does Consciousness Cause Behaviour?* (Cambridge, MA: MIT Press). An observation supporting such ideas is that electrical activity in the human cortex starts to rise a long time in advance of the initiation of a 'voluntary' movement, and considerably precedes the subject's sense of having made a decision to move: see Libet, B., Gleason, C. A., Wright, E. W. & Pearl, D. K. (1983). Time of conscious intention to act in relation to onset of cerebral activity (readiness-potential). The unconscious initiation of a freely voluntary act. *Brain* **106** 623–642.

APPENDIX

TECHNIQUES FOR STUDYING THE BRAIN

This appendix briefly outlines the main experimental techniques that can be used to discover the anatomical pathways within the brain, and how they relate to function. Some purely technical limitations are mentioned here, but the wider problems of functional interpretation are covered in Chapter 1 (p. 13).

The four main weapons at our disposal are the classical ones of path tracing, recording, stimulation and lesions.

Path tracing

Single-cell visualisation

Just looking with the naked eye at sections of the brain, one can make out the grosser nuclei and tracts; but a microscope is needed to elucidate the details of neuronal connections. Since neuronal tissue is virtually transparent, some kind of stain is needed; the problem then is that if we stain *all* the nerve cells, the brain is such a densely knotted structure that the whole thing will simply come out black and we shall be no better off: what is needed is a *selective* stain. One such is the *Golgi silver stain*, discovered 130 years ago by Camillo Golgi, which has the odd property that it is only taken up by a very small percentage (about one in a hundred) of the neurons in the tissue to which it is applied, apparently at random; those cells that do take it up do so completely, resulting in a complete and often very beautiful delineation of their dendritic and axonal microstructure. **Figure A.1** shows stellate cells from cerebral cortex stained by the Golgi method, and also using another substance with similar or possibly enhanced penetrating power, horseradish peroxidase (HRP). This kind of staining is capable of providing two distinct kinds of information:

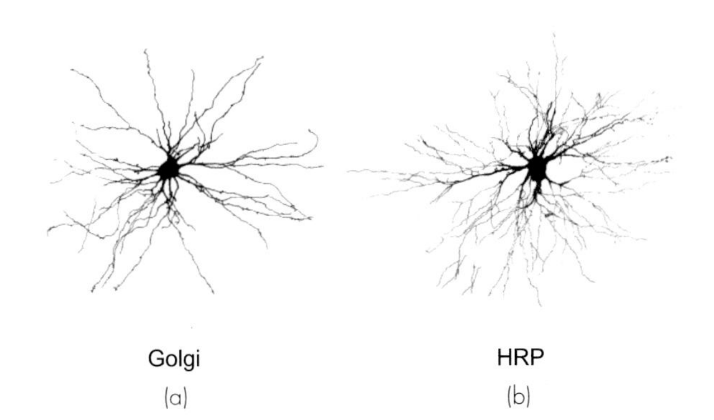

Figure A.1 Stellate cell from cerebral cortex, stained with Golgi method (a) and horseradish peroxidase (b).

- Detailed knowledge of a neuron's local morphology: knowing something about how spatial and temporal summation works, we can then make an informed guess about what sort of computational function the neuron is carrying out (p. 59). With modern computer models, these guesses can be relatively precise;
- Demonstration of the connections of neurons carrying information over long distances, by using serial sections (there is an example on p. 202). This can tell us with extreme precision exactly where neurons project to and the pattern of their innervation when they reach their destination: techniques that average over large numbers of neurons cannot do this.

Another technique is to inject a dye such as Procion Yellow into a neuron through a micropipette; the dye then diffuses into most parts of the cell but not outside it, and provides a good way of marking a cell whose electrical responses have previously been recorded with the same micropipette (**Fig. A.2**), since a series of different colours can be used.

Histochemical staining

Another method for selective staining that relates more closely to function is to use substances associated with particular neurotransmitters. They may be analogues to actual transmitters, such as serotonin, or they may be antibodies to enzymes associated with particular transmitters. An example of the latter is the tyrosine hydroxylase

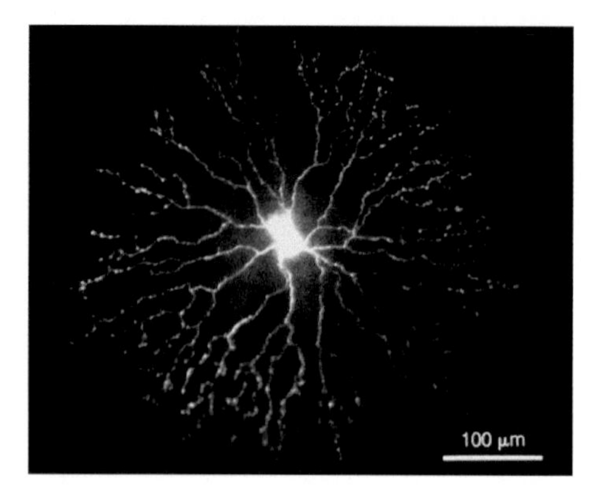

Figure A.2 'Starburst' amacrine cell from rabbit retina, injected with Lucifer Yellow.

found in neurons that use dopamine or noradrenaline. In addition, one may deploy labelled antibodies against neuropeptides themselves. In this way one may be able to identify homogeneous groups of cells within a nucleus.

Transport

Some marker substances are *transported* along axons, either from the cell body to the terminals (orthograde or anterograde) or in the opposite direction (retrograde) (**Fig. A.3**).

Labelled amino acids (tritiated leucine, for example) are taken up by cell bodies and transported towards the terminals, enabling one to identify the areas to which a particular nucleus projects. At least two transport mechanisms seem to be involved: one is fast (some 100–300 mm/day) and the other slow (1–10 mm/day). An attractive feature of this technique is that the transport may be transneuronal, the marker crossing the synapse to the next neuron along. **Figure A.4** shows a striking example of the use of this technique: dominance columns in monkey primary visual cortex are selectively stained after unilateral ocular injection of tritiated proline. HRP is an example of a substance showing retrograde transport. After extracellular injection at a particular site, it is taken up by axon terminals and carried back to the cell body. In this way one may identify the origin of efferents to a particular region

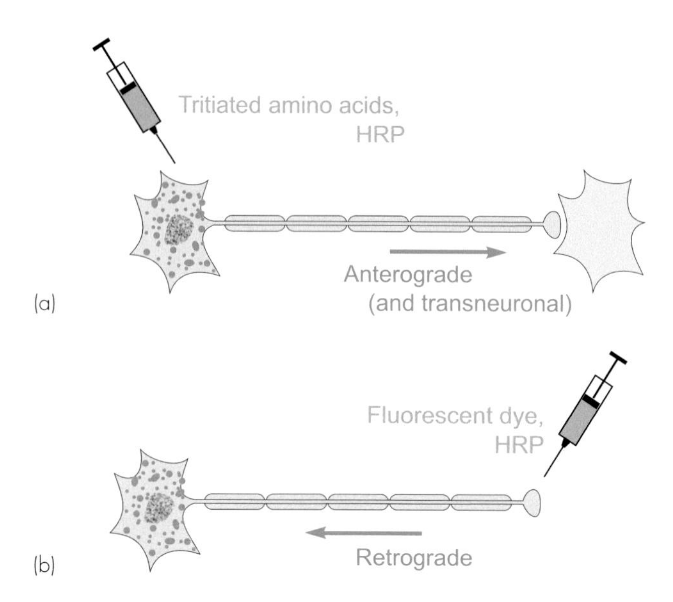

Figure A.3 Tracing connections by means of injected markers that are transported anterograde and possibly transneuronally (a) or retrograde (b). HRP, horseradish peroxidase.

of the brain, revealing for example subdivisions of nuclei according to their destinations.

Degeneration

An older, related, technique for tracing axonal pathways is to study the *degeneration* that results from injury to a nerve fibre. Two kinds of degeneration follow damage to an axon: *anterograde*, orthograde or Wallerian degeneration, distal to the cut; and *retrograde* degeneration, in the direction of the cell body. In the first case, one may use the Nauta stain that identifies certain of the degeneration products such as degenerating myelin sheaths and terminal boutons. In retrograde degeneration one may see various characteristic changes in the cell body (**Fig. A.5**): the nuclei and ribosomes show granule dispersion, which shows up with the Nissl stain.

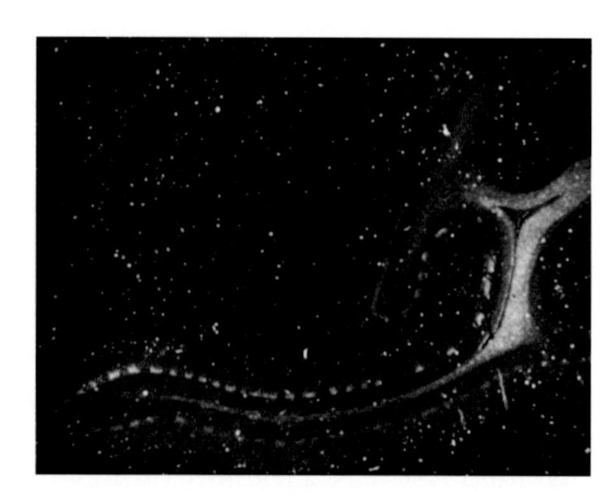

Figure A.4 Dominance columns in monkey visual cortex revealed by uniocular injection of tritiated proline.

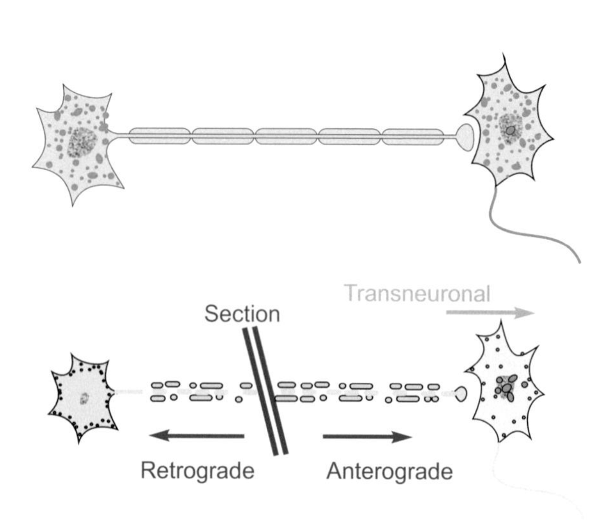

Figure A.5 Schematic representation of the types of degeneration that may follow injury to a neuron.

Figure A.6 Dominance columns in monkey visual cortex, delineated using 2DG while one eye was covered.

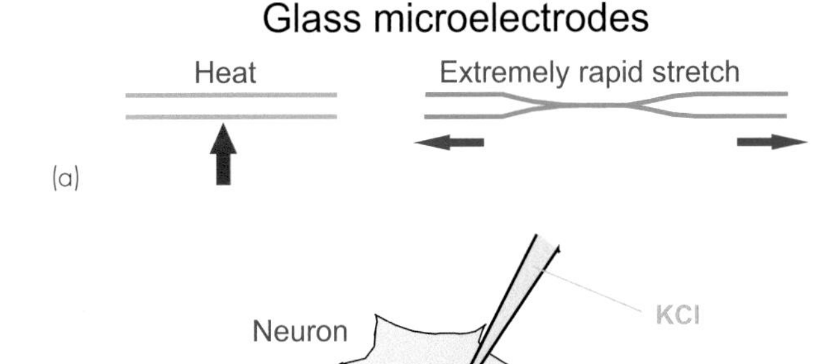

Figure A.7 Microelectrodes. (a) Common method of manufacture, by rapid drawing out of softened glass capillary tubing. (b) Microelectrode filled with potassium chloride and inserted in typical central horn cell.

Sometimes these degenerative changes may actually extend beyond the synapse to affect the next neuron along, and are then described as *transneuronal* anterograde or retrograde degeneration. In monkeys, for instance, removal of the eye results in shrinkage of the neurons in the thalamus with which the fibres from the eye make contact. Thus, by making a lesion in a particular nucleus, one may in principle trace both the afferent and efferent pathways associated with it, and in some cases one may identify the second-order cells with which these fibres synapse as well. A problem with degeneration studies, but not with HRP or labelled amino acids, is that they do not distinguish between fibres that genuinely begin or end in a particular region, and those *fibres of passage* that merely happen to pass through it.

Metabolic labelling

Local metabolic effects may also be used to pinpoint neural activity. One example is to treat an animal with radioactively labelled *2-deoxyglucose* (2DG), and then present it with a particular task, such as looking at a specific kind of visual stimulus. Those cells that are most active during that period take up the 2DG preferentially, and their localisation becomes apparent when the brain is subsequently sectioned (**Fig. A.6**): this technique has a resolution of some 50 μm.

Recording

Individual neurons

All neurons generate electrical currents when they are active, and these electrical effects may be picked up by means of *electrodes*. The choice here is between *intracellular microelectrodes* that are small enough to impale single nerve cells and record their activity in isolation from whatever else may be going on around them, and *extracellular* electrodes that record the external effects of currents generated by active cells, typically many of them at the same time. Intracellular electrodes often have the form of micropipettes, made by heating glass capillary tubing and then very rapidly drawing it out (**Fig. A.7**). By starting with fused pairs of tubes, double-barrelled electrodes can be made which enable more sophisticated measurements to be made, for instance the modification of response caused by injected currents or pharmacological substances (Chapter 3). Though 'micro', intracellular electrodes are not much smaller than the neurons themselves (**Fig. A.8**), and there is a danger of bias in making such recordings since populations of smaller cells may be missed entirely. Some workers have experimented with arrays of electrodes, which have the advantage that one may then be able to observe something of the *spatial pattern* of activity, which of course is what its all about: trying to find out how the brain works by recording from single neurons has been described as like trying to read a book by looking at just one letter on each page.

Paradoxically, by recording *extracellularly* with a large electrode one may then pick up activity of cells too small

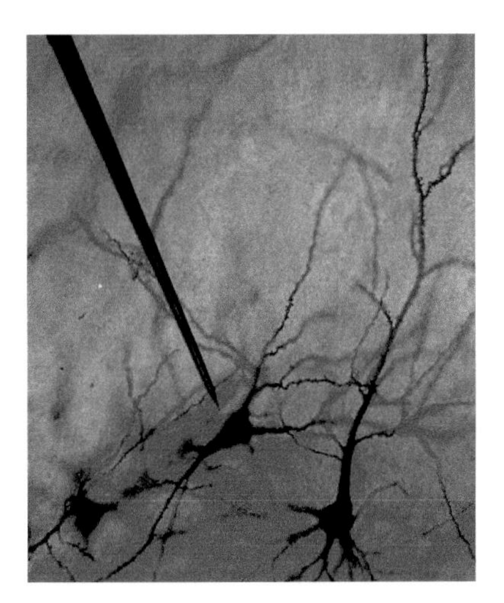

Figure A.8 Pyramidal cells, from cerebral cortex with an intracellular electrode, showing the relative sizes of each.

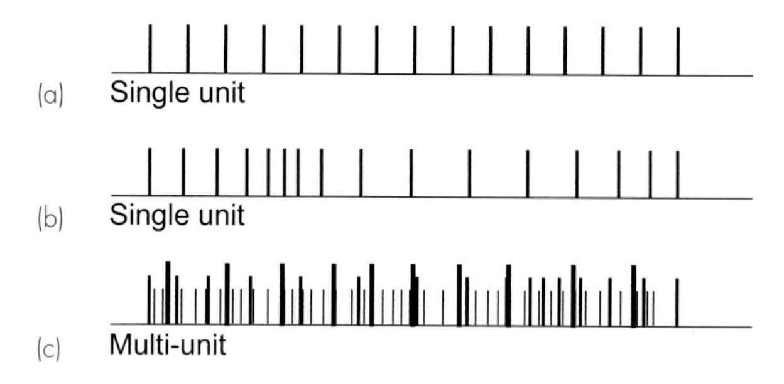

Figure A.9 Single and multi-unit recordings. (a and b) Schematic representation of responses from a single neuron at rest and undergoing stimulation. The action potentials are of essentially constant amplitude, with any changes in the intervals between spikes being relatively gradual. (c) In multi-unit recording, spikes are of different sizes (and often different shapes).

to be pierced by a microelectrode, and it is then possible to record from several cells at once (**Fig. A.9**), using a computer to sort out the different spikes from different neurons. Extracellular electrodes are also more stable than intracellular ones when chronically implanted in alert, normally behaving animals – an essential technique for studying higher aspects of brain function.

Electroencephalogram

In addition we may use *gross* electrodes that look at the average responses of many hundreds of neurons at once and are non-invasive. Just as electrodes on the skin can pick up the summed activity of the muscles below (the electromyogram), electrodes outside the skull can be used to record the average activity of very large numbers of cortical neurons over a wide area. An array of electrodes of this kind is used to measure the spontaneous *electroencephalogram* or EEG, discussed in Chapter 14 (**Fig. A.10**). But the same technique can also record *evoked* potentials from sensory stimulation: because they are very small in relation to the electrical noise created by the background activity of other cortical cells, it is necessary to repeat the stimulus many times, and average the evoked potential to get rid of the noise.

Magnetoencephalogram

The magnetoencephalogram (MEG) is similar to the EEG: it uses a sensor array around the head to respond to electrical activity at different points of the brain. However, in contrast to the EEG, the MEG measures the magnetic moments generated by the ionic fluxes underlying neuronal activity. An advantage over EEG is that the magnetic field is less susceptible to distortion than the electric field and spatial resolution may be better; an advantage over functional magnetic resonance imaging (fMRI; see p. 305) is that neuronal activity is directly measured, rather than the cerebrovascular response to activity, which is only indirectly related. MEG can generate a map of the functional organisation of the cortex and identify regions of abnormal electrical activity, hence offering a potential role in planning surgical treatment of epilepsy.

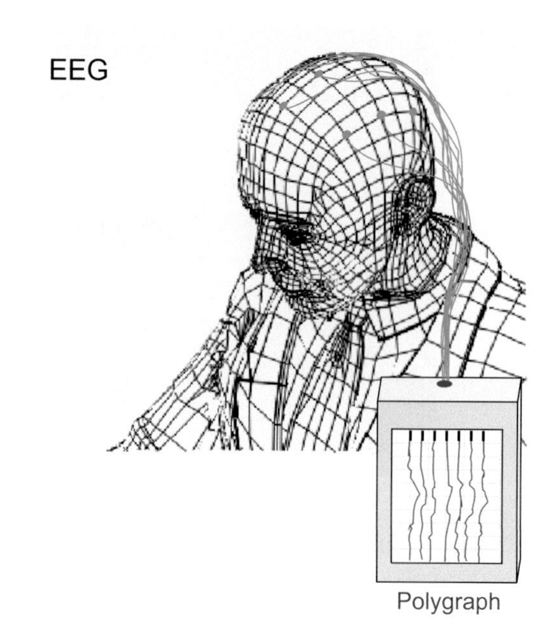

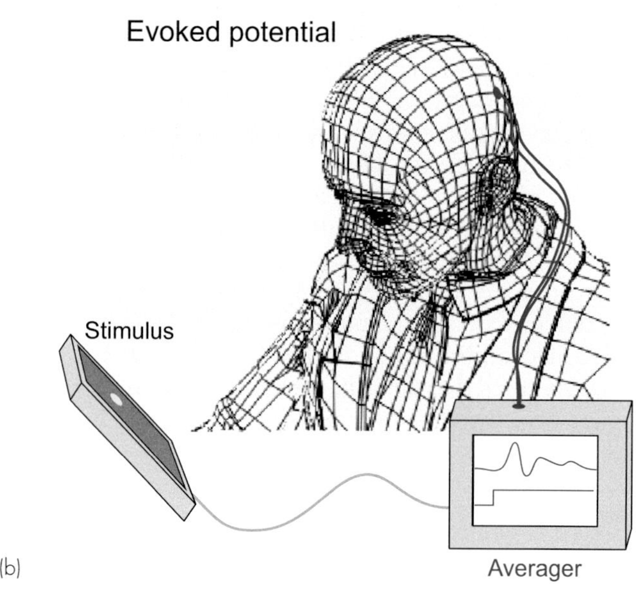

Figure A.10 Electroencephalography (EEG). (a) Simultaneous recording of spontaneous EEG with many electodes attached at defined locations on the scalp. (b) Recording evoked potentials: the stimulus is repeated many times, and the resultant activity averaged to eliminate background noise.

Brain scans

A popular and often colourful technique is the *brain scan*, of which there are various kinds.

Computed tomography (CT) is a computationally intensive technique whereby multiple X-ray images can be combined to build up a three-dimensional (3D) representation of body structures. CT is readily available in most hospitals, and is the 'scan of choice' in most medical emergencies, particularly head trauma or intracranial haemorrhage, being particularly adept at revealing cracks in the skull and blood where there shouldn't be blood. It is also logistically far more convenient than other types of neuroimaging. In a functional version of it, xenon CT, a subject breathes (non-radioactive) xenon, to the point of saturating the brain. During the subsequent washout, different tissues lose xenon at different rates, and this spatial variation can be detected in serial sections.

Positron emission tomography (PET) relies on the prior injection of radioactive markers to target specific cells and tissues. An example is shown on p. 268. While PET offers significantly greater specificity and superior serialisation of *in vivo* pathophysiological changes compared to MRI, the comparatively low resolution (5–10 mm) hampers visualisation of small areas.

Magnetic resonance imaging is even less invasive and, importantly, does not use ionising radiation. It depends on the fact that nuclei with an odd number of protons or neutrons possess a magnetic moment. Because of the magnetic moment, the application of a strong magnetic field causes a proportion of the nuclei to align with the magnetic field. This alignment is then perturbed by the transmission of a radiofrequency signal at the resonant frequency. When this signal is switched off the protons emit their own radiofrequency signal as they relax back into their original equilibrium states. Being the most abundant nucleus in organic tissue, the hydrogen nucleus (generating a 'proton signal') is used in nearly all nuclear MRI. The signal can be used to measure the proton density of different regions of tissue, and the rate of 'relaxation' gives further information about the chemical microenvironment being scanned. While X-rays, and hence CT images, are dependent on a single parameter to resolve different tissue planes (electron density), MRI images use multiple parameters (including spin density, relaxation times of vector alignment and phase coherence, spectral shifts) to create contrast between different tissues, greatly augmenting the ability to discriminate between them and identify pathology. For example, in **Figure A.11** note how much more clearly grey and white matter are delineated in the MRI than in the CT scan, as are the nuclei of the basal ganglia, the ventricles and the gyral contours. In addition, intravenously administered contrast agents can be used to resolve areas of increased vascularity. Lesions that can be completely invisible on CT, such as multiple sclerosis, early hypoxic brain injury, evolving stroke or herpetic encephalitis, can be clearly identified on MRI in time for emergent life-saving intervention. We can visualise brain anatomy with a resolution somewhat better than a millimetre, and in many ways MRI has revolutionised the practice of clinical neurology.

The principle of MRI can also be used to identify neural tracts, establishing routes of communication between various cortical and subcortical areas. In nerve axons the direction of diffusion of water molecules tends to run parallel to the bounding myelin sheaths and lipid membranes. This deviation from randomness can be picked up from the MR signal localising bundles of similarly orientated axons.

An extension of the basic technique, fMRI makes use of the fact that regions of neural activity augment their local blood flow. The resulting change in the amounts of oxygenated and deoxygenated haemoglobin alters the paramagnetic properties of the protons being scrutinised and the altered MR signal can thus be used to identify which areas of brain are active during certain activities; a computer looks at the difference in the image between one set of circumstances and another, for instance when doing a particular task as opposed to not doing it. Because the technique can be used on conscious humans, it can provide answers to questions such as which areas are differentially active when thinking of moving rather than actually moving. **Figure A.12** is an example of the complex interactions that can be studied in this way. It is from an obese subject: the red regions show significant differences in their activity between viewing images of

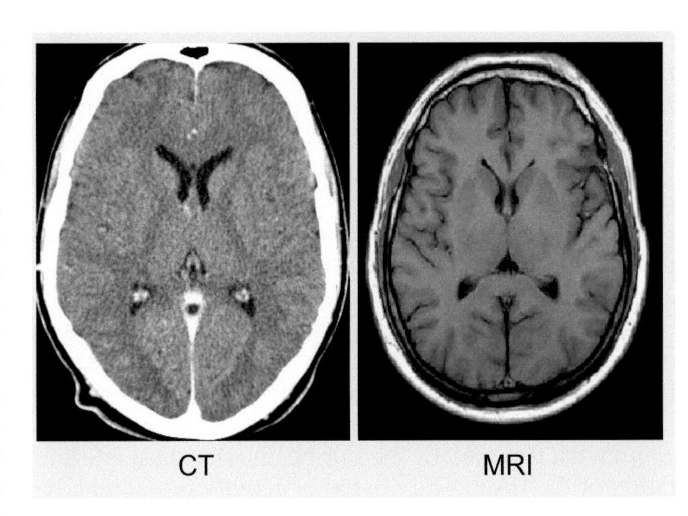

Figure A.11 A comparison between a CT and a T1-weighted MRI scan,

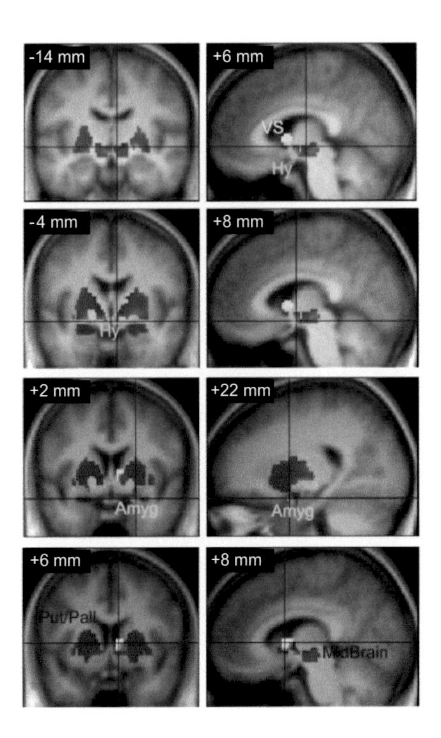

Figure A.12 Functional MRI (fMRI) can be used to visualise regions with increased blood flow, and thence increased neural activity (see text).

high-calorie and low-calorie foods, yellow indicates where these differences are greater when fasting, and green where they are reduced by the anti-obesity drug sibutramine.

However, the interpretation of fMRI is fraught with difficulties of various kinds. The reason that neural activity influences MRI is through the degree of oxygenation, but it is not obvious that this bears a simple relation to activity itself, rather than something like the difference between actual activity and the oxygen supply, determined by local blood flow. Because of over-simplistic and over-enthusiastic interpretation, the technique has acquired a poor reputation: sometimes it is described as the return of phrenology (cf. p. 15). Quite apart from the underlying uncertainty about what it is actually measuring, the statistical interpretation of the resultant patterns of activity is a somewhat subjective matter, dangerous in an area where apparently spectacular findings lead to career-enhancing headlines ('Scientists find the brain centre for happiness'). The fact remains that despite the huge sums spent on this extravagantly expensive technique, it is hard to think of anything very important that it has told us about brain function.[1]

Optical monitoring

Far more useful than recording from one or just a few arbitrarily selected neurons would be the ability to visualise the spatial pattern of activity across a 3D array of neural units. Calcium imaging allows investigators to do just that in living, and even awake, animals. Membrane-permeant indicator dye is injected through the skull into a targeted area of the brain, resulting in quite uniform staining of the area of interest; the dye fluoresces in response to calcium transients associated with single action potentials with high temporal resolution (up to 100 frames/sec) and spatial resolution approaching that of single cells. This fluorescence can be observed in the outer layers of cortex *in vivo* using two-photon laser scanning microscopy, generating a representation of activity in space and time across a functional neural network. Such dynamic mosaics of activity have been observed in rodent sensorimotor cortex during sensory-driven motor activity suggesting that shifting patterns of activity across neural networks drives such activity, rather than task-specific clumps of neurons. On the other hand, visualising neural activity in cat visual cortex has demonstrated discrete topographically related subgroups of neurons responding to similar visual attributes. Calcium imaging has also been used to analyse light- and olfactory-evoked responses. An exciting goal is the genetic encoding of fluorescent dyes, and the development of novel microscopic techniques allowing real-time imaging of neural activity throughout the entire brain. A similar technique is to stain neurons with a voltage-sensitive dye, though at present this is limited to tissue slices *in vitro*.[2]

Stimulation

Stimulation is a natural way to try to investigate the motor system, though for reasons discussed more fully in Chapter 1 and Chapter 9, it has not often proved a very helpful technique. Stimulation of single cells is usually inadequate to achieve any overt response at all, while simultaneous stimulation of large numbers of them with gross electrodes and heavy currents has proved in general to be too 'unphysiological' to give meaningful results. However, stimulation at one site while recording from another is often a good way to establish the existence of functional connecting pathways. Stimulation can be electrical, using intracellular or extracellular stimulation; or chemical: selective activation or inhibition of sub-sets of neurons within a nucleus may be achieved by application of pharmacological agonists, for instance muscimol mimicking γ-aminobutyric acid (GABA).

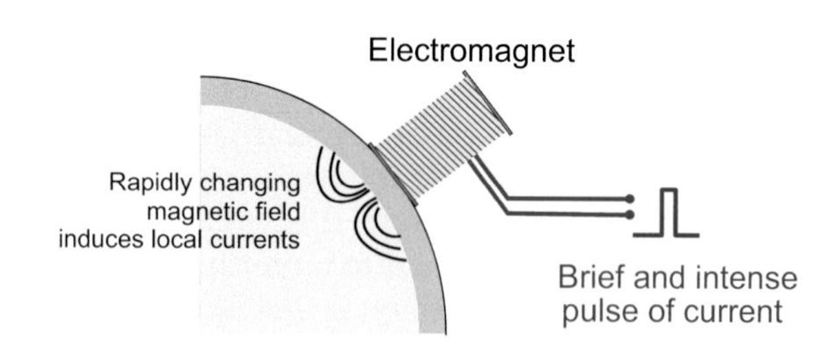

Figure A.13 Schematic representation of transcranial magnetic stimulation. A brief and very intense magnetic pulse is applied by means of an electromagnet placed on the scalp, inducing currents in the underlying neural tissue that stimulate neurons.

A relatively recent technique is *transcranial magnetic stimulation* (TMS). This is a non-invasive technique that can be used on conscious human subjects, and uses the principle that a change in magnetic field strength induces currents in a conducting medium. Since magnetic fields penetrate the skull without hindrance and are not in themselves painful, by placing a large electromagnet on the skull and passing through it a brief but very intense pulse of current, voltages are evoked in the underlying neurons that cause them to be stimulated (**Fig. A.13**). This technique does not have good spatial resolution, and is limited to neural structures such as cerebral cortex that are close to the surface; nevertheless, it has been used to good effect in studying various aspects of sensory and motor cortical function. In addition, by combining it with evoked potential measurements it can be used for tracing neural pathways in living human subjects.

Lesions

Apart from their use in demonstrating anatomical pathways through degeneration, lesions may also be used to try to associate particular functions with particular regions of the brain. Basically there are three ways of damaging the brain: natural pathology, or deliberate damage that may be irreversible or reversible.

Natural lesions

Pathological and developmental disorders are by far the least useful, for two reasons. First, although there may be a very visible lesion in one particular place, it may be due to an underlying condition that is causing invisible impairment in other places as well; for example, when CT scanning a comatose patient after a fall from a motorbike, the only visible lesion might be a discrete bleed in, for example, a frontal lobe. However, the most likely cause for persisting coma could actually be subtle diffuse axonal injury through the brainstem or a global hypoxic injury that occurred when bleeding from other injuries left the brain inadequately perfused – both of which might be invisible on an initial CT scan.

Perhaps the commonest type of neurological lesion dealt with by clinicians in the Western world is the ischaemic stroke, where occlusion of a terminal artery leads to death of the region of brain it supplied. In some cases this can lead to quite discrete areas of injury. Eliciting functional deficits allows the neurologist to predict the site of the lesion with some degree of accuracy, while the radiologist looking at the site of injury on a brain scan can predict, roughly, the pattern of neurological deficit found by the neurologist. What is much less precise, however, is relating the anatomical lesion and the functional loss quantitatively, particularly with regard to prognosis. Loss of almost an entire hemisphere can be succeeded by an astonishing degree of recovery, approaching functional independence. On the other hand, injuries which one might expect to cause relatively circumscribed deficits, such as isolated limb weakness, can be associated with profound, persistent unresponsiveness and global loss of normal electrical activity.

Ironically, the injury we are perhaps best able to prognosticate is perhaps the least specific, often practically invisible on an initial CT scan – global hypoxic brain injury (for instance after cardiac arrest). An absent pupil or corneal reflex, and absent motor response or limb extension only to pain 72 hours after the injury, has always reliably portended a dismal outcome.[3] However, recently even this clinical truism has been questioned.

Furthermore, apart from traumatic injuries, most conditions develop slowly, giving time for all kinds of adaptational compensation to occur. Consequently conclusions from studying neurological conditions such as parkinsonism can only be of the most general kind, though provided one does not try to read more into them than is justifiable, such broad-brush conclusions, if clearly formulated, can still be perfectly secure (see Chapter 12).

Experimental lesions

Deliberate local *irreversible* destruction can be done in many ways. Apart from the relatively crude traditional techniques of actual excision, or arresting blood supply, more localised destruction can be performed by using local heat, or by passing substantial currents through neural tissue that cause electrolytic lesions. All these techniques of course damage not only the neuronal cell bodies at a particular site, but also fibres that happen to pass through the region, which creates extra problems of interpretation. *Reversible* chemical lesions – the blocking of particular pathways by the local application of appropriate pharmacological agents, or with toxins – have the advantage of being more specific, as well as being free of some of the objections outlined above. For instance, excitotoxins such as ibotenic acid are specific for cell bodies and do not affect fibres of passage; other toxins are selective for neurons that use particular transmitters, e.g. 6-hydroxydopamine, which only affects noradrenergic and dopaminergic neurons. Better still is reversible block, which provides the experimental control of comparing behaviour during the block and after. Reversible blockage can be done by using pharmacological antagonists (for example bicuculline, blocking GABA), or local cooling by electrodes utilising the Peltier effect.

The future

Technological developments over the past decade or two have resulted in huge improvements to the tools at our disposal for investigating the brain, but there is still plenty of room for making them even better. What we would like is the ability to record the action potentials of many individual neurons simultaneously, anywhere in the brain, using apparatus that is portable and non-invasive. **Figure A.14** is a very approximate representation of the spatial and temporal resolution of some of the techniques used in brain research: in general it is the electrical techniques that provide good resolution, but at the expense of losing the bigger picture that comes from the various kinds of scan.

Splendid as all this technology is, in the end the single most important piece of kit is the human brain itself. Good science is not about providing answers so much as formulating questions, and there is a tendency for new techniques to encourage bandwagon, me-too research. Great discoveries are seldom the result of large teams deploying extremely expensive equipment; they occur when individuals permit themselves to think.

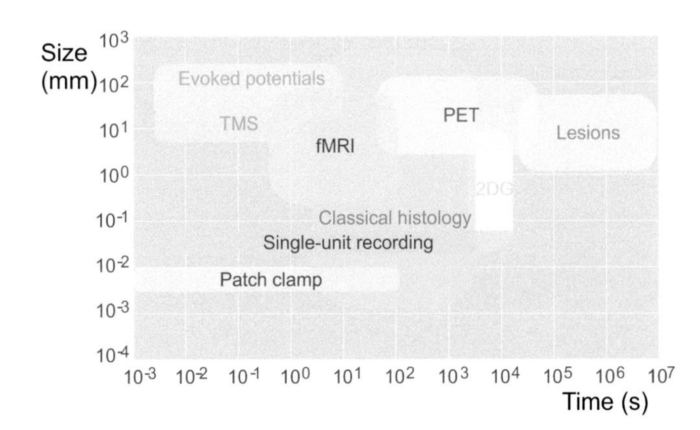

Figure A.14 Spatial and temporal resolution of some techniques used in brain research: highly approximate. 2DG, 2-deoxyglucose; fMRI, functional magnetic resonance imaging; PET, positron emission tomography; TMS, transcranial magnetic stimulation.

Notes

1. Brain scans – dynamic phrenology? See for instance Raichle, M. E. (1994). Visualizing the mind. *Scientific American* **April** 36–42.
2. Calcium and voltage imaging There is a useful review of calcium imaging in Garaschuk, O., Milos, R. J. & Grienberger, C. (2006). Pflugers Arch. *European Journal of Physiology* **453** 385–396; the use of voltage-sensitive dyes is described in Wu, J. Y. & Cohen, L. B. (1993). Fast multisite optical measurements of membrane potential. In W. T. Mason (ed.) *Fluorescent and Luminescent Probes for Biological Activity*. (London: Academic, pp. 389–404).
3. Predicting outcomes after hypoxic brain injury See Wijdicks, E. F. M, Hijdra, A., Young, G. B., Bassetti, C. L. & Wiebe, S. (2006). Practice parameter: Prediction of outcome in comatose survivors after cardiopulmonary resuscitation (an evidence-based review): Report of the Quality Standards Subcommittee of the American Academy of Neurology. *Neurology* **67** 203.

FIGURE REFERENCES

Chapter 1: The study of the brain

Buchsbaum, R. (1951) *Animals without Backbones.* (Harmondsworth: Penguin).

Chapter 2: Communication within neurons

Doyle, D. A., Cabral, J. M., Pfuetzner R. A., Kuo A., Gulbis J. M., Cohen S. L., Chait B. T. & MacKinnon R. (1998) The structure of the potassium channel: molecular basis of K+ conduction and selectivity. *Science* **280** 69–77.

Erlanger, J. & Gasser, H. S. (1938) *Electrical Signs of Nervous Activity.* (Philadelphia, PA: University of Pennsylvania Press).

Granit R., Kernell, D. & Shortess, G. K. (1963) Quantitative aspects of firing of mammalian motoneurons, caused by injected currents. *Journal of Physiology* **168** 911–931.

Hodgkin, A. L. (1938) The subthreshold potentials in a crustacean nerve fibre. *Proceedings of the Royal Society B* **126** 87–121

Hodgkin, A. L. (1958) Ionic movements and electrical activity in giant nerve fibres. *Proceedings of the Royal Society B* **148** 1–37. doi: https://doi.org/10.1098/rspb.1958.0001

Hodgkin, A. L. & Horowicz, P. (1959) The influence of potassium and chloride ions on the membrane potential of single muscle fibres. *Journal of Physiology* **148** 127–160.

Hodgkin, A. L. & Huxley, A. F. (1952a) The components of membrane conductance in the giant axon of *Loligo*. *Journal of Physiology* **116** 473–496.

Hodgkin, A. L. & Huxley, A. F. (1952b) A quantitative description of membrane current and its application to conduction and excitation in nerve. *Journal of Physiology* **117** 500–544. doi: 10.1113/jphysiol.1952.sp004764

Hodgkin, A. L. & Katz, B. (1949) The effect of sodium ions on the electrical activity of the giant axon of the squid. *Journal of Physiology* **108** 37–77.

Hodgkin, A. L. & Keynes. R. D. (1956) Experiments on the injection of substances into squid giant axons by means of a microsyringe. *Journal of Physiology* **131** 592–616.(With permission from John Wiley and Sons)

Huxley, A. F. (1959) Ion movements during nerve activity. *Annals of the New York Academy of Sciences* **81** 221–246.

Oshima, K. (1969) Studies of pyramidal tract cells. In H. H. Jasper, A. A. Ward & A. Pope (eds) *Basic Mechanisms of the Epilepsies*. (Boston, MA: Little, Brown).

Pattak, J. & Horn, R. (1982) Effect of N-bromoacetamide on single sodium channel currents in excised membrane segments. *Journal of General Physiology* **79** 333–351.doi: 10.1085/jgp.79.3.333

Rushton, W. A. H. (1951) A theory of the effects of fibre size in medullated nerve. *Journal of Physiology* **115** 101–122.

Chapter 3: Communication between neurons

Araki, T., Eccles, J. C. & Ito, M. (1960) Correlation of the inhibitory post-synaptic potential of motoneurons with the latency and time course of inhibition of monosynaptic reflexes. *Journal of Physiology* **154** 354–377.

Coombs, J. S., Eccles, J. C. & Fatt, P. (1955) Excitatory synaptic action in motoneurons. *Journal of Physiology* **130** 374–395. doi: 10.1113/jphysiol.1955.sp005413

Curtis, D. R. & Eccles, J. C. (1959) The time courses of excitatory and inhibitory synaptic actions. *Journal of Physiology* **145** 529–546.

Eccles, J. C. (1963) Presynaptic and postsynaptic inhibitory actions in the spinal cord. In G. Moruzzi (ed.) *Brain Mechanisms*. (Amsterdam: Elsevier).

Eccles, J.C. (1964) *The Physiology of Synapses*. (Cambridge: Cambridge University Press).

Eccles, J. C., Eccles, R. M. & Magni, F. (1961) Central inhibitory action attributable to presynaptic depolarization produced by muscle afferent volleys. *Journal of Physiology* **159** 147–166. https://doi.org/10.1113/jphysiol.1961.sp006798

Fatt P. & Katz B. (1950) Some observations on biological noise. Nature 166 597–598.

Heuser, J.E. (1977) Synaptic vesicle exocytosis revealed in quicj-frozen frog neuromuscular junction treated with 4-amino pyridine and given an electric shock. In Cowan, W.M., Ferrendelli, J.A. (eds.) *Society for Neuroscience Symposium Vol. 2: Approaches to the cell biology of neurons.* (Bethesda, MD: Society for Neuroscience), pp. 215–239.

Loewenstein, W. R. & Mendelsohn, M. (1965) Components of adaptation in a Pacinian corpuscle. *Journal of Physiology* **177** 377–397. https://doi.org/10.1113/jphysiol.1965.sp007598

Moczydlowkki, E. & Latorre, R. (1983) Gating kinetics of Ca2+-activated K+ channels from rat muscle incorporated into planar lipid bilayers. Journal of General Physiology, 82, 511–542. DOI: 10.1085/jgp.82.4.511

Mountcastle, V. B. (1965) The neural replication of sensory events. In J. C. Eccles (ed.) *Brain and Conscious Experience.* (Berlin: Springer).

Roll, J.P. & Vedel, J.P. (1982) Kinaesthetic role of muscle afferents in Man, studied by tendon vibration and microneurography. *Experimental Brain Research* **47** 177–190.

Chapter 4: Skin sense

Brodal, P. (1998) *Neurological Anatomy.* (New York, NY: Oxford University Press).

Brodmann, K. (1909) *Vergleichende Loakisationslehre des Grosshirnsrinde : ihren Prinzipien dargestellt auf Grund des Zellenbaues.* (Leipzig: Barth). https://wellcomecollection.org/works/vrnkkxtj

Halata, Z. (1975) The mechanoreceptors of mammalian skin. Ultrastructure and morphological classification. Advances in Anatomy, *Embryology and Cell Biology* **50** 1–77.

Kenshalo, D. R. (1976) Correlates of temperature sensitivity in man and monkey, a first approximation. In Y. Zotterman (ed.) *Sensory Functions of the Skin in Primates.* (Oxford: Oxford University Press).

Penfield, W. & Rasmussen, T. (1950) *The Cerebral Cortex of Man: a Clinical Study of Localisation of Function.* (New York, NY: Macmillan).

Chapter 5: Priprioception

Barker, D. (1948) The innervation of the muscle spindle. *Journal of Cell Science* **89** 143–186. https://doi.org/10.1242/jcs.s3-89.6.143

Boyd, I. A. & Roberts, T. D. M. (1953) Proprioceptive discharges from stretch-receptors in the knee-joint of the cat. *Journal of Physiology* **122** 38–58.

Matthews, P. B. C. (1964) Muscle spindles and their control. *Physiological Review* **44** 219–288. https://doi.org/10.1152/physrev.1964.44.2.219

Whitteridge, D. (1959) The effect of stimulation of intrafusal muscle fibres on sensitivity to stretch of extraocular muscle spindles. *Quarterly Journal of Experimental Physiology* **44** 385–393.

Chapter 6: Hearing

Dadson, R. S. & King, J. H. (1952) A determination of the normal threshold of hearing and its relation to the standardisation of audiometers. *Journal of Laryngology and Otolaryngology* **66** 366–378. https://doi.org/10.1017/S0022215100047812

Katsuki, Y. (1961) Neural mechanisms of auditory sensation in cats. In W. A. Rosenblith (ed.) *Sensory Communication.* (Boston, MA: MIT Press).

Spoendlin, H. (1968) Ultrastructure and peripheral innervation pattern of the receptor coding of the acoustic message. In A. V. S. de Renck & J. Knight (eds). *Hearing Mechanisms in Vertebrates.* (London: Churchill).

von Békésy, G. (1960) *Experiments in Hearing.* (New York, NY: McGraw-Hill).

Wood, A. (1930) *Sound Waves and their Uses.* (Glasgow: Blackie).

Chapter 7: Vision

Aguilar, M. & Stiles, W. S. (1954) Saturation of the rod mechanism at high levels of illumination. *Optica Acta* **1** 59–65. https://doi.org/10.1080/713818657

Baker, J., Gibson, A., Glickstein, M. & Stein, J. (1976) Visual cells in the pontine nuclei of the cat. *Journal of Physiology* **255** 415–433. 10.1113/jphysiol.1976.sp011287

Blakemore, C. B. & Rushton, W. A. H. (1965) The rod increment threshold during dark adaptation in normal and rod monochromat. *Journal of Physiology* **181** 629–640. 10.1113/jphysiol.1965.sp007787

Fisher, R. F. (1973) Presbyopia and the changes with age in the human crystalline lens. *Journal of Physiology* **228** 765–779.
Hubel, D. H. & Wiesel, T. N. (1965) Receptive fields and functional architecture in two nonstriate visual areas (18 and 19) of the cat. *Journal of Neurophysiology* **28** 229–289. 10.1152/jn.1965.28.2.229
Lamb, T. D. (1984) Effects of temperature changes on toad rod photocurrents. *Journal of Physiology* **346** 557–578. 10.1113/jphysiol.1984.sp015041
Rushton, W. A. H. & Powell, D. S. (1972) The rhodopsin content and the visual threshold of human rods. *Vision Research* **12** 1073–1081. https://doi.org/10.1016/0042-6989(72)90098-3
van Ness, F. L. & Bouman, M. A. (1967) Spatial modulation transfer in the human eye. *Journal of the Optical Society of America* **57** 401–406.

Chapter 8: Smell and taste

Carpenter, M.B. & Sutin, J. (1983) *Human Neuroanatomy.* (Baltimore, MD: Williams and Wilkins).
Firestein, S. & Werblin, F. (1989) Odor-Induced Membrane Currents in Vertebrate-Olfactory Receptor Neurons. *Science* **244** 79–82. DOI: 10.1126/science.2704991

Chapter 10: Local motor control

Brodal, P. (1998) *Neurological Anatomy.* (New York, NY: Oxford University Press).
Eldred, E., Granit, R. & Merton, P. A. (1953) Supraspinal control of the muscle spindles and its significance. *Journal of Physiology* **122** 498–523. 10.1113/jphysiol.1953.sp005017
Fitzgerald, M. J. T. (1985) *Neuroanatomy, Basic and Applied.* (London: Baillere Tindall).
Liddel, E. G. T. & Sherrington, C. S. (1924) Reflexes in response to stretch (myotatic reflexes). *Proceedings of the Royal Society B* **96** 212–242. https://doi.org/10.1098/rspb.1924.0023
Marsden, C. D., Merton, P. A. & Morton, H. B. (1972) Servo action in human voluntary movement. *Nature* **238** 140–143.
Matthews, P. B. C. (1972) *Mammalian Muscle Receptors and their Central Actions.* (London: Arnold).
Scheibel, M. E. & Scheibel, A. B. (1960) Spinal motor neurons, interneurons and Renshaw cells: a Golgi study. *Archives Italiennes de Biologie* **104** 328–353. https://doi.org/10.4449/aib.v104i3.2234
Vallbo, Å. B. (1971) Muscle spindle response at the onset of isometric voluntary contractions in Man: time difference between fusimotor and skeletomotor effects. *Journal of Physiology* **218** 405–438. 10.1113/jphysiol.1971.sp009625

Chapter 11: The control of posture

de Kleijn, A. (1921) Tonische Labyrinth und Halsreflexe auf die Augen. *Pflügers Archiv* **186** 82–97. https://doi.org/10.1007/BF01755737
Martin, J. P. (1967) *The Basal Ganglia and Posture.* (London: Pitman).
Rademaker, G. G. J. (1935) *Réactions labyrinthiques et équilibre; l'ataxie labyrinthique.* (Paris: Masson).
Witkin, H. A. (1949) Perception of body position and the position of the visual field. *Psychological Monographs* **302** (3) 1–46.

Chapter 12: Higher motor control

Penfield, W. & Rasmussen, T. (1950) The Cerebral Cortex of Man: a Clinical Study of Localisation of Function. (New York, NY: Macmillan).
Rosen, I. & Asanuma, H. (1972) Peripheral afferent inputs to the forelimb area of the monkey motor cortex: input-output relations. Experimental Brain Research 14 257–273. doi: 10.1007/BF00816162.
Woolsey, C. N. (1958) Organisation of somatic sensory and motor areas of the cerebral cortex. In H. F. Harlow and C. N. Woolsey (eds) Biological and Biochemical Bases of Behaviour. (Wisconsin, WI: University of Wisconsin Press).
Netter, F. H. (1962) Nervous System: The CIBA Collection of Medical Illustrations. (Basle: CIBA).
Cobb, S. (1946) Borderlands of Psychiatry. (Harvard, MA: Harvard University Press). Penfield, W. (1967) The Excitable Cortex in Conscious Man. (Liverpool: Liverpool University Press).
Freeman, W. & Watts, J.W. (1948) The thalamic projection to the frontal lobes. Research Publications of the Association for Nervous and Mental Diseases, 27, 200–209.

Chapter 13: Associatal cortex and memory

Cobb, S. (1946) *Borderlands of Psychiatry.* (Harvard, MA: Harvard University Press).
Cotterill, R. M. (1998) *Enchanted Looms: Conscious Networks in Brains and Computers.* (Cambridge, UK: Cambridge University Press).

Critchley, M. (1971) *The Parietal Lobes.* (London: Hafner).
Gazzaniga, M. S. (1967) The split brain in Man. *Scientific American* **217** 24–29.
Geschwind, N. & Levitsky, W. (1968) Human brain: left-right asymmetries in temporal speech region. *Science* **161** 186–187. DOI: 10.1126/science.161.3837.186
Lassen, N. A., Ingvar, D. H. & Skinhøj, E. (1978) Brain function and blood flow. *Scientific American* **239** 50–59. DOI: 10.1038/scientificamerican1078-62
Merzenich, M., Recanzone, G. H., Jenkins, W. M. & Nudo, R. J. (1990) How the brain functionally rewires itself. In M. A. Arbib & J. A. Robinson (eds) *Natural and Artificial Parallel Computation.* (Boston, MA: MIT Press).
Munoz, D. P. & Wurtz, R. H. (1995) Saccade-related activity in monkey superior colliculus. I. Characteristics of burst and build-up cells. *Journal of Neurophysiology* **73** 2313–2333. DOI: 10.1152/jn.1995.73.6.2313
Nauta W. J. H. & Feirtag M (1986) *Fundamental Neuroanatomy.* (New York, NY: W. H. Freeman).
Nolte, J. (1999) *The Human Brain.* (St. Louis, MO: Mosby).
Ochs, S. (1965) *Elements of Neurophysiology.* (New York, NY: Wiley).
Penfield, W. (1967) *The Excitable Cortex in Conscious Man.* (Liverpool: Liverpool University Press).
Zangwill, O. L. (1967) Speech and the minor hemisphere. *Acta Neurologica Belgica* **67** 1013–1020.

Chapter 14: Motivation and the control of behaviour

Aschoff, J. (1965) Circadian rhythms in man. *Science* **148** 1427–1432. DOI: 10.1126/science.148.3676.1427
Dement, W. & Kleitman, N. (1957) Cyclic variations in EEG during sleep and their relation to eye movements, body motility and dreaming. *Electroencephalography and Clinical Neurophysiology* **9** 673–690. DOI: 10.1016/0013-4694(57)90088-3
El Mansari, M., Sakai, K. and Jouvet, M. (1989) Unitary characteristics of presumptive cholinergic tegmental neurons during the sleep-waking cycle in freely moving cats. *Experimental Brain Research* **76** 519–529. doi: 10.1007/BF00248908
Engel, G. L. & Schmale, A. H. (1972) Conservation-withdrawal: a primary regulatory process for organismic homeostasis. In CIBA Symposium 8 (ed.) *Physiology, Emotion and Psychosomatic Illness.* (Amsterdam: Elsevier).
Fonberg, E. (1972) Control of emotional behaviour through the hypothalamus and amygdaloid complex. In CIBA Symposium 8 (ed.) *Physiology, Emotion and Psychosomatic Illness.* (Amsterdam:John Wiley and Sons)
Lothman, E. W. & Collins, R. C. (1990) Seizures and epilepsy. In A. L. Pearlman & R. C. Collins (eds) *Neurobiology of Disease.* (New York, NY: Oxford University Press), pp. 276–298.
Moruzzi, G. & Magoun, H. W. (1949) Brain stem reticular formation and activation of the EEG. *EEG Clinical Neurophysiology* **1** 455–473. https://doi.org/10.1016/0013-4694(49)90219-9
Roffwarg, H. P., Muzio, J. N. & Dement, W. C. (1966) Ontogenetic development of the human sleep-dream cycle. *Science* **152** 604–619. DOI: 10.1126/science.152.3722.604
Scheibel, M. E. & Scheibel, A. B. (1957) Structural substrates for integrative processes in the brainstem reticular core. In H. H. Jasper, L. D. Proctor, R. S. Knighton, W. C. Wothay and R. T. Costello (eds) *The Reticular Formation of the Brain.* (London: Churchill).
Wiener, S. I., Paul, C. A. & Eichenbaum, H. (1989) Spatial and behavioural correlates of hippocampal neuronal activity. *Journal of Neuroscience* **9** 2737–2763. https://doi.org/10.1523/JNEUROSCI.09-08-02737.1989

INDEX

Note: Page numbers followed by *f* and *t* indicate figures and tables, respectively. Clinical boxes are shown by *b*.